LIQUID–VAPOR PHASE-CHANGE PHENOMENA

Series in Chemical and Mechanical Engineering

G. F. Hewitt and C. L. Tien, *Editors*

Carey, Liquid–Vapor Phase-Change Phenomena: An Introduction to the Thermophysics of Vaporization and Condensation Processes in Heat Transfer Equipment

FORTHCOMING TITLES

Kemp, Pinch Technology and Process Integration
Nieuwenhuizen, Thermomechanics for Engineers

LIQUID–VAPOR PHASE-CHANGE PHENOMENA

An Introduction to the Thermophysics of Vaporization and Condensation Processes in Heat Transfer Equipment

Van P. Carey

Mechanical Engineering Department
University of California at Berkeley

HEMISPHERE PUBLISHING CORPORATION
A member of the Taylor & Francis Group
Washington Philadelphia London

LIQUID–VAPOR PHASE-CHANGE PHENOMENA: An Introduction to the Thermophysics of Vaporization and Condensation Processes in Heat Transfer Equipment

1 2 3 4 5 6 7 8 9 0   E B E B   9 8 7 6 5 4 3 2 1

This book was set in Times Roman by Edwards Brothers, Inc. The editors were Heather Jefferson and Lynne Lackenbach; and the production supervisor was Peggy M. Rote.
Cover design by Kathleen Ernst.
Printing and binding by Edwards Brothers, Inc.

A CIP catalog record for this book is available from the British Library.

Library of Congress Cataloging-in-Publication Data

Carey, V. P. (Van P.)
Liquid–vapor phase-change phenomena: an introduction to the thermophysics of vaporization and condensation processes in heat transfer equipment / Van P. Carey.
p. cm.—(Series in chemical and mechanical engineering)
Includes bibliographical references and index.
1. Heat exchangers. 2. Change of state (Physics) 3. Evaporation. 4. Condensation. I. Title. II. Series.
TJ263.C37 1992
621.402′5—dc20   91-25418
CIP

ISBN 0-89116-836-2 (case)
ISBN 1-56032-074-5 (paper)

To my daughter, Elizabeth Megan, and my son, Sean Wesley, whose curiosity is a constant reminder that eagerness to explore and the desire to understand are essential elements of the human spirit.

CONTENTS

PREFACE

This text was inspired by the need for instructional material for a graduate-level course on heat transfer with phase change taught on a yearly basis in mechanical engineering at the University of California at Berkeley. Several books published over the last 20 years have summarized the state-of-the-art in boiling and/or condensation phenomena. However, these texts invariably were less than ideal for instructional purposes because they focused almost entirely on the heat transfer and fluid mechanics aspects of boiling and condensation. They generally provided little, if any, treatment of the nonequilibrium thermodynamics and interfacial phenomena that frequently play central roles in such processes. In assembling this text, the goal was to provide a coherent presentation of the nonequilibrium thermodynamics and interfacial phenomena associated with vaporization and condensation processes, as well as the heat transfer and fluid flow mechanisms.

This book focuses on basic elements of condensation and vaporization processes. Those who work in the field know that the number of technical reports, papers, and books dealing with boiling and condensation processes is enormous. Coverage of all the work in these areas is clearly impossible within the limited space in a basic text. In the interest of conciseness, the tone of the presentation in this book is therefore illustrative rather than exhaustive. In most cases, the basic physical mechanisms associated with a particular phase-change phenomenon are described in detail, followed by a representative sample of the best models applicable to the circumstances of interest. Throughout the text, the importance of the basic phenomena to a wide variety of applications is discussed. However, space limitations precluded extensive discussion of special features that arise in some applications.

The sequence of material in this book was chosen to facilitate instruction at the advanced undergraduate or graduate level in mechanical or chemical engineering. The chapters in Part 1 of the book deal entirely with nonequilibrium thermodynamics and interfacial phenomena. If covered first, this material provides a useful foundation on which the later discussions of boiling and condensation phenomena can build. Part 2 covers boiling and condensation processes on the external surfaces of a body exposed to an extensive ambient. The material on internal flow boiling and condensation in Part 3 follows that in Part 2 because many of the concepts that apply to external condensation and boiling apply in a modified form to convective boiling or condensation in tubes.

Part 4 is a chapter that covers some additional special topics and applications. This material can be presented most efficiently after an understanding of the basic physics is attained from study of preceding chapters. A special effort has been made to incorporate material on the enhancement of boiling and condensation heat transfer, because engineers involved with such processes most often want to enhance the transport. The progressive flow of ideas provided by the book's structure should also make it useful to practicing engineers who wish to gain a further understanding of the thermophysics of vaporization and condensation processes through individual study.

As noted at the outset, this text evolved from material used to teach a graduate-level class in phase-change heat transfer at Berkeley. The author is indebted to the numerous students in that class who questioned and criticized the class notes that preceded this text. The author is also grateful to Professor John H. Lienhard, Professor Dennis O'Neal, and Professor Ralph Seban for their insightful comments on the early manuscript version of this text. An expression of appreciation is also due to the many investigators who have contributed to this area over the past 50 years. It is only through their combined efforts that a clear overview of this area is possible. Finally, the author wishes to express his thanks for the understanding and patience of his family during the many hours of work required to assemble the material in this text.

Van P. Carey
Berkeley, California

NOMENCLATURE

A	surface or cross-sectional area
A_f	fin area
A_o	tube open area
A_p	prime surface area
b	fin height
Bo	Bond number $[=g(\rho_l - \rho v)L_B^2/\sigma$ (where the length scale L_B depends on the circumstances of interest)]
	boiling number $(=q''/Gh_{lv})$
c_{pl}	liquid specific heat
c_{pv}	vapor specific heat
Co	convection number $\{=[(1 - x)/x]^{0.8}[\rho_v/\rho_l]^{0.5}\}$
d_d	bubble departure diameter
d_t	tube diameter
d_h	hydraulic diameter based on wetted perimeter
d_{hp}	hydraulic diameter based on heated perimeter
$(dP/dz)_{fr}$	frictional component of two-phase pressure gradient
$(dP/dz)_l$	pressure gradient for liquid flow alone through tube
$(dP/dz)_{lo}$	pressure gradient for entire flow as liquid through tube
$(dP/dz)_v$	pressure gradient for vapor flow alone through tube
D	tube diameter
D_C^*	binary diffusion coefficient for more volatile component
D_{AB}	binary diffusion coefficient for species A and B
E	mass fraction of liquid phase entrained in the core during annular flow
E''	rate of entrainment in mass of droplets per unit time per unit of wall area
E_{kin}	system kinetic energy
E_{pot}	system potential energy
f	bubble frequency
	friction factor

f_l	friction factor for liquid flowing alone in tube
f_v	friction factor for vapor flowing alone in tube
F	Helmholtz function $(= U - TS)$
	force
	Chen correlation parameter
F_{TD}	Taitel-Dukler flow regime parameter
Fr_{le}	Froude number $[= G^2/(\rho_l^2 gD)]$
g	gravitational acceleration
	specific Gibbs function
G	Gibbs function $(= H - TS)$
	mass flux through tube or channel
h	local heat transfer coefficient
$\bar{h}$	mean heat transfer coefficient
$\hat{h}$	specific enthalpy on per unit mass basis
h^*	mass transfer coefficient
h_l	heat transfer coefficient for the liquid phase flowing alone in the tube
h_{le}	heat transfer coefficient for entire flow as liquid
h_{lo}	heat transfer coefficient for entire flow as liquid
h_{lv}	latent heat of vaporization per unit mass
H_f	fin height
j_l	volume flux of liquid $[= G(1 - x)/\rho_l]$
j_v	volume flux of vapor $[Gx/\rho_v]$
J	flux of droplet or bubble embryos through size space
J^*	dimensionless droplet flux in size space
Ja	Jakob number $[= c_p \Delta T/h_{lv}$ (where the choices of c_p and ΔT depend on the circumstances of interest)]
k_B	Boltzmann constant $(= 1.3805 \times 10^{-23}$ J/K)
k_d	deposition coefficient in model of entrainment and deposition for annular flow
k_l	thermal conductivity of liquid
k_v	thermal conductivity of vapor
K_{TD}	Taitel-Dukler flow regime parameter
L	tube length
L_b	bubble or capillary length scale $\{= [\sigma/g(\rho_l - \rho_v)]^{1/2}\}$
L_f	fin length
m	mass of one molecule
$\dot{m}'$	mass flow rate of condensate in liquid film per unit width of surface
m''	mass flux
M	mass
$\bar{M}$	molecular weight
N_A	Avogadro's number $(= 6.02 \times 10^{26}$ molecules/kg mol)
N_l	number of liquid molecules per unit volume
N_n	number of embryos of n molecules at equilibrium per unit volume
P	pressure
P_c	critical pressure
P_l	ambient liquid pressure
$P_{pi}(T)$	saturation pressure of pure component i in mixture at temperature T

P_v	ambient vapor pressure
Pr_l	liquid Prandtl number
Pr_t	turbulent Prandtl number $(=\epsilon_M/\epsilon_H)$
Pr_v	vapor Prandtl number
q''	heat flux
q''_{cr}	critical heat flux
q''_{min}	minimum heat flux on pool boiling curve
q''_{mkc}	maximum heat flux limit dictated by kinetic theory for condensation
q''_{mkv}	maximum heat flux limit dictated by kinetic theory for vaporization
$\dot{q}$	total heat transfer rate
Q^*	dimensionless heat flux $\{=[4q''L/d_h(G/\rho_{\mathrm{in}})h_{lv}][(\rho_l - \rho_v)/\rho_v\rho_l]\}$
R	ideal gas constant on a per unit mass basis
	liquid jet radius
$\bar{R}$	universal gas constant $(=8.3144$ kJ/(kg mol K)
Re	Reynolds number
Re_F	film Reynolds number
Re_l	Reynolds number for liquid phase flowing alone $[=G(1-x)d_h/\mu_l]$
Re_{le}	Reynolds number for entire flow as liquid $(=Gd_h/\mu_l)$
Re_{lo}	Reynolds number for entire flow as liquid $(=Gd_h/\mu_l)$
Re_L	film Reynolds number $(=4\dot{m}'/\mu_l)$
Re_v	Reynolds number for vapor phase flowing alone $(=Gxd_h/\mu_v)$
s	specific entropy
	distance between fins in an offset fin matrix
S	entropy
	supersaturation ratio $[=(P_v)_{\mathrm{SSL}}/P_{\mathrm{sat}}(T_v)]$
	suppression factor in Chen correlation
	slip ratio $(=u_v/u_l)$
Sc	Schmidt number $(=\nu/D_{AB})$
Spl_s	spreading coefficient $[=-(\partial F/\partial A_s l)]$
St	Stanton number $(=h/Gc_p)$
Su	subcooling number $\{=[c_{pl}(T_{\mathrm{sat}} - T_{\mathrm{in}})/h_{lv}][(\rho_l - \rho_v)/\rho_v]\}$
T	temperature
T_c	critical temperature
T_i	interface temperature
T_{in}	fluid temperature at tube inlet
T_{sat}	saturation temperature
T_{TD}	Taitel-Dukler flow regime parameter
T_w	wall temperature
u	specific internal energy
	velocity component in the x direction
u_l	liquid mean downstream velocity in two-phase flow $[=G(1-x)/\rho_l(1-\alpha)]$
u_v	vapor mean downstream velocity in two-phase flow $(=Gx/\rho_v\alpha)$
U	internal energy
UA	overall conductance of a heat transfer device $(=\dot{q}$ divided by the driving temperature difference)
v	specific volume

	velocity component in the y direction
v_c	critical volume
V	volume
	velocity
w	velocity component in the z direction
$\bar{w}$	mean distance between fins
W	mass flow rate
x	coordinate (downstream coordinate for external flows)
	mass quality
x_a	actual ratio of vapor mass flow rate to total mass flow rate
x_{crit}	dryout quality
x_e	equilibrium quality
X	Martinelli parameter $\{=[(dP/dz)_l/(dP/dz)_v]^{1/2}\}$
X_l	mole fraction of more volatile component in liquid phase
X_{li}	mole fraction of component i in liquid phase
	mole fraction of more volatile component at the liquid–vapor interface
X_{tt}	Martinelli parameter for turbulent-turbulent flow
X_v	mole fraction of more volatile component in vapor phase
X_{vi}	mole fraction of more volatile component in vapor phase at the liquid–vapor interface
y	coordinate, surface normal coordinate for external flows
	twisted-tape insert ratio of length for 180° twist to tube inside diameter
y^+	dimensionless y coordinate $(=y\sqrt{\tau_0/\rho_l}/\nu_l)$
z	coordinate (downstream coordinate for tube flows)
α	wave number
	void fraction
α_c	critical wave number
α_T	thermal diffusivity $(=k/\rho c_p)$
α_{Tl}	thermal diffusivity of liquid
α_{Tv}	thermal diffusivity of vapor
β	frequency
β_f	volume fraction of liquid flowing in liquid film on tube wall
β_{max}	frequency of most rapidly growing disturbance
γ	multiplier in Baroczy correlation
δ	film thickness
δ^+	dimensionless film thickness $(=\delta\sqrt{\tau_0/\rho_l}/\nu_l)$
δ_f	fin thickness
δ_t	thermal boundary-layer thickness
ΔT_{vl}	temperature difference across liquid–vapor interface
ϵ	emissivity
ϵ_H	eddy diffusivity of heat for turbulent flow
ϵ_M	eddy diffusivity of momentum for turbulent flow
η_f	fin efficiency
θ	liquid contact angle
	angular coordinate
λ_c	critical wavelength
λ_D	most dangerous wavelength

μ	absolute viscosity
	chemical potential
μ_l	liquid viscosity
	liquid chemical potential
μ_v	vapor viscosity
	vapor chemical potential
ν_l	liquid kinematic viscosity
ν_v	vapor kinematic viscosity
ρ_l	liquid density
ρ_v	vapor density
σ	interfacial tension
$\hat{\sigma}$	accommodation coefficient
σ_{SB}	Stefan-Boltzmann constant ($=5.67 \times 10^{-8}$ W/m^2 K^4)
τ	shear stress
τ_i	shear stress at interface
τ_0	shear stress at wall
τ_w	shear stress at wall
ϕ_l	two-phase multiplier $\{=[(dP/dz)_{fr}/(dP/dz)_l]^{1/2}\}$
ϕ_{lo}	two-phase multiplier $\{=[(dP/dz)_{fr}/(dP/dz)_v]^{1/2}\}$
ϕ_v	two-phase multiplier $\{=[(dP/dz)_{fr}/(dP/dz)_v]^{1/2}\}$
Ω	vorticity
	angle between tube axis and horizontal

Subscripts

a	actual value
b	bulk
bp	bubble point
c	properties evaluated at the critical point
dp	dew point
ex	exit conditions
f	film
	fin
i	interface
in	inlet conditions
l	liquid
	corresponding to the liquid phase flowing alone
le	corresponding to the entire flow as liquid
	corresponding to liquid flow in equivalent separate cylinder
lo	corresponding to the entire flow as liquid
sat	corresponding to saturation conditions
SSL	supersaturation limit
v	vapor
	corresponding to the vapor phase flowing alone
ve	corresponding to vapor flow in equivalent separate cylinder
w	wall value
∞	far ambient conditions
0	wall value

PART ONE

THERMODYNAMIC AND MECHANICAL ASPECTS OF INTERFACIAL PHENOMENA AND PHASE TRANSITIONS

CHAPTER
ONE

INTRODUCTORY CONCEPTS

INTRODUCTION

Liquid–vapor phase-change processes play a vital role in many technological applications. The virtually isothermal heat transfer associated with boiling and condensation processes makes their inclusion in power and refrigeration cycles highly advantageous from a thermodynamic efficiency viewpoint. Liquid–vapor phase-change processes are also encountered in petroleum and chemical processing, liquefaction of nitrogen and other gases at cryogenic temperatures, and during evaporation or precipitation of water in the earth's atmosphere.

In addition, the high heat transfer coefficients associated with boiling and condensation have made the use of these processes increasingly attractive in the thermal control of compact devices that have high heat dissipation rates. Applications of this type include the use of boiling heat transfer to cool electronic components in mainframe computers and the use of compact evaporators and condensers for thermal control of aircraft avionics and spacecraft environments. Liquid–vapor phase-change processes are also of critical importance to nuclear power plant design, both because they are important in normal operating circumstances and because they dominate many of the accident scenarios that are studied in detail as part of design evaluation.

The heat transfer and fluid flow processes associated with liquid–vapor phase-change phenomena are typically among the more complex transport circumstances encountered in engineering applications. These processes may have all the complexity of single-phase convective transport (nonlinearities, transition to turbulence, three-dimensional or time-varying behavior) plus additional elements resulting from motion of the interface, nonequilibrium effects, or other complex dynamic interactions between the phases. Due to the highly complex nature of these processes, development of methods to predict the associated heat and mass transfer has often proved to be a formidable task. Nevertheless, the research efforts of numerous scientists over several decades have provided a fairly clear understanding of many aspects of vaporization and condensation processes in power and refrigeration systems. On the other hand, some elements of vaporization and condensation phenomena are not well understood, and research in these areas continues.

Before proceeding with the discussion of technical aspects of vaporization and condensation processes, a few comments about information sources and nomenclature are in order. There is a vast quantity of published information on liquid–vapor phase-change phenomena. Because such processes occur in a wide variety of applications, and because the thermodynamic, fluid mechanics, and heat transfer aspects of these processes appeal to different groups of investigators, technical papers on various aspects of vaporization and condensation processes are found in a number of journals, serial publications, and conference proceedings. English-language publications that frequently contain information on these topics include:

International Journal of Heat and Mass Transfer, published monthly by Pergamon Press

Journal of Heat Transfer, published quarterly by the American Society of Mechanical Engineers

AIChE Journal, published bimonthly by the American Institute of Chemical Engineers

Journal of Fluid Mechanics, published monthly by Cambridge University Press

International Journal of Multiphase Flow, published bimonthly by Pergamon Press

Heat Transfer Engineering, published quarterly by Hemisphere Publishing Corporation

Numerical Heat Transfer, Parts A and B, published quarterly by Hemisphere Publishing Corporation

ASHRAE Transactions, published annually by the American Society of Heating, Refrigerating and Air-Conditioning Engineers

Advances in Heat Transfer, currently published annually by Academic Press

Advances in Chemical Engineering, published at variable intervals by Academic Press

Experimental Heat Transfer, published quarterly by Hemisphere Publishing Corporation

Physiochemical Hydrodynamics, published bimonthly by Pergamon Press

Journal of Thermophysics and Heat Transfer, published monthly by the American Institute of Aeronautics and Astronautics

Experimental Thermal and Fluid Science, published quarterly by Elsevier Science Publishers

Proceedings of the International Heat Transfer Conferences, published every four years by Hemisphere Publishing Company

Journal of Physical Chemistry, published monthly by the American Chemical Society

Multiphase Science and Technology, published annually by Hemisphere Publishing Company

Nuclear Engineering and Design, published bimonthly by Elsevier Science Publishers

Nuclear Science and Engineering, published monthly by the American Nuclear Society

In addition to the sources listed above, useful information on liquid–vapor phase-change processes may also be obtained from government agency reports (e.g., those of the National Bureau of Standards or the Nuclear Regulatory Commission) and the reports of research laboratories (e.g., Sandia National Laboratory, Argonne National Laboratory, and the Electric Power Research Institute).

The physical diversity of the mechanisms involved in vaporization and condensation processes also makes selection of a consistent nomenclature a difficult problem. For example, symbols traditionally used for properties in thermodynamic analysis are commonly used to denote other physical quantities in fluid mechanics analysis or in analysis of heat exchanger performance. To avoid confusion, every effort has been made to make the definition of variables clear at the location in the text where they are introduced. In addition, a listing of the nomenclature for the text is provided in the back of the book to provide a quick means of checking variable definitions.

1.2 REVIEW OF FUNDAMENTAL THERMODYNAMIC PRINCIPLES

Thermodynamic considerations play a central role in the analysis of interfacial phenomena associated with phase-change processes. Consequently, a solid understanding of classical equilibrium thermodynamics is a necessary foundation for concepts presented in this text. The presentation of material presumes that the reader has been exposed to the basic principles of thermodynamics. A few of the main results will be mentioned here, which are particularly important to concepts developed later.

Specifically, the reader is assumed to be familiar with the first and second laws for a closed system,

$$\delta Q = dU + \delta W + dE_{\text{pot}} + dE_{\text{kin}} \tag{1.1}$$

$$dS \geq \frac{\delta Q}{T} \tag{1.2}$$

and the various versions of these equations for different system constraints. The reader is also assumed to be acquainted with the thermodynamic properties,

P pressure (kPa)
T temperature (°C or K)
V volume (m^3)
U internal energy (kJ)
H enthalpy, $U + PV$ (kJ)
S entropy (kJ/K)

In addition to these basic properties, the following additional properties will be useful in later sections:

G Gibbs function, $= H - TS$ (kJ)
F Helmholtz function, $= U - TS$ (kJ)
$\tilde{\mu}_i$ chemical potential, $= (\partial G/dN_i)_{P,T,N_{j\neq i}}$ (kJ/mol)

For the extensive properties defined above, V, U, H, S, G, and F, lowercase variables will be used to denote the corresponding specific properties on a per-unit mass basis (v, u, $\tilde{h}$, s, g, and f, respectively). Note also that G and F are sometimes referred to as the *Gibbs free energy* and the *Helmholtz free energy*, respectively. $\tilde{\mu}_i$ is the chemical potential associated with the ith component in a multicomponent mixture in which N_i is the number of moles of the ith component present. Because for a pure substance with a molecular weight $\overline{M}$ this implies that

$$\mu = \frac{\tilde{\mu}}{\overline{M}} = \frac{[\partial(\overline{M}g)/\partial N]_{T,P}}{\overline{M}} = g(T,P) \tag{1.3}$$

the chemical potential on a mass basis for a pure substance is equal to the specific Gibbs function, and they may be used interchangeably.

Equations of state will be seen, in later sections, to provide information that is essential to the understanding and analysis of phase transformations. The reader should understand the representation of equation-of-state data in tables and charts, and should be familiar with some of the commonly used analytical equations of state. These include the ideal gas law,

$$PV = \left(\frac{M}{\overline{M}}\right)\overline{R}T \tag{1.4}$$

where M is the mass of gas present, $\overline{M}$ is the molecular weight of the gas, and $\overline{R}$ is the universal gas constant. More complicated relations, such as the van der Waals equation of state,

$$P = \frac{(\overline{R}/\overline{M})T}{v - b} - \frac{a}{v^2} \tag{1.5a}$$

where

$$a = \frac{27(\overline{R}/\overline{M})^2 T_c^2}{64P_c} \tag{1.5b}$$

$$b = \frac{(\overline{R}/\overline{M})T_c}{8P_c} \tag{1.5c}$$

(and T_c and P_c are the critical temperature and pressure) are also sometimes useful in the analysis of phase-change processes.

Two additional results of classical thermodynamics will also be noted here, without proof, for future reference. The first is the *Clausius-Clapeyron equation*,

$$\left(\frac{dP}{dT}\right)_{\text{sat}} = \frac{s_{lv}}{v_{lv}} = \frac{h_{lv}}{Tv_{lv}} \tag{1.6}$$

The second result noted here is the *Gibbs-Duhem equation* for a multicomponent system with n_c components,

$$\sum_{i=1}^{n_c} x_i \, d\mu_i = -s \, dT + v \, dP \tag{1.7}$$

where x_i is the mass fraction of the ith component. For a pure substance, this reduces to

$$d\mu = dg = -s \, dT + v \, dP \tag{1.8}$$

Extensive discussions of the fundamental thermodynamic concepts described above are provided in several currently available texts [1.1–1.4]. Readers who are not familiar with these concepts are urged to consult one or more of these references.

1.3 CONDITIONS FOR EQUILIBRIUM

When a system or a portion of a system undergoes a phase change, the process is necessarily associated with a departure from equilibrium. However, because it is assumed that the system is moving toward an equilibrium state, the conditions corresponding to the equilibrium state are of considerable practical importance. Additional analytical tools for nonequilibrium circumstances will be developed in subsequent sections.

We first consider the conditions for equilibrium in a mixture of simple compressible substances in a thermally insulated, rigid container. For simple compressible substances, the only work mode is PdV work.

$$\delta W = PdV \tag{1.9}$$

Any spontaneous changes in the state of the mixture must satisfy the second law (1.2) and be consistent with constraints imposed by the container, which are

$$dV = 0 \quad \text{and} \quad dU = 0 \tag{1.10}$$

Combining the first and second laws, Eqs. (1.1) and (1.2), and using Eqs. (1.9) and (1.10), it is easily shown that spontaneous changes can occur only if

$$dS \geq 0 \tag{1.11}$$

Hence, all spontaneous changes increase the entropy of the system, and the equilibrium for a mixture of simple compressible substances held at constant volume and energy corresponds to conditions that maximize the entropy within the system constraints.

Alternate statements of the second-law extremum condition can be similarly developed for systems with different imposed constraints. Several of these alternate statements are summarized below:

System constraints	*Spontaneous internal processes correspond to:*	
T, V constant	$dF \leq 0$	(1.12)
P, T constant	$dG \leq 0$	(1.13)
S, V constant	$dU \leq 0$	(1.14)

Note that if a system is held at fixed pressure P_0 and temperature T_0, then G is equal to the *availability* Ψ, defined as

$$\Psi = U + P_0V - T_0S \tag{1.15}$$

Because $G = \Psi$, the above results imply that for a system with fixed P and T, a spontaneous process always decreases the availability of the system. Consequently, equilibrium corresponds to $d\Psi = 0$ for such a system, and stable equilibrium must correspond to a minimum value of Ψ. A further discussion of the thermodynamic significance of the availability function can be found in references [1.2] and [1.4.].

Thus, the second law requires that for a mixture of simple compressible substances, the equilibrium state is one that maximizes S or minimizes U, F, or G, depending on the imposed system constraints. A series of simple examples that demonstrate how these considerations come into play is shown in Fig. 1.1. We consider a system containing only pure water with both liquid and vapor phases present. In the usual manner, we designate the ratio of mass of water vapor to the total system mass as the quality, x.

In Fig. 1.1*a*, the volume and internal energy are assumed to be fixed. For this system, the amount of liquid and vapor present (i.e., the quality) at equilibrium will be that which maximizes the entropy of the system over all the possible x values, consistent with conservation of mass and the fixed values of V and U.

Figure 1.1*b* shows a similar system, except that in this case the volume and temperature of the system are fixed. With these constraints, the quality at equilibrium will be that which minimizes the Helmholtz free energy of the system over the possible x values, consistent with conservation of mass and the fixed values of V and T.

Before considering Fig. 1.1*c*, some digression is in order. For a system that contains n_c components and n_p phases, the *Gibbs phase rule* states that the number of independently variable intensive properties n_i is given by

$$n_i = n_c - n_p + 2 \tag{1.16}$$

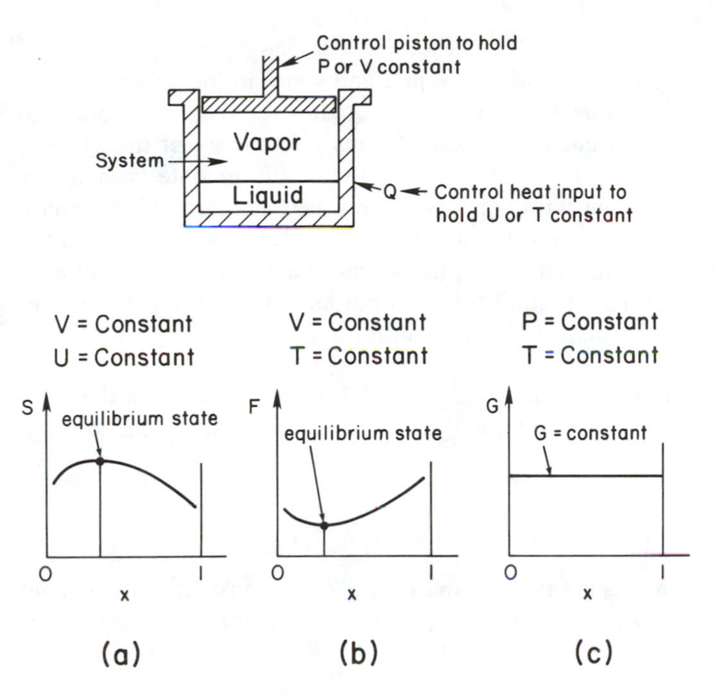

Figure 1.1 Comparison of equilibrium-state conditions for systems with different constraints.

The Gibbs phase rule is stated here without proof. Discussions of the reasons for its validity can be found in references [1.1–1.4].

For the systems shown in Fig. 1.1, $n_c = 1$ and $n_p = 2$, so $n_i = 1$ and there can be only one independently variable intensive parameter. In Fig. 1.1*c*, however, we have specified two intensive parameters, P and T. The Gibbs phase rule indicates that specification of both P and T for this system is redundant (i.e., they are not independent).

It is also known from basic thermodynamics that the number of independent variable properties that must be fixed to specify the state of a system is equal to the number of reversible work modes plus one. This is sometimes formalized in what is called the *state postulate* (see reference [1.1]). Because the systems in Fig. 1.1 contain a simple compressible substance, the only reversible work mode is PdV work, and two independent properties must be specified to fix the state of the system. But in Fig. 1.1*c*, P and T are not independent, as we have shown, and specifying P and T does not define an equilibrium state. How does this relate to the extremum principles described above?

To answer this question, we first consider the vaporization of a unit mass of saturated liquid at constant pressure (and temperature). For constant P and T, as shown above, the second law reduces to Eq. (1.13). Because the heat of vaporization is added reversibly at constant temperature, the equal sign applies in Eq. (1.13), resulting in

$$g_l = g_v \tag{1.17}$$

Thus, the specific Gibbs free energy of the liquid and vapor in the system in Fig. 1.1*c* are exactly the same. Hence, changing the quality (converting vapor into liquid or vice versa) does not change the total Gibbs free energy of the system, and no minimum in G exists with respect to x. An equilibrium state is therefore not defined, which is consistent with conclusions reached using the Gibbs phase rule and the state postulate. It should be noted that Eq. (1.13) still applies to the system shown in Fig. 1.1*c*. In fact, it applies at any value of quality and as a result, it alone does not define an equilibrium condition. An additional independent parameter must be specified to fix the equilibrium state.

Example 1.1 Determine the change in availability associated with the slow conversion of 1 kg of a pure saturated vapor to 1 kg of saturated liquid at constant pressure.

By definition,

$$\Delta\Psi = (U_2 + P_0V_2 - T_0S_2) - (U_1 + P_0V_1 - T_0S_1)$$

For condensation of saturated vapor, P and T are constant, so it is convenient to take $P_0 = P_1 = P_2$, $T_0 = T_1$. The relation for $\Delta\Psi$ can then be written as

$$\Delta\Psi = (U_2 + P_2V_2) - (U_1 + P_1V_1) - T_1(S_2 - S_1)$$
$$= H_2 - H_1 - T_1(S_2 - S_1)$$

From classical thermodynamics, the following relation is known to hold among the properties T, S, U, P, and V for a pure substance:

$$T\,dS = dU + P\,dV$$

Since P and T are constant during the vaporization process, this equation can be integrated to obtain

$$T_1(S_2 - S_1) = U_2 - U_1 + P_1(V_2 - V_1)$$

or, equivalently,

$$T_1(S_2 - S_1) = H_2 - H_1$$

Substituting this result into the last equation for $\Delta\Psi$ above yields

$$\Delta\Psi = 0$$

Thus the isothermal condensation of a pure vapor results in zero decrease in system availability. It can be similarly shown that vaporization of a saturated pure liquid also results in $\Delta\Psi = 0$.

1.4 PROPERTIES AT EQUILIBRIUM

To explore the constraints on system properties at equilibrium, we consider the closed system shown in Fig. 1.2*a*, which contains two phases I and II of a mixture

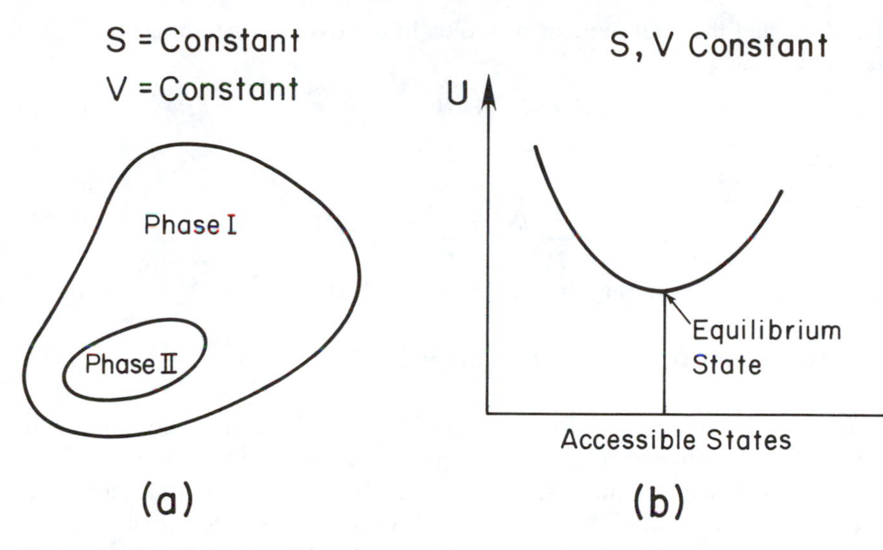

Figure 1.2 Equilibrium conditions in a two-phase system at fixed S and V.

of n_c pure simple compressible substances. For convenience, it is postulated that the system is constrained so that the overall entropy and volume remain constant regardless of internal changes. It was shown in the previous section that for a system constrained so that S and V are constant, the equilibrium state corresponds to conditions that minimize the internal energy over the range of conditions accessible to the system. This is shown schematically in Fig. 1.2*b*. All spontaneous processes within the system must satisfy Eq. (1.14): $dU \leq 0$ (i.e., they must result in a decrease in U).

The overall internal energy U for the system in Fig. 1.2*a* is equal to the sum of the contributions from the two phases I and II. Thus any differential change in U is given by

$$dU = dU^{\mathrm{I}} + dU^{\mathrm{II}} \tag{1.18}$$

At equilibrium, we must have $dU = 0$, so the equation

$$dU^{\mathrm{I}} + dU^{\mathrm{II}} = 0 \tag{1.19}$$

relates differential changes in U for the individual phases that preserve the overall equilibrium of the system. We further assume here that U for either of the phases can be determined as a function of S, V, and the number of moles of each component present, $N_i = M_i/\overline{M}_i$ (i.e., $U = U[S, V, N_i]$). Expanding the relation for U in a Taylor series for each of the two phases and substituting the resulting first-order relations for dU^I and dU^{II} into Eq. (1.19), we obtain

$$U_V^{\mathrm{I}}\, dV^{\mathrm{I}} + U_S^{\mathrm{I}}\, dS^{\mathrm{I}} + \sum_{i=1}^{n_c} U_{N_i}^{\mathrm{I}}\, dN_i^{\mathrm{I}} + U_V^{\mathrm{II}}\, dV^{\mathrm{II}} + U_S^{\mathrm{II}}\, dS^{\mathrm{II}} + \sum_{i=1}^{n_c} U_{N_i}^{\mathrm{II}}\, dN_i^{\mathrm{II}} = 0 \tag{1.20}$$

In the above equation and subsequent equations in this section, subscripts denote partial derivatives with respect to the subscripted variable; i.e., $U_V = \partial U/\partial V$, etc.

Because V, S, and the total number of moles in the overall system are fixed, we can also write that

$$dV = dV^{\mathrm{I}} + dV^{\mathrm{II}} = 0 \tag{1.21}$$

$$dS = dS^{\mathrm{I}} + dS^{\mathrm{II}} = 0 \tag{1.22}$$

$$dN_i = \sum_{i=1}^{n_c} dN_i^{\mathrm{I}} + \sum_{i=1}^{n_c} dN_i^{\mathrm{II}} = 0 \tag{1.23}$$

Substituting Eqs. (1.21) through (1.23) into Eq. (1.20) yields

$$(U_V^{\mathrm{I}} - U_V^{\mathrm{II}})\, dV^{\mathrm{I}} + (U_S^{\mathrm{I}} - U_S^{\mathrm{II}})\, dS^{\mathrm{I}} + \sum_{i=1}^{n_c} (U_{N_i}^{\mathrm{I}} - U_{N_i}^{\mathrm{II}})\, dN_i^{\mathrm{I}} = 0 \tag{1.24}$$

Because the two phases are actually composed of molecules that are colliding with each other, we expect that even at equilibrium the phases I and II will be subjected to perturbations of their volume, entropy, and number of moles (mass). However, we have shown that for equilibrium to exist in spite of these perturbations, Eq. (1.24) must be satisfied. If Eq. (1.24) must hold for all possible perturbations in V, S, and N_i, then it is necessary that

$$U_V^{\mathrm{I}} = U_V^{\mathrm{II}} \tag{1.25a}$$

$$U_S^{\mathrm{I}} = U_S^{\mathrm{II}} \tag{1.25b}$$

$$U_{N_i}^{\mathrm{I}} = U_{N_i}^{\mathrm{II}} \tag{1.25c}$$

But it can be shown using the Maxwell relations (see reference [1.2]) that

$$U_V = \left(\frac{\partial U}{\partial V}\right)_{S,N_i} = -P \tag{1.26}$$

$$U_S = \left(\frac{\partial U}{\partial S}\right)_{V,N_i} = T \tag{1.27}$$

$$U_{N_i} = \left(\frac{\partial U}{\partial N_i}\right)_{S,V} = \tilde{\mu}_i \tag{1.28}$$

Thus, using Eqs. (1.25*a*) through (1.28), it can be seen that at equilibrium we must have

$$P^{\mathrm{I}} = P^{\mathrm{II}} \tag{1.29}$$

$$T^{\mathrm{I}} = T^{\mathrm{II}} \tag{1.30}$$

$$\tilde{\mu}_i^{\mathrm{I}} = \tilde{\mu}_i^{\mathrm{II}} \quad \text{or} \quad \mu_i^{\mathrm{I}} = \mu_i^{\mathrm{II}} \tag{1.31}$$

Hence, necessary conditions for equilibrium are that the pressure, temperature, and chemical potential of each component must be the same in both phases of a pure substance in a closed system. When more than two phases are present, it can easily be shown that it is necessary that P, T, and μ_i be the same in all phases at equilibrium.

It should be noted that in the above discussion we have ignored any effects of interfacial tension at the interface between phases I and II. In many real systems, interfacial tension strongly effects the equilibrium conditions. In later chapters we will consider interfacial effects in more detail and develop concepts needed to understand and analyze their effects on vaporization and condensation processes.

1.5 REVIEW OF FUNDAMENTAL TRANSPORT PHENOMENA

Heat, mass, and momentum transfer within each of the individual phases generally play important roles in liquid–vapor phase-change phenomena. Consequently, knowledge of single-phase transport phenomena is vital to any effort to understand phase-change processes. The presentation of material in later chapters presumes that the reader has an understanding of fluid mechanics and heat transfer at least equal to that of a student who has completed a B.S. degree in mechanical or chemical engineering. A few of the fundamental aspects of single-phase transport, which are particularly relevant to phase-change processes considered later, will be briefly described here.

The reader is assumed to be familiar with Fourier's equation of heat conduction. For one-dimensional transfer of heat in a motionless homogeneous medium, Fourier's equation requires that

$$q'' = -k\left(\frac{dT}{dy}\right) \tag{1.32}$$

where y is the coordinate in the direction of heat flow, q'' is the heat flux in the y direction, and k is a property of the medium called the *thermal conductivity*. A description of the extension of Fourier's equation to more than one dimension, and its application to multidimensional and transient conduction processes, is described in most basic heat transfer texts and in references [1.5–1.7].

It is also assumed that the reader is familiar with what is usually called Fick's equation for mass diffusion in a binary system,

$$m''_A = -D_{AB}\left(\frac{dx_A}{dy}\right) \tag{1.33}$$

and the linear relation between shear stress and velocity gradient postulated for Newtonian fluids,

$$\tau_{yz} = -\mu\left(\frac{dw}{dy}\right) \tag{1.34}$$

Note that Eqs. (1.33) and (1.34) are the forms of these relations for one-dimensional transport of mass and momentum (in the y direction), respectively. In Eq. (1.33), m''_A is the mass flux of species A, x_A is the mass fraction of species A, and D_{AB} is called the *binary diffusion coefficient* of species A in a mixture of species

A and B. In Eq. (1.34), w is the flow velocity in the z direction (normal to y), τ_{yz} is the shear stress in the z direction, and μ is the absolute viscosity of the fluid.

Extension of Fick's equation (1.33) to more than one dimension and its application to steady and transient diffusion problems is analogous to similar treatments of Fourier's equation. For a detailed discussion of the analytical treatment of diffusion processes, see references [1.8–1.11]. Extension and application of the shear stress relation for Newtonian fluids to multidimensional and time-varying flows is described in a number of books on fluid mechanics and convective transport (see, e.g., references [1.8] and [1.11–1.13]).

The treatment of phase-change processes in later chapters also presumes that the reader is familiar with the basic equations for convective transport that reflect conservation of mass, momentum, energy, and species. In most vaporization and condensation processes, compressibility effects and viscous dissipation are small. Neglecting these effects and assuming constant fluid properties, for Cartesian coordinates with the z axis vertically up, the conservation equations for laminar convective transport become

$$\frac{\partial u}{\partial x} + \frac{\partial v}{\partial y} + \frac{\partial w}{\partial z} = 0 \tag{1.35}$$

$$\rho \frac{Du}{Dt} = - \frac{\partial P}{\partial x} + \mu \nabla^2 u \tag{1.36}$$

$$\rho \frac{Dv}{Dt} = - \frac{\partial P}{\partial y} + \mu \partial\nabla^2 v \tag{1.37}$$

$$\rho \frac{Dw}{Dt} = -\rho g - \frac{\partial P}{\partial z} + \mu \nabla^2 w \tag{1.38}$$

$$\rho c_p \frac{DT}{Dt} = k \nabla^2 T \tag{1.39}$$

$$\frac{Dx_A}{Dt} = D_A \nabla^2 x_A \tag{1.40}$$

where

$$\frac{D}{Dt} = \frac{\partial}{\partial t} + u \frac{\partial}{\partial x} + v \frac{\partial}{\partial y} + w \frac{\partial}{\partial z} \tag{1.41}$$

and

$$\nabla^2 = \frac{\partial^2}{\partial x^2} + \frac{\partial^2}{\partial y^2} + \frac{\partial^2}{\partial z^2} \tag{1.42}$$

In the above Eqs. (1.35) through (1.41), u, v, and w are velocities in the x, y, and z directions, respectively, and t is time. The variables μ and k are the viscosity and thermal conductivity defined previously, and ρ, c_p, x_A, and D_A are the fluid density, fluid specific heat, mass fraction of species A, and the diffusivity of species A, respectively. Equation (1.35) is the continuity or mass conservation equation, and Eqs. (1.36) through (1.38) are the momentum balance equations in the three coordinate directions. Conservation of energy and species A are enforced by Eqs. (1.39) and (1.40), respectively. Note that in deriving these convective transport equations, the fluid has been assumed to be Newtonian, and the Fourier and Fick equations for molecular diffusion of thermal energy and species A have been assumed to be valid.

As we shall later see, in some vaporization and condensation circumstances the flow is two-dimensional and conforms to what are known as the *boundary-layer approximations*. As an example, we consider the two-dimensional flow of a liquid film that is very thin in the y direction. The flow is primarily downward in the $-z$ direction with a v component of velocity that is generally small compared to w. Consistent with the usual boundary-layer approximations, downstream diffusion is neglected compared to convection and cross-stream diffusion, v is assumed to be much smaller than w, and the pressure is taken to be constant in the cross-stream direction y. The resulting two-dimensional boundary-layer equations are

$$\frac{\partial v}{\partial y} + \frac{\partial w}{\partial z} = 0 \tag{1.43}$$

$$\rho \frac{Dw}{Dt} = -\rho g - \frac{\partial P}{dz} + \mu \frac{\partial^2 w}{\partial y^2} \tag{1.44}$$

$$\rho c_p \frac{DT}{Dt} = k \frac{\partial^2 T}{\partial y^2} \tag{1.45}$$

$$\frac{Dx_A}{Dt} = D_A \frac{\partial^2 x_A}{\partial y^2} \tag{1.46}$$

where now D/Dt is given as

$$\frac{D}{Dt} = \frac{\partial}{\partial t} + v \frac{\partial}{\partial y} + w \frac{\partial}{\partial z} \tag{1.47}$$

If we further limit considerations to steady flow circumstances, Eqs. (1.43) through (1.47) reduce to

$$\frac{\partial v}{\partial y} + \frac{\partial w}{\partial z} = 0 \tag{1.48}$$

$$\rho\left(v \frac{\partial w}{\partial y} + w \frac{\partial w}{\partial z}\right) = -\rho g - \frac{\partial P}{dz} + \mu \frac{\partial^2 w}{dy^2} \tag{1.49}$$

$$\rho c_p\left(v \frac{\partial T}{dy} + w \frac{\partial T}{dz}\right) = k \frac{\partial^2 T}{dy^2} \tag{1.50}$$

$$v \frac{\partial x_A}{\partial y} + w \frac{\partial x_A}{\partial z} = D_A \frac{\partial^2 x_A}{\partial y^2} \tag{1.51}$$

In later chapters we will see that the steady boundary-layer equations (1.48) through (1.51), with appropriate boundary conditions, are a fairly accurate model of some film boiling and film condensation processes. Slightly different forms of the boundary-layer equations are appropriate for non-Cartesian geometries. A full discussion of these alternative forms can be found in reference [1.12].

The treatment of vaporization and condensation processes also assumes that the reader is familiar with the basic principles of radiation heat transfer. A surface is said to behave as a *blackbody* if it absorbs all the radiation incident upon it. A blackbody is also a perfect radiator that emits radiant energy from its surface at a rate q_r, given by

$$q_r = A_1 \sigma_{SB} T_1^4 \tag{1.52}$$

where A_1 is the surface area, T_1 is the absolute temperature of the body, and σ_{SB} is a constant, known as the *Stefan-Boltzmann constant,* whose value is

$$\sigma_{SB} = 5.67 \times 10^{-8}\ \mathrm{W/m^2}\ K^4 = 0.1714 \times 10^{-8}\ \mathrm{Btu/hr\ ft^2}\ R^4 \tag{1.53}$$

Equation (1.52) is sometimes referred to as the *Stefan-Boltzmann Law.*

If the blackbody radiates to an enclosure, as shown in Fig. 1.3, which is also a blackbody, the net rate of radiation heat transfer is given by

$$q_{\text{net}} = A_1 \sigma_{SB}(T_1^4 - T_2^4) \tag{1.54}$$

Of course, real bodies are not ideal radiators, but emit radiation at a rate lower than that of a blackbody. If, at each wavelength, a body emits a constant fraction ϵ_1 of the blackbody emission at the same temperature, it is said to be a *graybody.* The fraction ϵ_1 is called the *emittance* of the graybody. If the body inside the black enclosure in Fig. 1.3 were a graybody with an emittance of ϵ_1, the net rate

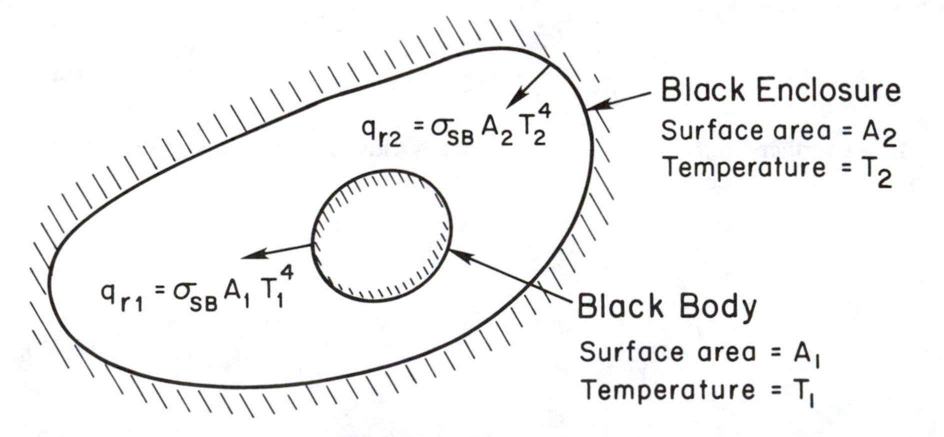

Figure 1.3 Radiative exchange between a blackbody and a surrounding black enclosure.

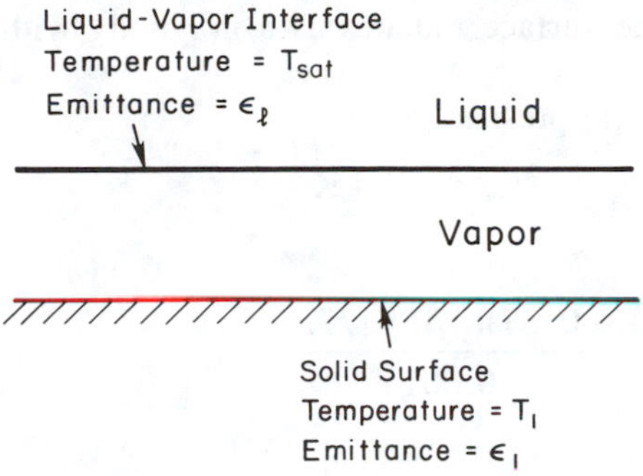

Liquid-Vapor Interface
Temperature = T_{sat}
Emittance = ϵ_ℓ

Liquid

Vapor

Solid Surface
Temperature = T_1
Emittance = ϵ_1

Figure 1.4 Radiative exchange across a vapor layer between a solid surface and a liquid surface.

of heat transfer between the graybody and the black enclosure is

$$q_{\text{net}} = A_1 \epsilon_1 \sigma_{SB} (T_1^4 - T_2^4) \tag{1.55}$$

A radiation heat transfer interaction that affects some film boiling processes is shown schematically in Fig. 1.4. A thin vapor layer exists between a liquid–vapor interface at the saturation temperature for the local pressure $T_{\text{sat}}(P)$ and a heated surface at a temperature $T_1 > T_{\text{sat}}$. If the interface and the solid surface are treated as infinite parallel gray surfaces, with emittances ϵ_l and ϵ_1, respectively, and the vapor is transparent to the radiation, it can be shown that the net radiation heat flux to the interface q''_{net} is given by

$$q''_{\text{net}} = \frac{\sigma_{SB}(T_1^4 - T_{\text{sat}}^4)}{\epsilon_1^{-1} + \epsilon_l^{-1} - 1} \tag{1.56}$$

The above relation will provide a reasonable estimate of radiation heat transfer during film boiling if the vapor does not significantly absorb or emit long-wavelength thermal radiation. If, as in the case of steam, the vapor does have absorption bands at wavelengths between 0.1 and 100 μm, it may be necessary to account for radiation interaction with the vapor to accurately predict the radiation heat transfer effect. A detailed discussion of radiative transport in absorbing and emitting gases can be found in references [1.14–1.17].

In addition to film boiling processes like that described above, radiation may also play an important role in the final stages of convective boiling in tubes. At very high mass quality, the wall of the tube may be dry while a mixture of liquid droplets and vapor flows through the tube. Radiative transfer of heat from the hot walls to the liquid droplets can be an important mechanism in the vaporization processes for such circumstances. Further discussion of vaporization processes under these conditions is provided in Chapter Twelve.

Example 1.2 During film boiling of liquid nitrogen at atmospheric pressure on a flat surface held at 270 K, the vapor film thickness δ is 1.0 mm. Assuming that heat is transferred across the vapor film by conduction and radiation, estimate the heat flux due to each mechanism. Assume that the interface radiates as a blackbody and the surface radiates as a graybody with an emmitance of 0.8.

For saturated nitrogen at atmospheric pressure,

$$k_v = 0.0137 \text{ W/m}^2 \text{ K} \qquad T_{\text{sat}} = 77.4 \text{ K}$$

For conduction,

$$q''_{\text{cond}} = \frac{k_v(T_w - T_{\text{sat}})}{\delta} = \frac{(0.0137)(270 - 77.4)}{0.001}$$
$$= 2639 \text{ W/m}^2$$

For radiation,

$$q''_{\text{rad}} = \frac{\sigma_{SB}(T_w^4 - T_{\text{sat}}^4)}{\epsilon_w^{-1} + \epsilon_l^{-1} - 1}$$
$$= \frac{(5.67 \times 10^{-8})[(270)^4 - (77.4)^4]}{(0.8)^{-1} - 1 + 1}$$
$$= 239 \text{ W/m}^2$$

Thus the radiation contribution is an order of magnitude lower than that due to conduction.

1.6 A MOLECULAR PERSPECTIVE ON LIQUID–VAPOR TRANSITIONS

Before considering further the macroscopic aspects of vaporization and condensation processes, it is useful to examine the nature of the liquid and vapor states, and the transition between them, from a molecular point of view. The relationship between the molecular behavior and macroscopic characteristics for the vapor phase can be understood, at least qualitatively, from the kinetic theory of gases. Basic elements of the kinetic theory are developed in Appendix I. Readers who are not familiar with at least the basic aspects of the kinetic theory of gases are advised to review this appendix before proceeding further in this section.

The force interaction between two molecules is usually characterized in terms of a potential function $\phi(r)$. The potential function $\phi(r)$ is defined as being the energy input (reversible work) required to bring two molecules, which are initially an infinite distance apart, to some finite spacing r. The variation of the potential function ϕ with r is, of course, dependent on the nature of the force interaction between the molecules.

The force fields that give rise to interaction forces between molecules vary widely in character. The fields can be attractive or repulsive in nature, and one type may or may not act independently of others. At very short range, two molecules generally exert a repulsive force on one another. This repulsion is a consequence of the interference of the electron orbits of one molecule with those of the other. Although the exact nature of these forces is not known for all molecules, it is generally assumed that the repulsive force increases rapidly as the spacing r decreases. Consequently, they are often represented by a repulsive potential of the form

$$\phi_R(r) = \frac{\lambda_0}{r^k} \qquad 9 \leqslant k \leqslant 15 \tag{1.57}$$

where λ_0 is a constant that varies depending on the type of molecule.

At somewhat larger distances, the forces acting between molecules generally fall into one of three categories: (1) electrostatic forces, (2) induction forces, or (3) dispersion forces. Electrostatic forces between molecules often arise because the molecules have a finite dipole moment (i.e., opposite sides of the molecule have opposite charges).

Common dipole molecules include water and the alcohols. Because the potential function rapidly becomes small as r increases, dipole interactions are generally significant only at very short range.

Induction forces arise when a permanently charged particle or dipole induces a dipole in a nearby neutral molecule. The strength of induction forces depends directly on how easily the initially neutral molecule is polarized. The potential function associated with the induction force interaction between a dipole molecule and a neutral molecule is inversely proportional to r^6.

Dispersion forces are a consequence of transient dipoles in nominally neutral molecules or atoms. Such transient dipoles are mutually induced in adjacent molecules as a result of instantaneous asymmetries in the electric field due to moving electrons in each molecule. This mechanism produces an attractive force between the molecules. The associated potential function for dispersion force interactions is proportional to r^{-6},

$$\phi_{\mathrm{DIS}} = -\frac{\lambda_{\mathrm{DIS}}\alpha_p^2}{r^6} \tag{1.58}$$

where α_p is the polarizability of the molecules and λ_{DIS} is a constant that varies with the type of molecule. Short-range attractive dispersion forces play an important role in the thermophysical behavior of virtually all liquids and vapors near saturation conditions.

For real molecules, all the force interaction mechanisms may come into play. Hence the overall potential function $\phi(r)$ is generally assumed to reflect the repulsive force behavior at small spacings and the attractive behavior due to electrostatic, induction, and/or dispersion forces at intermediate distances. Several model variations of $\phi(r)$ have been proposed that more or less conform to this general behavior. Perhaps the most well known of these models is the Lennard-Jones 6-12 potential,

$$\phi_{LJ}(r) = 4\epsilon\left[\left(\frac{r_0}{r}\right)^{12} - \left(\frac{r_0}{r}\right)^{6}\right] \tag{1.59}$$

which is plotted in Fig. 1.5. Appropriate values of the parameters ϵ and r_0 vary with the type of molecule.

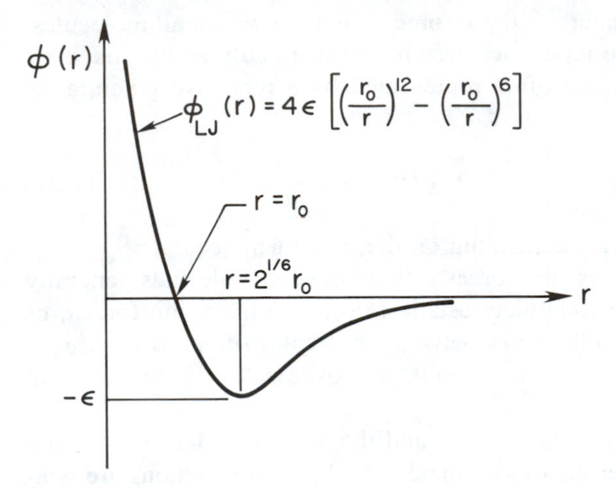

Figure 1.5 The Lennard-Jones 6-12 potential.

A study of all the implications of the Lennard-Jones potential and other models like it is beyond the scope of this book. The interested reader can find more information on this subject in references [1.18]–[1.20]. It is sufficient for our current purposes to note that this variation of the potential implies that to bring two molecules very far apart into closer proximity, we must remove energy. Conversely, if two molecules are close enough to feel attractive forces, but not so close that repulsive forces come into play, then energy must be supplied to increase the spacing of the molecules. Clearly, the energy exchanges associated with these processes are consistent with the input or removal of the latent heat of vaporization during vaporization (moving closely spaced molecules apart) or condensation (moving widely spaced molecules closer) processes.

In Appendix I, the following Maxwell-Boltzmann speed distribution for molecules in a gas is derived from the kinetic theory of gases:

$$dn_c = 4\pi n\left(\frac{m}{2\pi k_B T}\right)^{3/2} c^2 e^{-mc^2/2k_BT}\, dc \tag{1.60}$$

The relation $\epsilon = (\frac{1}{2})mc^2$ for the kinetic energy of a molecule can be inverted to obtain

$$c = \left(\frac{2\epsilon}{m}\right)^{1/2} \tag{1.61}$$

Differentiating this relation yields

$$dc = \left(\frac{1}{2m}\right)^{1/2_{\epsilon}1/2} d\epsilon \tag{1.62}$$

Substituting Eqs. (1.61) and (1.62) into (1.60), the following energy distribution is obtained:

$$dn_\epsilon = 2\pi n(\pi k_B T)^{-3/2}\epsilon^{1/2} e^{-\epsilon/k_BT}\, d\epsilon \tag{1.63}$$

In the above relation, dn_ϵ is interpreted as the number of molecules having kinetic energies between ϵ and $\epsilon + d\epsilon$.

It is often useful to know the fraction of the molecules in the gas that have energies exceeding a specified value ϵ^*. The number of molecules in the gas with energies above ϵ^*, $n(\epsilon^*)$, is given by

$$n(\epsilon^*) = \int_{\epsilon^*}^{\infty} dn_\epsilon \tag{1.64}$$

The fraction of the molecules with energies above ϵ^* is just $n(\epsilon^*)/n$. Substituting Eq. (1.63) for the integrand in Eq. (1.64), dividing both sides by n, and evaluating the integral yields

$$\frac{n(\epsilon^*)}{n} = \left(\frac{4\epsilon^*}{\pi k_B T}\right)^{1/2} e^{-\epsilon^*/k_B T} + \operatorname{erfc}\left(\sqrt{\frac{\epsilon^*}{k_B T}}\right) \tag{1.65}$$

When the threshold energy ϵ^* is much larger than $k_B T$, the complementary error function term will be very small and can be neglected. For such conditions, $n(\epsilon^*)/n$ is given by

$$\frac{n(\epsilon^*)}{n} = \left(\frac{4\epsilon^*}{\pi k_B T}\right)^{1/2} e^{-\epsilon^*/k_B T} \qquad (\epsilon^* \gg k_B T) \tag{1.66}$$

When the threshold energy ϵ^* is much larger than $k_B T$, the complementary error function term will be very small and can be neglected. For such conditions, $n(\epsilon^*)/n$ is given by

$$\frac{n(\epsilon^*)}{n} = \left(\frac{4\epsilon^*}{\pi k_B T}\right)^{1/2} e^{-\epsilon^*/k_B T} \qquad (\epsilon^* \gg k_B T) \tag{1.66}$$

Perhaps the most significant aspect of Eq. (1.66) is that it predicts that the fraction of molecules having energies above the threshold value ϵ^* increases rapidly with temperature. This behavior, which is characteristic of liquids as well as gases, plays an important role in determining chemical reaction rates and the equilibrium conditions in two-phase systems.

The equilibrium vapor pressure and its variation with temperature in a system containing saturated liquid and vapor are also consequences of the fact that the energy distribution among the molecules in the liquid and vapor phases is similar to the Maxwell-Boltzmann distribution. Because of the energy distribution, even at low temperatures, some fraction of the molecules in the liquid will have sufficient energy to escape the cohesive forces of other liquid molecules at the liquid–vapor interface. If the energy distribution in the liquid is similar to the Maxwell-Boltzmann result given by Eq. (1.66), the fraction capable of escaping in this manner will increase rapidly with temperature.

The rapid increase of the equilibrium vapor pressure suggested by these arguments is characteristic of most substances. This line of reasoning also suggests that, at a given temperature, a liquid with a small cohesive energy will have a higher vapor pressure than one with a large cohesive energy. Because the latent heat of vaporization is a macroscopic indicator of the cohesive energy of the liquid, it follows that, at the same temperature, a liquid with a high latent heat of vaporization should have a lower vapor pressure than a liquid with a smaller latent heat. This trend is also observed for most liquids.

Molecular-level characteristics are also reflected in the differences between the ideal gas law and the van der Waals equation. One defect in the ideal gas law $Pv = (\overline{R}/\overline{M})T$ is that it predicts that, for any finite pressure, the volume of the gas approaches zero as the absolute temperature approaches zero. Clearly this cannot be correct, because the molecules themselves occupy some finite volume regardless of the temperature. This could be corrected by adding a positive constant b to the volume predicted by the ideal gas relation,

$$v = b + \frac{\overline{R}T}{P\overline{M}} \tag{1.67}$$

whereupon the equation of state would become

$$P = \frac{\overline{R}T/\overline{M}}{v - b} \tag{1.68}$$

Another deficiency in the ideal gas law is that the pressure exerted on the walls of a container is always somewhat less than that predicted by the ideal gas relation. This deviation occurs because attractive forces between molecules tend to pull them together, thereby reducing the outward pressure force exerted against the container walls. The attractive force between any two volume elements in the gas is proportional to the number of molecules in each element. Assuming that the density of molecules is uniform throughout the gas, this pressure-reducing force will be proportional to the square of the density or, equivalently, $1/v^2$. To account for the attractive forces between molecules, a term proportional to $1/v^2$ is therefore subtracted from the right side of Eq. (1.68), which yields

$$P = \frac{\overline{R}T/\overline{M}}{v - b} - \frac{a}{v^2} \tag{1.69}$$

Equation (1.69) is in fact the van der Waals equation described earlier in this chapter. It is clear from this development that the constant b is associated with the volume occupied by the molecules themselves, and the a/v^2 term accounts for attractive forces between the molecules. These attractive forces, sometimes referred to as *van der Waals forces,* may in general be a combination of the dispersion, electrostatic, and induction forces described above.

Because the attractive forces represented by the a/v^2 term are the same forces that must be overcome to separate molecules that are initially closely spaced, the constant a is roughly proportional to the latent heat of vaporization of the liquid. Note also that the contribution of the a/v^2 term varies strongly with the density (and therefore the spacing) of the molecules. For low-density gases, v is large and the contribution of the a/v^2 is small. The van der Waals equation is also often assumed to be a valid representation for the liquid phase. For a liquid, the specific volume is very small and the contribution of the a/v^2 will be much larger than for the vapor phase of the same substance.

These arguments can be made more concrete by considering a specific example. It can easily be shown using density and molecular-weight data from Appendix II that, for saturated liquid water at atmospheric pressure (101 kPa), the mean volume occupied per molecule is about 3.1×10^{-29} m^3. If we envision the molecules as evenly spaced in a cubic lattice, the mean spacing is approximately the cube root of the mean volume, or 3.1×10^{-10} m apart. Because the diameter of a water molecule is about 2 Å (2.0×10^{-10} m), the center-to-center spacing of the molecules is only about 2.5 to 3 molecular radii in the saturated liquid.

For saturated water vapor at 101 kPa, the same line of reasoning suggests that the volume per molecule is 5.0×10^{-26} m^3 and the molecular spacing is 3.7×10^{-9} m or about 37 molecular radii. If r_0 in the Lennard-Jones potential shown in Fig. 1.5 is taken as being about one molecular radius, then it is clear from these numbers that attractive forces will be very small in the saturated vapor, but very important in the saturated liquid. This is completely consistent with the magnitude of these forces implied by the a/v^2 term in the van der Waals equation.

Example 1.3 Determine the deviation from the ideal gas law for saturated water vapor at atmospheric pressure using (*a*) tabulated data and (*b*) the van der Waals equation.

(*a*) From tables for saturated steam at 101 kPa, $v_v = 1.670$ m^3/kg, $T_{sat} = 100°$C, and

$$\frac{Pv}{(\overline{R}/\overline{M})T} = \frac{(101{,}000)(1.67)}{(8.314/18.0)(1000)(100 + 273)} = 0.981$$

(*b*) The van der Waals equation (1.5*a*) can be rearranged to obtain

$$\frac{Pv}{(\overline{R}/\overline{M})T} = \frac{v}{v - b} - \frac{a}{v(\overline{R}/\overline{M})T}$$

For water, evaluation of Eqs. (1.5*b*) and (1.5*c*) yields $a = 1.70 \times 10^3$ Pa m^6/kg^2 and $b = 1.69 \times 10^{-3}$ m^3/kg. Iteratively solving the above equation for the specified P and T values produces $v_v = 1.697$ m^3/kg. It can then be shown that

$$\frac{Pv}{(\overline{R}/\overline{M})T} = 0.995$$

Thus both methods indicate that there will be little deviation from the ideal gas predictions. In contrast, for saturated steam at 10 MPa, a similar calculation using tabulated data indicates that

$$\frac{Pv}{(\overline{R}/\overline{M})T} = 0.667$$

while the van der Waals equation predicts that

$$\frac{Pv}{(\overline{R}/\overline{M})T} = 0.820$$

At this elevated pressure, tabulated data suggest a significant deviation from ideal gas behavior. Because its accuracy is not good at this specific condition, the van der Waals equation predicts that there will be a smaller deviation from ideal gas behavior. While the van der Waals equation is qualitatively correct for most pure substances, its absolute accuracy may not be good in some cases.

For molecules in the center of a large body of liquid, the attractive forces from surrounding molecules are nominally spherically symmetric and hence they balance out to zero. Near a liquid–vapor interface, however, things are quite different. Within a few molecular diameters of the interface, molecules in the liquid must redistribute themselves to accommodate the lack of spherical symmetry in the molecular force interactions. This redistribution gives rise to interfacial phenomena, which are discussed in detail in the next chapter.

REFERENCES

1.1 Reynolds, W. C., and Perkins, H. C., *Engineering Thermodynamics,* 2nd ed., McGraw-Hill, New York, 1977.
1.2 Van Wylen, G., and Sonntag, R. E., *Introduction to Thermodynamics—Classical and Statistical,* 2nd ed., Wiley, New York, 1982.
1.3 Zemansky, M. W., and Dittman, R. H., *Heat and Thermodynamics,* 6th ed., McGraw-Hill, New York, 1981.
1.4 Howell, J. R., and Buckius, R. O., *Fundamentals of Engineering Thermodynamics,* McGraw-Hill, New York, 1987.
1.5 Carslaw, H. S., and Jaeger, J. C., *Conduction of Heat in Solids,* 2nd ed., Oxford University Press, Oxford, 1959.
1.6 Ozisik, M. N., *Heat Conduction,* Wiley, New York, 1980.
1.7 Arpaci, V. S., *Conduction Heat Transfer,* Addison-Wesley, Reading, MA, 1966.
1.8 Rohsenow, W. M., and Choi, H. Y., *Heat, Mass and Momentum Transfer,* Prentice-Hall, Englewood Cliffs, NJ, 1961.
1.9 Bird, R. B., Stewart, W. E., and Lightfoot, E. N., *Transport Phenomena,* chaps. 16–20, Wiley, New York, 1960.
1.10 Levich, V. G., *Physiochemical Hydrodynamics,* chaps. 1–3, Prentice-Hall, Englewood Cliffs, NJ, 1962.
1.11 Kays, W. M., and Crawford, M. E., *Convective Heat and Mass Transfer,* 2nd ed., McGraw-Hill, New York, 1980.
1.12 Schlichting, H., *Boundary Layer Theory,* 7th ed., McGraw-Hill, New York, 1979.
1.13 Panton, R. I., *Incompressible Flow,* Wiley, New York, 1984.
1.14 Siegel, R., and J. R. Howell, *Thermal Radiation Heat Transfer,* 2nd ed., Hemisphere, Washington, DC, 1981.
1.15 Sparrow, E. M., and Cess, R. D., *Radiation Heat Transfer,* 2nd ed., Hemisphere, Washington, DC, 1978.
1.16 Edwards, D. K., *Radiation Heat Transfer Notes,* Hemisphere, New York, 1981.
1.17 Tien, C. L., Thermal radiation properties of gases, in *Advances in Heat Transfer,* T. F. Irvine and J. P. Hartnett, eds., vol. 5, p. 253, Academic Press, New York, 1968.

1.18 Tien, C. L., and Lienhard, J., *Statistical Thermodynamics*, Hemisphere, New York, 1972.
1.19 McQuarrie, D. A., *Statistical Thermodynamics*, Harper & Row, New York, 1973.
1.20 Hirschfelder, J. O., Curtiss, C. F., and Bird, R. B., *Molecular Theory of Gases and Liquid*, Wiley, New York, 1954.

PROBLEMS

1.1 Using saturation property data from Appendix II for water, ammonia, and R-113, determine the volume and diameter of a spherical bubble created by vaporization of 0.1 mm^3 of liquid at atmospheric pressure. Repeat the calculations for the same fluids at a pressure equal to nine tenths of each fluid's critical pressure.

1.2 Film boiling occurs over an upward-facing flat horizontal surface in a pool of saturated water at atmospheric pressure. The system configuration is the same as shown in Fig. 1.4. Modeling the heat transfer across the layer as being due to conduction and radiation, estimate the fraction of the total heat transfer across the vapor layer that is due to each mechanism. The surface is held at a uniform temperature of 400°C and the mean vapor film thickness is 2 mm. Neglect any radiation interaction with the steam and treat the liquid–vapor interface as a blackbody. The surface emittance is 0.7.

1.3 Film boiling occurs over a hot sphere immersed in saturated liquid nitrogen at atmospheric pressure. At the bottom of the sphere, the local heat transfer coefficient is 7.5 W/m^2 K. Estimate the thickness of the vapor film at this location. Is radiation important for this system? Justify your answer quantitatively.

1.4 A thin film of liquid water forms on a cold vertical surface as saturated steam condenses on it at atmospheric pressure. The liquid in the film flows down the surface due to gravity as the condensation occurs. If the wall temperature is 60°C and the heat flux at a particular location is 50 kW/m^2, estimate the thickness of the liquid film at this location assuming that heat is transported across the film primarily by conduction. Are boundary-layer approximations appropriate for the liquid film flow? What must the local vapor velocity toward the surface be in order to maintain this rate of condensation?

1.5 The van der Waals constants are quoted as being $a = 1690$ Pa m^6/kg^2, $b = 1.75 \times 10^{-3}$ m^3/kg for water, and $a = 175$ Pa m^6/kg^2, $b = 1.40 \times 10^{-3}$ m^3/kg for nitrogen. How do these values compare to those computed using Eqs. (1.5*b*) and (1.5*c*)? (See Appendix II for values of T_c and P_c.) Based on the above values of the van der Waals constants, which of these two fluids would you expect to have the higher latent heat of vaporization? Do property data in Appendix II support your conclusion? Briefly explain your answers.

1.6 In a steam generator, 4 liters per second of water enters a flow passage at a pressure of 11 MPa and a temperature of 30°C and leaves as saturated vapor at 10 MPa. The flow channel is a round tube with an inside diameter of 3.0 cm. Determine the mean velocity of the flow at the inlet and exit of the tube. Also determine the overall heat transfer rate to the water. If this heat transfer is applied as a uniform flux of 24 MW/m^2 along the tube, determine the necessary length of the tube.

1.7 A rigid vessel with an internal volume of 0.014 m^3 contains 10 kg of water (vapor plus liquid) initially at 20°C. The vessel is then slowly heated. As a result of the heating process, will the liquid–vapor interface move upward toward the top of the vessel or downward toward the bottom? What if the vessel contains 1 kg instead of 10 kg? What would you see if you could view the contents of the vessel during the heating process and the vessel contained a mass equal to $0.014/v_c$, where v_c is the critical specific volume? Would you classify this as a condensation or vaporization process?

CHAPTER

TWO

INTERFACIAL TENSION

2.1 THE INTERFACIAL REGION

Macroscopic thermodynamic and fluid mechanics treatments of the boundary between two phases invariably assume a sharp discontinuity in density and/or composition across the boundary or interface. In considering the details of physical processes at the interface, however, it is important to recognize that rapid changes in fluid or material properties actually occur over a thin region centered about the interface. Consideration of molecular interactions in this interfacial region have led to a better understanding of several aspects of interfacial phenomena. Nevertheless, we will find it convenient to revert back to a two-dimensional surface treatment of the interface in many cases.

Anyone who has watched small bubbles rise in a carbonated beverage or a pot of boiling water has undoubtedly noted that the bubbles are almost perfectly spherical, as if an elastic membrane were present at the interface to pull the vapor into a spherical shape. From a thermodynamic point of view, this apparent *interfacial tension* may be interpreted in terms of energy stored in the molecules near the interface.

Figure 2.1 schematically shows the density distribution near a liquid–vapor interface. In the liquid, the density is lower in the interface region than in the bulk liquid phase. As described in Chapter One, because the molecules attract one another, energy must be supplied to move them apart. Hence the energy per molecule is greater in the interfacial region than in the bulk liquid. The system thus has an additional free energy per unit area of interface due to the presence of the interface.

There is also the more obvious mechanical interpretation of the interfacial

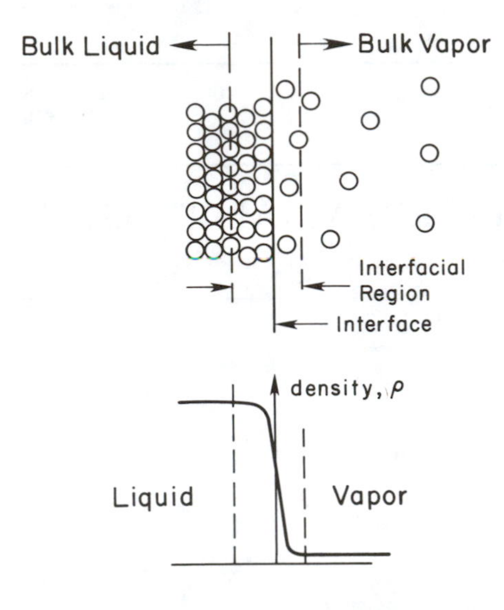

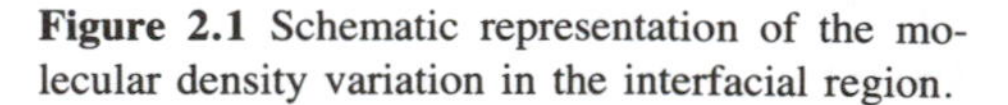
Figure 2.1 Schematic representation of the molecular density variation in the interfacial region.

tension as a force per unit length parallel to the interface (and perpendicular to the density gradient). This interpretation can also be viewed as being a consequence of attractive and repulsive interactions among molecules, which result from the molecular forces described in Section 1.6. In the bulk liquid, molecules are subject to forces of repulsion from their close neighbors and forces of attraction from all others. The repulsive forces are generally stronger, but because both types of forces act symmetrically in all directions, on the average, the resultant on each molecule is zero.

As indicated in Fig. 2.1, the mean spacing of the molecules in the liquid near the interface is greater than that in the bulk liquid. Note that the potential functions discussed in Section 1.6 imply that the close-range repulsive force (force being $-d\phi/dr$) varies more rapidly with spacing than the longer-range attractive forces. Hence, for a given molecule, this slightly increased spacing would significantly weaken the repulsive force it feels from its immediate neighbors, but would likely produce only a small change in the attractive force between it and more distant surrounding molecules. In the direction normal to the interface, this combination of effects would produce a force imbalance that would draw the molecules toward the bulk liquid. The mean spacing of the molecules in the direction normal to the interface could decrease slightly to establish a balance between long-range attractive forces and repulsion from close neighboring molecules.

In the directions parallel to the interface, however, the decrease in the repulsion force between molecules produced by the increased spacing does not create a force imbalance because of the radial symmetry of the force interactions. Consequently, there is no impetus to decrease the mean spacing in this direction. The decrease in repulsive forces between immediate neighbors, with little change in the longer-range attractive forces, may thus produce a net tension force among molecules in the interface region acting equally in all directions parallel to the interface. These arguments are clearly very crude. They ignore, for example, that, given the mobility of molecules in the liquid, it is difficult to envision how the liquid could maintain a surface structure in which the mean molecular spacing normal to the surface is different from that parallel to the surface. (See reference [2.1] for a further discussion of this matter.) Despite their crudeness, these arguments nevertheless imply that the existence of a net tension force on the molecules is a direct consequence of the increased mean molecular spacing in the interface region.

The intermolecular attractions that give rise to interfacial tension may result from several different types of molecular forces. These may include forces that are specific to particular types of molecules, such as the metallic bond or the hydrogen bond, as well as London dispersion forces, which exist in all types of matter and always give an attractive force between adjacent atoms or molecules no matter how dissimilar they are chemically. The London dispersion forces vary with the electrical properties of the substances involved and the distance between interacting elements, but they are independent of temperature.

Example 2.1 Assuming that the interfacial region between saturated liquid water and saturated water vapor at atmospheric pressure is three molecules thick, estimate the increase in molecular spacing necessary to account for the interfacial tension.

If we compare the specific volume of saturated liquid and vapor for water at 100°C, it is found that

$$\frac{v_v}{v_l} = \frac{1.673 \text{ m}^3/\text{kg}}{0.0010435 \text{ m}^3/\text{kg}} = 1603$$

This implies that the volume occupied per molecule in the vapor is about 1600 times that in the liquid. If the mean spacing between molecules is taken to be the cube root of this figure, then the ratio of the distances between molecules in the two phases L_v/L_l is

$$\frac{L_v}{L_l} = \left(\frac{v_v}{v_l}\right)^{1/3} = (1603)^{1/3} = 11.7$$

The volume per molecule in the liquid is equal to

$$\left(\frac{\text{vol}}{\text{molecule}}\right)_{\text{liq}} = \frac{v_l\overline{M}}{N_A} = \frac{(0.00104)(18)}{6.02 \times 10^{26}} = 3.11 \times 10^{-29} \text{ m}^3/\text{molecule}$$

If we take the cube root of this value as an estimate of the mean spacing of the molecules, we find that they are approximately 3.14×10^{-10} m apart. It follows that the thickness of the interfacial region δ_i and the number of molecules in it per unit of interfacial area are given by

$$\delta_i = 3.14 \times 10^{-10} \times 3 = 9.42 \times 10^{-10}\ \text{m}$$

$$\frac{\#\ \text{molecules}}{\text{area}} = \frac{\text{interfacial region vol./area}}{\text{vol./molecule}} = \frac{\delta_i}{\text{vol./molecule}}$$

$$= \frac{9.42 \times 10^{-10}}{3.11 \times 10^{-29}} = 3.03 \times 10^{19}\ \text{molecules/m}^2$$

The latent heat of vaporization will be interpreted as the energy required to increase the mean separation distance between molecules from the distance in the liquid to that (about 12 times larger) in the vapor. The energy required per molecule, e_{vap}, is given by

$$e_{\text{vap}} = h_{lv}\left(\frac{\overline{M}}{N_A}\right) = 2257\left(\frac{18}{6.02 \times 10^{26}}\right) = 6.75 \times 10^{-23}\ \text{kJ/molecule}$$

We will further make the very crude assumption that the energy required to increase the mean separation distance of the liquid molecules varies linearly with the separation distance up to the value for the vapor state. The energy required per molecule, e_i, to achieve an increase in the mean separation equal to ΔL_i in the interfacial region is then given by

$$e_i = e_{\text{evap}}\left(\frac{\Delta L_i}{\Delta L_{\text{evap}}}\right)$$

At a macroscopic level, the interfacial tension for saturated liquid water at atmospheric pressure is known to be 0.0588 N/m. The energy per molecule in the interfacial region must therefore equal

$$e_i = \frac{\sigma}{\#\ \text{molecules/area}} = \frac{0.0588}{3.03 \times 10^{19}} = 1.94 \times 10^{-21}\ \text{J/molecule}$$

Equating the above two relations for e_i, it follows that the increase in spacing, as a fraction of that associated with vaporization is

$$\frac{\Delta L_i}{\Delta L_{\text{evap}}} = \frac{e_i}{e_{\text{evap}}} = \frac{1.94 \times 10^{-21}}{6.75 \times 10^{-20}} = 0.029$$

As a fraction of actual molecular spacing on the bulk liquid, this becomes

$$\frac{\Delta L_i}{L_{\text{liq}}} = \frac{\Delta L_i}{\Delta L_{\text{evap}}}\left(\frac{L_{\text{vap}} - L_{\text{liq}}}{L_{\text{liq}}}\right) = \frac{\Delta L_i}{\Delta L_{\text{evap}}}\left(\frac{L_{\text{vap}}}{L_{\text{liq}}} - 1\right)$$

$$= 0.029(11.7 - 1) = 0.31$$

Although the arguments used here are very crude, they imply that an increase in the spacing of the liquid molecules in the interfacial region equal to about 30% of that in the bulk liquid would account for the macroscopically observed interfacial tension. They further indicate that the increase in the mean separation is only about 3% of the increase associated with vaporization of the liquid.

Fowkes [2.2] has shown that the surface tension can be considered as consisting of the sum of two parts: the part due to dispersion forces, σ_d, and the part due to specific forces such as metallic or hydrogen bonding, σ_s:

$$\sigma = \sigma_s + \sigma_d \tag{2.1}$$

In these terms, it is easy to understand why surface tensions of liquid metals are higher than those of hydrogen-bonded liquids such as water, which in turn are higher than those of nonpolar liquids such as pure hydrocarbons. Surface tension in nonpolar liquids is due entirely to dispersion forces. In hydrogen-bonded liquids there are contributions due to both dispersion forces and hydrogen bonding, generally resulting in slightly larger values of σ. For liquid metals, a large σ_s contribution results from the metallic bond attraction, which, when added to the dispersion-force contribution, results in even higher surface tension values.

The trends described above can be seen in the surface tension data shown in Table 2.1. The values are listed in the SI system units of millinewtons per meter (mN/m). Note that 1 mN/m is equal to 1 dyne/cm.

Both the thermodynamic and the mechanical interpretations of surface tension have their uses, as we will see in later sections. Regardless of its interpretation, the main effect of surface tension is that it causes the system to act to minimize its interfacial area. The effects of surface tension on the thermodynamic and mechanical aspects of liquid–vapor systems are discussed in more detail in the following sections.

Table 2.1 Values of surface tension for various liquids in contact with air or its own vapor at saturation

Liquid	Temperature (°C)	Surface tension (mN/m)
Silver (Ag)	1100	878
Mercury (Hg)	20	484
Hydrazine (N_2H_4)	25	91.5
Water (H_2O)	20	72.8
Ethylene glycol	20	48.4
Ammonia (NH_4)	−40	35.4
Carbon tetrachloride (CCl_4)	20	27.0
n-Butanol ($C_4H_{10}O$)	20	24.6
Acetone (CH_3COCH_3)	20	24.0
Ethanol (C_2H_6O)	20	22.8
Methanol (CH_4O)	20	22.6
R-113 (CCl_2FCClF_2)	26.7	19.0
R-11 (CCl_3F)	26.7	18.0
R-12 (CCl_2F_2)	26.7	8.9
Helium II (He_{II})	−271	0.32
Helium III (He_{III})	−271	0.069

2.2 THERMODYNAMIC ANALYSIS OF INTERFACIAL TENSION EFFECTS

The interfacial region is presumed to be bounded by two surfaces S_{I}^* and S_{II}^*, which are parallel to the interface surface S^* but are located in the corresponding bulk phases. As shown in Fig. 2.2, the density and internal energy actually vary continuously across the interfacial region. Note that because the transition between the bulk properties occurs over a finite thickness, the actual mass and internal energy of this region may differ from the values that would be calculated by assuming that the bulk phases I and II extended all the way to S^*. This difference is called the surface excess of the property. The surface excess mass $\Gamma_e^{S^*}$ is defined as

$$\Gamma_e^{S^*} = \int_{-y_{\mathrm{I}}}^{y_{\mathrm{II}}} \rho\, dy - \rho_{\mathrm{I}} y_{\mathrm{I}} - \rho_{\mathrm{II}} y_{\mathrm{II}} \tag{2.2}$$

and the surface excess internal energy $U_e^{S^*}$ is given by

$$U_e^{S^*} = \int_{-y_{\mathrm{I}}}^{y_{\mathrm{II}}} \rho u\, dy - \rho_{\mathrm{I}} u_{\mathrm{I}} y_{\mathrm{I}} - \rho_{\mathrm{II}} u_{\mathrm{II}} y_{\mathrm{II}} \tag{2.3}$$

Surface excess values of other thermodynamic properties (and mass concentrations in multicomponent systems) can be similarly defined. Note that these quantities have been given a superscript S^* because their values depend on the location of the surface S^*. A specific location may make either $\Gamma_e^{S^*}$ or $U_e^{S^*}$ zero, but in general they may be positive or negative. This concept facilitates a link between the ideal-

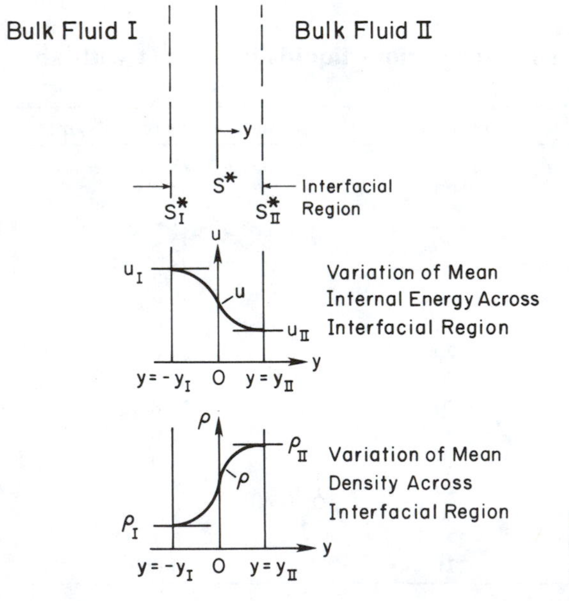

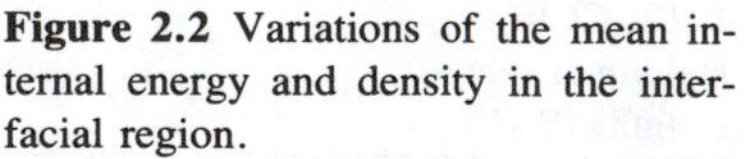
Figure 2.2 Variations of the mean internal energy and density in the interfacial region.

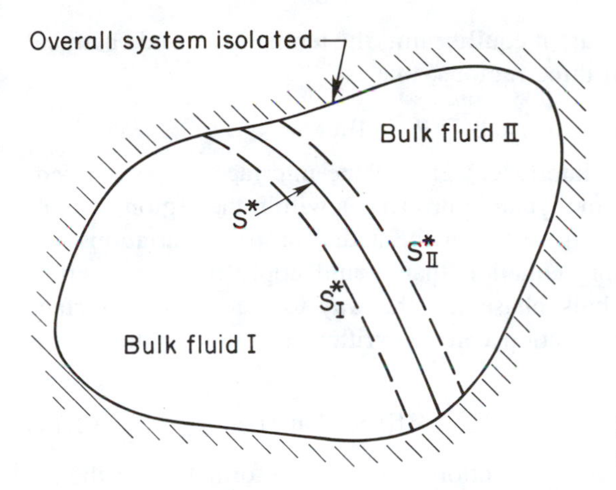

Figure 2.3 Model system for thermodynamic analysis of interfacial region.

ized concept of a two-dimensional interface surface and the actual three-dimensional character of the interfacial region.

To examine the relationships between properties in the interface region and corresponding ones in the surrounding bulk phases at equilibrium, we can proceed using a thermodynamic analysis similar to that presented in Section 1.3. The entire system consisting of the interface region (between S_{I}^* and S_{II}^*) and bulk fluids I and II in Fig. 2.3 is assumed to be completely isolated from the surroundings. If the shape of the interfacial region (and hence its volume) is assumed to be fixed, its internal energy may be taken to be a function only of its entropy and the number of moles present, $U = U(S,N)$. It follows that

$$dU = T\,dS + \mu\,dN \tag{2.4}$$

where

$$T = \left(\frac{\partial U}{\partial S}\right)_{N,\text{shape}} \qquad \mu = \left(\frac{\partial U}{\partial N}\right)_{S,\text{shape}} \tag{2.5}$$

We now consider possible internal perturbations that leave bulk fluid I unchanged while permitting exchange of energy or mass between the interfacial region and bulk fluid II. Because the overall system is isolated, we must have

$$dN_{\text{total}} = dN + dN_{\mathrm{II}} = 0 \tag{2.6}$$

$$dS_{\text{total}} = dS + dS_{\mathrm{II}} = 0 \tag{2.7}$$

$$dU_{\text{total}} = T\,dS + \mu\,dN + T_{\mathrm{II}}\,dS_{\mathrm{II}} + \mu_{\mathrm{II}}\,dN_{\mathrm{II}} = 0 \tag{2.8}$$

Combining Eqs. (2.6) through (2.8) yields

$$(T - T_{\mathrm{II}})\,dS + (\mu - \mu_{\mathrm{II}})\,dN = 0 \tag{2.9}$$

Because Eq. (2.9) must be satisfied for all possible perturbations of S and N, we must have $T = T_{\mathrm{II}}$ and $\mu = \mu_{\mathrm{II}}$. An identical set of arguments can be presented for bulk fluid I. The net result is that, at equilibrium, the temperature and chemical potential must be the same in all three regions; that is,

$$T = T_{\rm I} = T_{\rm II} \qquad \mu = \mu_{\rm I} = \mu_{\rm II} \tag{2.10}$$

Equation (2.4) applies to the interface region if the interface shape is fixed, regardless of the variation of thermodynamic properties within the region at equilibrium. We can, therefore, take Eq. (2.9) for the actual property variations and subtract from it the corresponding equation that would apply if the properties equaled those of the respective bulk phases all the way to the interface surface S^*. By definition, the resulting equation can be written in terms of the excess properties $U_e^{S^*}$, $S_e^{S^*}$, and $N_e^{S^*}$ as

$$dU_e^{S^*} = T\, dS_e^{S^*} + \mu\, dN_e^{S^*} \qquad \text{(fixed shape)} \tag{2.11}$$

If we now relax the fixed-shape restriction and allow deformation of the interface, both its area and its curvature can change. However, if the radii of curvature are much greater than the interfacial thickness, we might expect curvature effects on the internal energy, entropy, and the density in the interface region to be small. We do, however, want to account for the effect of any increase or decrease in the interface area resulting from deformation. Equation (2.11) is therefore modified as

$$dU_e^{S^*} = T\, dS_e^{S^*} + \mu\, dN_e^{S^*} + \sigma\, dA^{S^*} \tag{2.12}$$

where we have defined the interfacial tension σ as

$$\sigma = \left(\frac{\partial U_e^{S^*}}{\partial A^{S^*}}\right)_{S_e^{S^*},N_e^{S^*}} \tag{2.13}$$

to account for the effect of changing interface area on the energy balance.

For the interfacial region, the Helmholtz free energy F is given by

$$F = U - TS \tag{2.14}$$

Subtracting from Eq. (2.14) the corresponding equation that would apply if the two parts of the interfacial region were occupied by bulk fluids I and II yields an analogous relation for the surface excess free energy:

$$F_e^{S^*} = U_e^{S^*} - TS_e^{S^*} \tag{2.15}$$

Differentiating Eq. (2.15) and substituting Eq. (2.12) for $dU_e^{S^*}$ yields

$$dF_e^{S^*} = -S_e^{S^*}\, dT + \mu\, dN_e^{S^*} + \sigma\, dA_i \tag{2.16}$$

However, if we consider $F_e^{S^*} = F_e^{S^*}(T,N_e^{S^*},A_i)$ from a purely mathematical point of view, we can write

$$dF_e^{S^*} = \left(\frac{\partial F_e^{S^*}}{\partial T}\right)_{N_e^{S^*},A_i} dT + \left(\frac{\partial F_e^{S^*}}{\partial N_e^{S^*}}\right)_{T,A_i} dN_e^{S^*} + \left(\frac{\partial F_e^{S^*}}{\partial A_i}\right)_{N_e^{S^*},T} dA_i \tag{2.17}$$

Comparison of Eqs. (2.16) and (2.17) clearly indicates that the interfacial tension is related to the surface excess free energy as

$$\sigma = \left(\frac{\partial F_e^{S^*}}{\partial A_i} \right)_{N_e^{S^*},T} \tag{2.18}$$

Thus, σ is equal to the change in surface excess free energy produced by a unit increase in interface area, A_i.

We can extract an additional useful relation by considering a shift in the reference surface S^* uniformly toward region II by some small amount dy. The position of S^* is somewhat arbitrary because of the continuous property variations in the interface region. However, if our analysis is to yield meaningful results, the values of properties in the three regions in Fig. 2.3 should be independent of where S^* is located. In allowing S^* to shift by dy, we therefore expect that the overall free energy F for the system as a whole is unchanged, because no change occurs in its physical state. From basic thermodynamic relations, it can be shown that, for the bulk fluids,

$$dF = -S\,dT + \mu\,dN - P\,dV \tag{2.19}$$

Summing together the differential changes in F for the three regions in Fig. 2.3, using Eqs. (2.16) and (2.19) and requiring that the overall $dF = 0$, yields

$$\begin{aligned} dF = {} & -S_e^{S^*}\,dT + \mu\,dN_e^{S^*} + \sigma\,dA_i \\ & -S_{\mathrm{I}}\,dT + \mu\,dN_{\mathrm{I}} - P_{\mathrm{I}}\,dV_{\mathrm{I}} \\ & -S_{\mathrm{II}}\,dT + \mu\,dN_{\mathrm{II}} - P_{\mathrm{II}}\,dV_{\mathrm{II}} = 0 \end{aligned} \tag{2.20}$$

Because the total number of moles and total volume does not change, and there is no change of temperature,

$$dV = dV_{\mathrm{I}} + dV_{\mathrm{II}} = 0 \tag{2.21}$$

$$dN = dN_{\mathrm{I}} + dN_{\mathrm{II}} + dN_e^{S^*} = 0 \tag{2.22}$$

$$dT = 0 \tag{2.23}$$

and substitution of Eqs. (2.21) through (2.23) reduces Eq. (2.20) to

$$P_{\mathrm{I}} - P_{\mathrm{II}} = \sigma \frac{dA_i}{dV_{\mathrm{I}}} \tag{2.24}$$

If the surface S^* can be characterized by two radii of curvature measured in two perpendicular planes containing the local normal to S^*, as shown in Fig. 2.4, then the changes in A_i and V_{I} for a shift in S^* of dy are

$$dA_i = s_2\,ds_1 + s_1\,ds_2 \tag{2.25}$$

$$dV_{\mathrm{I}} = s_1 s_2\,dy \tag{2.26}$$

Using simple geometric relations between sides of similar triangles, it can be shown that

$$ds_1 = s_1 \frac{dy}{r_1} \qquad ds_2 = s_2 \frac{dy}{r_2} \tag{2.27}$$

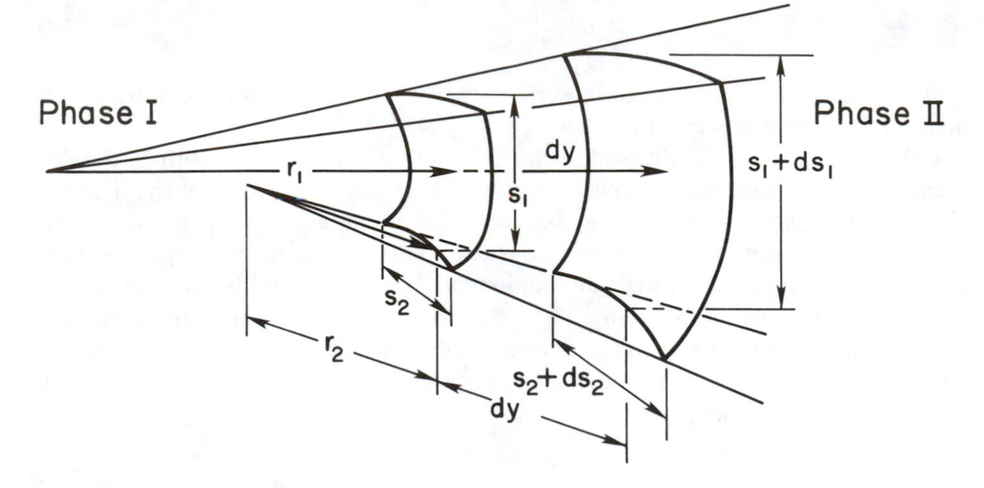

Figure 2.4 Geometry of interface between two phases.

Combining Eqs. (2.25) through (2.27) yields

$$\frac{dA_i}{dV_\mathrm{I}} = \frac{1}{r_1} + \frac{1}{r_2} \tag{2.28}$$

which can be substituted into Eq. (2.24) to obtain

$$P_\mathrm{I} - P_\mathrm{II} = \sigma\left(\frac{1}{r_1} + \frac{1}{r_2}\right) \tag{2.29}$$

Equation (2.29) is usually called the *Laplace* or the *Young-Laplace equation*. It relates the pressure difference across the interface to the interfacial tension and the geometry of the interface at equilibrium. As will be seen in later sections, this relation is a critical element in the thermodynamic and mechanical analysis of interfacial phenomena.

In this section, analysis of the interface as a region of finite thickness has demonstrated that the interfacial tension is related to the surface excess free energy by Eq. (2.18). This links the interfacial tension to the molecular properties that dictate the variation of the molecular spacing (density) and other properties across the interface region. The Young-Laplace equation, which is usually derived from force-balance considerations, was also derived from thermodynamic analysis of the interface region.

Finally, it should be noted that the analysis presented in this section has consistently considered an interface between two fluid phases of a pure substance (i.e., its liquid and vapor phases). This analysis can be extended to interfaces in multicomponent systems with only slight modifications. Further discussion of interfacial phenomena in multicomponent systems can be found in references [2.3] and [2.4].

2.3 DETERMINATION OF INTERFACE SHAPES AT EQUILIBRIUM

As might be expected, interfacial tension plays a major role in the determination of the shape of a liquid–vapor interface at equilibrium. The Young-Laplace equation (2.29) derived in the previous section is the principal mathematical tool used to predict the shape of the interface. Because this equation relates the interfacial tension, interface geometry, and pressure difference between the fluids at each point along the interface, it can be used with the equations of hydrostatics to compute the shape of a static interface. Alternatively, if the shape of the interface can be determined experimentally, the Young-Laplace equation can be used to infer the interfacial tension. Measurements of the interfacial tension will be described in more detail in the next section.

Evaluation of the radii of curvature in Eq. (2.29) for an arbitrary surface shape leads to a differential equation that can be quite difficult to solve. However, in many instances, the interface shape is sufficiently symmetric that the geometry can be handled without too much difficulty. The simplest case is the spherical drop or bubble, in which both r_1 and r_2 are equal to the radius r. Equation (2.29) then reduces to

$$P_{\mathrm{I}} - P_{\mathrm{II}} = P_{\text{inside}} - P_{\text{outside}} = \frac{2\sigma}{r} \tag{2.30}$$

A sign convention is adopted here such that the radii of curvature are always taken as positive, with the center of curvature on the concave side of the interface. Thus, in Eq. (2.30), the pressure inside the drop or bubble P_{I} exceeds the pressure outside P_{II} by an amount equal to $2\sigma/r$.

Another simple geometric configuration of interest is the sessile drop on a flat solid surface, as shown in Fig. 2.5. Determination of the interface shape requires an analysis that combines the equations of hydrostatics and the Young-Laplace equation with appropriate boundary conditions. Due to the hydrostatic heads, the pressure distributions in fluids I and II are given by

$$P_{\mathrm{I}} = P_{\mathrm{I}0} + \rho_{\mathrm{I}} g z \tag{2.31}$$

$$P_{\mathrm{II}} = P_{\mathrm{II}0} + \rho_{\mathrm{II}} g z \tag{2.32}$$

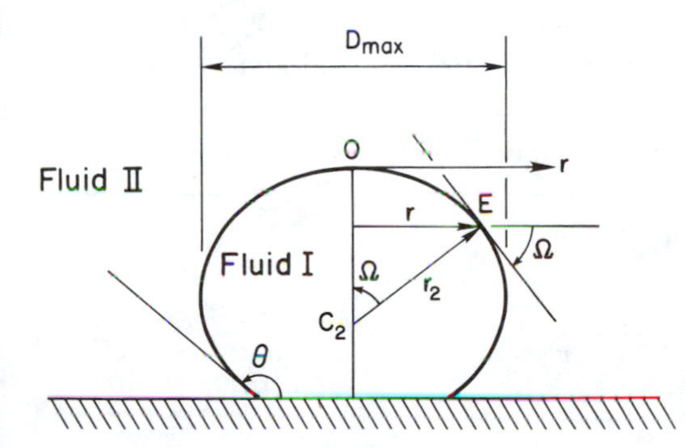

Figure 2.5 A sessile drop on a flat solid surface.

Note that in both fluids the reference pressure has been chosen as that at the drop apex (point 0) in Fig. 2.5. Cylindrical coordinates are used with the origin at point 0. The drop is assumed to be axisymmetric. Consequently, the radius of curvature r_0 at 0 is the same in all vertical planes passing through point 0, and from Eq. (2.30) we have

$$P_{I0} - P_{II0} = \frac{2\sigma}{r_0} \tag{2.33}$$

Combining Eq. (2.29) with Eqs. (2.31) through (2.33), the Young-Laplace equation for a general point E on the interface can be written as

$$\sigma\left(\frac{1}{r_1} + \frac{1}{r_2}\right) = \frac{2\sigma}{r_0} + (\rho_I - \rho_{II})gz \tag{2.34}$$

From simple geometric considerations, the principal radius of curvature r_1 in the vertical plane is given by

$$\frac{1}{r_1} = \frac{d^2z/dr^2}{[1 + (dz/dr)^2]^{3/2}} \tag{2.35}$$

The relation for r_2 is somewhat more difficult to derive, but it can be obtained with some manipulation. The radius of curvature r_2 is measured in a plane normal to the vertical plane of Fig. 2.5 but containing the local normal to the interface at E. The center of curvature for r_2 is at point C_2 in Fig. 2.5, where the normal meets the drop axis.

Example 2.2 Liquid droplets that condense to form fog in the atmosphere can be as small as 2 μm in diameter when they first form. Determine the pressure inside a droplet of this size at 20°C.

For a spherical droplet, the Young-Laplace equation (2.30) requires that

$$P_{\text{inside}} = P_{\text{outside}} + \frac{2\sigma}{r}$$

where r is the droplet radius. For water in contact with air at 20°C, $\sigma = 0.0728$ N/m and

$$P_{\text{inside}} = 101 \text{ kPa} + \frac{2(0.0728)}{1.0 \times 10^{-6}}(10^{-3}) = 243 \text{ kPa}$$

Hence the pressure inside the droplet is about 2.5 atm for these conditions.

From Fig. 2.5 it can be seen that

$$\sin \Omega = \frac{r}{r_2} \tag{2.36}$$

The trigonometric identity

$$1 + \cot^2 \Omega = \csc^2 \Omega \tag{2.37}$$

can be rearranged to obtain

$$\sin \Omega = \frac{\tan \Omega}{(1 + \tan^2 \Omega)^{1/2}} \tag{2.38}$$

Substituting Eq. (2.38) into Eq. (2.36) yields

$$\frac{r}{r_2} = \frac{\tan \Omega}{(1 + \tan^2 \Omega)^{1/2}} \tag{2.39}$$

As seen in Fig. 2.5, the normal to the interface at E makes an angle Ω with the horizontal. As a result, we can replace $\tan \Omega$ with dz/dr in Eq. (2.39), to obtain

$$\frac{1}{r_2} = \frac{dz/dr}{r[1 + (dz/dr)^2]^{1/2}} \tag{2.40}$$

Substituting Eqs. (2.35) and (2.40) into Eq. (2.34) and writing the resulting equation in terms of the dimensionless variables ξ and λ, defined as

$$\xi = \frac{z}{r_0} \qquad \lambda = \frac{r}{r_0} \tag{2.41}$$

we obtain

$$\frac{\xi''}{[1 + (\xi')^2]^{3/2}} + \frac{\xi'}{\lambda[1 + (\xi')^2]^{1/2}} = 2 + \text{Bo}\,\xi \tag{2.42}$$

where Bo is the Bond number, defined as

$$\text{Bo} = \frac{g(\rho_{\text{I}} - \rho_{\text{II}})r_0^2}{\sigma} \tag{2.43}$$

and the primes denote derivatives with respect to λ. For the second-order ordinary differential equation (2.42), two boundary conditions are required. As a consequence of the choice of coordinate systems and the symmetry of the sessile drop, it is required that

$$\xi = 0 \quad \text{and} \quad \xi' = 0 \qquad \text{at } \lambda = 0 \tag{2.44}$$

With the boundary conditions specified by Eq. (2.44), the system is closed and Eq. (2.42) can be solved to determine $\xi(\lambda)$ for specified Bo. In general, solving Eq. (2.42) analytically is virtually impossible, and a numerical approach has been used. Numerical solutions were first obtained by Bashforth and Adams [2.5] in the late 1800s. Complete and highly accurate tables giving ξ as a function of λ for specified Bo values are now available from more recent computer calculations (see Padday [2.6]). The results obtained for Bo = 20 are plotted in Fig. 2.6 as a function of the angle Ω.

It should be noted in Fig. 2.5 that θ is the *contact angle* of the interface with the solid surface. In liquid–vapor or liquid–gas systems, the convention is that

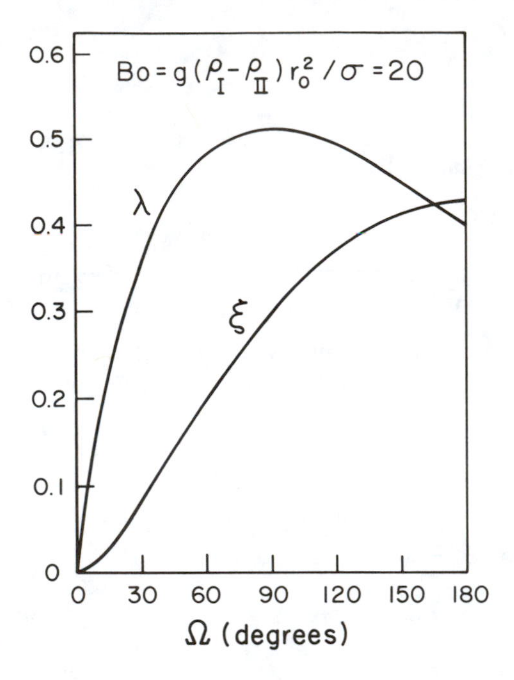

Figure 2.6 Variation of the dimensionless drop radius $\lambda = r/r_0$ and interface height $\xi = z/r_0$ as a function of angular position for a Bond number of 20.

the contact angle is the angle between the interface and the solid surface as measured through the liquid phase. Contact angles will be discussed in more detail in Chapter Three. It is sufficient here to acknowledge that the contact angle is a constant for the combination of fluids and solid surface in the system. In Fig. 2.5 it can also be seen that Ω starts at zero at the apex and increases as z increases until the interface contacts the surface. At the point where the interface contacts the surface, $\Omega = \theta$. Hence, in the computed results, the maximum value of ξ (where the interface stops) corresponds to $\Omega = \theta$.

Another circumstance of interest is the shape of a free liquid surface meeting a plane vertical wall. As shown in Fig. 2.7, if the liquid wets the wall (has a contact angle $\theta < 90°$), the liquid level will rise as the wall is approached, meeting the wall at its contact angle θ. For the two-dimensional configuration shown in Fig. 2.7, the principal radii of curvature are such that

$$\frac{1}{r_1} = \frac{d^2z/dy^2}{[1 + (dz/dy)^2]^{3/2}} \qquad \frac{1}{r_2} = 0 \tag{2.45}$$

Substituting these results into the Young-Laplace equation (2.29) yields

$$P_{\mathrm{I}} - P_{\mathrm{II}} = \frac{\sigma(d^2z/dy^2)}{[1 + (dz/dy)^2]^{3/2}} \tag{2.46}$$

The hydrostatic pressure variations in the two fluids are given by

$$P_{\mathrm{I}} = P_v = P_0 - \rho_v g z \tag{2.47a}$$

$$P_{\mathrm{II}} = P_l = P_0 - \rho_l g z \tag{2.47b}$$

where P_0 is the pressure at the interface far from the vertical wall. Substituting these equations into Eq. (2.46) and rearranging, we obtain

$$\frac{(\rho_l - \rho_v)gz}{\sigma} - [1 + (z')^2]^{-3/2} z'' = 0 \tag{2.48}$$

where the primes denote differentiation with respect to y. Multiplying Eq. (2.48) by z' and integrating gives

$$\frac{(\rho_l - \rho_v)gz^2}{2\sigma} + [1 + (z')^2]^{-1/2} = C_1 \tag{2.49}$$

Since z and $z' \to 0$ as $y \to \infty$, $C_1 = 1$. At $y = 0$, z' is given by the contact angle as

$$z'(0) = \cot \theta \tag{2.50}$$

Equation (2.49) with $C_1 = 1$ and Eq. (2.50) can be solved for $z(0)$. The resulting relation is

$$z_0 = z(0) = \left[\frac{2\sigma(1 - \sin \theta)}{(\rho_l - \rho_v)g}\right]^{1/2} \tag{2.51}$$

Note that z_0 is the height to which the liquid climbs at the vertical wall. Using Eq. (2.51) as a boundary condition, integration of Eq. (2.49) yields the following relation for the shape of the interface:

$$\frac{y}{L_c} = \cosh^{-1}\left(\frac{2L_c}{z}\right) - \cosh^{-1}\left(\frac{2L_c}{z_0}\right) + (4 + z_0^2/L_c^2)^{1/2} - (4 + z^2/L_c^2)^{1/2} \tag{2.52}$$

where

$$L_c = \left[\frac{\sigma}{(\rho_l - \rho_v)g}\right]^{1/2} \tag{2.53}$$

In a similar manner, the free surface of a liquid in small tubes and porous media will arise or fall to satisfy the Young-Laplace equation. This phenomenon

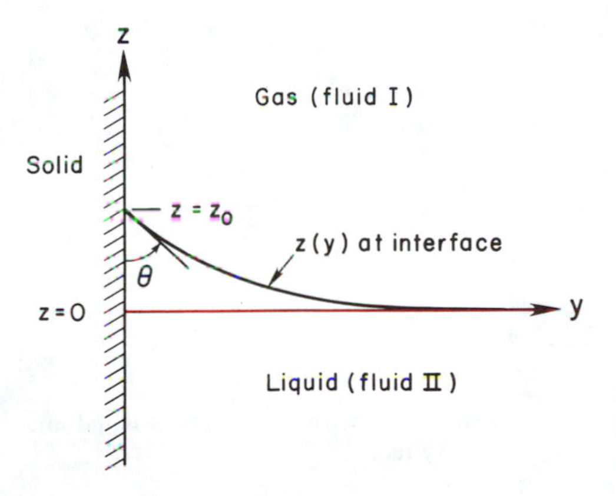

Figure 2.7 Rise height of a wetting liquid at the contact line with a solid vertical wall.

is known as *capillarity*. Consider the small tube of radius r_i shown in Fig. 2.8. The tube contains liquid with a free surface and is in contact with an extensive pool of liquid. The liquid is assumed to meet the wall at the contact angle $\theta < 90°$ (i.e., the liquid wets the wall). When $r_i \ll L_c$ given by Eq. (2.53), the radius of curvature of the interface is approximately uniform and equal to $r_i/\cos\theta$. For the small radius of curvature in this interface, the Young-Laplace equation requires a large jump in pressure across the interface. This pressure difference supports a considerable column of liquid against gravity.

Example 2.3 A cylindrical container is filled with saturated liquid R-12 and its vapor at 32°C. Determine the height to which the liquid will climb the vertical walls of the container if the contact angle with the walls is 5°.

At 32°C, the interfacial tension for saturated liquid R-12 liquid in contact with its vapor is 0.0077 N/m, and

$$\rho_v = 1.284 \text{ kg/m}^3 \qquad \rho_l = 44.8 \text{ kg/m}^3$$

The height to which the liquid climbs the vertical wall is given by Eq. (2.51):

$$z_0 = \left[\frac{2\sigma(1 - \sin\theta)}{(\rho_l - \rho_v)g}\right]^{1/2}$$

$$= \left\{\frac{2(0.0077)[1 - \sin(5°)]}{(44.8 - 1.284)(9.8)}\right\}^{1/2} = 0.0057 \text{ m} = 5.7 \text{ mm}$$

Combining the Young-Laplace equation with the hydrostatic pressure variation in the fluids, the condition for equilibrium is

$$P_{\text{I}} - P_{\text{II}} = (\rho_l - \rho_v)gz_i = 2\sigma\frac{\cos\theta}{r_i} \tag{2.54}$$

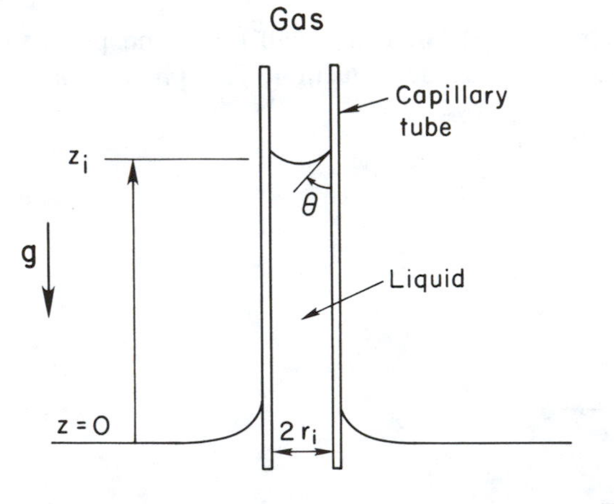

Figure 2.8 The rise of a wetting liquid in a capillary tube.

This equation can be solved for the equilibrium height of the liquid column, z_i:

$$z_i = 2\sigma \frac{\cos \theta}{(\rho_l - \rho_v) g r_i} \tag{2.55}$$

Equation (2.55) indicates that z_i can be very large if the tube diameter is very small. In a completely analogous way, porous materials such as blotting paper, bricks, or soil with very small pore sizes exert a strong "suction" on wetting fluids such as water. Note also that Eq. (2.55) correctly indicates that for a liquid that does not "wet" the tube walls, θ will be greater than 90° and z_i will be negative, corresponding to depression of the free surface in the tube below that outside the tube.

Example 2.4 Determine the capillary rise height for water at 20°C in a capillary tube with an inside diameter of 0.2 mm. The contact angle of the interface with the tube wall is 20°.

The surface tension and liquid density for water at 20°C are 0.0728 N/m and 998 kg/m^3, respectively. The rise height in the capillary tube is given by Eq. (2.55) as

$$z_i = \left[\frac{2\sigma \cos \theta}{(\rho_l - \rho_v) g r_i} \right]$$

For air at atmospheric pressure and 20°C,

$$\rho_{\text{air}} = \frac{P}{(\overline{R}/\overline{M})T} = \frac{101{,}000}{(8.314/28.97)(1{,}000)(273 + 20)} = 1.20 \text{ kg/m}^3$$

Substituting yields

$$z_i = \left[\frac{2(0.0728) \cos (10°)}{(998 - 1.20)(9.8)(0.0001)} \right] = 0.147 \text{ m} = 14.7 \text{ cm}$$

It is worth noting that the cos θ function is almost constant near $\theta = 0$. Consequently, the rise height is relatively insensitive to the exact wetting characteristics of the tube wall, provided that θ is small.

2.4 TEMPERATURE AND CONTAMINANT EFFECTS ON INTERFACIAL TENSION

For a pure liquid in contact with its vapor, the interfacial tension is a function of temperature. This dependence on temperature is somewhat expected, at least in the vicinity of the critical point, since as the critical temperature is approached (at saturation), the properties of the two fluids become identical and the interfacial tension must vanish. This implies that the surface tension will decrease with increasing temperature, vanishing at the critical temperature T_c. This is virtually always true, and is often the basis for interpolation schemes and curve-fit equations used to estimate the variation of surface tension with temperature.

For example, the surface tension of pure water in contact with its vapor can be computed from the relation [2.7]

$$\sigma = 235.8\left(1 - \frac{T_{sat}}{T_c}\right)^{1.256}\left[1 - 0.625\left(1 - \frac{T_{sat}}{T_c}\right)\right] \tag{2.56}$$

where both T_{sat} and T_c are in kelvins and the resulting σ values are in millinewtons per meter. This equation matches available data for water at temperatures between 0.01°C and 300°C to within ±0.5%. From 300°C to the critical point (374.15°C), it deviates from the data by no more than 13%.

As is the case for most liquids, the surface tension for water decreases almost linearly with temperature, as seen in Fig. 2.9. Consequently, curve-fit equations for surface tension data are invariably almost linear in nature. Jaspar [2.8] fit a linear relation of the form

$$\sigma = C_0 - C_1 T \tag{2.57}$$

to surface tension data for a wide variety of pure liquids over limited temperature ranges. In this dimensional equation, the temperature T is in degrees celsius. Values of the constants C_0 and C_1 for some substances are listed in Table 2.2.

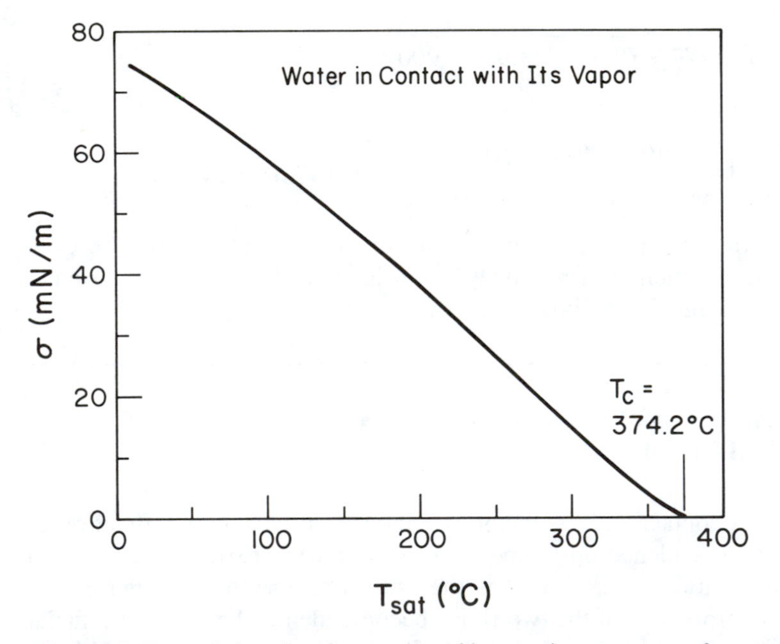

Figure 2.9 The variation of surface tension with saturation temperature for water in contact with its vapor.

Table 2.2 Constants for the linear surface tension relation proposed by Jasper [2.8] and values of the Reidel parameter [see Eq. (2.59)] for various substances (from Miller [2.11])

	Constants for Eq. (2.57)			
Substance	C_0 (mN/m)	C_1 (mN/m °C)	Temperature range (°C)	Reidel parameter
Acetone	26.26	0.112	25 to 50	7.30
Acetylene	3.42	0.1935	−90 to −50	
Argon	−34.28	0.2493	−189 to −181	
Butane	14.87	0.1206	−70 to 20	
n-Butyl alcohol	27.18	0.08983	10 to 100	8.91
Carbon tetrachloride	29.49	0.1224	15 to 105	6.75
Chlorine	19.87	0.1897	−80 to −30	
Ethyl alcohol	24.05	0.0832	10 to 70	8.98
Ethylene glycol	50.21	0.089	20 to 140	
Flourine	−16.10	0.1646	−202 to −188	
Heptane	22.10	0.0980	10 to 90	
Hydrazine	72.41	0.2407	15 to 40	
Hydrogen				4.74
Hydrogen peroxide	78.97	0.1549	2 to 20	
Isopropyl alcohol	22.90	0.0789	10 to 80	
Mercury	490.6	0.2049	5 to 200	
Methyl alcohol	24.00	0.0773	10 to 60	8.48
Nitrogen	−26.42	0.2265	−195 to −183	5.98
Octane	23.52	0.09509	10 to 120	
Oxygen	−33.72	0.2561	−202 to −184	5.92
Propane	9.22	0.0874	−90 to 10	
Sulfur dioxide	26.58	0.1948	−50 to 10	
Water	75.83	0.1477	10 to 100	7.39

Based on the principle of corresponding states, Brock and Bird [2.9] developed the following relation for estimating the surface tension of nonpolar liquids:

$$\sigma = P_c^{2/3} T_c^{1/3} (0.133 R_c - 0.281) \left[1 - \left(\frac{T}{T_c} \right) \right]^{11/9} \tag{2.58}$$

where R_c is the Reidel [2.10] parameter, defined as

$$R_c = \left\{ \frac{d[\ln P_{\text{sat}}(T)]}{d(\ln T)} \right\}_{T=T_c} \tag{2.59}$$

Values of R_c for various substances have been tabulated by Miller [2.11]. Some of the R_c values given by Miller [2.11] are listed in Table 2.2. Miller [2.11] also proposed the following empirical relation for R_c:

$$R_c = \frac{0.9076[1 + (T_b/T_c) \ln P_c]}{[1 - (T_b/T_c)]} \tag{2.60}$$

where T_b is the normal boiling point temperature of the liquid (at atmospheric pressure). In these relations, temperatures are in kelvins, P_c is in atmospheres, and σ is in millinewtons per meter.

Using Eqs. (2.58) and (2.60) together, it is possible to estimate the variation of surface tension with temperature from critical data and the normal boiling point. This predictive method is expected to be applicable to mechanically similar nonpolar molecules. A wide variety of simple organic compounds fit this description. This method is not expected to be accurate for very light molecules such as hydrogen and helium, highly polar inorganic substances such as water and ammonia, or associating substances such as alcohols, fused salts, and liquid metals.

In addition to temperature, the second major factor that affects surface tension is the presence of one or more other substances dissolved in the liquid. From a thermodynamic analysis of the liquid–vapor interfacial region for a binary mixture, in which species A is a solute that exhibits ideal mixture behavior in solvent B, it can be shown (see, e.g., reference [2.4]) that the surface excess mass of species A is related to the variation of σ with concentration as

$$\Gamma_A = -\left(\frac{x_A}{RT}\right)\left(\frac{\partial \sigma}{\partial x_A}\right)_T \tag{2.61}$$

In this equation, x_A is the concentration of the solute species A. It can be seen from Eq. (2.61) that if σ decreases as x_A increases, $(\partial \sigma / \partial x_A)$ is negative and Γ_A is positive, meaning that the solute tends to concentrate at the interface. Solutes for which this is true are called *surface-active materials*.

Some surface-active materials are so highly enriched at the interface that they are termed surface-active agents or *surfactants*. Because they concentrate so highly at the interface, the presence of materials of this type even in very low bulk concentrations may significantly alter the interfacial tension.

Equation (2.61) may also be interpreted in a converse manner (i.e., if for some reason species A accumulates at the interface), so that $\Gamma_A > 0$; then $(\partial \sigma / \partial x_A)_T$ is negative and the presence of the surface-active material decreases the surface tension. This is the situation when a typical soap is added to water. Because water is a highly polar material, polar molecules are readily accepted into its structure, whereas nonpolar entities, such as hydrocarbon chains, are not. Consequently, pure hydrocarbons are rather insoluble in water, while polar materials have considerable solubility in water.

With these observations, we can at least qualitatively interpret the effect of soap molecules on the surface tension of water. Typical soap molecules have a hydrocarbon chain and a polar group, as indicated schematically in Fig. 2.10. Given water's affinity for the polar group and aversion to the nonpolar end of the soap molecule, the system clearly would prefer a configuration in which the polar group remains in the water and the hydrocarbon end is removed from the water. As schematically indicated in Fig. 2.10, this preferred configuration can be achieved if the soap molecules take up positions in a monomolecular layer or *monolayer* at the interface between the water and the vapor or other fluid. Because this configuration is preferred, the soap molecules generally tend to concentrate at the

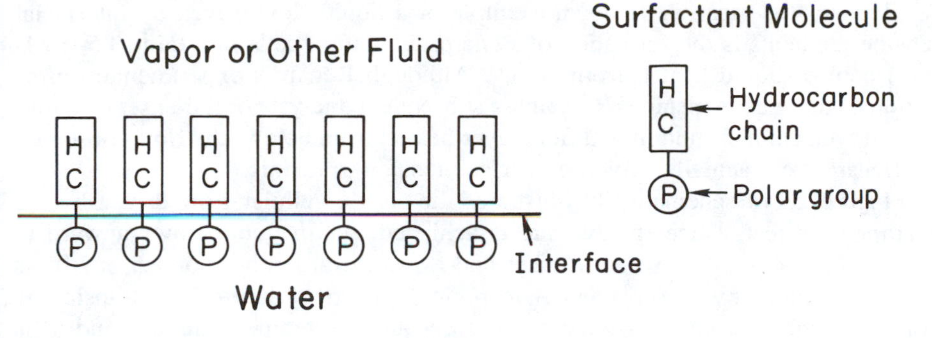

Figure 2.10 Orientation of surfactant molecules at the interface between water and another nonpolar fluid.

interface, in which case Eq. (2.61) implies that the interfacial tension will be reduced.

Many different types of polar groups exist, and they may be combined with many different hydrocarbon groups to form the type of surfactant molecule described here. Some of the more common simple polar groups are amine (NH_3^+), alcohol (OH), carboxylic acid (COOH), and sulfate (SO_4^-). A wide variety of straight or branched hydrocarbon chains may be linked to the polar group, making it possible to construct an almost infinite number of different surfactant molecules.

Because so many substances can act as surfactants in water, and because it takes only a minute amount of them to form a monolayer at the interface, it is particularly easy for the surface tension of water to be altered by trace contaminants. The surface tension can, in fact, be a useful indicator of the purity of water. The need for high purity to avoid surface contamination and a reduction in surface tension should be considered when planning or evaluating experiments involving interfacial transport phenomena.

Further information regarding methods to predict values of interfacial tension for pure substances and mixtures can be found in reference [2.12].

2.5 EFFECTS OF INTERFACIAL TENSION GRADIENTS

As noted in Section 2.4, the surface tension at the interface between a liquid and a vapor phase varies with the temperature and species concentrations in the liquid. Consequently, if the temperature or composition varies over the interface, the interfacial tension will also be nonuniform, with the result that liquid near the interface in regions of low surface tension will be pulled toward regions of higher surface tension. If the temperature or concentration nonuniformities are maintained somehow, a steady flow pattern may be established in which the pulling action at the interface resulting from the surface tension gradient is balanced locally by viscous shear stress in the liquid flow. Motion of the liquid due to surface tension gradients at the interface is sometimes called the *Marangoni effect*.

Perhaps the most well-known example of a liquid flow driven by interfacial tension gradients is the formation of Benard circulation cells in a thin (0.5- to 1-mm) pool of liquid heated from below. Although Benard's experimental observation of these cells inspired Rayleigh's subsequent analysis of the onset of buoyancy-driven cellular motion in a fluid layer heated from below, the flows observed by Benard were actually driven by surface tension gradients.

Figure 2.11*a* schematically illustrates the steady cellular flow driven by the Marangoni effect. Once steady state is achieved, warm liquid flows upward to the interface at point *A*, where it turns and flows toward either point *B* or *B'*. As the warm fluid flows from point *A* to point *B*, it cools due to heat transfer by convection to the cooler surrounding gas. Because the temperature at *B* and *B'* is less than at *A*, a surface tension gradient is maintained that drives the flow along the interface from *A* to *B* and from *A* to *B'*. The symmetric convergence of flow at points *B* and *B'* causes the fluid to turn downward toward the wall, whereupon it absorbs heat from the wall and repeats the cycle.

In some cases the cellular flow may be accompanied by significant deflection of the interface, as shown in Fig. 2.11*b*. Note that moving the interface at location *A* toward the surface may increase the temperature at *A*, while moving the interface at *B* away from the surface may reduce the temperature at *B*. Deflection of the interface in this manner thus has the effect of increasing the surface tension gradient between points *A* and *B*. In some systems the increased surface tension gradient will move fluid rapidly enough from *A* to *B* to maintain the deflected interface configuration.

For a thin liquid layer heated from below, the conditions for the onset of cellular motion can be predicted using linear stability analysis. The postulated initial condition of the liquid layer is schematically indicated in Fig. 2.12. Initially the liquid is motionless, and heat transfer from the wall to the interface is by

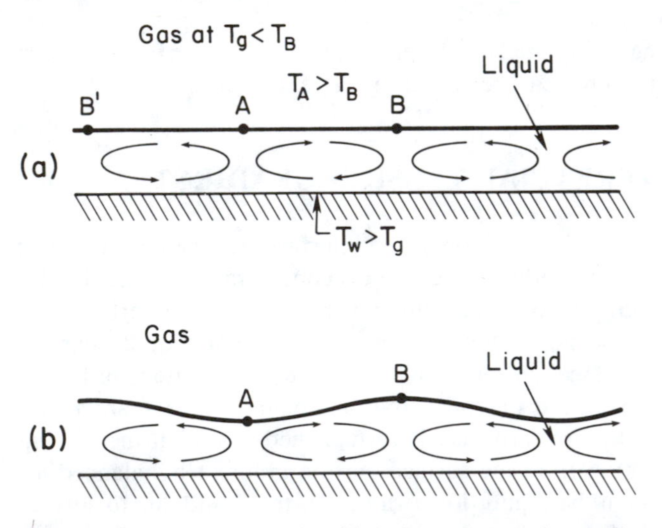

Figure 2.11 Cellular convection driven by surface tension gradients.

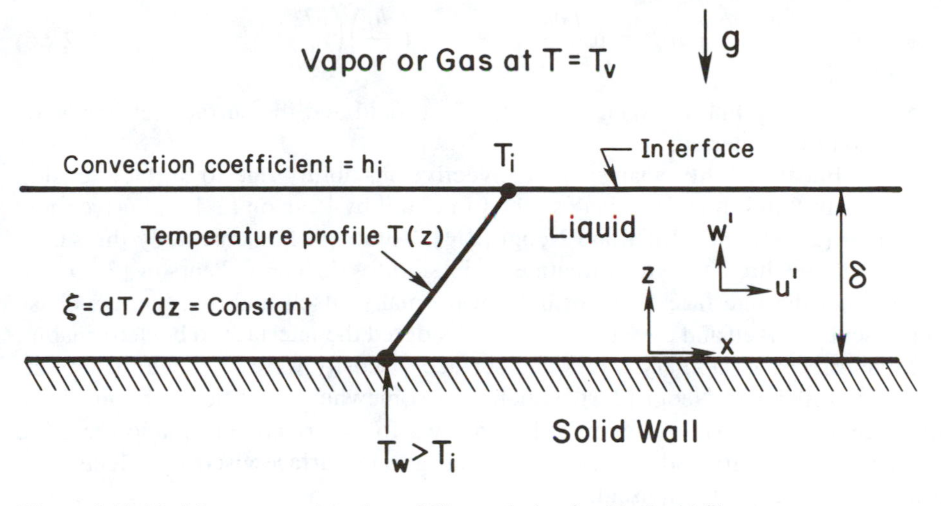

Figure 2.12 Model system used in the stability analysis of a thin liquid film.

conduction alone. The temperature profile in the liquid is linear, with a constant slope dT/dz equal to ξ. Heat is assumed to be transferred from the interface to the surrounding bulk vapor by convection in such a manner that the heat transfer coefficient $h_i = q''/(T_i - T_g)$ is constant and uniform over the interface.

Following the usual practice for linear stability analysis, the local temperature is assumed to be equal to the sum of the basic flow temperature $\overline{T}$ given by the initial linear profile, plus a small sinusoidal fluctuation T' representing a Fourier component of random disturbances in the system:

$$T = \overline{T} + T' = (T_w - \zeta z) + T' \tag{2.62}$$

Associated with the temperature perturbation are small perturbation velocities w' and u'. Because all base flow velocities for the initial state $\overline{w}$ and $\overline{u}$ are zero,

$$w = \overline{w} + w' = w' \qquad u = \overline{u} + u' = u' \tag{2.63}$$

The linear stability analysis proceeds by substituting Eqs. (2.62) and (2.63) into the governing continuity, momentum, and thermal energy transport equations, subtracting the corresponding base flow equations, and neglecting products of perturbation quantities. The resulting equations for the perturbation quantities T', w', and u' are solved with appropriate boundary conditions assuming that T' and w' are of the form

$$T' = \theta(z)e^{i\alpha x + \beta t} \tag{2.64}$$

$$w' = W(z)e^{i\alpha x + \beta t} \tag{2.65}$$

In the above relations, α is the wavenumber of the perturbation and the real and imaginary parts of β represent its amplification factor and temporal frequency. The boundary conditions at the interface ensure that the usual force, mass, momentum, and energy balances are enforced, and the additional condition

$$(\tau_{xz})_{z=\delta} = \mu_l \left(\frac{\partial u'}{\partial z} \right)_{z=\delta} = - \left(\frac{d\sigma}{dT} \right) \left(\frac{\partial T'}{\partial x} \right)_{z=\delta} \tag{2.66}$$

provides the key link between the liquid flow field and the surface tension variation along the interface.

The linear stability analysis of convective instability due to surface tension gradients in thin liquid films is described in detail by Pearson [2.13], Scriven and Sternling [2.14], and Miller and Neogi [2.4]. Although fundamentally the same, each of these three analytical treatments is slightly different. Pearson [2.13] assumed that the interface does not deform normal to itself and that the gas phase is inviscid. Scriven and Sternling [2.14] considered the interface to be deformable, and the effects of capillary waves and surface viscosity were included. The treatment in Miller and Neogi [2.4], which is a somewhat simplified version of the more general analysis of Smith [2.15], allows for interface deformation and the presence of capillary and gravity waves, but ignores surface viscosity effects, and assumes the gas to be inviscid.

In all of the previous studies noted above, both the real and imaginary parts of β were taken to be zero. The resulting stationary wave solutions, which do not amplify or damp with time, are therefore interpreted as being neutrally stable perturbations. The conditions for which these solutions are obtained are assumed to correspond to the onset of instability, which will ultimately lead to the transition to cellular convection.

Because details of the linear stability analysis are described in the references noted above [2.4, 2.13–2.15], a full description of the analysis is not presented here. Some of the results of these analyses are noteworthy, however. If the gas is assumed to be inviscid and surface viscosity effects are ignored, this type of linear stability analysis indicates that the neutral stability conditions can be specified in terms of a dimensionless wavenumber $\hat{\alpha}$, the Marangoni number Ma, the Biot number Bi, the Bond number Bo, and the "crispation number" Cr, defined as

$$\hat{\alpha} = \alpha\delta \tag{2.67}$$

$$\text{Ma} = \frac{\xi(d\sigma/dT)\delta^2}{\alpha_{Tl}\mu_l} \tag{2.68}$$

$$\text{Bi} = \frac{h_i\delta}{k_l} \tag{2.69}$$

$$\text{Bo} = \frac{(\rho_l - \rho_g)g\delta^2}{\sigma} \tag{2.70}$$

$$\text{Cr} = \frac{\mu_l\alpha_{Tl}}{\sigma\delta} \tag{2.71}$$

In the above expressions, k_l, μ_l, and α_{Tl} are the thermal conductivity, viscosity, and thermal diffusivity of the liquid, respectively. The linear stability analysis (see reference [2.4]) predicts that neutral (or marginal) stability corresponds to

$$\mathrm{Ma} = \frac{8\hat{\alpha}(\hat{\alpha} \cosh \hat{\alpha} + \mathrm{Bi} \sinh \hat{\alpha})(\hat{\alpha} - \sinh \hat{\alpha} \cosh \hat{\alpha})}{\hat{\alpha}^3 \cosh \hat{\alpha} - \sinh^3 \hat{\alpha} - (8\, \mathrm{Cr}\, \hat{\alpha}^5 \cosh \hat{\alpha})/(\mathrm{Bo} + \hat{\alpha}^2)} \tag{2.72}$$

The neutral stability curves predicted by Eq. (2.72) for various combinations of Bi, Cr, and Bo are shown in Fig. 2.13. For combinations of $\hat{\alpha}$ and Ma below the neutral curve, the system is stable; whereas above the curve the system is unstable. As illustrated in Fig. 2.13, an increase in the Biot number (sometimes referred to as a Nusselt number by earlier investigators) shifts the neutral curve upward, implying that the system is more stable. For thin liquid layers in real systems, the range of Bond numbers is small, and its variation typically has only a small effect on the location of the neutral curve.

Generally, the position of the neutral curve at small wavenumbers is very sensitive to the value of Cr, while at higher wavenumbers it is very insensitive to changes in Cr. The limit Cr $\rightarrow$ 0 corresponds to thick layers or high surface tension. For such conditions the interface deflection is small, vanishing in the limit Cr = 0. The early analysis of Pearson [2.13] considered conditions that correspond to the limit Cr $\rightarrow$ 0.

For real systems, however, the value of Cr is usually very small, but greater than zero. For example, if we consider a 0.5-mm-thick layer of water on a plate

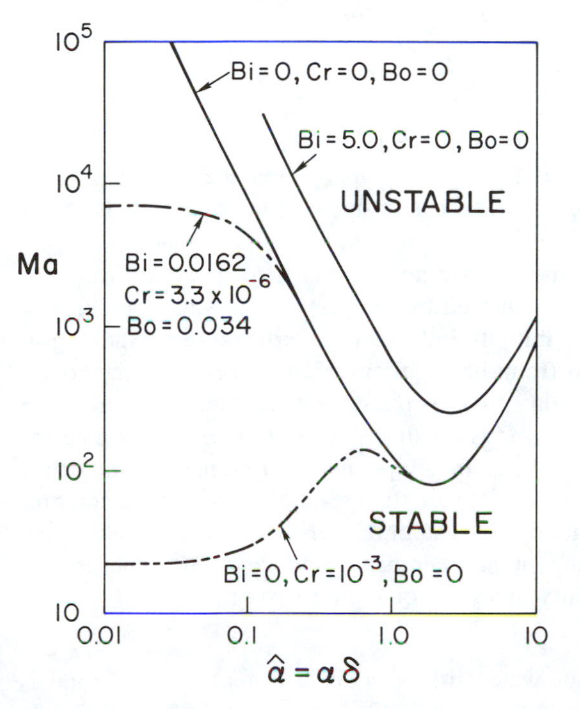

Figure 2.13 Stability plane for the onset of cellular motion as predicted using linear stability analysis.

held at 30°C and surrounded by air at 25°C and atmospheric pressure, the corresponding values of Cr and Bo are

$$\text{Cr} = 3.3 \times 10^{-6} \qquad \text{Bo} = 0.034$$

and if the heat transfer coefficient from the interface to the ambient air is 20 W/m^2 K (a typical value for natural convection), the value of the Biot number is

$$\text{Bi} = 0.0162$$

The neutral stability curve predicted by Eq. (2.72) for these values of Cr, Bo, and Bi is shown in Fig. 2.13. Even though the value of Cr is small, this system is much less stable to long-wavelength perturbations than would be the case if Cr = 0. The reason for this is that surface deflection is permitted for nonzero values of Cr. Long-wavelength deflection of the interface tends to make the system less stable by depressing high-temperature regions and elevating colder regions, as shown in Fig. 2.11*b*. For very long wavelengths, gravity has a stabilizing influence that causes the neutral stability curve to level off as $\hat{\alpha} \to 0$.

Because it is generally assumed that random disturbances contain Fourier components of all wavelengths, the system is expected to be unstable if any wavelength is unstable. Hence, the minimum value of Ma on the neutral stability curve is the critical value Ma_c, above which the system is likely to undergo a transition to cellular convection. The curves in Fig. 2.13 indicate that unless Cr exceeds about 10^{-4}, the critical Marangoni number is about 80 and the associated unstable dimensionless wavenumber $\hat{\alpha}_c$ is about 2, which corresponds to a wavelength λ_c of

$$\lambda_c = \frac{2\pi}{\alpha_c} = \frac{2\pi}{2/\delta} = \pi\delta$$

For the 0.5-mm-thick layer of water considered, this implies that the layer is unstable for temperature differences across the layer greater than about 0.12°C. Alternatively, it may be stated that the layer is unstable at heat flux levels above 145 W/m^2.

In real systems, the stability mechanism and cellular flow resulting from the Marangoni effect can be signficantly altered by the presence of small amounts of surface-active contaminants. Flow initially induced by interfacial temperature gradients will carry the contaminants from the high-temperature regions to the cooler regions. The resulting concentration of the surfactant at the cooler locations on the interface will reduce the surface tension there. This will tend to counteract the surface tension gradient resulting from temperature differences, which may weaken or suppress the cellular motion. Hence the presence of surfactant contaminants tends to enhance the stability of the system. Experimentally, instability may not be observed until the Marangoni number has greatly exceeded the critical value Ma_c predicted by the stability analysis for a pure system.

Example 2.5 A thin layer of water sits on a surface held at 100°C and is exposed to air at a bulk temperature of 20°C. The heat transfer coefficient between the interface and the bulk air is 8.0 $W/m^2\,°C$. Estimate the range of layer thicknesses for which Marangoni effects may become important.

Initially making the assumption that the Bond and Biot numbers are very small and the crispation number is small but finite, the critical Marangoni number is about 80. Taking the temperature gradient across the layer to be equal to $(T_w - T_i)/\delta$, the critical layer thickness δ_c is given by

$$\mathrm{Ma}_c = 80 = \frac{T_w - T_i}{\delta_c}\left(\frac{d\sigma}{dT}\right)\frac{\delta_c^2}{\alpha_{Tl}\mu_l}$$

Applying the resistance analogy for steady transport of heat from the surface to the bulk air, it can be shown that

$$T_w - T_i = \frac{\delta/k_l}{(1/h_c) + (\delta/k_l)}(T_w - T_{\mathrm{air}})$$

Substituting this relation into that above for the critical Marangoni number yields

$$\mathrm{Ma}_c = 80 = \left[\frac{T_w - T_{\mathrm{air}}}{(1/h_c) + (\delta_c/k_l)}\right]\left(\frac{d\sigma}{dT}\right)\frac{\delta_c^2}{\alpha_{Tl}\mu_l k_l}$$

For water at 60°C, $\alpha_{Tl} = 0.158 \times 10^{-6}\ m^2/s$, $k_l = 0.653$ W/m °C, $\mu_l = 4.67 \times 10^{-6}\ Ns/m^3$, $\sigma = 0.0644$ N/m, $d\sigma/dT = -1.79 \times 10^{-4}$ N/m °C. Substituting these property values and solving the above equation iteratively, the layer thickness corresponding to the critical Marangoni number is determined to be

$$\delta_c = 0.00026\ \mathrm{m} = 0.26\ \mathrm{mm}$$

Thus the system is unstable and Marangoni effects may become important for $\delta > 0.26$ mm. Using Eqs. (2.69) through (2.71), it can be shown that $\mathrm{Cr} = 4.41 \times 10^{-7}$, $\mathrm{Bi} = 3.19 \times 10^{-3}$, and $\mathrm{Bo} = 2.67 \times 10^{-6}$. These values are consistent with the initial assumptions regarding these parameters.

Besides being a driving mechanism for cellular convection, interfacial tension gradients are also responsible for other interfacial flow phenomena. In multicomponent systems, when one or more species is transferred across a liquid–vapor interface, an irregular, quasi-cellular flow pattern can result that is often referred to as *interfacial turbulence*. Although somewhat periodic in nature, flows of this type are generally much less regular than the Benard cells previously discussed. This irregular flow is driven by interfacial tension gradients resulting from interfacial concentration gradients.

Interfacial tension is typically a strong function of concentration. For evaporation from the interface of a multicomponent liquid, the variation of surface tension with concentration usually dominates over the variation with temperature, and, as a result, it usually plays a dominant role in the development of interfacial turbulence. Because heat and mass transfer at the interface may be greatly enhanced by the presence of interfacial turbulence, this phenomenon may be particularly important to the accurate prediction of heat and mass transfer in equipment where vaporization or condensation of multicomponent mixtures occurs. Further discussion of this phenomenon may be found in reference [2.16].

Marangoni flow can also arise during evaporation from the surface of liquid jets and at the interface of liquid drops or cylinders formed by mixing or agitation in an evaporating two-phase flow. Again, because surface tension usually varies strongly with concentration, the effects of Marangoni flow on the breakup of liquid jets and droplet breakoff from liquid films in two-phase flow may be particularly significant in multicomponent systems. The Marangoni effect may act to stabilize or destabilize the interface, depending on the nature of the multicomponent mixture. The role of the Marangoni effect in these circumstances is discussed in detail in reference [2.17].

The Marangoni effect also causes vapor bubbles in a liquid with an imposed temperature gradient to move toward the high-temperature region. Motion in this direction is thermodynamically favored because it reduces the interfacial free energy of the bubble. The liquid flow that causes the bubble motion toward the high-temperature region is driven by interfacial tension gradients.

Evaporation of a falling film of a binary liquid mixture on a heated vertical surface may also be affected by Marangoni flow if the more volatile component of the mixture has a higher surface tension. A transverse wavy perturbation of the film interface, like that shown in Fig. 2.14, would be expected to result in more rapid evaporation of the more volatile component in the thin portion of the film compared to that in the thick portion. This would produce a concentration difference between the thick and thin film regions, resulting in an interfacial tension gradient that draws liquid from the thin regions into the thick regions. Thus the resulting Marangoni flow promotes breakup of the film into rivulets. Conversely, the Marangoni effect opposes rivulet formation if the more volatile component has a lower surface tension. When it occurs, the transition from film flow

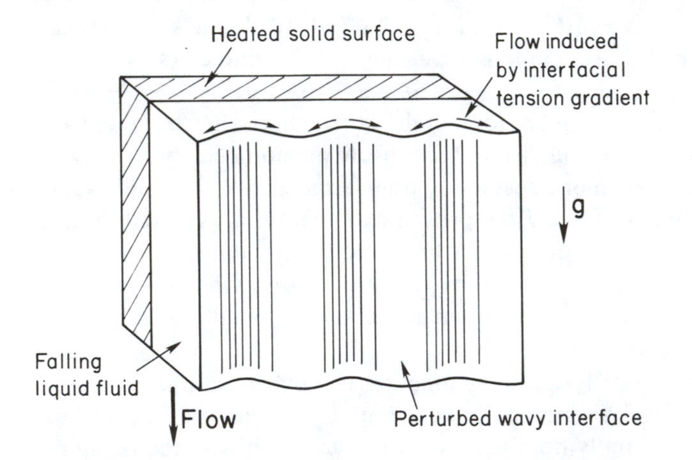

Figure 2.14 Lateral wavy perturbations that may lead to breakup of a falling film of a binary liquid mixture.

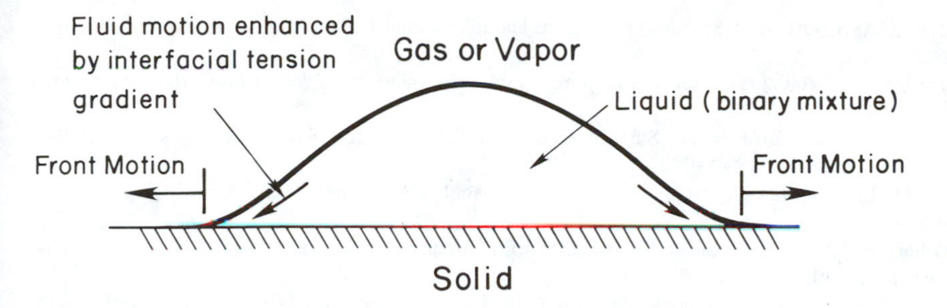

Figure 2.15 Schematic illustrating the possible effects of interfacial tension gradients on spreading of a liquid droplet on a solid surface.

to rivulets often produces a drastic drop in the heat transfer performance for these circumstances.

Interfacial tension gradients may also affect the spreading of an evaporating liquid droplet over a warm surface. We consider specifically a binary liquid mixture in which the more volatile component has a lower surface tension. As shown in Fig. 2.15, a thinner advancing layer of liquid forms at the perimeter of the droplet. Because it is thinner, evaporation of the more volatile component is more rapid than in the center of the droplet. The resulting concentration difference produces a surface tension gradient that enhances the motion of liquid from the bulk droplet to the advancing layer at its perimeter. The overall effect is to increase the rate of spreading of the liquid over the solid.

The formation of "wine tears" is also a consequence of the effect of Marangoni flow on the spreading of a binary liquid. Wine is, to a first approximation, a mixture of water and ethanol, with ethanol being the more volatile and having the lower surface tension of the two components. The rate of liquid spreading up the walls of the wine glass is enhanced by the Marangoni flow until enough liquid accumulates to form droplets or "tears." Spreading of liquids on solid surfaces is discussed in more detail in Chapter Three.

REFERENCES

2.1 Chappuis, J., Contact angles, in *Multiphase Science and Technology*, vol. 1, pp. 387–505, Hemisphere, New York, 1982.

2.2 Fowkes, F. M., Attractive forces at interfaces, in *Chemistry and Physics of Interfaces*, American Chemical Society, Washington, DC, 1965.

2.3 Adamson, A. W., *Physical Chemistry of Surfaces*, 4th ed., Wiley, New York, 1982.

2.4 Miller, C. A., and Neogi, P., *Interfacial Phenomena*, Marcel Dekker, New York, 1985.

2.5 Bashforth, F., and Adams, J. C., *An Attempt to Test the Theory of Capillary Action*, Cambridge University Press, Cambridge, 1893.

2.6 Padday, J. F., Theory of surface tension, in *Surface and Colloid Science*, vol. 1, pp. 39–151, E. Matijevic, Ed., Wiley, New York, 1969.

2.7 National Bureau of Standards, *Release of Surface Tension of Water Substance*, the International Association for the Properties of Steam, December 1976, available from the Executive Secretary, IAPS, Office of Standard Reference Data, National Bureau of Standards, Washington, DC.

2.8 Jasper, J. J., The surface tension of pure liquid compounds, *J. Phys. Chem. Ref. Data,* vol. 1, pp. 841–1010, 1972.

2.9 Brock, J. R., and Bird, R. B., Surface tension and the principle of corresponding states, *AIChE J.*, vol. 1, pp. 174–184, 1955.

2.10 Reidel, L., Eine neue universelle dampfdruckformel, *Chem. Ing. Tech.*, vol. 26, pp. 83–89, 1954.

2.11 Miller, D. G., On the reduced Frost-Kalkwarf vapor-pressure equation, *Ind. Eng. Chem. Fundamentals,* vol. 2, pp. 78–88, 1963.

2.12 Reid, R. C., Prausnitz, J. M., and Poling, B. E., *The Properties of Gases and Liquids,* 4th ed., McGraw-Hill, New York, 1987.

2.13 Pearson, J. R. A., On convection cells induced by surface tension, *J. Fluid Mech.*, vol. 4, pp. 489–500, 1958.

2.14 Scriven, L. E., and Sternling, C. V., On cellular convection driven by surface-tension gradients: Effects of mean surface tension and surface viscosity, *J. Fluid Mech.*, vol. 19, pp. 321–340, 1964.

2.15 Smith, K. A., On convective instability induced by surface-tension gradients, *J. Fluid Mech.*, vol. 24, pp. 401–414, 1960.

2.16 Sternling, C. V., and Scriven, L. E., Interfacial turbulence: Hydrodynamic instability and the Marangoni effect, *AIChE J.*, vol. 6, pp. 514–523, 1959.

2.17 Bainbridge, G. S., and Sawistowski, H., Surface tension effects in sieve plate distillation columns, *Chem. Eng. Sci.*, vol. 19, pp. 992–993, 1964.

PROBLEMS

2.1 A small quantity of liquid occupies the space near the contact point between two solid cylindrical rods of radius R as shown in Fig. P2.1. The liquid interface has a concave radius of curvature r_i, which becomes smaller as the volume of liquid present becomes smaller. For small values of the angle ϕ, derive a relation between the volume of liquid present and the force acting to hold the cylindrical rods against each other.

2.2 A small drop of liquid is squeezed between two flat plates to form a thin disk of height H and diameter D. The liquid wets the surfaces of the disk so that the perimeter of the disk has a concave radius of curvature in the plane normal to the disk as shown in Fig. P2.2. If D is much larger than H, derive a relation for the force acting to hold the plates together. Determine the magnitude of the force if the liquid is water at 20°C for D = 1.0 cm and H values of 1 mm, 0.1 mm, and 0.01 mm.

2.3 Determine the capillary depression of liquid mercury at 20°C in a capillary tube with an inside diameter of 0.5 mm. The contact angle of the mercury–air interface with the wall of the capillary tube is 100°.

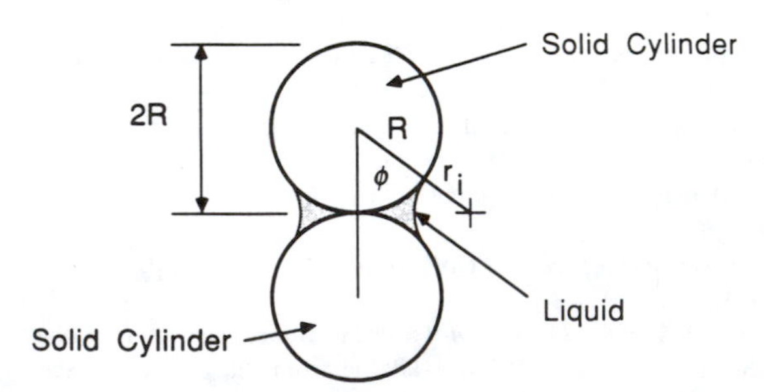

Figure P2.1

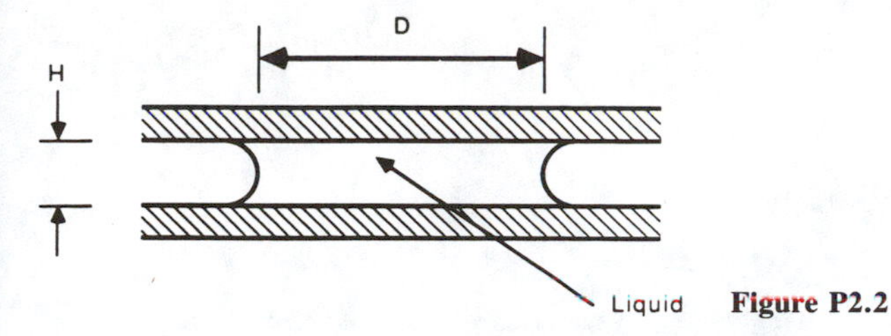

Figure P2.2

2.4 For water at 20°C in contact with air, use the results shown in Fig. 2.6 to determine and plot the variation of the sessile drop height with liquid contact angle for Bo = 20 and liquid contact angles between 5° and 90°.

2.5 Show that for a nonwetting liquid ($\theta > 90°$), the depression of the liquid–vapor interface at the side wall of the container is given by

$$z_0 = -\left[\frac{2\sigma(1 - \sin\theta)}{(\rho_l - \rho_v)g}\right]^{1/2}$$

2.6 A cylindrical vessel contains liquid (molten) silver at 1100°C. The contact angle of the silver–air interface with the vessel walls is 105°. Determine the depression of the silver–air interface at the vertical wall of the vessel. Note that for depression of the interface, Eq. (2.51) predicts the depression of the interface if the negative square root is taken as the physically realistic solution (see Problem 2.5).

2.7 A capillary tube with an inside diameter of 0.2 mm is heated so that a linear temperature gradient is maintained in its walls and in the fluid in the tube. The temperature is 20°C at the bottom of the tube, and the temperature rises linearly with height to 80°C at a height of 60 cm. The bottom of the tube is immersed in a pool of liquid water at 20°C and the top is open to the atmosphere, as is the surface of the pool. Using the fact that the variation of surface tension with temperature is given by Eq. (2.57) (with appropriate values of C_1 and C_2), find the rise height of the liquid in the capillary tube for these conditions.

2.8 For 1-mm-thick layers of methanol, liquid nitrogen, and liquid mercury at atmospheric pressure, use the results shown in Fig. 2.13 to estimate the temperature gradient that would have to be imposed to induce Marangoni instability. Use Table 2.2 and Appendix II to evaluate the physical properties in each case. The Biot number Bi may be taken as equal to zero.

2.9 A recommended means of accounting for the variation of surface tension with temperature is to use the following interpolation relation:

$$\sigma = \sigma_0\left[\frac{T_c - T}{T_c - T_0}\right]^{1.2}$$

Taking T_0 = 20°C and σ_0 = 0.0728 N/m, plot and compare the variation of σ with T predicted by this relation to that predicted by Eq. (2.56). Over the range $0°C < T < 374°C$, where is the largest difference?

CHAPTER

THREE

WETTING PHENOMENA AND CONTACT ANGLES

3.1 EQUILIBRIUM CONTACT ANGLES ON SMOOTH SURFACES

In most technological applications, a liquid–vapor phase change is accomplished by transferring energy through the walls of a container or channel into or out of a two-phase system. The vaporization or condensation process ultimately takes place at the liquid–vapor interface. However, in these circumstances the manner in which the liquid and vapor contact the solid walls through which the energy is transferred will strongly affect the resulting heat and mass transfer in the system. Consequently, the performance of heat transfer equipment in which vaporization or condensation occurs may strongly depend on the way the two phases contact the solid walls.

In everyday circumstances, it can be observed that the behavior of liquids in contact with solids may vary from one surface to another and with the type of liquid. If a small amount of liquid acetone is placed on a clean, flat aluminum surface, the liquid spreads out to form a thin film. If a small quantity of liquid water is placed on the same surface, a discrete drop is observed. Generally, liquids with weak affinities for a solid wall will collect themselves into beads, while those with high affinities for the solid will form films to maximize the liquid–solid contact area.

The affinity of liquids for solids is referred to as the *wettability* of the fluid. The general circumstance of a smooth axisymmetric bead of liquid in contact with a solid and surrounded by vapor is shown in Fig. 3.1. The solid surface is taken to be locally flat and is idealized as perfectly smooth. We will reconsider this

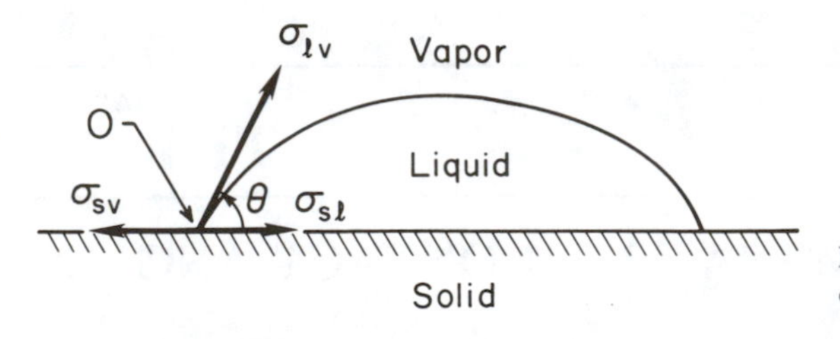

Figure 3.1 Interfacial tensions acting on a contact line.

idealization later in this chapter, but for the moment we adopt it as a plausible first approximation of a real surface.

The wettability of the liquid is quantified by the *contact angle* θ, defined as the angle between the liquid–vapor interface and the solid surface, measured through the liquid at the point 0 in Fig. 3.1 where all three phases meet. Note that for the real axisymmetric drop there is actually a line common to all three phases (the basal circle of the drop), called the *contact line*. A fixed quantity of liquid will spread more over the surface as θ decreases. In the limit of $\theta \to 0$, the liquid spreads over the entire available surface, forming a thin film.

The system shown in Fig. 3.1 has three interfaces: one between the vapor and liquid, another between the liquid and solid, and a third between the solid and vapor. For the liquid–vapor interface, σ_{lv} is the interfacial tension defined in Chapter Two. There are also interfacial tensions σ_{sl} and σ_{sv} associated with the solid–liquid and solid–vapor interfaces, respectively. At equilibrium, a force balance at point 0 in the horizontal direction requires that

$$\sigma_{sv} = \sigma_{sl} + \sigma_{lv} \cos \theta \tag{3.1}$$

The vertical force $\sigma_{lv} \sin \theta$ at point 0 must be balanced by a vertical reaction force in the solid. However, this force is usually so small, and the modulus of elasticity of the solid is usually so high, that there is no significant deformation of the solid surface.

Equation (3.1) can be rearranged to the form

$$\sigma_{lv} \cos \theta = \sigma_{sv} - \sigma_{sl} \tag{3.2}$$

which, although it is sometimes referred to as *Neumann's formula,* is more often called *Young's equation*.

We can gain a different perspective on this relation (3.2) by returning to the notion that the interfaces are actually interfacial regions, as shown in Fig. 3.2, and considering the thermodynamics of the system. We specifically consider a perturbation that results in a differential change in the positions of the interfacial surfaces S^*_{lv}, S^*_{sv}, and S^*_{sl} while the temperature and the volume of the individual phases in the system are constant. The resulting change in the total Helmholtz free energy is equal to the sum of the changes in the bulk phases and the interfacial regions:

$$dF = dF_v + dF_l + dF_s + dF_e^{lv} + dF_e^{sl} + dF_e^{sv} \tag{3.3}$$

Here, subscripts denote the bulk phase for the bulk phase properties and superscripts denote the interface for properties of the interfacial regions. In Chapter Two, it was shown that, for the interfacial region (see Eq. [2.16]),

$$dF_e^{S*} = -S_e^{S*}\,dT + \mu\,dN_e^{S*} + \sigma\,dA_i^{S*} \tag{3.4}$$

and for the bulk phases (see Eq. [2.19]),

$$dF = -S\,dT + \mu\,dN - P\,dV \tag{3.5}$$

Although Eq. (3.4) was originally developed specifically for a fluid–fluid interface, the arguments used in its development are equally valid for a solid–fluid interface. We will therefore use Eq. (3.4) to evaluate the dF terms in Eq. (3.3) for the solid–vapor and solid–liquid interfaces as well as for the liquid–vapor interface. If, in addition, we use Eq. (3.5) for the dF terms for the bulk phases, and simplify using the fact that $dT = 0$ and $dV = 0$ (because the temperature throughout the system is fixed), Eq. (3.3) becomes

$$\begin{aligned} dF = {} & \mu\,dN_v + \mu\,dN_l + \mu\,dN_s \\ & + \mu\,dN^{lv} + \mu\,dN^{sl} + \mu\,dN^{sv} \\ & + \sigma_{lv}\,dA_{lv} + \sigma_{sl}\,dA_{sl} + \sigma_{sv}\,dA_{sv} \end{aligned} \tag{3.6}$$

In deriving this result, we have also made use of the fact that, at equilibrium, μ must be the same in all the bulk and interfacial regions, as shown in Chapter Two.

Because the total number of moles in the system is unchanged, we must have

$$dN = dN_v + dN_l + dN_s + dN^{lv} + dN^{sl} + dN^{sv} = 0 \tag{3.7}$$

Substituting into Eq. (3.6) yields

$$dF = \sigma_{lv}\,dA_{lv} + \sigma_{sl}\,dA_{sl} + \sigma_{sv}\,dA_{sv} \tag{3.8}$$

For a droplet like that in Fig. 3.2, a shift in the interface surfaces that increases the area of the solid–liquid interface would result in an equal reduction in the solid–vapor interface area. This implies that

$$dA_{sv} = -\,dA_{sl} \tag{3.9}$$

If we further assume that the liquid–vapor interface is a spherical cap (which will

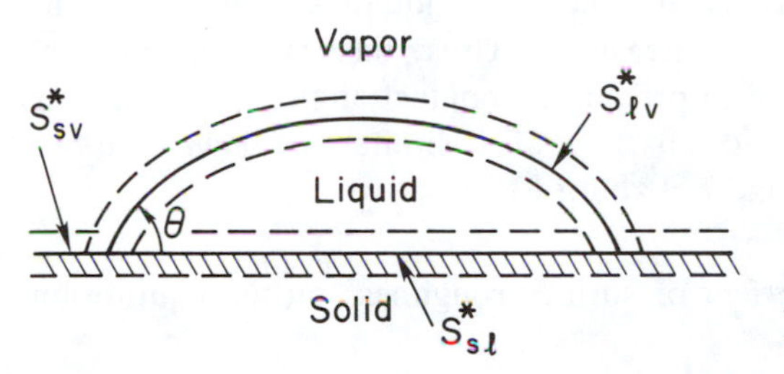

Figure 3.2 Interfacial regions associated with a liquid droplet on a solid surface.

be true if the effects of gravity body forces are small), then it can be shown from geometric considerations that

$$dA_{lv} = dA_{sl} \cos \theta \tag{3.10}$$

Substituting Eqs. (3.9) and (3.10) into Eq. (3.8) and rearranging yields

$$\frac{\partial F}{\partial A_{sl}} = \sigma_{lv} \cos \theta + \sigma_{sl} - \sigma_{sv} \tag{3.11}$$

Note that this relation gives the rate of change of the total free energy of the system with changing A_{sl} for a constant drop volume and constant temperature. As noted in Chapter One, for a system held at constant volume and temperature, equilibrium corresponds to a minimum in F. Thus, at equilibrium, F must be a minimum with respect to A_{sl}, that is, $\partial F/\partial A_{sl} = 0$. Hence, the right side of Eq. (3.11) must equal zero, which, when rearranged, yields Young's equation (3.2).

The above development demonstrates that Young's equation can be derived from a thermodynamic analysis of the interface regions and bulk phases in which the minimization of the free energy of system at equilibrium is imposed. In deriving Young's equation (3.2), we have ignored some aspects of the physical system. In particular, our treatment of the contact-line region omits some important features of the problem. In some systems, for example, there can be an adsorbed liquid film on the surface we have treated as the solid–vapor interface. In such cases, the presence of the adsorbed film affects the force balance and liquid–vapor interface shape at the contact line. Young's equation is valid when there is no adsorbed liquid film on the solid surface, but when one is present, Young's equation is not a completely accurate representation of the physics of the system.

In addition, it is assumed in the development presented above that the interfacial tensions are constant everywhere along their respective interfaces. It is known, however, that the interfacial tension very near the contact line may be different from the value far from the contact line. Fortunately, these differences are usually small, and the measured contact angle (for systems with no adsorbed film on the solid surface) usually agrees with that predicted by Young's equation.

Because the interfacial tensions are equilibrium properties, the contact angle defined by Young's equation is also an equilibrium property, sometimes called the *equilibrium contact angle*. Young's equation provides a useful framework for interpreting the observed shape of the liquid–vapor interface near its contact line with a solid surface. However, it is difficult to do much in the way of practical calculations with Young's equation, because the solid–liquid and solid–vapor interfacial tensions cannot ordinarily be measured. Thus, there are generally no available σ_{sl} and σ_{sv} data with which to predict the contact angle. More rigorous derivations of Young's equation and discussions of its limitations have been presented by Johnson [3.1] and Buff and Saltzburg [3.2].

Example 3.1 Determine the effect of surface roughness on the equilibrium contact angle.

For a rough horizontal surface, the ratio of the actual solid surface area per unit of area projected on a horizontal plane is defined as $\gamma = A_s/A_{hor}$. We again consider the system shown in Fig. 3.2, and assume that γ is a fixed constant. As for the case of a smooth surface, analysis of the free-energy change associated with a differential change in the positions of the interfaces S_{lv}, S_{sv}, and S_{sl} results in Eq. (3.8):

$$dF = \sigma_{lv}\, dA_{lv} + \sigma_{sl}\, dA_{sl} + \sigma_{sv}\, dA_{sv}$$

Assuming a spherical cap for the liquid–vapor interface, geometric considerations require that

$$dA_{sv} = -\, dA_{sl} \qquad dA_{lv} = (dA_{sl})_{hor} \cos\theta_R = dA_{sl}\frac{\cos\theta_R}{\gamma}$$

Combining these three relations yields

$$\frac{\partial F}{\partial A_{sl}} = \left(\frac{\sigma_{lv}}{\gamma}\right)\cos\theta_R + \sigma_{sl} - \sigma_{sv}$$

Since $\partial F/\partial A_{sl}$ must equal zero at equilibrium, setting the right-hand side to zero results in the relation

$$\cos\theta_R = \frac{\sigma_{sv} - \sigma_{sl}}{\sigma_{lv}}\gamma$$

Using Young's equation (3.2), this equation can be written as

$$\cos\theta_R = \gamma\cos\theta$$

where θ is the contact angle for a smooth surface. Because γ is greater than 1, $\cos\theta_R > \cos\theta$, which implies that θ_R is smaller than θ; that is, the contact angle on a rough surface is lower than on a smooth surface of the same material. Furthermore, using this equation and Young's equation (3.2), the equation derived above for $\partial F/\partial A_{sl}$ can be rearranged to obtain

$$\frac{\partial F}{\partial A_{sl}} = \frac{\sigma_{lv}}{\gamma}(\cos\theta_R - \gamma\cos\theta)$$

If $\gamma > \sec\theta$, the right-hand side is negative because $\cos\theta_R < 1$. This implies that $\partial F/\partial A_{sl}$ is negative. A decrease in the system free energy as the liquid spreads implies that such a process is thermodynamically favorable. This, in turn, implies that the liquid will spread over the rough surface spontaneously. This phenomenon is sometimes referred to as *wicking*.

3.2 WETTABILITY, COHESION, AND ADHESION

It can be seen in Fig. 3.1 that as the contact angle θ approaches 180°, if gravity body forces are small, the liquid drop becomes spherical, with only one point of contact with the solid. At the other extreme, as θ approaches zero, a drop with a fixed volume approaches a thin-film configuration. A liquid for which $\theta = 0$ is said to *completely wet* the solid surface. By convention, a liquid with a value of θ between 0° and 90° is termed a *wetting liquid*. For $90° < \theta < 180°$ the liquid is said to be *nonwetting*, and for $\theta = 180°$ it is *completely nonwetting*. Thus the contact angle is a direct index of the wettability of the liquid. Photographs illustrating the very different wetting characteristics of water and R-113 on Teflon (polytetrafluoroethylene) are shown in Fig. 3.3.

Note that in Young's equation (3.2), $|\cos \theta|$ cannot exceed 1. Hence, for an equilibrium contact angle to be established, it is necessary that $|(\sigma_{sv} - \sigma_{sl})/\sigma_{lv}| < 1$. Otherwise the interfacial tensions will be unable to achieve a force balance. If $(\sigma_{sv} - \sigma_{sl})/\sigma_{lv}$ is less than -1, σ_{sl} overcomes the combined horizontal components due to σ_{sv} and σ_{lv}. The contact line will be pulled toward the center of the drop ($\theta \rightarrow 180°$) until the contact disappears, and the liquid will be completely

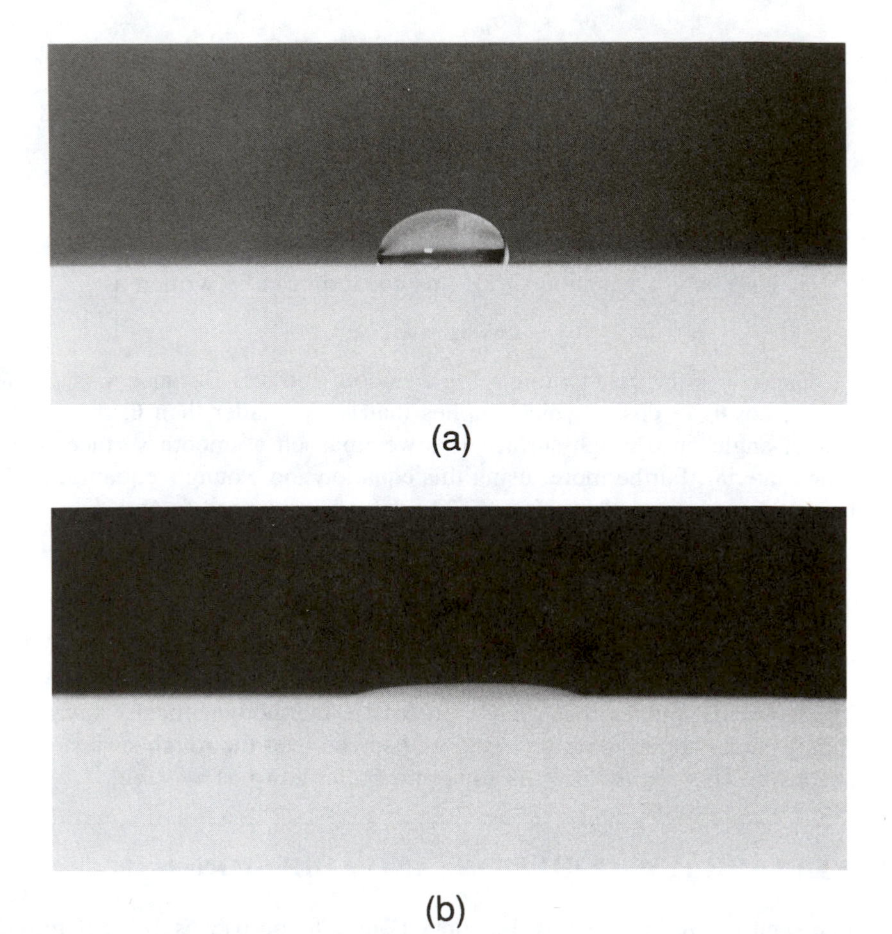

Figure 3.3 Profile shapes of liquid droplets on a solid Teflon surface: (*a*) water on Teflon, (*b*) R-113 on Teflon.

nonwetting. This virtually never happens for a droplet of liquid surrounded by vapor, but it could happen for a droplet of one liquid surrounded by another immiscible liquid.

Conversely, if $(\sigma_{sv} - \sigma_{sl})/\sigma_{lv}$ is greater than 1, σ_{sv} overcomes the combined horizontal force components due to σ_{sl} and σ_{lv}. The contact line will be pulled away from the center of the drop, spreading the liquid thinner and thinner until the surface of the solid is covered, or until the film becomes so thin that molecular interactions come into play. For a wetting liquid to spontaneously spread on a surface in this manner, the spreading process must result in a decrease in the free energy of the system.

For the sessile drop shown in Fig. 3.1, it was shown in Section 3.1 that

$$\frac{\partial F}{\partial A_{sl}} = \sigma_{sl} - \sigma_{sv} + \sigma_{lv} \cos \theta \tag{3.12}$$

If we define a *spreading coefficient* Sp_{ls} as

$$\mathrm{Sp}_{ls} = \sigma_{sv} - \sigma_{lv} - \sigma_{sl} \tag{3.13}$$

then Eq. (3.12) can be written as

$$\frac{\partial F}{\partial A_{sl}} = -[\mathrm{Sp}_{ls} + \sigma_{lv}(1 - \cos \theta)] \tag{3.14}$$

For a fully wetting liquid, $\theta = 0$ and Eq. (3.14) becomes

$$\frac{\partial F}{\partial A_{sl}} = -\ \mathrm{Sp}_{ls} \tag{3.15}$$

Equation (3.15) indicates the justification for the definition of the spreading coefficient given by Eq. (3.13). If the spreading coefficient Sp_{ls} is positive, then Eq. (3.15) indicates that the free energy decreases with increasing A_{sl}, as is necessary for spontaneous spreading of the liquid. If the liquid does not spread and establishes an equilibrium contact angle, then $\partial F/\partial A_{sl} = 0$. Substituting this result into Eq. (3.14) yields

$$\mathrm{Sp}_{ls} = -\sigma_{lv}(1 - \cos \theta) \tag{3.16}$$

which indicates that Sp_{ls} must be negative. Thus the spreading coefficient indicates the tendency of the liquid to wet and spread into a thin liquid film. A positive value of Sp_{ls} indicates that the liquid will wet and spontaneously spread into a thin film. A negative value of Sp_{ls} indicates that the liquid will partially wet the solid and establish an equilibrium contact angle. Although these results are theoretically satisfying, from a practical point of view it is generally difficult to evaluate Sp_{ls} from Eq. (3.13) for most materials because of the lack of data for the interfacial tensions σ_{sv} and σ_{sl}.

Consider now the cylindrical column formed by a liquid phase (l), in contact with a solid phase (s), which is surrounded by a low-density gaseous phase (g), as shown in Fig. 3.4*a*. If the column is torn apart at the l–s interface, as indicated in Fig. 3.4*b*, then the net reversible work required per unit of interface area is

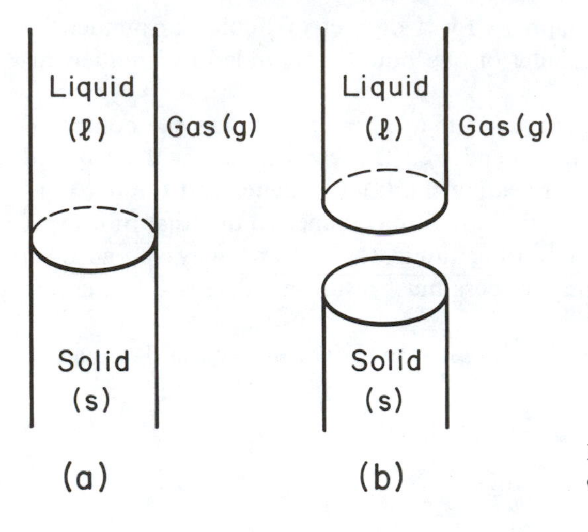

Figure 3.4 Separation of a solid–liquid column surrounded by a gas phase.

$$w_{sl} = \sigma_{lg} + \sigma_{sg} - \sigma_{sl} \tag{3.17}$$

Note that the terms on the right side are a direct result of the fact that, in tearing the column this way, we have created new liquid–vapor and solid–vapor interfaces and eliminated a solid–liquid interface. Assuming that we must do work on the system to create an interface and can retrieve work reversibly from destroyed interfaces, the right side of Eq. (3.17) follows immediately from the definitions of the interfacial tensions. The work w_{sl} defined by Eq. (3.17) is called the *work of adhesion,* because it is the minimum (reversible) work required to tear the liquid off the solid surface.

Instead of the lower portion being a different solid phase, suppose that the column torn in half were entirely the same liquid phase. Because we have created two new liquid–vapor interfaces without destroying an existing interface, the same arguments made above imply that the work required to break the column is

$$w_{ll} = 2\sigma_{lg} \tag{3.18}$$

Because w_{ll} is the work required to break internal bonds in the material, it is called the *work of cohesion*.

We can use a similar line of reasoning to analyze the spreading of a liquid on a solid. The spreading process can be idealized as schematically indicated in Fig. 3.5. In this idealized process, the liquid is sliced along the dotted lines in Fig. 3.5*a* and the cut-off portions are brought into contact with the solid to obtain the final configuration shown in Fig. 3.5*b*. In this manner the spreading process can be viewed as a sequence of creating and destroying interfaces. The process shown in Fig. 3.5 results in the creation of new liquid–gas and liquid–solid interface areas and the elimination of sections of solid–gas interface. The total work interaction for the processes (assuming that $\delta \ll \sqrt{A}$) is therefore given by

$$W_{sp} = \sigma_{lg}A + \sigma_{sl}A - \sigma_{sg}A \tag{3.19}$$

If we divide both sides of Eq. (3.19) by A and define $w_{sp} = W_{sp}/A$ as the work per unit area, Eq. (3.19) can be rearranged and combined with Eq. (3.13) to obtain

$$w_{sp} = \sigma_{lg} + \sigma_{sl} - \sigma_{sg} = -\,\mathrm{Sp}_{ls} \tag{3.20}$$

Thus the work interaction is negative (because Sp_{ls} is positive for spreading), implying that work could be extracted if we had the means to do so, and its value is equal to the spreading coefficient. Alternatively, the spreading coefficient can be interpreted as the reversible work interaction associated with the spreading process.

It can also be easily shown using Eqs. (3.17) through (3.20) that

$$\mathrm{Sp}_{ls} = -w_{sp} = w_{sl} - w_{ll} \tag{3.21}$$

Hence the spreading coefficient is also equal to the difference between the work of adhesion and the work of cohesion. In this sense it indicates the tendency of the liquid to adhere to the solid relative to its internal cohesive forces.

An additional useful result can be obtained by considering Eq. (3.17). If we generalize Eq. (3.17) to apply to any two solid or liquid phases a and b, and a low-density gas or vapor phase g, we obtain

$$w_{ab} = \sigma_{bg} + \sigma_{ag} - \sigma_{ab} \tag{3.22}$$

which can be rearranged to get

$$\sigma_{ab} = \sigma_{bg} + \sigma_{ag} - w_{ab} \tag{3.23}$$

This relation implies that the interfacial tension between a solid and a liquid or two liquid phases can be computed as the sum of their individual interfacial tensions while in contact with a third gas or vapor phase, minus the work of adhesion. If the two materials interact due to dispersion forces alone, it has been argued by

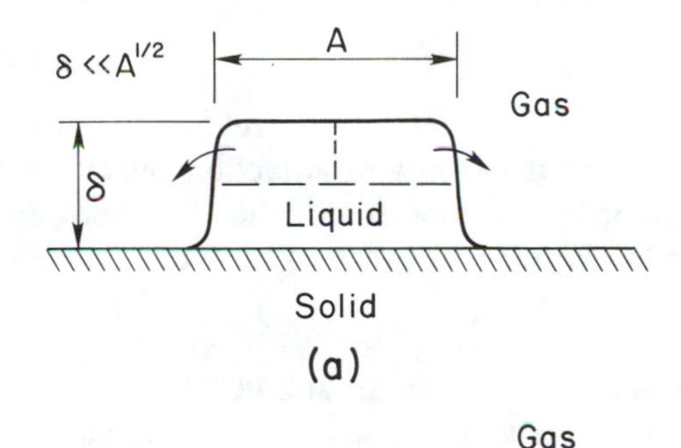

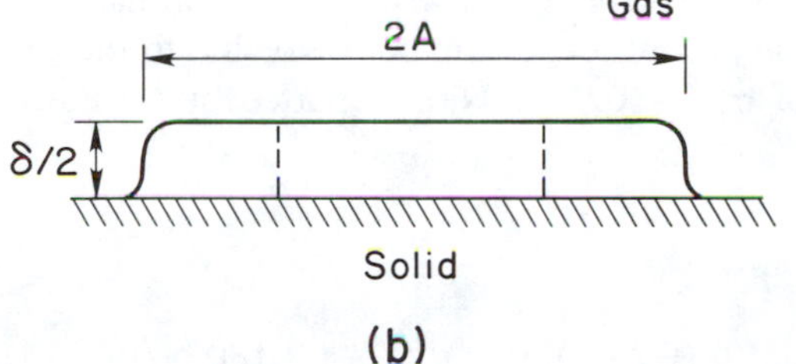

Figure 3.5 Idealized model of the spreading process on a solid surface. In this model, A is the area on the solid surface initially covered with liquid.

Fowkes (see reference [3.3] or [3.4]), based on the nature of these forces, that the work of adhesion is approximately given by

$$w_{ab,d} \cong 2(\sigma_{ag,d}\sigma_{bg,d})^{1/2} \tag{3.24}$$

If materials a and b are both nonpolar, the interaction at the interfaces will be due to dispersion forces alone, in which case $w_{ab,d} = w_{ab}$, $\sigma_{ag,d} = \sigma_{ag}$, and $\sigma_{bg,d} = \sigma_{bg}$. Substitution of these results with Eq. (3.24) into Eq. (3.23) yields

$$\sigma_{ab} = \sigma_{ag} + \sigma_{bg} - 2(\sigma_{ag}\sigma_{bg})^{1/2} \tag{3.25}$$

Equation (3.25) is a very useful result in that it can be used to calculate the interfacial tension between two immiscible liquid phases a and b if we know their individual interfacial tensions when in contact with a third gas or vapor phase g. Strictly speaking, the applicability of this relation is limited to three phases that are all nonpolar. However, Eq. (3.25) is usually also a good approximation for the case where either a or b is nonpolar. An example would be using surface tension values for water–air and kerosene–air interfaces to predict the interfacial tension for a kerosene–water interface. The large data base of interfacial tensions for liquids in contact with air can thus be used to compute interfacial tensions for an interface between two of the liquids.

Girifalco and Good [3.5–3.7] proposed the following relation for the interfacial tension of simple systems that is similar to, but more general than, Eq. (3.25):

$$\sigma_{ab} = \sigma_{ag} + \sigma_{bg} - 2\Phi(\sigma_{ag}\sigma_{bg})^{1/2} \tag{3.26}$$

In the above equation, Φ is a parameter characteristic of a given system. If a and b are liquids having spherical molecules of about equal size, and the interface is "regular," Φ is found experimentally to be approximately 1. If the molecules are not of equal size, then Φ may be approximated as

$$\Phi = 4(\hat{v}_a\hat{v}_b)^{1/3}(\hat{v}_a^{1/3} + \hat{v}_b^{1/3})^{-2} \tag{3.27}$$

where $\hat{v}_a$ and $\hat{v}_b$ are the molar volumes of the two phases. A "regular" interface is defined by these authors as one for which the energy of attraction between the unlike molecules is the geometric mean of the energies of attraction between pairs of like molecules. Further discussion of Eq. (3.26) can be found in references [3.5–3.7].

Example 3.2 For water and hexane in contact with air at 20°C, $\sigma_{wg} = 0.0728$ N/m and $\sigma_{hg} = 0.0184$ N/m, respectively. Use these data to estimate the interfacial tension between hexane and water. Compare this value to the experimentally determined value of $\sigma_{wh} = 0.0511$ N/m reported by Girifalco and Good [3.5].

Using Eq. (3.25),

$$\begin{aligned}\sigma_{wh} &= \sigma_{wg} + \sigma_{hg} - 2(\sigma_{wg}\sigma_{hg})^{1/2}\\ &= 0.0728 + 0.0184 - 2[(0.0728)(0.0184)]^{1/2} = 0.0180\ \text{N/m}\end{aligned}$$

Although Eq. (3.20) is written for a liquid spreading on a solid, it can also be applied to a liquid on a liquid. Rearranging this equation somewhat, the spreading coefficient for hexane on water can be computed as

$$\begin{aligned} Sp_{hw} &= \sigma_{wg} - \sigma_{hg} - \sigma_{wh} \\ &= 0.0728 - 0.0184 - 0.0511 = 0.0033 \text{ N/m} \end{aligned}$$

Because the spreading coefficient is positive, this suggests that the hexane will spontaneously spread over the surface of the water until it reaches the container boundary or forms a monolayer over the surface of the water. However, because the spreading coefficient is close to 0, this tendency is weak. Hexane is unlikely to form lens-shaped droplets. It is more likely to spread out into a film, but it may not aggressively cover the entire liquid surface.

3.3 EFFECT OF LIQUID SURFACE TENSION ON CONTACT ANGLE

In Section 3.2, it was argued that Eq. (3.24) is valid if *a* and *b* are substances for which interfacial tension effects are due to dispersion forces alone. For this category of substances, Eq. (3.24) can be combined with Eq. (3.23) and Young's equation (3.2) to obtain

$$\cos\theta = \frac{2}{\sigma_{bg}}(\sigma_{ag,d}\sigma_{bg,d})^{1/2} - 1 \tag{3.28}$$

Here we will specifically consider *a* to be a solid horizontal surface, with *b* corresponding to a liquid droplet and *g* a surrounding gas. The above equation can then be written as

$$\cos\theta = \frac{2}{\sigma_{lg}}(\sigma_{sg,d}\sigma_{lg,d})^{1/2} - 1 \tag{3.29}$$

If we pick a specific solid, so that $\sigma_{sg,d}$ is a fixed constant, and consider different nonpolar liquids for which $\sigma_{lg,d} = \sigma_{lg}$, then Eq. (3.29) indicates that the variation of θ with interfacial tension σ_{lg} is of the form

$$\cos\theta = 2\left(\frac{\sigma_{sg,d}}{\sigma_{lg}}\right)^{1/2} - 1 \tag{3.30}$$

As the surface tension tends to infinity, Eq. (3.30) indicates that $\cos\theta \to -1$ or $\theta \to 180°$, and the liquid is completely nonwetting. As σ_{lg} decreases, $\cos\theta$ increases until, at $\sigma_{lg} = \sigma_{sg,d}$, $\cos\theta = 1$, implying that $\theta = 0$. Liquids whose surface tension equals $\sigma_{sg,d}$ will completely wet the surface. For $\sigma_{lg} < \sigma_{sg,d}$, we must examine the spreading coefficient to determine the wetting behavior. Combining Eqs. (3.23) and (3.24), with *a*, *b*, and *c* taken to be the solid, liquid, and gas phases, and substituting into Eq. (3.13) to eliminate σ_{sl} yields

$$\mathrm{Sp}_{ls} = 2[(\sigma_{sg,d}\sigma_{lg,d})^{1/2} - \sigma_{lg}] \tag{3.31}$$

Using the fact that $\sigma_{lg,d} = \sigma_{lg}$ for the nonpolar liquids considered here, we obtain

$$\mathrm{Sp}_{ls} = 2\sigma_{lg}\left[\left(\frac{\sigma_{sg,d}}{\sigma_{lg}}\right)^{1/2} - 1\right] \tag{3.32}$$

It can be seen in Eq. (3.32) that for $\sigma_{lg} < \sigma_{sg,d}$, the spreading coefficient is positive, implying that the liquid spreads and wets the solid surface completely. Thus the relation between θ and σ_{lg} for nonpolar liquids is more correctly stated as

$$\cos\theta = \begin{cases} 2\left(\dfrac{\sigma_{sg,d}}{\sigma_{lg}}\right)^{1/2} - 1 & \text{for } \sigma_{lg} > \sigma_{sg,d} \\ 1 & \text{for } \sigma_{lg} \leqslant \sigma_{sg,d} \end{cases} \tag{3.33}$$

The variation of θ with σ_{lg} indicated by Eq. (3.33) is, in fact, observed in experimental data. Figure 3.6 shows, as an example, measured combinations of $\cos\theta$ and surface tension for various hydrocarbon liquids on Teflon (polytetrafluorethylene), which were obtained by Fox and Zisman [3.8]. It can be seen that the points distribute themselves in a fairly narrow band. The intersection of this band with the horizontal line at $\cos\theta = 1$ corresponds to $\sigma_{lg} = \sigma_{sg,d}$. This value of $\sigma_{lg} = \sigma_{sg,d}$ is called the *critical surface tension* of the solid.

For values of $\sigma_{lg} > \sigma_{sg,d}$ but close to $\sigma_{sg,d}$, the relation given by Eq. (3.33) may be expanded in a Taylor series about $\sigma_{lg} = \sigma_{sg,d}$. Retaining only the first term of the series yields

$$\cos\theta = 2 - \frac{\sigma_{lg}}{\sigma_{sg,d}} \qquad \text{for } \sigma_{lg} > \sigma_{sg,d},\ \frac{\sigma_{lg} - \sigma_{sg,d}}{\sigma_{sg,d}} \ll 1 \tag{3.34}$$

Thus, $\cos\theta$ varies linearly with σ_{lg} near $\sigma_{lg} = \sigma_{sg,d}$, which makes it possible to extrapolate data for $\sigma_{lg} > \sigma_{sg,d}$ linearly on this type of plot to determine the critical surface tension.

Example 3.3 Polytetrafluoroethylene (Teflon) is a low-energy surface having a critical surface tension of 0.018 N/m (see Fig. 3.6). Determine the degree to which the following fluids wet this material: liquid helium II at −271°C and saturated liquid methanol at 20°C and 160°C.

From Table 2.1, helium II at −271°C has a surface tension of 0.00032 N/m, which is far below the critical surface tension of Teflon. This implies, from Eq. (3.33), that $\cos\theta = 1$, from which it follows that $\theta = 0$ and the liquid fully wets the surface.

For methanol at 20°C, the interfacial tension is 0.0226 N/m, and from Eq. (3.33),

$$\theta = \cos^{-1}\left\{2\left(\frac{0.018}{0.0226}\right)^{1/2} - 1\right\} = 38.3°$$

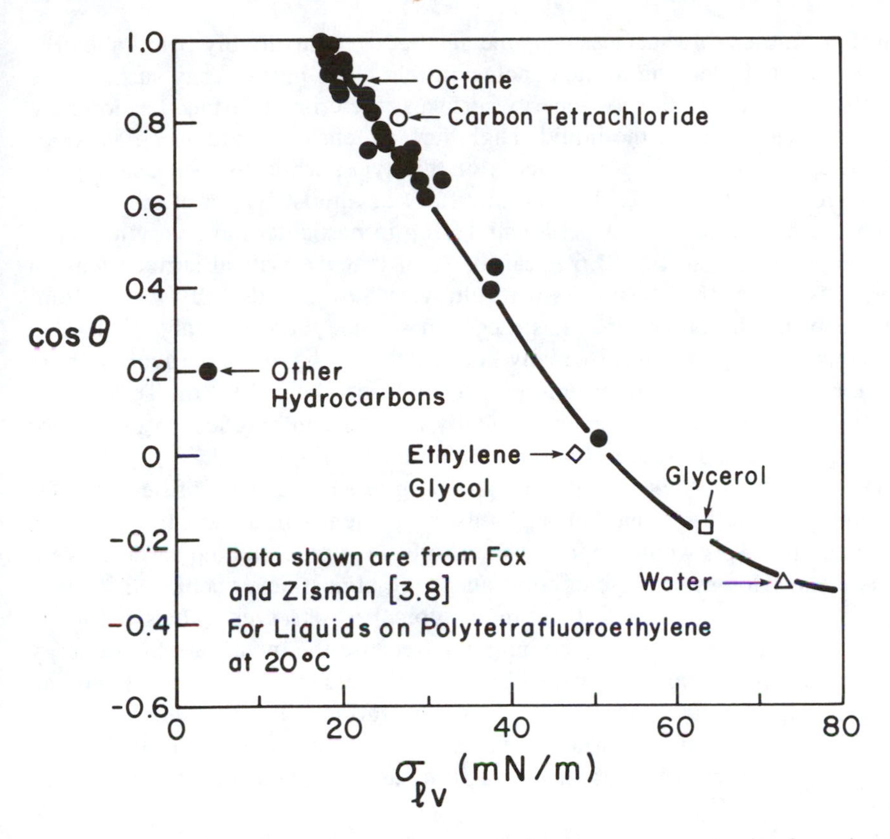

Figure 3.6 Plot of the observed contact θ versus liquid–vapor interfacial tension for various liquids on Teflon at 20°C. The solid curve is a best fit to the data.

For methanol at 160°C, the interfacial tension is 0.0069 N/m, which is less than the critical surface tension for the polytetrafluoroethylene. Hence, by the same logic as for the He_{II}, $\theta = 0$ and the liquid fully wets the surface. This demonstrates the often-observed fact that the wetting characteristics of the fluid varies with temperature. Generally, the liquid becomes more wetting as the temperature increases.

For nonpolar solids, $\sigma_{sg,d} = \sigma_{sg}$, and determination of the critical surface tension also indicates the interfacial tension of the solid. As noted in Section 3.2, this interfacial tension indicates the surface excess free energy per unit of surface area of the solid. Solid surfaces may be divided into two categories. There are high-energy surfaces having interfacial tensions of 500 mN/m or greater, and there are low-energy surfaces with interfacial tensions ranging from 15 to 40 mN/m. Glass and metals are high-energy surfaces, whereas hydrocarbons, polymers, and plastics are generally low-energy materials.

The surface tensions of nonmetallic liquids are typically between 15 and 75 mN/m. For high-energy surfaces, this means that σ_{lg} is invariably below the critical surface tension, leading to the conclusion that clean high-energy surfaces are almost always wettable. For low-energy surfaces, the critical surface tension may be above or below that of the liquid. High-surface-tension liquids such as water (see Table 2.1) may poorly wet surfaces of this type, while low-surface-tension liquids such as R-113 (see Table 2.1) may fully or almost fully wet the surface.

Fluorocarbon surfaces in general and Teflon in particular have very low critical surface tensions. In Fig. 3.6 it can be seen that the critical surface tension for Teflon is about 18 mN/m. As a result, very few liquids fully wet Teflon. Polyethylene, on the other hand, has a critical surface tension of about 31 mN/m and, consequently, it is wet by many liquids. Water, having a particularly high surface tension, has an equilibrium contact angle of about 108° on Teflon. Although higher contact angles are theoretically possible, in practice, liquid–vapor contact angles above 108° are almost never found.

The above results suggest that the high-energy metal surfaces of heat transfer equipment in which liquid and vapor phases are present will always be fully wetted by the liquid. This would be true if the fluids were pure and the surfaces were perfectly clean. However, these circumstances are almost never achieved in practice. A contact angle of about 20° is more typical of observed values for water on metal surfaces of heat transfer equipment. Because the metals are high-energy surfaces, they typically are also wetted by contaminants in the system, which may spread to form a thin layer over all or part of the surface. The working fluid then may not wet portions of the surface covered by a thin adsorbed film of the contaminant. The characteristics of such thin films are discussed further in the next sections.

3.4 ADSORPTION

As noted in Section 2.1, the variation of properties in the interfacial region between two phases is actually continuous, and the values of the properties in this region are generally different from those in the bulk phases. For a pure vapor in contact with a solid, attraction between solid and vapor molecules can thus increase the density of vapor molecules in the interface region near the solid surface. The excess vapor density above the bulk value is the amount said to be *adsorbed* onto the surface.

In a similar manner, if a mixture of gases is in contact with a solid, attraction between solid and gas molecules may increase the concentration of one or more of the component gases in the interfacial region above that in the bulk mixture. In this way, gases that have a particularly high affinity for molecules in the solid may be preferentially adsorbed onto the solid surface.

In general, *adsorption* is the retention at the interface of solid, liquid, or gas molecules, atoms, or ions by a solid or a liquid. Adsorption may affect the wetting of a liquid on a solid in at least two ways. First, we have already seen in Chapter Two that concentration of surfactant molecules at a liquid–vapor interface can reduce the interfacial tension. We have also seen in Section 3.3 that the liquid–vapor interfacial tension directly affects the contact angle. Hence substances adsorbed at the liquid–vapor interface can directly affect the wetting characteristics of the liquid.

In addition, adsorption of a substance onto the solid surface can alter the interfacial tension of the solid–vapor interface. The difference between the interfacial tension with and without the adsorbed species present is termed the *surface pressure* of the adsorbed material on the solid surface, π_S:

$$\pi_S = \sigma_{sv} - \sigma_{sv,a} \tag{3.35}$$

In this equation, $\sigma_{sv,a}$ is the interfacial tension with the absorbed substance present. It can be seen from Young's equation (3.2) or Eq. (3.33) that changing σ_{sv} will alter the contact angle. In fact, we can generalize Young's equation to include this effect as

$$\sigma_{lv} \cos \theta = (\sigma_{sv} - \pi_S) - \sigma_{sl} \tag{3.36}$$

In this equation, the π_S term is usually unimportant for nonwetting liquids. For these circumstances, σ_{sv} is small compared to σ_{sl} anyway (note that $\cos \theta < 0$ for $90° < \theta < 180°$), and the presence of adsorbed material usually just decreases it a bit more.

On the other hand, high-energy surfaces such as metals or glass have large values of σ_{sv}, which may be reduced appreciably by the presence of an adsorbed species. Equations (3.33) and (3.36) both indicate that this can lead to significant changes in the wetting angle for such cases.

Strictly speaking, the validity of Young's equation is suspect when additional adsorbed materials are present at one or more interfaces, because such effects are not considered in its derivation. Equation (3.36) is plausible if one considers the problem from a macroscopic force-balance viewpoint. However, this relation assumes that the sole effect of the adsorbed layer is to uniformly modify σ_{sv}, which may not be true in all cases.

The above discussion indicates that adsorbed materials on a solid metal surface may cause its wetting characteristics to deviate significantly from those for a perfectly clean surface. Hence for a system whose operation is sensitive to liquid wetting characteristics, exposure of the system to substances that may adsorb on its surfaces should be avoided if at all possible.

3.5 SPREAD THIN FILMS

In Section 3.3 it was noted that in some cases the spreading coefficient for a liquid on a solid, Sp_{ls}, is greater than 0, indicating that liquid brought into contact with the solid surface will spread into a thin film. From Eq. (3.32) it can be seen that Sp_{ls} will be positive if

$$\sigma_{sg,d} > \sigma_{lg} \tag{3.37}$$

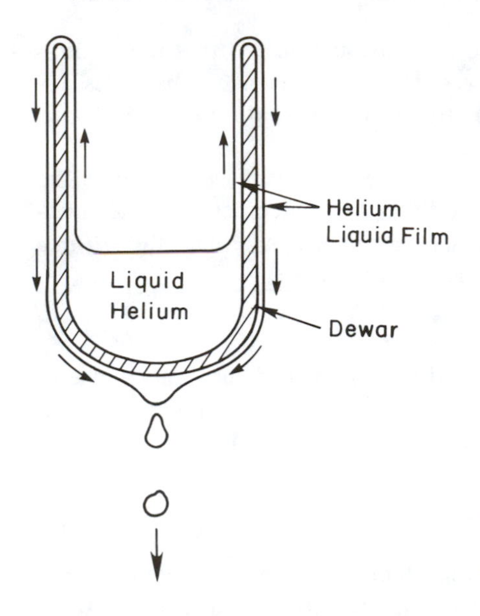

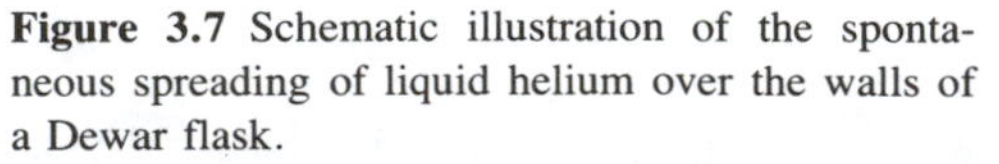

Figure 3.7 Schematic illustration of the spontaneous spreading of liquid helium over the walls of a Dewar flask.

Hence a spread thin film is more likely to be observed for low-surface-tension liquids on high-surface-energy solids, such as glass or metals. The most well-known examples of spread films do, in fact, correspond to these conditions.

Perhaps the most spectacular example of a spread film is the spreading of liquid helium. The surface tension of liquid helium varies with temperature, but is usually near or below 1 mN/m (see Table 2.1). A glass Dewar flask containing the liquid may have a surface free energy (σ_{sg}) on the order of several hundred millinewtons per meter. Thus the work of adhesion is much greater than the work of cohesion, and spreading will occur. The consequences of this strong tendency for spreading are shown in Fig. 3.7. The liquid helium spreads up the walls of the Dewar against gravity, over the rim of the opening at the top, and forms a thin film on the outside, which either evaporates or collects and drips off the bottom of the flask.

Low-viscosity silicone oils (e.g., Dow Corning 200 series) also spread spontaneously on most metal surfaces and many nonmetal surfaces. The surface tension of these fluids is typically near 18 mN/m, and consequently the work of cohesion, $w_{ll} = 2\sigma_{lg}$, is small compared to the work of adhesion on high-energy solids. Hence liquids of this type readily spread to form a thin film on such materials.

It should be noted, however, that adsorption of substances on the solid surface may sometimes inhibit the spreading of liquids over the surface. An adsorbed monolayer of a cationic surfactant can change a high-energy surface into a low-energy one, which can reduce the work of adhesion to the point that $\mathrm{Sp}_{lg} < 0$ and the liquid cannot spread. Surfactants of this type may be adsorbed selectively at specific locations, resulting in localized barriers to the spread of the otherwise wetting liquid.

To analyze more fully the spread thin film that results for these circumstances, we must first develop a new concept. Consider the system shown in Fig. 3.8, in which liquid is in contact with and fully wets the horizontal solid surface. A hemispherical housing with a trapped bubble of gas inside is brought into close proximity to the solid surface so that a thin film of liquid of thickness δ exists between the vapor and solid interfaces. The liquid does not wet the inside of the housing, and a tube connects the bubble inside the housing to a system that holds the pressure constant.

If the pressure inside the bubble is P_b and the liquid–gas interface is flat, then, at equilibrium, P_b must balance the liquid pressure across the interface in the liquid. If δ is large, then the pressure across the interface, P_f, will just equal the local ambient pressure in the liquid, $P_l = P_{atm} + \rho_l g z_h$. However, if the housing is brought very close to the solid surface, the pressure inside the bubble must not only balance the ambient liquid pressure P_l, it must also counteract the attractive forces between the liquid molecules and the solid surface, which otherwise would maintain a thicker film of liquid on the surface. When the film is very thin, these attractive forces act to pull liquid into the layer as if the pressure in the layer were reduced below the ambient pressure P_l by an amount P_d, which is known as the *disjoining pressure*. By convention, if the affinity of the liquid for the solid draws liquid into the film, P_d is taken to be negative.

For the circumstances shown in Fig. 3.8, the local liquid pressure P_l and the disjoining pressure effect (due to solid–liquid attraction) act in tandem to thicken

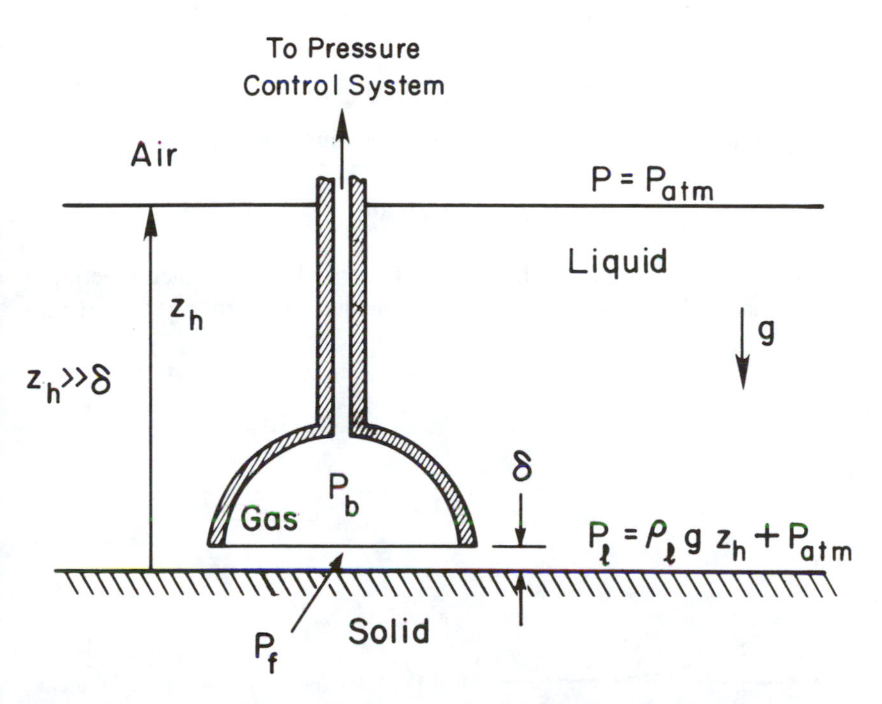

Figure 3.8 Model system considered in analysis of disjoining pressure effects.

the film. To maintain a thin liquid film, the pressure force in the bubble must balance both effects, which implies that, at equilibrium,

$$P_b = P_l + |P_d| = P_l - P_d = \rho_l g z_h + P_{\text{atm}} - P_d \tag{3.38}$$

Note in Eq. (3.38) that, because P_d is negative by convention, $-P_d$ is positive. Because the attractive forces between liquid molecules and those of the solid are expected to be stronger for molecules closer to the surface, the required disjoining pressure difference $-P_d$ to overcome them is expected to increase as δ gets smaller, as suggested in Fig. 3.9.

If we consider the process of slowly bringing the housing progressively closer to the solid wall while adjusting P_b to maintain equilibrium, the pressure required to thin the layer becomes continually larger until finally the last monolayer of the liquid phase is removed. The molecules of the vapor are then in contact with those of the solid to within an interfacial separation δ_0, which is less than the thickness of the liquid monolayer just removed. The work required to remove a unit area of the film in this manner, w_f, is given by

$$w_f = \int_{\delta=\delta_0}^{\delta=\infty} -P_d(\delta)\, d\delta \tag{3.39}$$

In forcing the spread layer of liquid away so a unit area is no longer wetted, we have just reversed the spreading process shown in Fig. 3.5 and described in Section 3.2. From this point of view, the work required to remove a unit area of the film w_f must be equal to the negative of the work of spreading, w_{sp}, given by Eq. (3.20):

$$w_f = -w_{sp} = \sigma_{sg} - \sigma_{lg} - \sigma_{sl} = \text{Sp}_{sl} \tag{3.40}$$

Hence the disjoining pressure is related to the spreading cofficient as

$$\text{Sp}_{sl} = \int_{\delta=\delta_0}^{\delta=\infty} -P_d(\delta)\, d\delta \tag{3.41}$$

In Section 2.3, we considered the shape of a free liquid surface meeting a vertical wall when the liquid only partially wets the surface, forming a finite con-

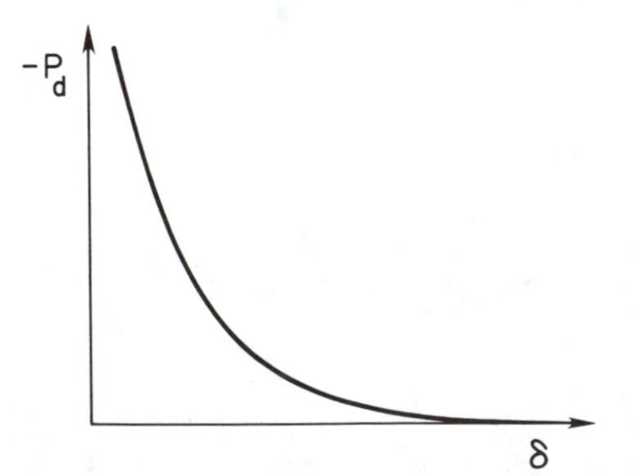

Figure 3.9 Expected variation of disjoining pressure with film thickness δ.

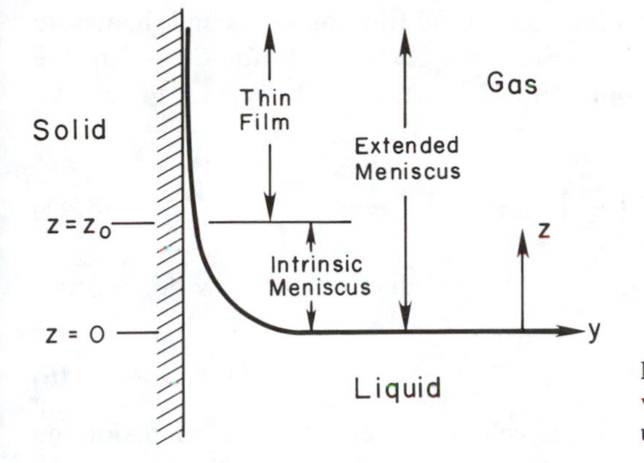

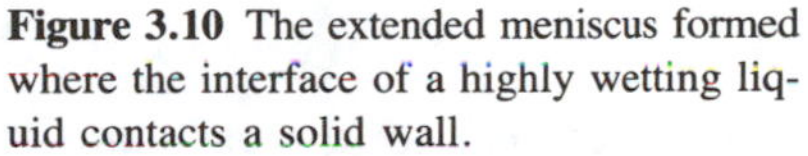
Figure 3.10 The extended meniscus formed where the interface of a highly wetting liquid contacts a solid wall.

tact angle. When the liquid fully wets and spontaneously spreads on the surface, the interface near the vertical wall differs in that there is a thin liquid film above the intrinsic meniscus, as indicated in Fig. 3.10. Here the *intrinsic meniscus* is defined as that portion of the meniscus profile that is dictated by the Young-Laplace equation (2.29), with the effects of disjoining pressure being negligible. The portion of the meniscus above the intrinsic meniscus will be referred to simply as the *thin film*. The thin film and intrinsic meniscus together will be termed the *extended meniscus*.

Here we will limit our attention to a two-dimensional extended meniscus system that is in equilibrium with no evaporation occurring. (In Section 8.3, characteristics of an evaporating extended meniscus will be examined.) In the intrinsic meniscus region, the interface profile is determined by solving Eq. (2.48) as described in Section 2.3. In solving Eq. (2.48), the requirement that $z \to 0$ as $y \to \infty$ is one boundary condition, and matching the profile to the thin film at the wall provides the other required boundary condition.

In the thin film, the interface profile is determined largely by the variation of the disjoining pressure. The hydrostatic pressure gradient in the liquid film is given by

$$\frac{dP_l}{dz} = -\rho_l g \tag{3.42}$$

At the lower limit of the thin film ($z = z_0$), the film is relatively thick and the pressure in the film equals the pressure in the surrounding vapor. Integrating Eq. (3.42) from this point to an arbitrary location z, we obtain

$$P_l - P_{l0} = P_d = -\rho_l(z - z_0)g \tag{3.43}$$

Hence the change in the hydrostatic pressure along the film is equal to the variation of the disjoining pressure. Differentiating Eq. (3.43) yields

$$\frac{dP_d}{dz} = -\rho_l g \tag{3.44}$$

The disjoining pressure is primarily a function of the film thickness and the nature of the liquid and the solid surface. Because, for a given solid–liquid system, the disjoining pressure is only a function of film thickness $y = \delta$, Eq. (3.44) can be written

$$\frac{dP_d}{d\delta}\left(\frac{d\delta}{dz}\right) = -\rho_l g \tag{3.45}$$

Following Potash and Wayner [3.9], we further assume a power-law dependence of P_d on δ:

$$P_d = -A\delta^{-B} \tag{3.46}$$

This functional form was found to agree well with measurements of disjoining pressure and film thickness obtained by Deryagin and Zorin [3.10].

Differentiating Eq. (3.46) with respect to δ and substituting into Eq. (3.45) yields

$$AB\delta^{-B-1}\left(\frac{d\delta}{dz}\right) = -\rho_l g \tag{3.47}$$

Rearranging Eq. (3.47) and integrating from the lower limit of the thin film z_0 to an arbitrary z location,

$$AB\int_{\delta_0}^{\delta} \delta^{-B-1}\, d\delta = -\rho_l g \int_{z_0}^{z} dz \tag{3.48}$$

Completion of the integration and solving for δ yields

$$\delta = \left[\delta_0^{-B} + \frac{\rho_l g(z - z_0)}{A}\right]^{-1/B} \tag{3.49}$$

Thus the disjoining pressure difference supports a thin liquid film against gravity above the intrinsic meniscus. The diminishing film thickness results in a gradient in the disjoining pressure that matches the hydrostatic pressure gradient. Additional thermodynamic aspects of the extended meniscus, and transport during evaporation of an extended meniscus, will be described in Section 8.3.

3.6 CONTACT-ANGLE HYSTERESIS

Equilibrium contact angles can usually be determined by simply photographing the contact location and measuring the angle on enlarged photographs. Data obtained in this manner generally indicate that the contact angle of the liquid varies depending on the motion history of the contact line. The nature of this phenomenon is schematically illustrated in Fig. 3.11.

Consider the process of partially immersing a thin slab of the solid into the liquid, moving it vertically downward very slowly, and stopping when the contact

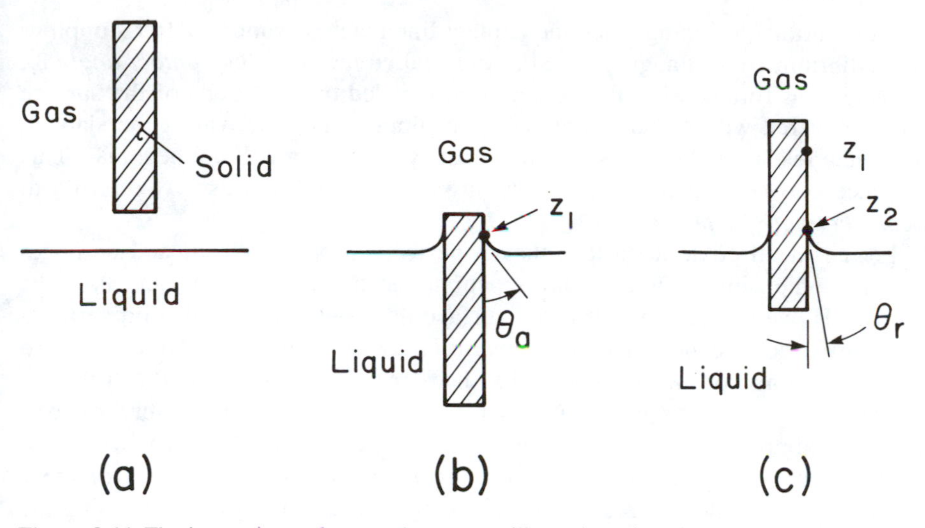

Figure 3.11 The immersion and removal sequence illustrating advancing and receding contact angles.

line reaches point z_1. The partially wetting liquid forms an equilibrium contact angle at point z_1, which we will designate as θ_a. Because this contact angle was established after advancing the contact line over dry solid, it is referred to as the *advancing contact angle*.

Example 3.4 Estimate the disjoining pressure for a layer thickness of 0.1 mm and 0.01 mm using the values of $A = 1.782$ Pa m^B and $B = 0.6$ given for carbon tetrachloride (CCl_4) and glass in reference [3.9]. How thick would the film be at a distance of 10 cm above a liquid pool of CCl_4 on the vertical walls of the container?

The disjoining pressure is determined from Eq. (3.46) as

$$P_d = -A\delta^{-B}$$

For $\delta = 0.10$ mm: $\quad P_d = -1.782(0.0001)^{-0.6} = -448$ Pa

For $\delta = 0.01$ mm: $\quad P_d = -1.782(0.00001)^{-0.6} = -1782$ Pa

For CCl_4, $\rho_l = 1590\ \text{kg/m}^3$. Using Eq. (3.49) and taking $\delta_0 = \infty$ for $z_0 = 0$ at the level of the free surface of the liquid–vapor interface, it follows that

$$\delta = \left(\frac{\rho_l g z}{A}\right)^{-1/B} = \left[\frac{1590(9.8)(0.1)}{1.782}\right]^{-1/0.6}$$

$$= 1.25 \times 10^{-5}\ \text{m} = 12.5\ \mu\text{m}$$

We now move the slab vertically upward, causing the contact line to move down the surface, stopping when the contact line reaches point z_2. After stopping, the equilibrium contact angle formed at z_2 is called the *receding contact angle* θ_r, because it was formed after the contact line receded over portions of the surface initially covered with liquid. After repeated dipping and removal of the slab, θ_a and θ_r may reach steady values, but for most systems they will not be equal. This difference between advancing and receding contact angles is usually referred to as *contact-angle hysteresis*.

For an idealized solid surface that is perfectly smooth, clean, and homogeneous in composition, there would appear to be no reason for θ_a and θ_r to be different. However, such an idealized surface does not exist. Real solid surfaces are never perfectly smooth; their composition may vary slightly with location; and molecules, atoms, or ions of other substances may be adsorbed on the surface.

Metals of the type used in heat transfer equipment (e.g., copper, brass, steel, aluminum) are particularly susceptible to these imperfections. Manufacturing processes always leave some degree of roughness on the surface. Alloys of these metals invariably have some distinct grain structure resulting from processing the material. Consequently, the surface is often a patchwork of two or more different grain types, which means that, on a microscopic scale, the surface is intrinsically heterogeneous. In addition, as noted in Section 3.4, clean metals are high-energy surfaces that easily adsorb thin films of many substances. As a result, metal surfaces are easily contaminated by substances in the environment, even if their concentrations are very low.

Contact-angle hysteresis is generally acknowledged to be a consequence of three factors: (1) surface inhomogeneity, (2) surface roughness, and (3) impurities on the surface. The effects of surface inhomogeneity and roughness can be more clearly understood by considering Fig. 3.12. Figure 3.12*a* schematically shows the behavior of advancing and receding two-dimensional liquid fronts on an idealized heterogeneous surface having alternating bands of a strongly wetted and poorly wetted material.

For an advancing liquid front, interface 1 in Fig. 3.12*a* establishes a contact angle appropriate to the poorly wetted material. However, when the contact line reaches the interface between the bands of different material, the interface changes to accommodate the smaller contact angle for the wetted material. This produces a curvature in the interface (see interface 2), which, because of the interfacial tension, lowers the pressure in the liquid near the solid surface. This lower pressure causes liquid to flow toward the contact line region, allowing the contact line to move rapidly across the strongly wetted material. At the next band of poorly wetted material, the contact line motion is slowed as the larger contact angle is reestablished. The tendency for the contact line to move rapidly across the strongly wetted surface means that if the front motion is stopped, the larger contact angle of the poorly wetted surface is more likely to be established as θ_a.

For the receding liquid front, interface 1 has established the low contact angle for the strongly wetted material that persists until the contact line reaches the boundary between the different materials. At this point the larger contact angle of the poorly wetted material is established, which produces curvature in the in-

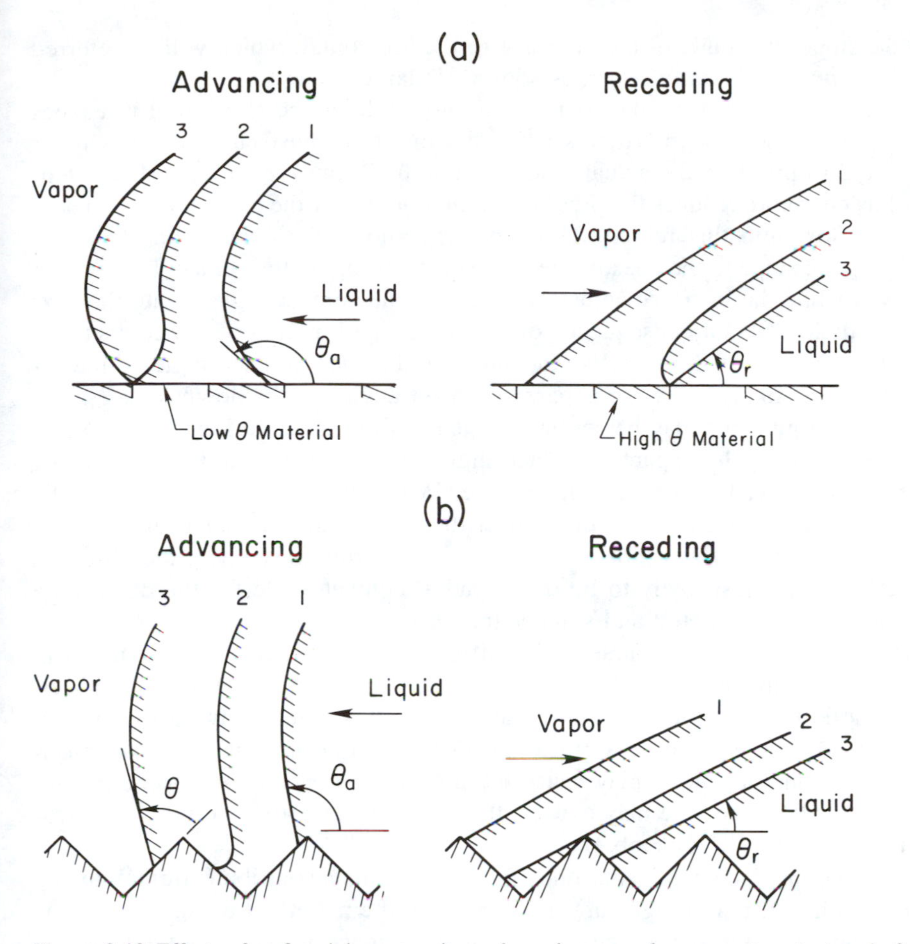

Figure 3.12 Effects of surface inhomogeneity and roughness on the apparent contact angle for advancing and receding liquid fronts.

terface near the contact line for receding interface 2 in Fig. 3.12*a*. Due to the interfacial tension, this curvature produces a rise in pressure in the liquid, which causes liquid to flow away from the contact line region, allowing the contact line to move rapidly across the poorly wetted material. At the next band of strongly wetted material, the motion of the contact line is slowed while the lower contact angle is established. In this case, the rapid motion of the contact line across the poorly wetted material means that if the front motion is stopped, the lower contact angle of the strongly wetted material is most likely to be established as θ_r.

Figure 3.12*b* schematically shows the behavior of advancing and receding liquid fronts on an idealized rough but homogeneous surface. The actual contact angle for liquid in contact with a flat section of the material θ is acute, indicating that the liquid actually moderately wets the surface. However, advancing interface 1 contacts a downward slope such that when a contact angle of θ is established with the slope, the angle of the interface to the horizontal, which will be referred to as the *apparent contact angle,* is somewhat larger than θ.

As the front advances, the contact line moves down the slope until it reaches the bottom of the groove. At this point the interface must curve as shown for interface 2 to preserve the actual contact angle θ. Because of the interfacial tension, this curvature reduces the pressure in the liquid near the contact line, causing liquid to flow into this region. This, in turn, allows the contact line to move rapidly up the slope to the apex of the groove. At the apex, the contact line motion is slowed while the interface adjusts to establish the contact angle θ with the next downward slope. As a consequence of the more rapid motion of the contact line along the upslope, if the front is stopped, the observed contact angle of the interface is most likely to be the apparent contact angle for the downslope θ_a.

Similar arguments may be applied to the receding fronts in Fig. 3.12*b*. Now, on the downslope, the apparent contact angle will be lower than θ if the contact angle θ is preserved on the sloping surface. When the interface (2) contacts the opposite wall of the groove, it does so at the apex, and the contact line simply transfers to the next downslope. As a result, if the front is stopped, the observed contact angle is most likely to be the apparent contact angle for the downslope θ_r, which is smaller than θ and smaller than θ_a.

Real surfaces are, of course, different from the idealized surfaces shown in Fig. 3.12. In general, the surface may be rough and have a nonsystematic variation of solid material properties over the surface. It is noteworthy, however, that consideration of the idealized surfaces in Fig. 3.12 indicates that both roughness and surface inhomogeneity may cause the advancing contact angle to be greater than the receding one. We may expect, then, that when both conditions are present, the same trend will be observed.

Adsorption of contaminants onto the surface may contribute further to the nonuniformity of its wetting characteristics, with the net effect being the same as the nonhomogeneous surface considered above. Hence it can be argued that all three of the mechanisms of contact-angle hysteresis mentioned earlier in this section can cause advancing contact angles to be larger than receding ones. This trend is, in fact, observed in contact-angle measurements.

Contact-angle hysteresis plays an important role in the behavior of liquid droplets on vertical or inclined surfaces and in small-diameter tubes. If a drop of liquid is placed on a horizontal solid surface, as shown in Fig. 3.13*a*, the resulting forward motion of the liquid front as the droplet contacts the surface and achieves an equilibrium shape leaves the droplet with the advancing contact angle all around its perimeter. If the surface is then rotated through 90°, the vertical components of the forces exerted on the liquid–vapor interface at the upper and lower contact lines effectively cancel, leaving the gravitational body force on the droplet initially unbalanced.

However, as indicated in Fig. 3.13*b*, as the droplet begins to move downward, the advancing contact angle is maintained near the bottom of the droplet, while the smaller, receding contact angle is established near the top. This differ-

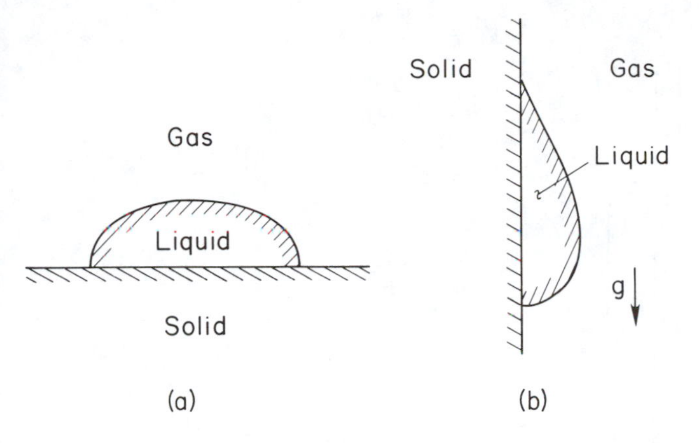

Figure 3.13 How contact hysteresis can allow a liquid droplet to resist downward motion when a horizontal surface is rotated to a vertical orientation.

ence between the top and bottom contact angles makes it possible for the droplet to adopt a shape that may support the weight of the liquid against gravity. Beginning at the top of the droplet and moving downward, the hydrostatic pressure in the liquid inside the droplet increases more rapidly than that in the surrounding gas because of the density difference between the phases.

Note in Fig. 3.13*b* that the interface radius of curvature is greater over the upper portion of the droplet and smallest near the bottom. As a result, the capillary pressure difference across the interface is largest at the bottom and smallest at the top of the droplet. The resulting variation of the surface tension forces over the interface may thus serve to balance the hydrostatic pressure difference across the interface, allowing the droplet to hold its position on the vertical wall against the force of gravity.

It should also be noted that a bubble on a vertical solid surface surrounded by liquid can similarly resist the upward buoyancy force on the vapor and remain fixed on the surface. For the bubble the mechanisms are virtually identical to those described above for the droplet surrounded by gas. In addition, differences in the contact angle and associated variations of the interface curvature can similarly allow the bubble or droplet to adopt an interface profile that resists drag forces resulting from motion of the surrounding fluid.

A similar phenomenon is exhibited by a droplet or slug of liquid in a small-diameter tube such as a soda straw. If a slug of liquid enters at the top of a vertical tube (open at both ends), gravity naturally tends to pull the liquid downward. However, as downward motion is initiated, the contact angles at the top and bottom of the slug must be the receding and advancing values, respectively, as indicated in Fig. 3.14. As a result, a smaller mean radius of curvature exists at the top of the slug than at the bottom end.

If the interface at each end of the liquid slug in Fig. 3.14 is idealized as a portion of a sphere, then the Young-Laplace equation requires that

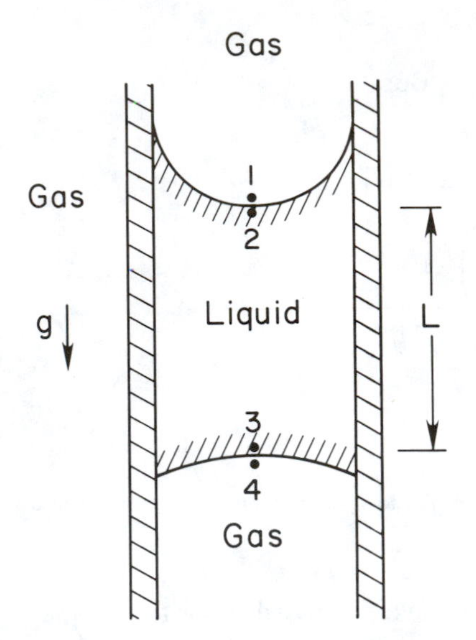

Figure 3.14 How contact-angle hysteresis can allow a liquid slug in a small vertical tube to resist downward motion.

$$P_1 - P_2 = \frac{2\sigma}{r_t} \tag{3.50}$$

$$P_4 - P_3 = \frac{2\sigma}{r_b} \tag{3.51}$$

where r_t and r_b are the radii of curvature at the top and bottom, respectively.

Because the tube is open at both ends, the hydrostatic pressure variations are such that

$$P_4 - P_1 = \rho_g g L \tag{3.52}$$

$$P_3 - P_2 = \rho_l g L \tag{3.53}$$

If all four equations (3.50) through (3.53) are satisfied, a force balance is achieved and the slug will remain fixed at its vertical location, despite the downward force of gravity on the slug. Combining these equations, it can be shown that a necessary condition for this to be true is

$$2\sigma\left(\frac{1}{r_t} - \frac{1}{r_b}\right) = (\rho_l - \rho_g)gL \tag{3.54}$$

This result implies that the length of a slug that can be supported against gravity depends directly on the difference between the radii of curvature at the upper and lower interfaces of the slug. Although the curvature of the interfaces most directly affects the force balance, contact-angle hysteresis is instrumental in establishing the different radii of curvature necessary to achieve equilibrium.

In preceding sections we have treated the contact angle as a well-defined constant property of a system. In most circumstances of interest we may still do so and obtain useful results. However, the results of this section indicate that, when evaluating the contact angle, we must be cognizant of its dependence on the prior motion of the contact line. Further discussion of wetting and contact angles may be found in references [3.11–3.13].

REFERENCES

3.1 Johnson, R. E., Jr., Conflicts between Gibbsian thermodynamics and recent treatments of interfacial energies in solid–liquid–vapor systems, *J. Phys. Chem.*, vol. 63, pp. 1655–1658, 1959.

3.2 Buff, F. P., and Saltzburg, H., Curved fluid interfaces. II. The generalized Neumann formula, *J. Chem. Phys.*, vol. 26, pp. 23–31, 1957.

3.3 Fowkes, F. M., Additivity of intermolecular forces at interfaces. I. Determination of the contribution to surface and interfacial tensions of dispersed forces in various liquids, *J. Phys. Chem.*, vol. 67, pp. 2538–2541, 1963.

3.4 Fowkes, F. M., Attractive forces at interfaces, in *Chemistry and Physics of Interfaces,* American Chemical Society, Washington, DC, 1965.

3.5 Girifalco, L. A., and Good, R. J., A theory for estimation of the surface and interfacial energies. I. Derivation and application to interfacial tension, *J. Phys. Chem.*, vol. 61, pp. 904–909, 1957.

3.6 Good, R. J., Girifalco, L. A., and Krause, G., A theory for estimation of the surface and interfacial energies. II. Application to surface thermodynamics of teflon and graphite. *J. Phys. Chem.*, vol. 62, pp. 1418–1421, 1958.

3.7 Good, R. J., and Girifalco, L. A., A theory for estimation of the surface and interfacial energies. III. Estimation of surface energies of solids from contact angle data, *J. Phys. Chem.*, vol. 64, pp. 561–565, 1960.

3.8 Fox, H. W., and Zisman, A., The spreading of liquids on low energy surfaces. I. Polytetrafluoroethylene, *J. Colloid Sci.*, vol. 5, pp. 514–531, 1950.

3.9 Potash, M., Jr., and Wayner, P. C., Jr., Evaporation from a two-dimensional extended meniscus, *Int. J. Heat Mass Transfer,* vol. 15, pp. 1851–1863, 1972.

3.10 Deryagin, B. V., and Zorin, A. M., Optical study of the adsorption and surface condensation of vapors in the vicinity of saturation on a smooth surface, *Proc. 2nd Int. Cong. on Surface Activity* (London), vol. 2, pp. 145–152, 1957.

3.11 Chappuis, J., Contact angles, in *Multiphase Science and Technology,* vol. 1, pp. 387–505, McGraw-Hill, New York, 1982.

3.12 Padday, J. F., Adhesion in a low-gravity environment, *Adhesion,* vol. 6, pp. 1–18, 1982.

3.13 Miller, C. A., and Neogi, P., *Interfacial Phenomena,* Marcel Dekker, New York, 1985.

PROBLEMS

3.1 Use Eq. (2.57) to predict the surface tension for an air–octane and an air–heptane interface at 20°C. Then use Eq. (3.25) to determine the interfacial tension for a water–octane and a water–heptane interface and compare the results with the following experimental data reported by Girifalco and Good [3.5]: $\sigma_{\text{water-octane}} = 0.0508$ N/m, $\sigma_{\text{water-heptane}} = 0.0502$ N/m.

3.2 At 20°C, to what degree will the following liquids wet Teflon: liquid mercury, ethylene glycol, acetone, and methanol? Briefly explain your answer.

3.3 For liquid carbon tetrachloride, use the liquid density quoted in Example 3.4 and assume that the molecules in the liquid are spaced in a rectangular array with a center-to-center distance of one molecular diameter to estimate the size of a cubic cell occupied by one molecule. Then use the

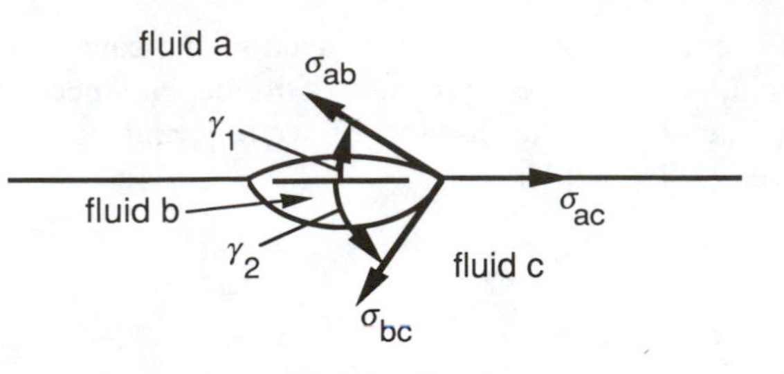

Figure P3.1

relation for $\delta(z)$ to estimate the height at which the thickness of the film just equals this dimension (i.e., the film is virtually a monolayer). (The molecular weight of carbon tetrachloride can be found in Appendix II.)

3.4 In Fig. 3.12, if the sawtooth pattern on the solid surface has an apex angle of 80° and the actual contact angle θ is 60°, what would be the advancing and receding contact angles?

3.5 At 20°C, the surface tension for water in contact with air is 0.0728 N/m. As indicated in Fig. 3.6, water poorly wets Teflon at this temperature. How does the wettability change if the water is at 60°C? At what temperature (if any) will liquid water fully wet Teflon?

3.6 Consider a capillary tube of radius R_t containing a liquid plug like that shown in Fig. 3.14. Determine a relation for the pressure drop required to initiate downward motion of the plug in terms of the liquid and vapor densities, the length of the slug L, and the advancing and receding contact angles θ_a and θ_r, respectively. You may assume that the gas–liquid interfaces are both portions of a spherical surface.

3.7 For water at 20°C and atmospheric pressure, if the advancing contact angle is 80° and the receding contact angle is 20°, estimate the length of a liquid slug that can be supported in air against gravity in a tube with an inside diameter of 1 mm. Assume that the interface surfaces are portions of spherical surfaces.

3.8 A slug of liquid water is supported against gravity inside a vertical tube with an inside diameter of 2 mm. For the slug, which is surrounded by water vapor at the same temperature, the advancing contact angle is 60° and the receding contact angle is 15°. Assuming that the contact angles do not change, determine and plot the variation of the length of the slug that can be supported with temperature for saturation temperatures between 100°C and the critical point.

3.9 A lens-shaped inclusion of fluid b can, under the right conditions, be trapped at the interface between two other fluids a and c, as shown in Fig. P3.1. Assuming that the bubble is radially symmetric, use force-balance requirements at the location where the three fluids meet to derive equations for the angles γ_1 and γ_2. Also derive two inequality constraints on the three interfacial tension values that must be satisfied if an equilibrium configuration of this type is to exist.

3.10 Determine the capillary rise height on the Teflon side wall of a container for the following liquids in air at 20°C: (*a*) water; (*b*) carbon tetrachloride; (*c*) acetone; and (*d*) ethanol. Use Eq. (2.51) and contact-angle values determined using Eq. (3.33) with information from Fig. 3.6 for your calculations. What would these rise heights be in a slowly spinning space station, where the effective gravity is only $0.02g$?

CHAPTER

FOUR

TRANSPORT EFFECTS AND DYNAMIC BEHAVIOR OF INTERFACES

4.1 TRANSPORT BOUNDARY CONDITIONS

In the preceding chapters, we have treated the interface between the two fluid phases to be a region over which the mean fluid properties vary continuously. Several important concepts and fundamental relations emerge from this view of the interface. At this juncture, we will depart from this viewpoint. Taking a more macroscopic perspective, we will now consider the interface to be a surface separating a liquid and its vapor.

To facilitate solution of the governing equations for heat, mass, and momentum transfer in the two fluids on either side of the interface, we must specify appropriate boundary conditions at the liquid–vapor interface. As a prelude to considering specific phase-change phenomena in later chapters, here we will formulate, in general terms, the interface conditions that will serve as boundary conditions for the transport equations in the adjacent phases.

At the interface, the system must satisfy the principles of conservation of mass, momentum, and energy. The transport of mass at the interface is schematically indicated in Fig. 4.1. In the liquid region, mass moves toward the interface with a velocity, $w_{l,n}$, with respect to a stationary observer. However, the interface is also moving with a velocity dZ_i/dt, so that the rate of liquid mass flow toward the control volume moving with the interface is $\rho_l \, (w_{l,n} - dZ_i/dt)$. In a similar manner, it can be argued that the rate of vapor mass flow rate out of the control volume moving with the interface is $\rho_v \, (w_{v,n} - dZ_i/dt)$. Note that we have arbitrarily adopted a sign convention whereby coordinates and velocities are positive to the right in Fig. 4.1.

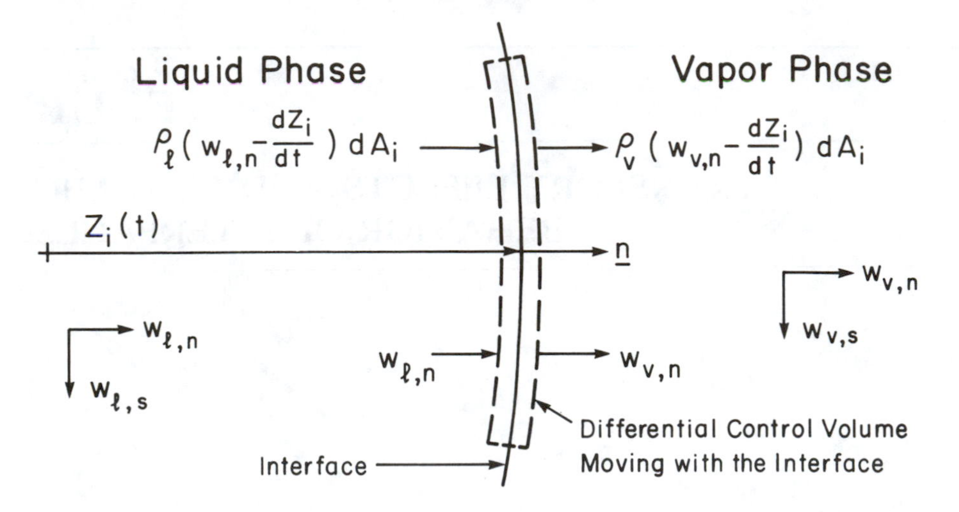

Figure 4.1 Mass fluxes across a liquid–vapor interface.

The control volume in Fig. 4.1 is assumed to be so thin that there can be negligible accumulation of mass within it. Conservation of mass for this control volume then requires that

$$\rho_l\left(w_{l,n} - \frac{dZ_i}{dt}\right) = \rho_v\left(w_{v,n} - \frac{dZ_i}{dt}\right) \tag{4.1}$$

which can be rearranged to obtain

$$\rho_l w_{l,n} - \rho_v w_{v,n} = (\rho_l - \rho_v)\frac{dZ_i}{dt} \tag{4.2}$$

Note in Eqs. (4.1) and (4.2) that the velocities $w_{l,n}$, $w_{v,n}$, and dZ_i/dt are all in the direction of the unit vector normal to the interface, **n**.

The transport of momentum normal to the interface and tangential to the interface are depicted in Figs. 4.2*a* and 4.2*b*, respectively. Because of the motion of the interface, momentum in the direction of the unit normal vector **n** is convected into the control volume moving with the interface at the relative velocity of the fluid with respect to the interface. Including the effects of pressure and surface tension forces, the force and momentum balance normal to the interface requires

$$P_l - P_v = \sigma\left(\frac{1}{r_1} + \frac{1}{r_2}\right) + \rho_v\left(w_{v,n} - \frac{dZ_i}{dt}\right)w_{v,n} - \rho_l\left(w_{l,n} - \frac{dZ_i}{dt}\right)w_{l,n} \tag{4.3}$$

where r_1 and r_2 are the principal radii of curvature of the interface surface. Consistent with our sign convention, r_1 and r_2 are positive if measured on the left (liquid) side of the interface and negative if measured on the right (vapor) side.

For some phase-change processes, the motion of the interface is limited by

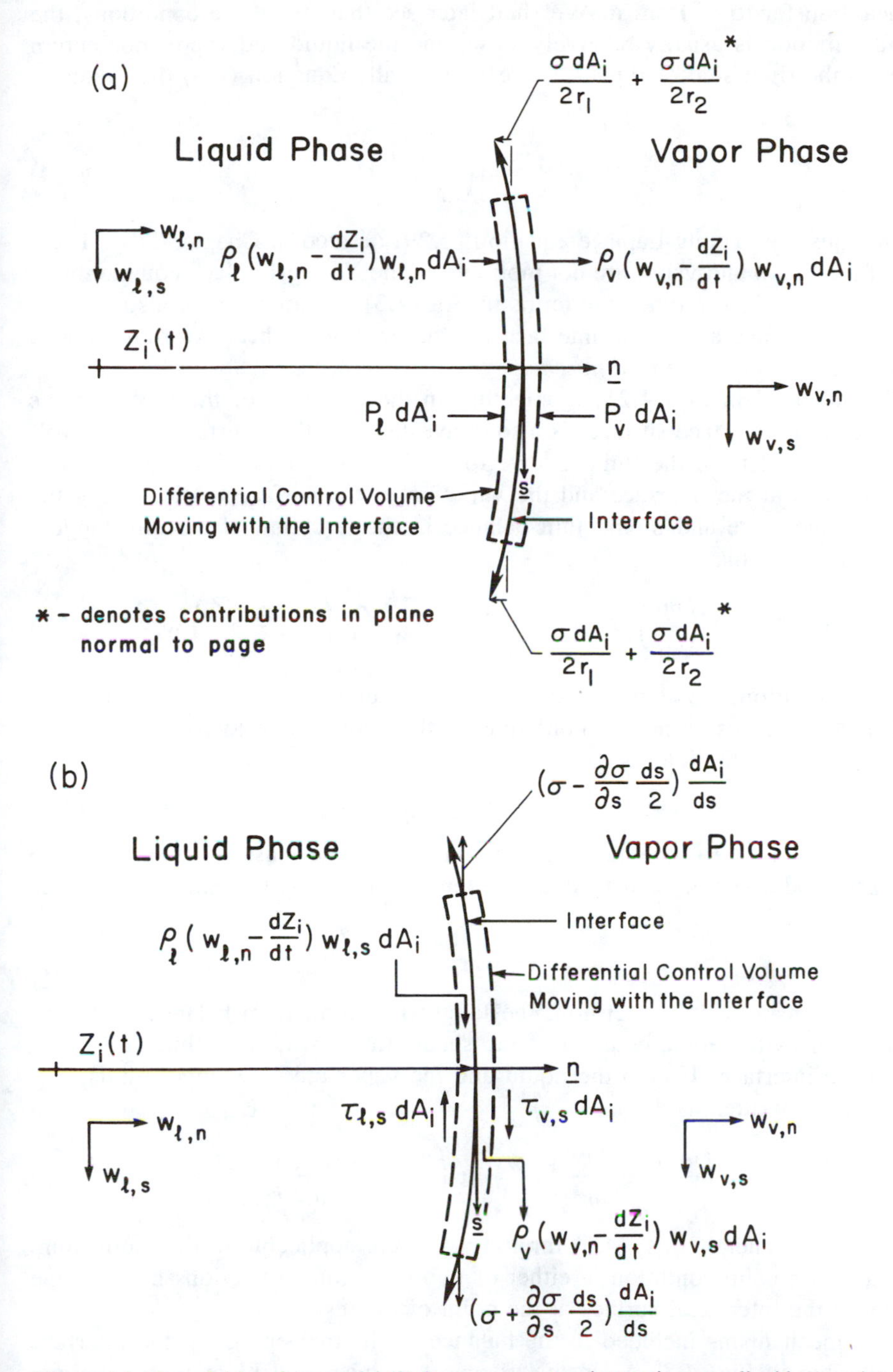

Figure 4.2 Normal (*a*) and tangential (*b*) force–momentum interactions at a liquid–vapor interface.

the heat transfer to or from it. We shall later see that, for these conditions, the interface motion is usually relatively slow, and the liquid and vapor momentum terms on the right side of Eq. (4.3) are very small. Equation (4.3) then reduces to

$$P_l - P_v = \sigma\left(\frac{1}{r_1} + \frac{1}{r_2}\right) \tag{4.4}$$

which is just the Young-Laplace equation (2.29) derived in Chapter Two. Thus, even if the liquid and vapor are not motionless, the Young-Laplace equation may still hold if the fluid momentum terms in Eq. (4.3) are small. Note also that the pressure difference across the interface for these circumstances is often referred to as the *capillary pressure difference*.

As indicated in Fig. 4.2*b*, momentum in the direction of the unit vector **s** (tangent to the interface surface) is also convected into the interface control volume at the velocity of the fluid relative to the interface. Including the effects of shear stresses at the interface and the variation of the surface tension along the interface, the force and momentum balance in the direction of the unit tangent vector **s** requires that

$$\tau_{l,s} - \tau_{v,s} - \left(\frac{\partial\sigma}{\partial s}\right) = \rho_l\left(w_{l,n} - \frac{dZ_i}{dt}\right) w_{l,s} - \rho_v\left(w_{v,n} - \frac{dZ_i}{dt}\right) w_{v,s} \tag{4.5}$$

If, in addition, we assume that the velocity fields vary continuously in each fluid, and we impose a no-slip condition on the tangential velocity components, it follows that, at the interface,

$$w_{l,s} = w_{v,s} \tag{4.6}$$

Substituting Eq. (4.6) together with the conservation of mass relation (4.1) into Eq. (4.5), and assuming constant surface tension ($\partial\sigma/\partial s = 0$), Eq. (4.5) reduces to

$$\tau_{l,s} = \tau_{v,s} \tag{4.7}$$

Thus, for constant surface tension, the tangential momentum balance at the interface reduces to simple equality of the shear stress in the two fluids on either side of the interface. If both the liquid and the vapor are Newtonian fluids, Eq. (4.7) may be written as

$$\mu_l\left(\frac{\partial w_{l,n}}{\partial s} + \frac{\partial w_{l,s}}{\partial z}\right)_{z=Z_i} = \mu_v\left(\frac{\partial w_{v,n}}{\partial s} + \frac{\partial w_{v,s}}{\partial z}\right)_{z=Z_i} \tag{4.8}$$

Note that, in general, Eqs. (4.5) through (4.8) are applicable to the momentum balance and no-slip condition in either of two orthogonal directions in the plane tangent to the interfacial surface at the point of interest.

The mechanisms included in the balance of thermal energy at the interface are indicated in Fig. 4.3. As in the transport of mass and momentum, thermal

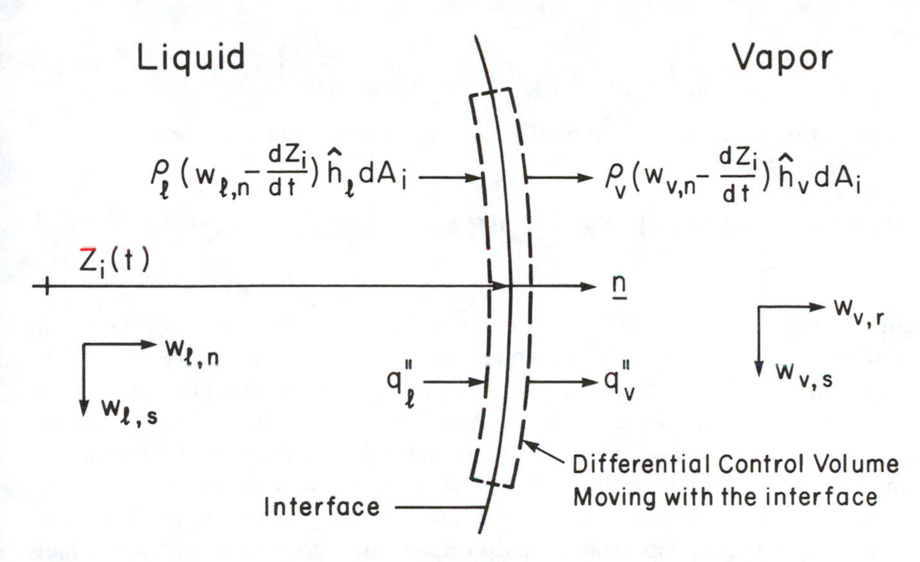

Figure 4.3 Energy transport across a liquid–vapor interface.

energy is convected into the control volume moving with the interface at the velocity of the fluid relative to the interface. Thermal energy can also be transported to or from the interface by Fourier conduction or, in the vapor, by radiative transport. The heat flux terms q_l'' and q_v'' include the transport due to these mechanisms. Conservation of thermal energy thus requires

$$q_v'' - q_l'' = \rho_l\left(w_{l,n} - \frac{dZ_i}{dt}\right)\hat{h}_l - \rho_v(w_{v,n} - \frac{dZ_i}{dt}\Bigg)\hat{h}_v \tag{4.9}$$

Substituting Eq. (4.1), Eq. (4.9) can be rearranged to obtain

$$q_l'' - q_v'' = \rho_l\left(w_{l,n} - \frac{dZ_i}{dt}\right)h_{l,v} \tag{4.10}$$

Note that, in obtaining Eq. (4.10), it has been assumed that local thermodynamic equilibrium exists at the interface so that $\hat{h}_l - \hat{h}_v = h_{lv}$. For the vapor and liquid phases of a pure substance, this assumption also implies a unique relation between the temperature and vapor pressure when momentum and interface curvature effects are sufficiently small that $P_l = P_v$, that is,

$$T_{\text{sat}} = T_{\text{sat}}(P_v) \tag{4.11}$$

When curvature effects are small, Eq. (4.11) is a necessary boundary condition for thermal transport in the vapor and liquid regions.

In many cases of practical interest, radiation effects in the vapor are small and the heat flux terms in Eq. (4.10) are due to Fourier conduction alone. Equation (4.10) may then be rewritten as

$$k_v\left(\frac{\partial T}{\partial z}\right)_{z=Z_i} - k_l\left(\frac{\partial T}{\partial z}\right)_{z=Z_i} = \rho_l\left(w_{l,n} - \frac{dZ_i}{\partial t}\right) h_{lv} \tag{4.12}$$

which can often be used as a thermal boundary condition at the interface.

4.2 KELVIN-HELMHOLTZ AND RAYLEIGH-TAYLOR INSTABILITIES

Instability associated with liquid–vapor interfaces can have a strong impact on the heat and mass transfer at the interface during phase-change processes. Often these instabilities cause a change in the morphology of the two-phase system at a particular set of transition conditions. Altering the interphase morphology invariably results in dramatic changes in heat and mass transport at the interface.

In this and subsequent sections of this chapter, we will examine several different instability mechanisms that (we will later see) play important roles in some commonly encountered vaporization and condensation processes. We begin here by considering the circumstances shown in Fig. 4.4. A vapor or gas phase is presumed to overlay a heavier liquid in a gravitational field that exerts a downward body force on the fluids.

Initially, the interface between the two phases is assumed to be a flat horizontal plane at $z = 0$. The liquid and vapor phases are moving with free-stream velocities $\bar{u}_l$ and $\bar{u}_v$, respectively, in the x direction parallel to the undisturbed interface. The object of our analysis here is to determine the range of conditions for which the interface is stable with respect to an arbitrary perturbation of the interface.

In general, determination of the time-dependent variation of the interface position in response to an arbitrary initial perturbation requires solution of a set of

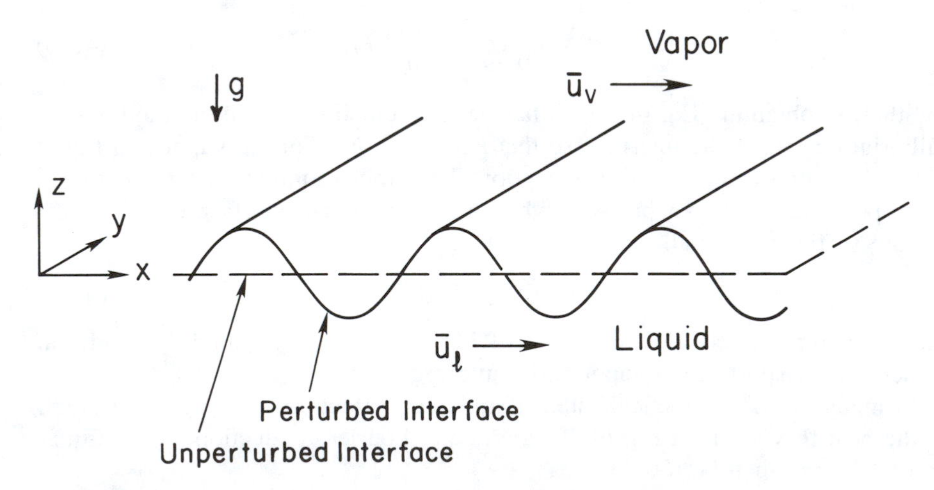

Figure 4.4 Perturbed interface considered in linear analysis of interface stability.

complex nonlinear equations with the associated boundary conditions. We can, however, simplify the analysis, while retaining most of the important physics, by adopting some appropriate idealizations.

As a first means of simplification, we will restrict the analysis to arbitrary disturbances $\delta(x,t)$ that vary in the x direction but not in the y direction. More general disturbances that vary with both x and y can be considered, at the expense of a more complicated analysis. However, because the method of analysis and the physical conclusions differ only slightly, we will consider simpler disturbances that vary only with x and t. In addition, we will also limit the analysis to circumstances in which viscous effects are negligible (i.e., the inviscid flow case). With these assumptions, the governing equations for the subsequent fluid motion in either of the two fluids are the two-dimensional forms of the laminar transport equations (1.35) through (1.38) presented in Chapter One with the viscosity μ set equal to zero.

$$\frac{\partial u}{\partial x} + \frac{\partial w}{\partial z} = 0 \tag{4.13}$$

$$\rho\left[\frac{\partial u}{\partial t} + u\left(\frac{\partial u}{\partial x}\right) + w\left(\frac{\partial u}{\partial z}\right)\right] = -\frac{\partial P}{\partial x} \tag{4.14}$$

$$\rho\left[\frac{\partial w}{\partial t} + u\left(\frac{\partial w}{\partial x}\right) + w\left(\frac{\partial w}{\partial z}\right)\right] = -\frac{\partial P}{\partial z} - \rho g \tag{4.15}$$

The velocities u and w and the pressure P are decomposed into base flow and perturbed components:

$$u = \overline{u} + u' \qquad w = \overline{w} + w' \qquad P = \overline{P} + P' \tag{4.16}$$

Because the governing equations (4.13) through (4.15) must hold for the unperturbed flow, the base flow quantities $\overline{u}$, $\overline{w}$, and $\overline{P}$ must satisfy these equations. Following the usual practice of linear stability analysis, Eqs. (4.16) for u, w, and P are substituted into Eqs. (4.13) through (4.15), products of perturbation (primed) quantities are neglected, and the corresponding equations for the base flow are subtracted. The resulting equations are further simplified using the facts that $\partial\overline{u}/\partial x = \partial\overline{u}/\partial z = \overline{w} = 0$ to obtain the following equations for the perturbation quantities:

$$\frac{\partial u'}{\partial x} + \frac{\partial w'}{\partial z} = 0 \tag{4.17}$$

$$\rho\left[\frac{\partial u'}{\partial t} + \overline{u}\left(\frac{\partial u'}{\partial x}\right)\right] = -\frac{\partial P'}{\partial x} \tag{4.18}$$

$$\rho\left[\frac{\partial w'}{\partial t} + \overline{u}\left(\frac{\partial w'}{\partial x}\right)\right] = -\frac{\partial P'}{\partial z} \tag{4.19}$$

Differentiating Eq. (4.18) with respect to x and Eq. (4.19) with respect to z, adding the resulting equations together, and substituting the continuity equation yields the Laplace equation for the perturbation pressure field:

$$\frac{\partial^2 P'}{\partial x^2} + \frac{\partial^2 P'}{\partial z^2} = 0 \tag{4.20}$$

Before the perturbation is applied, the base flow pressure variation in the two fluids is hydrostatic:

$$\overline{P}_v = P_0 - \rho_v g z \tag{4.21}$$

$$\overline{P}_l = P_0 - \rho_l g z \tag{4.22}$$

where P_0 is the pressure at the undisturbed (flat) interface. Any arbitrary two-dimensional initial disturbance can be represented as a sum of Fourier series components with contributions over a broad range of wavelengths. Because a wide variety of initial perturbations $\delta(x,0)$ may occur, all possible Fourier wavelengths are expected to be present, and the stability of the interface will depend on whether the amplitude of one or more of the Fourier waves will grow and cause the system to become unstable.

The above observations imply that the question of interface stability for these circumstances can be answered by considering the Fourier component waves themselves as initial perturbations,

$$\delta(x,0) = Ae^{i\alpha x} \tag{4.23}$$

where the disturbance wavelength is equal to $2\pi/\alpha$, and it is understood that only the real part of this relation is physically significant.

For the system considered here, it is plausible to expect that the spatial variations of δ, u', w', and P' with x will be similar to that for the original perturbation $\delta(x,0)$, and that their temporal variation will be oscillatory. We therefore postulate the following functional forms for the subsequent position of the interface δ and the perturbation quantities w' and P':

$$\delta(x,t) = Ae^{i\alpha x + \beta t} \tag{4.24}$$

$$w'(x,z,t) = \hat{w}(z)e^{i\alpha x + \beta t} \tag{4.25}$$

$$P'(x,z,t) = \hat{P}(z)e^{i\alpha x + \beta t} \tag{4.26}$$

Substituting Eq. (4.26) for P' into Eq. (4.20) yields the following equation for $\hat{P}$:

$$\frac{d^2\hat{P}}{dz^2} = \alpha^2\hat{P} \tag{4.27}$$

Solutions for $\hat{P}$ in the liquid and vapor regions, which satisfy Eq. (4.27) and the condition that $P' \rightarrow 0$ far from the interface, are

$$\hat{P}_v = a_v e^{-\alpha z} \tag{4.28}$$

$$\hat{P}_l = a_l e^{\alpha z} \tag{4.29}$$

Substituting Eqs. (4.25) and (4.26) into Eq. (4.19), we obtain

$$\hat{w}(z) = -[\rho(\beta + i\alpha\bar{u})]^{-1}\left(\frac{d\hat{P}}{dz}\right) \tag{4.30}$$

and using Eqs. (4.28) and (4.29) to evaluate $d\hat{P}/dz$ yields the following relations for $\hat{w}$ in the liquid and vapor regions:

$$\hat{w}_v(z) = \alpha a_v[\rho_v(\beta + i\alpha\bar{u}_v)]^{-1}\, e^{-\alpha z} \tag{4.31}$$

$$\hat{w}_l(z) = -\alpha a_l[\rho_l(\beta + i\alpha\bar{u}_l)]^{-1}\, e^{\alpha z} \tag{4.32}$$

Substituting Eq. (4.25) into the continuity equation (4.17), the resulting differential equation for u' can be integrated. Using the condition that $u' \to 0$ far from the interface, the following relation for u' is obtained:

$$u' = \frac{i}{\alpha}\left(\frac{d\hat{w}}{dz}\right)e^{i\alpha x + \beta t} \tag{4.33}$$

Direct substitution of the above relation for u' with Eqs. (4.25) and (4.26) for w' and P' verifies that Eq. (4.18) is identically satisfied.

Having established solution forms for u', w', and P' that satisfy the governing equations, we now apply appropriate boundary conditions at the interface. Note first that at a given x location, Eq. (4.24) indicates that the interface position oscillates sinusoidally with time. This produces a contribution to w' near the interface equal to $\partial\delta/\partial t$. In addition, for the fluid immediately adjacent to the interface to be convected in the x direction with a velocity $\bar{u}$ and yet follow the interface contour, at any instant of time w' must have a component equal to $\bar{u}\,\partial\delta/\partial x$. To satisfy both of these requirements, w' at the interface must be equal to

$$w'_{z\to 0} = \frac{\partial\delta}{\partial t} + \bar{u}\,\frac{\partial\delta}{dx} \tag{4.34}$$

Imposing this condition on both the vapor and liquid sides of the interface, substitution of Eqs. (4.24) and (4.25) with either Eq. (4.31) or (4.32) yields the following two relations:

$$\alpha a_v[\rho_v(\beta + i\alpha\bar{u}_v)]^{-1} = \beta A + i\alpha\bar{u}_v A \tag{4.35}$$

$$-\alpha a_l[\rho_l(\beta + i\alpha\bar{u}_l)]^{-1} = \beta A + i\alpha\bar{u}_l A \tag{4.36}$$

These relations can be solved for a_v and a_l in terms of A and the other parameters:

$$a_v = \frac{\rho_v}{\alpha}(\beta + i\alpha\bar{u}_v)^2 A \tag{4.37}$$

$$a_l = -\frac{\rho_l}{\alpha}(\beta + i\alpha\bar{u}_l)^2 A \tag{4.38}$$

For inviscid flow, the momentum balance tangential to the interface and continuity of the tangential velocity component at the interface cannot be imposed. However, the force and momentum balance in the direction normal to the interface must be enforced even for inviscid flow. Because the perturbation velocities normal to the interface are small, their contribution to the force and momentum balance normal to the interface is negligible. As described in Section 4.1, for such circumstances the force balance normal to the interface simply becomes the Young-Laplace equation:

$$P_l - P_v = \sigma\left(\frac{1}{r_1} + \frac{1}{r_2}\right) \tag{4.39}$$

For the system considered here,

$$\frac{1}{r_1} = -\frac{(\partial^2\delta/\partial x^2)}{[1 + (\partial\delta/\partial x)^2]^{3/2}} \qquad \frac{1}{r_2} = 0 \tag{4.40}$$

and

$$P_l = \overline{P}_l + P_l' \qquad P_v = \overline{P}_v + P_v' \tag{4.41}$$

The analysis may be continued by substituting the relations (4.40) and (4.41) into Eq. (4.39), using Eqs. (4.21) and (4.22) with $z = \delta$ to evaluate P_l and P_v, and using Eqs. (4.24), (4.26), (4.28), and (4.29) to evaluate P_l', P_v', and δ. Neglecting terms of order δ^2 compared to 1, we obtain the following relation from Eq. (4.39):

$$a_l - a_v = [(\rho_l - \rho_v)g + \sigma\alpha^2]A \tag{4.42}$$

Substituting Eqs. (4.37) and (4.38) for a_l and a_v into Eq. (4.42), A cancels out of the equation. After doing so, this equation can then be rearranged to get the following relation for β:

$$\beta = \pm\frac{\{\alpha^2\rho_l\rho_v(\overline{u}_l - \overline{u}_v)^2 - [\sigma\alpha^3 + (\rho_l - \rho_v)g\alpha](\rho_l + \rho_v)\}^{1/2}}{\rho_l + \rho_v} - \frac{i\alpha(\rho_l\overline{u}_l + \rho_v\overline{u}_v)}{\rho_l + \rho_v} \tag{4.43}$$

The relation (4.43) for β above actually provides much information about the type of system considered here. The fact that β always has an imaginary part as long as $\overline{u}_l$ or $\overline{u}_v$ is greater than zero implies that any horizontal velocity component in either fluid will produce waves at the interface. The amplitude of the perturbation will grow with time only if β has a positive real part. This can happen only if the sum of the two terms inside the braces is greater than zero. The second term in the braces, being negative, diminishes the sum. Thus surface tension and gravity tend to stabilize the interface.

Setting the term inside the braces to zero and rearranging, the condition for an unstable interface becomes

$$|\overline{u}_l - \overline{u}_v| > \left\{\frac{[\sigma\alpha + (\rho_l - \rho_v)g/\alpha](\rho_l + \rho_v)}{\rho_l\rho_v}\right\}^{1/2} \tag{4.44}$$

The term on the right side of the relation (4.44) above varies with the wave number of the disturbance α. It can be shown that the right side of this inequality has a minimum value at a critical wavenumber α_c equal to

$$\alpha_c = \left[\frac{(\rho_l - \rho_v)g}{\sigma}\right]^{1/2} \tag{4.45}$$

As a result, if $|\bar{u}_l - \bar{u}_v|$ is greater than the right side of the inequality (4.44) evaluated at $\alpha = \alpha_c$, there will be some range of disturbances that are unstable. Thus, if disturbances of all wavelengths are present in the system, the interface will be unstable when

$$|\bar{u}_l - \bar{u}_v| > u_c \tag{4.46}$$

where u_c is the critical velocity equal to the right side of relation (4.44) evaluated at α_c:

$$u_c = \left[\frac{2(\rho_l - \rho_v)}{\rho_l}\right]^{1/2}\left[\frac{\sigma(\rho_l - \rho_v)g}{\rho_v^2}\right]^{1/4} \tag{4.47}$$

This type of interface instability is referred to as *Kelvin-Helmholtz instability* or simply *Helmholtz instability*. If a specific disturbance wavelength λ is imposed on the system, the inequality (4.44) indicates that the interface will be unstable for

$$|\bar{u}_l - \bar{u}_v| > \left\{\frac{[2\pi\sigma/\lambda + (\rho_l - \rho_g)g\lambda/2\pi](\rho_l + \rho_v)}{\rho_l\rho_v}\right\}^{1/2} \tag{4.48}$$

Note also that when the interface is unstable, the oscillatory component of the disturbance is solely a result of the second term of Eq. (4.43), and the wave propagates in the x direction with a speed equal to $(\rho_l\bar{u}_l + \rho_v\bar{u}_v)/(\rho_l + \rho_v)$, which is between the fluid velocities $\bar{u}_l$ and $\bar{u}_v$

In addition to the system shown in Fig. 4.4, Eq. (4.43) also provides information on the stability of other similar systems. For example, if we change the sign on g, we get the dispersion relation for waves on an interface between a lower vapor region and an upper liquid region that have some relative velocity parallel to the interface. Gravity then acts to further destabilize the interface, with the interface being unstable for

$$|\bar{u}_l - \bar{u}_v| > \left\{\frac{[\sigma\alpha - (\rho_l - \rho_v)g/\alpha](\rho_l + \rho_v)}{\rho_l\rho_v}\right\}^{1/2} \tag{4.49}$$

Note that the right side of the above inequality now does not exhibit a minimum value as α varies. Hence if disturbances of all wavelengths are present, there will always be some long-wavelength (small-α) disturbances that will amplify.

If g is set to zero in Eq. (4.43), we get the dispersion relation for a vertical interface, which implies that it is unstable for

$$|\bar{u}_l - \bar{u}_v| > \left[\frac{\sigma\alpha(\rho_l + \rho_v)}{\rho_l\rho_v}\right]^{1/2} \tag{4.50}$$

Again, it is clear that if disturbances of all wavelengths are present, there will be some disturbances at small α and long wavelength that will amplify and cause the interface to be unstable.

If we set $\bar{u}_l = \bar{u}_v = 0$ and change the sign of g in Eq. (4.44), we obtain the following conditions for interface instability of a motionless liquid overlaying a motionless vapor region:

$$\alpha > \alpha_c = \left[\frac{(\rho_l - \rho_v)g}{\sigma}\right]^{1/2} \tag{4.51}$$

This condition is referred to as *Rayleigh-Taylor instability*. The critical wavelength $\lambda_c = 2\pi/\alpha_c$ corresponding to the critical wavenumber is

$$\lambda_c = 2\pi\left[\frac{\sigma}{(\rho_l - \rho_v)g}\right]^{1/2} \tag{4.52}$$

For a steam–water system at 100°C, λ_c is about 1.6 cm. Only perturbations having wavelenths greater than λ_c will grow, leading to instability of the interface. The stability of the interface can be assured by limiting the lateral extent of the interface. If the lateral dimensions of the interface are smaller than λ_c, no disturbances of wavelength greater than λ_c can arise, and the interface will be *stable*.

The value of β for Rayleigh-Taylor instability is obtained by setting $\bar{u}_l = \bar{u}_v = 0$ and changing the sign of g in Eq. (4.43). This yields the relation

$$\beta = \pm\left\{\frac{[(\rho_l - \rho_v)g\alpha - \sigma\alpha^3]}{\rho_l + \rho_v}\right\}^{1/2} \tag{4.53a}$$

If ρ_v is neglected compared to ρ_l, this relation can be written approximately as

$$\beta = \pm\left[\frac{(\rho_l - \rho_v)g\alpha}{\rho_l} - \frac{\sigma\alpha^3}{\rho_l}\right]^{1/2} \tag{4.53b}$$

From the form of the relation for the disturbance (see Eq. [4.24]) it is clear that the positive β value is equal to the fractional growth rate of the disturbance with time $(d\delta/dt)/\delta$. The variation of β with α given by Eq. (4.53*b*) indicates that there is a specific value of α where β is a maximum. Differentiating Eq. (4.53*b*) with respect to α and setting $d\beta/d\alpha = 0$, it can be shown that the maximum values of β and the corresponding α are given by

$$\alpha_{\max} = \left[\frac{(\rho_l - \rho_v)g}{3\sigma}\right]^{1/2} \tag{4.54}$$

$$\beta_{\max} = \left[\frac{4(\rho_l - \rho_v)^3 g^3}{27\sigma\rho_l^2}\right]^{1/4} \tag{4.55}$$

The disturbance wavelength corresponding to $\alpha_{\max}$ is often referred to as the *most dangerous wavelength* λ_D, given by

$$\lambda_D = 2\pi\left[\frac{3\sigma}{(\rho_l - \rho_v)g}\right]^{1/2} = \sqrt{3}\,\lambda_c \tag{4.56}$$

Because they grow most rapidly, disturbances with wavelengths near λ_D are expected to dominate the early stages of the instability where the linear theory applies. In real systems, the dominant disturbance wavelength observed experimentally is often close to λ_D, even though the linear analysis is not applicable to the larger interface deformations required to detect the instability.

Example 4.1 Determine the critical velocity for a stratified horizontal flow of saturated steam above liquid water at atmospheric pressure. What happens if the pressure is increased to 8.59 MPa?

For saturated steam and water at atmospheric pressure, $\rho_l = 958\ \text{kg/m}^3$, $\rho_v = 0.598\ \text{kg/m}^3$, $\sigma = 0.05878\ \text{N/m}$. Using Eq. (4.47),

$$u_c = \left[\frac{2(\rho_l - \rho_v)}{\rho_l}\right]^{1/2}\left[\frac{\sigma(\rho_l - \rho_v)g}{\rho_v^2}\right]^{1/4}$$

$$u_c = \left[\frac{2(958 - 0.598)}{958}\right]^{1/2}\left[\frac{0.05878(958 - 0.598)9.8}{(0.598)^2}\right]^{1/4} = 7.793\ \text{m/s}$$

At 8.59 MPa, $\rho_l = 712.5\ \text{kg/m}^3$, $\rho_v = 46.2\ \text{kg/m}^3$, $\sigma = 0.01439\ \text{N/m}$, and

$$u_c = \left[\frac{2(712.5 - 46.2)}{712.5}\right]^{1/2}\left[\frac{0.01439(712.5 - 46.2)9.8}{(46.2)^2}\right]^{1/4} = 0.627\ \text{m/s}$$

The density difference and surface tension that act to restore a perturbed interface to its initial flat configuration are weaker at high pressure, and the system becomes unstable at a lower relative velocity.

As noted at the beginning of this section, the linear stability analysis for two-dimensional waves presented here can be extended to three-dimensional waveforms. The procedure used to solve the three-dimensional problem is basically the same as for the two-dimensional one. Although the mathematical details are a bit more complex, solutions can be obtained for more general three-dimensional waveforms. In a very early study, however, Squire [4.1] showed for Rayleigh-Taylor instability that three-dimensional waves are generally more stable than two-dimensional waves for these circumstances. Hence two-dimensional waves are more likely to be the mechanism to initiate instability in systems of this type.

In addition, the results of a more recent study by Sernas [4.2] indicate that the most dangerous three-dimensional Rayleigh-Taylor waveform has peaks and valleys in a square grid pattern with the side of the square unit cell equal to the two-dimensional wavelength λ_D. Thus the two-dimensional wave analysis presented above provides a physically realistic indication of the stability criteria and the wavelength of the disturbance that is most rapidly amplified.

Strictly speaking, the analysis presented above applies only when the liquid and vapor depths are infinite. However, in later sections we will be interested in film boiling processes in which a vapor layer of finite depth exists under a liquid pool of virtually infinite depth. In a discussion of the paper by Hsieh [4.3], Dhir and Lienhard showed that for Rayleigh-Taylor instability, there was no effect of finite vapor depth on the most dangerous wavelength and its effect on the corresponding amplification rate was very small. Hence the results for infinite layer depths may be used for finite vapor layers with little loss in accuracy.

Although the analysis presented here ignores viscous effects, the conclusions based on the results of the analysis are usually valid for most liquid–vapor systems of interest in engineering systems. A further discussion of Kelvin-Helmholtz and Rayleigh-Taylor instabilities may be found in reference [4.4]. As we shall see, this type of instability also plays a role in some of the vaporization processes we shall consider in later sections.

Example 4.2 Determine the most dangerous wavelength for saturated liquid water at (*a*) 20°C and (*b*) 300°C in contact with its vapor, and (*c*) saturated liquid R-12 at 350°C in contact with its vapor.

(*a*) For saturated liquid water at 20°C, $\rho_l = 998\ \text{kg/m}^3$, $\rho_v = 0.0173\ \text{kg/m}^3$, $\sigma = 0.0728\ \text{N/m}$. It follows, using Eq. (4.56), that

$$\lambda_D = 2\pi\left[\frac{3\sigma}{(\rho_l - \rho_v)g}\right]^{1/2} = 2\pi\left[\frac{3(0.0728)}{(998. - 0.0173)9.8}\right]^{1/2} = 0.0297\ \text{m}$$

(*b*) Similarly, at 300°C, $\rho_l = 712\ \text{kg/m}^3$, $\rho_v = 46.2\ \text{kg/m}^3$, $\sigma = 0.0144$ N/m, and

$$\lambda_D = 2\pi\left[\frac{3(0.0144)}{(712. - 46.2)9.8}\right]^{1/2} = 0.0162\ \text{m}$$

(*c*) For saturated R-12 at 350 °C, $\rho_l = 1075\ \text{kg/m}^3$, $\rho_v = 136.4\ \text{kg/m}^3$, $\sigma = 0.0028\ \text{N/m}$, and

$$\lambda_D = 2\pi\left[\frac{3(0.0028)}{(1075. - 136.4)9.8}\right]^{1/2} = 0.0060\ \text{m}$$

Note that at low to moderate saturation pressures (away from the critical point), $\rho_l - \rho_v$ changes slowly as the temperature and pressure decrease, while σ increases about linearly with decreasing temperature (see Chapter Two), having the stronger effect on λ_D.

4.3 INTERFACE STABILITY OF LIQUID JETS

Careful observation of a jet of water issuing from a faucet or water hose reveals that although the air–water interface is cylindrical near the nozzle, farther downstream the jet invariably breaks up into droplets. The photograph in Fig. 4.5 illustrates this phenomenon.

The tendency of the liquid jet to break up into droplets is actually advantageous in some applications. When a jet of liquid fuel is injected into a combustion engine, the breakup of the jet increases the interfacial area, which, in turn, leads to more rapid vaporization of the fuel. A similar enhancement of vaporization results in the case of cooling ponds, where water sprayed into the air is cooled by its partial evaporation. On the other hand, the breakup phenomenon is a disadvantage for sprinkler systems, because loss of water by evaporation and the drift of small droplets is undesirable. In addition to these direct applications, the instability of liquid jets also provides a basis for understanding the breakdown of inverted annular flow into dispersed flow during convective film boiling in tubes.

In this section the stability of a thin liquid jet will be analyzed to try to determine the conditions that lead to breakup of the jet. We wish to consider a cylindrical jet of radius R_0 issuing from a nozzle at a uniform velocity u_l, as shown in Fig. 4.6*a*. To simplify the stability analysis, inviscid flow is assumed, gravity body forces are neglected, and the analysis will employ a coordinate system mov-

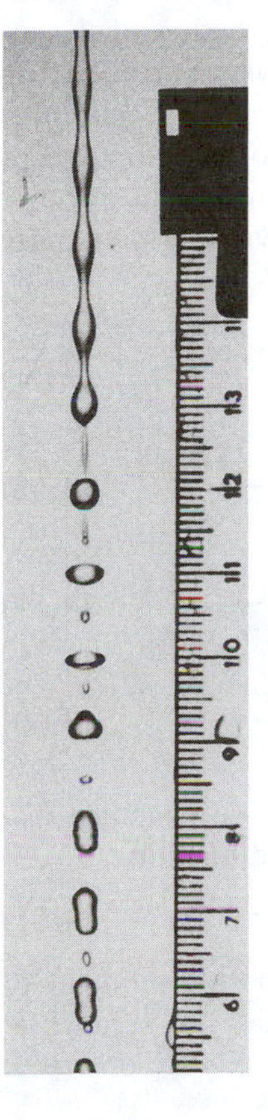

Figure 4.5 Breakup of a round water jet in air. The smallest divisions on the scale shown are 1 mm. *(From D. F. Rutland and G. J. Jameson, J. Fluid Mech., vol. 40, pp. 267–271, 1971, reproduced with permission, copyright © 1971, Cambridge University Press.)*

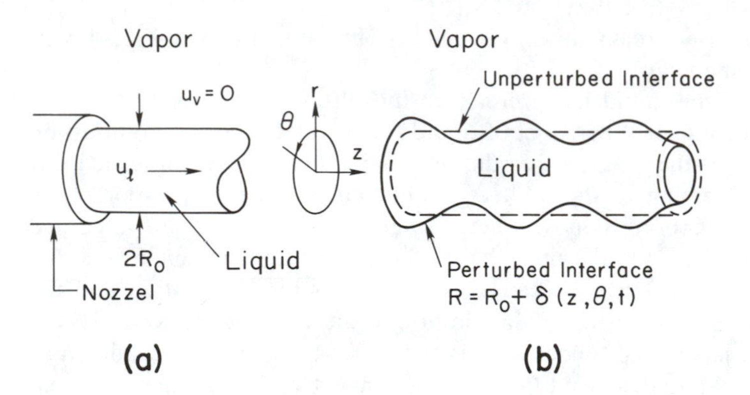

Figure 4.6 Model system considered in the analysis of the stability of a liquid jet.

ing with the liquid at a uniform velocity u_l. Hence in this coordinate system the unperturbed liquid velocity is zero.

With the idealizations noted above, the problem becomes that of analyzing the stability of the motionless liquid cylinder shown schematically in Fig. 4.6*b*. We specifically wish to know whether an arbitrary small perturbation of the interface will tend to grow in amplitude. As in the stability analyses considered in the previous sections, we will consider sinusoidal disturbances of the interface, because random disturbances in the real system are expected to be composed of Fourier components having wavelengths over virtually the entire spectrum.

For inviscid incompressible flow in the liquid jet, the governing continuity and momentum equations in cylindrical coordinates are

$$\frac{1}{r}\left[\frac{\partial(ru_r)}{\partial r}\right] + \frac{\partial u_z}{\partial z} + \frac{1}{r}\left(\frac{\partial u_\theta}{\partial r}\right) = 0 \tag{4.57}$$

$$\rho_l\left[\frac{\partial u_r}{\partial t} + u_r\left(\frac{\partial u_r}{\partial r}\right) + \frac{u_\theta}{r}\left(\frac{\partial u_r}{\partial \theta}\right) + u_z\left(\frac{\partial u_r}{\partial z}\right) - \frac{u_\theta^2}{r}\right] = -\frac{\partial P}{\partial r} \tag{4.58}$$

$$\rho_l\left[\frac{\partial u_\theta}{\partial t} + u_r\left(\frac{\partial u_\theta}{\partial r}\right) + \frac{u_\theta}{r}\left(\frac{\partial u_\theta}{\partial \theta}\right) + u_z\left(\frac{\partial u_\theta}{\partial z}\right) - \frac{u_r u_\theta}{r}\right] = -\frac{1}{r}\left(\frac{\partial P}{\partial \theta}\right) \tag{4.59}$$

$$\rho_l\left[\frac{\partial u_z}{\partial t} + u_r\left(\frac{\partial u_z}{\partial r}\right) + \frac{u_\theta}{r}\left(\frac{\partial u_z}{\partial \theta}\right) + u_z\left(\frac{\partial u_z}{\partial z}\right)\right] = -\frac{\partial P}{\partial z} \tag{4.60}$$

Following the usual procedure for linear stability analysis, we postulate that the pressure and velocities are the sum of steady plus perturbed quantities:

$$u_r = \overline{u}_r + u_r' \qquad u_\theta = \overline{u}_\theta + u_\theta' \qquad u_z = \overline{u}_z + u_z' \qquad P = \overline{P} + P' \tag{4.61}$$

Substituting these relations into Eqs. (4.57) through (4.60), ignoring products of primed quantities, and using the fact that the base flow velocities are zero ($\overline{u}_r = \overline{u}_\theta = \overline{u}_z = 0$), the following equations for the perturbation quantities are obtained:

$$\frac{1}{r}\left[\frac{\partial(ru_r')}{\partial r}\right] + \frac{\partial u_z'}{\partial z} + \frac{1}{r}\left(\frac{\partial u_\theta'}{\partial r}\right) = 0 \tag{4.62}$$

$$\rho_l \frac{\partial u_r'}{\partial t} = -\frac{\partial P'}{\partial r} \tag{4.63}$$

$$\rho_l \frac{\partial u_\theta'}{\partial t} = -\frac{1}{r}\left(\frac{\partial P'}{\partial \theta}\right) \tag{4.64}$$

$$\rho_l \frac{\partial u_z'}{\partial t} = -\frac{\partial P'}{\partial z} \tag{4.65}$$

Applying the operators $(1/r)\,\partial[r(\ )]/\partial r$, $(1/r)\,\partial/\partial\theta$, and $\partial/\partial z$ to Eqs. (4.63), (4.64), and (4.65), respectively, adding the resulting equations, and using the continuity equation (4.62) yields

$$\frac{1}{r}\left[\frac{\partial(r\,\partial P'/\partial r)}{\partial r}\right] + \frac{1}{r^2}\left(\frac{\partial^2 P'}{\partial \theta^2}\right) + \frac{\partial^2 P'}{\partial z^2} = 0 \tag{4.66}$$

where the left side of Eq. (4.66) is the expansion of $\nabla^2 P'$ in cylindrical coordinates.

For an initial interface perturbation of the form

$$R = R_0 + \delta_0 e^{i(m\theta + \alpha z)} \tag{4.67}$$

we postulate that the subsequent motion of the interface will result in variations of P' and u_r' of the form

$$P'(r,z,\theta,t) = \hat{P}(r)A(t)e^{i(m\theta+\alpha z)} \tag{4.68}$$

$$u_r'(r,z,\theta,t) = \hat{u}_r(r)A(t)e^{i(m\theta+\alpha z)} \tag{4.69}$$

In these equations, m can be any integer to allow for perturbations that vary sinusoidally around the circumference of the jet. Note that for $m = 0$ the disturbance perturbation is radially symmetric, with the radius of the jet varying sinusoidally with z, as schematically shown in Fig. 4.6*b*. Values of $m \neq 0$ correspond to more complicated, nonaxisymmetric perturbations.

Substituting the relation (4.68) for P' into Eq. (4.66) yields the following equation for $\hat{P}(r)$:

$$r^2 \frac{d^2\hat{P}}{dr^2} + r\frac{d\hat{P}}{dr} - (m^2 + \alpha^2 r^2)\hat{P} = 0 \tag{4.70}$$

For a specified value of m, the two independent solutions of this second-order linear differential equation are $I_m(\alpha r)$ and $K_m(\alpha r)$, where I_m and K_m are the modified Bessel functions (of order m) of the first and second kinds, respectively. In general, the solution may be a linear combination of $I_m(\alpha r)$ and $K_m(\alpha r)$:

$$\hat{P} = p_1 I_m(\alpha r) + p_2 K_m(\alpha r) \tag{4.71}$$

We note, however, that Bessel functions of the second kind are unbounded as $r \to 0$. To assure that the pressure field is finite throughout the liquid domain, we therefore take $p_2 = 0$.

$$\hat{P} = p_1 I_m(\alpha r) \tag{4.72}$$

Substituting the relations (4.68) and (4.69) for P' and u_r' into the r-direction momentum equation (4.63), we obtain

$$\frac{dA/dt}{A} = -\frac{(d\hat{P}/dr)}{\rho_l \hat{u}_r(r)} = \beta \tag{4.73}$$

Because the left-most term above is only a function of time and the center term is only a function of r, they can be equal for all t and r only if they both equal a constant, which we will denote as β. Setting the left-most term equal to β and solving for A yields

$$A(t) = a_0 e^{\beta t} \tag{4.74}$$

Similarly, using Eqs. (4.73) and (4.72) together with the identity

$$\frac{dI_m(x)}{dx} = I_{m+1}(x) + (m/x)I_m(x) \tag{4.75}$$

(see reference [4.5]) we obtain the following relation for $\hat{u}_r(r)$:

$$\hat{u}_r(r) = -\left(\frac{p_1\alpha}{\rho_l\beta}\right)\left[I_{m+1}(\alpha r) + \frac{m}{\alpha r} I_m(\alpha r)\right] \tag{4.76}$$

To obtain the dispersion relation between β and α, we now impose two boundary conditions. First, we require that the rate of change of the radius location of the interface must equal the radial velocity at the interface:

$$\frac{dR}{dt} = u_r'(R_0) \tag{4.77}$$

The variation of R with time is postulated to be of the form

$$R = R_0 + \delta \qquad \delta = \delta_0 e^{i(m\theta + \alpha z)} B(t) \tag{4.78}$$

where $B(0) = 1$. Substituting the relations (4.69), (4.76), and (4.78) into (4.77) and solving for $B(t)$ yields

$$B(t) = \left[\frac{\hat{u}_r(R_0)a_0}{\beta\delta_0}\right] e^{\beta t} \tag{4.79}$$

We further note that $B(0) = 1$ requires that

$$a_0 = \frac{\beta\delta_0}{\hat{u}_r(R_0)} \tag{4.80}$$

The second boundary condition imposed here is the force-momentum balance normal to the interface, which reduces to the Young-Laplace equation

$$(P_l - P_v)_{r=R_0} = \sigma\left(\frac{1}{r_1} + \frac{1}{r_2}\right) \tag{4.81}$$

Using the fact that $\delta/R_0 \ll 1$ for small perturbations, the following relations for the principal radii of curvature can be obtained from geometry considerations:

$$\frac{1}{r_1} = \frac{1 - \delta/R_r}{R_0} + \frac{m^2}{R_0^2} \tag{4.82}$$

$$\frac{1}{r_2} = -\frac{d^2\delta}{dz^2} = \alpha^2\delta \tag{4.83}$$

We further note that for the unperturbed case,

$$(\overline{P}_l - \overline{P}_v)_{r=R_0} = \frac{\sigma}{R_0} \tag{4.84}$$

Substituting $P_l = \overline{P}_l + P_l'$ together with Eqs. (4.68), (4.72), (4.78), and (4.82) through (4.84) into Eq. (4.81) and solving for β^2 yields

$$\beta^2 = -\left(\frac{\sigma\alpha^3}{\rho_l}\right)\left[\frac{1 + (m^2 - 1)}{\alpha^2 R_0^2}\right]\left[\frac{m}{\alpha R_0} + \frac{I_{m+1}(\alpha R_0)}{I_m(\alpha R_0)}\right] \tag{4.85}$$

Because the modified Bessel functions I_m are always positive, the right side of Eq. (4.145) is always negative for $m \geqslant 1$. This means that β is always a pure imaginary number, and the nonaxisymmetric disturbance perturbations for $m \geqslant 1$ do not grow in amplitude. This implies that the interface is stable for these nonaxisymmetric modes.

For $m = 0$, however, we see that β is positive if

$$\alpha > \alpha_c = \frac{1}{R_0} \tag{4.86}$$

This implies that the interface is unstable for axisymmetric perturbations with wavelengths $\lambda = 2\pi/\alpha$ greater than the circumference of the jet; that is,

$$\lambda > \lambda_c = 2\pi R_0 \tag{4.87}$$

In addition, by differentiating Eq. (4.85) with respect to α for $m = 0$ and setting the derivative equal to zero, it is possible to obtain an equation that can be solved for the value of $\alpha = \alpha_{max}$ that results in the maximum value of $\beta = \beta_{max}$. The resulting value of $\alpha = \alpha_{max}$ is the wavenumber of the disturbance that will be most rapidly amplified. Because of the presence of the Bessel functions in Eq. (4.85), explicit relations for α_{max} and β_{max} cannot be obtained. However, the equation for α_{max} can be solved numerically to yield

$$\alpha_{max} = \frac{0.70}{R_0} \tag{4.88}$$

$$\beta_{max} = 0.34\left(\frac{\sigma}{\rho_l R_0^3}\right)^{1/2} \tag{4.89}$$

These results were first obtained by Rayleigh [4.6].

Equation (4.88) implies that the fastest-growing disturbance has a wavelength λ_{max} equal to $9.0R_0$. It is interesting to note that λ_{max} depends only on the radius of the jet and not on surface tension, in spite of the fact that surface tension is the primary cause of the instability. On the other hand, the growth rate of the most rapidly amplified disturbance, which is proportional to $e^{\beta_{max}t}$, increases as σ increases.

If the Fourier component of a small perturbation of the interface at the jet nozzle with wavelength λ_{max} has an amplitude δ_0, the above analysis indicates that this disturbance component will grow most rapidly and may ultimately be responsible for the breakup of the jet. The distance downstream of the nozzle where breakup occurs, z_B for a jet injected with uniform velocity u_l, can be estimated to be

$$z_B = u_l t_B \tag{4.90}$$

where t_B is the time required for the disturbance amplitude to grow from δ_0 to R_0. If the disturbance grows as indicated by Eqs. (4.78) through (4.80) with $\beta = \beta_{max}$, then t_B is given by

$$t_B = \left(\frac{1}{\beta_{max}}\right) \ln\left(\frac{R_0}{\delta_0}\right) \tag{4.91}$$

Because of the logarithmic variation of t_B with R_0/δ_0, t_B is not strongly dependent on R_0/δ_0 at large values of R_0/δ_0. Hence if we take a reasonable estimate of R_0/δ_0, we can estimate t_B and z_B for a given system (see Example 4.3).

The above analysis specifically considers the temporal amplification of spatial wave distortions of the interface. Keller, Rubinow, and Tu [4.7] have also examined the spacial instability of liquid jets. Their analysis postulates wave distortions of the interface that oscillate with a fixed real frequency, and they examine the spacial growth rates of the waves as they propagate downstream from the nozzle.

The analysis of Keller et al. [4.7] indicates that the maximum growth rate obtained from the temporal analysis above is valid for high-velocity jets when

$$u_l\left(\frac{\rho_l R_0}{\sigma}\right)^{1/2} \gg 1 \tag{4.92}$$

Even when this condition is satisfied, the spacial stability analysis indicates that there are long-wavelength modes that amplify more rapidly than the axisymmetric ($m = 0$) mode, which amplifies most rapidly in the temporal analysis. Apparently, however, these long-wavelength modes cannot occur in a short jet, and the axisymmetric $m = 0$ mode of the temporal analysis dominates.

In many real circumstances, the jet velocity and liquid density are high enough and the surface tension is low enough that the condition (4.92) is satisfied, and the results of the temporal analysis are valid. When the jet velocity is so low that $u_l(\rho_l R_0/\sigma)^{1/2}$ is small compared to 1, the spatial instability analysis of Keller et al. [4.7] predicts results that are very different from those obtained in the temporal stability analysis presented above.

Example 4.3 A jet of liquid water at 20°C is injected into air at the same temperature. If the jet diameter is 0.5 mm and the liquid velocity is 3.0 m/s, estimate the distance downstream of the nozzle where the jet is likely to break up into droplets.

For water at 20°C, $\rho_l = 998$ kg/m^3 and $\sigma = 0.0728$ N/m. Using Eq. (4.89), we have

$$\beta_{max} = 0.34\left(\frac{\sigma}{\rho_l R_0^3}\right)^{1/2} = 0.34\left[\frac{0.0728}{998.(0.0005)^3}\right]^{1/2} = 259.7\ \text{s}^{-1}$$

If we take $\delta_0/R_0 = 10^{-3}$, Eq. (4.91) implies that

$$t_B = \left(\frac{1}{\beta_{max}}\right)\ln\left(\frac{R_0}{\delta_0}\right) = \left(\frac{1}{259.7}\right)\ln(10^3) = 0.0266\ \text{s}$$

and

$$z_B = u_l t_B = (3.0)(0.0266) = 0.080\ \text{m} = 8.0\ \text{cm}$$

If $\delta_0/R_0 = 10^{-4}$, a similar set of calculations yields $z_B = 10.6$ cm. This suggests that the jet will break up somewhere around 9 cm downstream of the nozzle. For these circumstances, $u_l(\rho_l R_0/\sigma)^{1/2}$ is equal to 7.9, implying that the growth rate obtained from the temporal analysis should be applicable to this system.

4.4 WAVES ON LIQUID FILMS

We will see in later sections that engineering applications often involve vaporization or condensation at the interface of a thin liquid film flowing over a solid surface. These films are typically thin compared to their lateral extent, but not so thin that disjoining pressure effects are important. The flow in films of this type is usually dominated by viscous, gravity, and surface tension effects.

In many instances, circumstances may give rise to interfacial waves on the thin liquid film. The presence of such waves may strongly affect the rate of vaporization or condensation at the surface of the film, because the waves increase the interfacial area and can enhance convective transport near the interface.

In Section 4.2 the Kelvin-Helmholtz stability of a liquid–vapor interface between two semiinfinite liquid and vapor regions was analyzed. It was shown in that analysis that any velocity component parallel to the interface in either fluid will produce waves at the interface. This suggests that any motion of an overlying vapor region parallel to the interface of a thin liquid film on a horizontal solid surface may produce small-amplitude waves. The analysis presented in Section 4.2 further suggests that if the horizontal vapor velocity exceeds a critical value, the amplitude of the waves may become very large, with a correspondingly large effect on interfacial heat and mass transfer.

Another circumstance of interest is that in which a thin liquid film surrounded by motionless vapor flows down a vertical or inclined surface due to gravity. This situation occurs during film condensation on a solid object surrounded by vapor, and in falling-film evaporators in which liquid is intentionally allowed to flow downward over heated surfaces in the evaporator to facilitate the vaporization process. In film flows of this type, three different flow regimes have been observed experimentally. Generally, these regimes correspond to different ranges of the film Reynolds number, Re_F, defined as

$$\mathrm{Re}_F = w_m \delta / \nu_l \tag{4.93}$$

where w_m is the mean liquid downstream velocity over the cross section of the film and δ is the film thickness.

For values of Re_F below about 30, the film flow is laminar and, if the inclination of the surface is constant, the film thickness is uniform. At Reynolds numbers Re_F between about 30 and 1500, a wavy flow regime is observed in which a wavy motion is superimposed on the forward motion of the film. At a value of Re_F of about 1500, the film undergoes a transition from laminar to turbulent flow.

To further investigate the existence of interfacial waves in film flows of this type, we will specifically consider the laminar downward-flowing thin liquid film shown schematically in Fig. 4.7. Because the film thickness is small and we wish

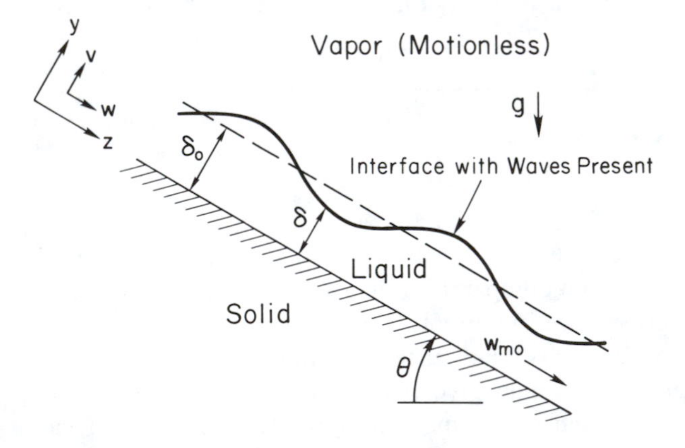

Figure 4.7 Model system considered in the analysis of waves on a liquid film flowing down an inclined surface.

to consider waves with wavelengths long compared to the film thickness (small wavenumbers), it is expected that velocity derivatives across the film (in the y direction) will be large compared to those in the z direction. Based on the usual boundary-layer approximations, we therefore assume that the fluid motion in the film is governed by the continuity equation and the boundary-layer form of the w momentum equation:

$$\frac{\partial v}{\partial y} + \frac{\partial w}{\partial z} = 0 \tag{4.94}$$

$$\frac{\partial w}{\partial t} + v\frac{\partial w}{\partial y} + w\frac{\partial w}{\partial z} = -\frac{1}{\rho_l}\left(\frac{\partial P}{\partial z}\right) + g \sin\theta + \nu_l \frac{\partial^2 w}{\partial y^2} \tag{4.95}$$

In the absence of any wave motion, steady, downward, fully developed flow exists in the liquid film, implying that the time and z derivatives and v are all zero. Noting that $\partial P/\partial z$ is just equal to the vapor hydrostatic gradient, the governing equation for w is then

$$\frac{(\rho_l - \rho_v)g \sin\theta}{\rho_l} + \nu_l \frac{d^2 w}{dy^2} = 0 \tag{4.96}$$

Using the boundary conditions

$$w = 0 \quad \text{at} \quad y = 0 \qquad \frac{\partial w}{\partial y} = 0 \quad \text{at} \quad y = \delta \tag{4.97}$$

integration of Eq. (4.96) yields the following expression for the w velocity profile:

$$w = \frac{(\rho_l - \rho_v)g \sin\theta(2y\delta - y^2)}{2\mu_l} \tag{4.98}$$

Defining the mean film velocity w_m as

$$w_m = \frac{1}{\delta}\int_0^\delta w(y)\, dy = \frac{(\rho_l - \rho_v)g\delta^2 \sin\theta}{3\mu_l} \tag{4.99}$$

we can rewrite Eq. (4.98) as

$$w = \frac{3w_m}{2}\left(\frac{2y}{\delta} - \frac{y^2}{\delta^2}\right) \tag{4.100}$$

We now will examine the unsteady flow that occurs when waves are present at the interface. For these circumstances, we retain the full time-dependent boundary-layer equation (4.95) for the w momentum balance. Assuming that the pressure variation in the liquid and vapor regions varies only hydrostatically, and that the Young-Laplace equation (4.4) holds at the interface, the pressure variation in the liquid film is given by

$$P_l = P_v + (\rho_l - \rho_v)g(\delta - y)\cos\theta - \sigma\frac{\partial^2 \delta}{\partial z^2} \tag{4.101}$$

Note that we are now allowing for the variation of δ with time and z location. Differentiating this equation with respect to z, we obtain

$$\frac{\partial P_l}{\partial z} = \frac{\partial P_v}{\partial z} + (\rho_l - \rho_v) g \cos\theta \frac{\partial \delta}{\partial z} - \sigma \frac{\partial^3 \delta}{\partial z^3} \tag{4.102}$$

Using the fact that $\partial P_v/dz$ is just equal to the hydrostatic gradient in the vapor,

$$\frac{\partial P_v}{dz} = \rho_v g \sin\theta \tag{4.103}$$

Eqs. (4.102) and (4.103) can be substituted into Eq. (4.95) to obtain

$$\frac{\partial w}{\partial t} + v \frac{\partial w}{\partial y} + w \frac{\partial w}{\partial z} = \left[\frac{(\rho_l - \rho_v) g}{\rho_l}\right]\left(\sin\theta - \cos\theta \frac{\partial \delta}{\partial z}\right) + \frac{\sigma}{\rho_l} \frac{\partial^3 \delta}{\partial z^3} + \nu_l \frac{\partial^2 w}{\partial y^2} \tag{4.104}$$

The analysis is continued by integrating both sides of this equation across the film from $y = 0$ to δ and using the continuity relation (4.94), integration by parts, Leibniz's rule, and the relation

$$w_{y=\delta} = \frac{3 w_m}{2}$$

to evaluate terms in the equation. After doing so, the w momentum relation becomes

$$(3w_m/2\delta) \frac{\partial \delta}{\partial t} + \left(\frac{3 w_m^2}{2\delta}\right) \frac{\partial \delta}{\partial z} = \left[\frac{(\rho_l - \rho_v) g}{\rho_l}\right]\left(\sin\theta - \cos\theta \frac{\partial \delta}{\partial z}\right) + \left(\frac{\sigma}{\rho_l}\right) \frac{\partial^3 \delta}{\partial z^3} - \frac{3 w_m \nu_l}{\delta^2} \tag{4.105}$$

By considering a control volume that extends across the film, it can be shown that conservation of mass requires that

$$\frac{\partial \delta}{\partial t} = -\frac{\partial (w_m \delta)}{\partial z} \tag{4.106}$$

Substituting Eq. (4.99) for w_m and carrying out the differentiation, Eq. (4.106) becomes

$$\frac{\partial \delta}{\partial t} = -3 w_m \frac{\partial \delta}{\partial z} \tag{4.107}$$

Substituting Eq. (4.107) into Eq. (4.105) reduces that equation to the form

$$\frac{3 w_m^2}{\delta} \frac{\partial \delta}{\partial z} = \left[\frac{(\rho_l - \rho_v) g}{\rho_l}\right]\left(\cos\theta \frac{\partial \delta}{\partial z} - \sin\theta\right) - \frac{\sigma}{\rho_l} \frac{\partial^3 \delta}{\partial z^3} + \frac{3 w_m \nu_l}{\delta^2} \tag{4.108}$$

Because w_m is a function of δ, we are left with a pair of nonlinear partial differential equations, (4.107) and (4.108), which must be solved together to determine $\delta(z,t)$. To avoid the difficulties of solving this nonlinear system, we postulate that the interface location δ is equal to its base flow value (with no waves present) δ_0 plus a small fluctuating component δ'. In addition, it is assumed that the perturbation of the interface is accompanied by perturbations of the local w velocity field, but that the mean mass flow rate, and hence the mean velocity in the film w_m, remain equal to the base flow values with no waves present.

$$\delta = \delta_0 + \delta' \qquad w_m = w_{m0} \tag{4.109}$$

We further note that from the definition of w_m in Eq. (4.99), the following relation must hold:

$$\left[\frac{(\rho_l - \rho_v)g}{\rho_l}\right] \sin\theta = \frac{3w_{m0}\nu_l}{\delta_0^2} \tag{4.110}$$

Substituting the relations (4.109) into (4.108), using Eq. (4.110) and the fact that $\partial\delta_0/\partial z = \partial w_{m0}/\partial z = 0$, and neglecting products of primed (fluctuating) quantities, we obtain

$$\frac{3w_{m0}^2}{\delta_0}\frac{\partial\delta'}{\partial z} = \left[\frac{(\rho_l - \rho_v)g}{\rho_l}\right] \cos\theta \frac{\partial\delta'}{\partial z} + \frac{\sigma}{\rho_l}\frac{\partial^3\delta'}{\partial z^3} \tag{4.111}$$

The fluctuating component δ' in the linearized equation above is postulated to be a two-dimensional sinusoidal wave of the form

$$\delta' = \delta^* e^{i\alpha(z-ct)} \tag{4.112}$$

where α is the wavenumber and c is the wave velocity. Substituting Eqs. (4.109) and (4.112) into Eq. (4.107), we find that $c = 3w_{m0}$. Thus the wave velocity is three times the mean base flow velocity. Substitution of Eq. (4.112) into Eq. (4.111) yields the following relation for the wavenumber α:

$$\alpha^2 = \frac{\rho_l}{\sigma}\left[\frac{(\rho_l - \rho_v)^2 g^2 \delta_0^3 \sin^2\theta}{3\mu_l^2} - \frac{(\rho_l - \rho_v)g\cos\theta}{\rho_l}\right] \tag{4.113}$$

If the right side of Eq. (4.113) is not greater than zero, α is a pure imaginary number and the solution for δ' is not a sinusoidal wave. It appears, therefore, that waves exist at the interface only if the right side of Eq. (4.113) is greater than zero. Rearranging this equation, it can be shown that this condition is met, and hence waves can exist for

$$\delta_0 > \left[\frac{3\mu_l^2\cos\theta}{(\rho_l - \rho_v)\rho_l g \sin^2\theta}\right]^{1/3} \tag{4.114}$$

Alternatively, if we combine Eqs. (4.110) and (4.114), we can write this criterion in terms of the film Reynolds number $\mathrm{Re}_F = \rho_l w_{m0}\delta_0/\mu_l$ simply as

$$\mathrm{Re}_F > \cot\theta \tag{4.115}$$

In light of the above results, we interpret the limiting case $\mathrm{Re}_F = \cot\theta$ as the boundary between a stable film with no waves (for $\mathrm{Re}_F < \cot\theta$) and conditions for which wavy flow may occur ($\mathrm{Re}_F > \cot\theta$).

It is interesting to note that the criterion for the appearance of wavy flow obtained from the approximate analysis above is consistent with the results of Benjamin's [4.8] more thorough analysis of wave formation in film flow down an inclined plane. The full stability analysis presented by Benjamin [4.8] predicts that very long waves (small α) become unstable (and hence amplify) for

$$\mathrm{Re}_F > \tfrac{5}{6}\cot\theta \tag{4.116}$$

In the case of a vertical surface, $\theta = 90°$ and both Eqs. (4.115) and (4.116) reduce to $\mathrm{Re}_F > 0$. This implies that conditions favor the presence of waves at the interface for any finite Reynolds number, no matter how small. At first glance, this appears to contradict the fact that waves are not observed experimentally at very low film Reynolds numbers.

This apparent contradiction is resolved by considering the amplification rate of wave amplitudes. For a vertical surface, Benjamin [4.8] also showed that for long-wavelength surface waves having the form given by Eq. (4.112), there is a specific wavelength that amplifies more rapidly than all others, and that the wavenumber and the imaginary component of the wave speed for this wavelength are

$$\alpha_{\max} = 1.12\delta_0^{-1}\left[\nu_l^{2/3}g^{1/6}\left(\frac{\sigma}{\rho_l}\right)^{-1/2}\right]\mathrm{Re}_F^{5/6} \tag{4.117}$$

$$\mathrm{Im}(c)_{\max} = 0.336w_{m0}\left[\nu_l^{2/3}g^{1/6}\left(\frac{\sigma}{\rho_l}\right)^{-1/2}\right]\mathrm{Re}_F^{11/6} \tag{4.118}$$

The amplitude of this wavelength must increase according to

$$\frac{|\delta|}{|\delta|_{t=0}} = \exp\left[\alpha_{\max}\,\mathrm{Im}(c)_{\max}\,t\right] \tag{4.119}$$

If we consider the amplification over the time interval required for the wave to travel 100 times the film thickness downstream, then $t = 100\delta_0/3w_{m0}$ and the amplification factor given by Eq. (4.119) becomes

$$\frac{|\delta|}{|\delta|_{t=0}} = \exp\left\{\left[12.5\nu_l^{4/3}g^{1/3}\left(\frac{\sigma}{\rho_l}\right)^{-1}\right]\mathrm{Re}_F^{8/3}\right\} \tag{4.120}$$

Generally, the term in square brackets in Eq. (4.120) is small. For water at 19°C under earth-normal gravity, it is 3.8×10^{-3}. Equation (4.120) is plotted for this value in Fig. 4.8. For these conditions, the amplification of the most rapidly amplified wavelength is extremely small for film Reynolds numbers below 6 and extremely large for $\mathrm{Re}_F > 13$. Hence, although wave disturbances are unstable for all $\mathrm{Re}_F > 0$, the growth rate is so low for $\mathrm{Re}_F < 6$ that waves may not be observed over the finite length of any real falling film for these conditions.

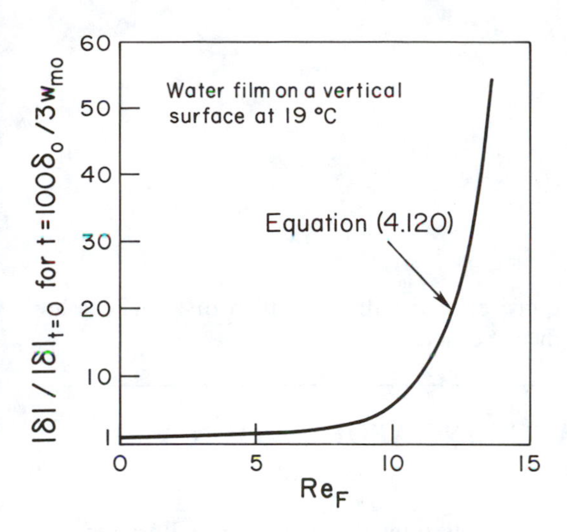

Figure 4.8 Wave amplitude growth predicted for a water film flowing down a vertical surface.

Furthermore, the very rapid increase in amplification rate over a narrow range of Re_F makes it likely that wavy flow will be observed at Reynolds numbers just above this range. This implies that although there is no critical Reynolds number in the usual sense (dividing wave-amplifying and wave-damping conditions), there is, nevertheless, a "quasi-critical" value of Re_F below which wavy flow is unlikely and above which it is very likely to occur. This is consistent with the experimentally observed behavior of low-Reynolds-number liquid films, as described at the beginning of this section.

Some additional aspects of wave motion on flat liquid–vapor interfaces in general, and thin liquid films in particular, are discussed in reference [4.9]. The effects of interfacial waves on evaporation and condensation processes involving liquid films will be explored in detail in later sections of this text.

Example 4.4 Estimate the quasi-critical value of $Re_F = \rho_l w_{m0} \delta_0 / \mu_l$ for a film flow of saturated liquid water at 300°C.

For saturated water at 300°C, $\rho_l = 712.5$ kg/m^3, $\nu_l = 0.127 \times 10^{-6}$ m^2/s, $\sigma = 0.01439$ N/m, and Eq. (4.120) requires that

$$\frac{|\delta|}{|\delta|_{t=0}} = \exp\left\{\left[12.5\, \nu_l^{4/3} g^{1/3}\left(\frac{\sigma}{\rho_l}\right)^{-1}\right] Re_F^{8/3}\right\}$$

$$= \exp\left\{\left[12.5(0.127 \times 10^{-6})^{4/3}(9.8)^{1/3}\left(\frac{0.01439}{712.5}\right)^{-1}\right] Re_F^{8/3}\right\}$$

The resulting variation of the disturbance amplification with Re_F is summarized in the following table:

Re_F	$\|\delta\|/\|\delta\|_{t=0}$
5	1.06
10	1.48
20	12.1
30	1570
35	6.63×10^4
40	7.66×10^6

Hence $Re_F = 35$ is about the quasi-critical value at which disturbances are expected to be amplified enough to be detectable.

4.5 INTERFACIAL RESISTANCE IN VAPORIZATION AND CONDENSATION PROCESSES

The extremely high heat transfer coefficients typically associated with vaporization and condensation processes make it possible to transfer thermal energy at high heat flux levels with relatively low driving temperature differences. The ability to handle high heat flux levels is particularly important in applications such as electronics cooling and power systems thermal control. A question of central importance in such applications is: "What is the highest heat flux possible in a given vaporization or condensation process?"

To explore this question we must consider the liquid–vapor interface at the molecular level. Because motion of vapor molecules in the vicinity of the interface plays a central role in heat flux limitations during vaporization and condensation processes, we will first examine some relevant aspects of the kinetic theory of gases.

One of the most useful results of the classical development of the kinetic theory of gases is the Maxwell velocity distribution:

$$\frac{dn_{uvw}}{n} = \left(\frac{m}{2\pi k_B T}\right)^{3/2} \exp\left[-\left(\frac{m}{2k_B T}\right)(u^2 + v^2 + w^2)\right] du\, dv\, dw \quad (4.121)$$

The left side of this equation is the fraction of the total number of molecules n with Cartesian velocities, u, v, and w in the ranges u to $u + du$, v to $v + dv$, and w to $w + dw$, respectively. In Eq. (4.121), m is the mass of one molecule, T is absolute temperature, and k_B is the Boltzmann constant, defined as

$$k_B = \frac{\overline{R}}{N_A} \quad (4.122)$$

where N_A is Avogadro's number. Fundamental aspects of the kinetic theory of gases and derivation of the Maxwell distribution are described in Appendix I. More detailed treatments of these subjects are also presented in most texts that treat statistical thermodynamics (see, e.g., references [4.10–4.12]).

Using the velocity distribution given by Eq. (4.121), it is possible to derive the following relation for the fraction of the molecules with speeds in the range c to $c + dc$ regardless of direction (see Appendix I):

$$\frac{dn_c}{n} = 4\pi\left(\frac{m}{2\pi k_B T}\right)^{3/2} c^2 e^{-mc^2/2k_BT}\, dc \tag{4.123}$$

As discussed in Appendix I, relations for the most probable speed c_{mp}, the mean speed $\langle c \rangle$, and the root-mean-square speed $\langle c^2 \rangle^{1/2}$ can be obtained from this speed distribution.

We now wish to determine the flux of molecules through an arbitrary plane for a gas that has a Maxwell velocity distribution. To do so, we consider the motion of molecules in the box shown in Fig. 4.9. We specifically wish to determine the number of molecules within the box that strike the shaded surface S_x^* per unit area and per unit time. A molecule with a given x component of velocity u and any v and w must lie within a distance $u\,\Delta t$ of the surface at the beginning of the time interval in order to pass through it. The fraction of the molecules having an x component of velocity between some value u and $u + du$ and having any v and w is found by integrating Eq. (4.121) over all possible v and w values:

$$\frac{dn_u}{n} = \int_v \int_w \frac{dn_{uvw}}{n} \tag{4.124}$$

Using Eq. (4.121) to evaluate the right side of Eq. (4.124), we obtain

$$\frac{dn_u}{n} = \left(\frac{m}{2\pi k_B T}\right)^{1/2} e^{-mu^2/2k_BT}\, du \tag{4.125}$$

The above observations indicate that only molecules within a fraction $u\,\Delta t/L_x$ of the total box volume will pass through the surface S_x^* in the time interval Δt. If the total number of molecules in the box is n, the number of molecules

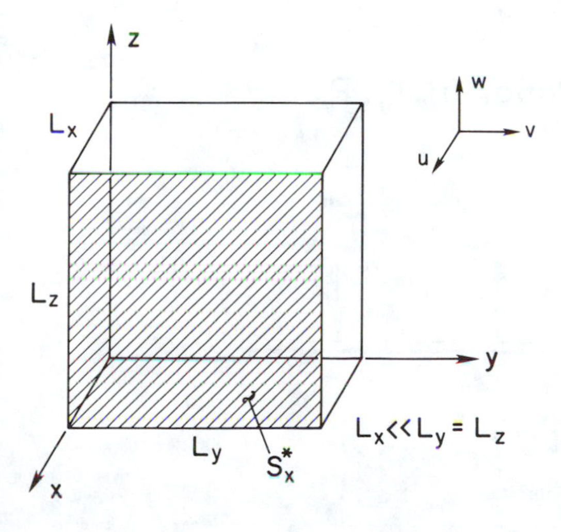

Figure 4.9 Geometry used in analysis of molecular flux through a surface.

with an x component of velocity equal to u is dn_u, as given by Eq. (4.125). The number of molecules with velocity u in the x direction that will pass through the surface S_x^* per unit area per unit time dj_u is therefore equal to

$$dj_u = \left(\frac{u\,\Delta t}{L_x}\right) dn_u \left(\frac{1}{L_z L_y}\right)\left(\frac{1}{\Delta t}\right) \tag{4.126}$$

Substituting Eq. (4.125) for dn_u and integrating the resulting expression over all possible values of u (0 to ∞), we obtain the following relation for the total rate at which molecules pass through surface S_x^* per unit area, j_n:

$$j_n = \left(\frac{1}{4}\right)\left(\frac{n}{V}\right)\left(\frac{8k_B T}{\pi m}\right)^{1/2} = \left(\frac{\overline{M}}{2\pi\overline{R}}\right)^{1/2} \frac{P}{mT^{1/2}} \tag{4.127}$$

where $\overline{M}$ is the molecular weight of the vapor. It is noteworthy that the flux of molecules j_n through the surface S_x^* is independent of the original box dimensions. The flux j_n must therefore be the same for any planar surface in the gas.

We can now use some of the above results from the kinetic theory of gases to interpret the motion of vapor molecules near a liquid–vapor interface. We now consider a plane surface in the vapor phase but immediately adjacent to the interface, as shown in Fig. 4.10. Even if no (net) vaporization or condensation occurs at the interface, a dynamic equilibrium is established in which molecules from the vapor phase that hit the interface and become part of the liquid phase are balanced (on the average) by an equal number of molecules that escape the liquid into the vapor region. When condensation occurs, the flux of vapor molecules joining the liquid must exceed the flux of liquid molecules escaping to the vapor phase. When vaporization occurs, the opposite must be true.

In considering evaporation or condensation from the interface, we will, at least initially, admit the possibility that the pressures and temperatures in the two phases may be different, as indicated in Fig. 4.10. For the system shown in this figure, the net mass flux across surface S_i^* immediately adjacent to the interface

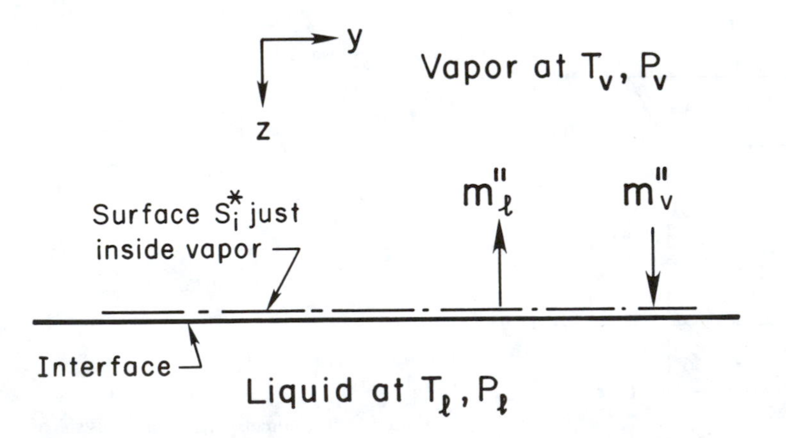

Figure 4.10 Mass fluxes at a liquid–vapor interface.

must be equal to the difference between the mass fluxes m_l'' and m_v'' passing through surface S_i^* in opposite directions just inside the vapor region:

$$m_{\text{net}}'' = m_v'' - m_l'' \tag{4.128}$$

To facilitate the evaluation of m_v'' and m_l'' in Eq. (4.128), it will be assumed here that the flux of molecules m_l'' is characterized by T_l and P_l, whereas m_v'' is characterized by T_v and P_v. With this assumption, it would appear that m_l'' and m_v'' can be obtained by multiplying the relation (4.127) by the mass of one molecule and evaluating the resulting expression at the appropriate T and P values. However, this approach is inadequate in two respects. First, Eq. (4.127) predicts the flux of molecules in a stationary gas, whereas in the system shown in Fig. 4.10, the vapor must have a bulk velocity $w = w_0$ in the z direction as a result of the phase change at the interface. The bulk velocity is toward the interface ($w_0 > 0$) for condensation and away from the interface ($w_0 < 0$) for vaporization.

By extending the analysis presented above, it can be shown from kinetic theory (see Schrage [4.13]) that when the gas moves normal to a planar surface with a speed w_0, the flux of molecules through the plane in the direction of the bulk motion is

$$j_{nw+} = \Gamma(a)\left(\frac{\overline{M}}{2\pi\overline{R}T}\right)^{1/2}\left(\frac{P}{m}\right) \tag{4.129a}$$

and the flux of molecules in the direction opposite to that of the bulk motion is

$$j_{nw-} = \Gamma(-a)\left(\frac{\overline{M}}{2\pi\overline{R}T}\right)^{1/2}\left(\frac{P}{m}\right) \tag{4.129b}$$

where

$$a = \frac{w_0}{(2\overline{R}T/\overline{M})^{1/2}} \tag{4.130}$$

and the factor Γ, which corrects for the effects of bulk gas motion, is given by one of the following relations:

$$\begin{aligned} \Gamma(a) &= \exp(a^2) + a\pi^{1/2}[1 + \operatorname{erf}(a)] \\ \Gamma(-a) &= \exp(a^2) - a\pi^{1/2}[1 - \operatorname{erf}(a)] \end{aligned} \tag{4.131}$$

The variations of $\Gamma(a)$ and $\Gamma(-a)$ with a are shown in Fig. 4.11.

Because the vapor in Fig. 4.10 is expected to be in bulk motion relative to the surface S_i^*, Eq. (4.129a) or (4.129b) must be used to compute m_v''. Following Schrage [4.13], we assume that the surface S_i^* is an infinitesimally small distance from the phase interface, so that there is no bulk motion effect on molecules emerging from the liquid and passing through the surface S_i^*. The mass flux m_l'' is therefore calculated using Eq. (4.127).

In addition to the effect of bulk vapor motion, it is clear that only a fraction $\hat{\sigma}_e$ of the molecules crossing the surface S_i^* in the negative z direction is actually

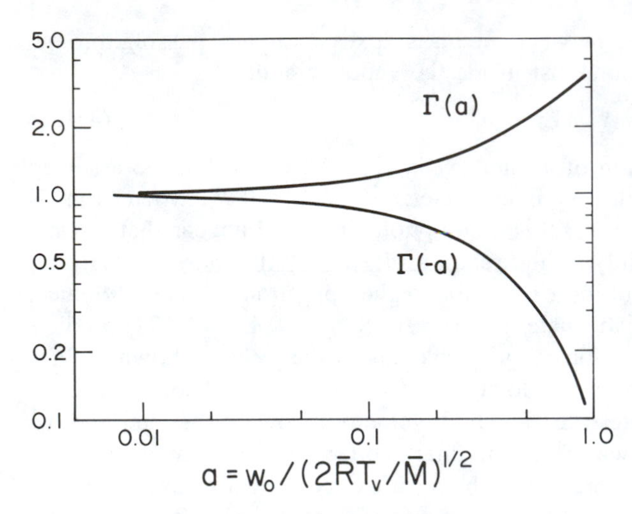

Figure 4.11 The $\Gamma(a)$ and $\Gamma(-a)$ functions developed in Schrage's [4.13] analysis of molecular fluxes at a liquid–vapor interface.

due to vaporization. The remaining fraction $1 - \hat{\sigma}_e$ is due to the "reflection" of vapor molecules that strike the interface but do not condense. The fraction of the molecules crossing the surface S_i^* in the positive z direction that condense and are not "reflected" is designated as $\hat{\sigma}_c$. If no phase change occurs at the interface, equilibrium requires that $\hat{\sigma}_e = \hat{\sigma}_c = \hat{\sigma}$. Usually $\hat{\sigma}_e$ and $\hat{\sigma}_c$ are assumed to be equal even for the dynamic case when phase change occurs at the interface, although the validity of this assumption is suspect. As a result, $\hat{\sigma}$ is sometimes referred to as a vaporization, evaporation, condensation, or accommodation coefficient. The term *accommodation coefficient* has been perhaps most commonly used for $\hat{\sigma}$ in recent years and will be adopted here.

Allowing for the bulk motion of the vapor and the role of the accommodation coefficient as described above, the portion of the mass flux m_v'' that actually enters the liquid phase upon striking the liquid, m_{vc}'' is given by

$$m_{vc}'' = \begin{cases} m\hat{\sigma}j_{nw+} & \text{for condensation} \\ m\hat{\sigma}j_{nw-} & \text{for vaporization} \end{cases} \tag{4.132}$$

and the portion of m_l'' that is due to molecules that actually emerged from the liquid phase, m_{le}'', is given by

$$m_{le}'' = m\hat{\sigma}j_n \tag{4.133}$$

The net mass flux to or from the interface as a result of the phase change m_i'' is just equal to the difference between m_{vc} and m_{le}'':

$$m_i'' = m_{vc}'' - m_{le}'' \tag{4.134}$$

Substituting Eqs. (4.132) and (4.133) into (4.134), evaluating j_{nw+} or j_{nw-} at P_v and T_v and j_n at P_l and T_l, and rearranging yields

$$m_i'' = \left(\frac{\overline{M}}{2\pi\overline{R}}\right)^{1/2}\left(\frac{\Gamma\hat{\sigma}P_v}{T_v^{1/2}} - \frac{\hat{\sigma}P_l}{T_l^{1/2}}\right) \tag{4.135}$$

Because the heat flux to the interface q_i'' must equal the net mass flux multiplied by the latent heat, a relation for q_i'' can be obtained directly from Eq. (4.135):

$$q_i'' = \hat{\sigma} h_{lv} \left(\frac{\overline{M}}{2\pi \overline{R}} \right)^{1/2} \left(\frac{\Gamma P_v}{T_v^{1/2}} - \frac{P_l}{T_l^{1/2}} \right) \tag{4.136}$$

It should be noted that because the rate of phase change is dictated by the heat flux, the resulting bulk velocity of the vapor is given by

$$w_0 = \frac{q_i''}{\rho_v h_{lv}} \tag{4.137}$$

which implies that

$$a = \frac{q_i''}{\rho_v h_{lv}} \left(\frac{2\overline{R} T_v}{\overline{M}} \right)^{1/2} \tag{4.138}$$

Because Γ is a function of a, Eq. (4.136) is actually an implicit relation for q_i''. If we assume that P_v and P_l are the saturation pressures corresponding to T_v and T_l, Eq. (4.136) provides an implicit relation between q_i'' and the temperatures T_v and T_l:

$$q_i'' = \hat{\sigma} h_{lv} \left(\frac{\overline{M}}{2\pi \overline{R}} \right)^{1/2} \left[\frac{\Gamma P_{\text{sat}}(T_v)}{T_v^{1/2}} - \frac{P_{\text{sat}}(T_l)}{T_l^{1/2}} \right] \tag{4.139}$$

For vaporization and condensation processes at high tempratures, a is often small. For example, for saturated pure water at 100°C condensing or vaporizing at a heat flux of 100 kW/m^2, a is 1.3×10^{-4}. On the other hand, the values of a for the vaporization or condensation of cryogenic liquids at low absolute temperatures may be much higher.

In the limit of small a, Eq. (4.131) is well approximated by

$$\Gamma = 1 + a\pi^{1/2} \tag{4.140}$$

Substituting this relation into Eq. (4.136) and using Eq. (4.138) with the ideal gas relation to evaluate ρ_v, after some manipulation, the following relation is obtained for q_i'':

$$q_i'' = \left[\frac{2\hat{\sigma}}{(2 - \hat{\sigma})} \right] h_{lv} \left(\frac{\overline{M}}{2\pi \overline{R}} \right)^{1/2} \left(\frac{P_v}{T_v^{1/2}} - \frac{P_l}{T_l^{1/2}} \right) \tag{4.141}$$

The associated relation for the interfacial mass flux (obtained by dividing the above equation by h_{lv}),

$$m_i'' = \left[\frac{2\hat{\sigma}}{(2 - \hat{\sigma})} \right] \left(\frac{\overline{M}}{2\pi \overline{R}} \right)^{1/2} \left(\frac{P_v}{T_v^{1/2}} - \frac{P_l}{T_l^{1/2}} \right) \tag{4.142}$$

was suggested by Silver and Simpson [4.14] in their study of condensing steam. In the Soviet literature, Eq. (4.142) has been referred to as the Kucherov-Rikenglaz equation [4.17].

An alternate form of the relation for q_i'' can be obtained if the relations

$$\Delta P_{vl} = P_v - P_l \quad \text{and} \quad \Delta T_{vl} = T_v - T_l \tag{4.143}$$

are substituted into Eq. (4.136) to eliminate P_l and T_l:

$$q_i'' = \hat{\sigma} h_{lv} \left(\frac{\overline{M}}{2\pi\overline{R}} \right)^{1/2} \left[\frac{\Gamma P_v}{T_v^{1/2}} - \frac{(P_v - \Delta P_{vl})}{(T_v - \Delta T_{vl})^{1/2}} \right] \tag{4.144}$$

To develop a more useful form of this equation, we expand the second term in square brackets in terms of $\Delta T_{vl}/T_v$, using the assumption that $\Delta P_{vl}/P_v \ll 1$ and $\Delta T_{vl}/T_v \ll 1$ so that terms of order $(\Delta P_{vl}/P_v)^2$ may be neglected. If we further assume that $P_v = P_{\text{sat}}(T_v)$ and $P_l = P_{\text{sat}}(T_l)$ and evaluate $\Delta P_{vl}/\Delta T_{vl}$ using the Clausius-Clapeyron equation (1.6), Eq. (4.144) becomes

$$q_i'' = \left(\frac{\hat{\sigma} h_{lv}^2}{T_v v_{lv}} \right) \left(\frac{\overline{M}}{2\pi\overline{R}T_v} \right)^{1/2} \left[\Gamma - \frac{P_v v_{lv}}{2h_{lv}} \right] \Delta T_{vl} \tag{4.145}$$

Defining the heat transfer coefficient for the interface, h_i, as

$$h_i = \frac{q_i''}{\Delta T_{vl}} \tag{4.146}$$

combining Eqs. (4.145) and (4.146) yields

$$h_i = \left(\frac{\hat{\sigma} h_{lv}^2}{T_v v_{lv}} \right) \left(\frac{\overline{M}}{2\pi\overline{R}T_v} \right)^{1/2} \left(\Gamma - \frac{P_v v_{lv}}{2h_{lv}} \right) \tag{4.147}$$

Because Γ is dependent on q_i'', Eq. (4.145) is an implicit relation for q_i'', and h_i as given by Eq. (4.147) depends on q_i''. For small a, the above analysis for small ΔP_{vl} and ΔT_{vl} can be applied in exactly the same manner, beginning with Eq. (4.141), yielding the following relations for q_i'' and h_i:

$$q_i'' = \left[\frac{2\hat{\sigma}}{(2 - \hat{\sigma})} \right] \left(\frac{h_{lv}^2}{T_v v_{lv}} \right) \left(\frac{\overline{M}}{2\pi\overline{R}T_v} \right)^{1/2} \left[1 - \frac{P_v v_{lv}}{2h_{lv}} \right] \Delta T_{vl} \tag{4.148}$$

$$h_i = \left[\frac{2\hat{\sigma}}{(2 - \hat{\sigma})} \right] \left(\frac{h_{lv}^2}{T_v v_{lv}} \right) \left(\frac{\overline{M}}{2\pi\overline{R}T_v} \right)^{1/2} \left(1 - \frac{P_v v_{lv}}{2h_{lv}} \right) \tag{4.149}$$

Although they are limited to low values of a, these equations provide explicit relations for q_i'' and h_i.

The above analysis of interfacial transport assumes that the fluxes of condensing and vaporizing molecules can be derived from kinetic theory for each flux direction separately and the results superimposed to obtain the net flux. The analysis fails to consider nonequilibrium interactions between the molecules leaving the interface and those approaching the interface, which must have different mean energy levels because of the difference between the liquid and vapor temperatures. Inclusion of these effects makes analysis of this problem considerably more difficult (see Wilhelm [4.16]).

Despite its approximate nature, the results of the above analysis provide a useful framework for assessing the effects of interfacial molecular transport on vaporization and condensation processes. It should be noted that the equations relating q_i'', h_i, and ΔT_{vl} derived above apply equally well to vaporization and condensation with the convention that q_i'' is positive for condensation and negative for vaporization.

Inspection of the relations derived above clearly indicates that the resulting q_i'' or h_i values depend directly on the value of the accommodation coefficient $\hat{\sigma}$. Quoted values of $\hat{\sigma}$ in the literature vary widely. Mills [4.17] suggested that molecular accommodation is less than perfect only when the interface is impure. Indeed, when great care is taken to ensure the purity of the test fluids, values of $\hat{\sigma}$ obtained from experiments are close to 1 (see, e.g., Volmer and Estermann [4.18]).

Because extreme purity is unlikely in most engineering systems, a value of σ less than 1 is expected. For condensing of mercury, Sukhatme and Rohsenow [4.19] found that their data implied $\hat{\sigma}$ values ranging from 0.37 to 0.61. Some of this variation was attributed to possible slight changes in the level of system contamination from test to test. They conclude that, for a particular system, with a relatively constant contamination level, a relatively constant $\hat{\sigma}$ value is attainable.

Accommodation coefficients for vaporization of a wide variety of substances have been compiled by Paul [4.20]. For liquid ethanol, methanol, n-propyl alcohol, and water, the reported values of $\hat{\sigma}$ range from 0.02 to 0.04. Paul [4.20] suggests that these low values may be due to the inadequacy of the kinematic theory model used to define $\hat{\sigma}$ for these materials. On the other hand, reported values of $\hat{\sigma}$ for benzene and carbon tetrachloride are near 1.

It appears, therefore, that values of $\hat{\sigma}$ less than 1 may result from either system contamination or deviation from the kinetic theory model used to define $\hat{\sigma}$. The value $\hat{\sigma} = 1$ still appears to be an upper bound for $\hat{\sigma}$, since no values above 1 have been reported. It is clear, however, that values reported in the literature should be regarded as typical values of $\hat{\sigma}$ rather than constants applicable to all systems.

It can be seen from Eq. (4.145) that the temperature difference across the interface is more likely to be significant at higher temperatures and heat flux levels in systems with high values of v_{lv} and low accommodation coefficients. These conditions may exist, for example, during condensation of pure water at 100°C. However, the high latent heat of water compensates for the other factors somewhat. At a heat flux of 100 kW/m^2 with $\hat{\sigma}$ taken to be 0.04, the values of h_i would be about 880 kW/m^2 K, resulting in a predicted temperature difference across the interface of only about 0.1°C.

The interfacial resistance may be particularly important in the condensation of liquid metals or other phase-change circumstances where overall system temperature differences are expected to be small because of very effective convective or conduction transport of heat within the liquid or vapor phases. In such instances, Eq. (4.147) or (4.149) may be used to provide an estimate of h_i that can be compared with heat transfer coefficients associated with other mechanisms in the system. If h_i is not large compared with other h values, the effect of interfacial resistance must be included in any performance calculations.

4.6 MAXIMUM FLUX LIMITATIONS

From the kinetic theory analysis described earlier in Section 4.5, it can also be argued that there is an upper limit to the heat flux that can be attained in a phase-change process. For a vaporization process, the maximum heat flux conceivable would result if molecules were emitted from the interface with perfect accommodation ($\hat{\sigma} = 1$) and no molecules were allowed to return to the liquid. Assuming that the vapor is a Maxwellian ideal gas, the heat flux for these idealized circumstances is equal to the flux of molecules j_N, given by Eq. (4.127), multiplied by the mass per molecule and the latent heat. The resulting expression for the maximum heat flux q''_{mkv} can be rearranged to obtain

$$q''_{mkv} = \rho_v h_{lv} \left(\frac{\overline{R} T_v}{2\pi \overline{M}} \right)^{1/2} \tag{4.150}$$

Schrage [4.15] has analyzed the maximum interfacial flux conditions in more detail. His results confirm that q''_{mkv} is proportional to the right side of Eq. (4.150). However, in his more detailed analysis, Schrage [4.15] found the proportionality constant to be 0.741, and he notes that the value of the proportionality constant depends on the velocity distribution function assumed at the interface. The right side of Eq. (4.150) is therefore best interpreted as a group that characterizes the upper limit of the heat flux, rather than an exact prediction of its value.

Figure 4.12 shows the variation of q''_{mkv} computed using Eq. (4.150) for R-12, R-113, ammonia, and water for saturation conditions spanning a wide range of reduced pressure P/P_c. It can be seen that, in general, q''_{mkv} increases with pressure up to approximately $P/P_c = 0.8$. Beyond this point, q''_{mkv} decreases as the critical pressure is approached, due to the fact that $h_{lv} \rightarrow 0$ as $P/P_c \rightarrow 1$. Water is seen to have the largest value of q''_{mkv} at a given reduced pressure for the fluids shown, largely because of its high latent heat.

Example 4.5 In the latter stages of saturated convective boiling in a vertical tube, a thin film of liquid water flows along the walls of the tube with mostly vapor flowing in the central core region. Consider such a case in which the film of liquid water is 0.1 mm thick and a heat flux of 200 kW/m^2 is applied to the tube wall. For a system pressure of 1 atm, estimate the interfacial heat transfer coefficient h_i for this condition. How does this compare to the heat transfer coefficient associated with conduction across the liquid film?

For water at atmospheric pressure, $\overline{M} = 18.0$ kg/(kg mol), $\rho_v = 0.598$ kg/m^3, $v_{lv} = 1.672$ m^3/kg, $h_{lv} = 2257$ kJ/kg, $\overline{R}/\overline{M} = 8314.4/18.0 = 462$ J/kg. It follows that

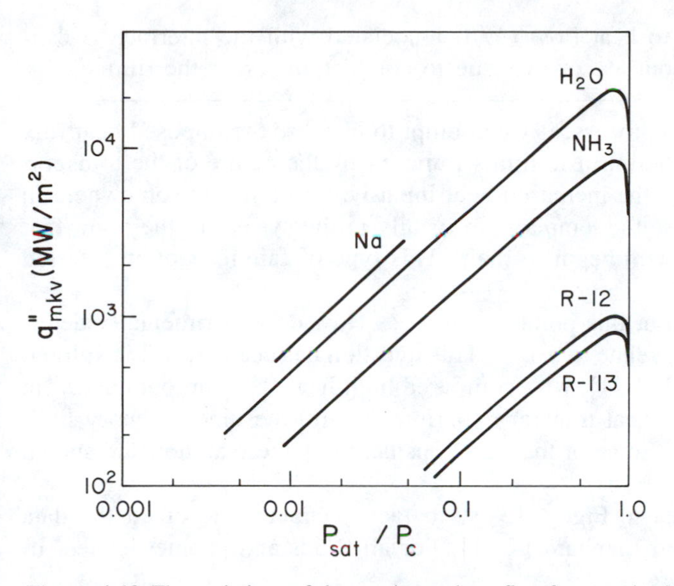

Figure 4.12 The variations of the maximum heat flux for vaporization with reduced pressure indicated by kinetic theory for different substances.

$$a = \frac{q_i''}{\rho_v h_{lv}(2\bar{R}T_v/\bar{M})^{1/2}} = \frac{200 \times 1000}{(0.598)(2257 \times 1000)(2 \times 462 \times 373)^{1/2}}$$

$$= 2.52 \times 10^{-4}$$

Because a is small, we can use Eq. (4.149):

$$h_i = \frac{2\hat{\sigma}}{2 - \hat{\sigma}}\left(\frac{h_{lv}^2}{T_v v_{lv}}\right)\left(\frac{\bar{M}}{2\pi\bar{R}T_v}\right)^{1/2}\left(1 - \frac{P_v v_{lv}}{2h_{lv}}\right)$$

$$= \frac{2\hat{\sigma}}{2 - \hat{\sigma}}\left[\frac{(2{,}257{,}000)^2}{(373)(1.672)}\right]\left[\frac{1}{2\pi(462)(373)}\right]^{1/2}\left[1 - \frac{(101{,}000)(1.672)}{2(2{,}257{,}000)}\right]$$

$$= \frac{2\hat{\sigma}}{2 - \hat{\sigma}}(7.55 \times 10^6)$$

Reported values of $\hat{\sigma}$ for water are in the range between 0.02 and 0.04. If we therefore take $\hat{\sigma} = 0.03$, we obtain

$$h_i = \frac{2(0.03)}{2 - 0.03}(7.55 \times 10^6) = 2.30 \times 10^5 \text{ W/m}^2\text{ K} = 230 \text{ kW/m}^2\text{ K}$$

For the liquid film, the conductivity $k_l = 0.681$ W/m K. Hence for conduction across the film,

$$h_c = \frac{k_l}{\delta} = \frac{0.681 \times 1000^{-1}}{0.0001} = 6.80 \text{ kW/m}^2\text{ K}$$

Thus the resistance to heat flow $(1/h)$ associated with the interface is estimated to be only about 3% of that due to conduction across the film.

In a real vaporization process, if we attempt to increase the imposed heat flux at the interface beyond the intrinsic limits imposed by the nature of the transport at the interface, generally the increased heat input will accumulate somewhere in the system. This can raise the temperature locally in the system to the point that solid elements of the system begin to melt. This type of failure is often referred to as *boiling burnout*.

An obvious question at this point would be: "How do experimentally determined burnout conditions relate to q''_{mkv}?" This question has been recently explored by Gambill and Lienhard [4.21], who compared high-heat-flux burnout data from a wide variety of boiling heat transfer experiments with the kinetic theory limit indicated by Eq. (4.150). Some of the data considered by these authors are shown in Fig. 4.13.

The trend in the data in Fig. 4.13 is, in fact, characteristic of all the data examined by Gambill and Lienhard [4.21]. For all fluids and geometries tested,

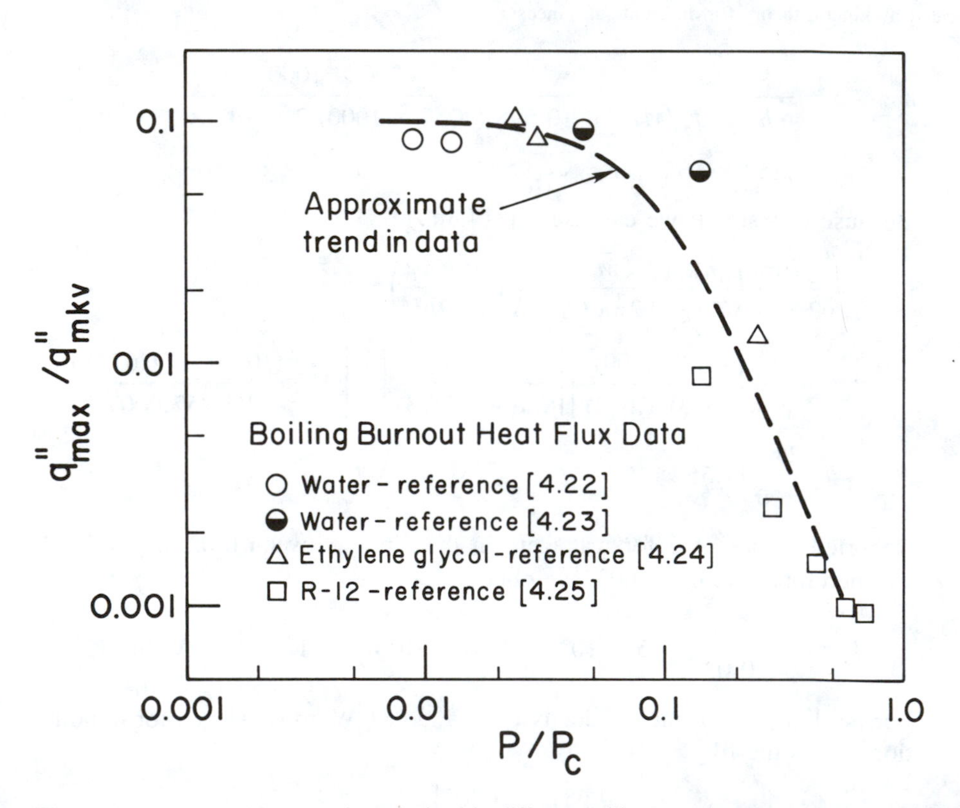

Figure 4.13 Measured burnout data obtained in different boiling experiments normalized with the corresponding maximum heat flux suggested by kinetic theory.

the burnout heat flux did not exceed 10% of q''_{mkv}. In addition, as P/P_c increased beyond 0.1, the ratio $q''_{\max}/q''_{mkv}$ invariably diminished.

The burnout heat flux is expected to be lower than q''_{mkv} for at least two reasons. First, it is clear that many molecules leaving the liquid at the interface will be returned to the liquid by molecular collisions. Hence the net molecular flux from the interface is likely to be significantly lower than that used to derive Eq. (4.150) for q''_{mkv}, resulting in a lower burnout flux. In addition, q''_{mkv} can be so large (see Fig. 4.12) in some cases that the wall temperature difference required to deliver even 10% of q''_{mkv} would result in wall temperatures that exceed the melting point of the material.

The decrease in $q''_{\max}/q''_{mkv}$ as P/P_c increases beyond 0.1 may be due, at least in part, to the wall temperature limitation problem noted above, and the increase of q''_{mkv} with reduced pressure for $0.1 < P/P_{cr} < 0.8$. In addition, this trend was also attributed, in part, to the tendency for more of the heat to find single-phase convection and/or conduction paths at higher pressures [4.21]. These observations suggest that the low range of $q''_{\max}/q''_{mkv}$ achieved to date may be due to the difficulty of experimentally imposing a sufficiently large heat flux and ensuring that it all produces vaporization. It may be possible, with a carefully designed experiment, to exceed $q''_{\max}/q''_{mkv} = 0.10$ somewhat, but currently available test data suggest that the practical limit to heat transfer by vaporization corresponds approximately to $q''_{\max} = 0.10q''_{mkv}$.

As noted by Schrage [4.13] and Gambill and Lienhard [4.21], the arguments underlying the prediction of q''_{mkv} apply equally well to condensation at the liquid–vapor interface. Strictly speaking, however, the equation (4.129*a*) for j_{Nw+} should be used to evaluate the maximum possible molecular flux to the interface, to account for the effect of the vapor bulk velocity on the velocity distribution function. Assuming perfect accommodation and that no molecules return to the vapor, we find, in a manner analogous to the vaporization case, that the maximum conceivable heat flux for condensation is given by

$$q''_{mkc} = \Gamma(a)\rho_v h_{lv}\left(\frac{\bar{R}T_v}{2\pi\bar{M}}\right)^{1/2} \tag{4.151}$$

As described above, Γ is a function of heat flux through a. As previously noted, however, Γ is close to 1 for very small values of a. If we substitute the small-a approximation for Γ given by Eq. (4.140) and use the ideal gas relation for ρ_v, we find that

$$q''_{mkc} = 2\rho_v h_{lv}\left(\frac{\bar{R}T_v}{2\pi\bar{M}}\right)^{1/2} \tag{4.152}$$

In addition, Schrage's [4.13] more detailed analysis predicts that

$$q''_{mkc} = 13.0\rho_v h_{lv}\left(\frac{\bar{R}T_v}{2\pi\bar{M}}\right)^{1/2} \tag{4.153}$$

Thus, as for the vaporization case, the maximum possible heat flux appears to be proportional to $\rho_v h_{lv}(\bar{R}T_v/2\pi\bar{M})^{1/2}$, but the proportionality constant depends on the model of the interfacial transport. It is interesting to note, however, that all three relations (4.151) through (4.153) predict that the proportionality constant is significantly greater than 1. This suggests that the limiting condensation heat flux may be somewhat higher than that for vaporization. However, Gambill and Lienhard [4.21] describe limited condensation data that suggest that the practical limit for q''_{max} for condensation is also near $0.1\rho_v h_{lv}(\bar{R}T_v/2\pi\bar{M})^{1/2}$.

The discussion in this section has focused on the interfacial transport limitations due to molecular transport in the vapor. In later sections, other mechanisms will be described that may also limit the rate of vaporization or condensation at the interface.

Example 4.6 Estimate the maximum possible heat flux associated with the vaporization of water and R-113 at atmospheric pressure.

Using the results of Schrage's [4.13] analysis, we have

$$q''_{mkv} = 0.741\rho_v h_{lv}\left(\frac{\bar{R}T_v}{2\pi\bar{M}}\right)^{1/2}$$

For water at atmospheric pressure, $\rho_v = 0.598$ kg/m³, $h_{lv} = 2257$ kJ/kg, $T_v = 373$ K, $\bar{R}/\bar{M} = 464$ J/kg K. Substituting yields

$$q''_{mkv} = 0.741(0.598)(2257)\left[\frac{462(373)}{2\pi}\right]^{1/2}$$

$$= 1.65 \times 10^5 \text{ kW/m}^2$$

It is interesting to note that the maximum heat flux attainable in nucleate pool boiling of water at atmospheric pressure is typically about 1.3×10^3 kW/m², which is far less than the limit determined above. For R-113 at atmospheric pressure, $\rho_v = 7.46$ kg/m³, $h_{lv} = 143.7$ kJ/kg, $T_v = 47.6°\text{C} = 321$ K, $\bar{R}/\bar{M} = 8314.4/187.4 = 44.4$ J/kg K, and

$$q''_{mkv} = 0.741(7.46)(143.7)\left[\frac{44.42(321)}{2\pi}\right]^{1/2}$$

$$= 3.78 \times 10^4 \text{ kW/m}^2$$

Thus, for R-113, q''_{mkv} is about a factor of 5 lower than for water. Note that this fluid's much lower h_{lv} value is somewhat offset by the higher vapor density for R-113.

REFERENCES

4.1 Squire, H. B., On the stability of three-dimensional distribution of viscous fluid between parallel walls, *Proc. Roy. Soc.*, London, vol. 142A, pp. 621–628, 1933.

4.2 Sernas, V., Minimum heat flux in film boiling—A three-dimensional model, *Proc. 2nd Canadian Congress of Applied Mechanics*, University of Waterloo, pp. 425–426, 1969.

4.3 Hsieh, D. Y., Effects of heat and mass transfer on Rayleigh-Taylor instability, *J. Basic Eng.*, vol. 94, pp. 156–162, 1972.
4.4 Chandrasekhar, S., *Hydrodynamic and Hydromagnetic Stability,* Clarendon Press, Oxford, 1961.
4.5 *CRC Standard Mathematical Tables,* 28th ed., CRC Press, Boca Raton, FL, 1986.
4.6 Rayleigh, Lord, On the instability of jets, *Proc. London Math. Soc.*, vol. 10, pp. 4–13, 1878.
4.7 Keller, J. B., Rubinow, S. I., and Tu, Y. O., Spatial instability of a jet, *Phys. Fluids,* vol. 16, pp. 2052–2055, 1973.
4.8 Benjamin, T. B., Wave formation in laminar flow down an inclined plane, *J. Fluid Mech.*, vol. 2, pp. 554–574, 1957.
4.9 Levich, V. G., *Physiochemical Hydrodynamics,* Prentice-Hall, Englewood Cliffs, NJ, 1962.
4.10 Sonntag, R. E., and Van Wylen, G., *Introduction to Thermodynamics—Classical and Statistical,* 2nd ed., Wiley, New York, 1987.
4.11 Knuth, E. L., *Introduction to Statistical Thermodynamics,* McGraw-Hill, New York, 1966.
4.12 Reif, F., *Fundamental of Statistical and Thermal Physics,* McGraw-Hill, New York, 1965.
4.13 Schrage, R. W., *A Theoretical Study of Interphase Mass Transfer,* Columbia University Press, New York, 1953.
4.14 Silver, R. S., and Simpson, H. C., The condensation of superheated steam, *Proc. of a conference held at the National Engineering Laboratory,* Glasgow, Scotland, 1961.
4.15 Kucherov, R. Y., and Rikenglaz, L. E., The problem of measuring the condensation coefficient, *Doklady Akad. Nauk. SSSR,* vol. 133, no. 5, pp. 1130–1131, 1960.
4.16 Wilhelm, D. J., Condensation of metal vapors mercury and kinetic theory of condensation, Argonne National Laboratory Report ANL-6948, 1964.
4.17 Mills, A. F., The condensation of steam at low pressures, Technical Report on NSF GP-2520, series no. 6, issue no. 39, Space Sciences Laboratory, University of California at Berkeley, 1965.
4.18 Volmer, M., and Estermann, J., Über den Verdampfungs-koeffizienten von Festem und Flüssigen Quecksilber, *Z. Physik,* vol. 7, pp. 1–12, 1921.
4.19 Sukhatme, S. P., and Rohsenow, W. M., Heat transfer during film condensation of a liquid metal vapor, *J. Heat Transfer,* vol. 88, pp. 19–28, 1966.
4.20 Paul, B., Compilation of evaporation coefficients, *ARS J.*, vol. 32, pp. 1321–1328, 1962.
4.21 Gambill, W. R., and Lienhard, J. H., An upper bound for the critical boiling heat flux, *Proc. 1987 ASME-JSME Thermal Engineering Joint Conf.*, vol. 3, pp. 621–626, 1987.
4.22 Gambill, W. R., Bundy, R. D., and Wamsbrough, R. W., Heat transfer, burnout, and pressure drop for water in swirl flow through tubes with internal twisted tapes, *Chem. Eng. Prog. Symp. Ser.*, vol. 57, no. 32, pp. 127–137, 1961.
4.23 Ornatskii, A. P., and Vinyarskii, L. S., Heat transfer crisis in a forced flow of underheated water in small-bore tubes, *High Temp.*, vol. 3, no. 3, pp. 400–406, 1965.
4.24 Gambill, W. R., and Bundy, R. D., High-flux heat-transfer characteristics of pure ethylene glycol in axial and swirl flow, *AIChE. J.*, vol. 9, pp. 55–59, 1963.
4.25 Katto, Y., and Shimizu, M., Upper limit of the CHF in the forced convection boiling on a heated disc with a small impinging jet, *J. Heat Transfer,* vol. 101, pp. 165–269, 1979.

PROBLEMS

4.1 At 20°C the surface tension of mercury in contact with air is 0.484 N/m and the liquid density is 12,800 kg/m^3. Determine the most dangerous wavelength for an air–mercury interface at 20°C.

4.2 In Chapter Two, it was argued that contact-angle hysteresis could cause a slug of liquid in a small-diameter vertical tube to be supported against gravity. This presumes, of course, that the interface is stable. Taylor instability can be avoided if the diameter of the tube is sufficiently small that disturbances larger than the critical wavelength can never exist (i.e., $D < \lambda_c$). Estimate the maximum tube diameter that can support a liquid slug for (*a*) water at 20°C ($\sigma = 0.0728$ N/m) and (*b*) saturated acetone at atmospheric pressure.

4.3 Determine the critical velocity difference for Helmholtz instability of a horizontal flow of saturated R-12 vapor above liquid R-12 at a pressure of 793 kPa.

4.4 Estimate the wind velocity necessary to cause water in a puddle to become Helmholtz unstable. Use $\sigma = 0.0728$ N/m for water at 20°C.

4.5 What effect does addition of a surfactant to water have on the critical velocity for Helmholtz instability? Explain briefly.

4.6 At 20°C the surface tension of an air–water interface is 0.0728 N/m and that for a synthetic silicon oil is 0.0350 N/m. Using Eq. (3.25), estimate the critical velocity difference for Helmholtz flow of silicon oil over water. The silicone oil has a specific gravity of 0.95.

4.7 A jet of liquid mercury at 20°C is injected into air at the same temperature. The jet diameter is 0.8 mm and the liquid velocity at the nozzle is 10 m/s. Estimate the distance downstream of the nozzle where the jet is likely to break up into droplets.

4.8 A jet of liquid R-113 with a diameter of 0.4 mm flows from a nozzle and impinges on a microelectronic component to cool it. If the jet velocity is 200 cm/s, estimate how close the nozzle must be to the target to ensure that the jet does not break up into droplets before striking the target. The R-113 in the jet is in contact with its own vapor at 20°C, at which the liquid density is 1575 kg/m^3 and the interfacial tension is 0.020 N/m.

4.9 A film of saturated liquid water at atmospheric pressure having a uniform thickness of 1.0 mm flows down a flat surface inclined at 30° relative to the horizontal. Estimate the mean film velocity at which waves are likely to be observed on the film interface.

4.10 A film of saturated liquid water at atmospheric pressure flows down a flat surface that is inclined at 15° to the horizontal. At the top of the surface, the liquid flows onto the surface with a thickness of 1 mm and a uniform velocity of 10 cm/s. The film flow is laminar and driven by gravity alone. After flowing down the surface for some distance, the film reaches a fully developed condition where its thickness and velocity field do not change. Are waves likely to be observed on this film? Justify your answer quantitatively, using an approximate integral analysis of the fully developed film flow.

4.11 Condensation of pure mercury vapor occurs on the interface of a film of liquid mercury on a cooled flat surface. The surface heat flux into the plate is 200 kW/m^2. Estimate the variation in the interfacial heat transfer coefficient associated with values of the accommodation coefficient between 0.3 and 0.6.

4.12 Estimate the interfacial heat transfer coefficient for evaporation of a thin film of saturated liquid ethanol at atmospheric pressure. The liquid film rests on a flat, solid surface to which a constant and uniform heat flux of 150 kW/m^2 is applied. The accommodation coefficient may be taken to be 0.03.

4.13 Film condensation of pure oxygen vapor on a cooled, flat surface occurs at atmospheric pressure. The heat flux removed at the solid wall is a constant and uniform value of 100 kW/m^2. Assuming an accommodation coefficient of 0.1, estimate the interfacial heat transfer coefficient for this process.

4.14 Water at 20°C is suddenly ejected from inside a spacecraft into the vacuum of space. Estimate an upper bound on the initial vapor flux rate from the liquid interface into space immediately after ejection. Also estimate the associated heat flux at the liquid surface to facilitate this vaporization. If this flux rate continues from a spherical droplet with a radius of 1 mm, estimate how long it would take for the droplet to begin freezing.

4.15 Estimate the maximum possible heat flux values associated with (*a*) the vaporization, and (*b*) the condensation of mercury at atmospheric pressure.

4.16 Estimate the maximum possible heat flux for R-12 vaporization at 333 kPa. Assume that this vaporization process was accomplished by transferring heat by conduction across a thin film of R-12 liquid. If the maximum possible driving temperature difference for such a process is the difference between the saturation temperature and the critical temperature for R-12, what is the maximum thickness that the liquid film can be?

CHAPTER

FIVE

PHASE STABILITY AND HOMOGENEOUS NUCLEATION

5.1 METASTABLE STATES AND PHASE STABILITY

In the development of classical thermodynamics, phase transitions are treated as if they occur as quasi-equilibrium processes at the equilibrium saturation conditions. As we will see in later sections, real phase transformations usually occur under nonequilibrium conditions. In real vaporization processes, some liquid in the system is almost always superheated above the equilibrium saturation temperature. Likewise, the initiation of condensation processes usually is achieved in real systems only after at least a portion of the vapor phase has been supercooled below its equilibrium saturation temperature.

In this section, we will first examine the stability of a closed system containing a pure substance that is not in thermodynamic equilibrium. Specifically, we wish to determine the conditions under which a portion of such a system is likely to undergo a change of phase. To do so, we consider an isolated closed system initially filled completely with phase I of a pure, simple, compressible substance. As schematically indicated in Fig. 5.1, a portion of this system is assumed to be separated from the rest by a diathermal, nonrigid, permeable membrane. The subsystem inside the membrane will be called region B, and the rest of the system will be called region A.

For the system as a whole, and each of the subsystems A and B shown in Fig. 5.1, U is taken to be a function of S, V, and the number of moles N. A small

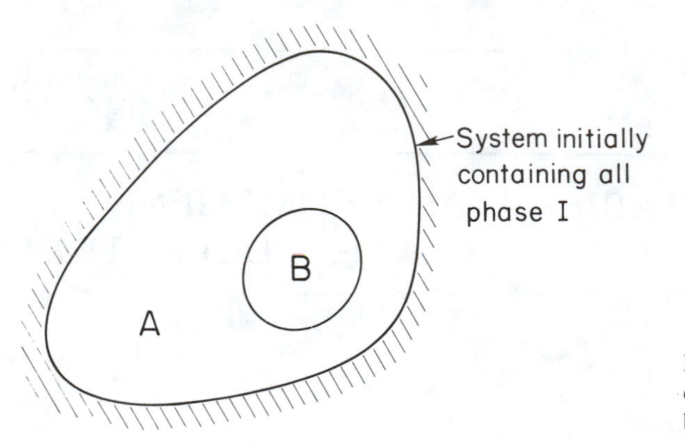

Figure 5.1 Model system considered in the analysis of phase stability.

change in U can then be written in terms of a Taylor series about the initial condition. Such a series can be expressed compactly as

$$\Delta U = \delta U + \left(\frac{1}{2!}\right)\delta^2 U + \left(\frac{1}{3!}\right)\delta^3 U + \ldots \tag{5.1}$$

where, if U is a function of three variables z_1, z_2, and z_3, δU, $\delta^2 U$, and $\delta^3 U$ are defined as

$$\delta U = \sum_{i=1}^{3} U_{z_i}\, dz_i \tag{5.2}$$

$$\delta^2 U = \sum_{i=1}^{3} \sum_{j=1}^{3} U_{z_i z_j}\, dz_i\, dz_j \tag{5.3}$$

$$\delta^3 U = \sum_{i=1}^{3} \sum_{j=1}^{3} \sum_{k=1}^{3} U_{z_i z_j z_k}\, dz_i\, dz_j\, dz_k \tag{5.4}$$

In Eqs. (5.2) through (5.4), the subscript on U denotes the partial derivative with respect to that variable (i.e., $U_{z_i} = \partial U/\partial z_i$, etc.). Equations (5.1) through (5.4) apply to the perturbations in U with z_1, z_2, and z_3 equal to S, V, and N. We will adopt the idealization that any quantity of phase II formed is in thermodynamic equilibrium with the surrounding phase I. It was noted in Section 1.3 that, for a system with fixed total S, V, and N, the system is in stable equilibrium only if U is at a local minimum. For two phases I and II of a pure, simple, compressible substance in a system with fixed total S, V, and N, it follows that, at equilibrium,

$$\begin{aligned} \delta U = \delta U^{\mathrm{I}} + \delta U^{\mathrm{II}} = U_V^{\mathrm{I}}\, dV^{\mathrm{I}} + U_S^{\mathrm{I}}\, dS^{\mathrm{I}} + U_N^{\mathrm{I}}\, dN^{\mathrm{I}} \\ + U_V^{\mathrm{II}}\, dV^{\mathrm{II}} + U_S^{\mathrm{II}}\, dS^{\mathrm{II}} + U_N^{\mathrm{II}}\, dN^{\mathrm{II}} = 0 \end{aligned} \tag{5.5}$$

To determine whether U is at a minimum, we must consider the higher-order terms in the Taylor series expansion for small changes of U about the equilibrium conditions. It is known from basic calculus that, for the Taylor series (5.1), a necessary and sufficient condition for minimum U is that $\delta^m U > 0$, where $\delta^m U$ is the lowest-order nonvanishing variation of U. Thus, for this system, the criteria for equilibrium and stability may be stated as

$$\delta U = 0 \qquad \text{criterion of equilibrium}$$

$$\left.\begin{array}{ll} \delta^m U > 0 & \text{for the smallest} \\ m = 2, 3, 4, \ldots & \text{at which } \delta^m U \neq 0 \end{array}\right\} \quad \text{criterion of stability}$$

The above criteria indicate that we must first consider $\delta^2 U$ to determine whether compatible perturbations of the two subsystems A and B may cause a shift to a new equilibrium by forming phase II in region B. For this system,

$$\begin{aligned} \delta^2 U &= \delta^2 U^A + \delta^2 U^B \\ &= U^A_{SS}(\delta S^A)^2 + U^A_{VV}(\delta V^A)^2 + U^A_{NN}(\delta N^A)^2 + 2U^A_{SN}\,\delta S^A\,\delta N^A \\ &\quad + 2U^A_{VN}\,\delta V^A\,\delta N^A + 2U^A_{VS}\,\delta V^A\,\delta S^A \\ &\quad + U^B_{SS}(\delta S^B)^2 + U^B_{VV}(\delta V^B)^2 + U^B_{NN}(\delta N^B)^2 + 2U^B_{SN}\,\delta S^B\,\delta N^B \\ &\quad + 2U^B_{VN}\,\delta V^B\,\delta N^B + 2U^B_{VS}\,\delta V^B\,\delta S^B \end{aligned} \tag{5.6}$$

where again the subscripts denote partial derivatives with respect to the subscripted variables.

Because $\delta^2 U$ is part of a Taylor series, the partial derivatives in it are evaluated at the initial condition, which is the same (phase I) in both regions A and B. The partial derivatives of intensive properties for A and B are therefore equal. As an example, consider

$$\left(\frac{\partial^2 u}{\partial s\,\partial v}\right)^A = \left(\frac{\partial^2 u}{\partial s\,\partial v}\right)^B \tag{5.7}$$

However, if this derivative is written in terms of the extensive properties of interest here, we find that

$$\frac{\partial^2 u}{\partial s\,\partial v} = \frac{\partial^2 (U/N)}{\partial (S/N)\,\partial (V/N)} = NU_{SV}$$

Hence Eq. (5.7) implies that

$$N^A U^A_{SV} = N^B U^B_{SV} \tag{5.8}$$

Similarly, it can be shown that

$$\begin{aligned} &N^A U^A_{SN} = N^B U^B_{SN} \qquad N^A U^A_{VN} = N^B U^B_{VN} \qquad N^A U^A_{VV} = N^B U^B_{VV} \\ &N^A U^A_{SS} = N^B U^B_{SS} \qquad N^A U^A_{NN} = N^B U^B_{NN} \end{aligned} \tag{5.9}$$

In addition, because the system is isolated with fixed S, V, and N, we know that

$$(\delta S^A)^2 = (\delta S^B)^2 \qquad (\delta V^A)^2 = (\delta V^B)^2 \qquad (\delta N^A)^2 = (\delta N^B)^2 \tag{5.10}$$

Multiplying the A terms in Eq. (5.6) by N^A/N^A and substituting Eqs. (5.8) through (5.10) to eliminate B terms yields

$$\delta^2 U = \left[\frac{(N^A + N^B)}{N^B}\right]\delta^2 U^* \tag{5.11}$$

where

$$\delta^2 U^* = U^A_{SS}(\delta S^A)^2 + U^A_{VV}(\delta V^A)^2 + U^A_{NN}(\delta N^A)^2 + 2U^A_{SN}\delta S^A \delta N^A + 2U^A_{SV}\,\delta S^A\,\delta V^A + 2U^A_{VN}\,dV^A\,dN^A \tag{5.12}$$

Note that, because the term in square brackets in Eq. (5.11) is always positive, the condition for stability $\delta^2 U > 0$ is satisfied if and only if $\delta^2 U^* > 0$. Note also that the right side of Eq. (5.12) contains only A terms. This implies that the stability depends only on the conditions of the original phase I in the system (which occupies region A). Because the stability is determined by conditions in the original phase I, the conditions for which $\delta^2 U > 0$ are called the criteria of *intrinsic stability* of the phase. The conditions under which these criteria are first violated, corresponding to $\delta^2 U = 0$, are called the *limits of intrinsic stability*.

Because the right side of Eq. (5.12) contains only A terms, for convenience, the A superscript will henceforth be dropped, with the understanding that all terms are evaluated at the initial conditions in A. Manipulating Eq. (5.12) to complete the squares for terms involving the cross derivatives yields

$$\delta^2 U^* = a_1\left[\delta S + \left(\frac{U_{SV}}{U_{SS}}\right)\delta V + \left(\frac{U_{SN}}{U_{SS}}\right)\delta N\right]^2 + \frac{a_2}{a_1}\left[\delta V + \left(\frac{U_{SS}U_{VN} - U_{SV}U_{SN}}{U_{SV}U_{VV} - U^2_{SV}}\right)\delta N\right]^2 + \frac{a_3}{a_1}(\delta N)^2 \tag{5.13}$$

where

$$a_1 \equiv |U_{SS}| = U_{SS} \tag{5.14}$$

$$a_2 \equiv \begin{vmatrix} U_{SS} & U_{SV} \\ U_{SV} & U_{VV} \end{vmatrix} \tag{5.15}$$

$$a_3 \equiv \begin{vmatrix} U_{SS} & U_{SV} & U_{SN} \\ U_{SV} & U_{VV} & U_{VN} \\ U_{SN} & U_{VN} & U_{NN} \end{vmatrix} \tag{5.16}$$

It can be shown using the Gibbs-Duhem equation (1.8), together with the Maxwell relations from classical thermodynamics, that $a_3 = 0$. Hence $\delta^2 U^*$ is positive definite if both a_1 and a_2 are greater than zero. It can easily be shown from basic thermodynamic relations that

$$a_1 = U_{SS} = \left(\frac{\partial T}{\partial S}\right)_{V,N} = \frac{T}{N\bar{c}_v} \tag{5.17}$$

where $\bar{c}_v$ is the molar specific heat at constant volume. Because T and N are always positive, it follows that a necessary condition for a_1 to be positive definite is that

$$\bar{c}_v > 0 \tag{5.18}$$

The above relation (5.18) is referred to as the *criterion of thermal stability*.

For $\delta^2 U^*$ to be positive definite, we also require that

$$\frac{a_2}{a_1} = U_{VV} - \frac{U_{SV}^2}{U_{SS}} > 0 \tag{5.19}$$

Using the thermodynamic relations

$$U_{VV} = -\left(\frac{\partial P}{\partial V}\right)_{S,N} \qquad U_{SV} = -\left(\frac{\partial P}{\partial S}\right)_{V,N} \tag{5.20}$$

and a form of the chain rule for $(\partial P/\partial V)_S$,

$$\left(\frac{\partial P}{\partial V}\right)_S = \left(\frac{\partial P}{\partial V}\right)_T + \left(\frac{\partial P}{\partial T}\right)_V \left(\frac{\partial T}{\partial V}\right)_S \tag{5.21}$$

the inequality (5.19) can be rewritten as

$$\frac{a_2}{a_1} = -\left(\frac{\partial P}{\partial V}\right)_T > 0 \tag{5.22}$$

This relation, which is usually written as

$$\left(\frac{\partial P}{\partial V}\right)_T < 0 \quad \text{or} \quad \left(\frac{\partial P}{\partial v}\right)_T < 0 \tag{5.23}$$

is called the *criterion for mechanical stability*. Because $\bar{c}_v$ is greater than zero for virtually all substances, the criterion of thermal stability is satisfied and Eq. (5.23) is a necessary and sufficient condition for the stability of a phase.

Although the above analysis has focused specifically on the stability of a phase of a pure substance, the analysis is readily extendable to multicomponent phases. The interested reader is referred to reference [5.1], which discusses phase stability of multicomponent systems in detail. It is worth noting, however, that extension of the above analysis to a binary system reveals that the inequalities (5.18) and (5.23), representing thermal and mechanical stability, are necessary but not sufficient conditions for stability. In addition, the following inequality must hold for the phase to be stable:

$$\left(\frac{\partial \mu_1}{\partial x_1}\right)_{T,P} \geqslant 0 \tag{5.24}$$

where μ_1 and x_1 are the chemical potential and mass fraction of component 1 in the binary mixture, respectively. The inequality (5.24) is sometimes called the *criterion of diffusional stability*. It should be noted, however, that this terminology is not meant to imply that phase stability of a binary system is somehow connected with the process of molecular diffusion. Assuming that the thermal and mechanical stability inequalities are satisfied, the inequality (5.24) is a necessary and sufficient condition for phase stability of a binary system.

Now that we have developed a criterion for stability, we will consider the role of this criterion in the behavior of a real system. It is generally known that liquid can be superheated and vapor can be supercooled (supersaturated) without a phase transformation occurring, even though the state should lie within the vapor dome at equilibrium. The expansion of steam in a nozzle, shown schematically in Fig. 5.2, is an example of a process in which supercooling of vapor may occur.

Superheated steam initially at state 1 expands irreversibly along a path approximated by the dotted line in the T–S diagram in Fig. 5.2*b*. Experimentally, condensation is almost never found to occur exactly at point 2 on the saturated vapor curve, where it should occur if thermodynamic equilibrium existed. Instead, condensation is observed to occur abruptly in what is usually called a *condensation shock* further downstream at point 3. Between points 2 and 3, the steam exists as a vapor below its equilibrium saturation temperature at the local pressure.

Vapor that is supercooled below its equilibrium saturation temperature and liquid that is superheated above its equilibrium saturation temperature exist in a nonequilibrium condition referred to as a *metastable state*. On a P–v diagram, metastable states for a pure substance can be visualized as lying on an isotherm that extends from the stable single-phase region into the vapor dome, as shown in Fig. 5.3.

If we begin with metastable liquid and increase v at constant T, we might expect that eventually the state would reach the metastable curve near the stable vapor condition and ultimately become stable vapor. This suggests that the locus of such states may form a continuous curve smoothly joining the isotherms in the stable single-phase regions. Such a curve is shown as the dotted line in Fig. 5.4.

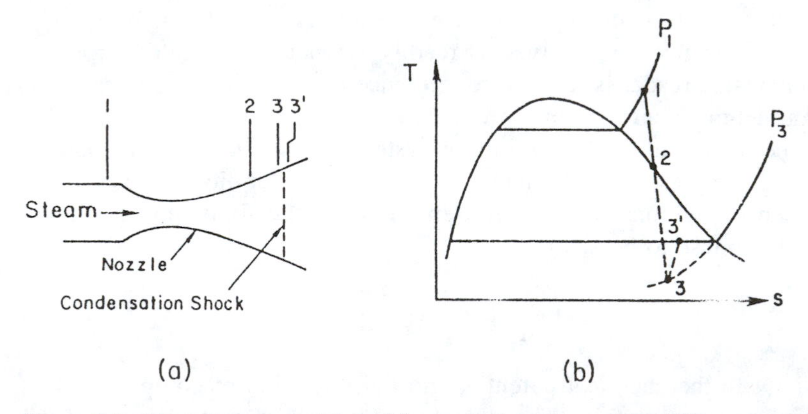

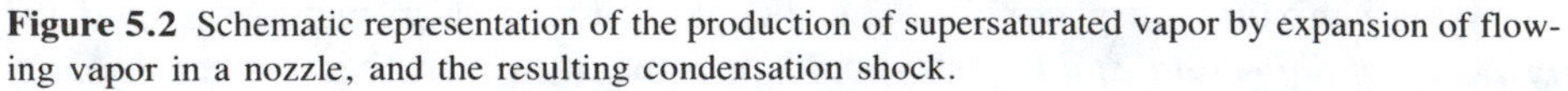
Figure 5.2 Schematic representation of the production of supersaturated vapor by expansion of flowing vapor in a nozzle, and the resulting condensation shock.

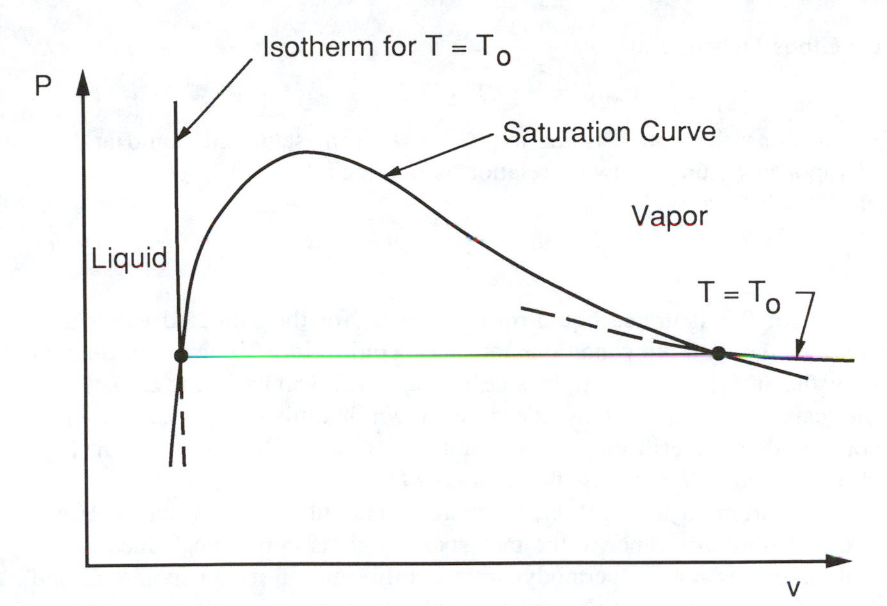

Figure 5.3 Behavior of isothermal lines near saturation.

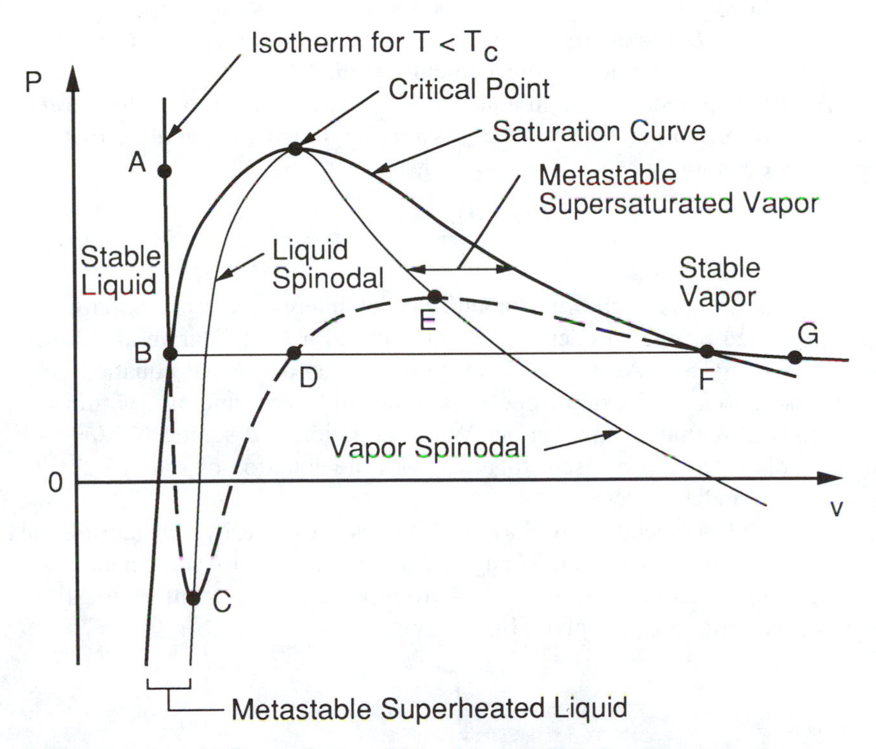

Figure 5.4 Spinodal lines and metastable regions on a *P–v* diagram.

If the Gibbs-Duhem equation

$$d\mu = -s\,dT + v\,dP \tag{5.25}$$

is integrated along the dotted isotherm ($dT = 0$) from saturated liquid at B to saturated vapor at F, the following relation is obtained:

$$\mu_F - \mu_B = \int_B^F v\,dP \tag{5.26}$$

However, μ_F for the saturated liquid must equal μ_B for the saturated vapor as a consequence of the required conditions for phase equilibrium discussed in Section 1.4. Hence the integral on the right side of Eq. (5.26) must equal zero. From elementary calculus considerations, it can be shown that this integral can be equal to zero only if the isotherm loops above and below line BDF, as shown in Fig. 5.4, and if the area $BCDB$ equals the area $DEFD$.

As shown earlier in this section, mechanical stability requires that $(\partial P/\partial v)_T < 0$. Note that liquid or vapor in the metastable regions is not mechanically unstable, although it is not in thermodynamic equilibrium. Between points C and E, however, $(\partial P/\partial v)_T$ is positive, and the stability criterion is violated. The location where $(\partial P/\partial v)_T$ changes from negative to positive is called the *limit of intrinsic stability* or the *spinodal limit*. The *spinodal curve* is the locus of spinodal limit points in the vapor dome. The region between the liquid and vapor spinodal curves shown in Fig. 5.4 is interpreted as being inaccessible to the system because these states violate the criterion of mechanical stability.

To predict the spinodal limit, an equation of state is needed to evaluate $(\partial P/\partial v)_T$ for nonequilibrium conditions. One possible approach is to assume that the van der Waals equation

$$P = \frac{(\overline{R}/\overline{M})T}{v - b} - \frac{a}{v^2} \tag{5.27}$$

is valid even for the nonequilibrium conditions of interest here. The shapes of isotherms predicted by the van der Waals equation on a P–v diagram are shown schematically in Fig. 5.5. As seen in this figure, the van der Waals equation does predict that isotherms will exhibit one maximum and one minimum within the vapor dome. Differentiating the van der Waals equation and setting $(\partial P/\partial v)_T = 0$ yields a relation that can be used, together with the equation of state (5.27), to determine the spinodal curves.

Another useful perspective on phase stability is provided by considering the variation of the chemical potential. We again integrate the Gibbs-Duhem equation (5.25) along the isotherm shown in Fig. 5.4 from point A to an arbitrary location. The resulting variation of μ is given by

$$\mu = \mu_A + \int_{P_A}^{P} v\,dP \tag{5.28}$$

Arbitrarily picking a datum value of μ at point A and using the van der Waals

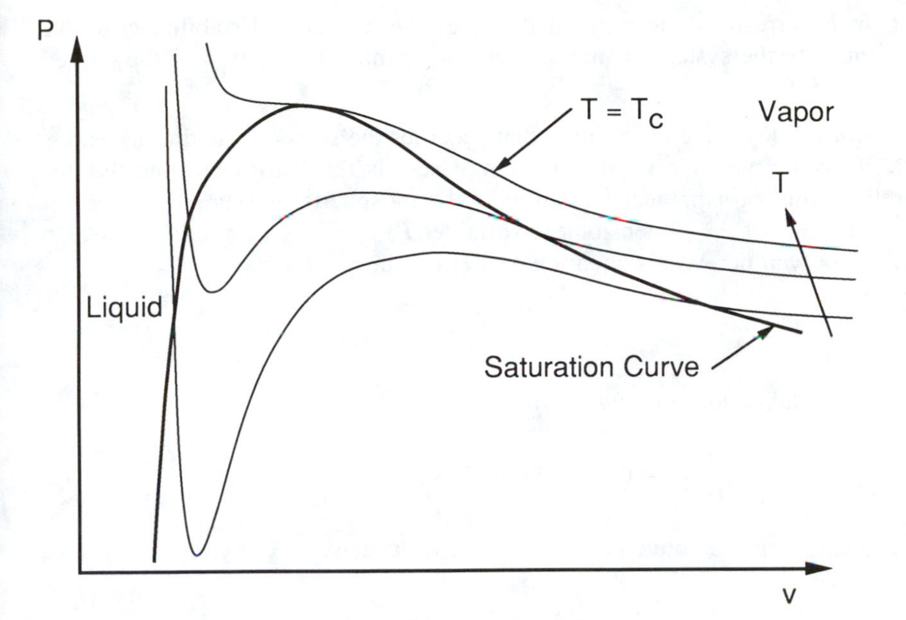

Figure 5.5 Isotherm shapes predicted by the van der Waals equation.

equation to evaluate the integral in Eq. (5.28) results in a variation of μ with P that looks like the curve shown in Fig. 5.6.

In Fig. 5.6, section AB of the curve corresponds to stable liquid, and section FG corresponds to stable vapor. Segments BC and EF of the curve correspond to metastable liquid and metastable vapor, respectively. The portion of the curve

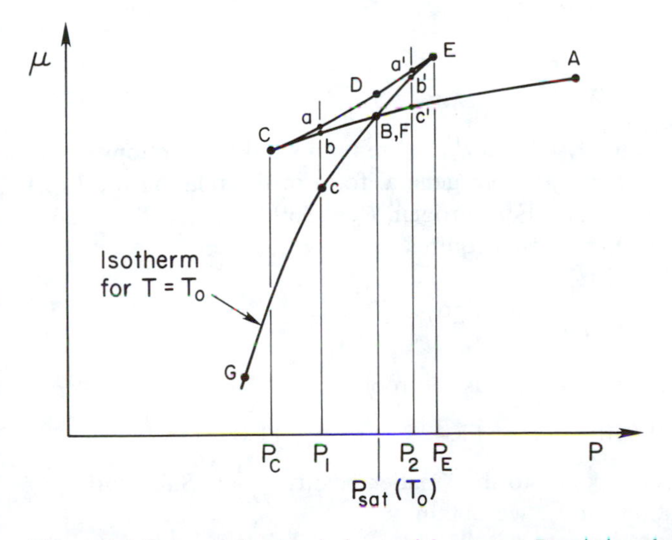

Figure 5.6 Shape of the chemical potential μ versus P variation along an isotherm as predicted by the van der Waals equation.

from C to E corresponds to states that violate the mechanical stability criterion, and presumably the system cannot access states along this section of the curve.

Example 5.1 Derive the relation that specifies the spinodal conditions as predicted by the van der Waals equation. Use this result to determine the theoretical limit of superheat for nitrogen at atmospheric pressure.

In terms of the dimensionless variables $P_r = P/P_c$, $v_r = v/v_c$, and $T_r = T/T_c$, the van der Waals equation (5.27) can be written as

$$P_r = \frac{8T_r}{3v_r - 1} - \frac{3}{v_r^2}$$

The spinodal condition requires that

$$\left(\frac{\partial P}{\partial v}\right)_T = 0 \quad \text{which implies that} \quad \left(\frac{\partial P_r}{\partial v_r}\right)_{T_r} = 0$$

Differentiating the above equation for P_r with respect to v_r yields

$$\frac{\partial P_r}{\partial v_r} = \frac{-24T_r}{(3v_r - 1)^2} + \frac{6}{v_r^3}$$

Equating this relation for $\partial P_r/\partial v_r$ to zero and solving for T_r yields

$$T_r = \frac{(3v_r - 1)^2}{4v_r^3}$$

This relation, in turn, can be back substituted into the dimensionless form of the van der Waals equation to eliminate T_r, which, after rearranging, results in the following relation for P_r:

$$P_r = \frac{3v_r - 2}{v_r^3}$$

The above relations for $T_r(v_r)$ and $P_r(v_r)$ are both valid descriptions of the spinodal condition. A somewhat more general form of the relation for $T_r(v_r)$ will be described in Section 5.4. For nitrogen, $P_c = 3396$ kPa and $T_c = 126.3$ K. The relation for $P_r(v_r)$ therefore requires

$$P_r = \frac{101}{3396} = 0.0297 = \frac{(3v_r - 2)}{v_r^3}$$

Iterative solution of this relation yields the roots

$$v_r = 0.670 \qquad v_r = 11.42 \qquad v_r = -11.09$$

The superheat limit corresponds to the smaller positive root. Substituting $v_r = 0.670$ into the relation for T_r, we obtain

$$T_r = \frac{(3v_r - 1)^2}{4v_r^3} = \frac{[3(0.670) - 1]^2}{4(0.670)^3} = 0.848$$

This implies that the superheat limit temperature, T_{SHL}, is

$$T_{SHL} = T_r \cdot T_c = 0.848(126.3) = 107 \text{ K}$$

For the temperature T_0 corresponding to the isotherm curve shown in Fig. 5.6, at $P = P_1 < P_{sat}(T_0)$ it can be seen that there are three possible states for the system, a, b, and c, which satisfy the van der Waals relation and hence lie on the isotherm curve. However, the system presumably cannot exist in state a because the stability criterion is not satisfied along segment CE of the curve. Both state b, metastable liquid, and state c, stable vapor, do not violate the criterion of mechanical stability, and hence are possible states for the system.

It was shown in Section 1.3 that, for a system containing a pure substance held at fixed T and P, equilibrium corresponds to the state that minimizes the Gibbs free energy and chemical potential of the system (note that $\mu = g$ for a pure substance). For a system at T_0 and P_1, equilibrium may be achieved for either state b or state c. For both states b and c, μ is a local minimum and the equilibrium is stable (in a local sense). It can be seen in Fig. 5.6, however, that the equilibrium μ for state c is lower than that at state b. Hence, given a sufficiently vigorous perturbation, the system would be inclined to spontaneously jump from metastable superheated liquid at b to stable vapor at state c. In many real systems the probability of a perturbation capable of causing a jump to state c is sufficiently low that the system may stay in state b almost indefinitely. If a system is initially in metastable state b and the pressure is reduced below P_C, the limit of intrinsic stability is exceeded, whereupon the system must jump to the stable vapor line.

The above comments regarding points a, b, and c apply in almost identical fashion to points a', b', and c' at $P = P_2 > P_{sat}(T_0)$. The system cannot exist in state a' because the stability criterion is violated. While both states b' and c' are possible stable equilibrium states, a system in state b' will spontaneously jump to c' if the right pertubation comes along, because of the lower value of μ at c'. Also, if a system is initially in a metastable state b' and the pressure is increased above P_E, the limit of intrinsic stability is exceeded and the system must jump to the stable liquid line.

It is worth emphasizing again that classical thermodynamic treatments usually assume that phase transitions occur at the equilibrium saturation conditions, which correspond to point B in Fig. 5.6. In real systems, however, a change in pressure at constant temperature would almost invariably carry the state point somewhat beyond point B, into the metastable range. As the system enters deeper into the metastable range, the likelihood that the system will undergo a phase transition generally increases. The spinodal limit is a barrier beyond which a phase change is virtually certain to occur. However, the actual conditions where a phase transformation occurs will depend on the probability that a suitable perturbation will arise to initiate the transition, which, in turn, is dependent on the molecular kinetics of the process. This will be discussed further in connection with homogeneous nucleation in the following sections.

5.2 THERMODYNAMIC ASPECTS OF HOMOGENEOUS NUCLEATION IN SUPERHEATED LIQUID

For a liquid heated at constant pressure above its equilibrium boiling temperature, the spinodal limit described in the previous section is a maximum upper limit on the superheat that results from thermodynamic considerations. For this reason it is sometimes called the *thermodynamic limit of superheat*. The spinodal limit for supersaturated (supercooled) vapor is similarly viewed as a thermodynamic limit.

Experimentally, bubble formation within superheated liquid, or droplet formation within supersaturated vapor, is generally observed to occur over a range of temperatures within the metastable range. Bubble nucleation completely within a superheated liquid is called *homogeneous nucleation*. The same term is used for droplet formation completely within a supercooled vapor. This type of process is to be contrasted with *heterogeneous nucleation,* which is nucleation at the interface between a metastable phase and another (usually solid) phase that it contacts. Heterogeneous nucleation will be discussed in detail in Chapter Six, while this section and those following will focus entirely on homogeneous nucleation phenomena.

Even in a thermodynamically stable liquid system, internal fluctuations take place. These localized transient deviations from the normal state include fluctuations of the local molecular density in the liquid. Far from the normal saturation conditions, such fluctuations in subcooled liquid are likely to be within limits consistent with the existence of a liquid phase. However, for states near the liquid saturation line in Fig. 5.4, density fluctuations in the liquid may exceed these limits, resulting in localized regions where the molecular density has been lowered almost to that of saturated vapor. Extreme fluctuations of this type thus give rise to small embryo bubbles of vapor within the liquid. Such fluctuations are sometimes referred to as *heterophase fluctuations*.

Given that embryo bubbles do form in liquid near the saturation curve, we now wish to determine whether such embryos are stable or unstable, and if unstable, whether they collapse or grow. To examine the stable equilibrium conditions for a small spherical bubble, we consider the system shown in Fig. 5.7*b*.

At equilibrium the temperature in the vapor and the liquid must be the same, and the chemical potential in the two phases must be equal.

$$\mu_l = \mu_{ve} \tag{5.29}$$

Because of the curvature of the interface, however, the pressures in the vapor bubble and the liquid are not equal, but are related through the Young-Laplace equation (2.30):

$$P_{ve} = P_l + \frac{2\sigma}{r_e} \tag{5.30}$$

Integrating the Gibbs-Duhem equation,

$$d\mu = -s\,dT + v\,dP$$

at constant temperature from $P = P_{sat}(T_l)$ to an arbitrary pressure P, we obtain

$$\mu - \mu_{sat} = \int_{P_{sat}(T_l)}^{P} v\,dP \tag{5.31}$$

For the vapor phase, we use the ideal gas law ($v = RT_l/P$) to evaluate the integral on the right side of Eq. (5.31), yielding the following relation for the chemical potential of the vapor and its equilibrium pressure P_{ve}:

$$\mu_{ve} = \mu_{sat,v} + RT_l \ln\left[\frac{P_{ve}}{P_{sat}(T_l)}\right] \tag{5.32}$$

For the liquid phase, Eq. (5.31) is again used, but this time, because the liquid is virtually incompressible, v is taken to be constant and equal to the value for saturated liquid at T_l, $v = v_l$. Evaluation of the integral in Eq. (5.31) for $P = P_l$ then yields

$$\mu_l = \mu_{sat,l} + v_l[P_l - P_{sat}(T_l)] \tag{5.33}$$

Substituting Eqs. (5.32) and (5.33) into the required equilibrium condition (5.29), using the fact that $\mu_{sat,v} = \mu_{sat,l}$ and rearranging, the following relation is obtained

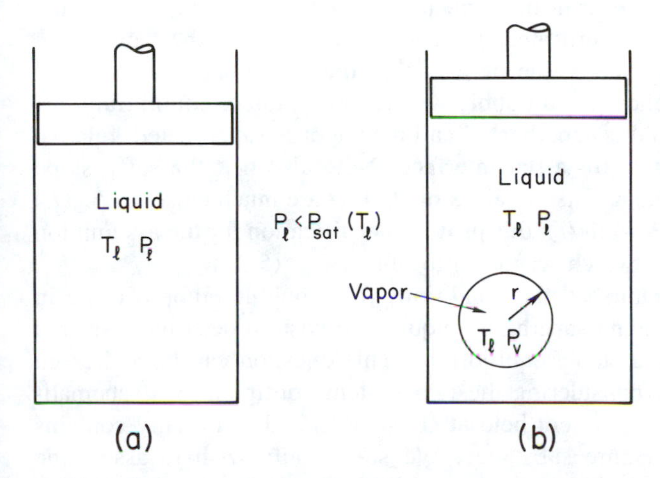

Figure 5.7 System model considered in the thermodynamic analysis of formation of an embryo vapor bubble.

for the vapor pressure inside the bubble at equilibrium:

$$P_{ve} = P_{\text{sat}}(T_l) \exp\left\{\frac{v_l[P_l - P_{\text{sat}}(T_l)]}{RT_l}\right\} \tag{5.34}$$

For superheated (metastable) liquid, P_l must be less than $P_{\text{sat}}(T_l)$, as can be seen from Fig. 5.4. Consequently, the exponential term in (5.34) is less than 1 and P_{ve} is less than $P_{\text{sat}}(T_l)$. Substituting Eq. (5.34) into Eq. (5.30), the following equation for r_e may be obtained:

$$r_e = \frac{2\sigma}{P_{\text{sat}}(T_l) \exp\{v_l[P_l - P_{\text{sat}}(T_l)]/RT_l\} - P_l} \tag{5.35}$$

Thus only bubbles having a radius $r = r_e$ given by Eq. (5.35) will be in equilibrium with the surrounding superheated liquid at T_l and P_l.

Alternatively, if Eq. (5.30) is substituted into Eq. (5.34) to eliminate P_l, the resulting relation for P_{ve} is

$$P_{ve} = P_{\text{sat}}(T_l) \exp\left\{\frac{v_l[P_{ve} - P_{\text{sat}}(T_l) - 2\sigma/r_e]}{RT_l}\right\} \tag{5.36}$$

In most cases $P_{ve} - P_{\text{sat}}(T_l)$ is small compared to $2\sigma/r_e$. Hence P_{ve} is well approximated for most systems by

$$P_{ve} = P_{\text{sat}}(T_l) \exp\left(\frac{-2v_l\sigma}{r_e RT_l}\right) \tag{5.37}$$

Because the exponential term in the above equation is less than 1, Eq. (5.37) implies, as did Eq. (5.34), that $P_{ve} < P_{\text{sat}}$. The situation regarding the state points of the vapor and liquid can be seen more clearly by considering Fig. 5.8. If the liquid state point is on the superheated liquid curve at point a, the vapor state point corresponding to an equal value of $\mu_{ve} = \mu_l$ must lie on the superheated vapor curve at point b. Hence, for a bubble with a finite radius, equilibrium can be achieved only if the liquid is superheated and the vapor is superheated, relative to the normal saturation state for a flat interface. Note also that the steep slope of the superheated vapor line results in values of P_{ve} that are much closer to $P_{\text{sat}}(T_l)$ than to P_l. Because $P_{ve} - P_l = 2\sigma/r_e$, this provides justification for the assumption that $P_{\text{sat}}(T_l) - P_{ve} \ll 2\sigma/r_e$, which was used to obtain Eq. (5.37).

Now that we have established the conditions for a bubble embryo to be in equilibrium with the surrounding superheated liquid, we wish to determine whether such an embryo can attain a stable equilibrium. This question can be addressed in an approximate way by considering the two system configurations schematically shown in Fig. 5.7. The system held at constant T_l and P_l initially contains all superheated liquid of a pure substance. We specifically wish to assess the stability of the system after the formation of a spherical embryo vapor bubble of radius r.

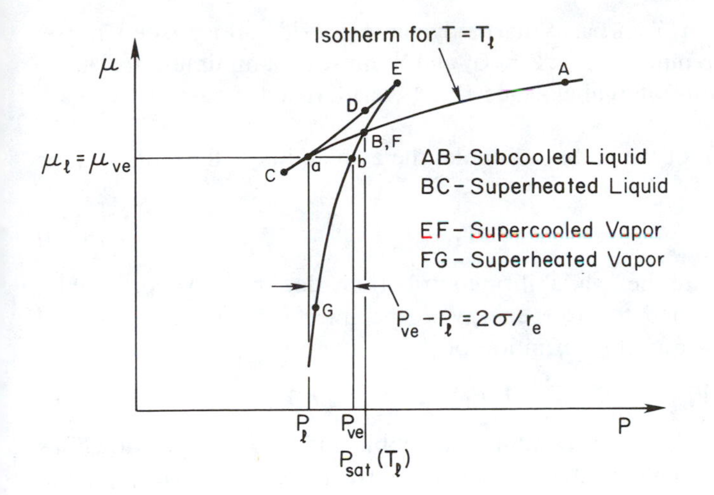

Figure 5.8 The liquid and vapor states for a vapor bubble in equilibrium with surrounding liquid.

Example 5.2 Determine the vapor pressure inside a water bubble in equilibrium with superheated liquid at 120°C and atmospheric pressure.

For water at 120°C, P_{sat} = 198.5 kPa, v_l = 0.001060 m^3/kg, σ = 0.0549 N/m, and $R = \overline{R}/\overline{M} = 8314.4/18 = 462$ J/kg. Substituting into Eq. (5.34),

$$P_{ve} = P_{sat}(T_l)\exp\left\{\frac{v_l[P_l - P_{sat}(T_l)]}{RT_l}\right\}$$

$$= 198.5\exp\left[\frac{0.001060[101{,}000 - 198{,}500]}{462(120 + 273)}\right] = 198.4 \text{ kPa}$$

The size of the bubble at equilibrium is given by Eq. (5.35) as

$$r_e = \frac{2\sigma}{P_{sat}(T_l)\exp\{v_l[P_l - P_{sat}(T_l)]/RT_l\} - P_l}$$

$$= \frac{2(0.0549)}{198{,}500\exp[0.001060(101{,}000 - 198{,}500)/462(120 + 273)] - 101{,}000}$$

$$= 1.13 \times 10^{-6} \text{ m}$$

In examining this problem, we will consider the availability function Ψ for the system, defined as

$$\Psi = U - T_l S + P_l V \tag{5.38}$$

This function is usually associated with the maximum (reversible) work that can be extracted from the system to bring it entirely to an equilibrium reference state at T_l and P_l. It is known from basic thermodynamic considerations (see Chapter One) that equilibrium requires that $d\Psi = 0$, and Ψ must be a minimum for stable equilibrium. Spontaneous internal changes always result in a decrease in the availability of the system.

After the formation of the embryo bubble, the availability is the sum of three components,

$$\Psi = \Psi_v + \Psi_l + \Psi_i \tag{5.39}$$

where Ψ_v, Ψ_l, and Ψ_i are the availability contributions due to the vapor bubble, the surrounding liquid, and the free energy in the interface (due to interfacial tension), respectively. From the definition of Ψ, Ψ_l is given by

$$\Psi_l = m_l(u_l - T_l s_l + P_l v_l) = m_l g_l(T_l, P_l) \tag{5.40}$$

where m_l is the mass of liquid surrounding the bubble and g_l is the specific Gibbs function for the liquid. The contribution due to interfacial tension is just equal to the work done to create the interface for a bubble of radius r:

$$\Psi_i = 4\pi r^2 \sigma \tag{5.41}$$

The vapor inside the bubble is at a temperature T_l equal to that of the surrounding liquid, but the pressure P_v must be different to satisfy the Young-Laplace equation $P_v = P_l + 2\sigma/r$. The availability for the vapor in the embryo is

$$\Psi_v = m_v(u_v - T_l s_v + P_l v_v)$$

which can be rearranged to obtain

$$\Psi_v = m_v[g_v(T_l, P_v) + (P_l - P_v)v_v] \tag{5.42}$$

where $g_v(T_l, P_v) = u_v - T_l s_v + P_v v_v$. Prior to formation of the embryo, the system is all superheated liquid and the availability is given by

$$\Psi_0 = (m_l + m_v)g_l(T_l, P_l) \tag{5.43}$$

In assessing the stability of the embryo bubble, we consider the change in the availability of the system associated with the formation of the embryo $\Delta\Psi$, defined as

$$\Delta\Psi = \Psi - \Psi_0 = \Psi_l + \Psi_v + \Psi_i - \Psi_0 \tag{5.44}$$

Substituting the relations (5.40) through (5.43) into (5.44) and rearranging, we obtain

$$\Delta\Psi = m_v[g_v(T_l, P_v) - g_l(T_l, P_l) + (P_l - P_v)v_v] + 4\pi r^2 \sigma \tag{5.45}$$

If the embryo is just the right size to be in equilibrium with the surrounding superheated liquid, then $P_v = P_{ve} = P_l + 2\sigma/r_e$ and the change in the availability function is equal to

$$\Delta\Psi_e = m_v\left[g_v(T_l, P_{ve}) - g_l(T_l, P_l) - \frac{2\sigma v_v}{r_e}\right] + 4\pi r_e^2 \sigma$$

For a pure substance, the chemical potential equals the specific Gibbs free energy, $\mu = g$. Because $\mu_l = g_l$ must equal $\mu_v = g_v$ at equilibrium, the g_l and g_v terms in the above equation cancel. Using the fact that $m_v v_v$ is just equal to the total volume of the embryo, the above relation for $\Delta\Psi_e$ then simplifies to

$$\Delta\Psi_e = \left(\frac{4}{3}\right)\pi r_e^2 \sigma \tag{5.46}$$

Because T_l and P_l are fixed, Eq. (5.45) indicates that $\Delta\Psi$ is a function of P_v, v_v, and r. If we assume that the vapor obeys the ideal gas law ($v_v = P_v/RT_l$) and the embryo is in mechanical equilibrium and satisfies the Young-Laplace equation ($P_v = P_l + 2\sigma/r$) regardless of size, then the dependency of $\Delta\Psi$ on P_v and v_v can be removed and $\Delta\Psi$ is only a function of r. With these idealizations, we can explore the stability of a vapor embryo with a radius near the equilibrium radius r_e by expanding $\Delta\Psi$ in a Taylor series about $r = r_e$:

$$\Delta\Psi = \Delta\Psi_e + \left(\frac{d\Delta\Psi}{dr}\right)_{r=r_e}(r - r_e) + \frac{1}{2}\left(\frac{d^2\Delta\Psi}{dr^2}\right)_{r=r_e}(r - r_e)^2 + \ldots \tag{5.47}$$

Using the assumptions described above together with the definition of g and the relation (5.45) for $\Delta\Psi$, it is a straightforward exercise to show that

$$\left(\frac{d\Delta\Psi}{dr}\right)_{r=r_e} = 0 \tag{5.48a}$$

$$\left(\frac{d^2\Delta\Psi}{dr^2}\right)_{r=r_e} = -\left(\frac{8\pi\sigma}{3}\right)\left(2 + \frac{P_l}{P_{ve}}\right) \tag{5.48b}$$

Substituting the above results into Eq. (5.47) and using Eq. (5.46) for $\Delta\Psi_e$ yields

$$\Delta\Psi = \frac{4}{3}\pi r_e^2\sigma - \frac{4\pi\sigma}{3}\left[2 + \frac{P_l}{P_{ve}}\right](r - r_e)^2 + \ldots \tag{5.49}$$

Note that the term in square brackets in Eq. (5.49) is approximately equal to 2, because $P_l \ll P_{ve}$.

The expansion (5.49) for $\Delta\Psi$ indicates that $\Delta\Psi$ has a local maximum at $r = r_e$, as indicated in Fig. 5.9. It can also be argued that in the limit $r \rightarrow \infty$, all the metastable liquid would undergo a transition to superheated vapor (the transition from b to c in Fig. 5.6), resulting in a decrease in μ and in the availability Ψ. Thus $\Delta\Psi$ must be negative for large values of r. Because we also know that $\Delta\Psi$ must approach zero as $r \rightarrow 0$, the overall variation of $\Delta\Psi$ with r must look like the solid-line representation of Eq. (5.49) plus the broken-line extensions shown in Fig. 5.9.

The variation of $\Delta\Psi$ with r shown in Fig. 5.9 has several noteworthy features. Because $\Delta\Psi$ must be at a minimum for a stable equilibrium, it is clear that an embryo of radius $r = r_e$ is in an unstable equilibrium. The loss of one molecule from the embryo decreases the radius into the range $0 < r < r_e$, where $d\Delta\Psi/dr$

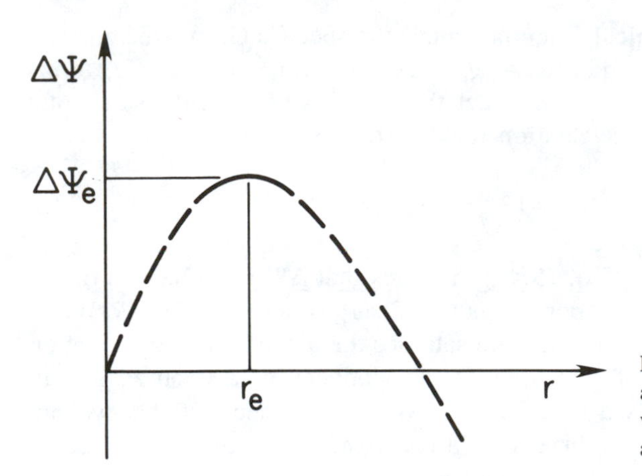

Figure 5.9 Variation of the system availability with bubble radius for a vapor bubble spontaneously formed in a superheated liquid.

is positive. Further reduction of r is expected in this range, because the system will favor spontaneous processes that decrease the availability. The end result is that, once entering this range of r, the embryo is likely to collapse completely.

If an embryo of radius $r = r_e$ gains one molecule and enters the range $r_e < r$, $\Delta\Psi$ decreases as r increases, and the embryo will spontaneously grow. Thus, if density fluctuations in a metastable liquid produce an embryo bubble of radius $r < r_e$, the bubble is likely to collapse. Conversely, if the embryo bubble has a radius greater than r_e, it will spontaneously grow, resulting in homogeneous nucleation of the vapor phase in the system.

In the above analysis, the assumption of mechanical equilibrium embodied in the use of the Young-Laplace equation is reasonable if the bubble radius readjusts rapidly to mechanical force variations. However, if the mechanical forces change rapidly and the liquid is too viscous to quickly readjust, then mechanical equilibrium may not be maintained. For such conditions, a better assumption would be that the values of the chemical potential in the two phases are equal. Use of this idealization, along with the others noted above, results in expansion for $\Delta\Psi$ that is identical to Eq. (5.49) except that the term in square brackets is simply equal to 3. Zeldovich [5.2] and Kagen [5.3] have shown that this modified form of Eq. (5.49) is more appropriate for cavitation processes. Regardless of whether chemical or mechanical equilibrium is assumed, the stability characteristics of the vapor embryos are qualitatively the same, and quantitatively only slightly different.

Example 5.3 Estimate the number of molecules needed to make up a bubble of critical size for superheated water at atmospheric pressure and 250°C.

The radius of a bubble of critical size is just that for a bubble in equilibrium, given by Eq. (5.35):

$$r_e = \frac{2\sigma}{P_{\text{sat}}(T_l)\exp\{v_l[P_l - P_{\text{sat}}(T_l)]/RT_l\} - P_l}$$

For water at 250°C, $P_{\text{sat}} = 3878$ kPa, $v_l = 0.001251$ m^3/kg, $\sigma = 0.0262$

N/m, and $R = \overline{R}/\overline{M} = 8314.4/18 = 462$ J/kg. Substituting into the above equation:

$$r_e = \frac{2(0.0262)}{3{,}978{,}000 \exp[0.001251(101{,}000 - 3{,}978{,}000)/462(250 + 273)] - 101{,}000}$$

$$= 1.38 \times 10^{-8} \text{ m}$$

The pressure inside the bubble is given by

$$P_v = P_l + \frac{2\sigma}{r_e} = 101{,}000 + \frac{2(0.0262)}{1.38 \times 10^{-8}} = 3{,}900{,}000 \text{ Pa}$$

and the specific volume of the gas is found from the ideal gas law:

$$v_v = \frac{RT}{P_v} = \frac{(462)(250 + 273)}{3{,}900{,}000} = 0.0620 \text{ m}^3/\text{kg}$$

The number of molecules in the bubble, n_e, is found as

$$n_e = \frac{V_d N_A}{v_v \overline{M}}$$

where N_A and $\overline{M}$ are Avogadro's number and the molecular weight, respectively, and the bubble volume V_d is equal to $4\pi r_e^3/3 = 1.10 \times 10^{-23}$ m^3. Substituting, it is found that

$$n_e = \frac{1.10 \times 10^{-23}(6.02 \times 10^{26})}{0.0620(18)} = 5940 \text{ molecules}$$

Thus, to form an embryo bubble of critical size requires about 6000 molecules—a relatively small number.

5.3 THE KINETIC LIMIT OF SUPERHEAT

The results of the analysis presented in the preceding section indicate that the likelihood that homogeneous nucleation will occur depends on the kinetics of the vapor embryo formation process. A vapor embryo of n vapor molecules is continuously gaining molecules due to evaporation and losing molecules due to condensation at the bubble interface. The difference between the rates of these two processes dictates whether the embryo will grow or decrease in size.

We consider a superheated liquid held at constant T_l and P_l in which heterophase fluctuations give rise to vapor embryos of various sizes. Let us first consider an idealized model of such a system in which an equilibrium distribution of embryo sizes is established. It is plausible to expect that the probability of finding an embryo of a particular size in such a system will vary with the change in the availability associated with the formation of the embryo. We postulate, therefore, that the number distribution of embryos for such an equilibrium system would be of the form

$$N_n = N_l \exp\left[-\frac{\Delta\Psi(r)}{k_B T_l}\right] \tag{5.50}$$

where N_n is the number of embryos of n molecules at equilibrium per unit volume, N_l is the number of liquid molecules per unit volume, and $\Delta\Psi(r)$ is the availability function previously defined. Note that r is related to the number of molecules n in the embryo by

$$\frac{\frac{4}{3}\pi r^3}{v_v} = nm = n\overline{M}/N_A \tag{5.51}$$

where m is the mass of one molecule, $\overline{M}$ is the molecular mass (kg/kg mol), and N_A is Avogadro's number (6.02×10^{26} molecules/kg mol). The number density of liquid molecules is given by $N_l = N_A/v_l\overline{M}$.

For an embryo of size n, j_{ne} is the number of molecules evaporating from the interface per unit time, per unit area, and j_{nc} is the number of molecules condensing at the interface per unit time, per unit area. An equilibrium number distribution can be established only if the following condition is satisfied:

$$N_n A_n j_{ne} = N_{n+1} A_{n+1} j_{(n+1)c} \tag{5.52}$$

In Eq. (5.52), A_n and A_{n+1} are the interfacial areas of embryos of sizes n and $n + 1$, respectively. This equality enforces the condition that the rate at which embryos of size n are converted to size $n + 1$ by evaporation is equal to the rate at which embryos of size $n + 1$ are converted to size n by condensation. As a result, the equilibrium number of embryos in each size category N_n is maintained, and there is no net flux of bubbles through the range of sizes (i.e., there is no steady stream of embryos that grow progressively in size).

For a real superheated liquid, the system is not expected to be in equilibrium, and Eq. (5.52) is not necessarily satisfied. Denoting the difference between the right and left sides of Eq. (5.52) as J_n, and the size distribution of embryos as N^* for the nonequilibrium circumstances, we can write that

$$J_n = N_n^* A_n j_{ne} - N_{n+1}^* A_{n+1} j_{(n+1)c} \tag{5.53}$$

Note that J_n is the excess number of embryos of size n that (due to evaporation) pass to size $n + 1$ over the number of size $n + 1$ that (due to condensation) pass to size n. Thus J_n is the net flux of the number of embryos in size space from n to $n + 1$. Using Eq. (5.52) to eliminate A_{n+1}, Eq. (5.53) can be rearranged to obtain

$$J_n = N_n A_n j_{ne}\left[\left(\frac{N_n^*}{N_n}\right) - \left(\frac{N_{n+1}^*}{N_{n+1}}\right)\right] \tag{5.54}$$

For an unsteady size distribution, the rate of change of the number of embryos in each category $\partial N_n^*/\partial t$ is given by

$$\frac{\partial N_n^*}{\partial t} = J_{n-1} - J_n \tag{5.55}$$

To facilitate further analysis of the kinetics of the embryo formation process, we will treat n as a continuous variable, whereupon Eqs. (5.54) and (5.55) can be rewritten as

$$J(n) = -N(n)A(n)j_e(n)\frac{\partial[N^*(n)/N(n)]}{\partial n} \tag{5.56}$$

$$\frac{\partial N^*(n)}{\partial t} = -\frac{\partial J(n)}{\partial n} \tag{5.57}$$

To simplify the analysis somewhat, we allow a steady flow of embryos through size space, but we insist that the number distribution of embryos is steady, so that $\partial N^*(n)/\partial t = 0$. From Eq. (5.57) it follows that

$$\frac{\partial J(n)}{\partial n} = 0 \tag{5.58}$$

which implies that

$$J(n) = \text{constant} = J \tag{5.59}$$

We know from the thermodynamic analysis presented above that embryos larger in size than the critical radius will spontaneously grow and allow the superheated liquid system to undergo a transition to an equilibrium state. Consequently, if we consider such a superheated system prior to such an event, there can be no embryos present with $r > r_e$. We therefore assume that

$$N^* \rightarrow 0 \qquad \text{at } n = n_0 \simeq n_e \tag{5.60}$$

where n_e is the number of molecules in an embryo of radius $r = r_e$ for the actual superheated liquid.

The qualitative variation of $N(n)$ is indicated in Fig. 5.10. Note that because $\Delta\Psi$ has a maximum at $r = r_e$, there is a corresponding minimum in N at n_e. The expected variation of N^* with n is also qualitatively shown in Fig. 5.10. N^* goes to zero at $n = n_0$, based on the reasoning described above. The number densities of very small embryos (small n) are so large that a departure from equilibrium will likely have little effect on their numbers. We therefore postulate that

$$\frac{N^*}{N} \rightarrow 1 \qquad \text{as } n \rightarrow 0 \tag{5.61}$$

as graphically suggested in Fig. 5.10.

Integrating Eq. (5.56) and using the condition (5.60), the following relation is obtained:

$$\frac{N^*(n)}{N(n)} = J\int_{n=n}^{n=n_0} [N(n)A(n)j_e(n)]^{-1}\, dn \tag{5.62}$$

which can be rearranged to yield

$$J = \left[\frac{N^*(n)}{N(n)}\right]\left\{\int_{n=n}^{n=n_0} [N(n)A(n)j_e(n)]^{-1}\, dn\right\}^{-1} \tag{5.63}$$

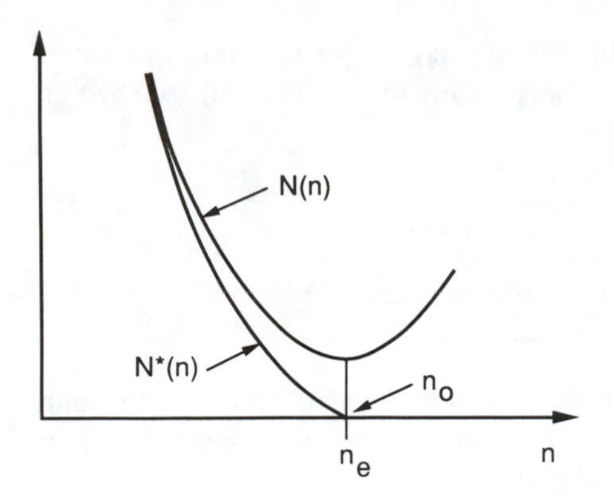

Figure 5.10 Qualitative variations of $N(n)$ and $N^*(n)$.

Because $N(n)$ possesses a sharp minimum at $n = n_e$, the integrand in Eq. (5.63) is significantly greater than zero only in the vicinity of $n = n_e$. We therefore take $j_e(n)$ to be equal to its value at $n = n_e$, and we use Eq. (4.129*a*), developed using kinetic theory of gases in Chapter Four, to evaluate $j_e(n_e)$:

$$j_e(n) \simeq j_e(n_e) = \frac{P_{ve}}{(2\pi m k_B T_l)^{1/2}} \tag{5.64}$$

Note that in obtaining the above approximation for $j_e(n_e)$, we have taken $\Gamma(a) = 1$ and rearranged Eq. (4.129*a*) slightly, writing it in terms of the Boltzmann constant $k_B = \overline{R}/N_A$.

In addition, because the integrand vanishes rapidly as $|n - n_e|$ becomes large, we can take the limits of integration to be 0 and ∞ with no significant loss of accuracy. Also, since J is a constant (independent of the value of n), we can elect to evaluate the right side of Eq. (5.63) at small n, where $N^*(n)/N(n) \rightarrow 1$.

Evaluating the right side of Eq. (5.63) as described above, we obtain the following simpler relation for J:

$$J = \left[\frac{P_{ve}}{(2\pi m k_B T_l)^{1/2}}\right]\left\{\int_0^\infty [N(n)A(n)]^{-1}\, dn\right\}^{-1} \tag{5.65}$$

The integral on the right side of Eq. (5.65) can be converted to an integral with respect to r using Eq. (5.50) for N and the relations

$$A = 4\pi r^2 \tag{5.66}$$

and

$$dn = \tfrac{4}{3}\pi r^2 \left(\frac{P_{ve}}{RT_l m}\right)\left(2 - \frac{P_l}{P_{ve}}\right) dr \tag{5.67}$$

where the second relation (5.67) is obtained by differentiating Eq. (5.51) and using the ideal gas law and the Young-Laplace equation to evaluate the dependence of v_v on r. After converting the integral in this manner, Eq. (5.65) becomes

$$J = \left(\frac{3N_l}{2 - P_l/P_{ve}} \right) \left(\frac{k_B T_l}{2\pi m} \right)^{1/2} \left\{ \int_0^\infty \exp\left[\frac{\Delta\Psi(r)}{k_B T_l} \right] dr \right\}^{-1} \tag{5.68}$$

Because of the maximum in $\Delta\Psi$ at $r = r_e$, the integrand takes on its greatest value at $r = r_e$ and rapidly becomes smaller as $|r - r_e|$ increases. Hence the value of the integral is dictated almost entirely by the variation of the integrand near $r = r_e$. The expansion (5.49) for $\Delta\Psi$ about $r = r_e$ is therefore adequate for evaluating the integral. We proceed by substituting the expansion for $\Delta\Psi$ into Eq. (5.68), changing the integration variable to $\xi = [4\pi\sigma(2 - P_l/P_{ve})/3k_B T_l]^{1/2}(r - r_e)$, and taking the lower integration limit in terms of ξ to be $-\infty$ (because the integrand goes rapidly to zero as $|\xi|$ becomes large). The integral can then be evaluated in closed form. The resulting equation for J is

$$J = N_l \left[\frac{6\sigma}{\pi m(2 - P_l/P_{ve})} \right]^{1/2} \exp\left(\frac{-4\pi r_e^2 \sigma}{3k_B T_l} \right) \tag{5.69}$$

Using the Young-Laplace equation to evaluate r_e and neglecting P_l/P_{ve} compared to 2, the relation for J can be written as

$$J = N_l \left(\frac{3\sigma}{\pi m} \right)^{1/2} \exp\left[\frac{-16\pi\sigma^3}{3k_B T_l (P_{ve} - P_l)^2} \right] \tag{5.70}$$

From Eq. (5.34), we can write

$$P_{ve} = \eta P_{sat}(T_l) \tag{5.71}$$

where

$$\eta = \exp\left\{ \frac{v_l [P_l - P_{sat}(T_l)]}{RT_l} \right\} \tag{5.72}$$

Substituting Eq. (5.71) into Eq. (5.70), we obtain

$$J = N_l \left(\frac{3\sigma}{\pi m} \right)^{1/2} \exp\left\{ \frac{-16\pi\sigma^3}{3k_B T_l [\eta P_{sat}(T_l) - P_l]^2} \right\} \tag{5.73}$$

The above equation for J is basically the same as that obtained by Kagen [5.3] and by Katz and Blander [5.4]. Note that J represents the rate at which embryo bubbles grow from n to $n + 1$ molecules per unit volume of liquid. Because in this steady-state solution J is constant for all n, it is also equal to the rate at which bubbles of critical size are generated. Hence, as J increases, the probability that a bubble will exceed critical size and grow spontaneously also increases.

Because most of the temperature-dependent quantities appear in the exponential term in Eq. (5.73), a slight change in T_l can have a very strong effect on the rate of nucleation. For typical circumstances, a change of 1°C can change J by as much as three or four orders of magnitude. Hence, although the embryo bubble formation rate increases continuously with temperature, the rate of nucleation changes so rapidly with temperature that we expect that there will exist a narrow range of temperature below which homogeneous nucleation does not occur, and above which it occurs almost immediately. The median temperature of this range is usually referred to as the *kinetic limit of superheat*.

Equation (5.73) can be written in laboratory units as

$$J = 1.44 \times 10^{40}\left(\frac{\rho_l^2\sigma}{\overline{M}^3}\right)^{1/2} \exp\left\{\frac{-1.213 \times 10^{24}\sigma^3}{T_l[\eta P_{\text{sat}}(T_l) - P_l]^2}\right\} \tag{5.74}$$

where J is in $m^{-3}\,s^{-1}$, P_{sat} and P_l are in Pa, T_l is in K, σ is in N/m, ρ_l is in kg/m^3, and $\overline{M}$ is in kg/kg mol. For a specific pure substance, if the physical properties are known and a suitable threshold value of J is chosen, then Eq. (5.74) can be solved iteratively to determine the limiting superheat temperature, $T_l = T_{SL}$.

The above analysis of the kinetics of the bubble formation process can also be made to apply to a cavitation process. It can be shown that doing so amounts to using the form of the expansion for $\Delta\Psi$ that is applicable to such processes (see Section 5.2) and taking the term in square brackets in Eq. (5.67) to be equal to 3. The resulting expression for J is identical to Eq. (5.73) or (5.74), except that the factor 3 inside the square brackets is replaced by a 2.

Example 5.4 Estimate the kinetic limit of superheat for water at 1000 kPa.

At 1000 kPa, $T_{\text{sat}} = 180°\text{C} + 273 = 453$ K. The kinetic limit of superheat is defined by Eqs. (5.72) and (5.74) with the threshold value of $J = 10^{12}$.

$$J = 1.44 \times 10^{40}\left[\frac{\rho_l^2\sigma}{\overline{M}^3}\right]^{1/2} \exp\left\{\frac{-1.213 \times 10^{24}\sigma^3}{T_l[\eta P_{\text{sat}}(T_l) - P_l]^2}\right\}$$

$$\eta = \exp\left[\frac{P_l - P_{\text{sat}}(T_l)}{\rho_l R T_l}\right]$$

Note that varying T_l causes ρ_l, $P_{\text{sat}}(T_l)$, and σ to vary in the above equation. Using values of ρ_l, $P_{\text{sat}}(T_l)$, and σ from the saturation table for water at various T_l values, the resulting values of $\log_{10}J$ are

T_l °(C)	$P_{\text{sat}}(T_l)$ (kPa)	ρ_l (kg/m^3)	σ (N/m)	$\log_{10}J$
300	8592	712	0.0144	−12.0
310	9869	691	0.0121	17.5
305	9231	702	0.0133	5.24

Interpolating between 305°C and 310°C to determine the value of T_l that corresponds to $\log_{10}J = 12$ yields $(T_l)_{SL} = 308°\text{C}$.

5.4 COMPARISON OF THEORETICAL AND MEASURED SUPERHEAT LIMITS

We now have two theoretical means of predicting the limiting superheat for metastable liquid: the thermodynamic or intrinsic limit conditions derived in Section

5.2, and the kinetic limits specified by the analysis in Section 5.3. We will now examine how each of these theoretical limits compares to experimental measurements.

The van der Waals and Berthelot equations of state can both be represented by the relation

$$(P + aT^{-\lambda}\overline{v}^{-2})(\overline{v} - b) = \overline{R}T \tag{5.75}$$

where $\lambda = 0$ corresponds to the van der Waals equation, and $\lambda = 1$ yields the Berthelot equation. If we further require that $(\partial P/\partial v)_T = 0$ and $(\partial^2 P/\partial v^2)_T = 0$ at the critical point, the coefficients a and b can be specified in terms of the critical properties, and the equation of state can be written as

$$(P_r + 3T_r^{-\lambda}v_r^{-2})(v_r - \tfrac{1}{3}) = (\tfrac{8}{3})T_r \tag{5.76}$$

where

$$P_r = \frac{P}{P_c} \qquad T_r = \frac{T}{T_c} \qquad v_r = \frac{v}{v_c} \tag{5.77}$$

In terms of reduced properties, the condition defining the spinodal limit $(\partial P/\partial v)_T = 0$ can be written as

$$\left(\frac{\partial P_r}{\partial v_r}\right)_{T_r} = 0 \tag{5.78}$$

Solving Eq. (5.76) for P_r, differentiating with respect to v_r, and setting the result equal to zero, the following relation is obtained for the spinodal limits:

$$(T_r)_s = \left[\frac{(3v_r - 1)^2}{4v_r^3}\right]^{1/(\lambda+1)} \tag{5.79}$$

Once a value of the exponent λ is specified, T_r can be computed for a given value of v_r using Eq. (5.79), and then Eq. (5.76) can be used to compute the corresponding P_r. The variation of T_r as a function of P_r along the spinodal curve determined in this manner is shown in Fig. 5.11 for $\lambda = 0$ and $\lambda = 1$, corresponding to the van der Waals and Berthelot equations, respectively. Also shown in Fig. 5.11 are measured superheat limit data summarized in reference [5.5] for various liquids at atmospheric pressure.

The spinodal curves in Fig. 5.11 imply that a liquid can generally be superheated to more than 80% of its critical temperature before the spinodal limit is reached. The thermodynamic limit of superheat computed from the Berthelot equation is consistently higher than that computed from the van der Waals equation at a given reduced pressure.

Experimental observations of the superheat limit temperature have typically been obtained using a test apparatus in which a drop of the liquid to be tested is injected through a capillary tube into the bottom of a vertical column filled with an immiscible liquid having a higher density and boiling point. Because of the density difference, the small drop rises slowly through the fluid in the column. The temperature in the column is controlled with external heaters so that the top

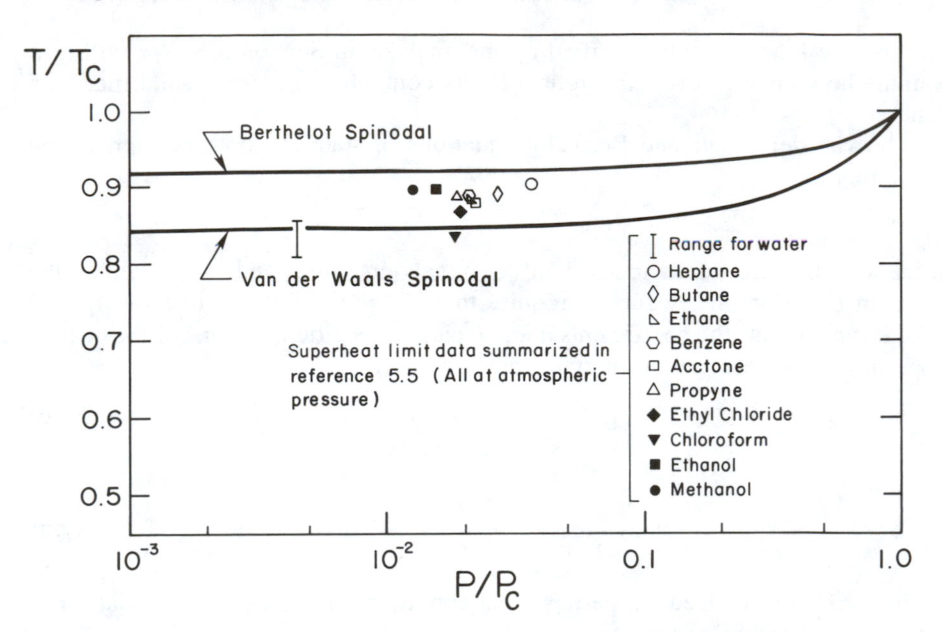

Figure 5.11 Comparison of measured superheat limit data with spinodal limit predictions.

is hotter than the bottom. The droplet is slowly heated as it rises, until homogeneous nucleation causes it to explode. Thermocouples at a number of locations indicate the temperature variation along the column, which in turn is used to determine the temperature of the droplet at the location where it is observed to explode.

The superheat limits determined in this manner for a number of fluids are plotted in Fig. 5.11. The experimentally observed superheat limits for a number of fluids exceed the spinodal limit predicted by the van der Waals equation, implying that this equation does not provide an accurate indication of the thermodynamic limit of superheat. The experimental data are all significantly below the spinodal limit predicted using the Berthelot equation. This is, however, consistent with the spinodal being an upper limit that cannot be exceeded. As a more accurate means of predicting experimentally observed superheat limits, Eberhart and Schnyders [5.6] suggest using Eqs. (5.76) and (5.79) with $\lambda = \frac{1}{2}$. The resulting superheat limit curve is halfway between the two curves shown in Fig. 5.11, and agrees fairly well with experimental data for a wide variety of fluids.

Equation (5.74) can be solved iteratively to determine the kinetic limit of superheat, if a threshold value of J corresponding to the onset of homogeneous nucleation is assumed. Blander and Katz [5.5] have compared the kinetic limit of superheat T_{SL} predicted in this manner for $J = 10^{12}$ m^{-3} s^{-1} with experimentally observed superheat limits at atmospheric pressure for a wide variety of hydrocarbon fluids. The intersection of Eq. (5.74) with this threshold value for several common fluids at atmospheric pressure is shown graphically in Fig. 5.12. A comparison of the measured superheat limits with those predicted by Eq. (5.74) with $J = 10^{12}$ is shown in Fig. 5.13 for some of the fluids considered by Blander and Katz [5.5].

The excellent agreement between the predicted and measured values of T_{SL} seen in Fig 5.13 was typical of the results for virtually all the organic fluids considered by Blander and Katz [5.5]. For water, however, the agreement is not quite as good. Equation (5.74) with $J = 10^{12}$ m^{-3} s^{-1} predicts a superheat limit of about 300°C for water at atmospheric pressure. Attempts to experimentally determine the superheat limit for water at atmospheric pressure by Blander et al. [5.7] and Apfel [5.8] produced measured values of T_{SL} between 250°C and 280°C.

While the discussion here has focused on systems with positive values of pressure, the results of these analyses can also be used to predict homogeneous nucleation in liquids subjected to negative pressure (i.e., in tension). Eberhart and Schnyders [5.6] found that the superheat limit data obtained by Apfel [5.8] at negative pressures fell between the van der Waals ($\lambda = 0$) and Berthelot ($\lambda = 1$) spinodal curves defined by Eqs. (5.76) and (5.79) for $P_r < 0$. Superheat limit data obtained by Skripov [5.9] and Avedisian [5.10] for pure liquids at high positive pressures are also consistent with the theoretical relations derived above.

The above analysis of the kinetics of the nucleation process in a superheated

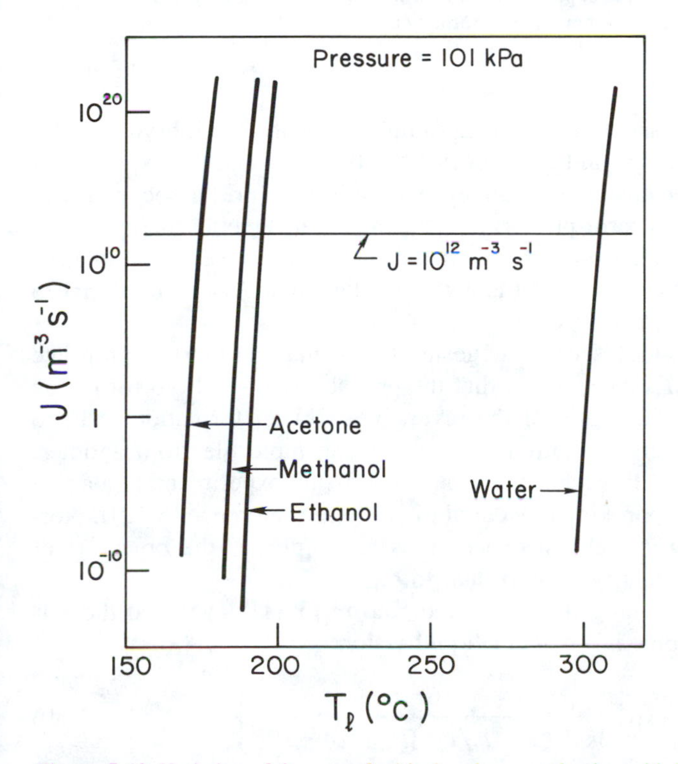

Figure 5.12 Variation of the rate of critical embryo production with liquid temperature predicted by analysis of the kinetics of vapor embryo formation for several fluids.

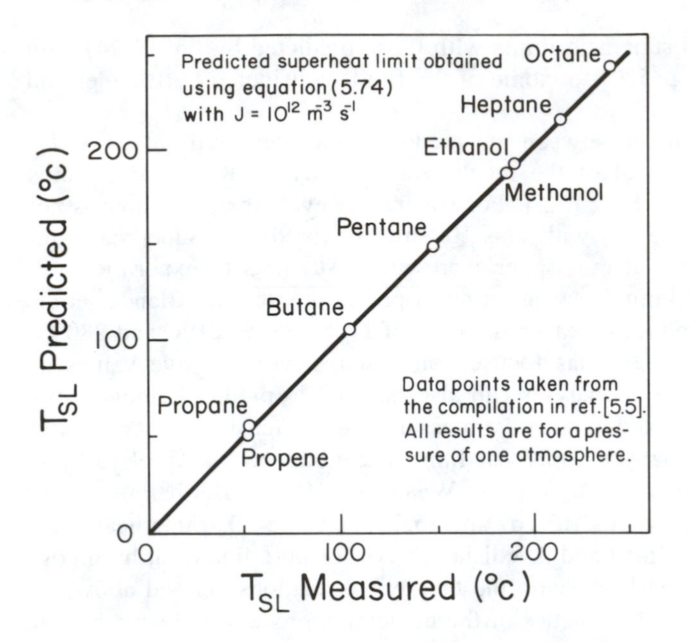

Figure 5.13 Comparison of measured superheat limit data with the superheat limit temperature predicted by analysis of the kinetics of vapor embryo formation.

liquid makes use of the assumption that the number density of embryo bubbles is proportional to $e^{-(\Delta\Psi/k_B T_l)}$, as indicated in Eq. (5.50). Note that the exponent of this term is the ratio of the change in availability (reversible work) associated with the embryo formation to a term proportional to the mean kinetic energy of the molecules. As a result of this hypothesis, which has become conventional in analyses of this type [5.2–5.4], the model analysis predicts that J is proportional to $e^{-(\Delta\Psi_e/k_B T_l)}$ (see Eqs. [5.46] and [5.69]).

Lienhard and Karimi [5.11] have suggested that it may be more appropriate to normalize $\Delta\Psi$ with $k_B T_c$. They argue that the probability of embryo formation ought to be dependent on the ratio of the reversible work of formation ($\Delta\Psi$) to the fixed value of the energy required to separate one molecule from another, which is denoted here as ϵ. It has been demonstrated using experimental data that for many substances ϵ is approximately equal to $\frac{3}{4}k_B T_c$ (see reference [5.12]). Normalizing $\Delta\Psi$ with $k_B T_c$ for such substances thus accomplishes the objective of normalizing $\Delta\Psi$ with a quantity proportional to ϵ.

Using this line of reasoning, Lienhard and Karimi [5.11] developed the following relation for the spinodal limit for liquid water:

$$10^{-5} = \exp\left\{\frac{-16\pi\sigma^3}{3k_B T_c(1 - v_l/v_v)^2[P_{\text{sat}}(T_l) - P_l]^2}\right\} \tag{5.80}$$

Equation (5.80) was found to closely match the spinodal limits obtained using isotherms fit to superheated liquid data at negative pressures. The spinodal limit for positive pressures generally corresponds to $T_{SL} > 0.8T_c$ (see Fig. 5.11). As a result, replacing $k_B T_l$ with $k_B T_c$ in Eq. (5.73) only slightly changes the argument of the exponential term. The choice of the threshold value of J could be altered to compensate, resulting in a relation that agrees as well with superheat data as Eq. (5.74). However, at negative pressures the superheat limit temperature may be substantially lower than T_c, and the choice of $k_B T_c$ over $k_B T_l$ as the quantity used to normalize $\Delta\Psi$ may have a greater impact on the predictive capabilities of the relation for the spinodal limit.

Treatment of homogeneous nucleation in multicomponent systems is beyond the scope of this text. It is worth noting, however, that the analytical techniques described here for pure liquids can also be extended to nucleation processes in multicomponent liquid mixtures. More information on nucleation in multicomponent systems can be obtained in references [5.5, 5.13–5.17].

The practical significance of homogeneous nucleation processes in liquids is usually related to the fact that, when it does occur, vapor is generated at an extremely rapid rate. In some instances, vapor explosions may occur, with disastrous consequences for nearby equipment and/or people. Perhaps the most well-known example is the vapor explosion that can result from a spill of cryogenically liquefied natural gas (LNG) on water. The explosive boiling resulting from homogeneous nucleation can generate shock waves much like a chemical explosion, with the same potentially destructive result. Further discussion of bubble growth after the initiation of nucleation is presented in Chapter Six.

5.5 THERMODYNAMIC ASPECTS OF HOMOGENEOUS NUCLEATION IN SUPERCOOLED VAPOR

The vapor phase of a pure substance may be brought to a supercooled or supersaturated state either by transferring heat through the walls of a containing structure or by rapidly dropping the pressure of the gas. For either type of process, once a saturated state is reached, further reduction in the system pressure or further removal of heat may initiate condensation of some vapor to liquid if nuclei of the liquid phase are present in the system. Often there are adsorbed molecules in a near-liquid state on the containment walls (see Section 3.4) or on dust particles suspended in the vapor that serve as nuclei. However, in the absence of these types of nuclei, the processes mentioned above may reduce the pressure or cool the vapor to a statepoint in the metastable vapor region of Fig. 5.4.

Heterophase fluctuations of the density of a metastable supersaturated vapor may result in the formation of an embryo droplet that may initiate the condensation process. Such a homogeneous nucleation process will, as in the case of vaporization, depend on the kinetics of the embryo formation process in the vapor. To explore these matters further, we will specifically consider the system shown in Fig. 5.14 containing a liquid droplet of radius r in equilibrium with a surrounding vapor held at fixed temperature T_v and pressure P_v.

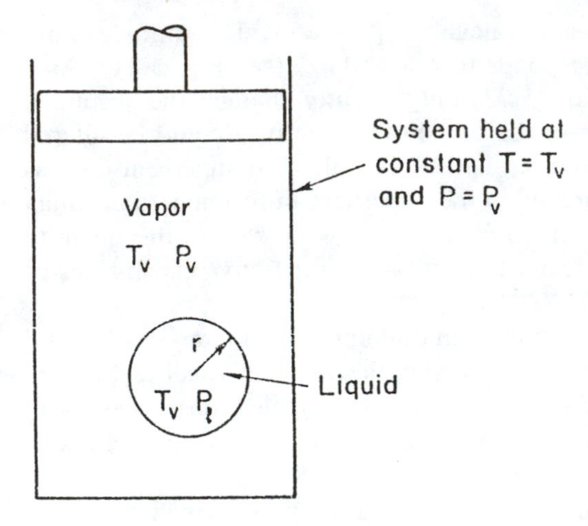

Figure 5.14 System model considered in the thermodynamic analysis of the formation of an embryo liquid droplet.

At equilibrium, the temperature and chemical potential in the vapor and the droplet must be equal:

$$\mu_v = \mu_l \tag{5.81}$$

The pressures in the two phases are related through the Young-Laplace equation (2.30):

$$P_{le} = P_v + \frac{2\sigma}{r_e} \tag{5.82}$$

Using the integrated form of the Gibbs-Duhem equation for a constant temperature process (derived in Section 5.2),

$$\mu - \mu_{\text{sat}} = \int_{P_{\text{sat}}(T_v)}^{P} v\, dP \tag{5.83}$$

We evaluate the integral on the right side using the ideal gas law ($v = RT_v/P$) for the vapor to obtain

$$\mu_{ve} = \mu_{\text{sat},v} + RT_v \ln\left[\frac{P_v}{P_{\text{sat}}(T_v)}\right] \tag{5.84}$$

For the liquid phase inside the droplet, the chemical potential can again be evaluated using Eq. (5.83). The liquid is taken to be incompressible, with v equal to its value for saturated liquid at T_v. With this assumption, evaluation of the integral in Eq. (5.83) for $P = P_{le}$ yields

$$\mu_{le} = \mu_{\text{sat},l} + v_l[P_{le} - P_{\text{sat}}(T_v)] \tag{5.85}$$

Equating the values of μ_{ve} and μ_{le} given by Eqs. (5.84) and (5.85) to satisfy Eq. (5.82), and using the fact that $\mu_{\text{sat},v} = \mu_{\text{sat},l}$, the following relation is obtained:

$$P_v = P_{sat}(T_v) \exp\left\{\frac{v_l[P_{le} - P_{sat}(T_v)]}{RT_v}\right\} \tag{5.86}$$

Comparison of the above relation with Eq. (5.34) reveals that the relation (5.86) for the equilibrium vapor pressure surrounding a small liquid droplet is identical to that for the pressure inside a small vapor bubble in equilibrium with a surrounding liquid pool. As seen in Fig. 5.15, if the vapor statepoint is on the metastable supercooled vapor curve at point a, the liquid state corresponding to equal μ must lie on the subcooled liquid line at point b. For a liquid droplet with finite radius, equilibrium can therefore be achieved only if the liquid is subcooled and the vapor is supersaturated, relative to the normal saturation state for a flat interface. Equation (5.86) also indicates that if P_v is greater than $P_{sat}(T_v)$, then P_{le} must also be greater than $P_{sat}(T_v)$, which is consistent with the statepoints indicated in Fig. 5.15.

Substituting Eq. (5.82) to eliminate P_{le}, Eq. (5.86) becomes

$$P_v = P_{sat}(T_v) \exp\left\{\frac{v_l[P_v - P_{sat}(T_v) + 2\sigma/r_e]}{RT_v}\right\} \tag{5.87}$$

In most instances, the steep slope of the supercooled vapor line in Fig. 5.15 results in values of P_v that are much closer to $P_{sat}(T_v)$ than to P_{le}. When this is the case, $P_v - P_{sat}(T_v)$ is small compared to $2\sigma/r_e = P_{le} - P_v$, and the relation (5.87) for

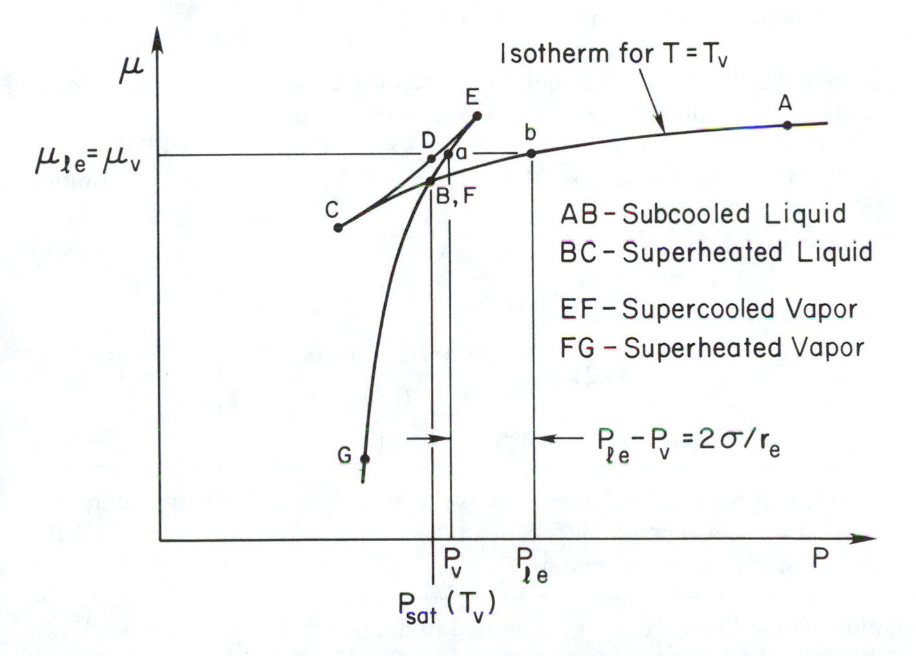

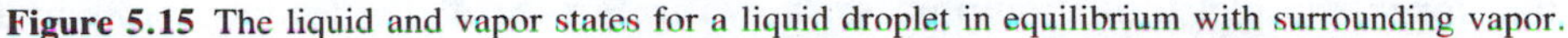

Figure 5.15 The liquid and vapor states for a liquid droplet in equilibrium with surrounding vapor.

P_v is well approximated by

$$P_v = P_{sat}(T_v) \exp\left(\frac{2v_l\sigma}{r_e RT_v}\right) \tag{5.88}$$

Equations (5.87) and (5.88) can also be inverted to solve for the equilibrium droplet radius for a given set of pressure and temperature conditions:

$$r_e = \frac{2\sigma}{(RT_v/v_l)\ln[P_v/P_{sat}(T_v)] - P_v + P_{sat}(T_v)} \tag{5.89}$$

$$r_e = \frac{2\sigma v_l}{RT_v \ln[P_v/P_{sat}(T_v)]} \qquad \text{for } P_v - P_{sat}(T_v) \ll \frac{2\sigma}{r_e} \tag{5.90}$$

The stability of an embryo droplet that forms in a system initially containing supercooled vapor can be analyzed in a manner almost identical to that presented in Section 5.2. Such an analysis again considers the change in the availability function $\Delta\Psi$ associated with the formation of an embryo droplet in a system held at constant temperature T_v and pressure P_v as indicated in Fig. 5.14. Proceeding in the manner described in Section 5.2, it can be shown that the change in the availability function associated with the formation of a droplet with just the right radius r_e to be in equilibrium with the surrounding vapor is given by

$$\Delta\Psi_e = \tfrac{4}{3}\pi r_e^2 \sigma \tag{5.91}$$

The above equation for $\Delta\Psi_e$ is, in fact, identical to the corresponding result obtained in the analysis of the embryo bubble.

Example 5.5 Determine the equilibrium vapor pressure for nitrogen vapor in equilibrium with droplets having a diameter of 0.5 μm at 80 K.

For saturated nitrogen at 80 K, $P_{sat} = 136.2$ kPa, $v_l = 0.001239$ m^3/kg, $\sigma = 0.00827$ N/m, and $R = \overline{R}/\overline{M} = 8314.4/28.0 = 296.9$ J/kg. Substituting into Eq. (5.88):

$$P_v = P_{sat}(T_v) \exp\left(\frac{2v_l\sigma}{r_e RT_v}\right)$$

$$= (136.2) \exp\left[\frac{2(0.001239)(0.00827)}{(0.5 \times 10^{-6})(296.9)(80)}\right]$$

$$= 136.2(1.0017) = 136.4 \text{ kPa}$$

Note that because of the very low surface tension at this temperature, the equilibrium vapor pressure differs by only a small amount from that for a flat interface at the same temperature.

Following the line of analysis described in Section 5.2, it can also be shown that for the droplet formation process shown in Fig. 5.14, the leading terms in a

Taylor series expansion for $\Delta\Psi$ about $r = r_e$ are

$$\Delta\Psi = \tfrac{4}{3}\pi r_e^2\sigma - 4\pi\sigma(r - r_e)^2 + \dots \tag{5.92}$$

These results imply that the variation of $\Delta\Psi$ with r is qualitatively similar to that shown in Fig. 5.9. The expansion (5.92) indicates that $\Delta\Psi$ has a local maximum at $r = r_e$. In the limit $r \to \infty$, all the metastable vapor would undergo a transition to subcooled liquid (the transition from b' to c' in Fig. 5.6), resulting in a decrease in μ and in the availability. Thus $\Delta\Psi$ must be negative for large r. Because $\Delta\Psi$ must go to zero as $r \to 0$, the overall variation of $\Delta\Psi$ with r must again look the curve indicated in Fig. 5.9.

The interpretation of the variation of $\Delta\Psi$ with r for the embryo droplet is also similar to that for the embryo bubble considered in Section 5.2. Because stable equilibrium corresponds to a minimum value of $\Delta\Psi$, it is clear that a droplet embryo of radius $r = r_e$ is in an unstable equilibrium. The loss of one molecule from the embryo will decrease its radius into the range $0 < r < r_e$, where $d\Delta\Psi/dr$ is positive and further reduction in size is favored due to the associated decrease in $\Delta\Psi$. Once entering this range of r, it is likely that the embryo will disappear.

If an embryo droplet of radius $r = r_e$ gains one molecule, it enters the range of $r > r_e$, where $\Delta\Psi$ decreases as r increases, which favors further growth of the droplet. Hence if density fluctuations in a metastable vapor result in the formation of an embryo droplet of radius $r < r_e$, the droplet is likely to lose molecules and disappear. If its radius is greater than r_e, it will grow spontaneously, initiating the condensation process in the vapor via homogeneous nucleation.

Example 5.6 Determine the critical radius of a water droplet for supersaturated steam at atmospheric pressure and a temperature of 80°C.

The critical radius is the equilibrium radius given by Eq. (5.89):

$$r_e = \frac{2\sigma}{(RT_v/v_l)\ln[P_v/P_{\text{sat}}(T_v)] - P_v + P_{\text{sat}}(T_v)}$$

For saturated water at $T_v = 80$°C, $P_{\text{sat}} = 47{,}400$ Pa, $v_l = 0.001029\ \text{m}^3/\text{kg}$, $\sigma = 0.0629$ N/m, and $R = \overline{R}/\overline{M} = 8314.4/18 = 462$ J/kg. Substituting into the above equation,

$$r_e = \frac{2(0.0626)}{[462(80 + 273)/0.001029]\ln(101{,}000/47{,}400) - 101{,}000 + 47{,}400}$$

$$= 1.04 \times 10^{-9}\ \text{m}$$

The number of molecules in the droplet n_e is found as

$$n_e = \frac{V_d N_A}{v_l \overline{M}}$$

where N_A and $\overline{M}$ are Avogadro's number and the molecular weight, respectively, and the droplet volume V_d is equal to $4\pi r_e^3/3 = 4.78 \times 10^{-27}\ \text{m}^3$.

Substituting, it is found that

$$n_e = \frac{4.78 \times 10^{-27}(6.02 \times 10^{26})}{0.001029(18)} = 155 \text{ molecules}$$

Thus, for these conditions, to form an embryo droplet of critical size requires about 155 molecules—a very small number.

5.6 THE KINETIC LIMIT OF SUPERSATURATION

As in the case of bubble formation in a superheated liquid, the likelihood of homogeneous droplet nucleation in a supersaturated vapor depends on the kinetics of the embryo formation process. In analyzing the kinetics, the droplet formation process is typically assumed to be analogous to the bubble formation process considered in Section 5.3. Using arguments virtually identical to those presented in Section 5.3 for bubble formation, the following relation is obtained for the net flux of the number of droplet embryos in size space, J:

$$J = \frac{N^*(n)}{N(n)} \left\{ \int_{n=n}^{n=n_0} [N(n)A(n)j_c(n)]^{-1} \, dn \right\}^{-1} \tag{5.93}$$

The above equation is, in fact, identical to Eq. (5.63). $N(n)$ is the equilibrium number distribution of droplet embryos for a supercooled vapor, postulated as being equal to

$$N(n) = N_v \exp\left[\frac{-\Delta\Psi(r)}{k_B T_v}\right] \tag{5.94}$$

where N_v is the number of vapor molecules per unit volume and $\Delta\Psi(r)$ is the availability function previously defined. The radius of the droplet r is related to the number of molecules n in the embryo as

$$\frac{\frac{4}{3}\pi r^3}{v_l} = nm = \frac{n\overline{M}}{N_A} \tag{5.95}$$

where, as before, m is the mass of one molecule, $\overline{M}$ is the molecular weight, and N_A is Avogadro's number.

$N^*(n)$ in Eq. (5.93) is the number distribution of droplet embryos in the supersaturated vapor for the actual nonequilibrium conditions. The surface area of an embryo is $A(n) = 4\pi r^2$, and $j_c(n)$ is the number of molecules condensing at the interface of a droplet of size n per unit time, per unit area.

As in the analysis of bubble embryo formation, the integrand in Eq. (5.93) rapidly approaches zero as $|n - n_e|$ or $|r - r_e|$ becomes large. We therefore approximate $\Delta\Psi$ for the droplet with the expansion about $r - r_e$ described in Section 5.5. In addition, we take $j_c(n)$ to be equal to its value at $n = n_e$ and use Eq. (4.129*a*), developed from the kinetic theory of gases in Chapter Four, to evaluate $j_c(n_e)$:

$$j_c(n) \simeq j_c(n_e) = \frac{P_v}{(2\pi m k_B T_v)^{1/2}} \tag{5.96}$$

In obtaining the approximate relation above for $j_c(n_e)$ we have taken $\Gamma(a) = 1$ and substituted $\overline{R} = k_B N_A$ in Eq. (4.129*a*). Use of this relation also implicitly assumes that the accommodation coefficient is unity.

For small n, we again assume that $N^*/N \rightarrow 1$, based on the argument that there will be so many very small embryos that departure from equilibrium has little effect on their numbers. The assumption that J is independent of time again leads to the conclusion that it is independent of n, and we therefore evaluate the right side of Eq. (5.93) at $n = 0$ so that N^*/N can be taken equal to 1. Substituting Eqs. (5.94) and (5.96) into Eq. (5.93) and using Eq. (5.95) to convert the integral in terms of n to an integral in terms of r, the following relation for J is obtained:

$$J = \left[\frac{N_v P_v v_l m}{(2m\pi k_B T_v)^{1/2}}\right]\left\{\int_0^\infty \exp\left[\frac{\Delta\Psi(r)}{k_B T_v}\right] dr\right\}^{-1} \tag{5.97}$$

Substituting the expansion (5.92) for $\Delta\Psi(r)$ in Eq. (5.97), changing the variable of integration to $\xi = (4\pi\sigma/k_B T_v)^{1/2}(r - r_e)$, and taking the lower limit of integration to be $-\infty$ (because the integrand rapidly becomes small as $|\xi|$ becomes large), the following relation is obtained for J:

$$J = \left(\frac{N_v P_v v_l}{k_B T_v}\right)\left(\frac{2\sigma m}{\pi}\right)^{1/2} \exp\left(\frac{-4\pi\sigma r_e^2}{3k_B T_v}\right) \tag{5.98}$$

using Eq. (5.90) to eliminate r_e, the above equation can be rearranged to obtain

$$J = \left(\frac{2\sigma\overline{M}}{\pi N_A}\right)^{1/2}\left(\frac{P_v}{k_B T_v}\right)^2 v_l \exp\left(\frac{-16\pi(\sigma/k_B T_v)^3(\overline{M}v_l/N_A)^2}{3\{\ln[P_v/P_{\text{sat}}(T_v)]\}^2}\right) \tag{5.99}$$

For a given pure substance with known physical properties, a threshold value of J can be chosen, corresponding to the onset of homogeneous nucleation, and Eq. (5.99) can be solved iteratively for the limiting supersaturation temperature at a given vapor pressure.

Supersaturation limit conditions determined experimentally have generally been found to be consistent with the theoretical relations developed in this chapter. Figure 5.16 shows a comparison between measured supersaturation limit data obtained by Frank and Hertz [5.18] and the vapor spinodal limits predicted by Eqs. (5.76) and (5.79) for the van der Waals ($\lambda = 0$) and Berthelot ($\lambda = 1$) equations of state. The data shown in Fig. 5.16 were obtained for methanol and ethanol using a diffusion cloud chamber.

It can be seen that the data in Fig. 5.16 are all well above the predicted spinodal limit curve for both the van der Waals and Berthelot equations. This trend is consistent with the interpretation of the spinodal as the ultimate stability limit for the metastable vapor. Of the two curves shown in Fig. 5.16, the Berthelot spinodal is somewhat closer to the actual supersaturation limits observed in the experiments of Frank and Hertz [5.18].

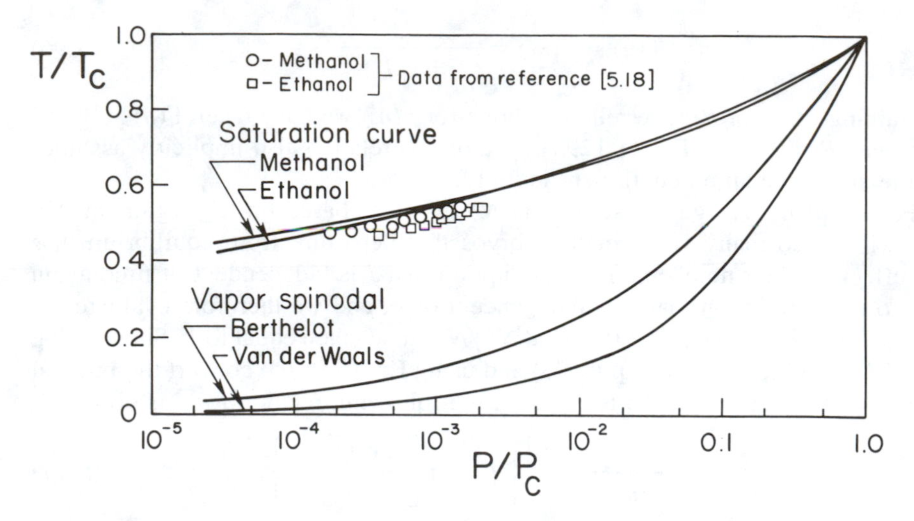

Figure 5.16 Comparison of measured supersaturation limit data with spinodal limit predictions.

In Fig. 5.17, the supersaturation limit data for a variety of fluids are compared with the corresponding limit predicted by Eq. (5.99) with J taken to be 10^6 m^{-3} s^{-1}. Note that the data are plotted in terms of the supersaturation ratio $(P_v)_{SSL}/P_{sat}(T_v)$, where $(P_v)_{SSL}$ is the vapor pressure at the supersaturation limit for $T = T_v$. As in the case of bubble nucleation, the choice of $J = 10^6$ m^{-3} s^{-1} for droplet nucleation was made to provide a best fit to available data. It is interesting that $J = 10^6$ m^{-3} s^{-1} provides a best fit to droplet nucleation data, whereas for bubble nucleation, $J = 10^{12}$ m^{-3} s^{-1} provides better agreement. The value of $(P_v)_{SSL}$ predicted by Eq. (5.99) varies weakly with J, so uncertainty in J does not strongly affect the accuracy of the predicted supersaturation limit.

As seen in Fig. 5.17, in general, the agreement between the experimentally determined S values and the theoretical predictions is fairly good. The fact that a single J value produced $(P_v)_{SSL}$ values that agree well with data for a wide variety of fluids argues strongly for the validity of this classical model analysis of the nucleation kinetics, and the resulting predictive Eq. (5.99). However, in a very recent investigation, Hung et al. [5.19] found systematic differences between measured nucleation rate data and the predictions of the classical model described above. Efforts have been made to develop an improved model of the nucleation kinetics (see, e.g., Reise et al. [5.21]). Hung et al. [5.19] found that the agreement between the model predictions and measured data for alternative models was not significantly better than that for the classical model.

To determine $(T_v)_{SSL}$ for a given vapor pressure P_v using Eq. (5.99) with $J = 10^6$ m^{-3} s^{-1} requires an iterative calculation because of the implicit temperature dependence of P_{sat} and σ. However, it can be shown that the following relation approximates Eq. (5.99) to a high degree of accuracy:

$$S = \frac{(P_v)_{SSL}}{P_{sat}(T_v)} = \exp\left\{\left[\frac{E^*(-\ln J^*)^{1/2}}{2(E^*)^{1/2} + (-\ln J^*)^{3/2}}\right]^{1/2}\right\} \qquad (5.100)$$

where

$$J^* = \frac{\overline{M}J}{N_A v_l}\left(\frac{\pi \overline{M}}{2\sigma N_A}\right)^{1/2}\left[\frac{RT_v}{P_{sat}(T_v)}\right] \tag{5.101}$$

$$E^* = \frac{16\pi\sigma^3 v_l^2}{3k_B R^2 T_v^3} \tag{5.102}$$

Equation (5.100) expresses $(P_v)_{SSL}/P_{sat}(T_v)$ as a function of the dimensionless variables J^* and E^*, which are functions only of T_v. Hence, if T_v is specified, the limiting supersaturation ratio $S = (P_v)_{SSL}/P_{sat}(T_v)$ can be computed explicitly using Eqs. (5.100) through (5.102). Values of S predicted by Eq. (5.100) agree with values obtained by iteratively solving Eq. (5.99) to within less than 0.5%. The variation of S with these parameters predicted by Eq. (5.100) is shown in Fig. 5.18 for ranges of J^* and E^* commonly encountered in real systems.

It is worth noting that the supersaturation limit data represented in Fig. 5.17 were obtained in experiments in which the supersaturated condition was carefully obtained in a diffusion or expansion chamber. Additional supersaturation limit

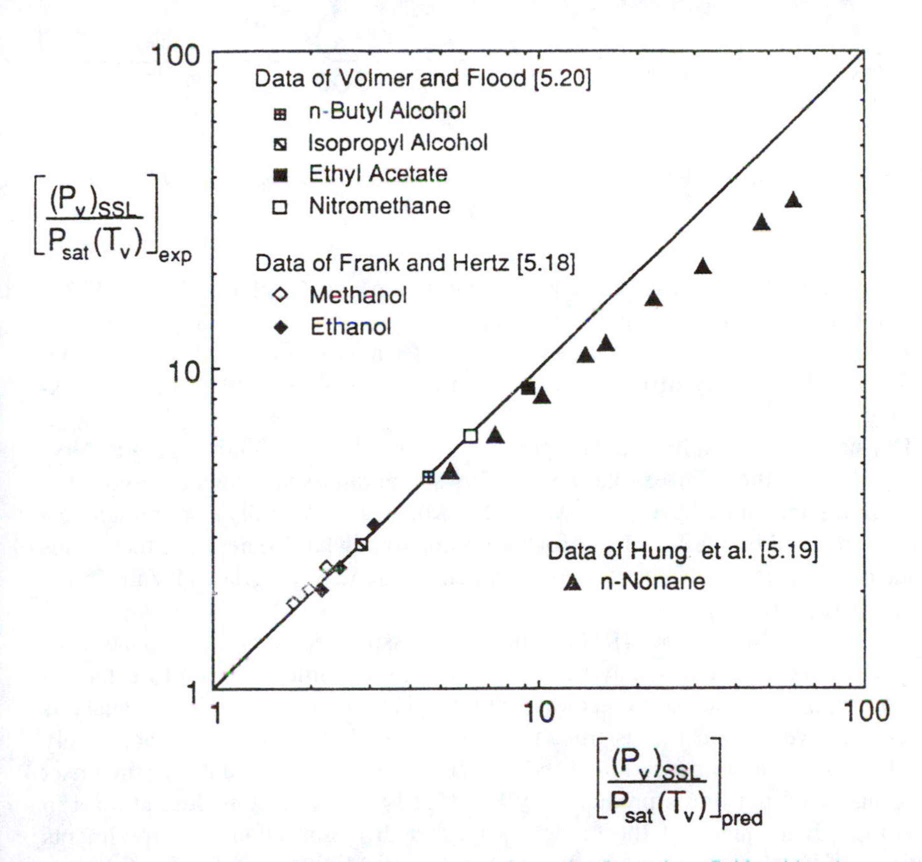

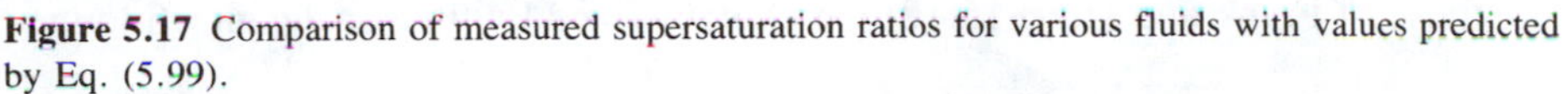
Figure 5.17 Comparison of measured supersaturation ratios for various fluids with values predicted by Eq. (5.99).

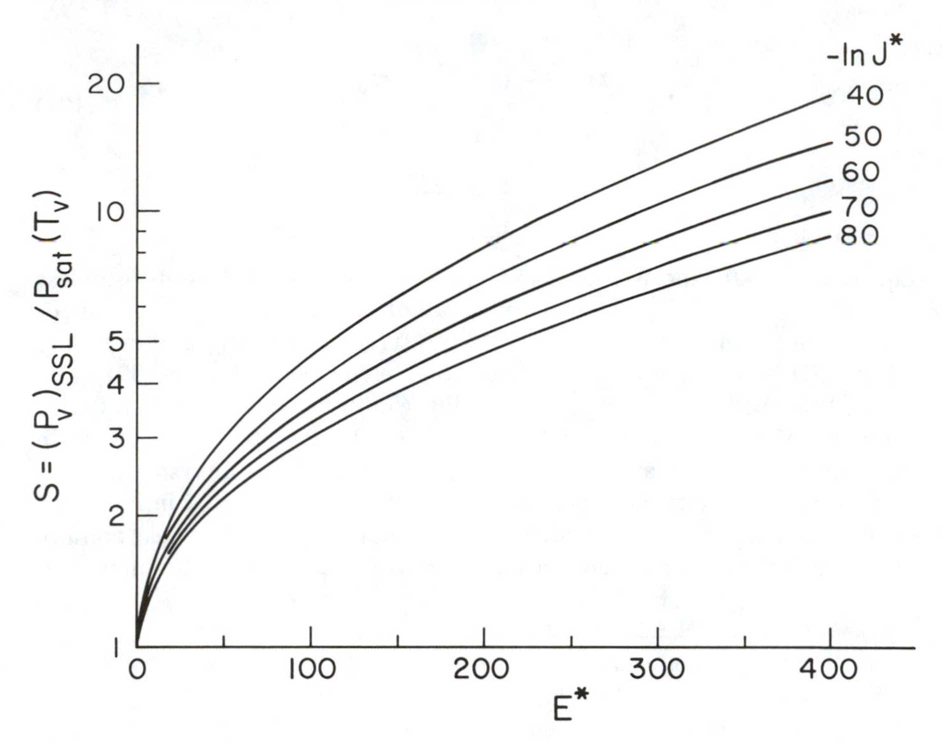

Figure 5.18 Supersaturation limit conditions plotted as a function of the parameters E^* and J^*.

data were also obtained by Yellot [5.22] for steam and Goglia and Van Wylen [5.23] for nitrogen gas undergoing high-speed expansion in diverging nozzles. It is generally acknowledged that the conditions at which a condensation shock occurs in such flows may differ significantly from those observed in nonflow expansions.

The supersaturation limit data reported by Yellot [5.22], Goglia and Van Wylen [5.23], and others for expanding gas flows generally scatter widely about the kinetic limits indicated by Eq. (5.99) or (5.100). Experimentally determined values of S obtained by Yellot [5.22] for steam are somewhat higher than the values predicted by Eq. (5.100), whereas for nitrogen, the data of Goglia and Van Wylen [5.23] are significantly lower.

As noted by Mason [5.24], the time scales associated with rapid expansions in supersonic wind tunnels may be comparable to the time required to establish the quasi-equilibrium embryo size distribution postulated to exist in the analysis described above. When this is true, this type of analysis may fail to accurately predict the onset of condensation in the tunnel. The scattering of data for the onset conditions about the limits predicted by Eq. (5.100) apparently is due, at least in part, to this inadequacy of the model. Note that diffusion chamber experiments like those of Frank and Hertz [5.18] are characterized by time scales much longer than the time needed to establish the embryo size distribution postulated in the analysis. Data from experiments of this type are therefore expected to agree well with the predictions of Eq. (5.100).

It is clear from the above discussion that it is, in fact, possible for a vapor to be in a highly supersaturated state without condensation occurring. This behavior can play a major role in many circumstances of practical interest, including cloud (fog) formation and precipitation in the atmosphere, expanding vapor flow in nozzles, and vapor-deposition manufacturing processes. As noted in Section 5.5, vapor can be maintained in a metastable supersaturated state only if there is not a liquid film on the solid surfaces of the system and only in the absence of dust particles or surface imperfections that can serve as nucleation sites. In most heat transfer equipment, however, it is difficult, if not impossible, to avoid the presence of these types of nucleation sites. As a result, condensation is often initiated by a heterogeneous nucleation process. This type of nucleation phenomena is considered in detail in Chapter Six.

Example 5.7 Estimate the kinetic limiting supersaturation vapor pressure for steam at a temperature of 20°C.

For saturated water at T_v = 20°C, P_{sat} = 2.36 kPa, v_l = 0.001002 m^3/kg, σ = 0.07278 N/m, and $R = \overline{R}/\overline{M}$ = 8314.4/18 = 462 J/kg. Using Eqs. (5.100) through (5.102),

$$J^* = \frac{\overline{M}J}{N_A v_l}\left[\frac{RT_v}{P_{sat}(T_v)}\right]^2\left(\frac{\pi\overline{M}}{2N_A\sigma}\right)^{1/2}$$

$$= \frac{18(10^6)}{6.02 \times 10^{26}(0.001002)}\left[\frac{462(20+273)}{2360}\right]^2\left[\frac{\pi(18)}{2(6.02 \times 10^{26})(0.07278)}\right]^{1/2}$$

$$= 7.87 \times 10^{-26}$$

from which it follows that

$$-\ln(J^*) = 57.803$$

$$E^* = \frac{16\pi\sigma^3 v_l^2}{3k_B R^2 T_v^3} = \frac{16\pi(0.07278)^3(0.001002)^2}{3(1.381 \times 10^{-23})(462)^2(20+273)^3} = 87.466$$

$$\frac{(P_v)_{SSL}}{P_{sat}(T_v)} = \exp\left\{\left[\frac{E^*(-\ln J^*)^{1/2}}{2(E^*)^{1/2} + (-\ln J^*)^{3/2}}\right]^{1/2}\right\}$$

$$= \exp\left\{\left[\frac{87.466(57.803)^{1/2}}{2(87.466)^{1/2} + (57.803)^{3/2}}\right]^{1/2}\right\} = 3.34$$

The supersaturation limit is therefore given as

$$(P_v)_{SSL} = 3.34P_{sat}(T_v) = 3.34(2.36) = 7.88 \text{ kPa}$$

REFERENCES

5.1 Modell, M., and Reid, R. C., *Thermodynamics and Its Applications,* 2nd Ed., Prentice-Hall, Englewood Cliffs, NJ, 1983.
5.2 Zeldovich, Y. B., On the theory of new phase formation: Cavitation, *Acta Physiochem. URSS,* vol. 18, pp. 1–22, 1943.
5.3 Kagen, Y., The kinetics of boiling of a pure liquid, *Russian J. Phys. Chem.,* vol. 34, pp. 42–46, 1960.
5.4 Katz, J. L., and Blander, M., Condensation and boiling: Corrections to homogeneous nucleation theory for nonideal gases, *J. Colloid Interface Sci.,* vol. 42, pp. 496–502, 1973.
5.5 Blander, M., and Katz, J. L., Bubble nucleation in liquids, *AIChE J.,* vol. 21, pp. 833–848, 1975.
5.6 Eberhart, J. G., and Schnyders, H. C., Application of the mechanical stability condition to the prediction of the limit of superheat for normal alkanes, ether and water, *J. Phys. Chem.,* vol. 77, pp. 2730–2736, 1973.
5.7 Blander, M., Hengstenberg, D., and Katz, J. L., Bubble nucleation in *n*-pentane, *n*-hexane, *n*-pentane + hexadecane mixtures, and water, *J. Phys. Chem.,* vol. 75, p. 3613, 1971.
5.8 Apfel, R. E., Vapor nucleation at a liquid–liquid interface, *J. Chem. Phys.,* vol. 54, p. 62, 1971.
5.9 Skripov, V. P., *Metastable Liquids,* Wiley, New York, NY, 1974.
5.10 Avedisian, C. T., Effect of pressure on bubble growth within liquid droplets at the superheat limit, *J. Heat Transfer,* vol. 104, pp. 750–757, 1982.
5.11 Lienhard, J. H., and Karimi, A., Homogeneous nucleation and the spinodal line, *J. Heat Transfer,* vol. 103, pp. 61–64, 1981.
5.12 McQuarrie, D. A., *Statistical Thermodynamics,* Harper & Row, New York, 1973.
5.13 Holden, B., and Katz, J. L., The homogeneous nucleation of bubbles in superheated binary mixtures, *AIChE J.,* vol. 24, pp. 260–267, 1978.
5.14 Ward, C. A., Balakrishnan, A., and Hooper, F. C., On the thermodynamics of nucleation in weak gas–liquid solutions, *J. Basic Eng.,* vol. 85, pp. 695–704, 1970.
5.15 Forest, T. W., and Ward, C. A., Effect of dissolved gas on the homogeneous nucleation pressure of a liquid, *J. Chem. Phys.,* vol. 66, pp. 2322–2330, 1977.
5.16 van Stralen, S., and Cole, R., *Boiling Phenomena,* vol. 1, chap. 3, Hemisphere, New York, 1979.
5.17 Avedisian, C. T., and Glassman, I., High pressure homogeneous nucleation of bubbles within superheated binary liquid mixtures, *J. Heat Transfer,* vol. 103, pp. 272–280, 1981.
5.18 Frank, J. P., and Hertz, H. G., Messung der kritischen Übersättigung von Dampfen mit der Diffusionsnebelkammer, *Z. Physik,* vol. 143, pp. 559–590, 1956.
5.19 Hung, C.-H., Krasnopler, M. J., and Katz, J. L., Condensation of a supersaturated vapor. VIII. The homogeneous nucleation of *n*-nonane, *J. Chem. Phys.,* vol. 90, pp. 1856–1865, 1989.
5.20 Volmer, M., and Flood, H., Tröpfchenbildung in Dampfen, *Z. Phys. Chem.,* vol. 170, pp. 273–285, 1934.
5.21 Reiss, H., Katz, J. L., and Cohen, E. R., *J. Chem. Phys.,* vol. 48, p. 5553, 1968.
5.22 Yellot, J. I., Supersaturated steam, *Trans. ASME,* vol. 56, pp. 411–430, 1934.
5.23 Goglia, G. L., and Van Wylen, G. J., Experimental determination of limit of supersaturation of nitrogen vapor expanding in a nozzle, *J. Heat Transfer,* vol. 83, pp. 27–32, 1961.
5.24 Mason, B. J., *The Physics of Clouds,* 2nd Ed., Oxford University Press, London, 1971.

PROBLEMS

5.1 Use the relation for the spinodal limits derived in Example 5.1 to predict the theoretical limit of superheat for liquid oxygen as a function of pressure. Plot your results for pressures between 1 atm and the critical point.

5.2 The pressure on a sample of saturated R-12 liquid is slowly decreased isothermally from an initial value of 500 kPa. Use the relation for the spinodal limits derived in Example 5.1 to predict the intrinsic limit of superheat at which the sample must begin to vaporize. How close to this limit do you expect the system to get before homogeneous nucleation actually begins? Explain briefly.

5.3 Using the relation for the spinodal limits derived in Example 5.1, determine the minimum temperature at which vapor is stable, and the maximum temperature at which liquid is stable for methane (CH_4) at atmospheric pressure. (For CH_4, T_c = 190.7 K, P_c = 45.8 atm.)

5.4 Determine the variation of the equilibrium bubble size with liquid superheat for water at atmospheric pressure. If the kinetic limit of superheat is 280°C, determine the smallest possible bubble that can exist in this system. How many molecules does such a bubble contain?

5.5 Determine the variation of the equilibrium bubble size with pressure for water superheated 20°C above its normal saturation temperature. Plot your results for pressure between 1 atm and the critical point.

5.6 Determine the vapor pressure inside a water bubble in equilibrium with superheated liquid at atmospheric pressure and 280°C. (Note that this is near the kinetic limit of superheat and is therefore about the smallest possible bubble size that can exist in such a system in equilibrium.)

5.7 Estimate the number of molecules needed to make up a bubble of critical size for superheated liquid nitrogen at atmospheric pressure and 100 K.

5.8 Use Eqs. (5.72) and (5.74) to estimate the kinetic limit of superheat for liquid nitrogen at atmospheric pressure. Compare your results with the spinodal limit prediction obtained in Example 5.1.

5.9 Use Eqs. (5.72) and (5.74) to estimate the kinetic limit of superheat for liquid oxygen at atmospheric pressure.

5.10 Determine the equilibrium vapor pressure for water vapor in equilibrium with droplets having a diameter of 0.5 μm at 250°C.

5.11 Mercury vapor at 630 K exists in equilibrirum with droplets of liquid mercury having a diameter of 2.0 μm. Determine the equilibrium vapor pressure in the system.

5.12 Droplets of liquid water in an atmospheric fog have a mean diameter of 10 μm. If the air temperature is 15°C, estimate the equilibrium partial pressure of water vapor in the air.

5.13 Determine the critical radius of a mercury droplet for supersaturated mercury vapor at atmospheric pressure and a temperature of 600 K.

5.14 In the last stages of the vaporization process inside tubes of an evaporator, the two-phase flow consists of 5-μm droplets of R-22 liquid entrained in the flowing vapor. If the vapor were in equilibrium with the droplets at a (vapor) pressure of 218 kPa, by how much would the equilibrium temperature differ from the standard (flat interface) saturation temperature?

5.15 Estimate the kinetic limiting supersaturation vapor pressure for mercury vapor at a temperature of 650 K.

5.16 Steam maintained at atmospheric pressure is cooled slowly below it normal saturation temperature of 100°C. Estimate the temperature corresponding to the kinetic limit of supersaturation for this process.

5.17 Nitrogen gas at atmospheric pressure is cooled slowly below its normal saturation temperature of 77.4 K. Estimate the temperature at which you would expect liquid droplets first to form in the system. Also determine the equilibrium size of the droplets that form at the limiting condition. Based on your computed size, how easy do you think it would be to see these droplets?

PART TWO

BOILING AND CONDENSATION NEAR IMMERSED BODIES

CHAPTER

SIX

HETEROGENEOUS NUCLEATION AND BUBBLE GROWTH IN LIQUIDS

6.1 HETEROGENEOUS NUCLEATION AT A SMOOTH INTERFACE

In many applications, vaporization of a liquid is made to occur by transferring heat through the solid walls of some containing structure. In such cases, the hottest liquid in the system will be in the region immediately adjacent to the wall. If enough heat is added to the system, the liquid near the wall may reach and slightly exceed the equilibrium saturation temperature. Since the temperature is highest right at the solid surface, the formation of a vapor embryo is most likely to occur at the solid–liquid interface. As noted in Chapter Five, formation of a vapor embryo at the interface between a metastable liquid and another solid phase is one type of heterogeneous nucleation.

The analysis described in Sections 5.2 and 5.3 for homogeneous nucleation within a metastable liquid phase can be extended to heterogeneous nucleation at the solid–liquid interface if the solid surface is idealized as being perfectly smooth. In general, the shape of a vapor embryo at the interface will be dictated by the contact angle and interfacial tension (see Section 2.3), together with the shape of the surface itself. If the solid surface is flat, the vapor embryo will have a profile shape like that shown in Fig. 6.1. The formation of such an embryo in a system held at constant temperature T_l and pressure P_l is schematically shown in Fig. 6.2.

If the embryo shape is idealized as being a portion of a sphere, the geometry dictates that the embryo volume V_v and the areas of the liquid–vapor (A_{lv}) and the solid–vapor (A_{sv}) interfaces are given by

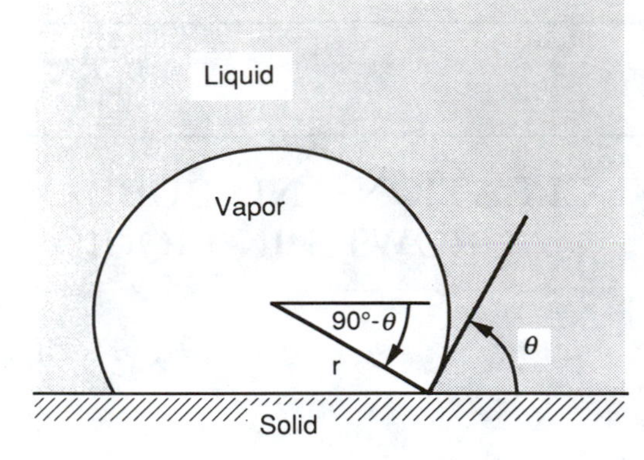

Figure 6.1 An embryo vapor bubble formed at an idealized liquid–solid interface.

$$V_v = \frac{\pi r^3}{3}(2 + 3\cos\theta - \cos^3\theta) \tag{6.1}$$

$$A_{lv} = 2\pi r^2(1 + \cos\theta) \tag{6.2}$$

$$A_{sv} = 2\pi r^2(1 - \cos^2\theta) \tag{6.3}$$

where θ is the contact angle and r is the spherical cap radius indicated in Fig. 6.1.

Initially, the system shown in Fig. 6.2 contains only superheated liquid, and

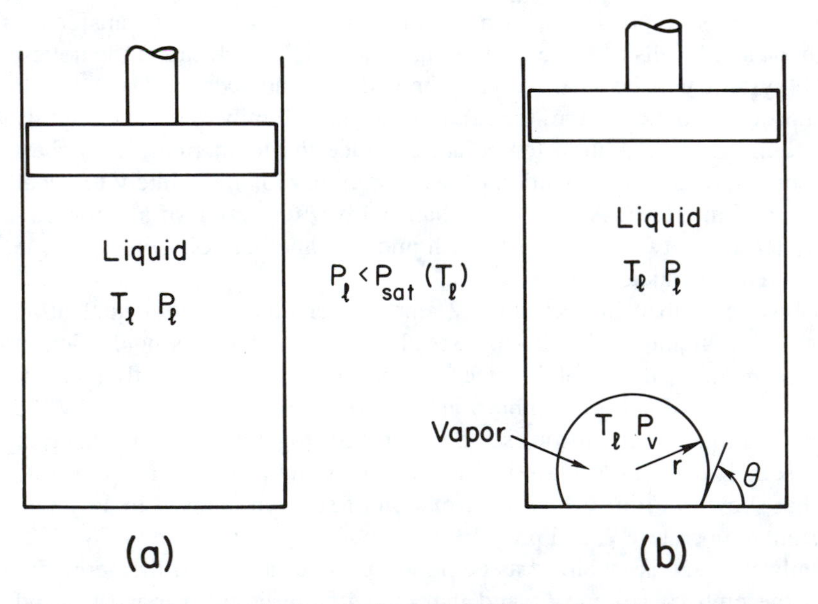

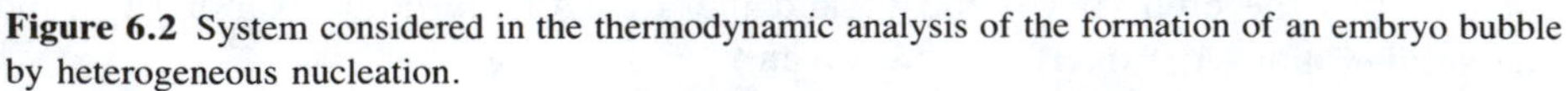
Figure 6.2 System considered in the thermodynamic analysis of the formation of an embryo bubble by heterogeneous nucleation.

the solid walls are fully wetted by the liquid. The initial availability Ψ_0 is therefore given by

$$\Psi_0 = m_T g_l(T_l, P_l) + (A_{sl})_i \sigma_{sl} \tag{6.4}$$

where m_T is the total mass in the system and, by definition, $g_l(T_l, P_l) = u_l - T_l s_l + P_l v_l$. After formation of the embryo, the total availability of the system Ψ is equal to the sum of contributions associated with the liquid (Ψ_l), vapor (Ψ_v), and interfacial (Ψ_i) regions:

$$\Psi = \Psi_l + \Psi_v + \Psi_i \tag{6.5}$$

The three terms on the right side of Eq. (6.5) are given by

$$\Psi_l = (m_T - m_v)g_l(T_l, P_l) \tag{6.6}$$

$$\Psi_v = m_v[g_v(T_l, P_v) + (P_l - P_v)v_v] \tag{6.7}$$

$$\Psi_i = \sigma_{lv} A_{lv} + \sigma_{sv} A_{sv} + (A_{sl})_f \sigma_{sl} \tag{6.8}$$

where m_V is the vapor mass and $g_v(T_l, P_v) = u_v - T_l s_v + P_v v_v$. We will also use the fact that

$$(A_{sl})_i - (A_{sl})_f = A_{sv} \tag{6.9}$$

and Young's equation (resulting from a force balance along the interline),

$$\sigma_{sv} - \sigma_{sl} = \sigma_{lv} \cos\theta \tag{6.10}$$

Combining Eqs. (6.2) through (6.10), the following relation is obtained for the change in the system availability $\Delta\Psi$ associated with appearance of the embryo:

$$\Delta\Psi = \Psi - \Psi_0 = m_v[g_v(T_l, P_v) - g_l(T_l, P_v)] + (P_l - P_v)V_v$$
$$+ 4\pi r^2 \sigma_{lv}[\tfrac{1}{2}(1 + \cos\theta) + \tfrac{1}{4}\cos\theta(1 - \cos^2\theta)] \tag{6.11}$$

If the embryo radius is exactly the right size ($r = r_e$) to be in thermodynamic equilibrium with the surrounding liquid, the g_v and g_l terms in Eq. (6.11) are equal, and the relation (6.11) for $\Delta\Psi$ reduces to

$$\Delta\Psi_e = \tfrac{4}{3}\pi r_e^2 \sigma_{lv}[\tfrac{1}{2} + \tfrac{3}{4}\cos\theta - \tfrac{1}{4}\cos^3\theta] \tag{6.12}$$

Because T_l and P_l are fixed, if we assume that the vapor obeys the ideal gas law and that the embryo is in mechanical equilibrium and satisfies the Young-Laplace equation $P_v = P_l + 2\sigma_{lv}/r$ regardless of size, the right side of Eq. (6.11) is only a function of r. Using these idealizations and appropriate thermodynamic relations, Eq. (6.11) can be used to derive the following Taylor series expansion for $\Delta\Psi$ about the equilibrium condition $r = r_e$:

$$\Delta\Psi = \tfrac{4}{3}\pi r_e^2 \sigma_{lv} F - \left(\frac{4\pi\sigma_{lv}F}{3}\right)\left(2 + \frac{P_l}{P_{ve}}\right)(r - r_e)^2 + \ldots \tag{6.13}$$

where

$$F = F(\theta) = \frac{2 + 3\cos\theta - \cos^3\theta}{4} \tag{6.14}$$

Equation (6.13) is identical to the corresponding expansion (5.49) for the homogeneous nucleation case, except that σ in Eq. (5.49) has been replaced with $\sigma_{lv}F$. It directly follows from the arguments described in Section 5.2 that the equilibrium condition corresponds to a maximum value of $\Delta\Psi$ and is a state of unstable equilibrium. The variation of $\Delta\Psi$ with r is expected to look like that for the homogeneous nucleation case (see Fig. 5.9), leading once again to the conclusion that embryos having a radius less than r_e spontaneously collapse, while those having a radius greater than r_e spontaneously grow (see Section 5.2).

As was done for homogeneous nucleation in Chapter Five, the expansion (6.13) for $\Delta\Psi$ can be used to determine the kinetic limit of superheat for the heterogeneous nucleation process illustrated in Fig. 6.2. The details of the analysis are virtually identical to that presented in Section 5.3 for homogeneous nucleation, and hence they will not be repeated here. However, there are two important differences worth noting. First, at the start of the kinetic limit analysis it is postulated that the equilibrium number density of embryos of n molecules per unit of interface area N_n is given by

$$N_n = N_l^{2/3} \exp\left[\frac{-\Delta\Psi(r)}{k_B T_l}\right] \tag{6.15}$$

where N_l is the number density of liquid molecules per unit volume and $\Delta\Psi(r)$ is the availability function previously defined. For the heterogeneous nucleation process considered here, only liquid molecules near the solid surface can participate in embryo bubble formation. The prefactor multiplying the exponential term in the relation (6.15) for N_n is therefore taken to be $N_l^{2/3}$, which is representative of the number of liquid molecules immediately adjacent to the solid surface per unit of surface area.

In addition, for the heterogeneous nucleation process considered here, the relationship between the number of molecules n in the embryo and its radius r is

$$n = \frac{N_A \pi r^3}{3\overline{M} v_v}(2 + 3\cos\theta - \cos^3\theta) \tag{6.16}$$

This relation differs from that used in the analysis of homogeneous nucleation because the embryo geometries are different.

Analysis of the kinetics of the heterogeneous nucleation process incorporates the modifications noted above and the expansion (6.13) for $\Delta\Psi(r)$ developed for this case. Otherwise the analysis is identical to that presented in Section 5.3 for homogeneous nucleation. Carrying the analysis to completion yields the following relation between the rate of embryo formation J ($\text{m}^{-2}\,\text{s}^{-1}$) and the system properties:

$$J = \frac{N_l^{2/3}(1 + \cos\theta)}{2F}\left(\frac{3F\sigma_{lv}}{\pi m}\right)^{1/2} \exp\left\{\frac{-16\pi F\sigma_{lv}^3}{3k_B T_l[\eta P_{sat}(T_l) - P_l]^2}\right\} \tag{6.17}$$

where

$$\eta = \exp\left\{\frac{v_l[P_l - P_{sat}(T_l)]}{RT_l}\right\} \tag{6.18}$$

and F is defined by Eq. (6.14). It should be noted that if θ is taken to be zero and $N_l^{2/3}$ is replaced by N_l, then Eq. (6.17) becomes identical to the expression (5.73) obtained in Section 5.3 for homogeneous bubble nucleation.

As in the homogeneous nucleation case, J is interpreted as the rate at which embryos of critical size are generated. As J increases, the probability that a bubble will exceed critical size and spontaneously grow becomes greater. If a threshold value of J is specified as corresponding to the onset of nucleation, the corresponding liquid temperature $T_l = T_{SL}$ can be determined from Eq. (6.17).

For superheated water at atmospheric pressure, the variation of J with liquid temperature predicted by Eq. (6.17) is shown in Fig. 6.3 for contact angles of 20° and 108°. A contact angle θ of 20° is typical of water on metal surfaces, whereas $\theta = 108°$ is the contact angle of water on Teflon. Also shown is the variation of J with T_l for homogeneous nucleation in superheated water predicted by Eq. (5.73) ($\theta = 0$).

As illustrated in Fig. 6.3, J increases rapidly with T_l for both the homogeneous and heterogeneous nucleation process. However, the curves shift as θ varies so that the rapid increase of J occurs over different temperature intervals for different values of θ. Relative to the homogeneous nucleation curve, for $\theta > 70°$ the curve shifts to lower temperatures, whereas for $\theta < 65°$ the curve shifts to higher temperatures. For $65 \leq \theta \leq 70$, the heterogeneous and homogeneous nu-

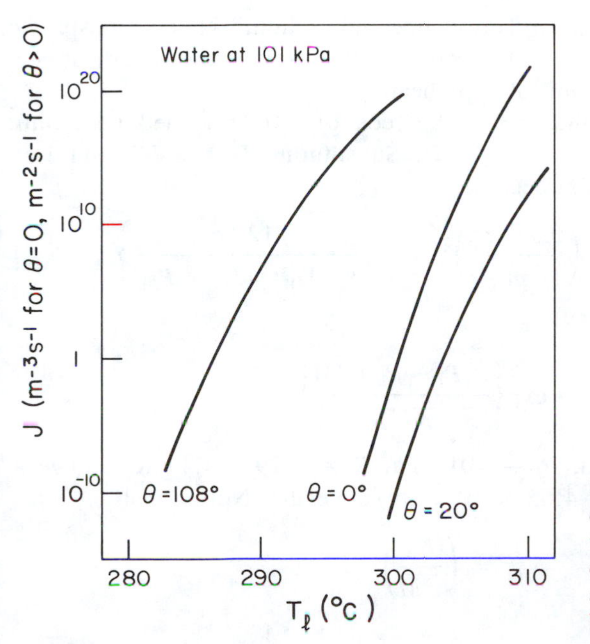

Figure 6.3 Variation of the rate of embryo formation at a water–solid interface with liquid temperature as predicted for different contact angles by analysis of the kinetics of embryo bubble formation.

cleation curves are at about the same location, implying that in a system of uniform temperature, the two modes are equally probable.

Analytical treatments of the kinetics of heterogeneous nucleation similar to that described above have also been developed for other interface geometries, such as spherical and conical cavities and spherical projections. In each case, the analysis must be modified to account for the manner in which the embryo geometry affects the equilibrium conditions and availability changes. Further description of this type of analysis for other geometries can be found in references [6.1]–[6.4]. An excellent summary of the results for different geometries is provided by Cole [6.5].

Jarvis et al. [6.6] have similarly analyzed heterogeneous nucleation at the interface between a volatile liquid (V) and a nonvolatile liquid (N), which are presumed to be immiscible. The profile of a vapor embryo formed at such an interface depends on the interfacial tension at the liquid–liquid and liquid–vapor interfaces. It has been shown by Jarvis et al. [6.6] that nucleation will occur at the interface only if the following conditions are both satisfied:

$$\sigma_{VN} > \sigma_N - \sigma_V \tag{6.19}$$

$$\sigma_{VN} > \sigma_V - \sigma_N \tag{6.20}$$

If Eq. (6.19) is not satisfied, the volatile liquid V spreads on the nonvolatile liquid N, and nucleation occurs entirely within the volatile liquid. If Eq. (6.20) is not satisfied, liquid N spreads on liquid V, and nucleation occurs completely within the nonvolatile liquid. This phenomena is usually referred to as *bubble-blowing*.

Example 6.1 A vessel with a Teflon liner holds liquid N_2 at atmospheric pressure. If the contact angle at the walls is 70°, estimate the superheat corresponding to the kinetic limit of superheat.

For $\theta = 70°$, $F = 0.25\,(2 + 3\cos\theta - \cos^3\theta) = 0.747$. The kinetic limit of superheat is specified by Eq. (6.17). Substituting $F = 0.747$ and $1 + \cos\theta = 1.342$, this equation becomes

$$J_{\theta=70} = 0.898 N_l^{2/3} \left(\frac{0.713\sigma_{lv}}{m} \right)^{1/2} \exp\left\{ \frac{-12.5\sigma_{lv}}{k_B T_l [\eta P_{\text{sat}}(T_l) - P_l]^2} \right\}$$

where

$$\eta = \exp\left\{ \frac{v_l[P_l - P_{\text{sat}}(T_l)]}{RT_l} \right\}$$

For N_2 liquid as specified, $P_l = 101$ kPa, $R = 0.297$ kJ/kg K, and $m = \overline{M}/N_A = 28.0/6.02 \times 10^{26} = 4.65 \times 10^{-26}$ kg/molecule. Noting that

$$N_l^{2/3} = \left(\frac{N_A}{v_l \overline{M}} \right)^{2/3}$$

and that v_l, σ_{lv}, and P_{sat} are functions of the temperature T_l, it is clear that variation of T_l causes these properties and $N_l^{2/3}$ to vary. Using the saturation tables for N_2 to evaluate v_l, σ_{lv}, and P_{sat} for the values of P_l, R, and m noted above and $k_B = 1.38 \times 10^{-23}$ J/K, the values of $J_{\theta=70}$ corresponding to different values of T_l can be determined. The values of $J_{\theta=70}$ obtained in this manner for several values of T_l are given in the following table:

T_l (K)	$P_{sat}(T_l)$ (kPa)	v_l (m^3/kg)	σ_{lv} (N/m)	$\log_{10} J_{\theta=70}$
105.	1083	0.00151	0.00279	−63.7
110.	1467	0.00160	0.00198	12.8
115.	1940	0.00171	0.00118	27.8

While an exact threshold J value corresponding to the kinetic limit of superheat is not known, Fig. 6.3 suggests that $J \simeq 10^{10}$ m^{-2} s^{-1} might be about right. Interpolating the results in the above table to obtain the value of T_l corresponding to $\log_{10} J_{\theta=70} = 10$ yields $(T_l)_{SL} \cong 109.1$ K.

When Eqs. (6.19) and (6.20) are satisfied, the analysis of Jarvis et al. [6.6] predicts the following relationship between J and the system properties:

$$J = N_{lv}^{2/3}\left(\frac{1 - Z_V}{2}\right)\left(\frac{3\sigma_V^{1/2}}{\pi m F_{ll}}\right) \exp\left\{\frac{-16\pi\sigma_V^3 F_{ll}}{3k_B T_l[\eta P_{sat}(T_l) - P_l]^2}\right\} \tag{6.21}$$

where

$$F_{ll} = \frac{1}{4}\left[(2 - 3Z_V + Z_N^3) + \left(\frac{\sigma_N}{\sigma_V}\right)^3 (2 - 3Z_N + Z_V^3)\right] \tag{6.22}$$

$$Z_V = \frac{\sigma_V^2 + \sigma_{VN}^2 - \sigma_N^2}{2\sigma_V \sigma_{VN}} \tag{6.23}$$

$$Z_N = \frac{\sigma_N^2 + \sigma_{VN}^2 - \sigma_V^2}{2\sigma_N \sigma_{VN}} \tag{6.24}$$

and η is given by Eq. (6.18).

Using Eq. (6.17), which was developed for heterogeneous nucleation on a flat, solid surface, it can be shown that for a given threshold J value, the superheat limit will decrease to the saturation temperature as $\theta \rightarrow 180°$. Decreasing superheat limit as $\theta \rightarrow 180°$ is also typical of the theoretical predictions for other solid surface geometries. However, in most real solid–liquid systems, the contact angle is virtually always significantly less than 180°. Even for water on Teflon, the contact angle is only 108°. Values above 108° are extremely rare, and contact angles for common liquids on metal surfaces are invariably much lower (see Chapter Three).

6.2 NUCLEATION FROM ENTRAPPED GAS OR VAPOR IN CAVITIES

For real systems with θ values significantly less than 180°, the analyses described in Section 6.1 indicate that heterogeneous nucleation of bubbles on flat or projecting solid surfaces requires superheat levels that are only slightly different from those required for homogeneous nucleation. Well-wetted cavities also require superheat approaching the homogeneous nucleation value to initiate boiling at a liquid–solid interface. Wall temperatures on the order of 300°C would be necessary to initiate boiling in water at atmospheric pressure if these analyses were correct.

Fortunately, the high wall superheat predicted by these analyses of heterogeneous nucleation is not usually observed experimentally. For boiling of water in a metal pan on a stove, wall superheats of no more than 10–15°C are typically necessary to initiate boiling once the water has reached saturation conditions. Initiation of nucleate boiling at these much lower superheat levels is attributed to the presence of trapped gas in narrow crevices in the solid surface.

Unlike the idealized smooth surfaces considered in the previously described models of heterogeneous nucleation, most real solid surfaces invariably contain natural or machine-formed pits, scratches, or other irregularities. The size of these imperfections may range from microscopic to macroscopic. If the surface is not completely wetted by the liquid, it is expected that many of the cavities will contain entrapped gas. Vaporization may occur at the liquid–gas interface in the cavity at relatively low temperatures.

The mechanism that causes the entrapment of vapor in cavities in the surface is schematically illustrated in Fig. 6.4*a*. For a liquid sheet passing over a gas-filled groove, the contact angle with the downslope tends to be maintained as liquid begins to fill the groove. Because the liquid front is convex, it is clear that the "nose" of the liquid front will strike the opposite wall before the contact line reaches the bottom of the groove if the contact angle is greater than the groove angle 2γ. The condition for entrapment of gas by the advancing liquid front can therefore be stated as

$$\theta_a > 2\gamma \tag{6.25}$$

It should be noted that in this case θ_a represents the advancing contact angle, which may be significantly greater than the static or equilibrium contact angle (see Chapter Three). Contact-angle hysteresis may thus serve to increase the tendency to trap vapor in surface cavities.

We next consider a receding liquid front passing over a liquid-filled groove, as shown in Fig. 6.4*b*. Arguments similar to those above lead to the conclusion that some liquid will be left behind in the groove if $\theta_r < 180° - 2\gamma$. Hence the condition for liquid entrapment in the groove may be stated as

$$\theta_r < 180° - 2\gamma \tag{6.26}$$

Contact-angle hysteresis may result in θ_r being much lower than the static or ad-

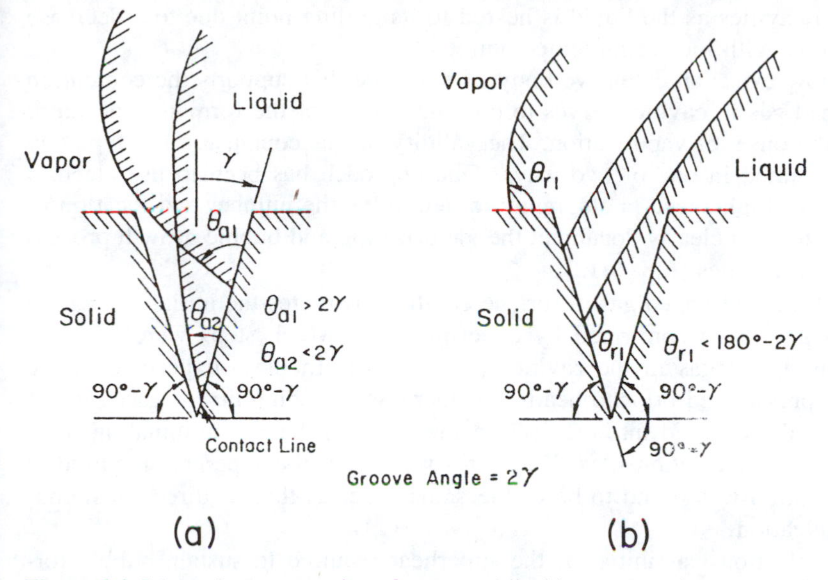

Figure 6.4 Schematic representation of vapor and liquid entrapment in a groove due to motion of the liquid front over the surface.

vancing contact angles. Thus, contact-angle hysteresis may also act to enhance liquid entrapment in surface cavities.

In an early study of liquid and vapor entrapment, Bankoff [6.7] noted that cavities may be classified into one of four categories: (1) cavities that trap gas only, (2) cavities that trap liquid only, (3) cavities that trap both liquid and vapor, and (4) cavities that trap neither liquid nor vapor. The category to which a particular cavity belongs depends on whether Eqs. (6.25) and (6.26) are satisfied. Contact-angle hysteresis enhances the probability that cavities that trap gas (category 1 or 3) will be present on the solid surface.

The above observations indicate that when the containment is initially filled with liquid, there is a high probability that gas will be trapped in some of the cavities on the surface of the containment wall. If the wall is heated and/or the liquid temperature is raised, vaporization of the liquid may occur first at the liquid–gas interface in cavities containing entrapped vapor.

The degree to which entrapped gas in surface cavities can act to initiate vaporization may be affected by the rate at which the gas in the cavity dissolves into the liquid and diffuses away from the gas–liquid interface. The population density of cavities that can act as vaporization initiators may therefore depend on the initial gas concentration in the liquid and the elapsed time between filling the system and the application of heat. Although dissolving of entrapped gas may deactivate a cavity, the opposite can also occur. Gas in a liquid saturated with dissolved gas at room temperature may come out of solution and form gas bubbles in additional cavities as the liquid is heated to its boiling point due to a decrease in its solubility with increasing temperature.

There now exists an extensive body of evidence that supports the contention that entrapped gas in cavities serves to provide nuclei for the formation of vapor bubbles at the onset of vaporization. The validity of this conclusion has typically been demonstrated in one of two ways. One approach has been to use electron micrographs or high-speed photography to determine the number and location of nucleation sites and clearly document the vaporization and bubble growth process within surface cavities [6.8–6.12].

The fact that entrapped gas in surface cavities facilitates the initial nucleation process has also been confirmed by experiments in which steps were taken to eliminate entrapped gas in the cavities [6.13–6.15]. In these experiments, the system was pressurized prior to heating to increase the solubility of gases in the liquid and to drive liquid into crevices in the surface, thereby eliminating most of the cavities as nucleation sites. When this was done, the superheat required to initiate nucleation was found to be of the same order as that required for homogeneous nucleation.

Once nucleation was initiated, the superheat required to sustain bubble formation typically dropped down to a much lower value. This drop apparently occurred because cavities in the surface refilled with vapor when nucleation commenced. Bubble formation could then be sustained at lower superheat levels because vaporization could occur at the liquid–vapor interface within these reactivated cavities.

Once a cavity becomes active, as it emits vapor bubbles, a portion of the entrapped gas is carried away with each bubble. For cavities with simple conical or groove geometries, this process can, over a few minutes or hours, remove the noncondensable gas from the cavity, leaving only vapor behind. Once this occurs, if the surface and surrounding liquid cools down, the vapor may condense completely, thereby deactivating the cavity.

Deactivation of nucleation sites in the manner described above may occur in heat transfer equipment because the duty cycle requires starting and stopping of the heat input or as a result of variations in the imposed pressure and temperature conditions. In some cases cavities may deactivate due to penetration of colder bulk liquid into the cavity as a bubble releases. The stability of active cavities subjected to this type of cooling after each bubble is released has been considered in detail by Bankoff [6.16]. Cooling as a result of liquid microlayer evaporation within the cavity, subsequent vapor condensation and liquid penetration into the cavity, and their effect on site stability have also been examined by Marto and Rohsenow [6.17].

The tendency for noncondensable gas to be replaced by vapor as bubbles are released varies with cavity geometry. Particularly noteworthy in this regard are reentrant cavities like that schematically represented in Fig. 6.5. Cavities of this type are typically very stable and may bubble for a very long period of time before the entrapped gas is completely eliminated.

The enhanced stability of reentrant cavities is linked to the fact that the in-

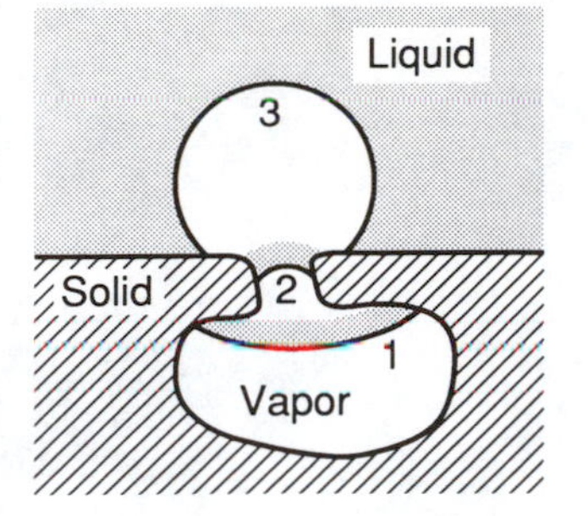

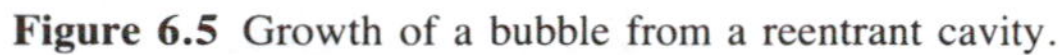
Figure 6.5 Growth of a bubble from a reentrant cavity.

terface curvature reverses when liquid penetrates the mouth of the cavity. For such circumstances, thermodynamic considerations dictate that the equilibrium saturation pressure of the vapor will be higher than the flat interface value at the same temperature (see Section 5.5). Consequently, substantial subcooling would be required to condense vapor in the reentrant cavity, making it more difficult to deactivate the site. Unfortunately, most naturally occurring activities are not reentrant.

It is useful at this point to consider the growth of a vapor embryo within a surface cavity. Although the geometry of real cavities is highly irregular, we will idealize the cavity as being conical with a mouth radius R and a cone angle of 2γ. In general, the principal radii of curvature may vary over the interface, as dictated by the Young-Laplace equation, the hydrostatic pressure gradient, and the contact angle (see Chapter Two). However, here we will idealize the interface profile as being a section of a sphere, characterized by a single radius of curvature, r. The manner in which the interface profile changes as the vapor embryo grows is illustrated in Fig. 6.6 for three different ranges of contact angle. The qualitative variation of R/r as the embryo grows and emerges from the cavity is indicated in Fig. 6.6*d*.

For a highly wetting liquid ($\theta < \gamma < 90°$), the interface radius of curvature starts at some initially small value and increases monotonically as the embryo grows up to and out of the mouth of the cavity (see Fig. 6.6*a*). Consequently, the ratio R/r starts at some finite value and decreases monotonically toward zero. When examining the interface motion, it is useful to consider the apparent contact angle θ_{ap} of the interface measured relative to a horizontal line through the liquid. For $\theta < \gamma < 90°$, the apparent contact angle is less than 90° both before and after the contact line emerges from the cavity, which facilitates the monotonic increase in the radius of curvature r.

It is unlikely that the bubble growth shown in Fig. 6.6*a* would actually occur in the idealized conical cavity considered here because, as described above, entrapment of gas in a "v"-shaped cavity will only occur if $\theta > 2\gamma$. However, if a nominally conical cavity had a small reentrant bottom, this growth pattern may still be obtained.

For contact angles in the range $2\gamma < \theta \leq 90°$, r first increases as the interface moves up to the mouth of the cavity (see Fig. 6.6*b*). However, as the contact

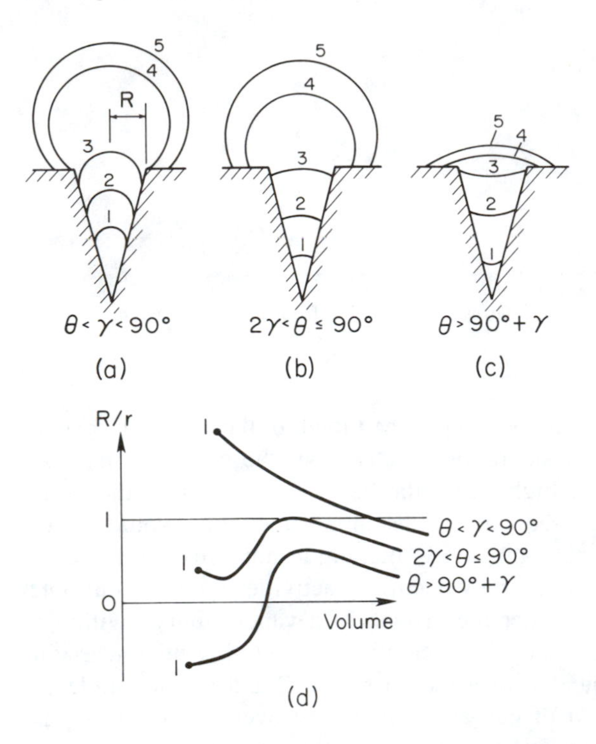

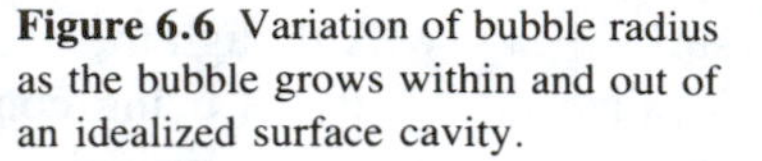
Figure 6.6 Variation of bubble radius as the bubble grows within and out of an idealized surface cavity.

line turns the corner at the mouth of the cavity, r decreases and then begins to increase again. This complicated r variation is associated with the fact that, as the interface emerges from the cavity, the apparent contact angle changes from a value greater than 90° inside the cavity to a value less than or equal to 90° outside. As a result, the R/r ratio decreases initially, then increases briefly, peaking as the interface emerges from the cavity, and then decreases again toward zero.

The interface may initially be concave at even higher contact angles, where $\theta > 90° + \gamma$. Such a circumstance is shown in Fig. 6.6*c*. The radius of curvature is then taken to be negative. As the embryo grows and the interface moves toward the mouth of the cavity, the interface becomes flatter and $1/r$ becomes less negative. As the contact line turns the corner at the mouth of the cavity, the curvature of the interface reverses and the ratio R/r changes sign. As the embryo grows further, R/r reaches a positive maximum and then begins to approach zero. The qualitative variation of R/r with embryo volume for each of the three cases described above is shown in Fig. 6.6*d*.

If we assume that the superheat must exceed the equilibrium value for growth of the embryo to continue, then for the embryo to grow beyond the mouth of the cavity, the system superheat must exceed the equilibrium value for the minimum interface radius. Making use of the assumption that $P_{ve} - P_l \cong P_{sat}(T_l) - P_l$ (see Section 5.2) and combining the Clausius-Clapeyron and Young-Laplace equations, the condition for the cavity to be active can therefore be stated as

$$T_l - T_{sat}(P_l) > \frac{2\sigma T_{sat} v_{lv}}{h_{lv} r_{min}} \tag{6.27}$$

The minimum of r corresponds to the maximum value of R/r in Fig. 6.6*d* as the embryo grows up to the mouth and out of the cavity. The maximum value of R/r occurs near $R/r = 1$ if the contact angle is large and the embryo becomes large enough that $R/r \leqslant 1$. For such circumstances, r_{min} can be approximated by R. The required superheat for the site to be active is then given by

$$T_l - T_{sat}(P_l) > \frac{2\sigma T_{sat}(P_l) v_{lv}}{h_{lv} R} \qquad \text{for } \frac{R}{r} \leqslant 1 \tag{6.28}$$

For small contact angles or moderate contact angles in cavities with a large mouth radius, the maximum value of R/r is larger than 2 and corresponds to the initial radius of the embryo. In such instances, a prediction of the initial embryo radius is needed before Eq. (6.27) can be used to determine the conditions necessary for the site to be active. Because fluid–solid combinations in heat transfer equipment often result in low to moderate contact angles, these circumstances may be commonly encountered in applications involving vaporization processes.

A model to predict the initial radius of the vapor embryo has been developed by Lorenz et al. [6.18]. In this model, the cavity is idealized as being conical, and during the filling process the liquid front is assumed to be planar as it passes over the cavity, preserving the contact angle with the downsloping wall, as indicated in Fig. 6.7*a*. The volume of vapor in the initial gas embryo was taken to be equal to the volume of vapor sealed into the cavity when the planar liquid front contacts the mouth of the cavity (see Fig. 6.7*a*). The radius of curvature is then geometrically determined for a section of the conical cavity bounded by a spherical cap that preserves the contact angle at the cavity wall and has a volume equal to that determined by the vapor-trapping model. This initial vapor embryo configuration is shown in Fig. 6.7*b*.

The radius of curvature for the initial embryo determined from the model described above is a function of the contact angle and the cone angle 2γ of the cavity. The variation of $r/R = \Omega$ with θ and 2γ computed by Lorenz et al. [6.18] using this model is summarized in Fig. 6.8. For given values of θ and 2γ, Ω can be determined from this figure and the conditions for the site to be active can be specified from Eq. (6.27) as

$$T_l - T_{sat}(P_l) > \frac{2\sigma T_{sat}(P_l) v_{lv}}{h_{lv} R \Omega(\theta, 2\gamma)} \qquad \text{for } \frac{R}{r} > 1 \tag{6.29}$$

For $\theta \leqslant 2\gamma$, the computed curves in Fig. 6.8 indicate that r/R is zero and no vapor is entrapped, which is consistent with the more general model of vapor entrapment described earlier. For a given cone angle 2γ, there is a maximum contact angle at which $\Omega = r/R = 1$. Above this maximum value of θ, $\Omega(\theta, 2\gamma)$ is taken to be unity, so that Eq. (6.29) becomes equivalent to Eq. (6.28).

In most real systems, cavities typically have very small half-angles γ, and the contact angle θ is invariably less than 90°. As a result, it is often found that

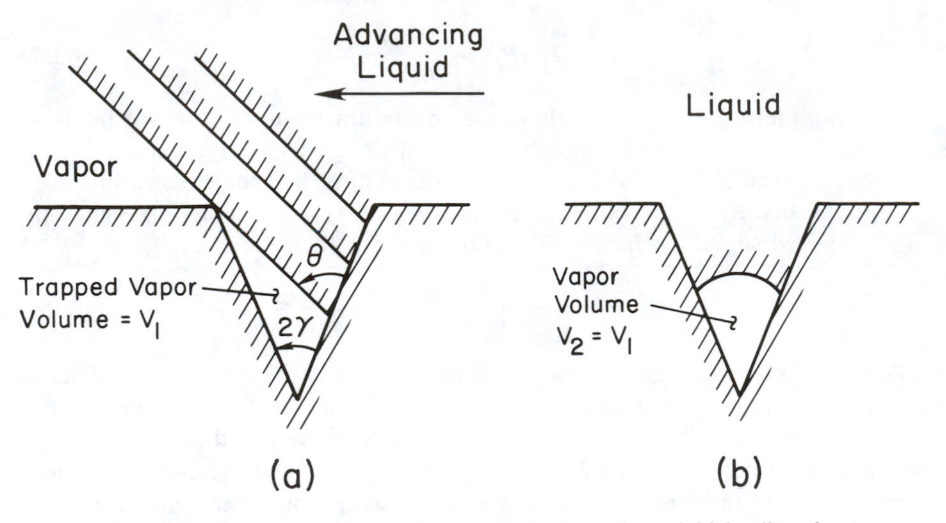

Figure 6.7 Idealized model of vapor trapping process used to estimate initial radius of embryo vapor bubble in the cavity.

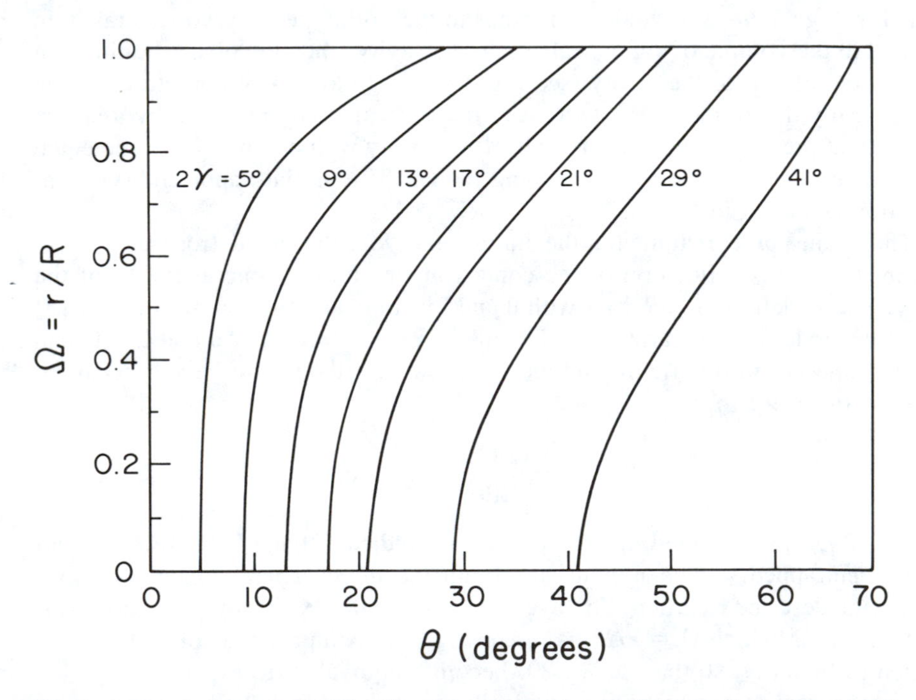

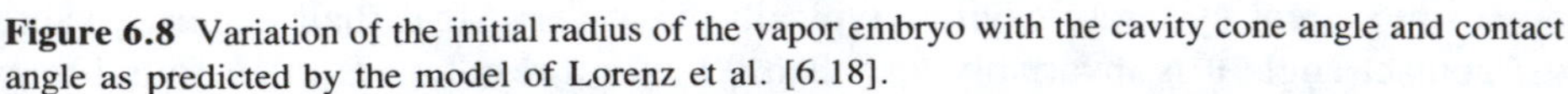

Figure 6.8 Variation of the initial radius of the vapor embryo with the cavity cone angle and contact angle as predicted by the model of Lorenz et al. [6.18].

$\gamma < \theta < 90°$ for cavities in real systems. The variation of R/r with embryo volume will then exhibit a maximum near $R/r = 1$, similar to the middle curve in Fig. 6.6*d*. In addition, the model results plotted in Fig. 6.8 indicate that if the cavity angle 2γ is small, the initial interface radius of curvature is nearly the same as or larger than the mouth radius for all but the smallest contact angles.

These observations suggest that in many real systems the minimum interface radius is equal to, or nearly the same as, the cavity mouth radius. This implies that the cavity mouth radius alone is sufficient to characterize the nucleation behavior of a given site. (See reference [6.19] for a further discussion of this idealization.) If that were true, the nucleation characteristics of a surface would be dictated primarily by the size distribution of potentially active cavities on the surface.

The above discussion clearly implies that the minimum interface radius during embryo growth is not always equal to the cavity mouth radius, but may vary with contact angle and cavity half-angle γ:

$$\frac{r_{\min}}{R} = \Omega(\theta, \gamma) \tag{6.30}$$

Hence, for a given surface having cavities with specified γ values, the associated values of $r_{\min}$ may depend on the contact angle θ.

Equation (6.27) implies that for a given level of imposed superheat, a cavity will be active if $r_{\min}$ is greater than a critical value r^*, given by

$$r^* = \frac{2\sigma T_{\text{sat}}(P_l) v_{lv}}{h_{lv}[T_l - T_{\text{sat}}(P_l)]} \tag{6.31}$$

Each cavity on a real surface has a specific $r_{\min}$ value dictated by its geometry and the contact angle θ. An idealized representation of the number distribution of cavities with respect to $r_{\min}$ for a typical surface is shown in Fig. 6.9. Only those cavities having $r_{\min}$ values greater than r^* will be active. The total number of active sites per unit area is therefore given by

$$n'_a = \int_{r^*}^{\infty} \left(\frac{dn'_c}{dr_{\min}} \right) dr_{\min}$$

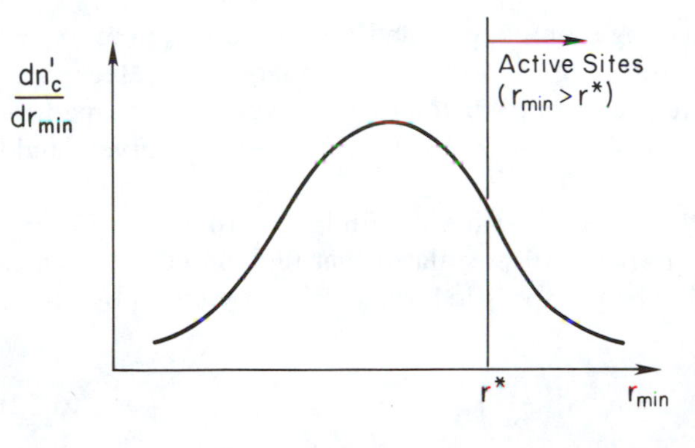

Figure 6.9 Idealized representation of the number distribution of cavities with respect to $r_{\min}$ for a typical surface.

As the wall superheat increases, r^* decreases and the number of active sites having $r_{\min}$ values greater than r^* increases.

Example 6.2 For water at atmospheric pressure, estimate the critical bubble radius r^* for liquid superheat levels of 2, 10, and 40°C.

For water at atmospheric pressure, $T_{sat} = 100°C$, $\sigma = 0.05878$ N/m, $v_{lv} = 1.672\ m^3/kg$, $h_{lv} = 2257$ kJ/kg. The critical radius r^* is determined by substituting these values into Eq. (6.31):

$$r^* = \frac{2\sigma T_{sat}(P_l)v_{lv}}{h_{lv}[T_l - T_{sat}(P_l)]}$$

$$= \frac{2(0.05878)(100 + 273)(1.672)}{(2257 \times 1000)[T_l - T_{sat}(P_l)]}$$

Substituting 2, 10, and 40 for $T_l - T_{sat}(P_l)$ yields

$T_l - T_{sat}(P_l)$ (°C)	r^* (μm)
2	16.20
10	3.25
40	0.81

The arguments described above suggest that the number of observed active sites per unit of surface area is a function primarily of θ and r^*. Experimental data obtained by Lorenz et al. [6.18] support this conclusion.

Data presented by Lorenz et al. [6.18] for boiling of four different liquids on a #240 (sandpaper) finish copper surface are shown in Fig. 6.10. The number of observed active sites n'_a were determined by visual counting at different superheat levels $[T_l - T_{sat}(P_l)]$. The measured superheat values were converted to r^* values using Eq. (6.31). The number of active sites increases rapidly as the superheat increases (and r^* decreases), for all fluids tested. The organic fluids all have a very low contact angle on copper. The discussion above implies that they should therefore have almost identical variations of n'_a with r^*. This is, in fact, observed in Fig. 6.10.

Water has a higher contact angle on copper, and we would therefore expect more effective vapor trapping, resulting in more sites with large $r_{\min}$ values. Relative to the organic fluids, more sites should therefore be active at lower superheat levels (higher r^*) and, as indicated in Fig. 6.10, the n'_a versus r^* curve should be shifted to the right.

The n'_a versus r^* distributions in the log-log plot in Fig. 6.10 are very nearly parallel straight lines. Mikic et al. [6.20] postulated that the number of cavities having a mouth radius greater than a specified value R^* obeys the power-law relation

$$n'_c = \left(\frac{R_s}{R^*}\right)^m \tag{6.32}$$

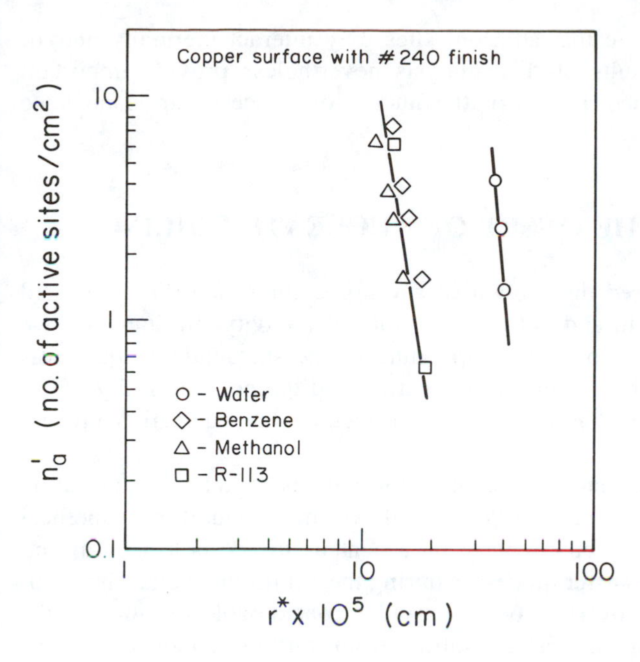

Figure 6.10 The number density of active sites observed by Lorenz et al. [6.18] for boiling of different liquids on a copper surface. (*Adapted with permission*).

where R_s and m are constants that depend on the surface geometry. If a single equivalent cone angle 2γ represents all surface cavities, then Eqs. (6.30) and (6.32) can be combined to obtain

$$n_c' = \left[\frac{R_s\Omega(\theta, 2\gamma)}{r_{\min}^*}\right]^m \tag{6.33}$$

This relation gives the number of sites per unit area having R values greater than R^*, or, equivalently, $r_{\min}$ values greater than $r_{\min}^*$ (because $r_{\min}$ is presumed to be proportional to R). The number density of active sites is equal to the number of cavities with $r_{\min}$ values greater than r^* given by Eq. (6.31). Replacing $r_{\min}^*$ by Eq. (6.31) for r^*, we therefore conclude that the number of active sites per unit area, n_a', is given by the relation

$$n_a' = \left\{\frac{R_s\Omega(\theta, 2\gamma)h_{lv}[T_l - T_{\text{sat}}(P_l)]}{2\sigma T_{\text{sat}}(P_l)v_{lv}}\right\}^m \tag{6.34}$$

The line of reasoning described above leads to the conclusion that the number of active sites varies proportional to $[T_l - T_{\text{sat}}(P_l)]^m$ for a given fluid and surface combination, which is consistent with the straight-line variation of n_a' with $T_l - T_{\text{sat}}(P_l)$ implied by the data in Fig. 6.10. The data in Fig. 6.10 further indicate that m is about the same for all fluids tested, confirming the original assumption that it depends only on the geometry of the solid surface.

Equation (6.34) appears to provide a means of characterizing the heterogeneous nucleation behavior of different solid–liquid combinations. However, this relation was developed by considering the behavior of a single isolated cavity. As a result, the accuracy of Eq. (6.34) becomes questionable when the nucleation site density becomes so large that adjacent sites may interact thermally and/or hydrodynamically. The results of this analysis nevertheless provide important physical insight into the mechanisms of the nucleation process during nucleate boiling at low superheat levels.

6.3 CRITERIA FOR THE ONSET OF NUCLEATE BOILING

In Section 6.2 we considered the growth of a vapor embryo and the associated motion of the interface up to and out of the mouth of a cavity. In doing so, we implicitly assumed that the superheat temperature of the surrounding liquid was uniform. In real systems, this is almost never true, and the nonuniformity of the temperature field very often has a significant effect on the nucleation process within surface cavities.

The steady cyclic growth and release of vapor bubbles at an active nucleation site is usually termed the *ebullition cycle*. In real systems, nonuniform superheat of the liquid surrounding an active cavity on a heated surface is a natural consequence of transient heat transfer processes during the ebullition cycle. The semi-theoretical model proposed by Hsu [6.21] provides considerable insight into the effects of nonuniform liquid superheat resulting from transient conduction in the liquid during the bubble growth and release process. It also provides an indication of the roles of subcooling, pressure, and physical properties in determining the range of surface cavity sizes that will be active.

The system considered in Hsu's model is schematically shown in Fig. 6.11. Initially, a small bubble embryo is assumed to exist at the mouth of the cavity. This bubble presumably was formed by residual vapor left behind after the release of the preceding bubble in an intermittent bubbling process. Cooler bulk liquid

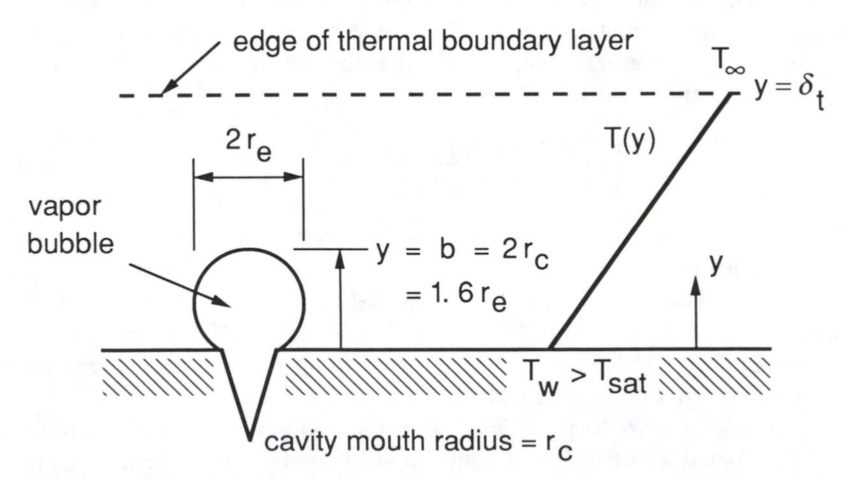

Figure 6.11 Model ebullition cycle.

has replaced the just-departed bubble. At this initial stage of the process, the fluid adjacent to the wall and the embryo bubble is idealized as being entirely at the bulk temperature T_∞. A period of time passes during which liquid adjacent to the wall is heated, and a thermal boundary layer (or penetration layer) grows near the wall as a result of transient conduction. This portion of the bubble growth and release cycle is often referred to as the *waiting period*. The treatment of the waiting period described by Hsu [6.21] is similar to that proposed in an earlier study by Hsu and Graham [6.22].

It is assumed in this analysis that convection and/or turbulence in the bulk fluid away from the wall limits the thickness of the thermal layer that develops as a result of transient conduction into the liquid near the wall. Although in real turbulent flows the eddy diffusivity generally increases with increasing distance from the wall, in this model, Hsu [6.21] assumed that a limiting thermal layer of thickness δ_t exists such that for $y < \delta_t$, transport of heat occurs by molecular diffusion alone; whereas for $y \geq \delta_t$, vigorous turbulent transport results in a uniform temperature of T_∞. Although δ_t in a real system may vary with time, we will assume here that δ_t is a fixed constant. We will also consider the wall to be at a constant and uniform temperature. The results for a uniform heat flux boundary condition were shown by Hsu [6.21] to be qualitatively the same. Heat transfer to the liquid during the waiting period is modeled as one-dimensional transient conduction, for which the governing equation may be written in the form

$$\frac{\partial \theta}{\partial t} = \alpha_l \left(\frac{\partial^2 \theta}{\partial y^2} \right) \tag{6.35}$$

where $\theta = T - T_\infty$. The appropriate boundary and initial conditions are

$$\theta = 0 \qquad \text{at } t = 0 \tag{6.36}$$

$$\theta = 0 \qquad \text{at } y = \delta_t \tag{6.37}$$

for $t > 0$,

$$\theta = \theta_w = T_w - T_\infty \qquad \text{at } y = 0 \tag{6.38}$$

Using elementary methods, the following solution for this system can be obtained (see, e.g., Carslaw and Jaeger [6.23]):

$$\frac{\theta}{\theta_w} = \frac{\delta_t - y}{\delta_t} + \frac{2}{\pi} \sum_{n=1}^{\infty} \frac{\cos n\pi}{n} \sin \left[n\pi \left(\frac{\delta_t - y}{\delta_t} \right) \right] e^{-n^2\pi^2(\alpha_l t/\delta_t^2)} \tag{6.39}$$

The qualitative behavior of the temperature profile predicted by this relation is shown in Fig. 6.12. This solution indicates that the region of heated liquid near the wall grows in thickness until its edge reaches $y = \delta_t$. The temperature profile then adjusts to ultimately establish a linear profile between the wall and $y = \delta_t$ at steady state.

The following relation between the equilibrium superheat and the bubble radius r_e can be derived by combining the Clausius-Clapeyron and Young-Laplace

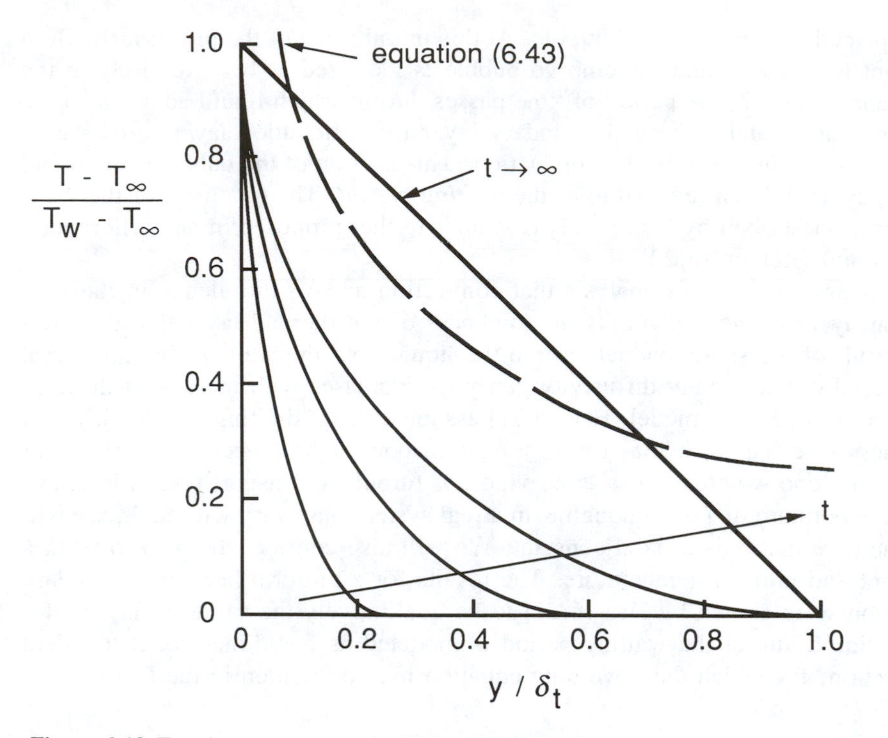

Figure 6.12 Transient temperature profile near surface.

equations and approximating $P_{ve} - P_l$ as $P_{sat}(T) - P_l$ (see Section 5.2):

$$T_{le} - T_{sat}(P_l) = \frac{2\sigma T_{sat}(P_l)}{\rho_v h_{lv} r_e} \tag{6.40}$$

Hsu [6.21] further postulated the following simple relations among the height of the embryo bubble b, the radius of the bubble embryo r_e, and the mouth radius of the cavity r_c:

$$b = 2r_c = 1.6r_e \tag{6.41}$$

The exact relations among these variables may, of course, depend on the fluids and the contact angle with the solid. These relations, although simplistic, are at least consistent with idealization of the embryo geometry as a truncated sphere.

Defining

$$\theta_b = T_{y=b} - T_\infty \qquad \theta_{sat} = T_{sat}(P_l) - T_\infty \tag{6.42}$$

and using the relations (6.41), Eq. (6.40) can be rearranged to obtain

$$\frac{\theta_b}{\theta_w} = \frac{3.2\sigma T_{sat}(P_l)}{\rho_v h_{lv} \delta_t \theta_w (1 - 2r_c/\delta_t)} \tag{6.43}$$

Hsu [6.21] postulated that the embryo bubble would grow and hence the cavity would be an active site if the equilibrium superheat was equaled or exceeded all around the perimeter of the embryo bubble. Because the temperature decreases with increasing y, this will be true if the temperature at $y = b$ is greater than the value of T_{le} specified by Eq. (6.40). Quantitatively, then, it may be stated that a site is active if

$$\frac{\theta}{\theta_w} \geqslant \frac{\theta_b}{\theta_w} \qquad \text{at } y = b = 2r_c \tag{6.44}$$

where θ_b/θ_w is given by Eq. (6.43). This condition can be interpreted graphically by considering Fig. 6.12. The broken curve in this figure represents the equilibrium superheat requirement at $y = b$ given by Eq. (6.43). Cavities having a given size r_c and corresponding b value will become active if, during the conduction transient, the value of θ/θ_w at $y/\delta_t = 2r_c/\delta_t$ exceeds the value of θ_b/θ_w determined from Eq. (6.43).

The highest temperature achieved at any location corresponds to the steady-state value indicated by the linear profile in Fig. 6.12. There may exist minimum and maximum values of $2r_c/\delta t$ between which the equilibrium superheat requirement at the corresponding value of $y/\delta_t = b/\delta_t$ is eventually exceeded. These limiting values are determined by the intersection of the linear steady-state profile with the broken curve representing the superheat requirement. Only those cavities on the surface for which the value of $2r_c/\delta_t$ is within these limits will be active nucleation sites. Specifically, a cavity is active if

$$\left(\frac{2r_c}{\delta_t}\right)_{\min} \leqslant 2r_c/\delta_t \leqslant \left(\frac{2r_c}{\delta_t}\right)_{\max} \tag{6.45}$$

The above-specified condition implies that there may be minimum ($r_{c,\min}$) and maximum ($r_{c,\max}$) values of r_c that define a range of active cavity sizes on the heated surface.

These results imply that for values of r_c above $r_{c,\max}$, the embryo bubble is sufficiently large that it protrudes beyond the superheated boundary layer, exposing the upper portion of the bubble to liquid that is below its equilibrium saturation temperature. Condensation of vapor may then occur at these locations, counteracting vaporization that occurs at portions of the bubble interface closer to the wall and preventing the bubble from growing large enough to release from the surface. Such a cavity is unable to sustain the ebullition and is therefore not an active site.

For values of r_c less than $r_{c,\min}$, the resulting embryo bubble is so small that the required equilibrium superheat cannot be supplied by the wall at the specified superheat level $T_w - T_{\text{sat}}(P_l)$. Thus cavities with r_c values less than $r_{c,\min}$ are inactive because the required superheat to make them active cannot be supplied by the system.

Mathematically, the values of $r_{c,\min}$ and $r_{c,\max}$ can be determined by substituting the linear steady-state relation for the temperature profile

$$\frac{\theta}{\theta_w} = \frac{\delta_t - y}{\delta_t} \tag{6.46}$$

into Eq. (6.43), setting $y = 2r_c$, and solving for r_c. The resulting equation can have two real solutions, corresponding to $r_{c,\min}$ and $r_{c,\max}$. The relation used to compute these values can be written compactly as

$$\left\{ \begin{matrix} r_{c,\max} \\ r_{c,\min} \end{matrix} \right\} = \frac{\delta_t}{4} \left[1 - \frac{\theta_{\text{sat}}}{\theta_w} \left\{ \begin{matrix} + \\ - \end{matrix} \right\} \sqrt{\left(1 - \frac{\theta_{\text{sat}}}{\theta_w}\right)^2 - \frac{12.8\sigma T_{\text{sat}}(P_l)}{\rho_v h_{lv} \delta_t \theta_w}} \right] \tag{6.47}$$

The plus and minus signs on the right side of this equation correspond to $r_{c,\max}$ and $r_{c,\min}$, respectively.

Equation (6.47) indicates that as the superheat θ_w decreases, eventually the collection of terms inside the square root will go to zero, whereupon $r_{c,\max}$ and $r_{c,\min}$ will be the same. For the superheat level at which this occurs, only one specific cavity size can be active. For values of θ_w below this value, the values of $r_{c,\min}$ and $r_{c,\max}$ computed from this relation are imaginary, implying that no cavities of any size will be active. The range of active sites predicted by Eq. (6.47) is indicated in Fig. 6.13 for water at atmospheric pressure and $\delta_t = 0.2$ mm. For the specified P_1 and δ_t values, a finite range of active cavity sizes exists at higher superheat levels. However, as θ_w decreases, the range of active sizes decreases in extent, vanishing completely at a certain threshold value of θ_w. This threshold value of θ_w is the minimum superheat necessary to initiate and sustain nucleate boiling.

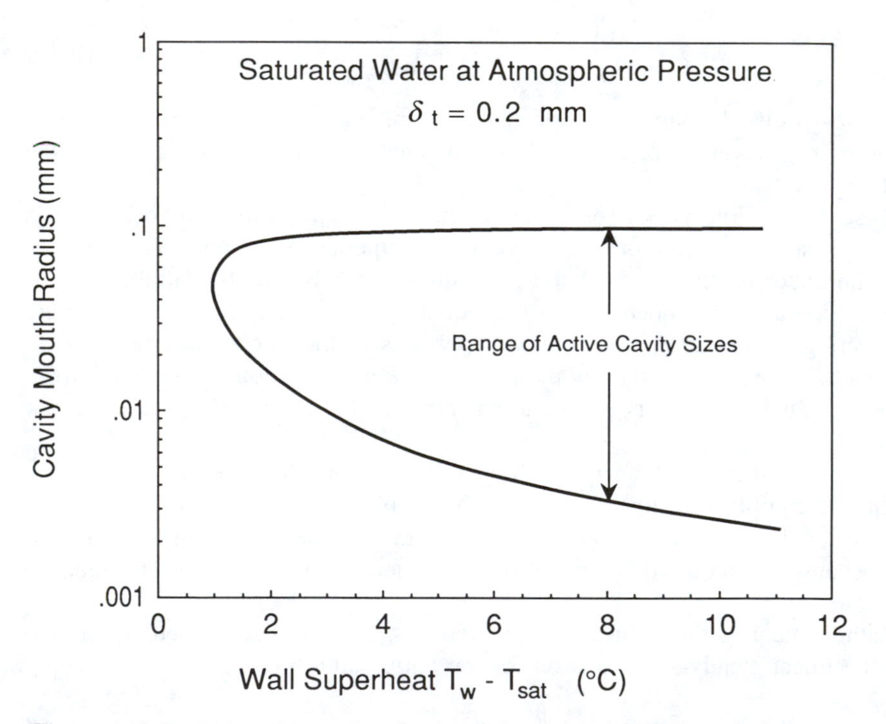

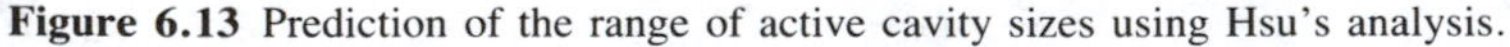
Figure 6.13 Prediction of the range of active cavity sizes using Hsu's analysis.

The arguments presented in the previous section indicate that a cavity is active if the minimum bubble radius during growth is greater than a limiting value dictated by the superheat of the surrounding liquid ($r_{min} > r^*$, given by Eq. [6.31]). If r_{min} is simply related to the cavity mouth radius, the arguments presented in the previous section imply that any cavity with a sufficiently large mouth will be active. Hsu's [6.21] analysis indicates that with a finite thermal boundary-layer thickness, there is, in fact, an upper bound to the range of cavity sizes that will be active.

The model developed by Hsu [6.21] thus predicts two important features of the heterogeneous nucleation on a heated wall:

1. A certain minimum value of wall superheat must be attained before any cavities on the surface will become active nucleation sites.
2. Above the superheat required to initiate nucleation, a finite range of cavities can become active sites. The extent of this range depends on the fluid properties, δ_t, and the subcooling (if any) of the bulk fluid.

Hsu [6.21] found that experimental observations of nucleate boiling reported by Clark et al. [6.9] were consistent with the trends predicted by this model analysis.

Example 6.3 For turbulent-flow forced convection of saturated liquid water at atmospheric pressure in a tube, the heat transfer coefficient at a particular location is 11.0 kW/m^2 °C. Estimate the required wall superheat to initiate nucleate boiling at this location. Also estimate the range of active cavity sizes at a wall superheat of 5°C.

For saturated liquid water, $\theta_{sat} = T_{sat} - T_\infty = 0$. Substituting the appropriate properties into Eq. (6.47) yields

$$\begin{Bmatrix} r_{c,\max} \\ r_{c,\min} \end{Bmatrix} = \frac{\delta_t}{4}\left(1 \begin{Bmatrix} + \\ - \end{Bmatrix} \sqrt{1 - \frac{2.08 \times 10^{-4}}{\delta_t \theta_w}}\right)$$

δ_t is approximated as $\delta_t = k_l/h = 0.681/(11.0 \times 10^3) = 6.09 \times 10^{-5}$ m. Setting the term inside the square-root sign equal to zero, it is found that

$$(\theta_w)_{onset} = 3.4°\text{C}$$

Since the onset of boiling occurs at $\theta_w = 3.4$°C, only a few widely spaced sites are expected to be active at $\theta_w = 5$°C and δ_t is likely to be close to that for turbulent single-phase convection. Substituting $\delta_t = 6.09 \times 10^{-5}$ m into the above relation with $\theta_w = 5$°C yields

$$r_{min} = 6.65 \times 10^{-6}\text{ m} \qquad r_{max} = 2.38 \times 10^{-5}\text{ m}$$

The significance of Hsu's [6.21] model is not so much its explicit predictive capacity, but the insight it provides into the mechanisms that affect the nucleation process. The model provides a basis for understanding the experimentally determined effects of subcooling, fluid properties, surface finish, and wall superheat on nucleate boiling. In particular, it is clear from the results of this model that the degree to which the distribution of cavity sizes on the surface overlaps the range of potentially active sizes will directly dictate the density of active sites observed at a given superheat level. The results also indicate that decreasing δ_t will increase the threshold level of θ_w required to initiate nucleation. This implies that enhancement of bulk convection (thereby reducing δ_t) will suppress nucleation. As will be seen in Chapter Twelve, this latter effect has important consequences in convective boiling.

6.4 BUBBLE GROWTH IN AN EXTENSIVE LIQUID POOL

Before considering further the bubble growth process during nucleate boiling at a heated surface, we will first consider the simpler case of bubble growth in an extensive uniformly superheated liquid. Many of the more complex features of bubble growth near a heated solid surface are absent for these circumstances. However, the spherical symmetry for this simpler case makes it possible to predict the nature of the bubble growth process using relatively simple analytical tools. In addition, because many of the fundamental mechanisms are the same, study of bubble growth in an extensive pool of liquid provides considerable insight into mechanisms that also play a role in bubble growth near a heated surface.

The physical circumstances of interest here are shown in Fig. 6.14. At any instant during the growth process, the interface is located at $r = R$ and is moving with a velocity dR/dt relative to the laboratory reference frame. The pressure and temperature are P_v and T_v inside the bubble and P_∞ and T_∞ in the surrounding liquid, respectively. When the bubble first forms, the interface temperature will be nearly equal to the superheated liquid temperature, and the vapor generated at the interface will be at a pressure nearly equal to $P_{sat}(T_\infty)$. As the liquid superheat near the interface is consumed to provide the latent heat of vaporization, the temperature at the interface will drop toward $T_{sat}(P_\infty)$. During the bubble growth pro-

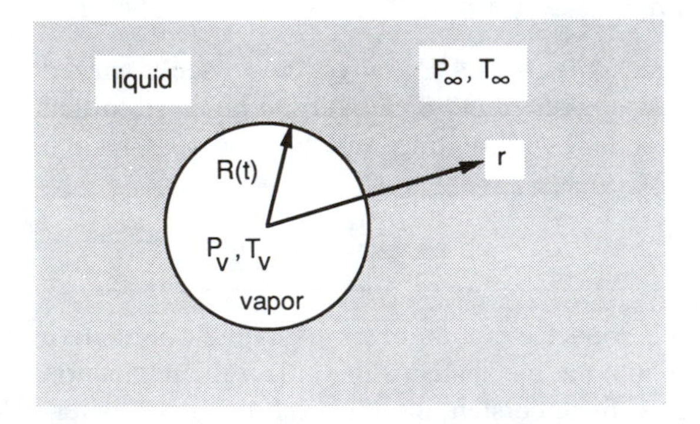

Figure 6.14 Bubble growing in superheated liquid.

cess, the capillary pressure difference across the interface decreases as the radius increases, and the pressure inside the bubble drops toward P_∞. During the growth process, P_v and T_v will therefore lie in the ranges

$$P_\infty \leqslant P_v \leqslant P_{sat}(T_\infty) \tag{6.48}$$

$$T_{sat}(P_\infty) \leqslant T_v \leqslant T_\infty \tag{6.49}$$

The rate of bubble growth at any instant of time during the growth process is dictated by three factors:

1. The fluid momentum and pressure difference interactions
2. The rate of heat transfer to the interface to supply the latent heat of vaporization
3. The thermodynamic constraint (consistent with the assumption of local thermodynamic equilibrium) that $P_v = P_{sat}(T_v)$

There are two limiting cases of the bubble growth process:

Inertia-controlled growth. In this regime P_v is near its maximum value. $P_{sat}(T_\infty)$, and $T_v \cong T_\infty$. Heat transfer to the interface is very fast and is not a limiting factor to growth. The growth rate is therefore governed by the momentum interaction between the bubble and the surrounding liquid (i.e., it is limited by how rapidly it can push back the surrounding liquid). These conditions usually exist during the initial stages of bubble growth, just after the embryo bubble forms and begins to grow.

Heat-transfer-controlled growth. In this regime, T_v is near its minimum value, $T_{sat}(P_\infty)$, and $P_v \cong P_\infty$. Growth is limited by the relatively slower transport of heat to the interface. As a result, the interface motion is slow compared to that for inertia-controlled growth, and the momentum transfer between the bubble and the surrounding liquid is not a limiting factor. These conditions generally correspond to the later stages of bubble growth when the bubble is larger and the liquid superheat near the interface has been significantly depleted.

The above limiting cases represent the initial and final stages of bubble growth. At intermediate times both heat transfer and liquid inertia effects come into play. This more general case is fairly complex to analyze. However, it is not too difficult to develop reasonably accurate analyses of the two limiting cases described above.

For inertia-controlled growth, the treatment originally proposed by Rayleigh [6.24] is particularly illuminating. For the incompressible, radially symmetric flow of liquid near the bubble, conservation of mass requires that

$$\frac{1}{r^2}\left[\frac{\partial(r^2 u)}{\partial r}\right] = 0 \qquad \text{where } u_{r=R} = \frac{dR}{dt} \tag{6.50}$$

Integrating the above equation from $r = R$ to an arbitrary location r yields

$$u = \frac{dR}{dt}\left(\frac{R}{r}\right)^2 \tag{6.51}$$

The total kinetic energy in the moving liquid surrounding the bubble $(KE)_l$ can be obtained from the above velocity variation as

$$(KE)_l = \left(\frac{\rho_l}{2}\right)\int_R^\infty u^2\, dV = 2\pi\rho_l\left(\frac{dR}{dt}\right)^2 R^3 \tag{6.52}$$

The net work W_l done against the surrounding liquid as the bubble grows from $R = 0$ to R is given by

$$W_l = \int_0^R P_{li}(4\pi R^2)\, dR - \frac{4}{3}\pi P_\infty R^3 \tag{6.53}$$

In this equation, P_{li} is the pressure at the interface in the liquid. The second term subtracts work done against the ambient surrounding liquid to accommodate the volume change of the bubble. The right side of Eq. (6.53) thus represents the net work on the liquid, which provides the kinetic energy of the liquid motion. Setting the expression (6.51) for $(KE)_l$ equal to the relation (6.53) for W_l yields

$$2\pi\rho_l\left(\frac{dR}{dt}\right)^2 R^2 = \int_0^R P_{li}(4\pi R^2)\, dR - \frac{4}{3}\pi P_\infty R^3 \tag{6.54}$$

We further note that P_{li} is related to the vapor pressure inside the bubble by the Young-Laplace equation,

$$P_{li} = P_v - \frac{2\sigma}{R} \tag{6.55}$$

Substituting the relation (6.55) for P_{li} into Eq. (6.54) and differentiating the entire equation with respect to R, the following equation is obtained:

$$R\frac{d^2R}{dt^2} + \frac{3}{2}\left(\frac{dR}{dt}\right)^2 = \frac{1}{\rho_l}\left(P_v - P_\infty - \frac{2\sigma}{R}\right) \tag{6.56}$$

This equation is sometimes referred to as the *Rayleigh equation* [6.24]. Although this equation was derived from a mechanical energy balance, it also can be interpreted as representing the force–momentum balance between the bubble and the surrounding liquid during the growth process.

For the inertia-controlled growth considered here, $2\sigma/R$ is usually much smaller than $P_v - P_\infty$. If $2\sigma/R$ is neglected compared to $P_v - P_\infty$, and the following linearized form of the Clapeyron equation is used to evaluate $P_v - P_\infty$,

$$P_v - P_\infty = \frac{\rho_v h_{lv}[T_v - T_{sat}(P_\infty)]}{T_{sat}(P_\infty)} \tag{6.57}$$

then Eq. (6.56) can be written as

$$R\frac{d^2R}{dt^2} + \frac{3}{2}\left(\frac{dR}{dt}\right)^2 = \frac{\rho_v}{\rho_l}[T_v - T_{\text{sat}}(P_\infty)]h_{lv} \tag{6.58}$$

This nonlinear differential equation has the solution

$$R(t) = \left\{\frac{2}{3}\left[\frac{T_\infty - T_{\text{sat}}(P_\infty)}{T_{\text{sat}}(P_\infty)}\right]\frac{h_{lv}\rho_v}{\rho_l}\right\}^{1/2} t \tag{6.59}$$

which satisfies the initial condition

$$R = 0 \qquad \text{at } t = 0 \tag{6.60}$$

This result implies that the radius increases linearly with time during the initial inertia-controlled stage of bubble growth.

The simplified analysis presented above obviously does not completely account for all mechanisms of momentum exchange during the bubble growth process. In particular, the effects of viscosity and the capillary pressure difference across the interface are neglected. Extended versions of Rayleigh's analysis that include more complete treatments of the momentum transport have been developed by Forster and Zuber [6.25], Plesset and Zwick [6.26], and Scriven [6.27].

In the latter stages of bubble growth, the rate at which heat is transported to the liquid–vapor interface of the bubble becomes the factor that limits the growth rate. During bubble growth, the governing equation for the transport of heat in the liquid surrounding the bubble is

$$\frac{\partial T}{\partial t} + u\frac{\partial T}{\partial r} = \left(\frac{\alpha_l}{r^2}\right)\frac{\partial}{\partial r}\left(r^2\frac{\partial T}{\partial r}\right) \tag{6.61}$$

As in the analysis of inertia-controlled growth analyzed above, it follows from continuity and the imposed boundary condition at the interface that the velocity field in the above equation is given by

$$u = \frac{dR}{dt}\left(\frac{R}{r}\right)^2 \tag{6.62}$$

The boundary and initial conditions for Eq. (6.61) are

$$\begin{aligned} T(r,0) &= T_\infty \\ T(R,t) &= T_{\text{sat}}(P_v) \\ T(\infty,t) &= T_\infty \end{aligned} \tag{6.63}$$

In addition, the requirement of conservation of mass and thermal energy at the interface can be stated in the form

$$k_l\frac{\partial T}{\partial r}(R,t) = \rho_v h_{lv}\frac{dR}{dt} \tag{6.64}$$

Equation (6.61) with conditions (6.63) and (6.64) have been solved exactly by Scriven [6.27] and approximately by Plesset and Zwick [6.26]. The latter investigators showed that for large Jakob number Ja, the solution took the relatively simple form

$$R(t) = 2C_R \sqrt{\alpha_l t} \tag{6.65}$$

where

$$C_R = \sqrt{\frac{3}{\pi}}\,\mathrm{Ja} \qquad \mathrm{Ja} = \frac{(T_\infty - T_{\mathrm{sat}})\rho_l c_{pl}}{\rho_v h_{lv}} \tag{6.66}$$

Thus, for heat-transfer-controlled growth, the bubble radius grows proportional to the square root of time for large values of the Jakob number.

The results of the analyses described above imply that for radially symmetric bubble growth in a superheated ilquid, the early stages of the bubble growth process are primarily inertia-controlled, and the long-time growth is heat-transfer-controlled. The complete bubble growth process must therefore be characterized by a smooth transition between these regimes. Based on this premise, Mikic et al. [6.28] used the asymptotic limiting behaviors described above to derive the following relation for the variation of the bubble radius with time:

$$R^+ = \frac{2}{3}\left[(t^+ + 1)^{3/2} - (t^+)^{3/2} - 1\right] \tag{6.67}$$

where

$$R^+ = \frac{RA}{B^2} \qquad t^+ = \frac{tA^2}{B^2} \tag{6.68}$$

$$A = \left\{\frac{2[T_\infty - T_{\mathrm{sat}}(P_\infty)]h_{lv}\rho_v}{\rho_l T_{\mathrm{sat}}(P_\infty)}\right\}^{1/2} \tag{6.69}$$

$$B = \left(\frac{12\alpha_l}{\pi}\right)^{1/2} \mathrm{Ja} \tag{6.70}$$

$$\mathrm{Ja} = \frac{[T_\infty - T_{\mathrm{sat}}(P_\infty)]c_{pl}\rho_l}{\rho_v h_{lv}} \tag{6.71}$$

Although Eq. (6.67) looks considerably more complicated that either of the limiting cases individually, it does reduce to Eq. (6.59) for small values of t^+ and Eq. (6.65) for large values of t^+. Mikic et al. [6.28] found good agreement between the $R^+(t^+)$ variation predicted by Eq. (6.67) and data obtained by Lien [6.29] for bubble growth in uniformly superheated liquid water over a range of ambient pressure levels.

At $t^+ = 1$, Eq. (6.67) predicts a value of $R^+ = 1$. Because R^+ increases monotonically with t^+, R^+ values much less than 1 correspond to $t^+ \ll 1$, indicating inertia-controlled growth. Likewise, R^+ values much greater than 1 correspond to $t^+ \gg 1$, indicating heat-transfer-controlled growth. Consequently, the magnitude of R^+ relative to 1 is an indication of the regime of bubble growth. For a bubble growing in an infinite ambient, this provides a convenient means of approximately determining the regime of bubble growth for a given bubble size R and ambient conditions.

While this formulation provides a convenient interpolation between the limiting bubble growth behaviors for large and small t^+, its accuracy at intermediate times is by no means guaranteed. In addition, because this formulation uses a linearized form of the Clapeyron equation to relate $P_v - P_\infty$ to $T_v - T_{sat}$, it becomes less accurate at high superheat levels. More recent studies by Theofanous and Patel [6.30] and Prosperetti and Plesset [6.31] have resulted in more refined treatments of the bubble growth process that address these inaccuracies.

Example 6.4 Assuming inertia-controlled growth, estimate the interface velocity of a 0.2-mm-diameter bubble growing in water at atmospheric pressure and 120°C.

For water at atmospheric pressure $T_{sat} = 100°\text{C}$, $h_{lv} = 2257 \text{ kJ/kg}$, $\rho_l = 958 \text{ kg/m}^3$, and $\rho_v = 0.598 \text{ kg/m}^3$. For inertia-controlled growth, Eq. (6.59) specifies that

$$R = \left\{ \frac{2}{3} \left[\frac{T_\infty - T_{sat}(P_\infty)}{T_{sat}(P_\infty)} \right] \frac{h_{lv}\rho_v}{\rho_l} \right\}^{1/2} t$$

It is clear from this relation that dR/dt is a constant, independent of time and bubble diameter. Thus,

$$\frac{dR}{dt} = \left\{ \frac{2}{3} \left[\frac{T_\infty - T_{sat}(P_\infty)}{T_{sat}(P_\infty)} \right] \frac{h_{lv}\rho_v}{\rho_l} \right\}^{1/2}$$

$$= \left[\frac{2}{3} \left(\frac{20}{100 + 273} \right) \frac{2257(1000)(0.598)}{958} \right]^{1/2}$$

$$= 7.10 \text{ m/s}$$

Example 6.5 For bubble growth in water at atmospheric pressure and 120°C, estimate the bubble size corresponding approximately to the transition between inertia-controlled and heat-transfer-controlled growth.

Since the transition corresponds approximately to $R^+ = 1$ or, equivalently,

$$R_{trans} = \frac{B^2}{A}$$

using the saturation properties for water at atmospheric pressure (see Example

6.4), it follows that

$$\mathrm{Ja} = \frac{[(T_\infty - T_{sat}(P_l)]c_{pl}\rho_l}{\rho_v h_{lv}} = \frac{20(4.22)(958)}{0.598(2257)} = 59.9$$

$$B = \left(\frac{12\alpha_l}{\pi}\right)^{1/2} \mathrm{Ja} = \left(\frac{12k_l}{\pi\rho_l c_{pl}}\right)^{1/2} \mathrm{Ja} = \left[\frac{12(0.681)}{\pi(958)(4220)}\right]^{1/2} (59.9)$$

$$= 0.0477 \text{ ms}^{-1/2}$$

$$A = \left\{\frac{2[T_\infty - T_{sat}(P_l)]h_{lv}\rho_v}{\rho_l T_{sat}(P_l)}\right\}^{1/2} = \left[\frac{2(20)(2{,}257{,}000)(0.598)}{958(100 + 273)}\right]^{1/2}$$

$$= 12.29 \text{ m/s}$$

$$R_{trans} = \frac{(0.0477)^2}{12.29} = 0.000185 \text{ m} = 0.185 \text{ mm}$$

6.5 BUBBLE GROWTH NEAR HEATED SURFACES

The growth of vapor bubbles in the thermal boundary-layer region near a superheated surface is more complex than the circumstances considered in Section 6.4 because of the lack of spherical symmetry and the nonuniformity of the temperature field in the surrounding liquid. Despite these significant differences, bubble growth near a superheated surface exhibits regimes of inertia-controlled and heat-transfer-controlled growth similar to those for spherical bubble growth in an infinite, uniformly superheated ambient.

The bubble growth process near a heated wall can be idealized as consisting of the sequence of stages indicated schematically in Fig. 6.15. After the departure of a bubble, liquid at the bulk fluid temperature T_∞ is brought into contact with the surface at temperature $T_w > T_{sat}(P_\infty)$. A brief period of time then elapses, during which transient conduction into the liquid occurs but no bubble growth takes place. This time interval is referred to as the *waiting period,* designated here as t_w.

Once bubble growth begins, the thermal energy needed to vaporize liquid at the interface comes, at least in part, from the liquid region adjacent to the bubble that was superheated during the waiting period. During the initial stage of bubble growth, the liquid immediately adjacent to the interface is highly superheated, and transfer of heat to the interface is not a limiting factor. As the bubble embryo bubble emerges from the nucleation site cavity, a rapid expansion is triggered as a result of the sudden increase in the radius of curvature of the bubble. The resulting rapid growth of the bubble is primarily resisted by the inertia of the liquid. For this inertia-controlled early stage of the bubble growth process, the bubble grows in a nearly hemispherical shape, as schematically shown in Fig. 6.15*c*. In this regime, as the bubble grows radially, a thin microlayer of liquid is left be-

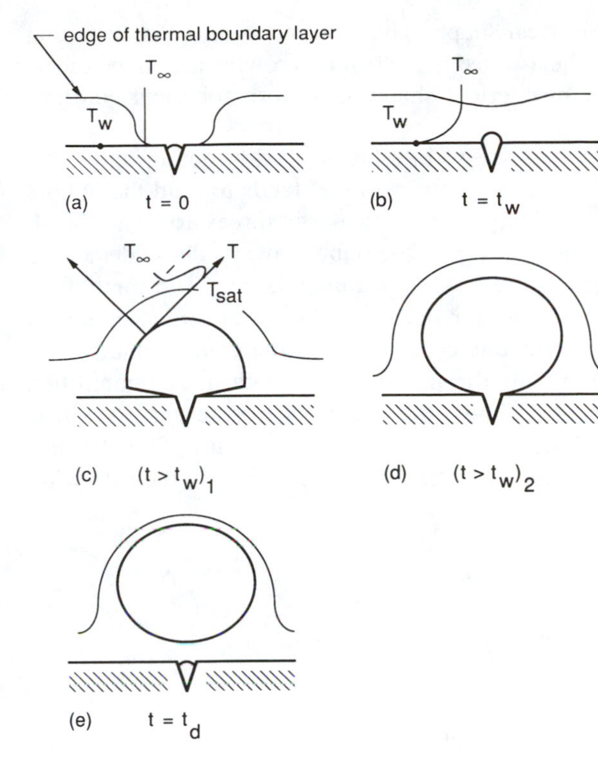

Figure 6.15 The waiting period and subsequent growth and release of a vapor bubble at an active cavity site.

tween the lower portion of the bubble interface and the heated wall (see Fig. 6.15*c*). This film, which is sometimes referred to as the *evaporation microlayer*, varies in thickness from nearly zero near the nucleation site cavity to a finite value at the edge of the hemispherical bubble. Heat is transferred across this film from the wall to the interface, directly vaporizing liquid at the interface. This film may completely evaporate near the cavity where nucleation began, significantly elevating the surface temperature there. When this occurs, the surface temperature may strongly fluctuate during the repeated growth and release of bubbles as the surface cyclically dries out and then rewets.

The liquid region adjacent to the interface is gradually depleted of its superheat as the bubble grows. This region is sometimes referred to as the *relaxation microlayer*. The nature of the temperature profile in this region at an intermediate stage of the bubble growth process is indicated in Fig. 6.15*c*. The interface is at the saturation temperature corresponding to the ambient pressure in the liquid. The liquid temperature increases with increasing distance from the interface, reaches a peak, and then decreases toward the ambient temperature. As growth continues, heat transfer to the interface may become a limiting factor, whereupon the bubble growth becomes heat-transfer-controlled.

If the bubble growth process does become heat-transfer-controlled, pressure and liquid inertia forces become relatively smaller, and surface tension then tends to pull the bubble into a more spherical shape. Thus, in undergoing the transition from inertia-controlled growth to heat-transfer-controlled growth, the shape of the bubble is transformed from a hemispherical shape to a more spherical configuration, as indicated in Fig. 6.15*d*.

Throughout the bubble growth process, interfacial tension acting along the contact line (where the interface meets the solid surface) tends to hold the bubble in place on the surface. Buoyancy, drag, lift, and/or inertia forces associated with motion of the surrounding fluid may act to pull the bubble away. These detaching forces generally become stronger as the bubble becomes larger. The bubble releases, as shown in Fig. 6.15*e*, when their net effect becomes large enough to overcome the retaining effect of surface tension forces at the contact line.

While the above description admits the possibility of both inertia-controlled and heat-transfer-controlled growth regimes, the occurrence or absence of either one depends on the conditions under which bubble growth occurs. Specifically, very rapid, inertia-controlled growth is more likely to be observed if the following conditions exist:

High wall superheat
High imposed heat flux
Highly polished surface having only very small cavities
Very low contact angle (highly wetting liquid)
Low latent heat of vaporization
Low system pressure (resulting in low vapor density)

The first four items on this list result in the buildup of high superheat levels during the waiting period. The last two items result in very rapid volumetric growth of the bubble once the growth process begins. The first item and the last two imply that inertia-controlled growth is likely for large values of the Jakob number Ja, defined in Eq. (6.71). The shape of the bubble is likely to be hemispherical when these conditions exist.

Conversely, heat-transfer-controlled growth of a bubble is more likely for:

Low wall superheat
Low imposed heat flux
A rough surface having many large and moderate-sized cavities
Moderate contact angle (moderately wetting liquid)
High latent heat of vaporization
Moderate to high system pressure

All of the conditions specified above result in slower bubble growth, which makes inertia effects smaller, or result in a stronger dependence of bubble growth rate on heat transfer to the interface. The more of these conditions that are met, the greater is the likelihood that heat-transfer-controlled growth will result.

Analytical treatments of heat-transfer-controlled bubble growth in the nonuniform temperature field near a superheated wall have been presented by Savic

[6.32], Griffith [6.33], Bankoff and Mikesell [6.34], Zuber [6.35], Han and Griffith [6.36], Cole and Shulman [6.37], van Stralen [6.38], and Mikic and Rohsenow [6.39]. The model of Mikic and Rohsenow [6.39] treats the heat-transfer-controlled growth as being governed by a one-dimensioinal transient conduction process that consists of two parts. During the waiting process, which begins at time $t = -t_w$, transient conduction in the liquid is postulated to satisfy the following well-known one-dimensional transport equation with appropriate boundary and initial conditions:

$$\frac{\partial T}{\partial t} = \alpha_l \left(\frac{\partial^2 T}{\partial y^2} \right) \tag{6.72}$$

$$T(y,-t_w) = T_\infty \tag{6.73}$$

$$T(\infty,t) = T_\infty \tag{6.74}$$

$$T(0,t) = T_w \qquad \text{for } -t_w \leqslant t < 0 \tag{6.75}$$

This model problem has a well-known conjugate error function solution that qualitatively varies as shown in Fig. 6.16*a*.

The second part of the overall model incorporates the idealization that conduction of heat to the bubble interface for $t > 0$ is again a one-dimensional transient process governed by Eq. (6.72). The solution is still subject to the far-field boundary condition (6.74). However, the boundary condition at $y = 0$ is taken to be the interface temperature $T_{sat}(P)$,

$$T(0,t) = T_{sat}(P_\infty) \qquad \text{for } t > 0 \tag{6.76}$$

and the initial condition is the temperature field that exists at the end of the first portion of the transient (at time $t = 0$). The variation of the temperature field during this second portion of the transient is shown qualitatively in Fig. 6.16*b*. For $t > 0$, Mikic and Rohsenow [6.39] obtained the following solution of the overall transient:

$$T(y,t) = T_\infty + (T_w - T_\infty) \operatorname{erfc}\left[\frac{y}{2\sqrt{\alpha_l (t + t_w)}} \right] - (T_w - T_{sat}) \operatorname{erfc}\left(\frac{y}{2\sqrt{\alpha_l t}} \right) \tag{6.77}$$

The above relation for the temperature field can be used to determine the rate of heat transfer and resulting vaporization rate at the bubble interface. Using the vaporization rate predicted in this manner, Mikic and Rohsenow [6.39] derived the following relation for the bubble radius as a function of time:

$$R(t) = \frac{2\,\text{Ja}\sqrt{3\pi\alpha_l t}}{\pi} \left\{ 1 - \frac{T_w - T_\infty}{T_w - T_{sat}} \left[\left(1 + \frac{t_w}{t}\right)^{1/2} - \left(\frac{t_w}{t}\right)^{1/2} \right] \right\} \tag{6.78}$$

where Ja is the Jakob number defined by Eq. (6.71). To evaluate the waiting time

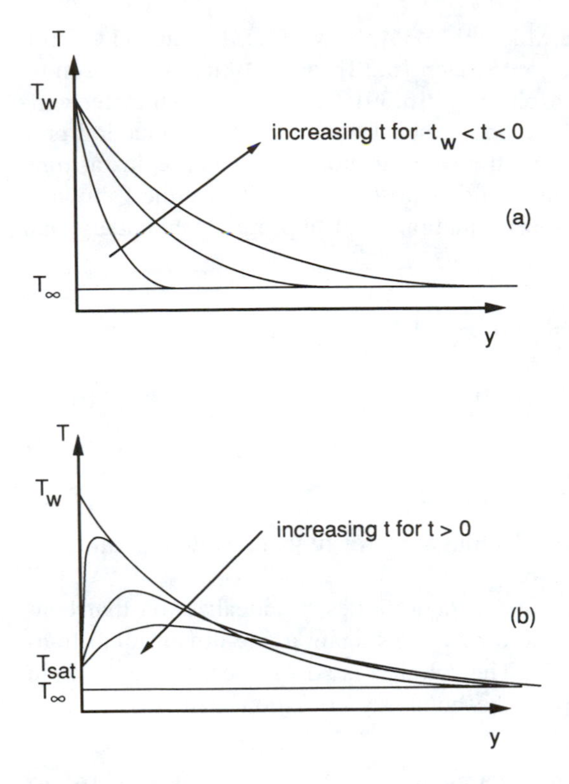

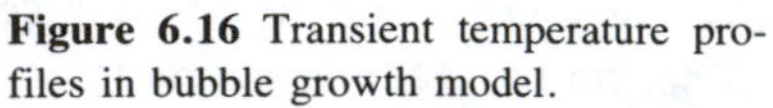
Figure 6.16 Transient temperature profiles in bubble growth model.

t_w, these investigators considered the temperature at the top of a hemispherical bubble of radius r_c covering a cavity of the same radius on the heated surface. They postulated that the temperature at this location in the nonuniform temperature field must exceed the equilibrium superheat for the bubble for growth to begin. The relation for the transient variation of the liquid temperature field for the waiting period was used to determine the time at which this condition was satisfied at a given $y = r_c$. This yielded the following relation for the waiting time:

$$t_w = \frac{1}{4\alpha_l}\left\{\frac{r_c}{\mathrm{erfc}^{-1}\left[\dfrac{T_{\mathrm{sat}} - T_\infty}{T_w - T_\infty} + \dfrac{2\sigma T_{\mathrm{sat}}(v_v - v_l)}{(T_w - T_\infty)h_{lv}r_c}\right]}\right\}^2 \tag{6.79}$$

Equations (6.78) with (6.79) and (6.71) provide a complete prediction of $R(t)$ for heat-transfer-controlled bubble growth near a superheated wall. Mikic and Rohsenow [6.39] reported good agreement between the predicted variation of R with t and observed variations in the experiments of Han and Griffith [6.36].

The $R(t)$ relation for heat-transfer-controlled growth obtained by Mikic and Rohsenow [6.39] will not be accurate when inertia effects are significant and hemispherical growth occurs. For such conditions, vapor production at the interface of the evaporation microlayer at the base of the bubble plays a major role and must be incorporated into the bubble growth model.

Analytical models of the transport in the evaporation microlayer have been proposed by Cooper and Lloyd [6.40] and van Stralen et al. [6.41]. For pure fluids at low pressures, the latent heat of vaporization is high and the density of the vapor is low. Because of these tendencies, at low pressures hemispherical growth may persist long enough to significantly deplete the relaxation microlayer adjacent to the dome of the bubble. Evaporation and hemispherical growth may then be sustained solely by vaporization of the evaporation microlayer between the base of the bubble and the wall.

van Stralen et al. [6.41] used the Pohlhausen [6.42] result for laminar boundary-layer heat transfer to estimate the initial thickness of the microlayer formed as the bubble grows. The Pohlhausen analysis predicts that, for laminar forced-convection boundary-layer flow, the local boundary layer thickness δ at a given downstream location x is given by

$$\frac{\delta}{x} = 3.012\left(\frac{Ux}{\nu}\right)^{-1/2} \mathrm{Pr}^{-1/3} \tag{6.80}$$

where U is the free-stream velocity. In the model proposed by van Stralen et al. [6.41], the initial thickness of the microlayer δ_{l0} at a given radial position is assumed to be given by Eq. (6.80) with δ replaced by δ_{l0}, x replaced by r, and U replaced by the interface velocity $\dot{R} = dR/dt$ at the instant it passes location r. The resulting expression for δ_{l0} is

$$\delta_{l0} = 3.012\left(\frac{\nu_l r}{\dot{R}}\right)^{1/2} \mathrm{Pr}_l^{-1/3} \tag{6.81}$$

If we further postulate that R is a power-law function of time t:

$$R = \gamma t^m \tag{6.82}$$

from which it follows that

$$\dot{R} = m\gamma t^{m-1} \tag{6.83}$$

Because $\dot{R}$ in Eq. (6.81) is evaluated at $r = R$,

$$\frac{r}{\dot{R}} = \frac{t}{m} \tag{6.84}$$

and

$$\delta_{l0} = 3.012\left(\frac{\nu_l t}{m}\right)^{1/2} \mathrm{Pr}_l^{-1/3} \tag{6.85}$$

Because this equation applies at $r = R = \gamma t^m$,

$$t = \left(\frac{r}{\gamma}\right)^{1/m} \tag{6.86}$$

which, when substituted into Eq. (6.85), yields

$$\delta_{l0} = 3.012 \, \mathrm{Pr}_l^{-1/3} \left(\frac{\nu_l}{m\gamma^{1/m}} \right)^{1/2} r^{1/2m} \tag{6.87}$$

Example 6.6 Using Eq. (6.79), estimate the waiting period associated with ebullition from a cavity with a mouth radius of 0.1 mm during boiling of saturated water at atmospheric pressure with a wall superheat of 20°C.

For saturated water at atmospheric pressure $T_{sat} = 100°C$, $v_l = 0.00104$ m³/kg, $v_v = 1.673$ m³/kg, $h_{lv} = 2257$ kJ/kg, $\sigma = 0.0588$ N/m, $\alpha_l = k_l/\rho_l c_{pl} = 5.16 \times 10^{-7}$ m²/s. Thus,

$$\frac{2\sigma T_{sat}(v_v - v_l)}{(T_w - T_\infty) h_{lv} r_c} = \frac{2(0.0588)(373)(1.673 - 0.001)}{20(2{,}257{,}000)(0.1 \times 10^{-3})}$$

$$= 0.0162$$

Substituting into Eq. (6.79) yields

$$t_w = \frac{1}{4(5.16 \times 10^{-7})} \left[\frac{0.1 \times 10^{-3}}{\mathrm{erfc}^{-1}(0 + 0.0162)} \right]^2$$

From standard math tables, $\mathrm{erfc}^{-1}(0.0162) = 1.70$, from which it follows that

$$t_w = 1.68 \times 10^{-3} \text{ s}$$

For growth of the hemispherical bubble solely as a result of evaporation of the microlayer under the base of the bubble, the following energy-balance relation must be satisfied:

$$\rho_v h_{lv} (2\pi R) \dot{R} = \int_0^R \frac{k_l (T_w - T_{sat})(2\pi r)}{\delta_{l0}} \, dr \tag{6.88}$$

Substituting Eq. (6.87) into Eq. (6.88) and using Eq. (6.82) to write the remaining R dependence in terms of t, the following equation is obtained:

$$\dot{R} = \left[\frac{\alpha_l \, \mathrm{Ja} \, \mathrm{Pr}_l^{1/3}}{3.012(2 - 1/2m)} \right] \left(\frac{m}{\nu_l} \right)^{1/2} t^{-1/2} \tag{6.89}$$

where Ja is the Jakob number defined by Eq. (6.71). Integrating and using the condition that $R = 0$ at $t = 0$ yields the following relation for $R(t)$:

$$R = \left[\frac{2\alpha_l \, \mathrm{Ja} \, \mathrm{Pr}_l^{1/3}}{3.012(2 - 1/2m)} \right] \left(\frac{m}{\nu_l} \right)^{1/2} t^{1/2} \tag{6.90}$$

Comparing this relation with Eq. (6.82), it is clear that $m = 1/2$, whereupon the above equation, after a bit of rearranging, becomes

$$R(t) = 0.470 \, \mathrm{Ja} \, \mathrm{Pr}_l^{-1/6} (\alpha_l t)^{1/2} \tag{6.91}$$

The above analysis suggests that for growth controlled by vaporization of the evaporation microlayer, the variation of the bubble radius with time is proportional to $t^{1/2}$, just as for heat-transfer-controlled growth, in which the latent heat is supplied to the bubble dome interface from the relaxation microlayer. As a consequence of this similarity, it is difficult to separate the effects of one mechanism from the other.

van Stralen et al. [6.41] also proposed the following relation for bubble growth in either the inertia-controlled or heat-transfer-controlled regime:

$$R(t) = \frac{R_1(t)R_2(t)}{R_1(t) + R_2(t)} \tag{6.92}$$

where

$$R_1(t) = 0.8165 \sqrt{\frac{\rho_v h_{lv}(T_w - T_{sat}) \exp[-(t/t_d)^{1/2}]}{\rho_l T_{sat}} t} \tag{6.93}$$

$$\begin{aligned} R_2(t) = {} & 1.9544\left\{b^* \exp\left[-\left(\frac{t}{t_d}\right)^{1/2}\right] + \frac{T_\infty - T_{sat}}{T_w - T_{sat}}\right\} \text{Ja}\,(\alpha_l t)^{1/2} \\ & + 0.3730\,\text{Pr}_l^{-1/6}\left\{\exp\left[-\left(\frac{t}{t_d}\right)^{1/2}\right]\right\}^{1/2} \text{Ja}\,(\alpha_l t)^{1/2} \end{aligned} \tag{6.94}$$

$$b^* = 1.3908 \frac{R_2(t_d)}{\text{Ja}\,(\alpha_l t)^{1/2}} - 0.1908\,\text{Pr}_l^{-1/6} \tag{6.95}$$

In the above relations, t_d is the time of departure of the bubble from the surface. Equation (6.92) is basically a superposition of the growth rate relations $R_1(t)$ for inertia-controlled growth and $R_2(t)$ for heat-transfer-controlled growth. $R_2(t)$ represents the combined effects of vapor generation at the bubble dome interface at the expense of the relaxation microlayer, and vaporization of the evaporation microlayer at the base of the bubble. The bubble growth parameter b^* accounts for the fact that only a portion of the vapor bubble dome may be in contact with superheated liquid. If the departure time t_d is known, b^* can be determined from Eqs. (6.94) and (6.95).

Equations (6.92) through (6.95) are actually the pure-liquid forms of equations given by van Stralen et al. [6.41], which apply to either pure liquids or binary systems. The more general forms of these relations can be obtained from reference [6.41]. In a companion study [6.43], these same investigators compared the predictions of these theoretically based relations with experimental bubble growth data for water over a range of temperatures. The agreement between the data and the predicted variation of R with t was generally found to be quite good.

The models of the ebullition cycle described above indicate that for a given cavity there will be a specific bubble size at which bubble release occurs and there will be a specific frequency at which bubbles are generated. These quantities, which parameterize the vapor generation process in a useful manner, are discussed in more detail in the next section.

6.6 BUBBLE DEPARTURE DIAMETER AND THE FREQUENCY OF BUBBLE RELEASE

The complete process of liquid heating, nucleation, bubble growth, and release, collectively referred to as the ebullition cycle, is the central mechanism of heat transfer from a superheated wall during nucleate boiling. Two features of this process that affect the rate of heat transfer during the ebullition cycle are the bubble diameter at departure, d_d, and the frequency, f, at which bubbles are generated and released.

The bubble growth rate analyses described in previous sections suggest that the departure diameter and frequency of release must be related. The inverse of the frequency, $\tau = 1/f$, which is the time period associated with the growth of each bubble, must equal the sum of the waiting period and the time required for the bubble to grow to its departure diameter:

$$\frac{1}{f} = \tau = t_w + t_{2R(t)=d_d} \tag{6.96}$$

The frequency of bubble release thus depends directly on how large the bubble must become for release to occur, and, as a consequence, on the rate at which the bubble can grow to this size.

The bubble diameter at release is primarily determined by the net effect of forces acting on the bubble as it grows on the surface. Interfacial tension acting along the contact line invariably acts to hold the bubble in place on the surface. Buoyancy is often a major player in the force balance, although its effect depends on the orientation of the surface with respect to the accelerating or gravitational body force vector. For an upward-facing horizontal surface, buoyancy directly acts to detach the bubble, whereas for a similar downward-facing surface, buoyancy acts to keep the bubble pressed against the wall. The effect of buoyancy will vary around the perimeter of a superheated horizontal cylinder.

If the bubble grows very rapidly, the inertia associated with the induced liquid flow field around the bubble may also tend to pull the bubble away from the surface. When the liquid adjacent to the surface has a bulk motion associated with it, drag and lift forces on the growing bubble may also act to detach the bubble from the surface. In addition, because the rate of bubble growth and the shape of the bubble (hemispherical or spherical) may affect the conditions for bubble release, the departure diameter may be affected by the wall superheat or heat flux, the contact angle θ, and the thermophysical properties of the liquid and vapor phases.

The departure diameter of bubbles during nucleate boiling has been the subject of numerous investigations over the past 60 years. In experimental studies, the departure diameter was typically determined from high-speed movies of the boiling process. Based on data obtained in this manner, a number of investigators have proposed correlation equations that can be used to predict the departure diameter of bubbles during nucleate boiling. A representative sample of correlations of this type is given in Table 6.1. In this table, many of the correlations are written in terms of the Bond number Bo, defined as

Table 6.1 Departure diameter correlations

Reference	Correlation	Eq.
(ref. [6.44])	$\mathrm{Bo}^{1/2} = 0.02080\theta$	(6.98)
	where θ is the contact angle in degrees	
(ref. [6.45])	$\mathrm{Bo}^{1/2} = \left[\dfrac{\sigma}{g(\rho_l - \rho_v)}\right]^{-1/6}\left[\dfrac{6k_l(T_w - T_{\mathrm{sat}})}{q''}\right]^{1/3}$	(6.99)
(ref. [6.46, 6.47])	$\mathrm{Bo}^{1/2} = \left[\dfrac{3\pi^2\rho_l\alpha_{Tl}^2 g^{1/2}(\rho_l - \rho_v)^{1/2}}{\sigma^{3/2}}\right]^{1/3}\mathrm{Ja}^{4/3}$	(6.100*a*)
	where $\mathrm{Ja} = \dfrac{\rho_l c_{pl}[T_w - T_{\mathrm{sat}}(P_\infty)]}{\rho_v h_{lv}}$	(6.100*b*)
(ref. [6.48, 6.49])	$\dfrac{d_d}{d_F} = -\dfrac{C}{d_F} + \sqrt{\dfrac{C^2}{d_F^2} + 1}$	(6.101)
	where $C = \left(\dfrac{6}{g}\right)\left(\dfrac{\rho_l}{\rho_l - \rho_v}\right)\left(\dfrac{\rho_v}{\rho_l}\right)^{0.4}\left(\dfrac{q''}{h_{lv}\rho_v}\right)$	(6.102)
	and d_F is the diameter calculated with Eq. (6.98)	
(ref. [6.50])	$\mathrm{Bo}^{1/2} = \dfrac{1000}{P}$ (P is pressure in mm Hg)	(6.103)
(ref. [6.51])	$\mathrm{Bo}^{1/2} = 0.04\ \mathrm{Ja}$	(6.104)
	where Ja is computed using Eq. (6.100*b*)	
(ref. [6.52])	$\mathrm{Bo}^{1/2} = C(\mathrm{Ja}^*)^{5/4}$	(6.105)
	where $\mathrm{Ja}^* = \dfrac{T_c c_{pl}\rho_l}{\rho_v h_{lv}}$	(6.106)
	$C = 1.5 \times 10^{-4}$ for water	
	$C = 4.65 \times 10^{-4}$ for fluids other than water	
(ref. [6.53])	$\dfrac{d_d}{d_1} = 1 + \dfrac{d_2}{d_1}$	(6.107)
	where	
	$d_1 = \dfrac{1.65d^*\sigma}{g(\rho_l - \rho_v)}$	(6.108)
	$d_2 = \left[\dfrac{15.6\rho_l}{g(\rho_l - \rho_v)}\right]^{1/3}\left[\dfrac{\beta_d k_l(T_w - T_{\mathrm{sat}})}{h_{lv}\rho_v}\right]^{2/3}$	(6.109)
	$d^* = 6.0 \times 10^{-3}$ mm	
	$\beta_d = 6.0$ for water, alcohol, and benzene	
(ref. [6.54])	$\mathrm{Bo}^{1/2} = 0.25(1 + 10^5K_1)^{1/2}$ for $K_1 < 0.06$	(6.110)
	where	
	$K_1 = \left(\dfrac{\mathrm{Ja}}{\mathrm{Pr}_l}\right)\left\{\left[\dfrac{g\rho_l(\rho_l - \rho_v)}{\mu_l^2}\right]\left[\dfrac{\sigma}{g(\rho_l - \rho_v)}\right]^{3/2}\right\}^{-1}$	(6.111)
(ref. [6.55])	$d_d\left(\dfrac{P_cM}{k_BT_c}\right)^{1/3} = 5.0 \times 10^5\left(\dfrac{P}{P_c}\right)^{-0.46}$	(6.112)
(ref. [6.56])	$\mathrm{Bo}^{1/2} = 0.19(1.8 + 10^5K_1)^{2/3}$	(6.113)
	where K_1 is given by Eq. (6.111)	

$$\text{Bo} = \frac{g(\rho_l - \rho_v)d_d^2}{\sigma} \tag{6.97}$$

This same dimensionless group is also sometimes referred to as the Eotvos number.

The first correlation given in Table 6.1 was developed by Fritz [6.44]. This relation reflects the assumption of a simple balance of surface tension forces and buoyancy at the instant of departure. The effect of the contact angle is taken into account in an empirical manner. Equation (6.99), proposed by Zuber [6.45], was developed from an analysis of the bubble growth in a nonuniform temperature field near a heated surface. The ratio $k_l(T_w - T_{sat})/q''$ on the right side of this equation is a length scale that is representative of the superheated thermal-layer thickness near the surface. Thus surface tension, buoyancy, and the size of the bubble relative to this layer thickness are represented in this correlation.

Equation (6.100) was proposed by Ruckenstein [6.46] (also cited in [6.47]). The form of this relation was obtained by considering the balance among buoyancy, drag, and surface tension forces. A more complicated relation (6.101) was developed by Borishansky and Fokin [6.48] (as cited in [6.49]) in an effort to account more accurately for the dynamic interaction between a growing bubble and fluid motion in the surrounding liquid.

Cole and Shulman [6.50] proposed a relation (6.103) in which $\text{Bo}^{1/2}$ is simply proportional to the inverse of the absolute pressure. This relation contrasts sharply with other relations listed in Table 6.1, in which the dimensionless departure diameter $\text{Bo}^{1/2}$ depends on a complex combination of physical properties. The success of Eq. (6.103) is apparently a result of the fact that $1000/P$ approximates the combined pressure dependence of the properties that appear in these other relations. In a subsequent study, Cole [6.51] proposed Eq. (6.104). This relation is an extension of Eq. (6.103) in the sense that the pressure term is taken into account by the inclusion of the vapor density in the Jakob number Ja.

Cole and Rohsenow [6.52] proposed Eq. (6.105) as an evolutionary improvement of Eq. (6.104). In Eq. (6.104) the wall superheat was replaced by the critical temperature T_c because experimental data contradicted the proportionality between wall superheat and departure diameter implied by Eq. (6.104).

Golorin et al. [6.53] later developed the correlation represented by Eqs. (6.107) through (6.109). This correlation includes the dynamic interaction between the growing bubble and the surrounding liquid as well as small-scale roughness of the heating surface. Kutateladze and Gogonin [6.54] found that they could correlate a large body of data from the literature with the correlation given by Eqs. (6.110) and (6.111), which contains only the Bond number and the dimensionless group K_1. The correlation given by Eq. (6.112) was developed by Borishanskiy et al. [6.55] based on thermodynamic similitude.

Jensen and Memmel [6.56] recently compared the correlations described above against available departure diameter data. As observed by Jensen et al. [6.56], there is considerable scatter in the available data from different references. Despite differences in the forms of the correlating equations, all of these correlations fit the data to some extent. This is undoubtedly a consequence of the scatter in the data and the empirical nature of the correlations. The correlation of Kutateladze and Gogonin [6.54] given by Eqs. (6.110) and (6.111) was found to provide the best overall fit to the data examined by Jensen and Memmel [6.56]. The average absolute deviation for this correlation was 45.4%. Jensen and Memmel [6.56] also proposed the correlation given by Eqs. (6.111) and (6.113) as an improved version of the correlation of Kutateladze and Gogonin [6.54]. This improved correlation fit the available data to an average absolute deviation of 44.4%.

The considerable scatter in the data about even the best correlation in Table 6.1 suggests that there may be considerable uncertainty in experimental data of this type. It also suggests that current understanding of factors that influence the bubble departure diameter is incomplete.

The frequency of bubble release depends directly on how large the bubble must become for release to occur and on the rate at which the bubble can grow to the release diameter. The frequency of release will therefore be a function of the departure diameter of the bubble and all the conditions and fluid properties that affect the waiting time and the bubble growth rate. As previously discussed, the size and nature of each cavity affects the nucleation and waiting-time behavior. This would seem to explain, at least partially, why visual observations of nucleate boiling generally indicate that individual sites emit bubbles with a nominally constant frequency, but the observed frequency varies from site to site. Although different cavities will bubble at different frequencies, it is useful to consider the mean bubbling frequency f associated with the boiling process for a given solid–liquid combination and imposed conditions.

There have been a number of attempts to develop relations to predict the mean bubble frequency for growth at a heated surface. The accuracy of some of the resulting relations is open to question, however. In a very early study, Jakob and Fritz [6.57] proposed the following relation for hydrogen and water vapor bubbles:

$$fd_d = 0.078 \tag{6.114}$$

In subsequent studies, Peebles and Garber [6.58] and Zuber [6.59, 6.60] also proposed relations that suggest that the bubble frequency f is inversely proportional to departure diameter d_d. Peebles and Garber [6.58] proposed the relation

$$fd_d = 1.18\left(\frac{t_g}{t_g + t_w}\right)\left[\frac{\sigma g(\rho_l - \rho_v)}{\rho_l^2}\right]^{1/4} \tag{6.115}$$

where t_w is the waiting time and t_g is the growth time to the departure diameter. Based on an analogy between the bubble release process and natural convection, Zuber [6.60] suggested the following relation:

$$fd_d = 0.59\left[\frac{\sigma g(\rho_l - \rho_v)}{\rho_l^2}\right]^{1/4} \tag{6.116}$$

Cole [6.61] showed that $1.18t_g/(t_g + t_w)$ can vary between 0.15 and 1.4, which implies that Eq. (6.116) is consistent with Eq. (6.115), but that the assumption of a constant multiplier of the term in square brackets for all possible circumstances is an oversimplification.

Ivey [6.62] argued that the f–d_d relation is dependent on the regime of bubble growth, with

$$f^2 d_d = \text{constant} \qquad \text{for dynamically (inertia) controlled growth}$$

$$f^{1/2} d_d = \text{constant} \qquad \text{for thermally (heat-transfer) controlled growth}$$

For the intermediate range between these limits, the exponent of f is postulated to change from 2 to $\frac{1}{2}$. This suggests that Eqs. (6.114) through (6.116) apply to conditions in the intermediate regime between heat-transfer- and inertia-controlled growth.

For the inertia-controlled regime, Cole [6.63] suggested the relation

$$f^2 d_d = \left(\frac{4}{3}\right) \frac{g(\rho_l - \rho_v)}{C_d \rho_l} \tag{6.117}$$

where C_d is a bubble drag coefficient, estimated to equal 1 for water at 1 atm.

Mikic and Rohsenow [6.39] used their model of the heat-transfer-controlled growth of a bubble in the nonuniform temperature field near a heated surface to evaluate the waiting and growth times and derived the following relation for the bubble frequency:

$$f^{1/2} d_d = \left(\frac{4}{\pi}\right) \text{Ja} \sqrt{3\pi\alpha_l} \left[\left(\frac{t_g}{t_w + t_g}\right)^{1/2} + \left(1 + \frac{t_g}{t_w + t_g}\right)^{1/2} - 1 \right] \tag{6.118}$$

These investigators further showed that for $0.15 < t_w/(t_w + t_g) < 0.8$, this expression is well approximated by the following simpler relation:

$$f^{1/2} d_d = 0.83 \, \text{Ja} \sqrt{\pi\alpha_l} \tag{6.119}$$

where Ja is the Jakob number defined by Eq. (6.71).

The relations described above for predicting the bubble frequency are based on theoretical arguments and a limited quantity of experimental data. Because their accuracy has not been extensively verified, and there are differences among the trends in different sets of data, they should be treated as being approximate. Despite their limited accuracy as predictive tools, these relations do provide useful insight into an important aspect of nucleate boiling heat transfer. The relationship between the bubbling frequency and the heat transfer performance during nucleate boiling is discussed further in Chapter Seven.

Example 6.7 Estimate the bubble departure diameter and bubbling frequency for saturated water at atmospheric pressure for a wall superheat of 20°C. Compare the period between bubbles with the waiting time determined in Example 6.6.

For saturated water at atmospheric pressure, $T_{sat} = 100°\text{C}$, $v_l = 0.00104$ m³/kg, $v_v = 1.673$ m³/kg, $h_{lv} = 2257$ kJ/kg, $\sigma = 0.0588$ N/m, $c_{pl} = 4.22$ kJ/kg K. Thus,

$$\text{Ja} = \frac{(T_w - T_{sat}) c_{pl} \rho_l}{\rho_v h_{lv}} = \frac{20(4.22)/0.00104}{(1/1.673)/(2257)} = 60.2$$

using Eq. (6.104), it follows that

$$\mathrm{Bo}^{1/2} = 0.04\ \mathrm{Ja} = 2.40$$

and using Eq. (6.97) it can be shown that

$$d_d = \left[\frac{\sigma\ \mathrm{Bo}}{g(\rho_l - \rho_v)}\right]^{1/2} = \left[\frac{0.0588}{9.8(1/0.00104 - 1/1.673)}\right]^{1/2} (2.40)$$

$$= 6.00 \times 10^{-3}\ \mathrm{m} = 6.0\ \mathrm{mm}$$

Using Eq. (6.116),

$$fd_d = 0.59\left[\frac{\sigma g(\rho_l - \rho_v)}{\rho_l^2}\right]^{1/4}$$

$$= 0.59\left[\frac{0.0588(9.8)(1/0.00104 - 1/1.673)}{(1/0.00104)^2}\right]^{1/4} = 0.156\ \mathrm{m/s}$$

Substituting the value d_d obtained above,

$$f = \frac{0.156}{d_d} = \frac{0.156}{0.006} = 26.1\ \mathrm{s}^{-1}$$

The period τ_b between bubble releases is therefore $\tau_b = 1/f = 1/26.1 = 0.038$ s, which is significantly longer than the 0.0017 s waiting period determined in Example 6.6.

REFERENCES

6.1 Volmer, M., Über Keimbildung und Keimwiskung als Spezial fälle der heterogenen Katalyse, *Z. Electrochem.*, vol. 35, pp. 555–561, 1929.
6.2 Fisher, J. C., The fracture of liquids, *J. Appl. Phys.*, vol. 19, pp. 1062–1067, 1948.
6.3 Turnbull, D., Kinetics of heterogeneous nucleation, *J. Chem. Phys.*, vol. 18, pp. 198–203, 1950.
6.4 Bankoff, S. G., Ebullition from solid surfaces in the absence of a pre-existing gaseous phase, *Trans. ASME*, vol. 79, pp. 735–740, 1957.
6.5 Cole, R., Boiling nucleation, *Adv. Heat Transfer*, vol. 10, pp. 85–167, 1974.
6.6 Jarvis, T. J., Donahue, M. D., and Katz, J. L., *J. Colloid Interface Sci.*, vol. 50, p. 359, 1975.
6.7 Bankoff, S. G., *AIChE J.*, vol. 4, p. 24, 1958.
6.8 Corty, C., and Faust, A. S., *Chem. Eng. Prog. Symp. Ser.*, vol. 51, no. 17, p. 1, 1955.
6.9 Clark, H. B., Strenge, P. S., and Westwater, *Chem. Eng. Prog. Symp. Ser.*, vol. 55, no. 29, p. 103, 1959.
6.10 Cornwell, K., *Lett. Heat Mass Transfer*, vol. 4, p. 63, 1977.
6.11 Wei, C. C., and Preckshot, G. W., *Chem. Eng. Sci.*, vol. 19, p. 838, 1964.
6.12 Kosky, P. G., *Int. J. Heat Mass Transfer*, vol. 11, p. 929, 1968.
6.13 Harvey, E. N., McElroy, W. D., and Whiteley, A. H., *J. Appl. Phys.*, vol. 18, p. 162, 1947.
6.14 Knapp, R. T., *Trans. ASME*, vol. 80, p. 1315, 1958.
6.15 Sabersky, R. H., and Gates, C. W., Jr., *Jet Propul.*, Vol. 25, p. 67, 1955.
6.16 Bankoff, S. G., *Chem. Eng. Prog. Symp. Ser.*, vol. 55, no. 29, p. 87, 1959.
6.17 Marto, P. J., and Rohsenow, W. M., *J. Heat Transfer*, vol. 91, p. 315, 1966.
6.18 Lorenz, J. J., Mikic, B. B., and Rohsenow, W. M., *Proc. Fifth Int. Heat Transfer Conf.*, vol IV, p. 35, 1974.

6.19 Griffith, P., and Wallis, J. D., The role of surface conditions in nucleate boiling, *Chem. Eng. Prog. Symp. Ser.,* vol. 561, no. 30, pp. 49–63, 1960.

6.20 Mikic, B. B., and Rohsenow, W. M., *J. Heat Transfer,* vol. 91, pp. 245–250, 1969.

6.21 Hsu, Y. Y., On the size range of active nucleation cavities on a heating surface, *J. Heat Transfer,* vol. 84, pp. 207–213, 1962.

6.22 Hsu, Y. Y., and Graham, R. W., An analytical and experimental study of the thermal boundary layer and the ebullition cycle in nucleate boiling, NASA TN-D-594, 1961.

6.23 Carslaw, H. S., and Jaeger, J. C., *Conduction of Heat in Solids,* 2nd ed., Oxford University Press, Oxford, 1978.

6.24 Rayleigh, Lord, On the pressure developed in a liquid during the collapse of a spherical cavity, *Phil. Mag.,* vol., 34, pp. 94–98, 1917 (also in *Scientific Papers,* vol. 6, Cambridge University Press, Cambridge, 1920).

6.25 Forster, H. K., and Zuber, N., Dynamics of vapor bubbles and boiling heat transfer, *AIChE J.,* vol. 1, pp. 531–535, 1955.

6.26 Plesset, M. S., and Zwick, S. A., The growth of vapor bubbles in superheated liquids. *J. Appl. Phys.,* vol. 25, pp. 493–500, 1954.

6.27 Scriven, L. E., On the dynamics of phase growth, *Chem. Eng. Sci.,* vol. 90, pp. 1–13, 1959.

6.28 Mikic, B. B., Rohsenow, W. M., and Griffith, P., On bubble growth rates, *Int. J. Heat Mass Transfer,* vol. 13, pp. 657–666, 1970.

6.29 Lien, Y., Bubble growth rates at reduced pressure, Sc.D. thesis, Mechanical Engineering Dept., M.I.T., 1969.

6.30 Theofanous, T. G., and Patel, P. D., Universal relations for bubble growth, *Int. J. Heat Mass Transfer,* vol. 2, pp. 83–98, 1976.

6.31 Prosperetti, A., and Plesset, M. S., Vapour-bubble growth in a superheated liquid, *J. Fluid Mech.,* vol. 85, pp. 349–368, 1978.

6.32 Griffith, P., Bubble growth rates in boiling, *J. Heat Transfer,* vol. 80, pp. 721–726, 1958.

6.33 Savic, P., Discussion on bubble growth rates in boiling, *J. Heat Transfer,* vol. 80, pp. 726–728, 1958.

6.34 Bankoff, S. G., and Mikesell, R. D., Growth of bubbles in a liquid of initially non-uniform temperature, ASME paper no. 58-A-105, ASME, New York, 1958.

6.35 Zuber, N., Vapor bubbles in non-uniform temperature field, *Int. J. Heat Mass Transfer,* vol. 2, pp. 83–98, 1961.

6.36 Han, C. Y., and Griffith, P., The mechanism of heat transfer in nucleate pool boiling, Parts I and II, *Int. J. Heat Mass Transfer,* vol. 8, pp. 887–914, 1965.

6.37 Cole, R., and Shulman, H. L., Bubble growth rates of high Jakob numbers, *Int. J. Heat Mass Transfer,* vol. 9, pp. 1377–1398, 1966.

6.38 van Stralen, J. D., The mechanism of nucleate boiling in pure liquids and in a binary mixture, Parts I and II, *Int. J. Heat Mass Transfer,* vol. 9, pp. 995–1046, 1966.

6.39 Mikic, B. B., and Rohsenow, W. M., Bubble growth rates in non-uniform temperature field, *Prog. Heat Mass Transfer,* vol. II, pp. 283–292, 1969.

6.40 Cooper, M. G., and Lloyd, A. J. P., The microlayer in nucleate pool boiling, *Int. J. Heat Mass Transfer,* vol. 12, pp. 895–913, 1969.

6.41 van Stralen, S. J. D., Sohal, M. S., Cole, R., and Sluyter, W. M., Bubble growth rates in pure and binary systems: Combined effect of relaxation and evaporation microlayers, *Int. J. Heat Mass Transfer,* vol. 18, pp. 453–467, 1975.

6.42 Pohlhausen, E., Der Warmeaustauch Zwishen Festen Korpern und Flussigkeiten mit kleiner Treibung und Warmeleitung, *Z. Angew. Math. Mech.,* vol. 1, pp. 115–121, 1921.

6.43 van Stralen, S. J. D., Cole, R., Sluyter, W. M., and Sohal, M. S., Bubble growth rates in nucleate boiling of water at subatmospheric pressures, *Int. J. Heat Mass Transfer,* vol. 18, pp. 655–669, 1975.

6.44 Fritz, W., Berechnung des Maximalvolume von Dampfblasen, *Phys. Z.,* vol. 36, pp. 379–388, 1935.

6.45 Zuber, N., Hydrodynamic aspects of boiling heat transfer, U.S. AEC report AECU 4439, June, 1959.

6.46 Ruckenstein, E., Physical model for nucleate boiling heat transfer from a horizontal surface, *Bul. Institutului Politech. Bucaresti*, vol. 33, no. 3, pp. 79–88, 1961; *Appl. Mech. Rev.*, vol. 16, Rev. 6055, 1963.
6.47 Zuber, N., Recent trends in boiling heat transfer research. Part I: Nucleate pool boiling, *Appl. Mech. Rev.*, vol. 17, pp. 663–672, 1964.
6.48 Borishansky, V. M., and Fokin, F. S., Heat transfer and hydrodynamics in steam generators, *Trudy TsKTI*, vol. 62, p. 1, 1963.
6.49 Voloshko, A. A., and Vurgaft, A. V., Dynamics of vapor-bubble breakoff under free-convection boiling conditions, *Heat Transfer—Sov. Res.*, vol. 2, no. 6, pp. 136–141, 1970.
6.50 Cole, R., and Shulman, H. L., Bubble departure diameters at subatmospheric pressures, *Chem. Eng. Prog. Symp. Ser.*, vol. 62, no. 64, pp. 6–16, 1966.
6.51 Cole, R., Bubble frequencies and departure volumes at subatmospheric pressures, *AIChE J.*, vol. 13, pp. 779–783, 1967.
6.52 Cole, R., and Rohsenow, W. M., Correlation of bubble departure diameters for boiling of saturated liquids, *Chem. Eng. Prog. Symp. Ser.*, vol. 65, no. 92, pp. 211–213, 1968.
6.53 Golorin, V. S., Kol'chugin, B. A., and Zakharova, E. A., Investigation of the mechanism of nucleate boiling of ethyl alcohol and benzene by means of high-speed motion-picture photography, *Heat Transfer—Sov. Res.*, vol. 10, no. 4, pp. 79–98, 1978.
6.54 Kutateladze, S. S., and Gogonin, I. I., Growth rate and detachment diameter of a vapor bubble in free convection boiling of a saturated liquids, *High Temperature*, vol. 17, pp. 667–671, 1979.
6.55 Borishanskiy, V. M., Danilova, G. N., Gotovskiy, M. A., Borishanskiy, A. V., Danilova, G. P., and Kupriyanova, A. V., Correlation of Data on Heat Transfer in, and Elementary Characteristics of the Nucleate Boiling Mechanism, *Heat Transfer—Sov. Res.*, vol. 13, no. 1, pp. 100–116, 1981.
6.56 Jensen, M. K., and Memmel, G. J., Evaluation of Bubble Departure Diameter Correlations, *Proc. Eighth Int. Heat Transf. Conf.*, vol. 4, pp. 1907–1912, 1986.
6.57 Jakob, M., and Fritz, W., *Frosh. Geb. Ingenieurwes*, vol. 2, p. 434, 1931.
6.58 Peebles, F. N., and Garber, H. J., Studies on motion of gas bubbles in liquids, *Chem. Eng. Prog.*, vol. 49, pp. 88–97, 1953.
6.59 Zuber, N., Hydrodynamic aspects of boiling heat transfer, Ph.D. thesis, UCLA, USAEC Report AECU-4439, 1959.
6.60 Zuber, N., Nucleate boiling—the region of isolated bubbles—similarity with natural convection, *Int. J. Heat Mass Transfer*, vol. 6, pp. 53–65, 1963.
6.61 Cole, R., Frequency and departure diameter at sub-atmospheric pressures, *AIChE Journal*, vol. 13, pp. 779–783, 1967.
6.62 Ivey, H. J., Relationships between bubble frequency, departure diameter and rise velocity in nucleate boiling, *Int. J. Heat Mass Transfer*, vol. 10, pp. 1023–1040, 1967.
6.63 Cole, R., Photographic study of boiling in the region of critical heat flux, *AIChE J.*, vol. 6, pp. 533–542, 1960.

PROBLEMS

6.1 A vessel made of stainless steel holds liquid acetone at atmospheric pressure. If the contact angle at the walls is 10°, estimate the wall superheat corresponding to the kinetic limit of superheat. Use $J = 10^{10}\ \mathrm{m^{-2}\ s^{-1}}$ as a threshold value of J corresponding to the superheat limit.

6.2 Determine and plot the variation of the critical bubble radius r^* with pressure for water between atmospheric pressure and the critical pressure for a wall superheat of 10°C. Based on your results, how do you expect nucleation behavior to be affected by an increase in system pressure? Explain briefly.

6.3 For a thermal boundary-layer thickness of 5.0×10^{-6} m, use Hsu's analysis to determine the size range of active nucleation site cavities for saturated R-12 at 200 kPa flowing over a wall at a temperature of 10°C.

6.4 Using Hsu's [6.21] analysis, compute for saturated water the variation of the range of active nucleation site sizes as a function of pressure for a thermal boundary-layer thickness of 3.0×10^{-6} m and a wall superheat of 15°C. Plot the results for pressures between atmospheric pressure and the critical point.

6.5 For forced-convection flow of liquid nitrogen at 360 kPa over a surface held at 100 K, the thermal boundary layer is estimated to be 2×10^{-5} m thick. Using the results of Hsu's [6.21] analysis, determine the size range of active nucleation sites on the surface for (*a*) saturated liquid and (*b*) liquid at a bulk temperature of 85 K.

6.6 For water at atmospheric pressure, use the results of Hsu's [6.21] analysis to plot the variation of the range of active cavity sizes with bulk subcooling for a thermal boundary-layer thickness of 2.0×10^{-6} m and a wall superheat of 15°C.

6.7 Using the relation $\delta = k_l/h$ to estimate the thermal boundary-layer thickness, determine the minimum value of the heat transfer coefficient h that will ensure that nucleate boiling does not occur on a surface held at 67°C over which saturated liquid R-12 is flowing. The pressure in the system is 1145 kPa.

6.8 For R-12 liquid at 528 kPa superheated to a bulk temperature of 300 K, estimate the time required for a bubble to grow from essentially zero radius to a radius of 1 mm.

6.9 A specific bubble radius R_{trans} corresponds approximately to the transition between the inertia-controlled growth regime and the heat-transfer-controlled growth regime. Determine and plot the variation of R_{trans} with pressure for water superheated above its normal saturation temperature by 10°C. (*Hint:* Assume that R_{trans} corresponds to $R^+ \simeq 1$.)

6.10 Using the results of the Mikic and Rohsenow [6.39] model discussed in Section 6.5, determine the waiting period associated with ebullition from a cavity with a mouth radius of 0.1 mm during boiling of saturated liquid nitrogen at atmospheric pressure. Compare your result to that obtained for boiling of saturated liquid mercury from a cavity of the same size at atmospheric pressure.

6.11 Use Eqs. (6.100) and (6.116) to estimate the bubble departure diameter and bubble frequency for boiling of saturated liquid nitrogen at atmospheric pressure with a wall superheat of 10°C. Repeat the calculations for wall superheats of 2, 5, and 20°C, and plot the variations of these quantities with superheat.

6.12 The frequency–bubble diameter product is a lower bound for the upward velocity of bubbles departing the surface. For water boiling at atmospheric pressure on a wall superheated by 70°C, compute the fd_d product and d_d using Eqs. (6.116) and (6.104). Compare the results with the rise velocity v_r for spherical bubbles of diameter d_d predicted by the relation $v_r = 1.79 \sqrt{g d_d(\rho_l - \rho_v)/\rho_l}$. (This relation is valid for bubbles modeled as solid spheres at Reynolds numbers between 1000 and 10^5.)

6.13 A polished copper surface is immersed in a horizontal position in saturated liquid water at a pressure of 571 kPa. The liquid flows over the surface with a free-stream velocity of 1.0 m/s. The surface is 0.2 m long in the direction of flow and is held at a constant and uniform temperature of 435 K. Cavities and a mouth radius of 0.03 mm have been added over the entire surface. Because of the polishing, naturally occurring cavities on the surface have mouth radii that are less than 5.0×10^{-7} m. For single-phase forced convection under these conditions, the thermal boundary-layer thickness is given by

$$\delta_t = 24.9 x^{0.2} \left(\frac{\mu_l}{\rho_l u_\infty} \right)^{0.8}$$

where x is the distance downstream of the leading edge of the surface. (*a*) At the locations $x = 0.05$ m, 0.1 m, and 0.2 m, are active nucleation sites present or absent? Justify your answer quantitatively. (*b*) With the system initially at the conditions specified above, the pressure is suddenly decreased to atmospheric pressure. Describe the changes in the nucleation behavior that you expect to observed immediately after the drop in pressure. (*c*) Estimate the maximum growth rate dR/dt for bubble growth at the surface immediately following the drop in pressure described in part (*b*).

CHAPTER

SEVEN

POOL BOILING

7.1 REGIMES OF POOL BOILING

Boiling at the surface of a body immersed in an extensive pool of motionless liquid is generally referred to as *pool boiling*. This type of boiling process is encountered in a number of applications, including metallurgical quenching processes, flooded tube-and-shell evaporators (with boiling on the shell side), immersion cooling of electronic components, and boiling of water in a pot on the burner of a stove. The nature of the pool boiling process varies considerably depending on the conditions at which boiling occurs. The level of heat flux, the thermophysical properties of the liquid and vapor, the surface material and finish, and the physical size of the heated surface all may have an effect on the boiling process.

The regimes of pool boiling are most easily understood in terms of the so-called boiling curve: a plot of heat flux q'' versus wall superheat, $T_w - T_{sat}$, for the circumstances of interest. Most of the features of the classical pool boiling curve were determined in the early investigations of pool boiling conducted by Nukiyama [7.1], Jakob and Linke [7.2], and Drew and Mueller [7.3]. Strictly speaking, the classical pool boiling curve defined by the work of these and other investigators applies to well-wetted surfaces for which the characteristic physical dimension L is large compared to the bubble or capillary length scale L_b, defined as

$$L_b = \sqrt{\frac{\sigma}{g(\rho_l - \rho_v)}}$$

The discussion in this section is limited to pool boiling of wetting liquids on surfaces with dimensions large compared to L_b. In subsequent sections, features of the boiling curve when the liquid poorly wets the surface or when L/L_b is not large will be examined more closely.

For the purposes of this discussion, we will assume that the ambient liquid surrounding the immersed body is at the saturation temperature for the ambient pressure. If the surface temperature of the immersed body is controlled and slowly increased, the boiling curve will look similar to that shown in Fig. 7.1. The regimes of pool boiling encountered for a horizontal flat surface as its temperature is increased are schematically indicated in Fig. 7.2. The lateral extent of the surface is presumed to be much larger than L_b. As described in Chapter Six, at very low wall superheat levels, no nucleation sites may be active and heat may be transferred from the surface to the ambient liquid by natural convection alone. The heat transfer coefficient associated with natural convection is relatively low, and q'' increases slowly with $T_w - T_{sat}$.

Eventually the superheat becomes large enough to initiate nucleation at some of the cavities on the surface. This onset of nucleate boiling (ONB) condition occurs at point c in Fig. 7.1. Because we have postulated that the temperature of the surface is controlled, the sudden appearance of this added heat transfer mech-

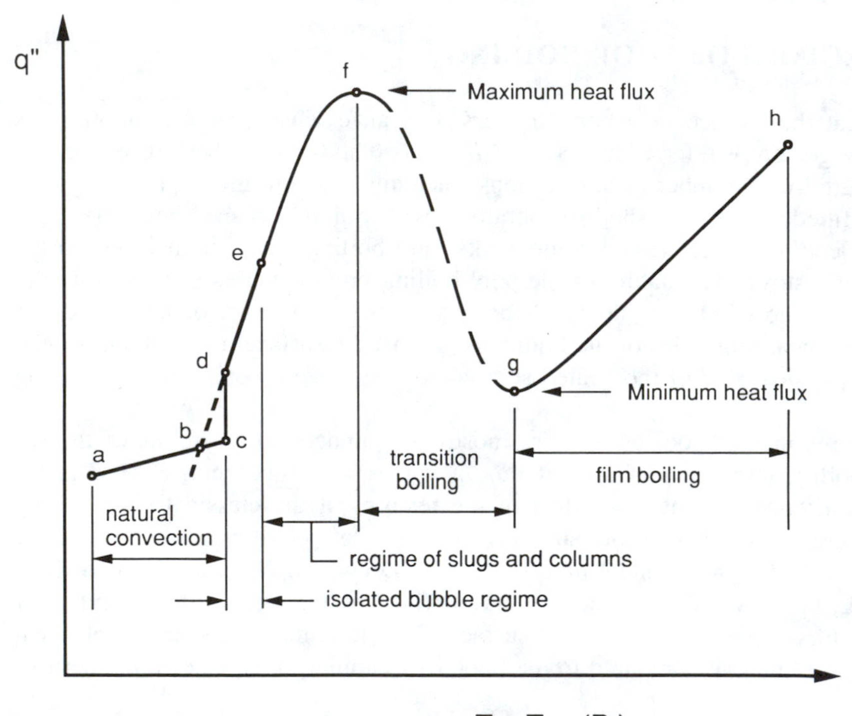

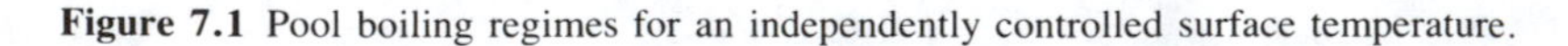

Figure 7.1 Pool boiling regimes for an independently controlled surface temperature.

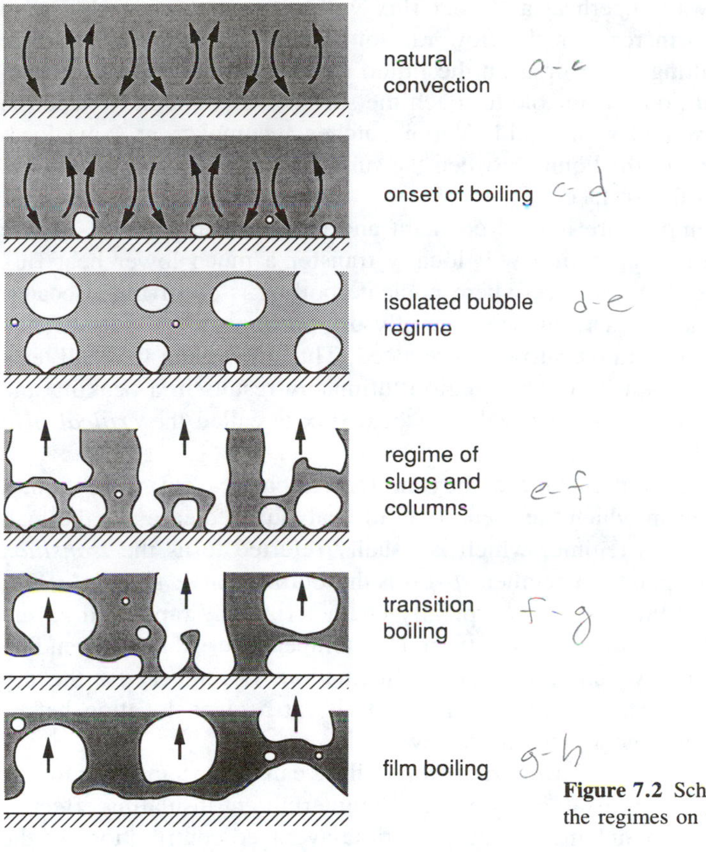

Figure 7.2 Schematic representation of the regimes on the pool boiling curve.

anism does not change the surface temperature, but it does increase the heat flux, with the result that the system operating point moves vertically upward from point c to point d in Fig. 7.1.

Once nucleate boiling is initiated, any further increase in wall temperature causes the system operating point to move upward along section d–f of the curve in Fig. 7.1. This portion of the curve corresponds to the *nucleate boiling regime*. The active sites are few and widely separated at low wall superheat levels. This range of conditions, corresponding to segment d–e of the curve, is sometimes referred to as the *isolated bubble regime*.

With increasing surface superheat, more and more sites become active, and the bubble frequency at each site generally increases. Eventually, the active sites are spaced so closely that bubbles from adjacent sites merge together during the final stages of growth and release. Vapor is being produced so rapidly that bubbles merging together form columns of vapor slugs that rise upward in the liquid pool toward its free surface. This higher range of wall superheat, corresponding to segment e–f of the boiling curve in Fig. 7.1, is referred to as the *regime of slugs and columns*.

Increasing the wall superheat and heat flux within the regime of slugs and columns produces an increase in the flow rate of vapor away from the surface. Eventually, the resulting vapor drag on the liquid moving toward the surface becomes so severe that liquid is unable to reach the surface fast enough to keep the surface completely wetted with liquid. Vapor patches accumulate at some locations, and evaporation of the liquid between the surface and some of these patches dries out portions of the surface.

If the surface temperature is held constant and uniform, dry portions of the surface covered with a vapor film will locally transfer a much lower heat flux than wetted portions of the surface where nucleate boiling is occurring. Because of the reduction in heat flux from intermittently dry portions of the surface, the mean overall heat flux from the surface is reduced. Thus increasing the wall temperature within the slugs and columns region ultimately results in a peaking and rollover of the heat flux. The peak value of heat flux is called the *critical heat flux* (CHF), designated as point f in Fig. 7.1.

If the wall temperature is increased beyond the critical heat flux condition, a regime is encountered in which the mean overall heat flux decreases as the wall superheat increases. This regime, which is usually referred to as the *transition boiling regime,* corresponds to segment f–g on the boiling curve shown in Fig. 7.1. The transition boiling regime is typically characterized by rapid and severe fluctuations in the local surface heat flux and/or temperature values (depending on the imposed boundary condition). These fluctuations occur because the dry regions are generally unstable, existing momentarily at a given location before collapsing and allowing the surface to be rewetted.

The vapor film generated during transition boiling can be sustained for longer intervals at higher wall temperatures. Because the intermittent insulating effect of the vapor blanketing is maintained longer, the time-averaged contributions of the blanketed locations to the overall mean heat flux are reduced. The mean heat flux from the surface thus decreases as the wall superheat is increased in the transition regime. As this trend continues, eventually a point is reached at which the surface is hot enough to sustain a stable vapor film on the surface for an indefinite period of time. The entire surface then becomes blanketed with a vapor film, thus making the transition to the *film boiling regime*. This transition occurs at point g in Fig. 7.1.

Within the film boiling regime, the heat flux monotonically increases as the superheat increases. This trend is a consequence of the increased conduction and/or convection transport due to the increased driving temperature difference across the vapor film $[T_w - T_{sat}(P_l)]$. Radiative transport across the vapor layer may also become important at higher wall temperatures.

In general, once a surface is heated to a superheat level in the film boiling regime, if the surface temperature is slowly decreased, the system will progress through each of the regimes described above in reverse order. Experimental evidence indicates, however, that the path of the boiling curve may differ significantly from that observed for increasing wall superheat. Differences between the increasing and decreasing superheat curves may arise in the transition boiling regime and at the transition between nucleate boiling and natural convection, as indicated in Fig. 7.3.

Experimental evidence summarized by Witte and Lienhard [7.4] implies that the path of the transition boiling curve is determined, to a large degree, by the wetting characteristics of the liquid on the solid surface. For a given wall superheat level in the transition boiling regime, a higher heat flux is generally obtained if the liquid wets the surface than if it poorly wets the surface. During the intermittent contact of liquid with the surface, apparently the liquid spreads and transports heat more effectively from the wall if it wets the surface better.

When transition boiling is initiated by increasing the wall superheat beyond the critical heat flux condition, the observed liquid contact angle is usually near the receding contact angle θ_r for the system. If transition boiling results from decreasing the wall superheat to the point that film boiling collapses, the observed liquid contact angle for liquid intermittently touching the surface is usually near the advancing contact angle θ_a for the system. For systems in which θ_a is significantly larger than θ_r, the transition boiling curves obtained for decreasing and increasing wall superheat may therefore be quite different. The transition boiling curve for decreasing wall superheat may be significantly below that for increasing

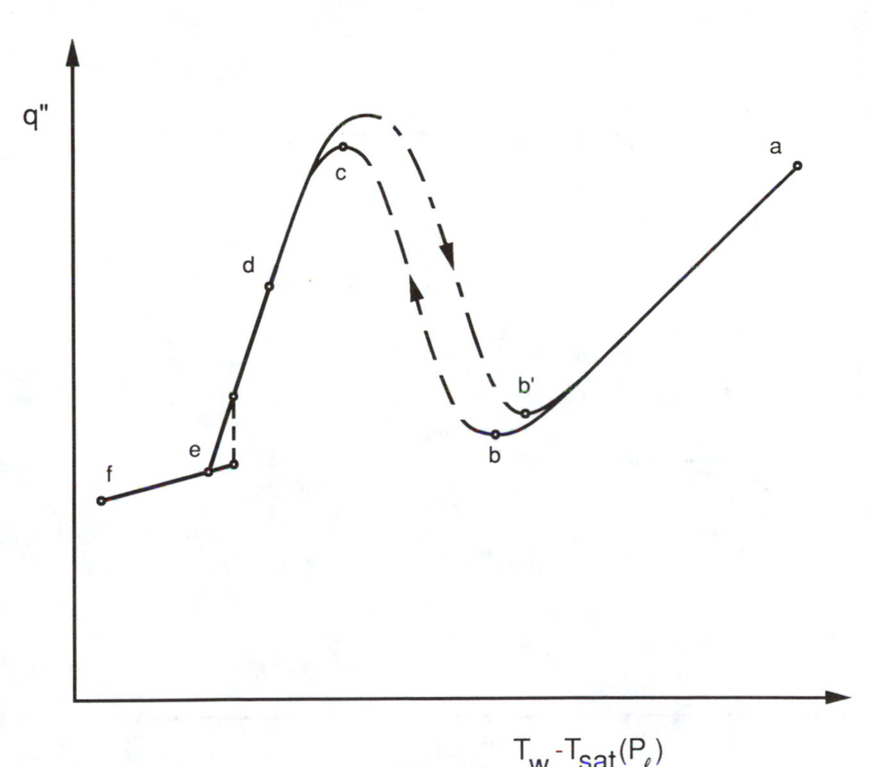

Figure 7.3 Pool boiling curve exhibiting two transition curves.

hysterisis

superheat for such circumstances, as illustrated in Fig. 7.3. These trends are discussed in more detail in Section 7.7.

The boiling curves in Fig. 7.3 also differ near the onset of nucleate boiling. For increasing wall superheat, nucleation is not initiated until the onset condition is reached, and then the system jumps upward to the nucleate boiling curve (see Fig. 7.1). Once activated, nucleation sites may remain active below the superheat required for onset. As a result, the boiling curve for decreasing wall superheat often simply follows the nucleate boiling curve downward, merging into the natural convection curve near the location where they cross. Thus the boiling curve for rising and dropping wall temperature may differ substantially near the onset condition.

The boiling curve for a surface subjected to a uniform and controlled heat flux generally takes on a different character. If the applied surface heat flux is slowly increased, the boiling curve typically looks like that shown in Fig. 7.4. The natural convection, onset of boiling, and nucleate boiling regimes are essentially the same as those observed for the temperature-controlled case shown in Fig. 7.1. One difference between the temperature-controlled and heat-flux-controlled boundary conditions is that, at the onset condition, the system state jumps horizontally to the nucleate boiling curve for the heat-flux-controlled case. Another

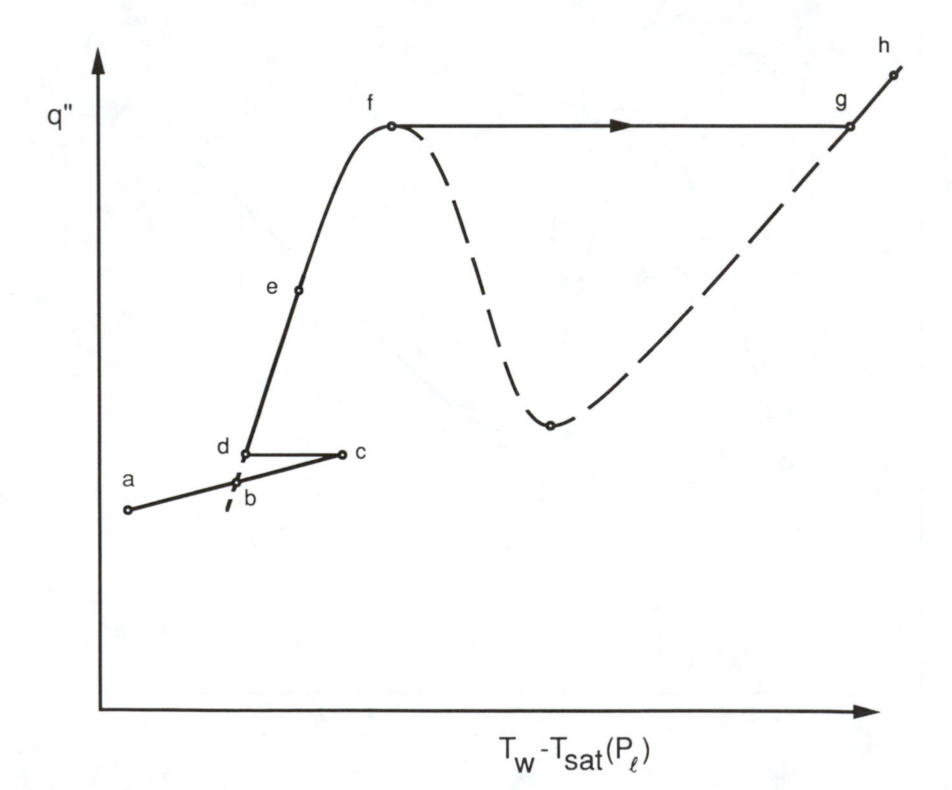

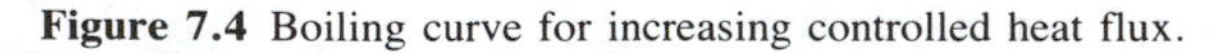
Figure 7.4 Boiling curve for increasing controlled heat flux.

difference is that when the heat flux is increased beyond the critical heat flux, the surface temperature must jump to a much higher temperature on the film boiling curve to deliver the increased heat flux. As a result, the boiling curve jumps from point *f* to point *g* in Fig. 7.4 before further increase in heat flux can be accommodated. Hence the system never encounters the transition regime observed when the surface temperature is controlled.

For an electrically heated surface, the rise in temperature associated with the jump from nucleate to film boiling at the critical heat flux is very often large enough to melt component materials and burn out the component. As a result, the critical heat flux is often referred to as the *burnout heat flux* to acknowledge the potentially damaging effects of applying this heat flux level to components cooled by nucleate boiling. Once the jump to film boiling has been made, any further increase in applied heat flux increases the wall superheat, and the system follows basically the same film boiling curve as in the temperature-controlled case.

For a surface at a very high heat flux, already in the film boiling regime, if the heat flux is slowly reduced, the system generally tracks down the film boiling curve to point *b* in Fig. 7.5, which corresponds to the minimum heat flux that can sustain stable film boiling. The system operating point must then jump to the nucleate boiling curve at point *c* before further reduction in heat flux can be ac-

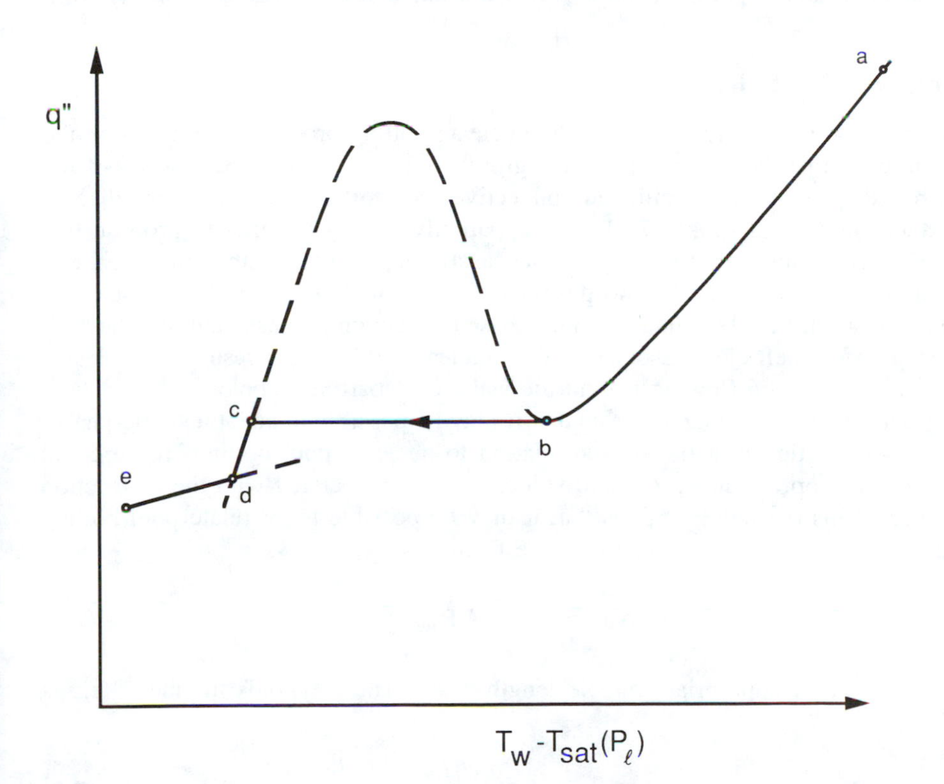

Figure 7.5 Boiling curve for decreasing controlled heat flux.

commodated. This sequence again bypasses the transition boiling regime. Further reduction then typically follows the nucleate boiling curve and the natural-convection curve as for the temperature-controlled case previously described.

The discussion in this section has been primarily at an overview level. Even from this limited discussion, however, it can be seen that the mechanisms of pool boiling are complex. The following sections of this chapter will attempt to describe these mechanisms in more detail.

7.2 MODELS OF TRANSPORT DURING NUCLEATE BOILING

The importance of nucleate boiling in a wide variety of applications has provided the incentive for numerous investigations of its basic mechanisms over the past 60 years. A substantial number of such efforts has been devoted to understanding and modeling of the transport during the nucleate boiling process. A complete account of all attempts to model the heat transfer and fluid motion during nucleate boiling is beyond the scope of this text. A few of the modeling efforts are described here, however, to illustrate the different types of models that have been considered, and to point out models that illuminate the physics particularly well.

Rohsenow's Model

Many of the very early models of the nucleate boiling process were based on the assumption that the process of bubble growth and release induced motions of the surrounding liquid that facilitated convective transport of heat from the adjacent surface. Jakob and Linke [7.2] were apparently among the first to propose this line of reasoning. Perhaps the most successful application of this approach was made by Rohsenow [7.5], who postulated that heat flows from the surface first to the adjacent liquid, as in any single-phase convection process, and that the high heat transfer coefficient associated with nucleate boiling is a result of local agitation due to liquid flowing behind the wake of departing bubbles.

The above reasoning suggests that it may be possible to adapt a single-phase, forced-convection heat transfer correlation to nucleate pool boiling, if we could specify the appropriate length and velocity scales associated with the convection process. This reasoning implies that it may be possible to correlate pool boiling heat transfer data with a relation of the form

$$\mathrm{Nu}_b = \frac{hL_b}{k_l} = A\,\mathrm{Re}_b^n\,\mathrm{Pr}_l^m \tag{7.1}$$

where L_b is an appropriate bubble length scale. The Reynolds number, Re_b, is given by

$$\mathrm{Re}_b = \frac{\rho_v U_b L_b}{\mu_l} \tag{7.2}$$

where U_b is an appropriate velocity scale. Rohsenow [7.5] took the length scale to be the bubble departure diameter d_d and the vapor superficial velocity, defined as

$$L_b = d_d = C_b\theta\left[\frac{2\sigma}{g(\rho_l - \rho_v)}\right]^{1/2} \tag{7.3}$$

$$U_b = \frac{q''}{\rho_v h_{lv}} \tag{7.4}$$

where θ is the contact angle and C_b is a constant specific to the system. By using the vapor superficial velocity and vapor density in the numerator of the right side of Eq. (7.2), the bubble Reynolds number so defined can be interpreted as a ratio of vapor inertia to liquid viscous forces.

Experimental data indicate that the effect of subcooling on heat transfer generally disappears rapidly with increasing heat flux. The heat transfer coefficient in the bubble Nusselt number was therefore defined as

$$h = \frac{q''}{[T_w - T_{sat}(P_l)]} \tag{7.5}$$

where T_w is the surface temperature and P_l is the ambient pressure. The relationship between the Nusselt number and the Reynolds and Prandtl numbers is postulated to be of the form

$$\mathrm{Nu}_b = A\,\mathrm{Re}_b^{(1-r)}\,\mathrm{Pr}_l^{(1-s)} \tag{7.6}$$

Substitution of Eqs. (7.2) through (7.5) into this relation and rearranging yields

$$\frac{q''}{\mu_l h_{lv}}\left[\frac{\sigma}{g(\rho_l - \rho_v)}\right]^{1/2} = \left(\frac{1}{C_{sf}}\right)^{1/r} \mathrm{Pr}_l^{-s/r}\left[\frac{c_{pl}[T_w - T_{sat}(P_l)]}{h_{lv}}\right]^{1/r} \tag{7.7}$$

where

$$C_{sf} = \frac{\sqrt{2}\,C_b\theta}{A} \tag{7.8}$$

Equation (7.7) is of the form of the well-known Rohsenow correlation [7.5] for pool boiling heat transfer. Recommended values of the constants in this equation are described in the next section.

Microconvection Models

Reasoning similar to that followed by Rohsenow [7.5] was used by Forster and Zuber [7.6] in the development of their microconvection model. They also postulated that the heat transfer could be represented with a correlation of the form given by Eq. (7.1). In evaluating the length and velocity scales, however, they made use of bubble growth relations that they had developed in an earlier investigation [7.7]. For heat-transfer-controlled growth, they showed that the bubble growth radius R and the velocity of the interface are given by

$$R = \mathrm{Ja}(\pi\alpha_l t)^{1/2} \tag{7.9}$$

$$\dot{R} = \mathrm{Ja}\left(\frac{\pi\alpha_l}{4t}\right)^{1/2} \tag{7.10}$$

where

$$\mathrm{Ja} = \frac{[T_\infty - T_{\mathrm{sat}}(P_\infty)]c_{pl}\rho_l}{\rho_v h_{lv}} \tag{7.11}$$

Using $2R$ as the length scale and $\dot{R}$ as the velocity scale, the bubble Reynolds number and Nusselt number are

$$\mathrm{Re}_b = \frac{2R\rho_l\dot{R}}{\mu_l} = \pi\ \mathrm{Ja}^2\ \mathrm{Pr}_l^{-1} \tag{7.12}$$

$$\mathrm{Nu}_b = \frac{q''(2R)}{(T_w - T_l)k_l} \tag{7.13}$$

where T_∞ in the Ja definition (Eq. [7.11]) is replaced by T_w. Equation (7.12) indicates that the bubble Reynolds number is independent of bubble radius. If this Reynolds number is interpreted as an indicator of the level of agitation in the liquid, the above expression for Re_b suggests that rapidly growing small bubbles produce about the same agitation of the surrounding liquid as larger bubbles that grow more slowly.

Direct determination of a characteristic length scale from Eq. (7.9) is not possible, because the radius varies with time. Based on theoretical arguments, Forster and Zuber [7.6] proposed the following definition for the length scale $2R$:

$$2R = R_c\left[\left(\frac{\rho_l\dot{R}^2}{2\sigma/R_c}\right)\left(\frac{R^2}{R_c^2}\right)\right]^{1/4} \tag{7.14}$$

$$= \mathrm{Ja}\left\{\frac{4\pi^2\alpha_l^2\rho_l^2\sigma^2}{[P_{\mathrm{sat}}(T_w) - P_l]^3}\right\}^{1/4} \tag{7.15}$$

where

$$R_c = \frac{2\sigma}{P_{\mathrm{sat}}(T_w) - P_l} \tag{7.16}$$

Using this characteristic length scale in the Nusselt and Reynolds numbers defined in Eqs. (7.12) and (7.13), Forster and Zuber [7.6] found that pool-boiling heat transfer data could be correlated using the forced-convection relation (7.1) with $A = 0.0015$, $n = 0.62$, and $m = 0.33$.

Unlike the Rohsenow model described above, the heat transfer coefficient in the Nusselt number defined by Eq. (7.13) is based on $T_w - T_l$. This suggests that the difference between the wall and liquid bulk temperatures is the temperature difference that characterizes the driving potential for heat transfer. Data for saturated pool boiling, where $T_l = T_{sat}(P_l)$, generally support this conclusion. However, experimental data for nucleate pool boiling generally correlate better in terms of the wall superheat, $T_w - T_{sat}(P_l)$, even when the liquid pool is significantly subcooled below the saturation temperature. This is particularly surprising because the agitation mechanism for subcooled boiling may be considerably different from that for saturated boiling. In saturated boiling the bubbles will simply grow and release, whereas bubbles in subcooled boiling tend to collapse as they rise after release, and some bubbles may grow only until they extend far enough into the subcooled ambient for condensation to cause them to collapse again.

As noted by Forster and Greif [7.7], the microconvection model implies that the heat flux for subcooled pool boiling should be a function of the temperature difference between the wall and the bulk fluid. This is not observed in experimental data, which suggests that the microconvection model outlined above is not a completely accurate model of the mechanism. Experiments and subsequent analysis by Gunther and Kreith [7.8] indicate that in many instances the latent heat of vaporization associated with bubble formation during subcooled nucleate boiling represents only a small fraction of the total heat transfer rate from the surface. This implies that convective transport must play a dominant role in nucleate boiling heat transfer.

Vapor–Liquid Exchange Models

The vapor–liquid exchange model proposed by Forster and Greif [7.7] provides a means of explaining the seemingly contradictory trends noted above. This model postulates that bubbles act as microscopic pumps, which draw cold ambient fluid to the surface as the bubble releases or collapses. As a new bubble grows, the pumping action pushes heated liquid from the near-wall region out into the cooler ambient. Forster and Greif [7.7] further assumed that each bubble pumps a quantity of liquid equal to its volume at release. Based on this postulated behavior, if the bubble grows hemispherically to a maximum radius R_{max}, and the bubble frequency is f, the sensible contribution to the heat transfer rate from the surface at a given site q_s is given approximately by

$$q_s = \rho_l c_{pl} \left(\frac{2\pi}{3} \right) R_{max}^3 \left(\frac{T_w + T_l}{2} - T_l \right) f \tag{7.17}$$

This relation incorporates the assumption that the mean temperature of the warm liquid pumped away from the surface is equal to the average of the wall and bulk temperatures. Equation (7.17) can be rearranged slightly to obtain

$$q_s = \rho_l c_{pl} \left(\frac{2\pi}{3} \right) R_{max}^3 \left(\frac{1}{2} \right) (T_w - T_l) f \tag{7.18}$$

If n_a' is the density of active nucleation sites on the surface, and natural convection between the sites is neglected, the mean heat flux from the surface is given by

$$q'' = \rho_l c_{pl}\left(\frac{2\pi}{3}\right)R_{\max}^3\left(\frac{1}{2}\right)(T_w - T_l)fn_a' \tag{7.19}$$

Equation (7.18) appears to imply a strong dependence of heat transfer rate on $T_w - T_l$. However, Forster and Greif [7.7] demonstrated, using experimental data, that the right side of Eq. (7.18) is, in fact, largely independent of subcooling. As the subcooling is increased, the temperature difference $T_w - T_l$ obviously increases. But increasing the subcooling also was shown to decrease the maximum radius $R_{\max}$ and increase the bubble frequency f. The net result is that these three effects approximately compensate for each other, resulting in very little change in the right side of Eq. (7.18). A change of over 300% in subcooling was found to produce only about a 20% change in the heat transfer rate per nucleation site.

The question then arises: even if the heat transfer per site is insensitive to subcooling, why should the total heat flux from the surface depend primarily on wall superheat? One reason is that both $R_{\max}$ and n_a' in Eq. (7.19) increase with increasing wall superheat. As described in Chapter Six, higher superheat will produce more rapid bubble growth, causing the bubble to achieve a larger radius before releasing or collapsing. Likewise, more (and smaller) sites can become active as the wall superheat increases and the critical radius gets smaller. As a result of these trends, the heat flux varies with superheat even though its dependence on subcooling is weak.

Using the vapor–liquid exchange concept developed for subcooled boiling, Han and Griffith [7.9] developed a more detailed model of saturated boiling. A similar model was subsequently used by Mikic and Rohsenow [7.10] to derive a correlation for saturated pool-boiling heat transfer. Their model postulated that, in the isolated bubble regime, the surface can be envisioned as consisting of regions of natural convection between active nucleation sites, and regions immediately adjacent to active sites that are cooled by microconvection resulting from the liquid–vapor exchange mechanism. Microconvection regions were postulated to correspond to an area of influence within a circle around each active site with a radius equal to twice the bubble departure radius.

After departure of a bubble, colder fluid at $T_{\text{sat}}(P_l)$ is brought into contact with the surface in the region of influence, initiating transient conduction of heat into the liquid. The actual conduction process was modeled as the simple one-dimensional transient conduction process into a semi-infinite medium considered in the bubble growth model of Mikic and Rohsenow [7.11]. This model problem and its solution are described in Section 6.5. Using the conjugate error function solution to this problem (Eq. [6.77]) to evaluate the heat flux from the surface in the region of influence q''_{mc} yields

$$q''_{mc} = -k_l\left(\frac{\partial T}{\partial y}\right)_{y=0} = \frac{k_l[T_w - T_{\text{sat}}(P_l)]}{(\pi\alpha_l t)^{1/2}} \tag{7.20}$$

This relation gives the instantaneous heat flux, which varies with time over the period of one bubble cycle. This time period equals $1/f$, where f is the bubble frequency. Taking an integral average of the heat flux over this time interval yields

$$\bar{q}''_{mc} = f\int_0^{1/f} q''_{mc}\,dt = f\int_0^{1/f}\left\{\frac{k_l[T_w - T_{\text{sat}}(P_l)]}{(\pi\alpha_l t)^{1/2}}\right\}dt \tag{7.21a}$$

$$= \left(\frac{4k_l\rho_l c_{pl}}{\pi}\right)^{1/2}[T_w - T_{\text{sat}}(P_l)]f^{1/2} \tag{7.21b}$$

The mean heat flux density over the entire surface is expected to be equal to the area weighted average of the microconvection and natural-convection contribution:

$$q'' = \left(\frac{A_{mc}}{A_{\text{tot}}}\right)\bar{q}''_{mc} + \left(\frac{A_{nc}}{A_{\text{tot}}}\right)q''_{nc} \tag{7.22}$$

The natural-convection heat flux is expected to be so small that its contribution may be neglected. Doing so reduces Eq. (7.22) to

$$q'' = \left(\frac{A_{mc}}{A_{\text{tot}}}\right)\bar{q}''_{mc} = \frac{\pi d_d^2}{4}n'_a\bar{q}''_{mc} \tag{7.23}$$

where $\pi d_d^2/4$ is the area of the region of influence per site and n'_a is the number of active sites per unit area of surface. Combining this equation with Eq. (7.21b), we find that

$$q'' = 2(\pi k_l\rho_l c_{pl})^{1/2}f^{1/2}d_d^2 n'_a[T_w - T_{\text{sat}}(P_l)] \tag{7.24}$$

Examination of Eq. (7.24) reveals one of the fundamental difficulties in developing heat transfer relations for saturated (and subcooled) nucleate boiling: The dependence of heat flux on wall superheat can be specified only if the variations of bubble frequency f, departure diameter d_d, and active site density n'_a with superheat is known. Mikic and Rohsenow [7.10] used the following relations to evaluate the departure diameter and frequency:

$$d_d = A_1\left[\frac{\sigma}{g(\rho_l - \rho_v)}\right]^{1/2}\left(\frac{\rho_l c_{pl}T_{\text{sat}}}{\rho_v h_{lv}}\right)^{5/4} \tag{7.25}$$

$$fd_d = A_2\left[\frac{\sigma g(\rho_l - \rho_v)}{\rho_v^2}\right]^{1/4} \tag{7.26}$$

where $A_1 = 1.5 \times 10^{-4}$ for water and 4.65×10^{-4} for other fluids, and $A_2 = 0.6$. These values of A_1 and A_2 were recommended in references [7.12] and [7.13]. For n'_a, Mikic and Rohsenow [7.10] argued that the number of active nucleation sites should vary as a power-law function of cavity mouth radius R:

$$n'_a = A_3\left(\frac{R_0}{R}\right)^m \tag{7.27}$$

Equation (7.27) is interpreted as meaning that n_a' is the number of active nucleation sites per unit area having a mouth radius greater than R. R_0 is the threshold radius at which n_a' is A_3 sites per unit area. As described in Section 6.2, for real systems that have cavities with small half-angles and contact angles less than 90°, the minimum radius of curvature of a bubble growing in a cavity is essentially equal to the cavity mouth radius. This implies that the cavity mouth radius alone characterizes the nucleation behavior of a given site. (See reference [7.14] for a further discussion of this idealization.) It was argued on this basis in Section 6.2 that the minimum superheat for the bubble to grow beyond the mouth of the cavity is

$$[T_w - T_{sat}(P_l)]_{\text{min for growth}} = \frac{2\sigma T_{sat}(P_l) v_{lv}}{h_{lv} R} \tag{7.28}$$

If the superheat is fixed at all cavities, the smallest active cavity radius R_{min} for a given superheat is therefore given by

$$R_{min} = \frac{2\sigma T_{sat}(P_l) v_{lv}}{h_{lv}[T_w - T_{sat}(P_l)]} \tag{7.29}$$

If we neglect v_l compared to v_v, this relation can be written as

$$R_{min} = \frac{2\sigma T_{sat}(P_l)}{\rho_v h_{lv}[T_w - T_{sat}(P_l)]} \tag{7.30}$$

Mikic and Rohsenow [7.10] substituted this relation for R_{min} for R in Eq. (7.27) to obtain

$$n_a' = A_3 \left[R_0 \frac{\rho_v h_{lv}}{2\sigma T_{sat}(P_l)} \right]^m [T_w - T_{sat}(P_l)]^m \tag{7.31}$$

Substituting Eqs. (7.25), (7.26), and (7.31) into Eq. (7.24), a relation is obtained for the mean surface heat flux q'' that can be written in the form

$$q^* = \frac{\overline{q}''}{\mu_l h_{lv}} \left[\frac{\sigma}{g(\rho_l - \rho_v)} \right]^{1/2} = B[\phi(T_w - T_{sat})]^{m+1} \tag{7.32}$$

where

$$B = A_1^{3/2} A_2^{1/2} A_3 \left(\frac{2\pi^{1/2}}{g^{9/8}} \right) \left[\frac{R_0}{2} \right]^m \tag{7.33}$$

and

$$\phi = \left[\frac{k_l^{1/2} \rho_l^{17/8} c_{pl}^{19/8} h_{lv}^{m-23/8} \rho_v^{m-15/8}}{\mu_l (\rho_l - \rho_v)^{9/8} \sigma^{m-11/8} T_{sat}^{m-15/8}} \right]^{1/(m+1)} \tag{7.34}$$

Equation (7.32) has basically the same form as Rohsenow's [7.5] original correlation as specified in Eq. (7.7). The advantage of Eq. (7.32) is that the constants now have some physical significance in terms of the properties of the fluids and the cavity size distribution. If these system characteristics are known, A_1, A_2, A_3, R_0, and m are determined and Eq. (7.32) would specify the nucleate boiling curve.

Mikic and Rohsenow [7.10] found that nucleate boiling data for several different systems were well correlated in terms of q^* and $(T_w - T_{sat})$, although the values of B and m varied from one system to the next. The data of Addoms [7.15] for water are plotted in terms of these variables in Fig. 7.6.

Natural-Convection Analogy Model

Zuber [7.16] proposed a model of the microconvective effect during nucleate boiling based on an analogy between nucleate boiling and single-phase turbulent natural convection. Beginning with the form of a heat transfer correlation for turbulent natural convection, Zuber [7.16] substituted appropriate length, velocity, and time scales into the correlation and evaluated the fluid density as the mean density of the two-phase system, accounting for the vapor void fraction at the surface. The resulting correlation has the form

$$\frac{q''L}{k_l} = C_0 \left\{ \frac{gL^3}{\nu_l \alpha_l} \left[\beta(T_w - T_{sat}) + \frac{\pi}{6} n_a' d_d^2 \frac{d_d f(\rho_l - \rho_v)}{\rho_l u_T} \right] \right\}^{1/3} \tag{7.35}$$

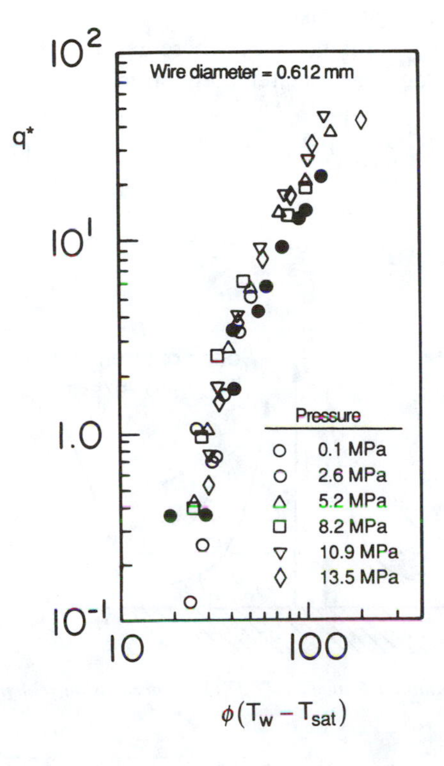

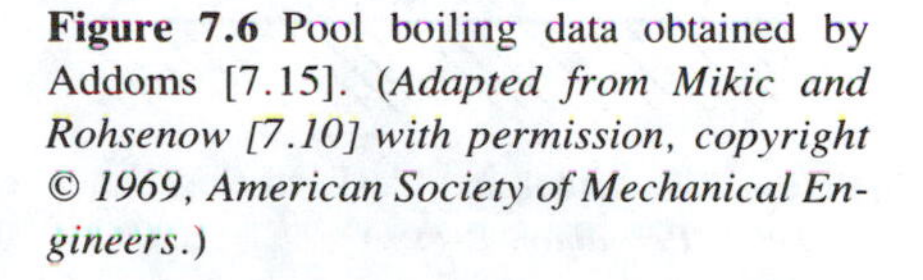
Figure 7.6 Pool boiling data obtained by Addoms [7.15]. (*Adapted from Mikic and Rohsenow [7.10] with permission, copyright © 1969, American Society of Mechanical Engineers.*)

Note in this relation that the length scale L cancels out. C_0 is an undetermined constant. In the square brackets on the right side, the second terms can be evaluated only if the active site density n_a', bubble departure diameter d_d, bubble frequency f, and (terminal) rise velocity of the vapor bubbles u_T are known. Using experimental data and appropriate correlations, Zuber [7.16] demonstrated that Eq. (7.35) is consistent with nucleate boiling heat transfer data in the isolated bubble regime, supporting the idea of an analogy between nucleate boiling and turbulent natural convection.

Inverted Stagnation Flow Model

Tien [7.16] noted the analogy between the induced flow associated with a rising bubble column and an inverted stagnation flow, as shown in Fig. 7.7. For laminar axisymmetric stagnation flow it is known that

$$\frac{hr}{k_l} = 1.32\left(\frac{u_\infty r}{\nu_l}\right)^{0.5} \mathrm{Pr}_l^{0.33} \tag{7.36}$$

where

$$u_\infty = ar \tag{7.37}$$

with a being a constant and r being the radial distance from the center of the axisymmetric flow. For these conditions, the thermal boundary-layer thickness $\delta_t = k_l/h$ is constant (independent of r). The heat transfer coefficient associated with an active site is evaluated over a circle of diameter s, where s is the spacing between sites. Replacing r with $s/2$ in the heat transfer correlation yields

$$h = 1.32\,\frac{k_l}{s}\left(\frac{as^2}{\nu_l}\right)^{0.5} \mathrm{Pr}_l^{0.33} \tag{7.38}$$

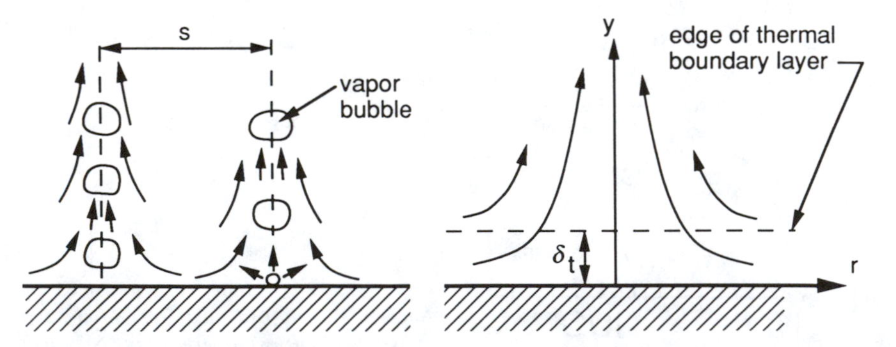

Figure 7.7 Actual and idealized flow fields. (*Adapted from Tien [7.17] with permission, copyright © 1962, Pergamon Press.*)

If the density of active sites n'_a is taken as being equal to s^{-2}, this relation can be written as

$$h = 1.32 k_l (n'_a)^{0.5} \left(\frac{a}{n'_a \nu_l} \right)^{0.5} \mathrm{Pr}_l^{0.33} \tag{7.39}$$

Based on the active-site density data of Yamagata et al. [7.18] and Gaertner and Westwater [7.19], Tien argued that $a/n'_a\, \nu_l$ is about equal to 2,150 and approximately constant. Substituting this value in Eq. (7.39), it is found that

$$h = \frac{q''}{T_w - T_{\mathrm{sat}}} = 61.3\, \mathrm{Pr}_l^{0.33}\, k_l (n'_a)^{0.5} \tag{7.40}$$

As seen in Fig. 7.8, this relation is in general agreement with data at moderate to high active site densities for a wide variety of fluids. This model is yet another demonstration that single-phase convection heat transfer concepts can, in some instances, be extended to nucleate boiling.

Latent Heat and Microlayer Evaporation Effects

As intricate as they are, the models described so far only account for the microconvective effects of bubble growth and release. This type of model is plausible for circumstances in which microconvective effects dominate. For some conditions, however, the latent heat associated with the vapor bubble and/or evapo-

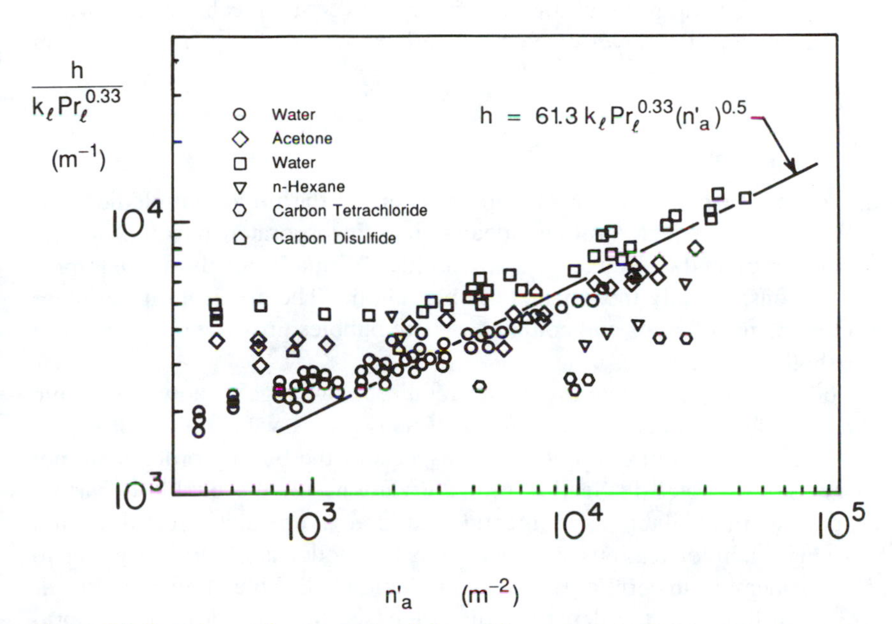

Figure 7.8 Experimentally observed variation of the nucleate boiling heat transfer coefficient with nucleation site density. (*Adapted from Tien [7.17] with permission, copyright © 1962, Pergamon Press.*)

ration of a liquid microlayer under the bubble may also play an important role. Analytical models similar to that of Mikic and Rohsenow [7.10] have also been developed that attempt to account for these effects.

van Stralen [7.20] developed a theoretical treatment of nucleate boiling heat transfer for pure and binary liquids that accounted for both microconvective and latent heat transfer using a superposition approach. This model treats the microconvective effect in a manner similar to the models of Han and Griffith [7.9] and Mikic and Rohsenow [7.10], and considers the total heat transfer rate to equal the sum of heat flow due to the microconvection and latent heat mechanisms.

Microconvection due to bubble agitation appears to account for most of the heat transfer at low heat flux densities in the isolated bubble regime. However, experimental data suggest that microconvection does not completely account for the total heat flow at higher heat flux levels, where closer spacing of active sites results in interference that reduces the bubble's ability to convect superheated liquid into the ambient. More of the superheat energy then supplies the latent heat of vaporization to the bubble, resulting in a larger fraction of the total energy flow being carried away from the surface by this mechanism. van Stralen [7.20] presents comparisons with experimental data that appear to confirm the superposition behavior postulated for these circumstances.

As described in Chapter Six, at relatively low pressures, bubble growth during the ebullition cycle may be in the inertia-controlled regime, resulting in hemispherical bubble growth. For such circumstances, microlayer evaporation may also contribute significantly to the total heat transfer from the surface. A model that incorporates the natural-convection and microconvection mechanisms, as well as a treatment of the microlayer evaporation effect, was developed by Judd and Hwang [7.21].

General Observations

While the models described so far all capture some of the important elements of the transport during nucleate boiling, idealizations incorporated into the models limit their accuracy and/or range of applicability. Virtually all the models proposed to date consider only the isolated bubble regime. The applicability of these models to the regime of slugs and columns, where bubbles growing on the surface interact, is highly suspect.

The modeling efforts to date invariably require knowledge of how the bubble frequency, departure diameter, and density of active sites vary with wall superheat. Despite many years of research, these aspects of the boiling process are not well understood. As noted in Section 6.6, correlations for the bubble departure diameter, at best, fit available experimental data to an average absolute deviation of 44%. Using such correlations in a heat transfer model is obviously going to introduce considerable uncertainty into its predictions. Bubble frequency correlations are often linked to the departure diameter (see Section 6.6). Hence, predictions of the bubble frequency are likely to suffer from the same high uncertainty level as the departure diameter. In addition, most of the models described so far do not account for the fact that the bubble frequency relation should change depending on whether inertia-controlled or heat-transfer-controlled growth is expected (again, see Section 6.6).

Similar difficulties arise in attempting to incorporate in these models an appropriate prediction of the active nucleation site density n'_a. As discussed in Chapter Six, the number of active sites depends in a complicated way on the surface morphology, the thermophysical properties of the fluids, the imposed pressure and flow conditions, and the wetting characteristics of the liquid–surface combination. The simplistic relations between n'_a and the wall superheat adopted in most models cannot account for the complex dependence of n'_a on all these factors. The inability of such simple relations to account for factors other than wall superheat may be the source of some of the scatter in experimental data when compared with model predictions. This undoubtedly is the source of some of the scatter in the data shown in Fig. 7.8.

Further complicating this issue is the fact that different models suggest different dependencies of q'' on n'_a and $T_w - T_{sat}$. The modeling efforts described so far suggest that, overall, the nucleate boiling heat flux exhibits a power-law dependence on the nucleation site density and wall superheat,

$$q'' \propto (n'_a)^a (T_w - T_{sat})^b \tag{7.41}$$

The microconvection model of Mikic and Rohsenow [7.10] implies a linear dependence of q'' on n'_a. Equation (7.33), developed from Zuber's [7.16] natural-convection analogy model, implies that q'' is proportional to $(n'_a)^{1/3}$, and Tien's [7.17] model indicates that q'' is proportional to $(n'_a)^{1/2}$. The relations for all three of these models imply an almost linear variation of q'' with $T_w - T_{sat}$.

Several experimental studies, such as those of Gaertner and Westwater [7.19] and Heled et al. [7.22], also indicate power-law dependencies of q'' on nucleation site density and superheat. These studies suggest that a lies somewhere between 0.3 and 0.5 and b is between 1.0 and 1.8. This appears to support the values of a suggested by the models of Zuber [7.16] and Tien [7.17]. Mikic and Rohsenow [7.10] noted, however, that very seldom in these experimental studies was only n'_a changed, without simultaneously changing the wall superheat, bubble departure diameter, and/or the bubble frequency. Equation (7.24), developed in their analysis, implies that q'' is linearly proportional to n'_a, provided that all other parameters on the right side of Eq. (7.24) are held constant. It is usually not possible to distinguish the influence of departure diameter and bubble frequency on heat flux from experimental data. Consequently, the results of these experimental studies do not allow a definitive evaluation of the validity of Eq. (7.24).

The exact values of the exponents in the power-law relation (7.41) continue to be a subject of investigation and debate. However, for many common systems, the variation of n'_a (and any other proportionality factors not indicated in Eq. [7.42]) with wall superheat is usually such that q'' varies in an overall manner about proportional to $(T_w - T_{sat})^3$. It seems likely that this fortuitous situation is a consequence of the fact that common manufacturing practices used on metals, glass, and other frequently used materials produce cavity distributions that lie within specific bounds. The probability of the distribution for a given surface being within these limits is usually high, and as a result, the increase in the number of active sites with increasing wall superheat is likely to be comparable to that for other surfaces subjected to similar manufacturing methods.

The models of nucleate boiling heat transfer described here are, at best, crude idealizations. The development of better models has been hindered by an inability to accurately predict some basic features of the boiling process such as the number density of active sites, and the mean bubble frequency and departure diameter. Fortunately, the nearly identical power-law dependence of heat flux on superheat for many wall material-and-liquid combinations makes it possible to develop relatively simple nucleate boiling heat transfer correlations that can be used with fair accuracy for a wide variety of systems. Correlations of this type are described in more detail in the next section.

7.3 CORRELATION OF NUCLEATE BOILING HEAT TRANSFER DATA

As noted in previous sections of this chapter, nucleate pool boiling heat transfer data have been obtained for a number of pure substances at different system pressure levels. Correlations of such data have typically been used as a tool to predict nucleate boiling heat transfer in engineering systems and heat exchangers. Many investigators have proposed methods of correlating data of this type—so many, in fact, that a complete discussion of them all could easily fill a major portion of this chapter.

In this section, four specific correlation methods will be described in detail: (1) Rohsenow's correlation [7.5], (2) the Forster-Zuber correlation [7.6], (3) the Borishansky-Mostinski correlation [7.23, 7.24], and (4) the correlations of Stephan and Abdelsalam [7.25]. Other correlations have been proposed by Jakob and Linke [7.2], Fritz [7.26], Kutateladze [7.27], Kruzhilin [7.28], Forster and Greif [7.7], Levy [7.29], Alad'yev [7.30], Labuntsov [7.31], Katuteladze et al. [7.32], Lienhard [7.33], and Mikic and Rohsenow [7.10].

The first four correlations are discussed in detail here because either they are widely used (as is the case for the first two) or they are convenient, yet seem to provide reasonably reliable results. These correlations were selected because they are representative of the spectrum of different logical approaches used to correlate data of this type. Rohsenow's correlation [7.5] and the Forster-Zuber correlation [7.6] are based on analogies with forced-convection processes. In contrast, the Borishanski-Mostinski correlation [7.23, 7.24] is based on the principle of corresponding states, and the correlations proposed by Stephan and Abdelsalam [7.25] were obtained from dimensional analysis and regression fits of available data.

Before further discussing correlation techniques, two aspects of the interpretation of such correlations are worth noting. First, because the subcooling of the liquid pool has virtually no effect on the resulting heat transfer rate (see Section 7.2), the pool boiling correlations are generally regarded as being valid for both subcooled and saturated nucleate boiling. Second, it has also been observed that a pool boiling heat transfer correlation developed for one heated surface geometry in one specific orientation often works reasonably well for other geometries and/or other orientations. Hence, although a correlation was developed for a specific geometry and orientation, it may often be used, at least as a good approximation, for others as well.

The effects of surface orientation on nucleate boiling heat transfer can be assessed somewhat more directly by examining the results of a recent study by Nishikawa et al. [7.34]. These investigators obtained pool boiling heat transfer data for a flat heated surface in an extensive liquid pool. The test surface was specifically constructed so that it could be rotated to change the orientation of the surface with respect to the gravity vector. The resulting boiling curves for water at atmospheric pressure for different surface orientations are shown in Fig. 7.9. The data in this figure indicate that above a certain threshold heat flux level the pool boiling curves for all orientations are virtually identical. Below this threshold value the curves differ significantly for the different surface orientations. This threshold value corresponds approximately to the transition from the isolated bubble regime to the regime of slugs and columns.

This transition is sometimes referred to as the Moissis-Berenson transition [7.35]. Based on arguments that can be traced back to Zuber [7.16], these investigators developed a semiempirical model that predicts that bubbles releasing from adjacent sites on an upward-facing flat plate will touch one another, merging into vapor columns, when the heat flux reaches the level given by

$$q''_{\mathrm{MB}} = 0.11\rho_v h_{lv}\theta^{1/2}\left(\frac{\sigma g}{\rho_l - \rho_v}\right)^{1/4} \tag{7.42}$$

where θ is the liquid contact angle. This relation was found to agree well with visual observations of the transition between the isolated bubble regime and the regime of slugs and columns. A corresponding relation for this transition on horizontal cylinders has been obtained by Bhattacharya and Lienhard [7.36]. Leinhard [7.37] showed that the transition suggested in the data of Nishikawa et al. [7.34] is consistent with that indicated by Eq. (7.42), if the contact angle is between 35° and 85° (see Fig. 7.9).

These results imply that surface orientation in the slugs and columns regime has little impact on the pool boiling heat transfer performance. However, in the isolated bubble regime, at lower heat flux, surface orientation appears to affect the boiling curve in a systematic way. Interestingly enough, the upward-facing surface resulted in the highest superheat for a given heat flux, whereas the surface superheat was the lowest in the downward-facing position.

At lower heat flux levels, the enhancement of heat transfer for the downward-facing orientations would appear to be due, at least in part, to two effects. First, the natural-convection boundary layer for a downward-facing surface is thicker than for upward-facing or vertical surface orientations. Hsu's analysis, described in Section 6.3, implies that the wall superheat required to initiate nucleate boiling will be lower for the thicker thermal boundary layer associated with the downward-facing surface.

In addition, when a bubble grows and releases from an inclined or horizontal downward-facing surface, the bubble must travel along the surface to its lateral edge before escaping to the ambient. This sweeping of the surface may serve to enhance heat transfer by facilitating vaporization of the liquid film between the interface of the bubble and the wall as it moves along the surface. The enhancement provided by this mechanism, together with the earlier onset of boiling due to the mechanism described above, may account for the upward shift in the nucleate boiling curve for a downward-facing surface relative to that for a vertical or upward-facing surface. Further discussion of the mechanisms that may be responsible for these trends can be found in references [7.38]–[7.40].

The mechanisms that give rise to the trends in the boiling curves in Fig. 7.9

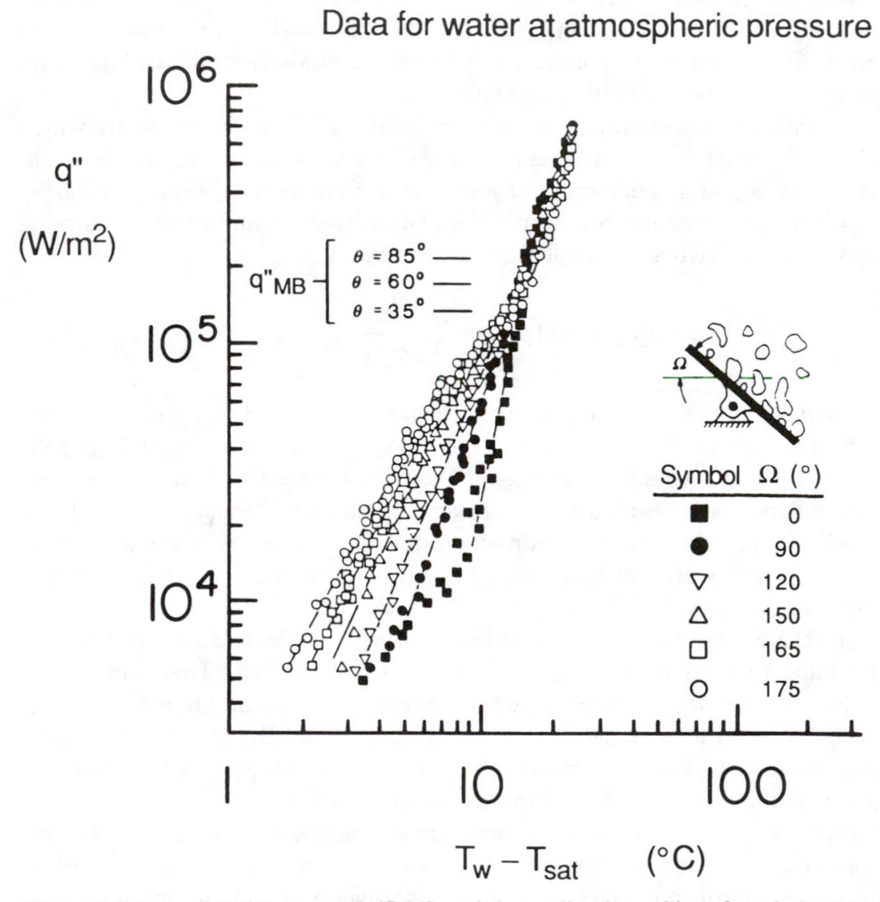

Figure 7.9 Data of Nishikawa et al. [7.34], and the predicted transitions from isolated bubbles to slugs and columns. (*Adapted from Lienhard [7.37] with permission, copyright © 1985, American Society of Mechanical Engineers.*)

Table 7.1 Values of C_{sf} in the Rohsenow correlation (Eq. [7.43]) for different liquid–surface combinations

Liquid–surface combination	C_{sf}
Water on Teflon pitted stainless steel	0.0058
Water on scored copper	0.0068
Water on ground and polished stainless steel	0.0080
Water on emery polished copper	0.0128
Water on chemically etched stainless steel	0.0133
Water on mechanically polished stainless steel	0.0132
Water on emery polished, paraffin-treated copper	0.0147
n-Pentane on lapped copper	0.0049
n-Pentane on emery polished nickel	0.0127
n-Pentane on emery polished copper	0.0154
Carbon tetrachloride on emery polished copper	0.0070

Source: Data from [7.41].

are not fully understood at this time. These results clearly imply, however, that at low heat flux levels, nucleate pool boiling correlations are not universally applicable to all surface geometries, as is often assumed. At higher heat flux levels, the general applicability is expected to be better.

Having taken heed of the above warning, we will now consider in detail the four correlations noted earlier in this section. As described in the previous section, Rohsenow's [7.5] correlation is of the form

$$\frac{q''}{\mu_l h_{lv}}\left[\frac{\sigma}{g(\rho_l-\rho_v)}\right]^{1/2} = \left(\frac{1}{C_{sf}}\right)^{1/r} \mathrm{Pr}_l^{-s/r}\left\{\frac{c_{pl}[T_w - T_{\mathrm{sat}}(P_l)]}{h_{lv}}\right\}^{1/r} \tag{7.43}$$

Originally, values of $r = 0.33$ and $s = 1.7$ were recommended for this correlation. Subsequently, Rohsenow recommended that, for water only, s be changed to 1.0. Values of C_{sf} recommended for different liquid–solid combinations are listed in Table 7.1. These values were tabulated by Vachon et al. [7.41] based on fits to pool boiling data available in the literature.

Additional values of C_{sf} have been obtained for other liquid–surface combinations from subcooled forced-convective boiling data. Following a superposition model proposed by Rohsenow [7.42], the total heat transfer rate was postulated as being equal to the sum of a forced-convection contribution and a nucleate boiling contribution, with the latter given by the Rohsenow correlation. Subtracting the forced-convective contribution from the total, values of C_{sf} have been determined that best fit the implied nucleate boiling contribution in the data of Rohsenow and Clarke [7.43], Kreith and Sommerfield [7.44], Piret and Isbin [7.45], and Bergles and Rohsenow [7.46]. These values of C_{sf} are listed in Table 7.2. The accuracy of C_{sf} values obtained in this manner is thus limited by the accuracy of the superposition method. This type of model will be discussed further in Chapter Twelve.

The values of C_{sf} listed in Tables 7.1 and 7.2 are useful indicators of typical values. However, it is generally recommended that, whenever possible, an ex-

Table 7.2 Values of C_{sf} in the Rohsenow correlation (Eq. [7.43]) for different liquid–surface combinations

Liquid–surface combination	C_{sf}
Water on nickel (vertical tube)	0.006
Water on stainless steel (horizontal tube)	0.015
Water on stainless steel (horizontal tube)	0.020
Water on copper (vertical tube)	0.013
Carbon tetrachloride on copper (vertical tube)	0.013
Isopropyl alcohol (vertical tube)	0.0022
n-Butyl alcohol on copper (vertical tube)	0.0030

Source: Data from [7.43]–[7.46].

periment be conducted to determine the appropriate value of C_{sf} for the particular solid–liquid combination of interest. If this is not possible, and the combination is not listed in Table 7.1 or 7.2, a value of $C_{sf} = 0.013$ is recommended as a first approximation.

Another frequently quoted relation is the Forster-Zuber correlation (7.6), which can be written as

$$q'' = 0.00122\left(\frac{k_l^{0.79} c_{pl}^{0.45} \rho_l^{0.49}}{\sigma^{0.5} \mu_l^{0.29} h_{lv}^{0.24} \rho_v^{0.24}}\right)[T_w - T_{sat}(P_l)]^{1.24}\, \Delta P_{sat} \tag{7.44}$$

where ΔP_{sat} is the difference in saturation pressure corresponding to a difference in saturation temperature equal to the wall superheat $T_w - T_{sat}(P_l)$. There are several possible combinations of property units that could be used in this relation. One set that does work is $k_l \sim$ kW/m °C, $c_{pl} \sim$ kJ/kg °C, $\rho_l \sim$ kg/m^3, $\rho_v \sim$ kg/m^3, $T_w - T_{sat}(P_l) \sim$ °C, $\Delta P_{sat} \sim$ Pa, $\sigma \sim$ N/m, $\mu_l \sim$ Ns/m^3, and $h_{lv} \sim$ kJ/kg with the resulting value of q'' in kW/m^2.

Based on thermodynamic similitude, Borishansky [7.23] proposed a correlation that can be written as

$$q'' = (A^*)^{3.33}[T_w - T_{sat}(P_l)]^{3.33}[F(P_{r})]^{3.33} \tag{7.45}$$

where q'' is in W/m^2 and $F(P_r)$ is a function of the reduced pressure $P_r = P/P_c$. For this correlation, Mostinski [7.24] proposed the following relations for A^* and $F(P_r)$:

$$A^* = 0.1011 P_c^{0.69} \text{ (with } P_c \text{ in bar)} \tag{7.46}$$

$$F(P_r) = 1.8P_r^{0.17} + 4P_r^{1.2} + 10P_r^{10} \tag{7.47}$$

$$P_r = \frac{P}{P_c} \tag{7.48}$$

In a more recent study, Stephan and Abdelsalam [7.25] proposed the following correlations based on dimensional analysis and optimal fits to experimental data:

For water:

$$q'' = \{C_1[T_w - T_{sat}(P_l)]\}^{1/0.327} \tag{7.49}$$

For hydrocarbons:

$$q'' = \{C_2[T_w - T_{sat}(P_l)]\}^{1/0.330} \tag{7.50}$$

For cryogenic fluids:

$$q'' = \{C_3(\rho c_p k)_c^{0.117}[T_w - T_{sat}(P_l)]\}^{1/0.376} \tag{7.51}$$

For refrigerants:

$$q'' = \{C_4[T_w - T_{sat}(P_l)]\}^{1/0.255} \tag{7.52}$$

Values of the constants C_1 through C_4 for materials of the indicated types are shown in Figs. 7.10 through 7.13. In the above relations the units to be used are kg/m^3 for ρ, kJ/kg °C for c_p, W/m °C for k, °C for $T_w - T_{sat}(P_l)$, and W/m^2 for q''.

For cryogenic fluids $(\rho c_p k)_c$ represents the indicated properties of the heated surface or the surface cover material, which were found by Stephan and Abdelsalam [7.25] to affect the heat transfer performance. The authors also noted that,

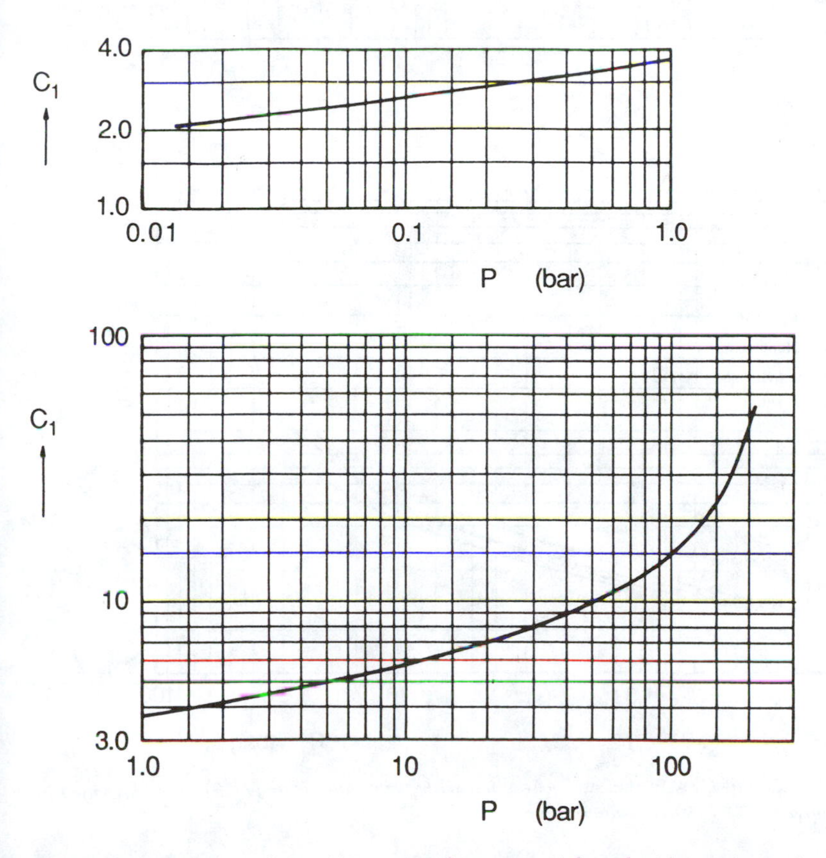

Figure 7.10 Variation of C_1 with pressure for water. (*Adapted with permission from [7.25]; plots provided by K. Stephan, copyright © 1980, Pergamon Press.*)

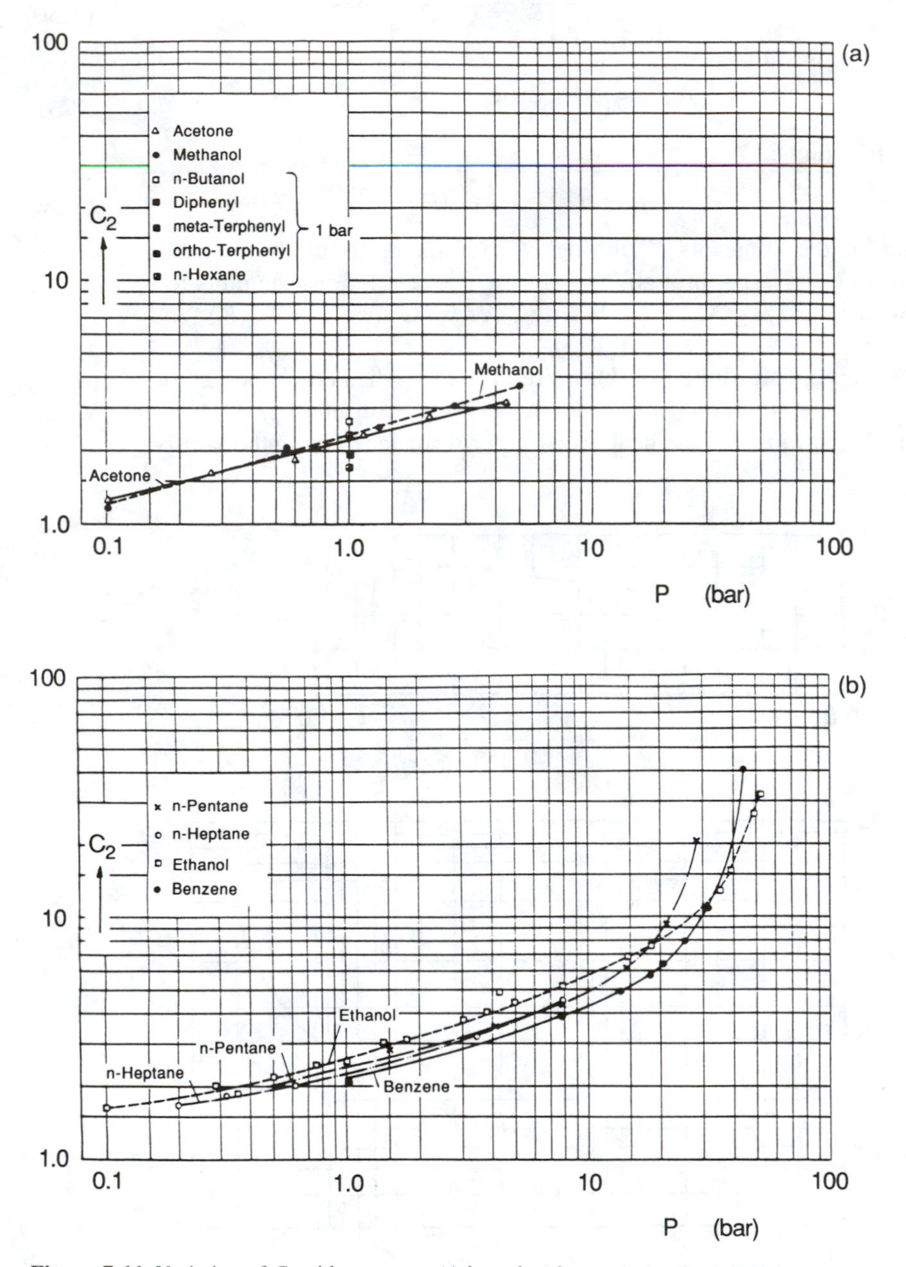

Figure 7.11 Variation of C_2 with pressure. (*Adapted with permission from [7.25]; plots provided by K. Stephan, copyright © 1980, Pergamon Press.*)

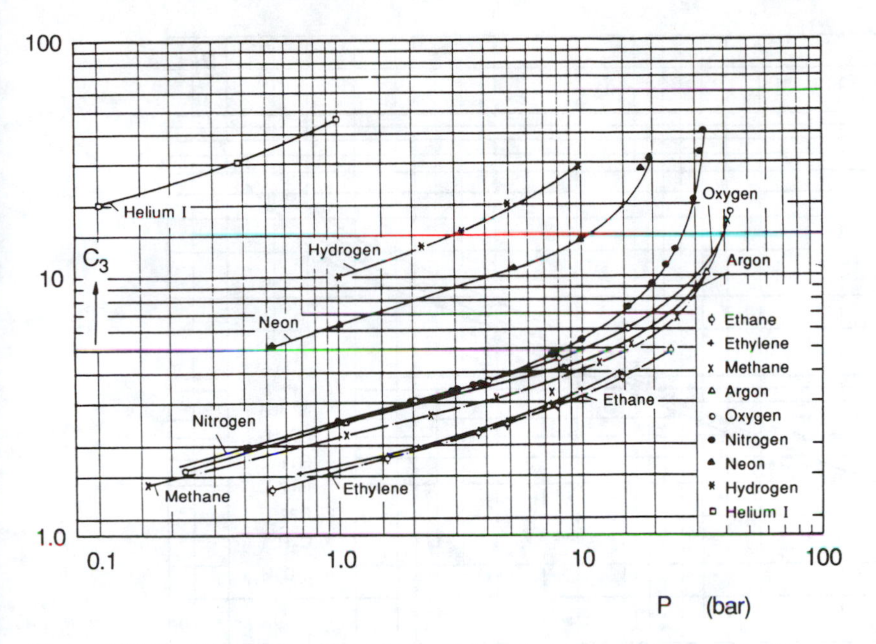

Figure 7.12 Variation of C_3 with pressure. (*Adapted with permission from [7.25]; plot provided by K. Stephan, copyright © 1980, Pergamon Press.*)

in developing these equations, the mean surface roughness R_s was assumed to be 1 μm. For other roughness values in the range $0.1 \leqslant R_s \leqslant 10$ μm, they suggest that the heat transfer coefficient predicted by the above relations be corrected by multiplying it by $R_s^{0.133}$, where R_s is the mean surface roughness in micrometers.

The pool boiling curves for water at atmospheric pressure predicted by the four correlation methods described here are shown in Fig. 7.14. For Rohsenow's correlation, C_{sf} was taken to be 0.013. Although the four curves in this figure are somewhat different, the overall variation is within the typical scatter in nucleate pool boiling heat transfer data (see Fig. 7.6). The heat flux predicted by these correlations may vary by a factor of 2 or more for a given wall superheat, whereas the difference in the wall superheat predicted for a specified heat flux is at most only a few degrees.

Example 7.1 Compare the heat flux values predicted by the Rohsenow correlation and the Stephan and Abdelsalam correlation for nucleate pool boiling of *n*-butyl alcohol (*n*-butanol) on a copper surface at atmospheric pressure for a wall superheat of 10°C.

From the saturation table for *n*-butanol for $P_{sat} = 101$ kPa:

$$\rho_l = 712 \text{ kg/m}^3 \qquad \rho_v = 2.30 \text{ kg/m}^3 \qquad h_{lv} = 591.3 \text{ kJ/kg}$$

$$c_{pl} = 3.20 \text{ kJ/kg K} \qquad \mu_l = 4.04 \times 10^{-4} \text{ Ns/m}^2$$

$$k_l = 0.127 \text{ W/m K} \qquad \text{Pr}_l = 10.3 \qquad \sigma = 0.0171 \text{ N/m}$$

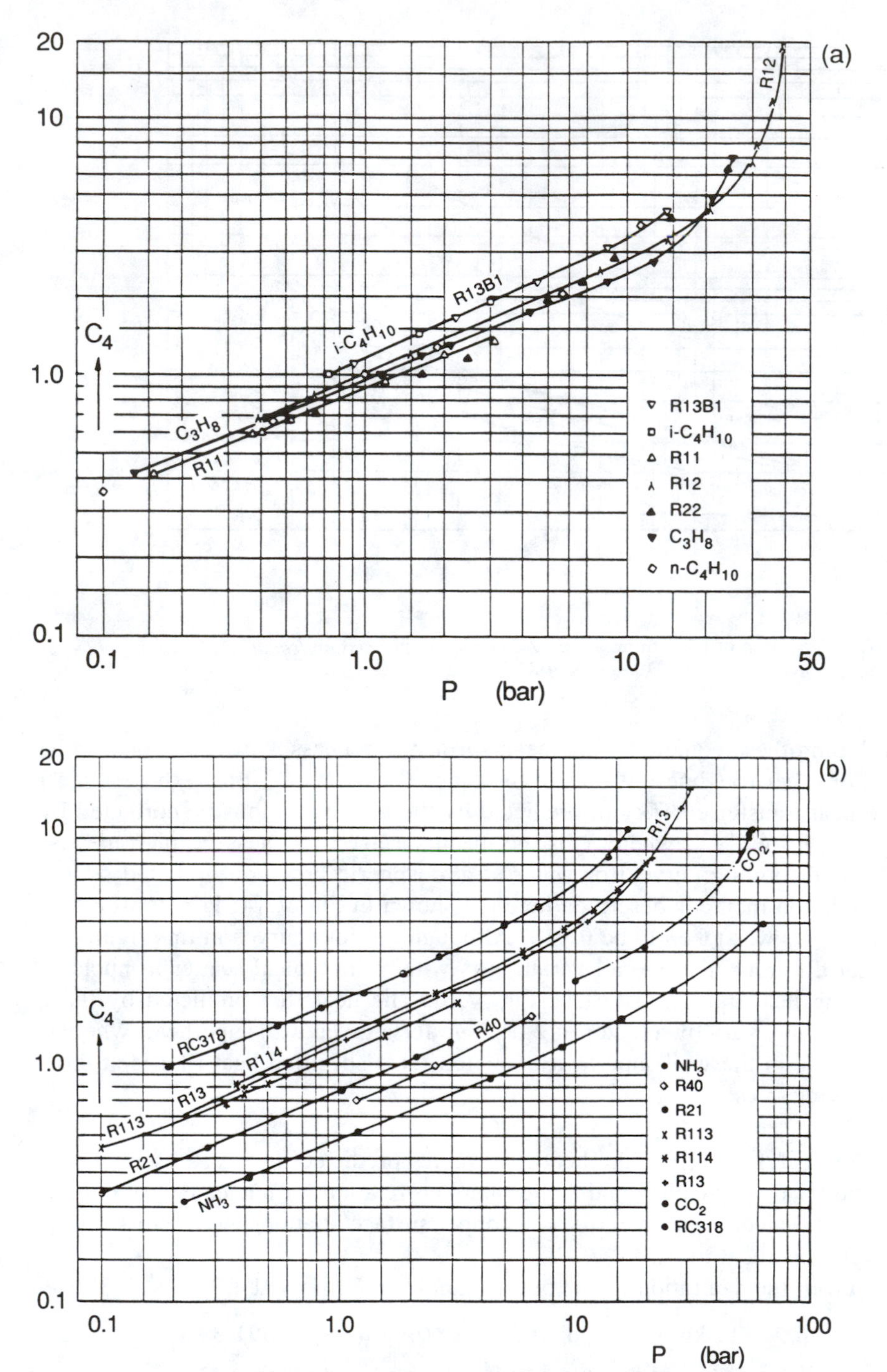

Figure 7.13 Variation of C_4 with pressure. (*Adapted with permission from [7.25]; plots provided by K. Stephan, copyright © 1980, Pergamon Press.*)

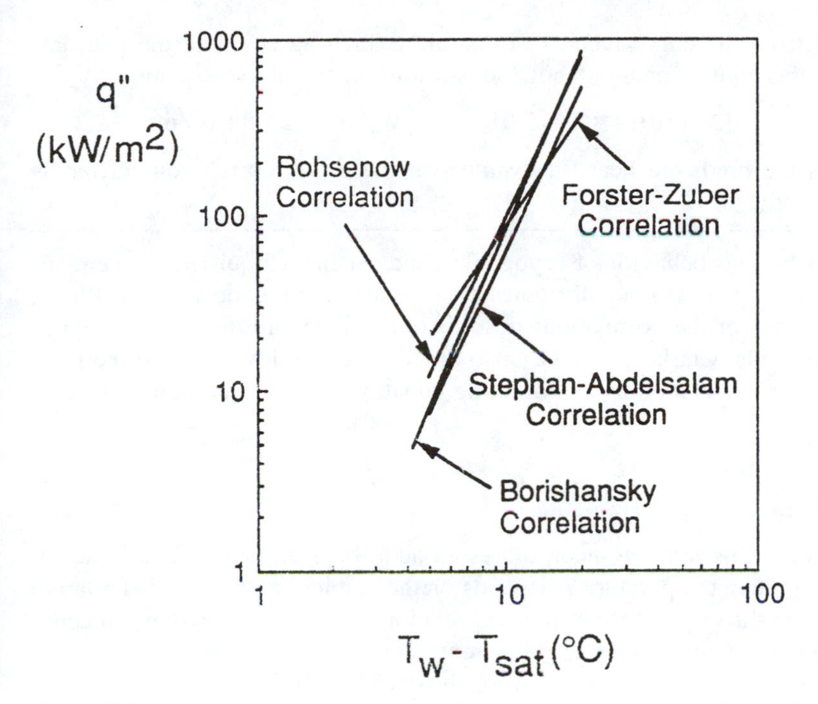

Figure 7.14 Comparison of the nucleate boiling curves predicted by the four indicated correlations for water at atmospheric pressure.

The Rohsenow correlation for a fluid other than water can be written in the form

$$q'' = \mu_l h_{lv} \left[\frac{g(\rho_l - \rho_v)}{\sigma}\right]^{1/2} \left[\frac{c_{pl}(T_w - T_{sat})}{C_{sf} h_{lv}}\right]^{3.0} \mathrm{Pr}_l^{-5.15}$$

Using $C_{sf} = 0.0030$ for n-butanol/copper from Table 7.2 and substituting the physical properties,

$$q'' = (4.40 \times 10^{-4})(591.3)\left[\frac{9.8(712 - 23)}{0.0171}\right]^{1/2}$$

$$\cdot \left[\frac{3.20(10)}{(0.0030)(591.3)}\right]^{3.0} (10.3)^{-5.15}$$

$$= 5.42 \text{ kW/m}^2$$

For hydrocarbons, the appropriate form of the Stephan and Abdelsalam correlation is

$$q'' = [C_2(T_w - T_{sat})]^{1/0.330}$$

From Fig. 7.11, the C_2 value for n-butanol at 1.0 bar is $C_2 = 2.7$. This was obtained from the curve labeled "Ethanol," which agrees with the one data point on this figure for *n*-butanol. Substituting in the above equation,

$$q'' = [2.7(10)]^{1/0.330} = 2.18 \times 10^4 \text{ W/m}^2 = 21.8 \text{ kW/m}^2$$

Thus the predicted heat flux values for these two correlations differ by about a factor of 4.

The pool boiling behavior of cryogenic fluids such as liquified oxygen, nitrogen, and hydrogen is generally consistent with that of noncryogenic fluids. Appropriate forms of the correlations described above should therefore be applicable to these fluids. Clarke [7.47] has also proposed the following modified form of the Rohsenow correlation as a correlating equation for nucleate boiling of cryogenic liquids:

$$\frac{q''}{\mu_l h_{lv}}\left[\frac{\sigma}{g(\rho_l - \rho_v)}\right]^{1/2} = 3.25 \times 10^5 \left\{\frac{c_{pl}[T_w - T_{\text{sat}}(P_l)]}{h_{lv}\,\text{Pr}_l^{1.8}}\left(\frac{T}{T_c}\right)^{1.8}\right\}^{1/2.89} \tag{7.53}$$

The T/T_c term in this relation incorporates an additional pressure effect, since at saturation the system temperature T depends on the ambient pressure. This relation also differs from the original Rohsenow correlation in that C_{sf} is fixed at a specific value and the exponents r and s are different.

Some aspects of nucleate pool boiling in liquid metals are significantly different from the corresponding behavior of nonmetallic fluids. Because of the high conductivity of the liquid, heat can be transferred to the vapor–liquid interface very rapidly during bubble growth. Bubbles therefore grow rapidly, resulting in inertia-controlled growth over most of the growth process.

Although bubble growth times are shorter, the waiting period between bubbles in liquid metals is typically considerably longer. When a bubble departs and cold liquid metal is brought into contact with the wall, locally, the surface temperature will drop more than for a nonmetal under comparable circumstances, because the nonmetal has a much lower value of $k_l \rho_l c_{pl}$ than the liquid metal. This localized drop in the surface temperature lengthens the waiting period, resulting in much lower bubble frequencies.

In addition, the low contact angles of liquid metals on solid metal surfaces often result in large cavities being fully wetted and inactive as nucleation sites. As a result, a high wall superheat may be required to initiate boiling at smaller cavity sites. Once bubbling is initiated, the inrush of cold liquid as the bubble releases may drop the temperature of the surface so low that the cavity cannot sustain the boiling process. A relatively long waiting period must then pass before the wall and adjacent fluid build up enough superheat to initiate boiling again. At high superheat, the bubbles grow almost explosively, generating a shock wave in the liquid that makes an audible sound. The resulting intermittent violent growth of vapor bubbles in liquid metals is often referred to as *bumping*.

The following correlation for pool boiling heat transfer has been proposed by Subbotin et al. [7.48] as a best fit to experimental data for potassium, sodium, and cesium:

$$\frac{(q'')^{1/3}}{T_w - T_{\text{sat}}(P_l)} = C_s\left(\frac{k_l h_{lv}\rho_l}{\sigma T^2}\right)^{1/3}\left(\frac{P_l}{P_c}\right)^s \tag{7.54}$$

where

$$C_s = 8.0,\ s = 0.45 \qquad \text{for } \frac{P_l}{P_c} < 0.001 \tag{7.55a}$$

$$C_s = 1.0,\ s = 0.15 \qquad \text{for } \frac{P_l}{P_c} \geqslant 0.001 \tag{7.55b}$$

Further discussion of the special features of pool boiling in liquid metals may be found in Dwyer's [7.49] comprehensive book on the subject.

Example 7.2 Using Clark's correlation and the Stephan and Abdelsalam correlation, determine the heat flux for nucleate boiling of liquid nitrogen at atmospheric pressure on a copper surface for a wall superheat of 5°C.

For saturated nitrogen at atmospheric pressure, $T_{\text{sat}} = 77.4$ K, $\rho_l = 807.1$ kg/m^3, $\rho_v = 4.62$ kg/m^3, $h_{lv} = 197.6$ kJ/kg, $c_{pl} = 2.06$ kJ/kg K, $\mu_l = 1.63 \times 10^{-4}$ Ns/m^2, $k_l = 0.137$ W/m K, $\text{Pr}_l = 2.46$, $\sigma = 0.00885$ N/m, and $T_c = 126.3$ K. Clarke's correlation can be written in the form

$$q'' = 3.25 \times 10^5 \mu_l h_{lv}\left[\frac{g(\rho_l - \rho_v)}{\sigma}\right]^{1/2}\left[\frac{c_{pl}(T_w - T_{\text{sat}})}{h_{lv}\,\text{Pr}_l^{1.8}}\left(\frac{T}{T_c}\right)^{1.8}\right]^{2.89}$$

Substituting the above properties and using $T = (1/2)(T_w + T_{\text{sat}}) = 79.9$ K,

$$\begin{aligned} q'' &= 3.25 \times 10^5(1.63 \times 10^{-4})(197.6) \\ &\quad \times \left[\frac{9.8(807 - 4.6)}{0.00885}\right]^{1/2}\left[\frac{2.06(5)}{197.6(2.46)^{1.8}}\left(\frac{79.9}{126.3}\right)^{1.8}\right]^{2.89} \\ &= 1.61 \text{ kW/m}^2 \end{aligned}$$

For a copper surface,

$$(\rho c_p k)_c = 8954\,\frac{\text{kg}}{\text{m}^3} \times 0.384\,\frac{\text{kJ}}{\text{kg K}} \times 398\,\frac{W}{\text{m K}} = 1.37 \times 10^6\,\frac{\text{W kJ}}{\text{m}^4\,\text{K}^2}$$

For cryogenic liquids, the Stephan and Abdelsalam correlation is

$$q'' = [C_3(\rho c_p k)_c^{0.117}(T_w - T_{\text{sat}})]^{1/0.376}$$

From Fig. 7.12 at 1.0 bar pressure, $C_3 = 0.26$. Substituting,

$$\begin{aligned} q'' &= [2.6(1.37 \times 10^6)^{0.117}(5)]^{1/0.376} \\ &= 7.45 \times 10^4 \text{ W/m}^2 = 74.5 \text{ kW/m}^2 \end{aligned}$$

Thus these two correlations predict widely different heat flux levels for these conditions.

7.4 MAXIMUM HEAT FLUX CONDITIONS

As noted in Section 7.1, the pool boiling curve generally exhibits a maximum or critical heat flux (CHF) at the transition between nucleate and transition boiling. This peak value is the maximum level of heat flux from the surface that the system can provide in a non-film-boiling mode at a given pressure. This maximum can be approached either by reducing the surface temperature in the transition boiling regime or by increasing the heat flux or wall superheat in the nucleate boiling regime (see Fig. 7.1).

The mechanism responsible for the critical heat flux has been the subject of considerable investigation and debate over the past five decades. Four different mechanisms have been postulated as the cause of the critical heat flux phenomena in pool boiling:

1. As the heat flux increases, bubbles generated at the surface coalesce, as shown in Fig. 7.15*a*, to form vapor columns in which there are liquid droplets that fall back to the surface (assuming it is horizontal). Vapor blanketing of the surface occurs when the vapor velocity becomes high enough to carry the droplets away from the surface against gravity.
2. As the nucleation site density increases, a critical bubble packing is eventually reached that inhibits liquid flow to the surface to such a degree that a vapor blanket forms over portions of the surface.
3. As the heat flux increases, bubbles generated at the surface coalesce, as shown in Fig. 7.15*a*, to form vapor columns or jets. The CHF condition occurs when Helmholtz instability of the large vapor jets leaving the surface distorts the jets, blocking liquid flow to portions of the heated surface. Continued vaporization of liquid at locations on the surface that are starved of replacement liquid then leads to formation of a vapor blanket over part or all of the surface.
4. At high heat flux levels small vapor jets connected with individual nucleation sites carry vapor to a large vapor bubble that is fed by a number of the small jets, as shown in Fig. 7.15*b*. The jets carry vapor through a thin film of liquid that exists under the larger slugs of vapor. The thickness of this liquid film must be sufficiently small that the jets are not Helmholtz unstable. Hence it is postulated that the layer thickness must be proportional to the Helmholtz-unstable wavelength of the small jets. The large slugs of vapor "hover" over the surface until they accumulate enough vapor to escape. It is postulated that the CHF condition is attained when the liquid film under the bubble evaporates completely during the "hovering" time interval needed for the bubble to grow large enough to escape.

Each of these models of the CHF condition is based on a scenario that results in vapor blanketing of portions of the surface, with the added thermal resistance of the vapor blanket reducing the heat flux for a given wall superheat.

Mechanism 1 was proposed by Kutateladze [7.50] based on an analogy between the CHF condition and flooding phenomena in process equipment. The

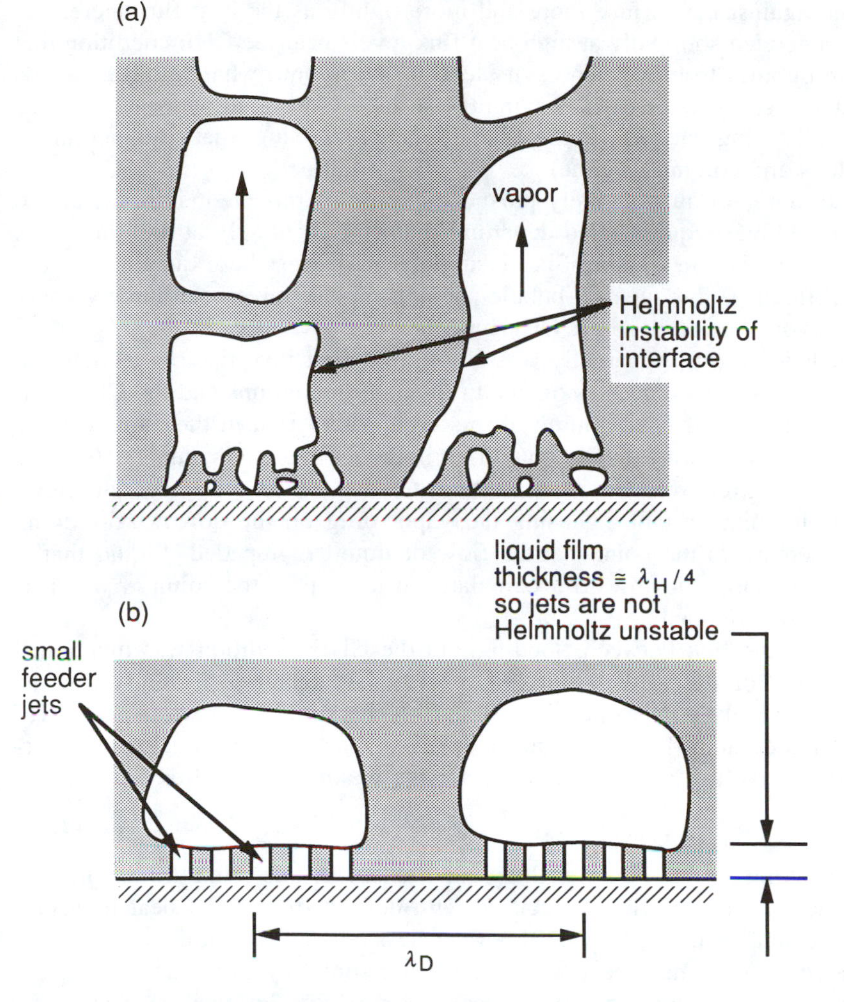

Figure 7.15 Postulated Helmholtz instability CHF mechanisms.

bubble packing mechanism 2 was the basis of early CHF models proposed by Rohsenow and Griffith [7.51] and Chang and Snyder [7.52]. The Helmholtz-instability mechanism described as 3 was first incorporated into a formal model by Zuber [7.53] for a flat horizontal surface. This type of modeling was subsequently refined and extended to other geometries by Lienhard and co-workers [7.54]–[7.58]. Model 4, which focuses on the layer of liquid under the large vapor bubbles, has been recently proposed by Haramura and Katto [7.59].

There is some logic to the idea that packing of bubbles at the surface could ultimately shut off the liquid flow to the surface and induce vapor blanketing over portions of the surface. However, this line of reasoning has two pitfalls. First, visual evidence does not support this model's basic premise that round bubbles are packed against the surface more and more tightly as the heat flux increases. Vapor is generated so rapidly at high heat flux levels near the CHF condition that successive bubbles from a given cavity tend to merge into what would better be described as a small jet of vapor leaving the surface. The small vapor jets formed at an upward-facing flat surface immediately merge into slugs that rise in columns (in the slugs and columns regime).

Furthermore, to quantitatively pursue this type of model, it is necessary to predict the bubble frequency and departure diameter accurately at high heat flux levels. As described in Chapter Six, accurately predicting these quantities is extremely difficult. Consequently, bubble-packing models have been largely abandoned in favor of hydrodynamic CHF models.

Kutateladze [7.50] apparently was among the first investigators to note the similarity between flooding phenomena in distillation columns and the CHF condition in pool boiling. In a column of this type, vapor rich in the more volatile components flows upward, while liquid rich in the less volatile components flows downward. If the relative velocity of the two streams becomes too large, the flows become Helmholtz unstable, causing the vapor drag on the downward-moving liquid to increase to the point that the flow of liquid is impeded. Liquid that is then unable to move downward accumulates at the top of the column, which is then said to be "flooded."

Although the link between flooding and the CHF condition was mentioned in a paper by Bonilla and Perry in 1941 [7.53], Kutateladze apparently was the first to aggressively pursue the idea. Based on the similarity between the CHF condition and column flooding, Kutateladze [7.52] used dimensional analysis arguments to derive the following relation for the maximum heat flux:

$$q''_{\max} = C_K \rho_v^{1/2} h_{lv} [g(\rho_l - \rho_v)\sigma]^{1/4} \tag{7.56}$$

Strictly speaking, the flooding analogy used to obtain this relation is applicable only to one-dimensional flow associated with boiling from a flat heated surface of infinite extent. Kutateladze [7.50] nevertheless concluded that C_K was equal to 0.131 based on maximum heat flux data for horizontal cylinders and other configurations different than the infinite surface for which this relation was developed.

The interface stability analysis developed by Taylor [7.60] in 1950 was a key factor in the subsequent development of more detailed hydrodynamic models of the CHF mechanism. Chang [7.61] appears to have been the first to suggest a link between Taylor wave motion and pool boiling processes. Chang's observations, together with arguments in the Soviet literature linking flooding phenomena to the CHF condition, apparently influenced Zuber [7.53] to include Taylor wave motion and Helmholtz instability as key elements in his model of the CHF mechanism. (See Chapter Four for a detailed discussion of Taylor and Helmholtz instabilities.)

Zuber's model analysis, as refined by Lienhard and Dhir [7.54], is based on the following idealizations:

1. The critical heat flux is attained when the interface of the large vapor columns leaving the surface becomes Helmholtz unstable.
2. The columns leave the surface in a rectangular array as shown in Fig. 7.16. The centerline spacing of the columns coincides with nodes of the most dangerous wavelength associated with the two-dimensional wave pattern for Taylor instability of the horizontal interface between a semiinfinite liquid region above a layer of vapor.
3. The column radius is equal to $\lambda_D/4$, where λ_D is the spacing of the columns as predicted by Taylor instability analysis (see Section 4.2).
4. The Helmholtz-unstable wavelength imposed on the columns is equal to the Taylor wave node spacing λ_D.

As described in Section 4.3, the critical Helmholtz velocity u_c for vertical vapor and liquid flow (from Eq. [4.50]) is given by

$$u_c = |\bar{u}_l - \bar{u}_v| = \left[\frac{\sigma\alpha(\rho_l + \rho_v)}{\rho_l\rho_v}\right]^{1/2} \tag{7.57}$$

Assuming that $\rho_l \gg \rho_v$ and substituting $2\pi/\lambda$ for the wavenumber α, this relation can be written as

$$u_c = \left(\frac{2\pi\sigma}{\rho_v\lambda}\right)^{1/2} \tag{7.58}$$

Because the downward liquid velocity is much smaller than the upward vapor velocity, due to the large density difference between the phases, u_c is essentially given by

$$u_c = \frac{q''_{\max}}{\rho_v h_{lv}}\left(\frac{A_{\text{surf}}}{A_{\text{col}}}\right) \tag{7.59}$$

For the unit cell of the model column array shown in Fig. 7.16, the surface-to-column area ratio is given by

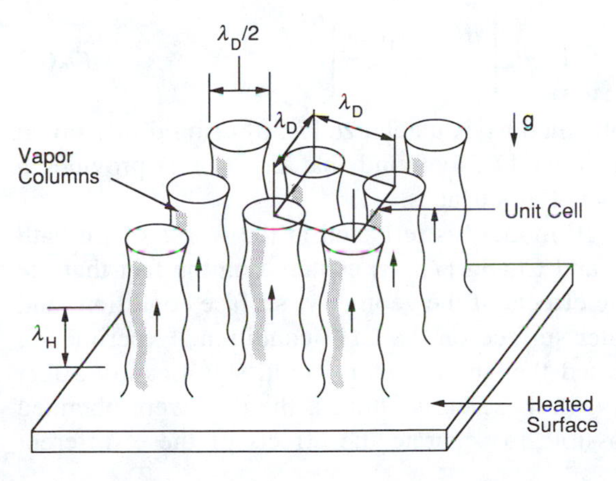

Figure 7.16 Vapor column spacing in the Zuber critical heat flux model.

$$\frac{A_{surf}}{A_{col}} = \frac{\lambda_D^2}{\pi(\lambda_D/4)^2} = \frac{16}{\pi} \tag{7.60}$$

Combining Eqs. (7.58) through (7.60), substituting

$$\lambda_D = 2\pi \left[\frac{3\sigma}{(\rho_l - \rho_v)g} \right]^{1/2} \tag{7.61}$$

for λ_D (Eq. [4.56]), and solving for q''_{max} yields

$$q''_{max} = \frac{\pi}{16(3)^{1/4}} \rho_v h_{lv} \left[\frac{\sigma(\rho_l - \rho_v)g}{\rho_v^2} \right]^{1/4} \tag{7.62}$$

$$= 0.149 \rho_v h_{lv} \left[\frac{\sigma(\rho_l - \rho_v)g}{\rho_v^2} \right]^{1/4} \tag{7.63}$$

The analysis described above basically follows Zuber's original model, except for the assumption that the Helmholtz-unstable wavelength is equal to λ_D, which was later proposed by Lienhard and Dhir [7.54]. The assumption that the Helmholtz-unstable wavelength is equal to the Taylor-unstable wavelength is perhaps the least transparent of the idealizations in this model. However, this is a plausible position if we consider approaching the maximum heat flux from the transition boiling side (see Fig. 7.1).

For film boiling on an upward-facing surface, release of vapor bubbles is expected to occur at the nodes of waves having the most dangerous wavelength λ_D, since waves of this wavelength grow most rapidly (this is discussed further in the next section). The existence of vapor bubble columns with a similar spacing is also expected in the transition regime near the minimum heat flux, since the departure from film boiling is small. If the spacing of the jets does not significantly change as the heat flux increases, then the assumption that this spacing exists near q''_{max} is reasonable.

Zuber's original model resulted in a relation for q''_{max} identical to Eq. (7.63) except that the constant on the right side was $\pi/24 = 0.131$:

$$q''_{max,Z} = 0.131 \rho_v h_{lv} \left[\frac{\sigma(\rho_l - \rho_v)g}{\rho_v^2} \right]^{1/4} \tag{7.64}$$

This relation is identical to that obtained by Kutateladze [7.50] using dimensional analysis. However, Lienhard and Dhir [7.54] found that Eq. (7.63) provides a better fit to available data for large, flat surfaces.

Initial criticism of Zuber's CHF model, as reflected in the works of Bernath [7.62], Costello and Frea [7.63], and Chang [7.64], centered on the fact that the model did not account for possible effects of the geometry, surface condition, and wetting characteristics of the heater surface on the CHF condition. Experimental data available at that time suggested that these factors could significantly affect the CHF condition. Unfortunately, in these early studies, the data were obtained in such a way that it was impossible to separate the effects of these different influences.

The role of heater geometry in hydrodynamic models of the CHF condition has been clarified largely by the systematic studies of Lienhard and co-workers [7.55]–[7.58]. These investigators have adapted Zuber's model for the critical heat flux mechanism to saturated pool boiling for square and round heated surfaces of finite size, horizontal cylinders, horizontal ribbons, and spheres. For each of these configurations, the heated surface is finite in extent in at least one dimension. Hence, the length scale characterizing the size of the heater becomes an important parameter. Correlations for the critical heat flux for these finite-sized surfaces have typically been written in the form

$$\frac{q''_{max}}{q''_{max,Z}} = f\left(\frac{L}{L_b}\right) \tag{7.65}$$

where L_b is the bubble (capillary) length scale, defined as

$$L_b = \sqrt{\frac{\sigma}{g(\rho_l - \rho_v)}} \tag{7.66}$$

and $q''_{max,Z}$ is the Zuber maximum heat flux, defined by Eq. (7.64). Since $L_b = \lambda_D/2\pi\sqrt{3}$, the ratio L/L_b indicates the size of the heater relative to the expected spacing of the vapor columns carrying vapor away from the surface near the critical condition. For heaters of finite size, variation of the value of this dimensionless group (i.e., the Bond number) is expected to significantly alter the CHF condition, particularly if its value is near or below 1.

CHF correlations and the range of applicability suggested by the developers are given in Table 7.3 for a number of finite heater configurations. The improved q''_{max} correlation for the infinite flat plate developed by Lienhard and Dhir [7.54] can also be written in the form of Eq. (7.65): $q''_{max}/q''_{max,Z} = 1.14$.

The correlations given in Table 7.3 apply to saturated pool boiling near bodies with smooth surfaces in a motionless liquid ambient. The relations in this table generally agree with experimental data to within about ±20%, which is about the level of scatter in data of this type. The relations for an isolated flat surface, a sphere, and a cylinder are plotted in Fig. 7.17. The success of this type of correlation methodology suggests that the Helmholtz instability mechanisms associated with these different heater configurations have common features, but do differ from one type of geometry to another.

More recently, Lienhard and Hasan [7.65] proposed an alternative analytical approach to predicting the pool boiling CHF condition that retains the vapor column instability mechanism of the Zuber model. These investigators used the mechanical energy stability criterion as a key element of the model. This criterion is a special case of the more general thermodynamic requirement for stable equilibrium. In Chapter One, it was shown that, for a simple compressible system, the system is in stable equilibrium if, for all internal fluctuations,

$$(\Delta U)_{S,V} \leq 0 \tag{7.67}$$

where U denotes the system internal energy and the subscripts denote changes at

Table 7.3 Correlations for the maximum pool boiling heat flux

Geometry	Correlation	Range of applicability	Reference
Infinite, heated flat plate	$\dfrac{q''_{max}}{q''_{max,Z}} = 1.14$	$\dfrac{L}{L_b} > 30$	[7.55]
Small heater of width or diameter L with vertical side walls	$\dfrac{q''_{max}}{q''_{max,Z}} = \dfrac{1.14\lambda_D^2}{A_{heater}}$	$9 < \dfrac{L}{L_b} < 20$	[7.55]
Horizontal cylinder of radius R	$\dfrac{q''_{max}}{q''_{max,Z}} = 0.89 + 2.27 \exp\left(-3.44 \sqrt{\dfrac{R}{L_b}}\right)$	$\dfrac{R}{L_b} > 0.15$	[7.56]
Large horizontal cylinder of radius R	$\dfrac{q''_{max}}{q''_{max,Z}} = 0.90$	$\dfrac{R}{L_b} > 1.2$	[7.57]
Small horizontal cylinder of radius R	$\dfrac{q''_{max}}{q''_{max,Z}} = 0.94\left(\dfrac{R}{L_b}\right)^{-1/4}$	$0.15 \leq \dfrac{R}{L_b} \leq 1.2$	[7.57]
Large sphere of radius R	$\dfrac{q''_{max}}{q''_{max,Z}} = 0.84$	$4.26 \leq \dfrac{R}{L_b}$	[7.58]
Small sphere of radius R	$\dfrac{q''_{max}}{q''_{max,Z}} = 1.734\left(\dfrac{R}{L_b}\right)^{-1/2}$	$0.15 \leq \dfrac{R}{L_b} \leq 4.26$	[7.58]
Small horizontal ribbon oriented vertically with side height H—both sides heated	$\dfrac{q''_{max}}{q''_{max,Z}} = 1.18\left(\dfrac{H}{L_b}\right)^{-1/4}$	$0.15 \leq \dfrac{H}{L_b} \leq 2.96$	[7.57]
Small horizontal ribbon oriented vertically with side height H—back side insulated	$\dfrac{q''_{max}}{q''_{max,Z}} = 1.4\left(\dfrac{H}{L_b}\right)^{-1/4}$	$0.15 \leq \dfrac{H}{L_b} \leq 5.86$	[7.57]
Small, slender, horizontal cylindrical body of arbitrary cross section with transverse perimeter L_p	$\dfrac{q''_{max}}{q''_{max,Z}} = 1.4\left(\dfrac{L_p}{L_b}\right)^{-1/4}$	$0.15 \leq \dfrac{L_p}{L_b} \leq 5.86$	[7.57]
Small bluff body with characteristic dimension L	$\dfrac{q''_{max}}{q''_{max,Z}} = C_0\left(\dfrac{L}{L_b}\right)^{-1/2}$	Large $\dfrac{L}{L_b}$	[7.57]

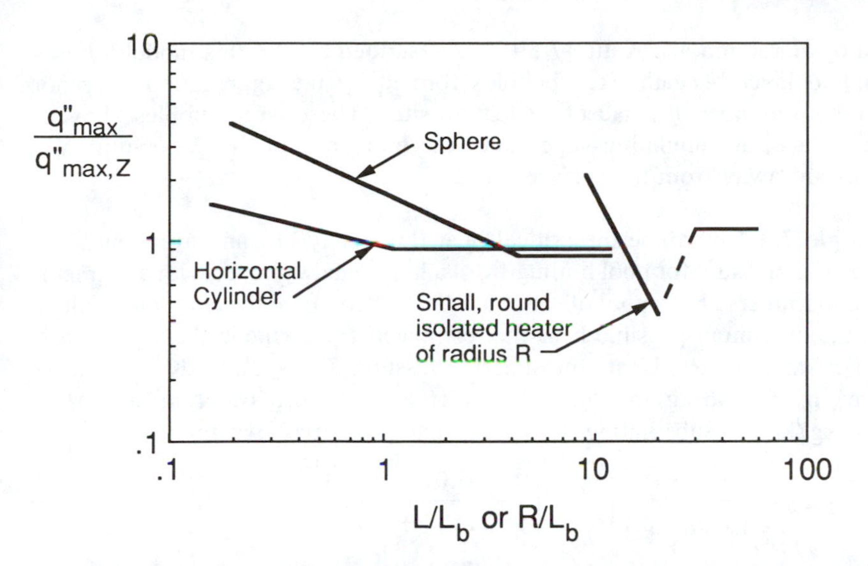

Figure 7.17 Variation of the critical heat flux with the characteristic dimension of the heated surface.

constant entropy and volume. If the arguments presented in Chapter One are broadened to include mechanical energy exchanges, the condition for equilibrium can be stated as

$$(\Delta E)_{S,V} \leq 0 \tag{7.68}$$

where E denotes the system energy. With respect to the boiling process, this condition is interpreted as meaning that the vapor-removal process in the columns (or wakes) adjacent to the surface remains stable as long as the net mechanical energy transfer to the system is negative.

Lienhard and Hasan [7.65] postulated that violation of the mechanical energy stability criterion corresponded to the maximum (critical) heat flux condition. Specifically, this condition was violated when the rate of vapor kinetic energy into the column system shown in Fig. 7.16 just exceeds the capillary energy given up when the columns break up into bubbles. These investigators equated expressions for the rate of vapor kinetic energy addition and the rate of capillary energy removal from the bubble column to obtain a relation for the vapor flux from the surface at the critical heat flux condition.

By adopting relations between the geometry of the vapor column array and the Taylor-unstable wavelength similar to those in the original Zuber analysis, Lienhard and Hasan [7.65] showed that the results of this type of analysis are consistent with those of the Zuber-Helmholtz instability model. They also showed that using this approach eliminates the need for information regarding the Helmholtz-unstable wavelength, but instead requires empirical information or assumptions regarding the bubble departure diameter when the vapor column breaks up.

The fourth model listed at the beginning of this section has recently been proposed by Haramura and Katto [7.59]. As described above, this model focuses on the liquid layer beneath large bubbles formed by the aggregation of vapor generated by a number of adjacent nucleation sites. These large bubbles "hover" over the surface, accumulating vapor until the buoyancy of the vapor pulls the bubble upward away from the surface.

Example 7.3 Determine the critical heat flux for R-113 and nitrogen at atmospheric pressure for pool boiling from a horizontal cylinder with a diameter of 5 mm immersed in a pool of saturated liquid. Compare the computed values with the maximum possible heat flux estimated from kinetic theory.

For saturated R-113 at atmospheric pressure, $T_{sat} = 320.7$ K, $\rho_l = 1507$ kg/m^3, $\rho_v = 7.46$ kg/m^3, $h_{lv} = 143.8$ kJ/kg, $\sigma = 0.0169$ N/m, and $\overline{M} = 187.4$ kg/kmol. Substituting the appropriate properties, we find that

$$L_b = \left[\frac{\sigma}{g(\rho_l - \rho_v)}\right]^{1/2} = \left[\frac{0.0169}{9.8(1507 - 7.5)}\right]^{1/2} = 0.00107 \text{ m}$$

It follows that $R/L_b = (5/2)/1.07 = 2.33$. Because this ratio is greater than 1.2, the large radius relation $q''_{max}/q''_{max,Z} = 0.9$ applies (see Table 7.3). Thus,

$$q''_{max} = 0.9(0.131)\rho_v h_{lv}\left[\frac{\sigma(\rho_l - \rho_v)g}{\rho_v^2}\right]^{1/4}$$

$$= 0.9(0.131)(7.46)(143.8)\left[\frac{0.0169(1507 - 7.5)9.8}{(7.46)^2}\right]^{1/4}$$

$$= 183.8 \text{ kW/m}^2$$

For saturated N_2 at atmospheric pressure, $T_{sat} = 77.4$ K, $\rho_l = 807.1$ kg/m^3, $\rho_v = 4.62$ kg/m^3, $h_{lv} = 197.6$ kJ/kg, $\sigma = 0.00885$ N/m, and $\overline{M} = 28.02$ kg/kmol. Computing L_b as before:

$$L_b = \left[\frac{0.00885}{9.8(807.1 - 4.6)}\right]^{1/2} = 0.00106 \text{ m}$$

Because $R/L_b = 2.5/1.06 = 2.36$ is again greater than 1.2, we again use the large-cylinder relation for q''_{max}:

$$q''_{max} = 0.9(0.131)(4.62)(197.6)\left[\frac{0.00885(807.1 - 4.6)9.8}{(4.62)}\right]^{1/4}$$

$$= 144.6 \text{ kW/m}^2$$

In Section 4.7 it was argued that an estimate of the maximum possible vaporization heat flux is provided by the relation

$$q''_{mkv} = 0.741\rho_v h_{lv}\left(\frac{\overline{R}T_v}{2\pi\overline{M}}\right)^{1/2}$$

where $\bar{R}$ = 8314.4 J/kmol is the universal gas constant. It was shown in Example 4.6 that, for R-113 at atmospheric pressure,

$$q''_{mkv} = 3.78 \times 10^4 \text{ kW/m}^2,$$

which is about 200 times more than the q''_{max} value determined above. Substituting appropriate properties for nitrogen:

$$q''_{mkv} = 0.741(4.62)(197.6)\left[\frac{(8314.4)(77.4)}{2\pi(28.02)}\right]^{1/2}$$
$$= 4.09 \times 10^4 \text{ kW/m}^2$$

Thus the value of q''_{max} for pool boiling of nitrogen is also several hundred times lower than the maximum possible flux at the interface.

A key element of the Haramura and Katto model is the argument that the thickness of the liquid film must be smaller than the Helmholtz-unstable wavelength for the vapor jets to ensure that they are not Helmholtz unstable. Specifically, these investigators assumed that the layer thickness is equal to $\lambda_H/4$. It was further postulated that the large bubbles are spaced in a rectangular array with a centerline spacing equal to λ_D, and that the critical heat flux condition occurs when the liquid film under the large bubbles evaporates completely over the "hovering" time interval during which the bubble grows before release. Based on these arguments (for an infinite flat surface), Haramura and Katto [7.59] derived the following relation for the critical heat flux:

$$\frac{q''_{max}}{\rho_v^{1/2} h_{lv}[\sigma(\rho_l - \rho_v)g]^{1/4}} = \frac{0.131 q''_{max}}{q''_{max,Z}}$$
$$= \left(\frac{\pi^4}{2^{11}3^2}\right)^{1/16}\left(\frac{A_v}{A_w}\right)^{5/8}\left(1 - \frac{A_v}{A_w}\right)\left\{\frac{(\rho_l/\rho_v) - 1}{[(11/16)(\rho_l/\rho_v) + 1]^{3/5}}\right\}^{5/16} \quad (7.69)$$

where A_v/A_w is the ratio of the cross-sectional area of the vapor stems to the heater surface area. They further showed that this relation is in good agreement with Zuber's CHF relation (7.62) if the ratio A_v/A_w is given by

$$\frac{A_v}{A_w} = 0.584\left(\frac{\rho_v}{\rho_l}\right)^{0.2} \quad (7.70)$$

Haramura and Katto [7.59] also demonstrated that this model could be extended in a straightforward manner to saturated pool boiling from other geometries and to forced-convection boiling from several different geometries.

The extent to which the hydrodynamic theories have led to useful correlations for the critical heat flux is impressive. Yet despite their success, hydrodynamic models have been openly questioned on theoretical grounds and because predictions of correlations developed from the models do not completely agree with trends in experimental data. The original Zuber model and the more recent model of Haramura and Katto [7.59] both contain a number of idealizations that are, at best, weakly justifiable.

For example, there is no clear justification for why the radius of the vapor columns in the Zuber model should be a constant fraction of λ_D independent of the heat flux. Furthermore, the assumption that the vapor column spacing for film boiling persists throughout the transition boiling regime is plausible, but difficult to strongly justify.

The model of Haramura and Katto [7.59] postulates the existence of a thin liquid sublayer adjacent to the surface that is replenished only after the large bubble covering it departs. This feature of their model has been questioned because large vapor bubbles in boiling systems have been visually observed to be separated from the sublayer by long larger vapor jets, suggesting that the liquid layer under the large bubbles could be continuously replenished. This visual evidence contradicting the postulated morphology of the system leaves serious doubts about the correctness of the CHF mechanism embodied in this model. The good agreement of the resulting correlation with data could simply be a consequence of the fact that it is dimensionally consistent with the correlation obtained from the Zuber model.

In addition, the CHF data presented by Berensen [7.66], Dhir and Liaw [7.67] and others indicate that surface wetting characteristics can have a significant effect on the CHF condition. In the extreme case of a liquid boiling on a completely nonwetted surface, the nucleate boiling regime and maximum heat flux condition both vanish. (See Section 8.1 for further discussion of this phenomenon.) Neither the Zuber model nor the Haramura and Katto model [7.59] accounts for the influence of contact angle on the CHF condition.

Despite the questionable aspects of the model proposed by Haramura and Katto [7.59], their idea that Helmholtz instability might actually occur on the feeder jets near the surface could be the key to including a weak surface influence in an improved hydrodynamic model. Dhir and Liaw [7.67] have, in fact, developed a hydrodynamic model for the CHF condition on a vertical plate in which the shape of the feeder jets near the surface is influenced by the contact angle. Although hydrodynamic considerations will continue to play a central role in our understanding of the CHF transition, there clearly exists a need to develop models of the CHF mechanism that account for both hydrodynamic and surface effects. Further discussion of the development and limitations of hydrodynamic CHF theories can be found in references [7.68] and [7.69].

The available experimental evidence indicates that the CHF models described above will provide reasonably accurate predictions of the CHF condition for water, hydrocarbons, cryogenic liquids, and halogenated refrigerants. However, for boiling in liquid metals, the experimentally determined CHF values can be from two to four times the values predicted by the relations given in Table 7.3. The higher CHF values for liquid metals apparently are due to the stronger conduction-convection mechanism in addition to the vapor transport mechanism considered in the above models.

Example 7.4 Determine the variation of the critical heat flux with pressure for pool boiling from a horizontal cylinder with a diameter of 5 mm immersed in a pool of saturated water.

Using properties from the saturation tables and either the small- or large-diameter relation for q''_{max} given in Table 7.3, the values of q''_{max} obtained at different pressures are summarized below. Formulas used in the calculations are

$$L_b = \left[\frac{\sigma}{g(\rho_l - \rho_v)}\right]^{1/2}$$

$$q''_{max,Z} = 0.131\rho_v h_{lv}\left[\frac{\sigma(\rho_l - \rho_v)g}{\rho_v^2}\right]^{1/4}$$

$$q''_{max} = 0.90 q''_{max,Z} \qquad \left(\frac{R}{L_b} > 1.2\right)$$

$$q''_{max} = 0.94(R/L_b)^{-1/4} q''_{max,Z} \qquad \left(0.15 \leq \frac{R}{L_b} \leq 1.2\right)$$

P (kPa)	$\rho_l - \rho_v$ (kg/m^3)	σ (N/m)	h_{lv} (kJ/kg)	ρ_v (kg/m^3)	R/L_b	q''_{max} (kW/m^2)
101	958	0.0588	2,257	0.598	0.999	1,042
1,003	882	0.0423	2,015	5.16	1.130	2,390
3,348	797	0.0286	1,765	16.8	1.306	3,297
8,592	666	0.0144	1,404	46.2	1.683	3,503
14,608	518	0.00571	1,027	92.8	2.357	2,706
18,674	384	0.00203	719	144.0	3.404	1,879

As can be seen in the table, q''_{max} first increases and then decreases as pressure increases toward the critical point, peaking at about 8.6 MPa. This trend is a consequence of the competing effects of decreasing ρ_l, σ, and h_{lv} and increasing ρ_v as the saturation pressure increases.

The stronger conduction-convection contribution is directly a consequence of the high thermal conductivity of the liquid metals. For other liquids with much lower conductivities, the conduction-convection mechanisms is sufficiently small that it can be neglected. For liquid metals, this implies that the critical heat flux $q''_{max,lm}$ might best be predicted by a relation of the form

$$q''_{max,lm} = q''_{c,c} + q''_{max} \tag{7.71}$$

where q''_{max} is the critical heat flux predicted by one of the correlations described above, and $q''_{c,c}$ is an additional term that accounts for the conduction-convection effect. In fact, Noyes and Lurie [7.70] have correlated CHF data for pool boiling of sodium with the relation

$$q''_{\max,\mathrm{lm}} = q''_{c,c} + 0.16\rho_v^{1/2} h_{lv}(\sigma\rho_l g)^{1/4} \tag{7.72}$$

where

$$q''_{c,c} = 1260\ \mathrm{kW/m^2} \tag{7.73}$$

The second term in this relation differs from the Zuber model relation (7.64) only in that 0.131 has been replaced by 0.16 and ρ_v has been neglected compared to ρ_l. This relation was found to fit CHF data for sodium to within ±20%. Further discussion of the CHF phenomena in liquid metals can be found in reference [7.49].

Corresponding-states correlations for the pool boiling CHF condition have also been proposed by Soviet investigators (see, e.g., [7.71]) and, more recently, by Sharan et al. [7.72]. As noted by Sharan et al. [7.72], corresponding-states correlations can be established whenever $q''_{\max}$ depends only on thermodynamic properties and not on surface characteristics or transport properties. Over a limited range of conditions, the agreement of such correlations with experimental data is generally comparable to that for the correlations given in Table 7.3. In cases where it is appropriate, the use of corresponding-states correlations may be preferred because of their mathematical simplicity.

7.5 MINIMUM HEAT FLUX CONDITIONS

As shown in Fig. 7.1, the boundary between the transition boiling regime and the film boiling regime corresponds to a minimum in the heat flux-versus-superheat curve. This condition is referred to as the *minimum heat flux condition,* and sometimes, for reasons that will be described later, it is referred to as the *Leidenfrost point*. It is clear from the nature of the transition and film boiling regimes (see Section 7.1) that the minimum heat flux corresponds approximately to the lowest heat flux that will sustain stable film boiling.

For an infinite, flat (upward-facing) heated surface, vapor generated at the interface during stable film boiling is released as bubbles at the nodes of a standing two-dimensional Taylor wave pattern. Chang [7.73] apparently was the first to propose a link between the Taylor-instability wave behavior and film boiling processes. Subsequently, Zuber [7.53] postulated that the minimum heat flux condition occurs when vapor is not produced rapidly enough to compensate for the normal collapse rate of the film. These considerations suggest that a relation of the following form can be written for the minimum heat flux $q''_{\min}$:

$$q''_{\min} = e_b n''_b f_{\min} \tag{7.74}$$

where e_b = energy per bubble
n''_b = number of bubbles released per unit area, per release cycle
$f_{\min}$ = minimum number of cycles per second (consistent with the Taylor mechanism) to just compensate for the normal collapse rate

Assuming that the radius of the bubbles released at the nodes of the Taylor wave is equal to $\lambda_D/4$, the latent heat energy per bubble is given by

$$e_b = \frac{4\pi}{3}\left(\frac{\lambda_D}{4}\right)^3 \rho_v h_{lv} \tag{7.75}$$

Because Taylor instability analysis indicates that the "most dangerous" wave will amplify most rapidly (see Section 4.2), Zuber postulated that the standing wave established as the vapor release mechanism is, in fact, this most dangerous wave pattern. Bubbles are released at the nodes and antinodes of this wave pattern, as shown in Fig. 7.18. For this postulated behavior, one bubble is released per unit cell, per half-cycle of the standing wave (see Fig. 7.18). It follows directly that

$$n_b'' = \frac{2}{\lambda_D^2} \tag{7.76}$$

At the minimum heat flux, Zuber [7.53] argued that the frequency of release f_{min} could not be slower than that due to the Taylor instability mechanism alone.

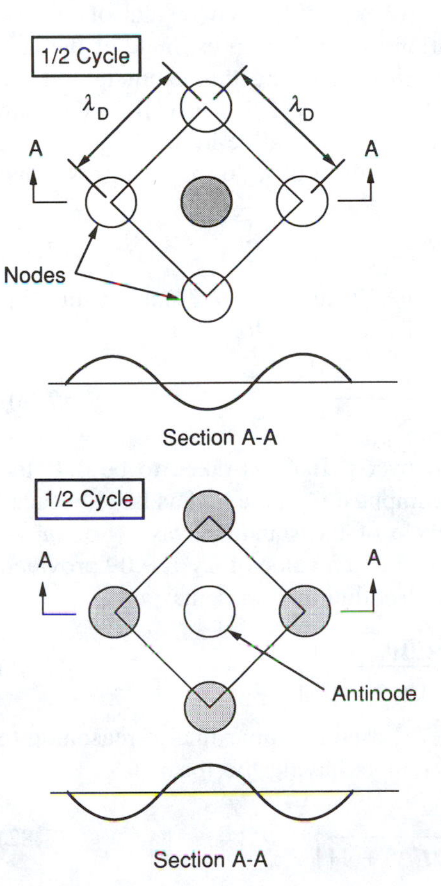

Figure 7.18 Schematic of the vapor release cycle during film boiling from a horizontal heated surface facing upward.

Waves that grow large enough to generate bubble release are expected to be those with the maximum possible Taylor wave growth rate. The general interfacial stability analysis described in Section 4.2 applies to Taylor instability if $\bar{u}_v$ and $\bar{u}_l$ are set to zero and g is replaced by $-g$ so the analysis applies to a more dense liquid overlaying a less dense vapor. Incorporating these modifications, the dispersion relation (4.43) becomes

$$\beta = \left[\frac{(\rho_l - \rho_v)g\alpha - \sigma\alpha^3}{\rho_l + \rho_v}\right]^{1/2} \tag{7.77}$$

where $\alpha = 2\pi/\lambda$ is the wavenumber. To obtain $\beta_{\max}$, we evaluate β at $\lambda = \lambda_D$, which yields, after substituting Eq. (7.61) for λ_D,

$$\beta_{\max} = \left[\frac{4(\rho_l - \rho_v)^3 g^3}{27(\rho_l + \rho_v)^2\sigma}\right]^{1/4} \tag{7.78}$$

Because it is obtained from linear stability analysis of small disturbances, this relation for $\beta_{\max}$ will accurately predict the temporal growth of the wave amplitude only in the early stages of growth. Experimental evidence indicates that the growth rate increases as the amplitude increases. Zuber [7.53] used results of experimental studies of interface wave amplification to develop an estimate of the amplification rate in the later stages of wave growth. Using this estimate and the results of linear stability theory, Zuber [7.53] concluded that near the minimum heat flux condition, the bubble release frequency $f_{\min}$ should be in the range $0.4\beta_{\max} < f_{\min} < 0.43\beta_{\max}$. This line of reasoning is crude, but it strongly suggests that $f_{\min}$ is proportional to $\beta_{\max}$:

$$f_{\min} = C_1\beta_{\max} \tag{7.79}$$

Substituting Eqs. (7.75), (7.76), (7.78), and (7.79) into Eq. (7.74) and using Eq. (7.61) to evaluate λ_D, the following equation is obtained for $q''_{\min}$:

$$q''_{\min} = C_2\rho_v h_{lv}\left[\frac{g\sigma(\rho_l - \rho_v)}{(\rho_l + \rho_v)^2}\right]^{1/4} \tag{7.80}$$

where $C_1(\pi^2/12)(4/3)^{1/4}$ has been replaced by C_2. If C_1 is taken to be 0.4, the resulting value of C_2 is 0.35. Zuber [7.53] computed C_2 to be half as large because he counted only one bubble release per cycle of the standing wave (i.e., $n''_b = 1/\lambda_D^2$). Somewhat later, Berenson [7.74] found that a value of $C_2 = 0.09$ provided a better fit to $q''_{\min}$ data determined from pool boiling experiments:

$$q''_{\min,B} = 0.09\rho_v h_{lv}\left[\frac{g\sigma(\rho_l - \rho_v)}{(\rho_l + \rho_v)^2}\right]^{1/4} \tag{7.81}$$

More recently, Lienhard and Wong [7.75] used a similar line of reasoning to develop a $q''_{\min}$ correlation for horizontal cylinders having the form

$$q''_{\min} = C_3\left\{\frac{18}{(R/L_b)^2[2(R/L_b)^2 + 1]}\right\}^{1/4} q''_{\min,B} \tag{7.82}$$

where R is the cylinder radius and $q''_{\min,B}$ is given by Eq. (7.81). As a best fit to experimental data, they recommend 0.515 for the value of C_3. They note, however, that the best-fit value of C_3 may vary depending on the end mounting of the cylinder, which, in some instances, was found to have a significant effect on the condition at which film boiling will collapse.

For spheres, Gunnerson and Cronenberg [7.76] have developed a complex correlation for $q''_{\min}$ that includes the effects of intermittent liquid contact with portions of the surface, as well as the Taylor wave release mechanism. Their correlation also includes the effects of vapor superheat and ambient pool subcooling on $q''_{\min}$.

Example 7.5 Estimate the values of $q''_{\min}$ for pool boiling of water and R-113 on a flat, infinite surface at atmospheric pressure.

For water at atmospheric pressure, $\rho_v = 0.597$ kg/m^3, $\rho_l = 958$ kg/m^3, $h_{lv} = 2257$ kJ/kg, and $\sigma = 0.0589$ N/m. Using Eq. (7.81),

$$q''_{\min} = 0.09\rho_v h_{lv}\left[\frac{g\sigma(\rho_l - \rho_v)}{(\rho_l + \rho_v)^2}\right]^{1/4}$$

$$= 0.09(0.597)(2257)\left[\frac{9.8(0.0589)(958)}{(959)^2}\right]^{1/4} = 19.0 \text{ kW/m}^2$$

For R-113 at atmospheric pressure, $\rho_v = 7.46$ kg/m^3, $\rho_l = 1507$ kg/m^3, $h_{lv} = 143.8$ kJ/kg, and $\sigma = 0.0169$ N/m. Substituting into the above equation for $q''_{\min}$ yields

$$q''_{\min} = 0.09(7.46)(143.8)\left[\frac{9.8(0.0169)(1500)}{(1514)^2}\right]^{1/4}$$

$$= 9.85 \text{ kWm}^2$$

7.6 FILM BOILING

Boundary-Layer Analysis

As described in the first section of this chapter, at very high wall superheat levels, a layer of vapor completely blankets the heated surface. In general, transport of heat across the vapor film from the wall to the interface is accomplished by convection, conduction, and radiation. The radiation contribution may depend on the nature of the solid surface. However, because liquid does not contact the surface, nucleation in surface cavities is not part of the transport process. Consequently, when the radiation effect is small, the heat transfer for film boiling is independent of the material properties and finish of the surface. This feature, together with the relatively simple morphology of the interface, allows the application of modified boundary-layer techniques to the analysis of film boiling transport.

Perhaps the simplest pool boiling circumstance that can be modeled with boundary-layer analysis is laminar film boiling from a heated flat vertical surface in a motionless pool of saturated liquid. For the purposes of the following analysis, the coordinates and corresponding velocity components indicated in Fig. 7.19 are adopted here. If the usual boundary-layer approximations are adopted, and the vapor flow is assumed to be laminar and two-dimensional, the governing equations for transport in the vapor film become

$$\frac{\partial u}{\partial x} + \frac{\partial v}{\partial y} = 0 \tag{7.83}$$

$$u\frac{\partial u}{\partial x} + v\frac{\partial u}{\partial y} = \nu_v \frac{\partial^2 u}{\partial y^2} + \frac{g(\rho_l - \rho_v)}{\rho_v} \tag{7.84}$$

$$u\frac{\partial T}{\partial x} + v\frac{\partial T}{\partial y} = \alpha_{T,v}\frac{\partial^2 T}{\partial y^2} \tag{7.85}$$

where ν_v and $\alpha_{T,v}$ are the kinematic viscosity and thermal diffusivity of the vapor, respectively. In this analysis, transport of heat across the vapor film by radiation will be neglected.

For the liquid adjacent to the vapor film, the temperature is assumed to be equal to $T_{sat}(P_l)$ everywhere, since the liquid in the ambient pool is at the saturation temperature. Consistent with the usual boundary-layer approximations, the motion of the liquid is assumed to be governed by the following forms of the continuity and u-momentum equations:

$$\frac{\partial u}{\partial x} + \frac{\partial v}{\partial y} = 0 \tag{7.86}$$

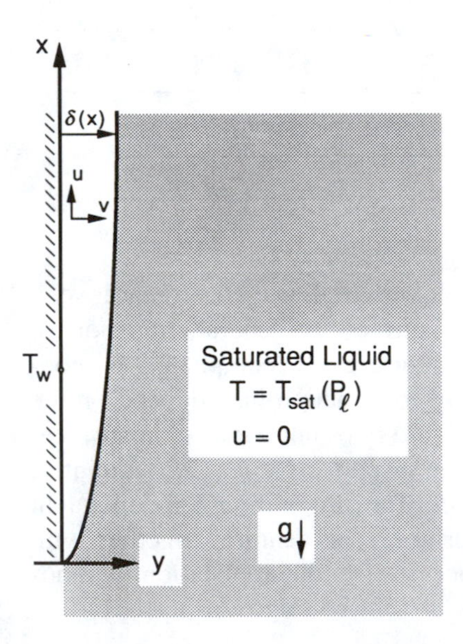

Figure 7.19 Coordinates for boundary-layer analysis of film boiling on a vertical surface.

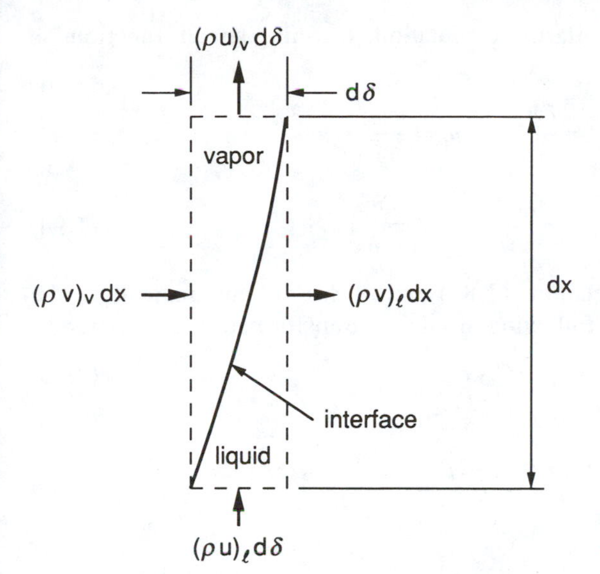

Figure 7.20 Mass balance at the liquid–vapor interface.

$$u \frac{\partial u}{\partial x} + v \frac{\partial u}{\partial y} = \nu_l \frac{\partial^2 u}{\partial y^2} \tag{7.87}$$

The boundary conditions at the heated wall ($y = 0$), at the liquid–vapor interface ($y = \delta$), and in the far ambient ($y \to \infty$) are
At $y = 0$:

$$u = v = 0 \qquad T = T_w \tag{7.88}$$

At $y = \delta$:

$$u_v = u_l \qquad \left(\mu \frac{\partial u}{\partial y}\right)_v = \left(\mu \frac{\partial u}{\partial y}\right)_l \tag{7.89}$$

$$\left(\rho u \frac{d\delta}{dx} - \rho v\right)_v = \left(\rho u \frac{d\delta}{dx} - \rho v\right)_l \tag{7.90}$$

$$T = T_{\text{sat}}(P_l) \tag{7.91}$$

as $y \to \infty$:

$$u \to 0 \tag{7.92}$$

At the wall, the boundary conditions are a consequence of the no-slip condition and the isothermal wall specification. The relations (7.89) and (7.91) that apply for $y = \delta$ specify continuity of the u velocity, shear stress, and the temperature profile across the interface. Equation (7.90) imposes mass conservation on a different control volume at the curved interface, as indicated in Fig. 7.20.

Following the analysis presented by Koh [7.77], these equations and boundary conditions can be cast into a similarity formulation. Defining stream functions in the liquid and vapor so that

$$u_v = \frac{\partial \psi_v}{\partial y} \qquad u_l = \frac{\partial \psi_l}{\partial y} \tag{7.93}$$

$$v_v = -\frac{\partial \psi_v}{\partial x} \qquad v_l = -\frac{\partial \psi_l}{\partial x} \tag{7.94}$$

it follows that the continuity relations (7.83) and (7.86) are automatically satisfied. For the vapor region, the following similarity transformation is imposed:

$$\eta = C_v y x^{-1/4} \tag{7.95}$$

$$\psi_v = 4\nu_v C_v x^{3/4} f(\eta) \tag{7.96}$$

$$\phi = \frac{T - T_{\text{sat}}}{T_w - T_{\text{sat}}} \tag{7.97}$$

where

$$C_v = \left[\left(\frac{g}{4\nu_v^2}\right)\frac{\rho_l - \rho_v}{\rho_v}\right]^{1/4} \tag{7.98}$$

The velocity components are related to the similarity variables as

$$u_v = 4C_v^2 \nu_v x^{1/2} f'(\eta) \tag{7.99}$$

$$v_v = C_v \nu_v x^{-1/4}[\eta f'(\eta) - 3f(\eta)] \tag{7.100}$$

where the primes denote differentiation with respect to η.

In terms of the similarity variables, the governing momentum and energy equations in the vapor layer become

$$f''' + 3ff'' - 2(f')^2 + 1 = 0 \tag{7.101}$$

$$\phi'' + 3\,\text{Pr}_v\, f\phi' = 0 \tag{7.102}$$

A similar transformation is applied to the liquid-region variables:

$$\xi = C_l y x^{-1/4} \tag{7.103}$$

$$\psi_l = 4\nu_l C_l x^{3/4} F(\xi) \tag{7.104}$$

where

$$C_l = \left[\left(\frac{g}{4\nu_l^2}\right)\frac{\rho_l - \rho_v}{\rho_v}\right]^{1/4} \tag{7.105}$$

This converts the momentum equation for the liquid to the form

$$F''' + 3FF'' - 2(F')^2 = 0 \tag{7.106}$$

where the primes denote derivatives with respect to ξ.

In terms of the similarity variables, the boundary conditions become
At $\eta = 0$:

$$f(0) = f'(0) = 0 \qquad \phi(0) = 1 \tag{7.107}$$

At $\eta = \eta_\delta$ or $\xi = \xi_\delta$:

$$F(\xi_\delta) = \left[\frac{\rho_v \mu_v}{\rho_l \mu_l}\right]^{1/2} f(\eta_\delta) \tag{7.108}$$

$$F'(\xi_\delta) = f'(\eta_\delta) \tag{7.109}$$

$$F''(\xi_\delta) = \left[\frac{\rho_v \mu_v}{\rho_l \mu_l}\right]^{1/2} f''(\eta_\delta) \tag{7.110}$$

$$\phi(\eta_\delta) = 0 \tag{7.111}$$

at $\xi \to \infty$:

$$F' \to 0 \tag{7.112}$$

Examination of the mathematical problem posed by Eqs. (7.101), (7.102), and (7.106) with boundary conditions (7.107) through (7.112) reveals that this is an eighth-order system of nonlinear ordinary differential equations with eight boundary conditions. The system is therefore closed if we can specify the location of the interface $y = \delta$. However, the location of the interface is not known a priori. Because the interface location is dictated by the transport, the location of the interface must be determined as part of the solution process.

The additional relation needed to allow determination of the interface location and solution of the transport equations is obtained from an energy and mass balance at the interface, which can be written approximately as

$$\frac{k_v}{h_{lv}} \left(\frac{\partial T}{\partial y}\right)_{y=\delta} = \frac{d}{dx} \int_0^\delta \rho_v u_v \, dy \tag{7.113}$$

This relation simply equates the rate of vapor generation at the interface due to vaporization (left side) to the rate of change of vapor flow in the layer (right side) with downstream distance. This approximate formulation neglects the sensible thermal energy convected in the vapor layer, assuming that all heat transferred into the system goes into the latent heat of vaporization. This is a valid assumption provided that $c_{pv}(T_w - T_{sat})/h_{lv}$ is small.

In terms of the similarity variables defined above, Eq. (7.113) can be written as

$$-\frac{3f(\eta_\delta)}{\phi'(\eta_\delta)} = \frac{c_{pv}(T_w - T_{sat})}{h_{lv}\,\mathrm{Pr}_v} \tag{7.114}$$

where, as before, η_δ is η evaluated at $y = \delta$. The problem is now completely closed mathematically. For a given set of physical boundary conditions, solution of this system of equations can be achieved using the following iterative procedure:

1. A value of η_δ is guessed.
2. Values of $f''(0)$ and $\phi'(0)$ are guessed.
3. Equations (7.101) and (7.102) are numerically integrated from $\eta = 0$ to $\eta = \eta_\delta$.
4. Equations (7.108) through (7.111) are used to generate boundary conditions for $F(\xi_\delta)$, $F'(\xi_\delta)$, and $F''(\xi_\delta)$. Equation (7.106) is then integrated numerically from $\xi = \xi_\delta$ to $\xi = \xi_e$, where ξ_e is sufficiently large to approximate $\xi \to \infty$.
5. If the conditions $\phi(\eta_\delta) = 0$ and $F'(\xi_e) = 0$ are satisfied to an acceptable level of accuracy, the solutions for $f'(\eta)$, $\phi(\eta)$, and $F'(\xi)$ (for the specified η_δ) are complete and the scheme proceeds to step 6. If these conditions are not satisfied, new values of $f''(0)$ and $\phi'(0)$ are guessed and the procedure returns to step 3.
6. $f(\eta_\delta)$ and $\phi'(\eta_\delta)$ from the converged numerical solution are used to evaluate the left side of Eq. (7.114). If Eq. (7.114) is not satisfied, a new value of η_δ is guessed and the procedure returns to step 2. If the left side equals the right side, the solution completely satisfies the imposed physical conditions and is therefore complete.

Dimensionless temperature and u-velocity profiles computed using this similarity formulation by Koh [7.77] for $[\rho_v\mu_v/\rho_l\mu_l]^{1/2} = 0.01$, $\mathrm{Pr}_v = 1.0$, and various values of $c_{pv}(T_w - T_{sat})/h_{lv}$ are shown in Fig. 7.21. Beginning with the primitive relation

$$h = \frac{q''}{T_w - T_{sat}} = \frac{-k_v}{T_w - T_{sat}}\left(\frac{\partial T}{\partial y}\right)_{y=0} \tag{7.115}$$

writing $\partial T/\partial y$ in terms of ϕ and rearranging yields

$$\mathrm{Nu}_x = \frac{hx}{k_v} = [-\phi'(0)]\left[\frac{g(\rho_l - \rho_v)x^3}{4\rho_v\nu_v^2}\right]^{1/4} \tag{7.116}$$

For a given set of physical conditions, the value of $\phi'(0)$ can be computed using the scheme described above and then the variation of the Nusselt number or heat transfer coefficient along the surface can be determined from Eq. (7.116).

Although additional complications may arise in more complex surface geometries, or when the liquid pool is subcooled, the simple film boiling circumstance considered above exhibits the key features that are common to all film boiling processes. The computational scheme required to solve the governing transport equations for such processes is always complicated by the nonlinearity of the equations, the complexity of the boundary conditions, and the requirement that the interface position be found simultaneously while solving the transport equations.

Radiation Effects

Despite its complexity, the film boiling model described above is highly idealized, and several physical effects have been ignored. In particular, the transport of heat

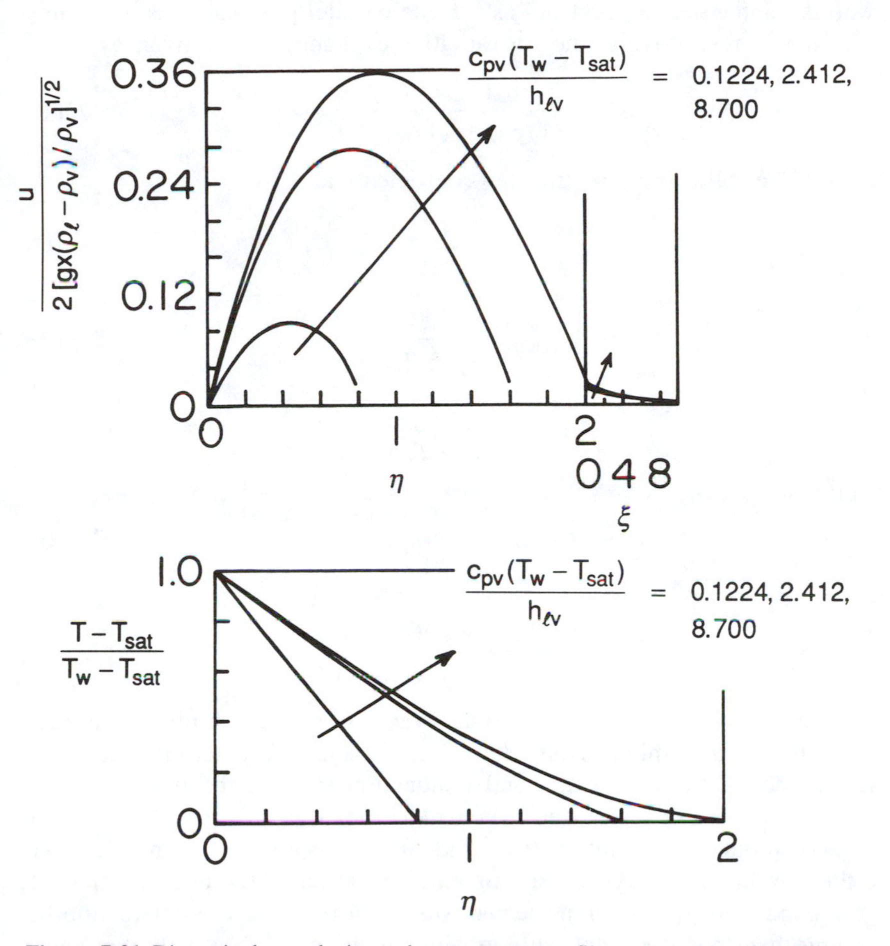

Figure 7.21 Dimensionless velocity and temperature profiles computed using the similarity formulation for film boiling on an isothermal vertical surface. (*Adapted from Koh [7.77] with permission, copyright © 1962, American Society of Mechanical Engineers.*)

by radiation from the surface to the interface has not been considered. If the vapor in the film absorbs and emits radiation at infrared wavelengths, a detailed treatment of the radiation interaction with the vapor may be necessary to accurately predict the film boiling heat transfer.

Fortunately, in most cases the vapor is virtually transparent to infrared radiation emitted by the wall and the interface. For thin vapor films, the radiation transport can then be modeled with reasonable accuracy as the radiation exchange between two parallel plates. The radiative transfer is often treated as if it acts independently and in parallel with the convection/conduction mechanism. The total heat flux from the surface would then be equal to the sum of the convection/conduction heat flux $q''_{\text{con,act}}$ and a radiation contribution q''_{rad}:

$$q'' = q''_{\mathrm{con,act}} + q''_{\mathrm{rad}} \tag{7.117}$$

If the wall and interface are modeled as infinite parallel gray surfaces with emissivities ϵ_w and ϵ_i, respectively, the net radiative exchange q''_{rad} is given by

$$q''_{\mathrm{rad}} = \frac{\sigma_{SB}(T_w^4 - T_{\mathrm{sat}}^4)}{(1/\varepsilon_w) + (1/\varepsilon_i) - 1} \tag{7.118}$$

Defining the following heat transfer coefficients as

$$h = \frac{q''}{T_w - T_{\mathrm{sat}}} \tag{7.119a}$$

$$h_{\mathrm{con,act}} = \frac{q''_{\mathrm{con,act}}}{T_w - T_{\mathrm{sat}}} \tag{7.119b}$$

$$h_{\mathrm{rad}} = \frac{q''_{\mathrm{rad}}}{T_w - T_{\mathrm{sat}}} \tag{7.119c}$$

Eq. (7.117) implies that

$$h = h_{\mathrm{con,act}} + h_{\mathrm{rad}} \tag{7.120a}$$

where, from Eq. (7.118),

$$h_{\mathrm{rad}} = \frac{\sigma_{SB}(T_w^2 + T_{\mathrm{sat}}^2)(T_w + T_{\mathrm{sat}})}{(1/\varepsilon_w) + (1/\varepsilon_i) - 1} \tag{7.120b}$$

This line of reasoning suggests than an overall h can be computed using Eq. (7.120*a*) with h_{rad} determined from Eq. (7.120*b*) and h_{con} determined from an analytical model of the convection and conduction such as the boundary-layer analysis described above. However, when radiation is significant, the increased rate of vapor production resulting from radiative transport to the interface will tend to thicken the vapor layer, resulting in a reduction of the heat transfer rate due to the conduction/convection mechanism. As a result of this interaction between the mechanisms, the simple superposition relation (7.120*a*) will overpredict the overall h for the film boiling process if a model for convection/conduction transport alone is used to predict $h_{\mathrm{con,act}}$.

Based on arguments about the interaction between the radiation and convection mechanisms, Bromley [7.78] recommended the following relation for the overall heat transfer coefficient h when both mechanisms are present:

$$h = h_{\mathrm{con}}\left(\frac{h_{\mathrm{con}}}{h}\right)^{1/3} + h_{\mathrm{rad}} \tag{7.121}$$

where h_{con} is the heat transfer coefficient associated with the convective effect in the absence of radiative effects. For a flat, vertical surface, Lubin [7.79] has shown that the above relation can be justified as follows. If convective effects are neglected in the u-momentum equation (7.84), the remaining terms in this equation can be integrated from $y = 0$ to $y = \delta$ to obtain

$$u = \frac{g(\rho_l - \rho_v)}{\mu_v}\left(\delta y - \frac{y^2}{2}\right) \tag{7.122}$$

Assuming that the convection terms in the energy equation are also negligible, integration of the remaining conduction term across the film yields

$$-\left(\frac{\partial T}{\partial y}\right)_{y=\delta} = \frac{T_w - T_{sat}}{\delta} \tag{7.123}$$

Neglecting the sensible heating of the vapor, conservation of mass and energy within a differential control volume in the vapor film further requires that

$$-k_v\left(\frac{\partial T}{\partial y}\right)_{y=\delta} + h_{rad}(T_w - T_{sat}) = \rho_v h_{lv}\frac{d}{dx}\int_0^\delta u_v\,dy \tag{7.124}$$

Substituting Eqs. (7.122) and (7.123) into (7.124), integrating, and differentiating as indicated yields

$$\frac{\delta^3\,d\delta}{(1 + h_{rad}\delta/k_v)} = \left[\frac{k_v\mu_v(T_w - T_{sat})}{g\rho_v(\rho_l - \rho_v)h_{lv}}\right]dx \tag{7.125}$$

Lubin [7.79] argued that with the initial condition $\delta = 0$ at $x = 0$, the solution of the above equation for $h_{rad}\,\delta/k_v \ll 1$ is well approximated by the relation

$$\frac{\delta^4}{(1 + h_{rad}\delta/k_v)} = \left[\frac{4k_v\mu_v(T_w - T_{sat})x}{g\rho_v(\rho_l - \rho_v)h_{lv}}\right] \tag{7.126}$$

For the linear temperature profile indicated by Eq. (7.123), it follows that

$$h_{con,act} = \frac{k_v}{\delta} \tag{7.127}$$

From Eqs. (7.126) and (7.127), it can be seen that, in the absence of a radiation effect ($h_{rad} = 0$),

$$h_{con} = \frac{k_v}{\delta_{con}} = \left[\frac{k_v^3 g\rho_v(\rho_l - \rho_v)h_{lv}}{4\mu_v(T_w - T_{sat})x}\right]^{1/4} \tag{7.128}$$

Combining Eqs. (7.126) through (7.128) and Eq. (7.120*a*), the following relation is obtained after some manipulation:

$$\frac{h}{h_{con}} = \left(\frac{h_{con}}{h}\right)^{1/3} + \frac{h_{rad}}{h_{con}} \tag{7.129}$$

This relation is identical to Eq. (7.121), thus justifying Bromley's more qualitative derivation. This relation was also verified by Sparrow [7.80], who found that calculation of the total surface heat transfer using Eq. (7.121) agrees within a few percent with the results obtained by including radiation in a solution of the boundary-layer equations.

One drawback of Eq. (7.121) is that h is not easily determined from it because it cannot be explicitly solved for h. To avoid an iterative determination of h, for $h_{rad} < h_{con}$, Bromley [7.78] recommended the following explicit relation, which approximates Eq. (7.121) to within 5%:

$$h = h_{con} + \frac{3}{4} h_{rad} \tag{7.130}$$

For large values of h_{rad}, Bromley [7.78] recommended the relation

$$h = h_{con} + h_{rad}\left\{\frac{3}{4} + \frac{1}{4}\left(\frac{h_{rad}}{h_{con}}\right)\left[\frac{1}{2.62 + (h_{rad}/h_{con})}\right]\right\} \tag{7.131}$$

This more general approximation of Eq. (7.121) was recommended for values of h_{rad}/h_{con} between 0 and 10.

It should be noted that the strong temperature dependence of the radiation component implies a rapidly increasing radiation contribution to the total film boiling heat flux as the wall superheat increases. The resulting effect on the pool boiling curve is indicated in Fig. 7.22.

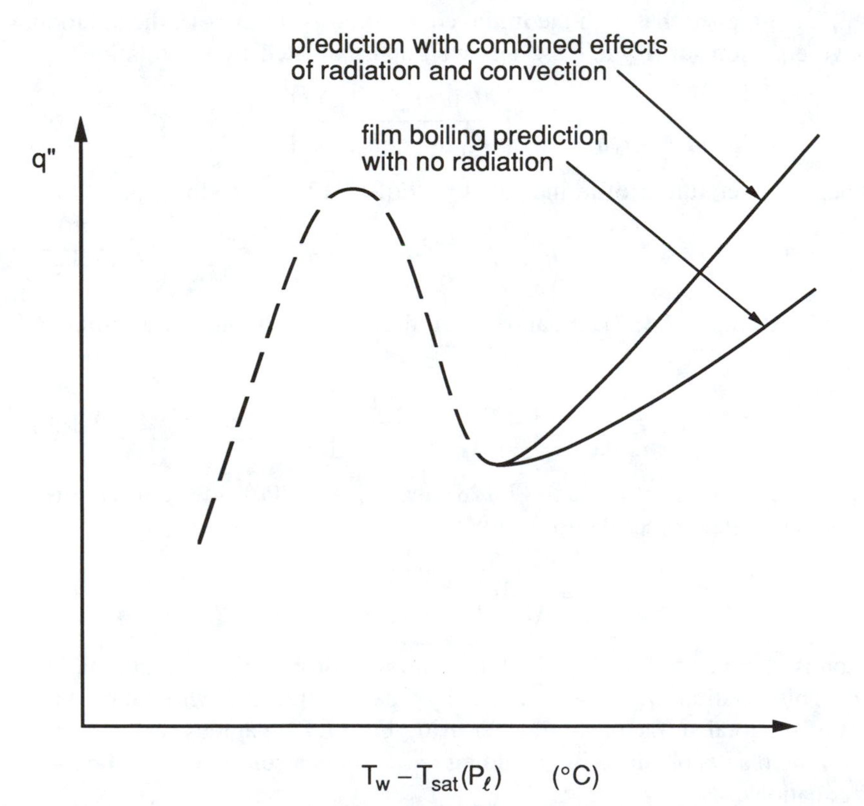

Figure 7.22 Effect of radiation on the film boiling portion of the boiling curve.

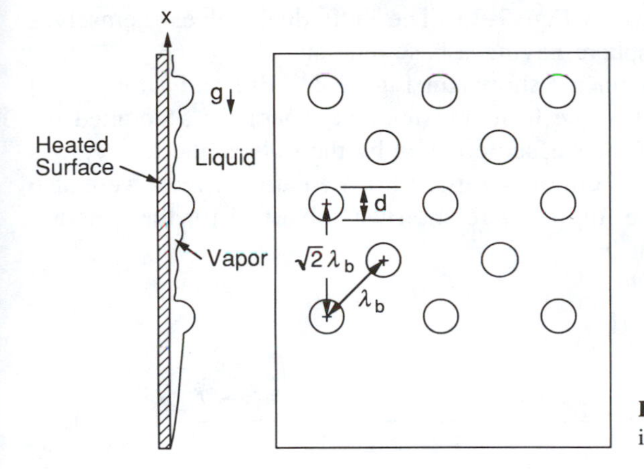

Figure 7.23 Bulge pattern postulated in film boiling model.

Interfacial Waves and Turbulence

Even with a correction for radiation effects, the similarity analysis for saturated laminar film boiling over a vertical surface is very often an accurate representation of the transport over only a small portion of the vertical surface. Generally, the vapor flow is laminar and the interface is smooth only near the leading edge of the surface. With increasing distance downstream, the flow rate of vapor in the film increases and the film thickness increases. Eventually the velocity of the vapor relative to the adjacent liquid becomes large enough that the interface becomes Helmholtz unstable and the interface begins to show small-amplitude capillary waves. Despite the presence of such waves, the flow in the vapor film often remains essentially laminar.

The effects of large-scale bulges or bubbles on transport during film boiling on a vertical surface has been the subject of increasing interest in recent years. The significance of these bulges to the heat transfer process was recognized in an early investigation by Greitzer and Abernathy [7.81]. If the flow in the film becomes turbulent, most of the resistance to heat transfer will be in the viscous sublayer, and the presence of the bulges may not significantly affect the transport of heat across the layer. However, Bui and Dhir [7.82] note that for film boiling of common fluids on vertical surfaces of modest extent, the vapor film Reynolds number may be too low to justify the assumption of turbulent flow. The presence of the bulges may then significantly affect the heat transfer.

Bui and Dhir [7.82] proposed an idealized model of laminar film boiling over a vertical surface with bulges present in the interface. The features of this idealized model are indicated schematically in Fig. 7.23. The region very near the lower edge of the surface is presumed to be largely free of bulges. The bulges are presumed to appear only after the vapor velocity becomes high enough that the interface becomes Helmholtz unstable. Over the portion of the surface where they exist, the bulges are idealized as being spaced in a rectangular array with a characteristics spacing λ_b, as shown in Fig. 7.23. The individual bulges themselves are modeled as sections of a sphere having a base diameter d_b.

Based on the idealized geometry shown in Fig. 7.23, Bui and Dhir [7.82] developed an analytical model of the heat transfer that separately accounted for contributions of the regions of the surface covered by the bulges and the regions between the bulges covered with a thinner vapor film. Radiation effects were also included. From this model, the following relation was obtained for the time-averaged heat transfer coefficient $\langle h \rangle$:

$$\langle h \rangle = C\bar{h}_f\left(1 - \frac{\pi d_b^2}{4\lambda_b^2}\right) + 0.3\left(\frac{k_v}{\bar{\delta}}\right)\left(\frac{\pi d_b^2}{4\lambda_b^2}\right) + \frac{\sigma_{SB}}{[(1/\varepsilon_w) + (1/\varepsilon_l) - 1]}\left[\frac{T_w^4 - T_{\text{sat}}^4}{T_w - T_{\text{sat}}}\right] \tag{7.132}$$

The first term on the right side of this relation represents the contribution due to heat transfer across the thin vapor film in the regions between the bulges. On the right side of the equation, $\bar{h}_f$ is the average heat transfer coefficient over the region between the bulges,

$$\bar{h}_f = \frac{1}{L}\int_0^L h(x)\,dx \tag{7.133}$$

where L is the effective flow length determined from the postulated geometry of the bulge pattern. The local value of $h(x)$ was taken to equal k_v/δ, where δ is the local thickness of the thin film between the bulges. From an integral analysis of the transport across the thin vapor film, Bui and Dhir [7.82] obtained the following relation for the local thickness of the thin film:

$$\frac{\delta g^{1/3}}{\nu_v^{2/3}} = \sqrt{2}\,[(U_i^{*2} + 4x^*)^{1/2} - U_i^*]^{1/2} \tag{7.134}$$

where

$$U_i^* = \frac{U_i \mu_v}{g^{1/3}(\rho_l - \rho_v)\nu_v^{4/3}} \tag{7.135}$$

$$x^* = \frac{xg^{1/3}(\rho_l - \rho_v)^2}{\nu_v^{2/3}\rho_v^2}\left(\frac{(\rho_l - \rho_v)c_{pv}(T_w - T_{\text{sat}})}{\rho_v h_{lv}\,\text{Pr}_v}\right) \tag{7.136}$$

In Eq. (7.135), U_i is the velocity at the interface, for which Bui and Dhir [7.82] derived the following relation:

$$U_i = 0.37 d_b\left[\frac{(\rho_l - \rho_v)g}{\rho_l \nu_l^{1/2}}\right]^{2/3} \tag{7.137}$$

The constant C in Eq. (7.132) was chosen to account for the effect of interface ripples on the thin film heat transfer. A value of 1.15 was recommended for this constant. In using these relations to predict the heat transfer coefficient, λ_b is postulated to be approximately equal to the most dangerous Helmholtz-unstable wavelength. A value of 0.5 for d_b/λ_b was justified on the basis of simple geometric arguments, which allowed d_b to be computed from the value of λ_b.

The second term on the right side of Eq. (7.132) represents the small but significant contribution of heat transfer across the vapor at the bulge locations. The constant factor 0.3 in this term is an estimate of the ratio of the actual heat transfer at these locations to that which would occur if the vapor thickness were equal to the mean value of the thickness of the vapor film around the perimeter of the bulge $\bar{\delta}$. The value of $\bar{\delta}$ was determined from the variation of the film thickness predicted by the analysis of the transport across the film between the bulges, as described above.

The final term on the right side of Eq. (7.132) represents the radiation contribution to the overall heat transfer coefficient, treating this effect as radiant transport between two infinite parallel plates. In this term, ϵ_w and ϵ_l are the emissivities of the heated wall and the liquid–vapor interface, respectively.

The predictions of this model were found to agree well with data and visual observations for film boiling of water over a vertical surface 10.3 cm tall. Specifically, the model not only agrees well quantitatively with the data, but it predicts that the heat transfer coefficient is independent of downstream location for these conditions, which is consistent with the trend in the data. Because of its realistic treatment of the effects of the highly irregular morphology of the film, this model is a promising development in the effort to better understand film boiling heat transfer processes. However, the model needs to be tested more extensively before it can be widely used with confidence.

For film boiling on a long vertical surface, when a sufficiently high vapor film Reynolds number is reached, transition to turbulent flow occurs and the interface generally becomes more wavy, as indicated in Fig. 7.24. Hsu and Westwater [7.83] estimated the condition for the onset of transition to turbulent flow as

$$\mathrm{Re}_\delta = \frac{\rho_v \delta u_{y=\delta}}{\mu_v} = 100 \tag{7.138}$$

where $u_{y=\delta}$ is the local vapor u-velocity at the interface ($y = \delta$). With increasing height above the onset of transition location, the film thickness continues to increase, the vapor flow becomes more fully turbulent, and the interfacial waves increase in wavelength, eventually becoming unstable. When this occurs, the interfacial waves may roll up and "break," releasing vapor bubbles into the adjacent liquid.

If film boiling occurs on a long vertical surface, the turbulent film boiling process described above may exist over a major portion of the surface. Modeling of this complex transport process is difficult at best. However, a model of the transport for these circumstances has been developed by Hsu and Westwater [7.83].

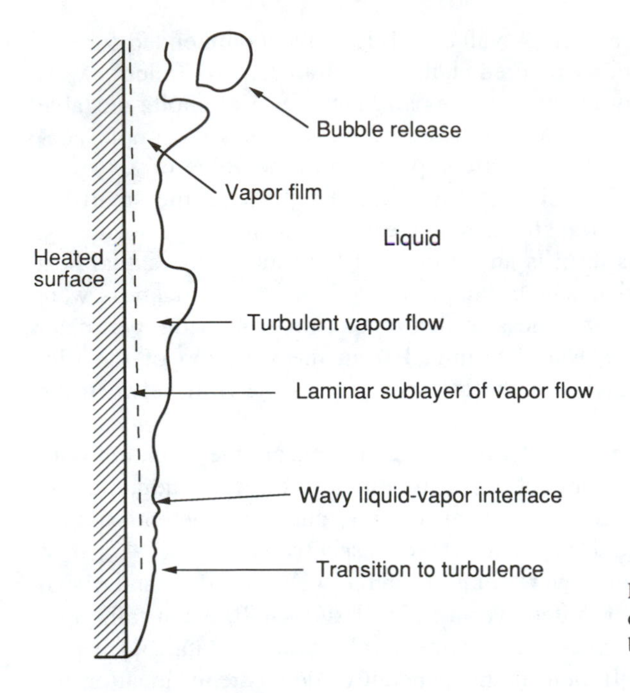

Figure 7.24 Schematic of phenomena associated with turbulent film boiling.

Based on this model, a complicated expression for the heat transfer coefficient was derived that accounts separately for the contributions associated with the laminar and turbulent portions of the flow.

Hsu and Westwater [7.84] also experimentally investigated turbulent film boiling from vertical cylinders. Based on available data of this type, they recommended the following empirical correlation for the mean film boiling heat transfer coefficient for the cylinder:

$$h_{\text{con}} \left[\frac{\mu_v^2}{k_v^3 \rho_v (\rho_l - \rho_v) g} \right]^{1/3} = 0.002 \, \text{Re}_v^{0.6} \tag{7.139}$$

$$\text{Re}_v = \frac{4\dot{m}_v}{\pi D \mu_v} \tag{7.140}$$

In computing Re_v, the mass flow rate of vapor at the upper end of the cylinder can be calculated as $\dot{m}_v = q'' \pi DL / h_{lv}$ if the heat flux q'' and the cylinder diameter D and length L are known. This correlation, which applies for conditions where radiation effects are negligible, was recommended for Reynolds number values in the range $800 \leq \text{Re}_v \leq 5000$.

An analytical model of turbulent film boiling on vertical surfaces was also developed by Suryanarayana and Merte [7.85]. Their model is based on turbulent boundary-layer theory and includes an enhancement factor to account for the interfacial oscillations and distortion.

Correlations Developed from Integral Analysis

The boundary-layer analysis of laminar film boiling near immersed bodies can also be done in a more approximate integral formulation. Such an analysis for a heated sphere is described in Example 7.6. Bromley's [7.78] analysis of this type for a horizontal cylinder of diameter D yielded the following relation for the mean heat transfer coefficient $\bar{h}_{con}$ for the cylinder:

$$\bar{h}_{con} = \frac{k_v}{\delta} = C_0 \left[\frac{k_v^3 g \rho_v (\rho_l - \rho_v) h'_{lv}}{\mu_v (T_w - T_{sat}) D} \right]^{1/4} \tag{7.141}$$

where

$$h'_{lv} = h_{lv} \left[1.0 + \frac{0.40 c_{pv} (T_w - T_{sat})}{h_{lv}} \right] \tag{7.142}$$

The constant C_0 was determined to be 0.512 if the adjacent liquid was taken to be virtually stationary. If the liquid at the interface exerts zero shear on the vapor, C_0 equals 0.724. Bromley [7.78] proposed a mean value of $C_0 = 0.62$ for average heat transfer coefficients on horizontal round tubes. The appearance of h'_{lv} in Eq. (7.141) results from the inclusion of sensible heating effects in the vapor film.

It should be noted that this result is expected to be valid only as long as the tube diameter D is large compared with the film thickness. The correlation covers the diameter range that is most important for common practical applications. However, it is not valid for very large diameters or for very small wires. The effect of diameter has been examined in detail by Breen and Westwater [7.87]. The effect of cylinder diameter can be best assessed in terms of the ratio λ_D/D, where λ_D is the Taylor most dangerous wavelength defined by Eq. (7.61). When the diameter of the cylinder is large compared to the most dangerous wavelength, a Taylor instability mechanism similar to that for a large, flat, horizontal surface controls vapor release from the cylinder during film boiling, suggesting a departure from the Bromley theory.

Example 7.6 Determine a relation for the mean heat transfer coefficient for laminar film boiling on an isothermal sphere of radius R. Use an integral boundary-layer analysis and neglect radiation effects.

In terms of the x,y coordinates and u,v velocities indicated in Fig. 7.25, the boundary-layer equations governing the flow in the vapor film and in the adjacent liquid are

$$\left.\begin{aligned} \frac{\partial (ur)}{\partial x} + \frac{\partial (vr)}{\partial y} &= 0 \\ u\frac{\partial u}{\partial x} + v\frac{\partial u}{\partial y} &= g\frac{(\rho_l - \rho_v)}{\rho_v}\sin\left(\frac{x}{R}\right) + \nu_v \frac{\partial^2 u}{\partial y^2} \\ u\frac{\partial T}{\partial x} + v\frac{\partial T}{\partial y} &= \alpha_v \frac{\partial^2 T}{\partial y^2} \end{aligned}\right\} \quad \text{vapor}$$

$$\left.\begin{aligned} \frac{\partial(ur)}{\partial x} + \frac{\partial(vr)}{\partial y} &= 0 \\ u\frac{\partial u}{\partial x} + v\frac{\partial u}{\partial y} &= \nu_l \frac{\partial^2 u}{\partial y^2} \end{aligned}\right\} \quad \text{liquid}$$

Boundary conditions for the above equations are
At $y = 0$:

$$u = v = 0 \qquad T = T_w$$

At $y \to \infty$:

$$u \to 0$$

At $y = \delta$:

$$u_v = u_l \qquad \mu_l\left(\frac{\partial u}{\partial y}\right)_l = \mu_v\left(\frac{\partial u}{\partial y}\right)_v \qquad T = T_{sat}$$

Neglecting the sensible energy due to superheating the vapor (compared to the latent heat), an energy balance on the vapor film requires that

$$k_v\, dA_s\left[-\left(\frac{\partial T}{\partial y}\right)_{y=\delta}\right] = h_{lv}\rho_v \frac{d}{dx}\left(\int_0^{A_\delta} u dA_\delta\right) dx \tag{7.6.1}$$

Continuity at the liquid–vapor interface also requires

$$\rho_l\left(u\frac{d\delta}{dx} - v\right)_l = \rho_v\left(u\frac{d\delta}{dx} - v\right)_v$$

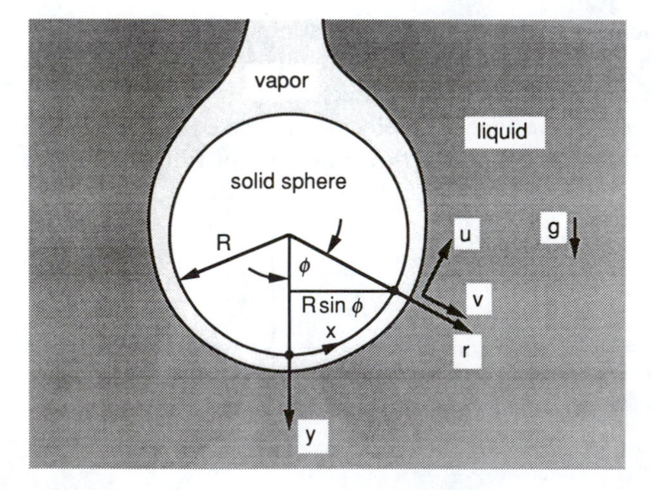

Figure 7.25 Film boiling from a sphere.

In this analysis, the inertia and convection terms are neglected in the governing equations and it is assumed that $\mu_v/\mu_l \cong 0$. The latter assumption together with the interface boundary condition implies that

$$\left(\frac{\partial u}{\partial y}\right)_l = \frac{\mu_v}{\mu_l}\left(\frac{\partial u}{\partial y}\right)_v \cong 0 \qquad \text{at } y = \delta$$

Neglecting the inertia terms, the liquid momentum equation becomes

$$\frac{\partial^2 u}{\partial y^2} = 0$$

Integration of this equation yields $u_l = Ay + B$. Application of the above condition at $y = \delta$ implies that $A = 0$. Also, requiring that $u_l \to 0$ as $y \to \infty$ implies that $B = 0$. Thus $u_l = 0$ everywhere in the liquid. Neglecting the convection terms, the vapor energy equation is simply

$$\frac{\partial^2 T}{\partial y^2} = 0$$

Integrating twice and applying the temperature boundary conditions at $y = 0$ and $y = \delta$ yields the linear temperature profile,

$$\frac{T - T_{\text{sat}}}{T_w - T_{\text{sat}}} = 1 - \frac{y}{\delta}$$

Likewise, neglecting the inertia terms in the vapor momentum equation, it reduces to

$$\frac{g(\rho_l - \rho_v)}{\rho_v}\sin\left(\frac{x}{R}\right) + \nu_v \frac{\partial^2 u}{\partial y^2} = 0$$

Assuming the following velocity profile, which satisfies the boundary conditions $u = 0$ at $y = 0$ and $u = u_l = 0$ at $y = \delta$,

$$\frac{u}{\bar{u}} = 6\left(\frac{y}{\delta} - \frac{y^2}{\delta^2}\right)$$

where $\bar{u}$ is the mean u-velocity, defined as

$$\bar{u} = \frac{1}{\delta}\int_0^\delta u\,dy$$

substituting into the momentum equation yields

$$\bar{u} = \frac{\delta^2 g}{12\nu_v}\left(\frac{\rho_l - \rho_v}{\rho_v}\right)\sin\left(\frac{x}{R}\right)$$

In the energy balance relation it can be shown from simple geometry considerations that

$$dA_s = 2\pi R \sin\left(\frac{x}{R}\right) dx = 2\pi R^2 \sin\phi \, d\phi$$

$$dA_\delta = 2\pi r(x)\, dy \qquad r(x) = R \sin\left(\frac{x}{R}\right) = R \sin\phi$$

where dA_s is a differential element of surface area on the sphere and dA_δ is a differential element of cross-sectional area across the film. It follows that

$$\int_0^{A_\delta} u \, dA_\delta = \int_0^\delta 2\pi r(x) \left[6\bar{u}\left(\frac{y}{\delta} - \frac{y^2}{\delta^2}\right)\right] dy = 2\pi r(x)\bar{u}\delta$$

$$-\left(\frac{\partial T}{\partial y}\right)_{y=\delta} = \frac{T_w - T_{sat}}{\delta}$$

Substitution of these results into the energy balance relation (7.6.1) yields, after some rearranging,

$$\frac{96 \text{ Ja}}{\text{Ra}} = \frac{\delta/R}{\sin\phi} \frac{d}{d\phi}\left(\frac{\delta^3}{R^3}\sin^2\phi\right) \tag{7.6.2}$$

where

$$\text{Ra} = \frac{gD^3(\rho_l - \rho_v)}{\nu_v \alpha_v \rho_v} \qquad D = 2R \qquad \text{Ja} = \frac{c_{pv}(T_w - T_{sat})}{h_{lv}}$$

Equation (7.6.2) can be solved by elementary methods using the condition that $d\delta/d\phi = 0$ at $\phi = 0$, to obtain

$$\frac{\delta}{R} = 2\left(\frac{8 \text{ Ja}}{\text{Ra}}\right)^{1/4} \left(\frac{\int_0^\phi \sin^{5/3}\bar{\phi}\, d\bar{\phi}}{\sin^{8/3}\phi}\right)^{1/4}$$

Noting that the local surface heat flux is given by $q'' = k_v(T_w - T_{sat})/\delta$, the total heat transfer from the sphere is given by

$$q = \int_0^\pi \frac{k_v(T_w - T_{sat})}{\delta(\phi)} 2\pi R^2 \sin\phi \, d\phi = \frac{\pi}{4} D k_v (T_w - T_{sat}) \left(\frac{2 \text{ Ra}}{\text{Ja}}\right)^{1/4} \int_0^\pi f_s(\phi)\, d\phi$$

where

$$f_s(\phi) = \sin^{5/3}\phi \left(\int_0^\phi \sin^{5/3}\bar{\phi}\, d\bar{\phi}\right)^{-1/4}$$

Evaluating the integral of $f_s(\phi)$ and rearranging yields the relation

$$\text{Nu}_D = \frac{\bar{h}D}{k_v} = \frac{qD}{\pi D^2 (T_w - T_{sat}) k_v} = 0.586 \left(\frac{\text{Ra}}{\text{Ja}}\right)^{1/4}$$

Note that this analysis does not account for separation of the wake, waviness of the interface, or transition to turbulence. Frederking and Clark [7.86] recommended the following relation as a better fit to data for film boiling of liquid nitrogen:

$$\mathrm{Nu}_D = 0.14 \left[\frac{\mathrm{Ra}\left(1 + \frac{1}{2}\,\mathrm{Ja}\right)}{\mathrm{Ja}} \right]^{1/3}$$

Breen and Westwater [7.87] did, in fact, observe that data deviate from Bromley's theoretical prediction for $\lambda_D/D < 0.8$. For $0.8 < \lambda_D/D < 8$, the portion of the interface over the lower part of the tube was smooth, the vapor flow was generally laminar, and the heat transfer data generally agreed with the Bromley correlation given by Eqs. (7.141) and (7.142). For $\lambda_D/D > 8$ (small diameters), the heat flux became almost independent of diameter. Heat transfer for such conditions again deviated from the Bromley theory, apparently being determined by the bubble size and release mechanism only.

As a best fit to available experimental data for a broad spectrum of cylinder sizes, Breen and Westwater [7.87] proposed the following empirical correlation for film boiling on horizontal cylinders:

$$\overline{h}_{\mathrm{con}} = \frac{[0.59 + 0.069(\lambda_D/D)]F}{(\lambda_D)^{1/4}} \tag{7.143}$$

where

$$F = \left[\frac{k_v^3 g \rho_v (\rho_l - \rho_v) h'_{lv}}{\mu_v (T_w - T_{\mathrm{sat}})} \right]^{1/4} \tag{7.144}$$

$$h'_{lv} = h_{lv} \left[1.0 + \frac{0.34 c_{pv} (T_w - T_{\mathrm{sat}})}{h_{lv}} \right] \tag{7.145}$$

This correlation is recommended for any λ_D/D value. Alternatively, one of the following relations can be used, depending on the specific value of λ_D/D:
For $\lambda_D/D < 0.8$:

$$\overline{h}_{\mathrm{con}} = 0.6 F \lambda_D^{-1/4} \tag{7.146}$$

For $0.8 < \lambda_D/D < 8$:

$$\overline{h}_{\mathrm{con}} = 0.6 F D^{-1/4} \tag{7.147}$$

For $8 < \lambda_D/D$:

$$\overline{h}_{\mathrm{con}} = 0.16 \left(\frac{\lambda_D}{D} \right)^{0.83} F \lambda_D^{-1/4} \tag{7.148}$$

Large Horizontal Surfaces

Heat transfer during film boiling from a large, flat, horizontal heated surface is dictated, to a large degree, by the vapor release mechanism, which determines the mean vapor film thickness. Chang [7.73] proposed a model of the film boiling process on a horizontal surface that linked the wave action associated with Taylor instability to the breakup of the interface associated with the vapor release process. This model predicts a heat transfer relation of the form

$$h_{\text{con}} = C_1 \left[\frac{k_v^3 g(\rho_l - \rho_v)}{\mu_v \alpha_0} \right]^{1/3} \tag{7.149}$$

where

$$\alpha_0 = \frac{k_v(T_w - T_{\text{sat}})}{2h_{lv}\rho_v} \tag{7.150}$$

Chang [7.73] proposed 0.43 as a value of C_1 that provided good agreement with available data.

The vapor escape model used by Zuber to analyze the minimum heat flux condition (see Section 7.5) has been extended by Berenson [7.74] to model steady film boiling from a flat horizontal surface. From observations of the boiling process, Berenson [7.74] again postulated that bubbles were released at the nodes of the most dangerous Taylor wave, with the nodes spaced in a rectangular array, as shown in Fig. 7.18.

The conceptual model on which Berenson's analysis is based is shown in Fig. 7.26. For the postulated spacing of the bubble release locations, the unit cell shown in Fig. 7.18 has an area equal to $\lambda_D^2/2$. For a radially symmetric unit cell of the same area, $r_2 = (2\pi)^{-1/2}\lambda_D$. Vapor is generated and flows with a radial velocity u_v from a distance $r_2 = (2\pi)^{-1/2}\lambda_D$ to the bubble, which is characterized by the radius $r = r_1$. Based on experimental observation of bubble departure diameters, Berenson argued that r_1 is given approximately by

$$r_1 = 2.35 \left[\frac{\sigma}{g(\rho_l - \rho_v)} \right]^{1/2} \tag{7.151}$$

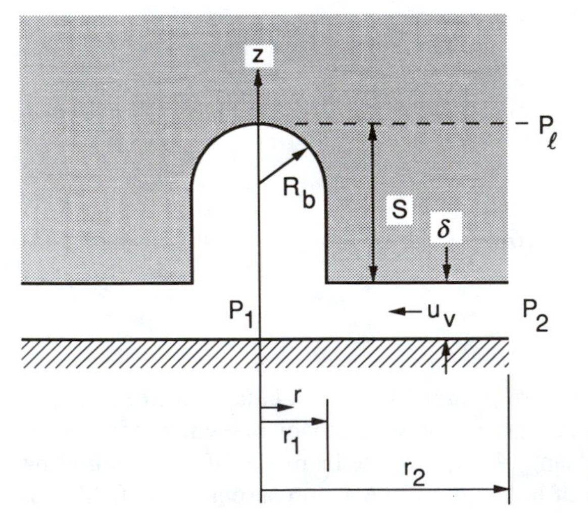

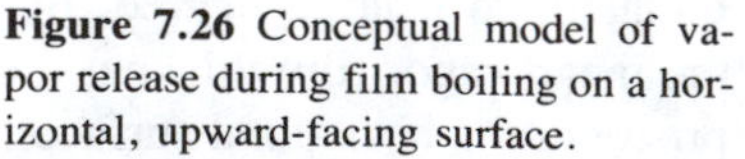
Figure 7.26 Conceptual model of vapor release during film boiling on a horizontal, upward-facing surface.

Conservation of mass requires that the mass flow rate of vapor $\dot{m}_v$ toward the bubble at any location between r_1 and r_2 be given by

$$\dot{m}_v = \rho_v(2\pi r)\delta u_v \tag{7.152}$$

Assuming that heat is transferred across the thin portion of the film by conduction alone, an energy balance also requires that the vapor mass flow rate satisfy

$$\dot{m}_v h'_{lv} = \pi\left(\frac{k_v}{\delta}\right)(r_2^2 - r^2)(T_w - T_{\text{sat}}) \tag{7.153}$$

where

$$h'_{lv} = h_{lv} + 0.5c_{pv}(T_w - T_{\text{sat}}) \tag{7.154}$$

The use of h'_{lv} as defined above approximately accounts for the sensible heating of the vapor film.

Combining Eqs. (7.151) through (7.153), the following relation is obtained for the vapor velocity:

$$u_v = \left[\frac{k_v(T_w - T_{\text{sat}})}{\rho_v h'_{lv}\delta^2}\right]\frac{(\lambda_D^2 - 2\pi r^2)}{4\pi r} \tag{7.155}$$

The approximate momentum balance for the inward flow of vapor in the film requires that

$$dP_v = \frac{b\mu_v u_v}{\delta^2}\,dr \tag{7.156}$$

where $b = 12$ if the liquid surface is taken to be stationary, and $b = 3$ if the shear stress is zero at the interface. Substituting Eq. (7.155) into Eq. (7.156) yields

$$dP_v = \left[\frac{b\mu_v k_v(T_w - T_{\text{sat}})}{\rho_v h'_{lv}\delta^4}\right]\left(\frac{\lambda_D^2 - 2\pi r^2}{4\pi r}\right)dr \tag{7.157}$$

Integrating this relation from r_1 to r_2 and using Eq. (7.60) to evaluate λ_D yields

$$P_2 - P_1 = 1.36\left[\frac{8b\mu_v k_v(T_w - T_{\text{sat}})}{\pi\rho_v h'_{lv}\delta^4}\right]\left[\frac{\sigma}{g(\rho_l - \rho_v)}\right] \tag{7.158}$$

At a height $z = S + \delta$ above the surface, the local liquid pressure P_l is a constant at all horizontal locations. In the vapor near the surface, the pressures P_1 at $r = r_1$ and P_2 at $r = r_2$ are given by

$$P_1 = P_l + \rho_v g S + \frac{2\sigma}{R_b} \tag{7.159}$$

$$P_2 = P_l + \rho_l g S \tag{7.160}$$

The last term on the right side of Eq. (7.159) accounts for the capillary pressure difference across the curved interface of the bubble. Combining Eqs. (7.159) and (7.160), we find that

$$P_2 - P_1 = (\rho_l - \rho_v)gS - \frac{2\sigma}{R_b} \tag{7.161}$$

In this equation, S and R_b were evaluated using empirical relations derived from bubble release data published by Borishansky [7.88]

$$S = 1.36R_b = 3.2\left[\frac{\sigma}{g(\rho_l - \rho_v)}\right]^{1/2} \tag{7.162}$$

Combining Eqs. (7.158) and (7.161) to eliminate $P_2 - P_1$ and using Eq. (7.162), the following equation for δ is obtained:

$$\delta = \left\{1.36\left[\frac{1.09b\mu_v k_v(T_w - T_{sat})}{g\rho_v(\rho_l - \rho_v)h'_{lv}}\right]\left[\frac{\sigma}{g(\rho_l - \rho_v)}\right]^{1/2}\right\}^{1/4} \tag{7.163}$$

Assuming that heat is transferred across the film by conduction only, the heat transfer coefficient is given by $h_{con} = k_v/\delta$. It follows directly from Eq. (7.163) that

$$h_{con} = C_2\left\{\left[\frac{k_v^3 g\rho_v(\rho_l - \rho_v)h'_{lv}}{\mu_v(T_w - T_{sat})}\right]\left[\frac{g(\rho_l - \rho_v)}{\sigma}\right]^{1/2}\right\}^{1/4} \tag{7.164}$$

where the numerical constants and b have been absorbed into C_2. The form of this relation can be obtained from Eq. (7.128) derived for a vertical surface, if λ_D is substituted for the characteristic length x. A value of 0.425 was proposed for C_2 because it correlated Berenson's [7.74] data for the minimum film boiling heat flux within $\pm 10\%$.

More recently, Klimenko [7.89] has extended the models described above for laminar film boiling on a horizontal flat plate to circumstances for which turbulent transport occurs in the vapor film. He noted that the previous predictions of laminar film boiling heat transfer from a horizontal flat surface developed by Berenson [7.74] and others have typically been of the form

$$\text{Nu}_\lambda = C_3\left[\frac{g\lambda_D^3(\rho_l - \rho_v)\,\text{Pr}_v\, h'_{lv}}{\nu_v c_{pv}(T_w - T_{sat})}\right]^m \tag{7.165}$$

where

$$\text{Nu}_\lambda = \frac{q''\lambda_D}{k_v(T_w - T_{sat})} \tag{7.166}$$

$$h'_{lv} = h_{lv}(1 + C_4\,\text{Ja}) \tag{7.167}$$

$$\text{Ja} = \frac{c_{pv}(T_w - T_{sat})}{h_{lv}} \tag{7.168}$$

In the above expressions, C_3, m, and C_4 are constants that vary somewhat from one correlation to the next.

Klimenko [7.89] proposed (as did Berenson [7.74]) that heat is transferred from the surface to the interface across a forced-convection flow of vapor in the film, with the vapor flow being driven by hydrostatic pressure differences generated by the bubble release process. He argued that the vapor flow would be turbulent if

$$R = \frac{g\lambda_D^3}{(3)^{3/2}\nu_v^{2/3}}\left(\frac{\rho_l - \rho_v}{\rho_v}\right) > 10^8 \tag{7.169}$$

For laminar flow, Klimenko [7.89] proposed the correlation

$$\text{Nu}_\lambda = 0.19\sqrt{3}\,R^{1/3}\,\text{Pr}_v^{1/3}\,f_1(\text{Ja}) \tag{7.170}$$

where

$$f_1 = 1 \qquad \text{for Ja} \geqslant 0.714 \tag{7.171a}$$

$$= 0.89\,\text{Ja}^{-1/3} \qquad \text{for Ja} < 0.714 \tag{7.171b}$$

Note that the function f_1 results in a more complex variation of the heat transfer coefficient with wall superheat than is predicted by Berenson's correlation (7.164). However, this variation agrees better with currently-available data than Eq. (7.164).

For turbulent flow, using Reynolds analogy arguments, Klimenko [7.89] developed the following correlation for the film boiling heat transfer coefficient:

$$\text{Nu}_\lambda = 0.0086\sqrt{3}\,R^{1/2}\,\text{Pr}_v^{1/3}\,f_2(\text{Ja}) \tag{7.172}$$

where

$$f_2 = 1 \qquad \text{for Ja} \geqslant 0.5 \tag{7.173a}$$

$$= 0.71\,\text{Ja}^{-1/2} \qquad \text{for Ja} < 0.5 \tag{7.173b}$$

Ramilison and Lienhard [7.90] found that the trends in their data for film boiling on a flat, horizontal surface and Berenson's [7.74] earlier data are well represented by Klimenko's turbulent film boiling correlation for Ja > 0.5. However, they replaced the constant 0.0086 in that correlation with the following specific values for each liquid: 0.0057 for R-113, 0.0066 for acetone, and 0.0154 for benzene.

Example 7.7 Use Berenson's correlation (7.164) together with the $q''_{\min}$ values predicted in Example 7.5 to estimate the minimum wall superheat required to maintain stable film boiling on an upward-facing infinite horizontal surface for pool boiling of water and R-113 at atmospheric pressure.

Berensen's correlation can be written in the form

$$\frac{q}{T_w - T_{\text{sat}}} = 0.425\left\{\left[\frac{k_v^3 g\rho_v(\rho_l - \rho_v)h'_{lv}}{\mu_v(T_w - T_{\text{sat}})}\right]\left[\frac{g(\rho_l - \rho_v)}{\sigma}\right]^{1/2}\right\}^{1/4}$$

where

$$h'_{lv} = h_{lv} + 0.5c_{pv}(T_w - T_{\text{sat}})$$

Because c_{pv} is low (2.034 kJ/kg K for water and 0.665 kJ/kg K for R-113 at atmospheric pressure), as a first approximation we will take $h'_{lv} = h_{lv}$. It is then possible to rearrange the above equation to solve for $T_w - T_{sat}$, yielding

$$T_w = T_{sat} = 3.130 \left\{ \frac{(q'')^4 \mu_v}{k_v^3 g \rho_v (\rho_l - \rho_v) h_{lv}} \left[\frac{\sigma}{g(\rho_l - \rho_v)} \right]^{1/2} \right\}^{1/3}$$

For water at atmospheric pressure (taking the properties at saturation), $k_v = 0.0249$ W/m K, $\mu_v = 1.21 \times 10^{-5}$ Ns/m^2. The other needed properties are given in Example 7.5. Substituting in the above relation, using $q'' = q''_{min}$ from Example 7.5,

$$T_w - T_{sat}$$

$$= 3.130 \left\{ \frac{(19.0 \times 10^3)^4 (1.21 \times 10^{-5})}{(0.0249)^3 (9.8)(0.598)(958)(2257 \times 10^3)} \left[\frac{0.0589}{9.8(958)} \right]^{1/2} \right\}^{1/3}$$

$$= 85.2°\text{C}$$

For R-113 at atmospheric pressure (again using the saturation properties), $k_v = 0.00866$ W/m K, $\mu_v = 1.08 \times 10^{-5}$ Ns/m^2. Substituting again these properties and the q''_{min} value from Example 7.5, we obtain

$$T_w - T_{sat}$$

$$= 3.130 \left\{ \frac{(9.85 \times 10^3)^4 (1.08 \times 10^{-5})}{(0.00866)^3 (9.8)(7.46)(1500)(143.8 \times 1000)} \left[\frac{0.0169}{9.8(1500)} \right]^{1/2} \right\}^{1/3}$$

$$= 68.9°\text{C}$$

Note that these results imply that for water,

$$\frac{0.5 c_{pv}(T_w - T_{sat})}{h_{lv}} = \frac{0.5(2.034)(85.2)}{2257} = 0.038$$

while for R-113,

$$\frac{0.5 c_{pv}(T_w - T_{sat})}{h_{lv}} = \frac{0.5(0.665)(68.9)}{143.8} = 0.159$$

Since $T_w - T_{sat}$ is proportional to $(h'_{lv})^{1/3}$, the fractional error in $T_w - T_{sat}$ associated with taking $h'_{lv} = h_{lv}$ is only one third of the values of $0.5 c_{pv}(T_w - T_{sat})/h_{lv}$ indicated above for each case. Thus, $h'_{lv} = h_{lv}$ is quite good for water, but a bit less accurate for R-113. Because of the large temperature difference across the vapor film, it would also be more accurate to repeat the calculations with the transport properties evaluated at the mean film temperature.

Corrections for Sensible Heat and Convective Effects

A number of the correlations for the film boiling heat transfer coefficient described in this section make use of an effective latent heat term h'_{lv}. The definition of this term typically has the general form

$$h'_{lv} = h_{lv}(1 + C_L \,\mathrm{Ja}) \tag{7.174a}$$

where Ja is the Jakob number, defined as

$$\mathrm{Ja} = \frac{c_{pv}(T_w - T_{\mathrm{sat}})}{h_{lv}} \tag{7.174b}$$

The recommended value of the constant C_L varies from one correlation to another. Heat transfer correlations involving a corrected latent heat term evolve most directly from integral boundary-layer treatments of laminar film boiling heat transfer. For laminar film boiling on a vertical flat plate, the integral analysis in this section led to the following equation for the heat transfer coefficient in the absence of radiation:

$$h_{\mathrm{con}} = \frac{k_v}{\delta_{\mathrm{con}}} = \left[\frac{k_v^3 g \rho_v(\rho_l - \rho_v)h_{lv}}{4\mu_v(T_w - T_{\mathrm{sat}})x}\right]^{1/4} \tag{7.128}$$

No correction to the latent heat appears in this relation. However, in deriving this relation, terms accounting for superheating of the vapor in the energy-balance equation (7.124) were neglected, as were the convective transport terms in the momentum and energy equations. Inclusion of the convective terms in the momentum and energy transport equations adds considerable complexity to the analysis. However, a term accounting for superheating of the vapor in the energy balance can be handled without much difficulty. Doing so and neglecting radiation converts Eq. (7.124) to the form

$$-k_v\left(\frac{\partial T}{\partial y}\right)_{y=\delta} = \frac{d}{dx}\int_0^\delta \rho_v u_v[h_{lv} + c_{pv}(T - T_{\mathrm{sat}})]\, dy \tag{7.175}$$

It is left as an exercise for the reader to show that proceeding with the integral analysis described above results in the following equation for h:

$$h_{\mathrm{con}} = \frac{k_v}{\delta_{\mathrm{con}}} = \left[\frac{k_v^3 g \rho_v(\rho_l - \rho_v)h'_{lv}}{4\mu_v(T_w - T_{\mathrm{sat}})x}\right]^{1/4} \tag{7.176}$$

where h'_{lv} as defined by Eq. (7.174a) with $C_L = \frac{3}{8}$. Including the convective transport terms in the analysis tends to further increase C_L. Bromley [7.78] suggested a value of 0.5 for C_L. Because of the high wall temperatures associated with some film boiling processes, this correction can significantly affect the predicted heat transfer coefficient. As a result, this type of correction to the latent heat is also usually found in correlations for film boiling heat transfer on cylinders, spheres, and flat horizontal surfaces.

For laminar film boiling on a vertical surface, Sadasivan and Lienhard [7.91] have examined this issue in some detail. These investigators determined the variation of C_L necessary to make Eq. (7.176) match the predictions of Koh's [7.77] similarity solution described above. In computing the similarity solutions, they took $(\rho\mu)_v/(\rho\mu)_l$ to be zero, which implies that the upward vapor velocity is essentially zero at the interface. Computed results for Prandtl numbers between 0.6 and 1000 and Ja values up to 0.8 indicated that C_L was virtually independent of

Ja, but did vary with Prandtl number. Based on these results, they recommended that C_L be evaluated using the following relation, which closely fit their computed variation of C_L:

$$C_L = 0.968 - \frac{0.163}{\text{Pr}_v} \tag{7.177}$$

For a vapor with a Prandtl number of 1.0, this relation indicates that C_L should be 0.81, which is well above the value of $\frac{3}{8}$ obtained from the approximate integral analysis. Sadasivan and Lienhard [7.91] note that this relation will not be accurate when $(\rho\mu)_v$ is not small compared to $(\rho\mu)_l$ (i.e., at high pressures) or when bulk motion of the liquid exerts significant shear on the interface. Within the appropriate constraints, they argued that the inaccuracy for non-flat-plate geometries should be minimal, implying that this relation for C_L is applicable to cylinders and spherical bodies as well.

Further discussion of film boiling heat transfer can be found in the review articles by Jordan [7.92] and Bressler [7.93].

7.7 TRANSITION BOILING

As described in Section 7.1, the range of wall superheat levels between the critical heat flux and the minimum heat flux conditions on the boiling curve is usually termed the transition boiling regime. Transition boiling has traditionally been interpreted as a combination of nucleate and film boiling alternately occurring over the heated surface. Compared with the other regimes of pool boiling discussed in this chapter, relatively fewer investigations of transition boiling have been conducted. Much of the transition boiling data that is available has been obtained in transient quenching experiments.

Quenching studies typically have been initiated by heating a solid body to a high temperature and suddenly immersing the body into a pool of liquid. The body temperature is typically high enough that film boiling occurs at its surface immediately after immersion. As the body cools, its surface temperature drops to the point that the boiling process shifts to the transition boiling regime, followed by transitions to nucleate boiling and finally to natural convection.

The heat flux versus wall superheat for this transient process would typically traverse a curve similar to *a–b–c–d–e–f* in Fig. 7.3. Transition boiling heat transfer data obtained in this manner correspond to the transition boiling curve obtained when the transition region is entered from the film boiling side. Based on analysis of available data, Witte and Lienhard [7.4] concluded that two distinctly different transition boiling curves are possible for a given liquid and heater surface combination. The data suggest that the two different transition boiling curves correspond to the extreme limits of surface wettability—one corresponding to conditions where the liquid wets the surface (low contact angle), and the other corresponding to circumstances for which the liquid poorly wets the surface (higher contact angle). Witte and Lienhard [7.4] also note that, in quenching of a hot body, a transition from a higher advancing contact angle to a lower receding one can occur abruptly, causing a dramatic jump in the transition boiling heat flux.

The transition boiling conditions obtained by increasing the surface temperature beyond that for the critical heat flux are generally presumed to correspond to a well-wetted surface condition. For these circumstances, when liquid contacts the surface as a dry patch rewets, the low contact angle apparently facilitates a rapid spread of liquid over the surface, resulting in vigorous nucleation, which tends to blow liquid away from the surface. The slugs and columns structure tends to persist beyond the critical heat flux, with vapor supplied to the columns mainly by the vigorous nucleate boiling occurring when a dry patch is rewetted. Because it is an extension of the nucleate boiling curve beyond the critical heat flux, the q''-versus-superheat curve for these conditions is sometimes referred to as the *nucleate transition boiling* curve.

Beginning in the film boiling regime and decreasing the wall superheat, eventually the surface temperature drops to the point that vapor is not produced rapidly enough to sustain a stable vapor layer. The breakdown of the film at these high wall temperatures is postulated to result from liquid being brought into contact with the hot surface by interfacial wave action. The contact occurs over small areas near the nodes of waves, as indicated schematically in Fig. 7.27.

The liquid that contacts the solid surface during this breakdown is expected to exhibit the advancing contact angle θ_a for the system along the contact line (see Chapter Three). If θ_a is large, the liquid contacting the surface may spread very little, with the result that the wetted area over which nucleate boiling can occur is small. For such circumstances, the heat transfer from the surface may be only slightly enhanced above that for film boiling alone.

The effect of surface wettability on the departure from film boiling is indicated by the data in Figure 7.28. The data in this figure were obtained by Ramilison and Lienhard [7.90] for pool boiling of *n*-pentane. The data represent the pool boiling heat transfer obtained for three different surfaces as the wall temperature was gradually decreased beginning in the film boiling regime. In tests conducted prior to the boiling experiments, the Teflon was found to have the

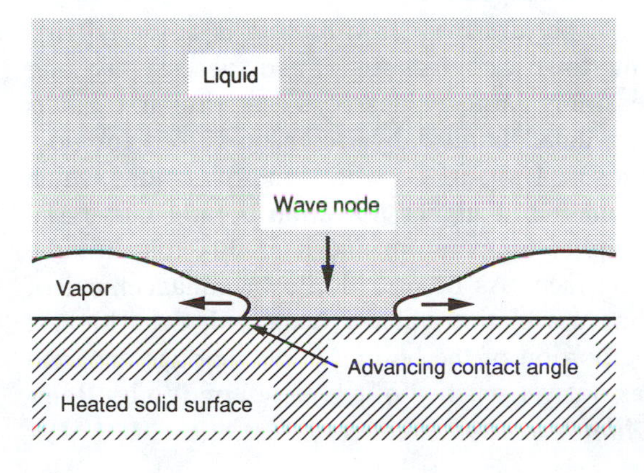

Figure 7.27 Onset of transition boiling due to breakdown of the vapor film at node locations in the Taylor wave motion of the interface.

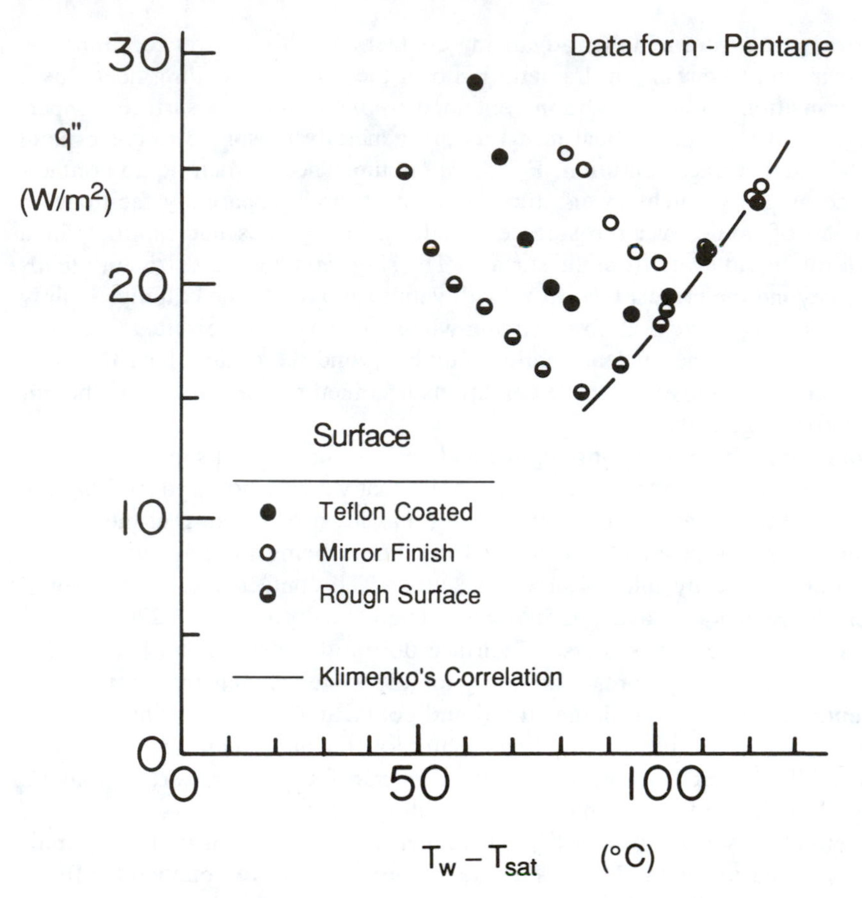

Figure 7.28 Transition and film boiling data for different surface conditions. Also shown is the film boiling prediction of Klimenko's correlation. (*Adapted from Ramilison and Lienhard [7.90] with permission, copyright © 1987, American Society of Mechanical Engineers.*)

lowest advancing contact angle of 30°, while the mirror and rough surfaces had higher measured values of 40° and 50°, respectively.

As the superheat is decreased, departure from film boiling is observed first in Fig. 7.28 for the surface with the lowest contact angle, and successively later departures are observed for increasingly large values of θ_a. A similar trend is evident in Fig. 7.29, which shows data obtained by Berenson [7.66] for pool boiling of pentane on a copper surface. The surface with some oxidation, which is better wetted by the liquid, resulted in a higher transition boiling curve and departure from film boiling at a higher temperature and heat flux than for the more poorly wetted clean copper surface. As described above, enhancement of the heat flux above the film boiling correlation value is expected to be highest for the best wetting condition. The portion of the boiling curve that has departed from the film boiling curve but has not yet reached $q''_{\min}$ was referred to by Ramilison and Lienhard [7.90] as the *film transition regime*.

The trends described above imply that if a system exhibits significant contact-angle hysteresis (i.e., $\theta_a > \theta_r$), the transition boiling curve obtained for increasing wall superheat may be very different from that obtained for decreasing superheat. For increasing superheat, the nucleate transition boiling is encountered first. For such conditions, the liquid–solid contact angle is generally near the receding contact angle θ_r. This lower contact angle apparently persists as the wall superheat is increased, resulting in the higher transition boiling curve associated with better-wetting liquids.

For decreasing superheat the film transition boiling regime is encountered first, and, as noted above, the liquid–solid contact is initially near the advancing contact-angle value for the system. If the advancing contact angle is significantly higher than the receding value, then the transition boiling curve for decreasing superheat will, at least initially, be lower than that for increasing superheat, as indicated in Fig. 7.30. If the contact angle remains near θ_a as the wall temperature drops, the entire transition boiling curve may be lower than for the increasing superheat case, as indicated in Fig. 7.30. However, if the contact angle suddenly changes to a value near the receding contact angle, the heat flux may suddenly rise, and the system may jump to the upper curve in Fig. 7.30. The inability to control the contact angle may thus be responsible for considerable variability and/or scatter in transition pool boiling data.

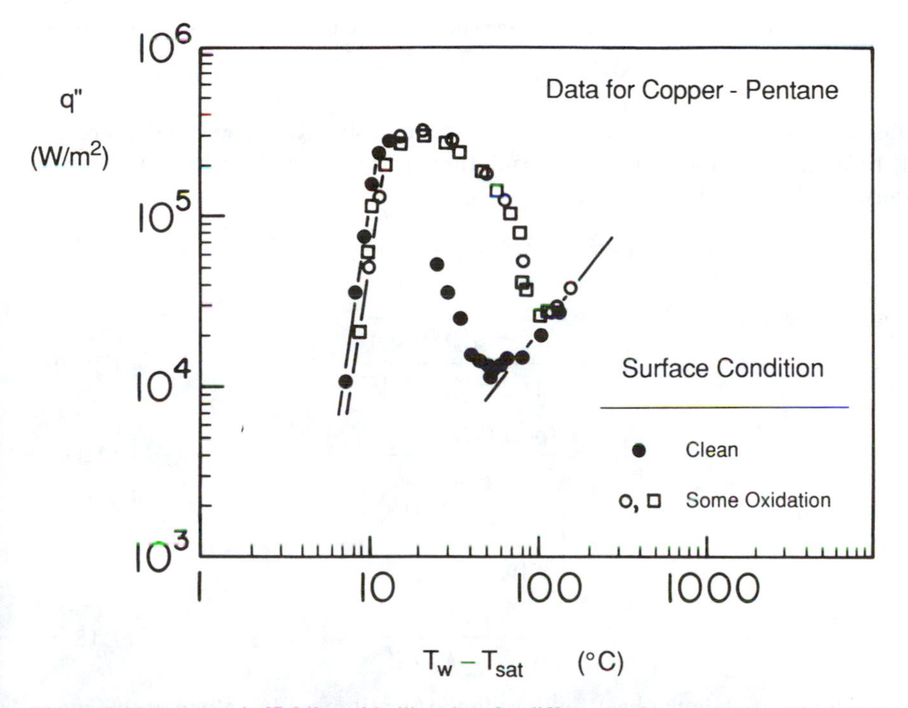

Figure 7.29 Berenson's [7.94] pool boiling data for different surface conditions. The solid lines are best fits to the data in the nucleate boiling and film boiling regimes.

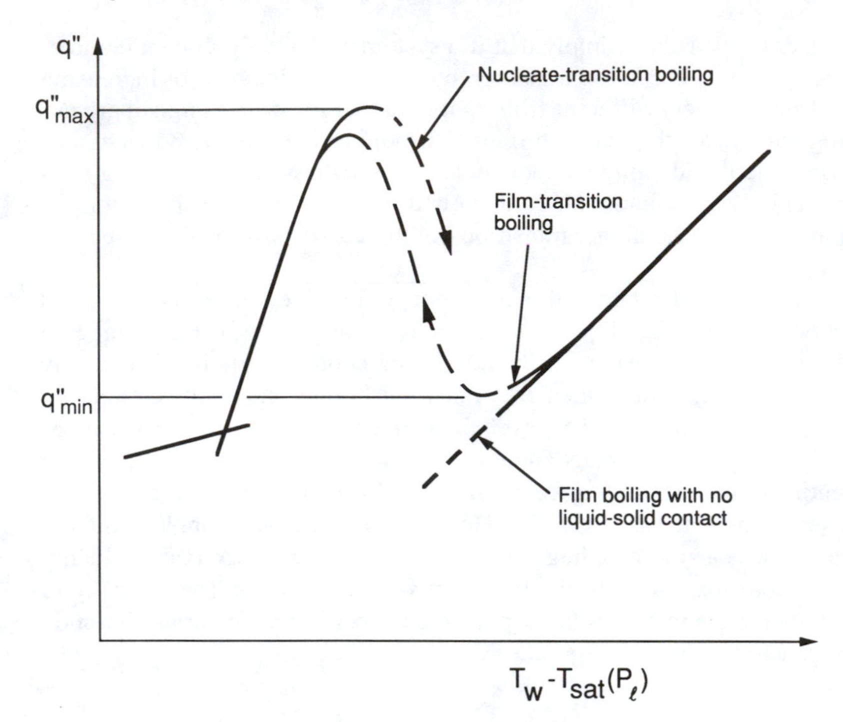

Figure 7.30 Relative locations of the nucleate transition and film transition portions of the boiling curves.

Based on a model of the heat transfer mechanism, Ramilison and Lienhard [7.90] recommended the following correlation for predicting the heat transfer in the transition film boiling regime:

$$\text{Bi}^* = 3.74 \times 10^{-6}(\text{Ja}^*)^2 K \tag{7.178}$$

where

$$\text{Bi}^* = \frac{(q'' - q''_{fb})\sqrt{\alpha_h \tau}}{k_h[T_w - T_{\text{sat}}(P_l)]K} \tag{7.179}$$

$$\text{Ja}^* = \frac{(\rho c_p)_h(T_{dfb} - T_w)}{\rho_v h_{lv}} \tag{7.180}$$

$$\tau = \left[\frac{\sigma}{g^3(\rho_l - \rho_v)}\right]^{1/4} \tag{7.181a}$$

$$K = \frac{k_l/\alpha_l^{1/2}}{k_l/\alpha_l^{1/2} + k_h/\alpha_h^{1/2}} \tag{7.181b}$$

In these relations, q''_{fb} is the heat flux predicted for film boiling alone at the specified wall superheat, τ is the characteristic period of the Taylor wave action at the interface, and T_{dfb} is the wall temperature at which the boiling process departs from film boiling due to the onset of liquid contact with the surface. The variables k_h and α_h are the thermal conductivity and thermal diffusivity of the heated surface, respectively. Ramilison and Lienhard [7.90] gave a graphical correlation for determining T_{dfb}, which is well approximated by the relation

$$\frac{T_{dfb} - T_{sat}}{T_{hn} - T_{sat}} = 0.97 \exp(-0.00060\theta_a^{1.8}) \tag{7.182}$$

where θ_a is the advancing contact angle in degrees and T_{hn} is the homogeneous nucleation temperature. The latter can be determined using the corresponding-states correlation proposed by Lienhard [7.94]:

$$T_{hn} = \left[0.932 + 0.077\left(\frac{T_{sat}}{T_c}\right)^9\right]T_c \tag{7.183}$$

where T_c is the critical temperature of the coolant. The above correlation was found to provide a good fit to transition boiling heat transfer data obtained by Ramilison and Lienhard [7.90] and Berenson [7.66], which spanned a wide variety of fluid and surface combinations.

The variation of the transition boiling curve with liquid contact angle was also observed in the experimental data obtained by Chowdhury and Winterton [7.95] for vertical cylinders and Bui and Dhir [7.96] for a vertical flat surface. In both cases the transition boiling curve was found to shift upward as the liquid became more wetting. Bui and Dhir [7.96] also indicated that, for the transient cooling typical of quenching processes, the maximum heat fluxes were as much as 60% lower than the maximum steady-state heat fluxes for the same system. This reduction apparently is a direct consequence of the lower transition boiling curve that results when transition boiling is entered from the film boiling side.

In a recent study, Dhir and Liaw [7.67] modeled transition boiling from a vertical surface as a rectangular array of vapor stems carrying vapor across a thin region of wall-dominated flow near the surface. Based on this model, they proposed that the heat flux in the transition boiling regime be predicted by adding together the volume-fraction-weighted contributions to the total heat transfer associated with the dry and wet areas on the surface:

$$q'' = h(T_w - T_{sat}) = \bar{h}_l(1 - \alpha_w)(T_w - T_{sat}) + \bar{h}_v\alpha_w(T_w - T_{sat}) \tag{7.184}$$

where α_w is the void fraction (vapor volume fraction) at the wall. The heat transfer coefficient in the dry region was taken to be that given by a correlation for film boiling heat transfer from a vertical surface developed by Bui and Dhir [7.82]:

$$\bar{h}_v = 0.47\left\{\left[\frac{k_v^3 g\rho_v(\rho_l - \rho_v)h_{lv}}{\mu_v(T_w - T_{sat})}\right]\left[\frac{g(\rho_l - \rho_v)}{\sigma}\right]^{1/2}\right\}^{1/4} \tag{7.185}$$

The heat transfer to the wet portions of the surface was idealized as being simple conduction from the heated surface to the liquid–vapor interface near the base of the vapor stem. From an analysis of this conduction problem, Dhir and Liaw [7.67] derived the following relation for $\bar{h}_l$:

$$\bar{h}_l = \frac{2}{L^2(1-\alpha_w)} \int_0^{\pi/4} Ck_l(L \sec \psi + D_w)\, d\psi \tag{7.186}$$

where

$$C = \sum_{n=1}^{\infty} \frac{2 \sin^2(\lambda_n b)}{\lambda_n b + \sin(\lambda_n b)\cos(\lambda_n b)} \tag{7.187}$$

$$b = \frac{L \sec \psi - D_w}{2} \tag{7.188}$$

and the λ_n values are roots of the equation

$$\lambda_n b \tanh(\lambda_n b) = \theta \tag{7.189}$$

where θ is the liquid contact angle.

Although the computation is complex, in principle, the above relations can be used to determine $\bar{h}_l$ if the wall void fraction α_w, the center-to-center spacing of the vapor stems L, and the diameter of the vapor stems D_w at the wall are specified. Dhir and Liaw [7.67] argued that L should equal the nucleation site spacing for nucleate boiling near the maximum heat flux, and they used measured void fraction data to evaluate α_w. D_w can then be determined from the following relation, which follows from the geometry of the vapor stem model:

$$\alpha_w = \frac{\pi D_w^2}{4L^2} \tag{7.190}$$

Dhir and Liaw [7.67] found good agreement between measured heat transfer data and the predictions of the above model. Although many aspects of the model are highly idealized, it does appear to provide a useful theoretical framework for correlation transition boiling data.

REFERENCES

7.1 Nukiyama, S., The maximum and minimum values of heat Q transmitted from metal to boiling water under atmospheric pressure, *J. Jap. Soc. Mech. Eng.*, vol. 37, pp. 367–374, 1934 (translated in *Int. J. Heat Mass Transfer,* vol. 9, pp. 1419–1433, 1966).

7.2 Jakob, M., and Linke, W., *Phys. Z.*, vol. 36, p. 267, 1935.

7.3 Drew, T. B., and Mueller, C., Boiling, *Trans. AIChE,* Vol. 33, p. 449, 1937.

7.4 Witte, L. C., and Lienhard, J. H., On the existence of two "transition" boiling curves, *Int. J. Heat Mass Transfer,* vol. 25, pp. 771–779, 1982.

7.5 Rohsenow, W. M., A method of correlating heat transfer data for surface boiling of liquids, *Trans. ASME,* vol. 84, p. 969, 1962.

7.6 Forster, H. K., and Zuber, N., Dynamics of vapor bubbles and boiling heat transfer, *AIChE J.*, vol. 1, p. 531, 1955.

7.7 Forster, H. K., and Greif, R., Heat transfer to a boiling liquid—mechanisms and correlations, *J. Heat Transfer*, vol. 81, p. 45, 1959.
7.8 Gunther, F. C., and Kreith, F., Photographic study of bubble formation in heat transfer to subcooled water, *Prog. Rept.* 4-120, Jet Propulsion Lab., California Institute of Technology, Pasadena, CA, March 1956.
7.9 Han, C.-Y., and Griffith, P., *Int. J. Heat Mass Transfer*, vol. 8, p. 887, 1965.
7.10 Mikic, B. B., and Rohsenow, W. M., A new correlation of pool boiling data including the effect of heating surface characteristics, *J. Heat Transfer*, vol. 91, p. 245, 1969.
7.11 Mikic, B. B., and Rohsenow, W. M., Bubble growth rates in non-uniform temperature field, *Prog. in Heat Mass Transfer*, vol. II, pp. 283–292, 1969.
7.12 Cole, R., and Rohsenow, W. M., *Chem. Eng. Prog. Symp. Ser.*, vol. 65, no. 92, p. 211, 1965.
7.13 Cole, R., *AIChE J.*, vol. 13, p. 779, 1967.
7.14 Griffith, P., and Wallis, J. D., The role of surface conditions in nucleate boiling, *Chem. Eng. Prog. Symp. Ser.*, vol. 56, no. 30, pp. 49–63, 1960.
7.15 Addoms, J. N., Heat transfer at high rates to water boiling outside cylinders, D.Sc. thesis, Chem. Eng. Dept., MIT, Cambridge, MA, June 1948.
7.16 Zuber, N., *Int. J. Heat Mass Transfer*, vol. 6, p. 53, 1963.
7.17 Tien, C.-L., A hydrodynamic model for nucleate pool boiling, *Int. J. Heat Mass Transfer*, vol. 5, pp. 533–540, 1962.
7.18 Yamagata, K., Kirano, F., Nishikawa, K., and Matsuoka, H., Nucleate boiling of water on the horizontal heating surface, *Mem. Fac. Eng. Kyushu Univ.*, vol. 15, no. 1, pp. 97–163, 1955.
7.19 Gaertner, R. F., and Westwater, J. W., Population of active sites in nucleate boiling heat transfer, *Chem. Eng. Prog. Symp. Ser.*, vol. 55, no. 30, pp. 39–48, 1959.
7.20 van Stralen, S. J. D., *Chem. Eng. Sci.*, vol. 25, p. 149, 1970.
7.21 Judd, R. L., and Hwang, K. S., *J. Heat Transfer*, vol. 98, p. 630, 1976.
7.22 Heled, Y., Ricklis, J., and Orell, A., Pool boiling from large arrays of artificial nucleation sites, *Int. J. Heat Mass Transfer*, vol. 13, pp. 503–516, 1970.
7.23 Borishansky, V. M., Correlation of the effect of pressure on the critical heat flux and heat transfer rates using the theory of thermodynamic similarity, In *Problems of Heat Transfer and Hydraulics of Two-Phase Media*, Pergamon Press, New York, pp. 16–37, 1969.
7.24 Mostinski, I. L., *Teploenergetika*, vol. 4, p. 63, 1963 (English abstract in *Br. Chem. Eng.*, vol. 8, p. 580, 1963).
7.25 Stephan, K., and Abdelsalam, M., Heat-transfer correlations for natural convection boiling, *Int. J. Heat Mass Transfer*, vol. 23, pp. 73–87, 1980.
7.26 Fritz, W., *Z. d. Ver deutsch. Ing., Beiheft "Verfahrenstechnik,"* no. 5, p. 149, 1937.
7.27 Kutateladze, S. S., Heat transfer during condensation and boiling, 2nd ed. (in Russian), State Scientific and Technical Publishing House of Literature on Machinery, 1952 (also in translated form as AEC-TR-3770).
7.28 Kruzhilin, G. N. Generalization of experimental data on heat transfer during boiling of a liquid with natural convection (in Russian), *Izvestiya AN SSSR, OTN (News of the Academy of Sciences of the USSR, Division of Technical Sciences)*, no. 5, 1949.
7.29 Levy, S., A generalized correlation of boiling heat transfer, *J. Heat Transfer*, vol. 81, pp. 37–42, 1959.
7.30 Alad'yev, I. T., Heat transfer to liquids boiling in tubes and pools, *Teploenergetika*, vol. 10, p. 57, 1969.
7.31 Labuntsov, D. A., General relationships for heat transfer during nucleate boiling of liquids, *Teploenergetika*, vol. 7, p. 76, 1966.
7.32 Kutateladze, S. S., Leonte'ev, A. I., and Kirdyashkin, A. G., Theory of heat transfer in nucleate boiling, *Inzh. Fiz. Zh.*, vol. 8, p. 7, 1965.
7.33 Lienhard, J. H., A semi-rational nucleate boiling heat flux correlation, *Int. J. Heat Mass Transfer*, vol. 6, pp. 215–219, 1963.
7.34 Nishikawa, K., Fujita, Y., Uchida, S., and Ohta, H., Effect of heating surface orientation on

nucleate boiling heat transfer, *Proc. ASME-JSME Thermal Eng. Joint Conf.*, Honolulu, March 20–24, 1983, ASME, New York, vol. 1, pp. 129–136, 1983.

7.35 Moissis, R., and Berenson, P. J., On the hydrodynamic transitions in nucleate boiling, *J. Heat Transfer,* vol. 85, pp. 221–229, 1963.

7.36 Bhattacharya, A., and Lienhard, J. H., Hydrodynamic transition in electrolysis, *J. Basic Eng.*, vol. 94, pp. 804–810, 1972.

7.37 Lienhard, J. H., On the two regimes of nucleate boiling, *J. Heat Transfer,* vol. 107, pp. 262–264, 1985.

7.38 Dhir, V. K., and Tung, V. X., A thermal model for fully developed nucleate boiling of saturated liquids, in *Collected Papers in Heat Transfer 1988–Volume Two,* K. T. Yang, Ed., ASME HTD vol. 104, pp. 153–164, 1988.

7.39 Tong, W., Simon, T. W., and Bar-Cohen, A., A bubble sweeping heat transfer mechanism for low flux boiling on downward-facing inclined surfaces, in *Collected Papers in Heat Transfer 1988—Volume Two,* K. T. Yang, Ed., ASME HTD vol. 104, pp. 173–178, 1988.

7.40 Merte, H., Jr., Combined roles of buoyancy and orientation in nucleate pool boiling, in *Collected Papers in Heat Transfer 1988—Volume Two,* K. T. Yang, Ed., ASME HTD vol. 104, pp. 179–186, 1988.

7.41 Vachon, R. I., Nix, G. H., and Tanger, G. E., Evaluation of constants for the Rohsenow pool-boiling correlation, *J. Heat Transfer,* vol. 90, pp. 239–247, 1968.

7.42 Rohsenow, W. M., Heat transfer with evaporation, *Heat Transfer—A Symposium Held at the University of Michigan During the Summer of 1952,* University of Michigan Press, Ann Arbor, MI, pp. 101–150, 1953.

7.43 Rohsenow, W. M., and Clarke, J. A., Heat transfer and pressure drop data for high heat flux densities to water at high subcritical pressures, *1951 Heat Transfer and Fluid Mechanics Inst.*, Stanford University Press, Stanford, CA, 1951.

7.44 Kreith, F., and Sommerfield, M., Heat transfer to water at high flux densities with and without surface boiling, *Trans. ASME,* vol. 71, pp. 805–815, 1949.

7.45 Piret, E. L., and Isbin, H. S., Two-phase heat transfer in natural circulation evaporators, *Chem. Eng. Prog. Symp. Ser.*, vol. 50, no. 6, p. 305, 1953.

7.46 Bergles, A. E., and Rohsenow, W. M., The determination of forced-convection surface-boiling heat transfer, *J. Heat Transfer,* vol. 86, pp. 365–372, 1964.

7.47 Clarke, J. A., in *Cryogenic Technology,* R. W. Vance, Ed., Wiley, New York, chap. 5, 1963.

7.48 Subbotin, V. I., Sorokin, D. N., and Kudryavtsev, R. R., Generalized relationship for heat transfer in developed boiling of alkali metals, *Atomic Energy,* vol. 29, p. 45, 1970.

7.49 Dwyer, O. E., *Boiling Liquid Metal Heat Transfer,* American Nuclear Society, Hinsdale, IL, 1976.

7.50 Kutateladze, S. S., On the transition to film boiling under natural convection, *Kotloturbostroenie,* no. 3, p. 10, 1948.

7.51 Rohsenow, W. M., and Griffith, P., Correlation of maximum heat transfer data for boiling of saturated liquids, *Chem. Eng. Prog. Symp. Ser.*, vol. 52, no. 18, p. 47, 1956.

7.52 Chang, Y. P., and Snyder, N. W., Heat transfer in saturated boiling, *Chem. Eng. Prog. Symp. Ser.*, vol. 56, no. 30, pp. 25–38, 1960.

7.53 Zuber, N., Hydrodynamic aspects of boiling heat transfer, AEC Rep., AECU-4439, June 1959.

7.54 Lienhard, J. H., and Dhir, V. K., Extended hydrodynamic theory of the peak and minimum pool boiling heat fluxes, NASA CR-2270, July 1973.

7.55 Lienhard, J. H., Dhir, V. K., and Riherd, D. M., Peak pool boiling heat flux measurements on finite horizontal flat plates, *J. Heat Transfer,* vol. 95, pp. 477–482, 1973.

7.56 Sun, K. H., and Lienhard, J. H., The peak pool boiling heat flux on horizontal cylinders, *Int. J. Heat Mass Transfer,* vol. 13, pp. 1425–1439, 1970.

7.57 Lienhard, J. H., and Dhir, V. K., Hydrodynamic prediction of peak pool-boiling heat fluxes from finite bodies, *J. Heat Transfer,* vol. 95, pp. 152–158, 1973.

7.58 Ded, J. S., and Lienhard, J. H., The peak pool boiling heat flux from a sphere, *AIChE J.*, vol. 18, pp. 337–342, 1972.

7.59 Haramura, Y., and Katto, Y., A new hydrodynamic model of the critical heat flux, applicable widely to both pool and forced convective boiling on submerged bodies in saturated liquids, *Int. J. Heat Mass Transfer,* vol. 26, pp. 389–399, 1983.

7.60 Taylor, G. I., The instability of liquid surfaces when accelerated in a direction perpendicular to their planes. I, *Proc. Roy. Soc. London, Ser. A,* vol. 201, pp. 192–196, 1950.

7.61 Chang, Y. P., A theoretical analysis of heat transfer in natural convection and in boiling, *Trans. ASME,* vol. 79, p. 1501, 1957.

7.62 Bernath, L., A theory of local burnout and its application to existing data, *AIChE Symp. Ser.,* vol. 56, no. 30, p. 59, 1959.

7.63 Costello, C. P., and Frea, W. J., A salient non-hydrodynamic effect on pool boiling burnout of small semi-cylindrical heaters, AIChE Preprint No. 15, 6th Natl. Heat Transfer Conf., Boston, August 11–14, 1963.

7.64 Chang, Y. P., Some possible critical conditions in nucleate boiling, *J. Heat Transfer,* vol. 85, pp. 90–100, 1963.

7.65 Lienhard, J. H., and Hasan, M. M., On predicting boiling burnout with the mechanical energy stability criterion, *J. Heat Transfer,* vol. 101, pp. 276–279, 1979.

7.66 Berenson, P. J., Transition boiling heat transfer from a horizontal surface, *AIChE Paper 18, ASME-AIChE Heat Transfer Conf.,* Buffalo, NY (also MIT Heat Transfer Lab. Tech. Report 17), 1960.

7.67 Dhir, V. K., and Liaw, S. P., Framework for a unified model for nucleate and transition pool boiling, in *Radiation, Phase Change Heat Transfer and Thermal Systems,* ASME HTD vol. 81, pp. 51–58, 1987.

7.68 Leinhard, J. H., and Witte, L. C., A historical review of the hydrodynamic theory of boiling, *Rev. Chem. Eng.,* vol. 3, pp. 187–277, 1985.

7.69 Lienhard, J. H., Things we don't known about boiling heat transfer: 1988, *Int. Commun. Heat Mass Transfer,* vol. 15, pp. 401–428, 1988.

7.70 Noyes, R. C., and Lurie, H., Boiling sodium heat transfer, *Proc. 3d Int. Heat Transfer Conf.,* Chicago, IL, vol. 5, p. 92, 1966.

7.71 Borishansky, V. M., Novikov, I. I., and Kutateladze, S. S., Use of thermodynamic similarity in generalizing experimental data of heat transfer, Paper No. 56, Int. Heat Transfer Conf., Univ. of Colorado, Boulder, CO, 1961, *Int. Dev. in Heat Transfer,* ASME, New York, pp. 475–482, 1963.

7.72 Sharan, A., Lienhard, J. H., and Kaul, R., Corresponding states correlations for pool and flow boiling burnout, *J. Heat Transfer,* vol. 107, 392–397, 1985.

7.73 Chang, Y. P., Wave theory of heat transfer in film boiling, *J. Heat Transfer,* vol. 81, p. 112, 1959.

7.74 Berenson, P. J., Film boiling heat transfer from a horizontal surface, *J. Heat Transfer,* vol. 83, p. 351, 1961.

7.75 Lienhard, J. H., and Wong, P. T. Y., The dominant unstable wavelength and minimum heat flux during film boiling on a horizontal cylinder, *J. Heat Transfer,* vol. 86, pp. 220–226, 1964.

7.76 Gunnerson, F. S., and Cronenberg, A. W., On the minimum film boiling conditions for spherical geometries, *J. Heat Transfer,* vol. 102, pp. 335–341, 1980.

7.77 Koh, J. C. Y., Analysis of film boiling on vertical surfaces, *J. Heat Transfer,* vol. 84, p. 55, 1962.

7.78 Bromley, J. A., Heat transfer in stable film boiling, *Chem. Eng. Prog.,* vol. 46, no. 5, pp. 221–227, 1950.

7.79 Lubin, B. T., Analytical derivation for total heat transfer coefficient in stable film boiling from vertical plate, *J. Heat Transfer,* vol. 91, pp. 452–453, 1969.

7.80 Sparrow, E. M., The effect of radiation on film-boiling heat transfer, *Int. J. Heat Mass Transfer,* vol. 7, pp. 229–238, 1964.

7.81 Greitzer, E. M., and Abernathy, F. H., Film boiling on vertical surfaces, *Int. J. Heat Mass Transfer,* vol. 15, pp. 475–491, 1972.

7.82 Bui, T. D., and Dhir, V. K., Film boiling heat transfer on an isothermal vertical surface, *J. Heat Transfer,* vol. 107, pp. 764–771, 1985.

7.83 Hsu, Y. Y., and Westwater, J. W., Approximate theory for film boiling on vertical surfaces, *Chem. Eng. Prog. Symp. Ser.*, vol. 56, no. 30, p. 15, 1960.
7.84 Hsu, Y. Y., and Westwater, J. W., *AIChE J.*, vol. 4, p. 58, 1958.
7.85 Suryanarayana, N. V., and Merte, H., Film boiling on vertical surfaces, *J. Heat Transfer,* vol. 94, pp. 377–384, 1972.
7.86 Frederking, T. H. K., and Clark, J. A., Natural convection film boiling on a sphere, *Adv. Cryogen. Eng.*, vol. 8, pp. 501–506, 1963.
7.87 Breen, B. P., and Westwater, J. W., Effect of diameter of horizontal tubes on film heat transfer, *Chem. Eng. Prog.*, vol. 58, no. 7, p. 67, 1962.
7.88 Borishansky, V. M., in *Problems of Heat Transfer During a Change of State,* S. S. Kutateladze, Ed. (in translated form as AEC-tr-3405), 1959.
7.89 Klimenko, V. V., Film boiling on a horizontal plate—New correlation, *Int. J. Heat Mass Transfer,* vol. 24, pp. 69–79, 1981.
7.90 Ramilison, J. M., and Lienhard, J. H., Transition boiling heat transfer and the film transition regime, *J. Heat Transfer,* vol. 109, pp. 746–752, 1987.
7.91 Sadasivan, P., and Lienhard, J. H., Sensible heat correction in laminar film boiling and condensation, *J. Heat Transfer,* vol. 109, pp. 545–546, 1987.
7.92 Jordan, D. P., Film and transition boiling, *Adv. Heat Transfer,* pp. 55–128, 1968.
7.93 Bressler, R. G., A review of physical models and heat transfer correlations for free convection film boiling, *Adv. Cryogen. Eng.*, vol. 17, pp. 382–406, 1972.
7.94 Lienhard, J. H., Corresponding states correlations for the spinoidal and homogeneous nucleation temperatures, *J. Heat Transfer,* vol. 104, pp. 379–381, 1982.
7.95 Chowdhury, S. K. R., and Winterton, R. H. S., Surface effects in pool boiling, *Int. J. Heat Mass Transfer,* vol. 28, pp. 1881–1889, 1985.
7.96 Bui, T. D., and Dhir, V. K., Transition boiling heat transfer on a vertical surface, *J. Heat Transfer,* vol. 107, pp. 756–763, 1985.

PROBLEMS

7.1 Compute the heat flux values predicted by the Forster-Zuber and Borishansky nucleate boiling correlations for *n*-butanol at atmospheric pressure for a wall superheat of 10°C. Compare the results with those in Example 7.1 for the Rohsenow and Stephan-Abdelsalam correlations.

7.2 Use the Forster-Zuber correlation to predict the heat flux for nucleate boiling of liquid nitrogen at atmospheric pressure at a wall superheat of 5°C. Compare the result with those for the Clark and Stephan-Abdelsalam correlations given in Example 7.2.

7.3 You are asked to design an enhanced pool boiling heat transfer surface made of copper with artificial nucleation sites. This surface must transfer as high a heat flux as possible to a saturated liquid pool at atmospheric pressure and a wall superheat of 5°C. To estimate mouth size for the cavities, proceed as follows: (*a*) Use the Rohsenow correlation for a rough copper surface to estimate q'' at the designated operating conditions. Then estimate δ_t as $\delta_t = k_l(T_w - T_{sat})/q''$. (*b*) Use the results of Hsu's analysis described in Section 6.3 to estimate the range of cavity mouth radii that will be active for these conditions. This provides an estimate for the mouth size of the artificial cavities on the enhanced surface.

7.4 Pool boiling of liquid mercury at atmospheric pressure occurs on an upward-facing horizontal surface immersed in saturated liquid mercury. The wall superheat for the surface is 10°C. Determine the resulting heat flux using (*a*) the Forster-Zuber correlation and (*b*) the correlation proposed by Subbotin et al. [7.48].

7.5 Determine the heat flux corresponding to the Moissis-Berenson transition for water boiling at atmospheric pressure, and at pressures of 571, 2185, 6124, and 14044 kPa. The heated surface is upward facing and infinite in extent, and the contact angle is 20°. Also, compute the maximum heat flux condition at these same pressures. From the computed results, determine and plot the fraction of the maximum heat flux corresponding to the Moissis-Berenson transition as a function of reduced pressure (P/P_c). What do you conclude about the effect of pressure on this transition?

7.6 For a horizontal cylinder with a diameter of 2 mm, determine the critical heat flux for pool boiling of methanol at a pressure of 1 atm. Using the appropriate form of the Stephan-Abdelsalam correlation, estimate the wall temperature corresponding to the critical heat flux condition.

7.7 Estimate the critical heat flux for pool boiling from a 1-mm-diameter sphere immersed in saturated liquid oxygen at atmospheric pressure. By how much is the value estimated to change if the sphere diameter is increased to 5 mm?

7.8 For pressures between 1 atm and the critical point, determine and plot the variation of the maximum heat flux with pressure for pool boiling from an infinite upward-facing horizontal surface immersed in saturated liquid nitrogen. Estimate the pressure corresponding to the highest maximum heat flux value for these conditions.

7.9 For pressures between 1 atm and the critical point, determine and plot the variation of the minimum heat flux with pressure for pool boiling from an infinite upward-facing horizontal surface immersed in saturated liquid water.

7.10 For values of cylinder radius between 1 mm and 10 mm, determine and plot the variation of the minimum heat flux for pool boiling from a horizontal cylinder immersed in saturated liquid nitrogen at atmospheric pressure.

7.11 For pressures between 1 atm and the critical point, determine and plot the variation of the minimum heat flux with pressure for pool boiling from an infinite upward-facing horizontal surface immersed in saturated liquid mercury.

7.12 A vertical flat surface held at 190°C is immersed in saturated liquid water at atmospheric pressure. Assuming that laminar film boiling occurs at the surface, use Bromley's superposition technique to estimate the total heat flux from the surface due to the combined effects of convection and radiation. Assume that the wall radiates as a graybody with an emittance of 0.9 and that the interface radiates as a blackbody. (Neglect any radiation interaction with water vapor in the film.)

7.13 (*a*) For an infinite, upward-facing, flat, horizontal surface, determine the minimum heat flux for pool boiling of saturated R-22 at 619 kPa. (*b*) Use Berenson's correlation for laminar film boiling, with the radiation effect included, to estimate the wall superheat corresponding to the minimum heat flux condition determined in part (*a*). Assume that the wall radiates as a graybody with an emmitance of 0.85 and the interface acts like a blackbody.

7.14 For laminar, natural-convection film boiling over an isothermal horizontal cylinder, use an integral analysis to derive an expression for the local heat transfer coefficient near the bottom of the cylinder. Assume that the interface is smooth, neglect radiation effects, and take all thermophysical properties to be constant. Compare your results to the correlation represented in Eqs. (7.141) and (7.142). (*Hint*: Follow the analysis presented for the sphere in Example 7.6. Make similar idealizations, but allow for the differences in geometry.)

7.15 Use an integral analysis of laminar, natural-convection film boiling over a vertical, isothermal, flat surface to show that, when radiation from the surface dominates over conduction and convection, the film thickness varies with downstream location as

$$\delta = \left[\frac{3\mu_v h_{rad}(T_w - T_{sat})x}{\rho_v g(\rho_l - \rho_v)h'_{lv}}\right]^{1/3}$$

where

$$h'_{lv} = h_{lv}\left(1 + \frac{3}{8}\,\mathrm{Ja}\right) \qquad \mathrm{Ja} = \frac{c_{pv}(T_w - T_{sat})}{h_{lv}}$$

and

$$h_{rad} = \frac{\sigma_{SB}(T_w^2 + T_{sat}^2)(T_w + T_{sat})}{(1/\varepsilon_w) + (1/\varepsilon_i) - 1}$$

7.16 Film boiling of liquid nitrogen at atmospheric pressure occurs over an infinite, upward-facing flat plate held at 120 K. For these circumstances, determine whether the flow is laminar or turbulent, and use the appropriate form of Klimenko's correlation to predict the heat transfer coefficient.

7.17 Use Berenson's correlation (7.164) together with the q''_{min} value predicted by Eq. (7.81) to estimate the minimum wall superheat required to maintain stable film boiling on an upward-facing infinite horizontal surface for pool boiling of liquid nitrogen at atmospheric pressure.

CHAPTER

EIGHT

OTHER ASPECTS OF BOILING AND EVAPORATION IN AN EXTENSIVE AMBIENT

8.1 ADDITIONAL PARAMETRIC EFFECTS ON POOL BOILING

Subcooling

Subcooling of the ambient pool below saturation generally shifts the boiling curve as shown in Fig. 8.1. The natural-convection portion of the boiling curve will shift upward because the driving temperature difference increases with increasing subcooling. As noted in Section 7.2, subcooling has little, if any, effect on nucleate boiling heat transfer. As a result, the nucleate boiling portion of the boiling curve usually changes very little with increasing subcooling. Saturated nucleate boiling correlations may therefore be used with reasonable accuracy, in most cases, even for subcooled conditions.

The maximum heat flux is strongly influenced by subcooling. The vapor leaving the region near the heated surface will tend to condense as it rises through the subcooled pool, making it easier for liquid to flow toward the surface. This suggests that a higher heat flux can be attained before the critical Helmholtz-unstable condition is reached. This trend is, in fact, observed experimentally.

Kutateladze [8.1] apparently was the first to derive an equation for the critical heat flux that accounts for the effect of liquid subcooling. He postulated that the critical heat flux for subcooled conditions should exceed that for saturation by the amount of heat needed to bring the subcooled liquid approaching the surface to

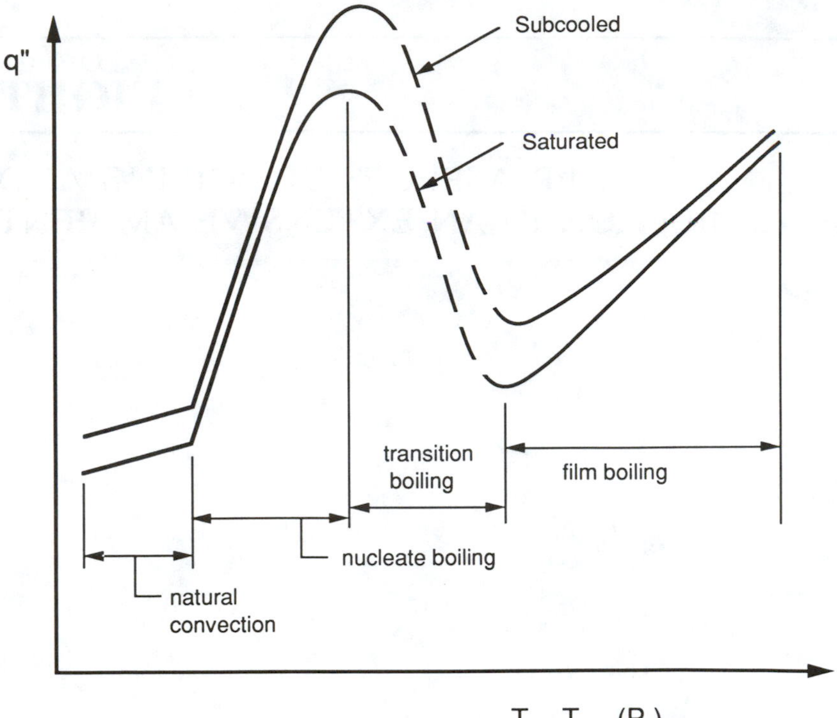

Figure 8.1 The effects of subcooling on the boiling curve.

the saturation temperature. Based on this line of reasoning, he proposed the correlation

$$q''_{\max} = 0.16\rho_v h_{lv}\left[\frac{g\sigma(\rho_l - \rho_v)}{\rho_v^2}\right]^{1/4}\left[1 + C_0\left(\frac{\rho_l}{\rho_v}\right)^m\left(\frac{c_{pl}(T_{\text{sat}} - T_l)}{h_{lv}}\right)\right] \tag{8.1}$$

where T_l is the bulk liquid temperature. Values of the constants C_0 and m equal to 0.065 and 0.8, respectively, were recommended as a best fit to available data.

Somewhat later, Ivey and Morris [8.2] recommended the same correlation, but proposed constant values $C_0 = 0.1$ and $m = 0.75$. Zuber, Tribus, and Westwater [8.3] proposed the relation

$$q''_{\max} = 0.16\rho_v h_{lv}\left[\frac{g\sigma(\rho_l - \rho_v)}{\rho_v^2}\right]^{1/4} \times \left\{1 + \frac{(5.32/\rho_v h_{lv})[g(\rho_l - \rho_v)/\sigma]^{1/4}(k_l c_{pl}\rho_l)^{1/2}(T_{\text{sat}} - T_l)}{[g\sigma(\rho_l - \rho_v)/\rho_v^2]^{1/8}}\right\} \tag{8.2}$$

Transition boiling is also strongly affected by subcooling. In general, the transition boiling curve is expected to shift upward as the subcooling increases, but at the present time there is virtually no quantitative information on its effect.

Film boiling heat transfer is also enhanced by increased subcooling of the liquid pool. However, the enhancement tends to diminish as the heat flux and wall superheat increase, because convection of sensible heat into the subcooled liquid pool becomes progressively smaller relative to the energy convected away in the vapor. Boundary-layer analyses of film boiling in a subcooled liquid pool have been presented by Sparrow and Cess [8.4] and Frederking and Hopenfeld [8.5].

Example 8.1 Compute q''_{max} using Kutateladze's [8.1] correlation and the correlation of Zuber et al. [8.3] for water boiling on an infinite horizontal surface at atmospheric pressure and subcooling levels of 10°C and 50°C.

For saturated water at atmospheric pressure, $\rho_l = 958$ kg/m^3, $\rho_v = 0.597$ kg/m^3, $h_{lv} = 2257$ kJ/kg, $c_{pl} = 4.22$ kJ/kg K, $k_l = 0.679$ W/m K, and $\sigma = 0.0589$ N/m. With the recommended C_0 and m values, Kutateladze's [8.1] correlation is

$$q''_{max} = 0.16\,\rho_v h_{lv}\left[\frac{g\sigma(\rho_l - \rho_v)}{\rho_v^2}\right]^{1/4} \times \left\{1 + 0.065\left(\frac{\rho_l}{\rho_v}\right)^{0.8}\left[\frac{c_{pl}(T_{sat} - T_l)}{h_{lv}}\right]\right\}$$

Substituting for 10°C subcooling,

$$q''_{max} = 0.16(0.597)(2257)\left[\frac{9.8(0.0589)(958 - 0.597)}{(0.597)^2}\right]^{1/4}$$

$$\times \left\{1 + 0.065\left(\frac{958}{0.597}\right)^{0.8}\frac{(4.22)(10)}{2257}\right\} = 1953 \text{ kW/m}^2$$

Repeating the calculation for $T_{sat} - T_l = 50$°C yields

$$q''_{max} = 4353 \text{ kW/m}^2$$

The correlation of Zuber et al. [8.3] can be written as

$$q''_{max} = 0.16\,\rho_v h_{lv}\left[\frac{g\sigma(\rho_l - \rho_v)}{\rho_v^2}\right]^{1/4}(1 + \xi)$$

where

$$\xi = \frac{5.32[g(\rho_l - \rho_v)/\sigma]^{1/4}(k_l c_{pl}\rho_l)^{1/2}(T_{sat} - T_l)}{\rho_v h_{lv}[g\sigma(\rho_l - \rho_v)/\rho_v^2]^{1/8}}$$

Substituting for 10°C subcooling,

$$\xi = \frac{5.32[9.8(957)/0.0589]^{1/4}[(0.679)(4220)(958)]^{1/2}(10)}{(0.597)(2257000)[9.8(0.0589)(957)/(0.597)^2]^{1/8}}$$

$$= 0.522$$

$$q''_{max} = 0.16(0.597)(2257)\left[\frac{9.8(0.0589)(957)}{(0.597)^2}\right]^{1/4} (1 + 0.522)$$

$$= 2059 \text{ kW/m}^2$$

Repeating the calculation for $T_{sat} - T_l = 50°$ yields

$$q''_{max} = 4884 \text{ kW/m}^2$$

It can be seen that the predicted q''_{max} values for these correlations differ only slightly. Increasing the subcooling produces a substantial increase in q''_{max}.

Forced Convection

The introduction of a forced-convection effect generally increases the single-phase heat transfer coefficient above that for natural convection alone, resulting in an upward shift in the single-phase portion of the curve. The associated reduction in the thermal boundary-layer thickness may also suppress the onset of nucleation until a higher wall superheat is attained, as indicated schematically in Fig. 8.2.

In most instances, nucleate boiling is such a strong heat transfer mechanism that introduction of a forced-convection effect has little effect on the nucleate boiling portion of the boiling curve. The exception would be systems with weak nucleate boiling (at low wall superheat levels) and/or very strong convective effects. When both mechanisms are important and the amount of vapor generated is small, Bergles and Rohsenow [8.6] recommend the following empirical relation for predicting the heat flux:

$$q'' = q''_{fc}\sqrt{1 + \frac{q''_{pb}}{q''_{fc}}\left(1 - \frac{q''_i}{q''_{pb}}\right)^2} \tag{8.3}$$

where q''_{fc} is the heat flux for single-phase liquid forced convection alone for the heater at the specified wall temperature, q''_{pb} is the heat flux for pool boiling alone for the surface at the actual wall superheat, and q''_i is the heat flux for pool boiling at the threshold superheat where nucleate boiling just begins. The wall superheat for incipient boiling can be estimated using Hsu's analysis (see Section 6.2) or from empirical correlations. While this is a useful interpolation scheme, it is strictly applicable to subcooled flow or low-quality flow, where the amount of vapor generated is small.

As indicated in Fig. 8.2, the maximum pool boiling heat flux is generally strongly increased by the addition of a forced-convection effect. For parallel forced flow of saturated liquid over a flat plate, Haramura and Katto [8.7] recommend the correlation

$$\frac{q''_{max}}{Gh_{lv}} = 0.175\left(\frac{\rho_v}{\rho_l}\right)^{0.467}\left(\frac{\sigma\rho_l}{G^2L}\right)^{1/3} \tag{8.4}$$

where L is the length of the plate in the flow direction and G is the bulk mass

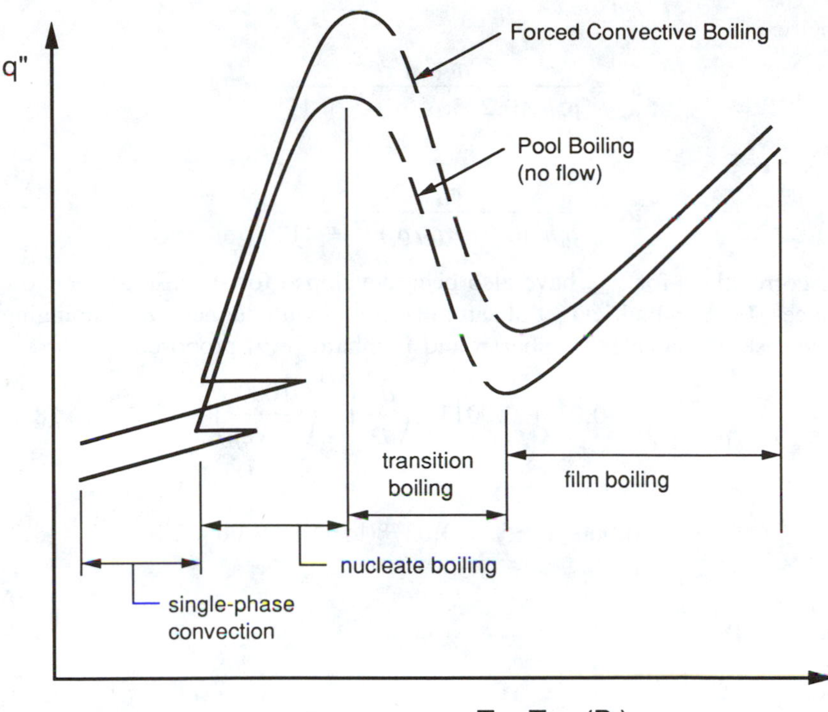

Figure 8.2 The effects of forced convection on the boiling curve.

flux, $G = \rho_l u_l$, u_l being the bulk flow velocity parallel to the plate. For a cylinder of diameter d in a saturated liquid cross flow, these investigators recommended

$$\frac{q''_{max}}{Gh_{lv}} = 0.151 \left(\frac{\rho_v}{\rho_l}\right)^{0.467} \left(\frac{\sigma \rho_l}{G^2 d}\right)^{1/3} \tag{8.5}$$

For cross flow over a cylinder at high velocities, Lienhard and Eichhorn [8.8] recommended the relation

$$\frac{q''_{max}}{\rho_l u_\infty h_{lv}} = \frac{1}{\pi}\left[\frac{1}{169}\left(\frac{\rho_v}{\rho_l}\right)^{1/4} + \frac{1}{19.2}\left(\frac{\rho_v}{\rho_l}\right)^{1/6}\left(\frac{\sigma}{\rho_l u_\infty^2 d}\right)^{1/3}\right] \tag{8.6a}$$

while at low velocities they recommended

$$\frac{q''_{max}}{\rho_l u_\infty h_{lv}} = \frac{1}{\pi}\left[\left(\frac{\rho_v}{\rho_l}\right) + (4)^{1/3}\left(\frac{\rho_v}{\rho_l}\right)^{2/3}\left(\frac{\sigma}{\rho_l u_\infty^2 d}\right)^{1/3}\right] \tag{8.6b}$$

In these relations, u_∞ is the cross-flow velocity and d is the diameter of the cylinder. The high- and low-velocity regimes were specified as

Low velocity:

$$u_\infty \leq \frac{\pi q''_{max}}{\rho_v h_{lv}[0.275(\rho_l/\rho_v)^{1/2} + 1]}$$

High velocity:

$$u_\infty > \frac{\pi q''_{max}}{\rho_v h_{lv}[0.275(\rho_l/\rho_v)^{1/2} + 1]}$$

Other correlations for q''_{max} have also been developed for specific forced-flow circumstances. For a cylindrical jet of saturated liquid with diameter d_{jet} impinging on a heated disk of diameter D, Sharan and Lienhard [8.9] proposed

$$\frac{q''_{max}}{\rho_v h_{lv} u_{jet}} = (0.21 + 0.0017\gamma)\left(\frac{d_{jet}}{D}\right)^{1/3}\left(\frac{1000}{\gamma\,\mathrm{We}_D}\right)^{A} \tag{8.7}$$

where

$$A = 0.486 + 0.06052 \ln \gamma - 0.0378(\ln \gamma)^2 + 0.00362(\ln \gamma)^3 \tag{8.8}$$

$$\gamma = \frac{\rho_l}{\rho_v} \tag{8.9}$$

$$\mathrm{We}_D = \frac{\rho_l u_{jet}^2 D}{\sigma} \tag{8.10}$$

and u_{jet} is the mean velocity of the jet.

For a submerged jet of saturated liquid flowing parallel to a small rectangular heater, Katto and Kurata [8.10] recommended the correlation

$$\frac{q''_{max}}{G_l h_{lv}} = 0.186\left(\frac{\rho_v}{\rho_l}\right)^{0.559}\left(\frac{\sigma\rho_l}{G_l^2 L}\right)^{0.264} \tag{8.11}$$

where L is the length of the heater and $G_l = \rho_l u_{jet}$. Critical heat flux data for jets of water and R-113 with velocities ranging from 1.25 to 10 m/s were correlated with this relation.

Example 8.2 Determine q''_{max} for flow of saturated R-113 at atmospheric pressure over a flat plate 5 mm long (in the flow direction) for free-stream flow rates of 0.5, 5.0, and 50 m/s.

Equation (8.4), recommended by Haramura and Katto [8.7], can be written as

$$q''_{max} = 0.175\rho_l U h_{lv}\left(\frac{\rho_v}{\rho_l}\right)^{0.467}\left(\frac{\sigma}{\rho_l U^2 L}\right)^{1/3}$$

For saturated R-113 at atmospheric pressure, $\rho_l = 1507$ kg/m^3, $\rho_v = 7.46$ kg/m^3, $h_{lv} = 143.8$ kJ/kg, and $\sigma = 0.0169$ N/m. Substituting these properties with $L = 0.005$ m for $U = 0.5$ m/s yields

$$q''_{max} = 0.175(1507)(0.5)(143.8)\left(\frac{7.46}{1507}\right)^{0.467}\left[\frac{0.0169}{1507(0.5)^2(0.005)}\right]^{1/3}$$

$$= 330\ \text{kW/m}^2$$

Repeating the calculation for $U = 5.0$ and 50 m/s yields

$$q''_{max} = 712\ \text{kW/m}^2 \qquad \text{for } U = 5.0\ \text{m/s}$$

$$q''_{max} = 1534\ \text{kW/m}^2 \qquad \text{for } U = 50\ \text{m/s}$$

For a disk heater of diameter D flush mounted in a rectangular-flow channel, Yagov and Puzin [8.11] found that data for R-12 flow velocities ranging from 0.5 to 12.5 m/s are well represented by the correlation

$$\frac{q''_{max}}{G_l h_{lv}} = 0.66\left(\frac{\rho_v}{\rho_l}\right)^{0.604}\left(\frac{\sigma\rho_l}{G_l^2 D}\right)^{0.415} \tag{8.12}$$

For subcooled boiling from an isolated small square heater element on the wall of a channel, Mudawwar and Maddox [8.12] recommend the following correlation for the critical heat flux:

$$q_m^* = 0.161 We^{-8/23} \tag{8.13}$$

where

$$q_m^* = \frac{q''_{max}/\rho_l U_l h_{lv}}{(\rho_v/\rho_l)^{15/23}(L/d_h)^{1/23}(1 + \text{Ja}_s)^{7/23}[1 + 0.021\ \text{Ja}_s(\rho_v/\rho_l)]^{16/23}} \tag{8.14}$$

$$\text{We} = \frac{\sigma}{\rho_l U^2 L} \tag{8.15}$$

$$\text{Ja}_s = \frac{c_{pl}(T_{sat} - T_l)}{h_{lv}} \tag{8.16}$$

In these expressions L is the heater length in the flow direction, d_h is the channel hydraulic diameter, and U is the mean liquid velocity in the channel. Data for boiling of FC-72 liquid coolant were well correlated by this set of relations.

Example 8.3 R-12 flows in a channel at a mean velocity of 1.0 m/s over a flush-mounted heater with a diameter of 5 mm. For saturated liquid, use the correlation of Yagov and Puzin [8.11] to determine the variation of q''_{max} with pressure for pressures between 1 atm and the critical point.

The correlation of Yagov and Puzin [8.11] can be written as

$$q''_{max} = 0.66\rho_l u_m h_{lv}\left(\frac{\rho_v}{\rho_l}\right)^{0.604}\left(\frac{\sigma}{\rho_l u_m^2 D}\right)^{0.415}$$

The properties ρ_l, ρ_v, σ, and h_{lv} must be determined from the saturation table at each pressure. At atmospheric pressure,

$$q''_{max} = 0.66(1486)(1.0)(168.3)\left(\frac{6.33}{1486}\right)^{0.604}\left[\frac{0.0155}{(1486)(1.0)^2(0.005)}\right]^{0.415}$$

$$= 471 \text{ kW/m}^2$$

The properties and results for other pressures are

P_l (kPa)	ρ_l (kg/m^3)	ρ_v (kg/m^3)	h_{lv} (kJ/kg)	σ (N/m)	q''_{max} (kW/m^2)
101	1486	6.33	168.3	0.0155	471
333	1388	19.2	154.7	0.0114	746
793	1284	44.8	137.7	0.0077	943
1602	1157	94.6	114.0	0.0042	956
2907	970	203	75.8	0.0013	621

It can be seen that q''_{max} first increases and then decreases with increasing pressure, peaking at a pressure of about 1600 kPa.

The addition of a forced-convection effect is generally expected to enhance transition boiling heat transfer relative to that for pool boiling under similar conditions. However, at the present time there is very little information regarding the effects of forced convection on transition boiling.

Forced-convection film boiling over immersed bodies will result in a higher heat flux for a given wall superheat level than for ordinary pool boiling under the same conditions. Experimental data for forced-convection film boiling over cylinders have been reported by Bromley, LeRoy, and Robbers [8.13] and Yilmaz and Westwater [8.14]. Bromley et al. [8.13] recommended the following relation for film boiling heat transfer during forced flow of saturated liquid over a cylinder:

$$q'' = C_0\left[\frac{k_v(T_w - T_{sat})\rho_v h'_{lv} u_\infty}{D}\right]^{1/2} \tag{8.17}$$

where

$$h'_{lv} = h_{lv}\left[1 + 0.5\frac{c_{pv}(T_w - T_{sat})}{h_{lv}}\right] \tag{8.18}$$

A value of $C_0 = 2.70$ was found to provide a best fit to experimental data. Witte [8.15] recommended the same relation for saturated forced-convection film boiling over a sphere, but recommended 2.98 as a better value of C_0 for spherical bodies.

In a recent study, Orozco and Witte [8.16] obtained additional heat transfer data for forced-convection film boiling over a sphere, and found good agreement between their data and predictions of an integral boundary-layer analysis. Analytical treatments of forced-convection film boiling have also been developed by Motte and Bromley [8.17], Cess and Sparrow [8.18, 8.19], and Cess [8.20].

Example 8.4 Use the correlation of Bromley et al. [8.13] to predict the heat flux for film boiling over a cylinder with a diameter of 4 mm in a cross flow of saturated liquid nitrogen at atmospheric pressure. Compute the heat flux for a wall superheat of 40°C with cross-flow velocity values of 0.5, 2.0, and 5.0 m/s.

For saturated nitrogen at atmospheric pressure, $\rho_v = 4.62$ kg/m^3, $c_{pv} =$ 1.12 kJ/kg K, $h_{lv} = 197.6$ kJ/kg, and $k_v = 0.00754$ W/m K. It follows that

$$h'_{lv} = h_{lv}\left[1 + \frac{0.5c_{pv}(T_w - T_{\text{sat}})}{h_{lv}}\right]$$

$$= (197.6)\left[1 + \frac{0.5(1.12)(40)}{197.6}\right] = 220.0 \text{ kJ/kg}$$

The correlation (8.17) proposed by Bromley et al. [8.13] is

$$q'' = 2.70\left[\frac{k_v(T_w - T_{\text{sat}})\rho_v h'_{lv} u_\infty}{D}\right]^{1/2}$$

Substituting for $u_\infty = 0.5$ m/s yields

$$q'' = 2.70\left[\frac{(0.00754)(40)(4.62)(220{,}000)(0.5)}{0.004}\right]^{1/2}$$

$$= 16{,}700 \text{ W/m}^2 = 16.7 \text{ kW/m}^2$$

Repeating the calculation for $u_\infty = 2.0$ and 5.0 m/s results in

$$q'' = 33.4 \text{ kW/m}^2 \qquad \text{for } u_\infty = 2.0 \text{ m/s}$$

$$q'' = 52.8 \text{ kW/m}^2 \qquad \text{for } u_\infty = 5.0 \text{ m/s}$$

Size and Wettability of the Surface

The classical boiling curves described in Section 7.1 are characteristic of heater surfaces that satisfy two conditions: (1) They must be at least partially wetted by the liquid in the surrounding pool, and (2) the characteristic dimension of the heater L must be large compared with the capillary length scale $L_b = \sqrt{\sigma/g(\rho_l - \rho_v)}$. If the surface does not satisfy these conditions, the resulting boiling curve can be very different from the classical curves described in Chapter Seven.

If the liquid does not wet the heated surface, vapor and/or air will be trapped in virtually every cavity on the surface when the body is immersed. Because of the abundance of vapor-filled cavities, vaporization is initiated immediately when the surface temperature begins to exceed the saturation temperature. If the liquid does not wet the surface, vapor produced at one location on the surface will displace liquid adjacent to the surface and spread laterally to form a vapor blanket over the surface. Thus, once boiling is initiated, the boiling process immediately enters the film boiling regime. The nucleate boiling regime, critical heat flux condition, and transition boiling regimes are not observed for a nonwetting liquid.

The pool boiling curve for a nonwetting liquid is schematically presented in Fig. 8.3. The surface heat flux increases monotonically with superheat, eventually merging with the "classical" film boiling curve for the particular fluid and surface geometry at high superheat levels. This behavior is consistent with the reduction in $q''_{\min}$ with increasing contact angle described in Section 7.7. In the limit of $\theta = 180°$, reduction of $q''_{\min}$ to zero is equivalent to the entire boiling curve being in the film boiling regime.

Circumstances in which the liquid does not wet the heated surface are not usually encountered in common applications. This type of condition can arise, however, when boiling of water occurs on a heated surface that has adsorbed or has been coated with a material that is hydrophobic. It can also occur if a high-surface-tension fluid such as mercury is vaporized on a low-energy surface such as Teflon. As discussed in Chapter Three, the very high interfacial tension will result in a contact angle that is so high that it is practically nonwetting on Teflon. For the more common circumstances of water or organics boiling on typical metal surfaces in heat transfer equipment, the liquid does wet the surface and the immediate jump to film boiling indicated in Fig. 8.3 does not occur.

The "classical" boiling curves discussed in Section 7.1 also may be significantly altered if the characteristic length scale of the heated surface is small compared to the capillary length scale $L_b = \sqrt{\sigma/g(\rho_l - \rho_v)}$. The ebullition cycle associated with nucleation boiling was considered in Chapter Six under the implicit assumption that the extent of the heated surface was large compared to the departure diameter of the bubble, d_d. The departure diameter correlations discussed in Chapter Six indicate that d_d/L_b is typically of order 1 in most systems.

A heated surface that is small compared to L_b is small compared to the expected departure diameter of bubbles that may grow from active cavities on the

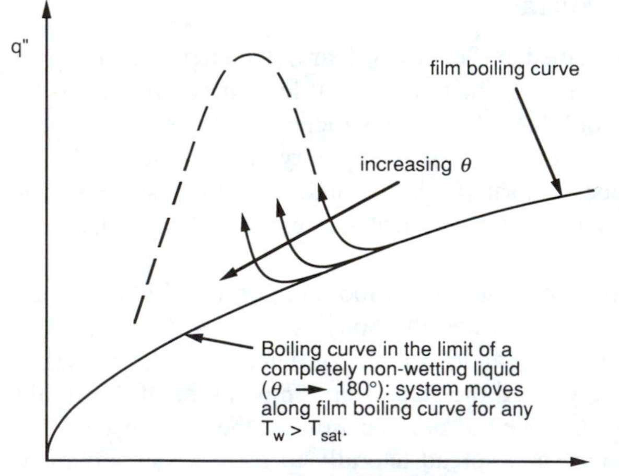

Figure 8.3 Boiling curve in the limit of a completely nonwetting liquid. (Note that this is a linear plot of q'' versus superheat.)

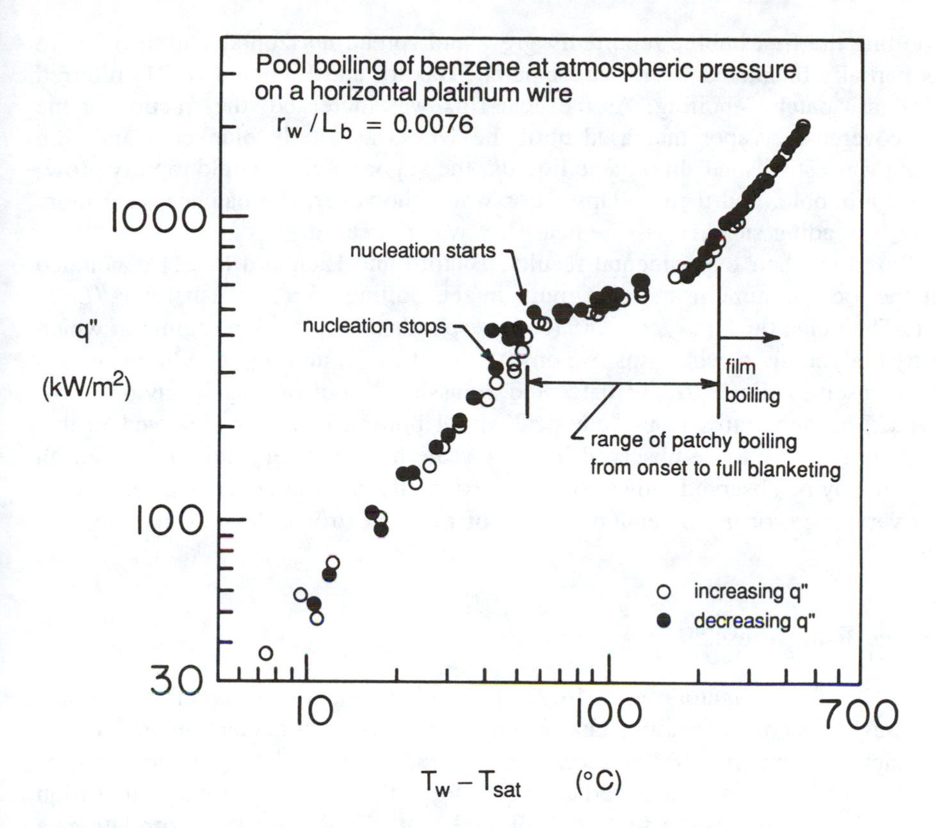

Figure 8.4 Pool boiling data obtained by Bakhru and Lienhard [8.21] for a horizontal wire with a diameter much smaller than L_b. (*Adapted with permission, copyright © 1972, Pergamon Press.*)

surface. For such a surface, a growing bubble may completely cover the surface of the heater, evaporating all liquid in contact with the surface and thereby inducing a transition to film boiling. Thus, for very small heaters, the possibility exists that the onset of boiling may initiate a film-type boiling process. The "classical" nucleate boiling, transition boiling, and the maximum heat flux behavior may be absent for small heaters, or if present, they may be much different in character.

The departure from "classical" boiling behavior discussed above has been observed in experiments conducted by Bakhru and Lienhard [8.21]. These investigators determined the pool boiling curves for very small-diameter horizontal wires in a variety of fluids. An example of their results is shown in Fig. 8.4. The data in this figure were obtained for boiling of benzene at atmospheric pressure on a wire with a radius r_w of 12.7×10^{-6} m. For these conditions the ratio r_w/L_b is 0.0076. The boiling curve indicated by the data in Fig. 8.4 is obviously markedly different from the "classical" pool boiling curves described in Chapter Seven. The regimes of nucleate boiling and transition boiling are absent, and q'' increases monotonically without passing through a maximum and minimum. At the onset of boiling the first bubble reportedly grew and spread horizontally until the wire was partially blanketed with a vapor patch. Bakhru and Lienhard [8.21] referred to this as "patchy" boiling. As the heat flux was increased, the fraction of the wire covered by vapor increased until the wire was totally blanketed and film boiling was established. In organic liquids, the vapor patches would rapidly grow, release a bubble, and then collapse. For water, however, the patches were more stable, spreading steadily, as the heat flux was increased.

Based on their experimental results, Bakhru and Lienhard [8.21] concluded that the local minimum and maximum in the boiling curve vanish for $r_w/L_b \leq 0.01$. The range $0.01 < r_w/L_b < 0.15$ corresponds to a transition regime in which the hydrodynamic mechanisms responsible for the q''_{max} and q''_{min} conditions establish themselves. For $r_w/L_b \geq 0.15$, the "classical" pool boiling behavior is observed. The departures from "classical" pool boiling behavior observed in this study are unlikely to be observed in large-scale heat transfer equipment used on earth. It may be observed, however, in microgravity environments in space, where L_b is very large, or in immersion cooling of microelectronic devices that are very tiny.

Surface Roughness

A number of investigators (e.g., [8.22, 8.23]) have examined the effects of surface roughness on pool boiling data. From the discussion of nucleation phenomena in Chapter Seven, rougher surfaces are generally expected to provide a higher heat flux for a given wall superheat because of the generally higher nucleation site density. It can also be seen in Table 7.1 that, for Rohsenow's correlation, a rougher surface finish generally corresponds to a lower value of C_{sf}, which has the effect of shifting the nucleate pool boiling curve to the left on a q''-versus-$(T_w – T_{\text{sat}})$ plot. This again implies a higher heat flux at a given superheat level for a rough surface relative to a smoother one.

In general, surface finish may also affect transition boiling because it affects the wettability of the surface by the liquid. Although measured data suggest some dependence, the effect of surface roughness on the critical heat flux and minimum heat flux values associated with the boiling curve is generally weak. The effect of surface roughness on film boiling is usually small. It can, however, have an effect on the radiation transport, since it may affect the emissivity of the surface.

Other Factors

The technical literature contains a number of publications that report the results of experimental studies of additional factors that affect pool boiling heat transfer. These include the effects of system pressure [8.24, 8.25], surface aging [8.26], surface coatings or deposits [8.23, 8.27, 8.28], the presence of particulates [8.29, 8.30], dissolved gases [8.31], hysteresis [8.32], agitation [8.33], gravity [8.34, 8.35] and dissolved lubricating oil [8.36–8.39].

8.2 THE LEIDENFROST PHENOMENON

Simply defined, the *Leidenfrost phenomenon* refers to the film boiling of small liquid masses on a hot surface. When a small droplet of liquid is brought into the proximity of a highly superheated surface, the liquid may vaporize so rapidly that the production of vapor on the side of the droplet facing the surface establishes a pressure field that acts to repel the droplet away from the surface. If the droplet is above a solid surface, the repelling force may just balance gravity, allowing the droplet to hover over the surface on a film of vapor, as shown in Fig. 8.5.

Heat is transferred from the hot surface to the droplet by conduction through the vapor film and by radiation. For small liquid masses, interfacial tension acts to pull the mass into a spheroidal shape. Consequently, the term *spheroidal state* is sometimes used to describe liquid in these circumstances.

This phenomenon is named after J. G. Leidenfrost, a German medical doctor who observed the film boiling of water droplets on a red-hot spoon. His report of his observations [8.40] document what is generally acknowledge to be the first experimental investigation of boiling phenomena.

Interest in the Leidenfrost phenomenon stems from the fact that impingement of liquid droplets on hot surfaces can arise in a number of technological applications. These include spray cooling of hot metal during metallurgical processing, the design of quick-response steam generators that spray liquid on a hot surface, film cooling of a rocket nozzle, post-dryout mist flow heat transfer in evaporators, vaporization of fuel droplets in fuel-injected engines, and reflooding of a nuclear reactor core after a loss-of-coolant accident. Of central interest in these applications is whether an impinging droplet will contact the surface, wetting it and vaporizing via nucleate or transition boiling, or whether the droplet will undergo film boiling, hovering over the surface or being repelled by it. If the droplet does hover over the surface, the heat transfer rate from the surface and the time required to vaporize the droplet are of major interest.

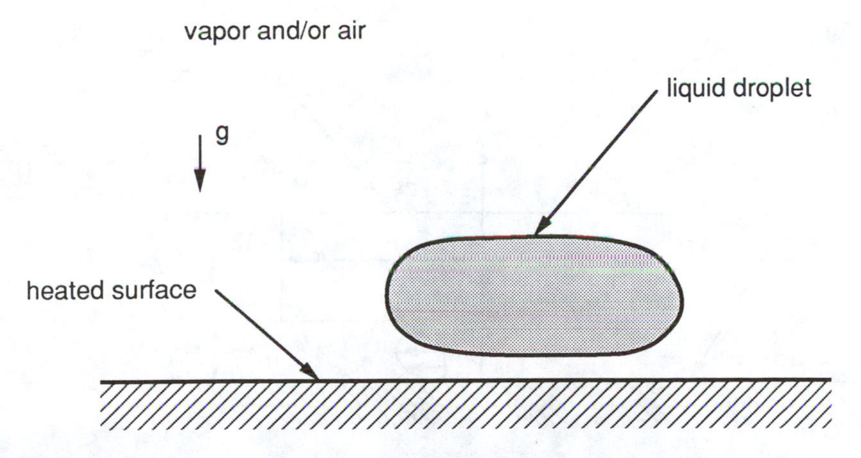

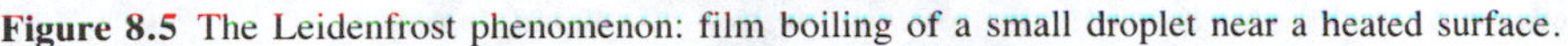

Figure 8.5 The Leidenfrost phenomenon: film boiling of a small droplet near a heated surface.

The mechanisms of the Leidenfrost phenomenon can be understood more clearly by considering the idealized model of the process indicated schematically in Fig. 8.6. For the purposes of determining the heat transfer, the droplet is modeled as a section of a cylinder of radius R and height H. The laminar vapor flow in the thin film between the droplet and the surface is governed by the following continuity and u-momentum equations:

$$\frac{\partial(ur)}{\partial r} + \frac{\partial(vr)}{\partial z} = 0 \tag{8.19}$$

$$\rho_v u \frac{\partial u}{\partial r} + \rho_v v \frac{\partial u}{\partial z} = -\frac{dP}{dr} + \mu_v \left(\frac{\partial^2 u}{\partial z^2}\right) \tag{8.20}$$

Consistent with typical thin-film approximations, the pressure P is taken to be independent of z, and the r derivatives of u are neglected compared to z derivatives. Boundary conditions for these equations are taken to be

$$u(r, 0) = 0 \qquad u(r, \delta) = 0 \qquad v(r, 0) = 0 \tag{8.21}$$

$$T(r, 0) = T_w \qquad T(r, \delta) = T_b \tag{8.22}$$

$$h_{lv} \rho_v v_{z=\delta} = -k_v \left(\frac{\partial T}{\partial z}\right)_{z=\delta} \tag{8.23}$$

$$\left(\frac{\partial P}{\partial r}\right)_{r=0} = 0 \qquad P(R, z) = P_s \tag{8.24}$$

Integrating Eqs. (8.19) and (8.20) from $z = 0$ to $z = \delta$ and using boundary conditions (8.21) yields the following forms of the continuity and u-momentum equations:

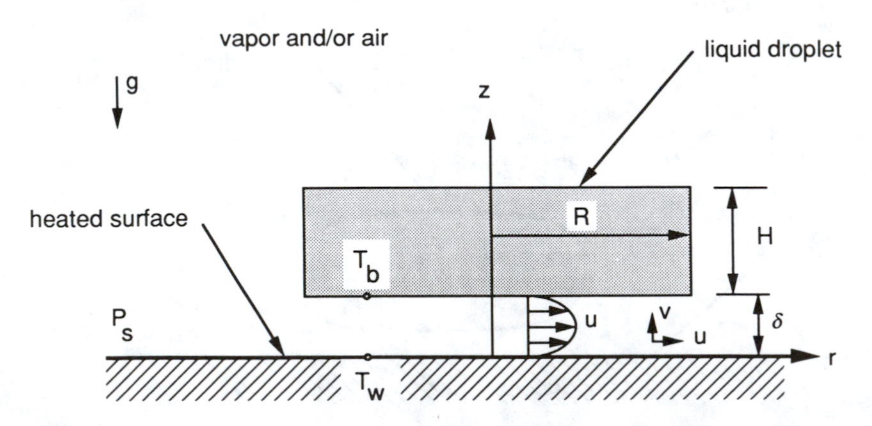

Figure 8.6 Idealized model of Leidenfrost vaporization of a liquid droplet.

$$\frac{\partial}{\partial r}\int_0^{\delta}(ur)\,dz + rv_{z=\delta} = 0 \tag{8.25}$$

$$\int_0^{\delta}\left[\frac{1}{r}\frac{\partial}{\partial r}(\rho_v r u^2)\right]dz = -\delta\left(\frac{\partial P}{\partial r}\right) + \mu_v\left(\frac{\partial u}{\partial z}\right)_{z=\delta} - \mu_v\left(\frac{\partial u}{\partial z}\right)_{z=0} \tag{8.26}$$

Assuming that transport of thermal energy across the vapor film is by conduction alone, the temperature profile is linear for the specified boundary conditions and given by

$$\frac{T - T_b}{T_w - T_b} = 1 - \frac{z}{\delta} \tag{8.27}$$

We will also assume the existence of a parabolic velocity profile,

$$\frac{u}{u_m} = 6\left(\frac{z}{\delta} - \frac{z^2}{\delta^2}\right) \qquad u_m = \frac{1}{\delta}\int_0^{\delta} u\,dz \tag{8.28}$$

Using Eq. (8.27), boundary condition (8.23) can be written as

$$h_{lv}\rho_v v_{z=\delta} = -k_v\left(\frac{\partial T}{\partial z}\right)_{z=\delta} = \frac{k_v(T_w - T_b)}{\delta} \tag{8.29}$$

Combining the continuity relation (8.25) with Eqs. (8.28) and (8.29), and using the fact that $u_m = 0$ at $r = 0$ yields the following relation for u_m:

$$u_m = \frac{\alpha_v \,\mathrm{Ja}\, r}{2\delta^2} \tag{8.30}$$

where

$$\mathrm{Ja} = \frac{c_{pv}(T_w - T_b)}{h_{lv}} \tag{8.31}$$

Substituting the parabolic profile given by Eq. (8.28) into the integral momentum equation (8.26) yields the following differential equation for the radial pressure variation under the droplet:

$$\frac{dP}{dr} = -\left(\frac{6\mu_v\alpha_v\,\mathrm{Ja} + 9\rho_v\alpha_v^2\,\mathrm{Ja}^2/2}{\delta^4}\right)r \tag{8.32}$$

Integrating this equation and applying the boundary conditions (8.24), we obtain the following relation for the pressure variation:

$$P - P_s = \frac{3\rho_v\nu_v^2}{\delta^4}\left(\frac{\mathrm{Ja}}{\mathrm{Pr}_v}\right)\left[1 + \frac{3}{4}\left(\frac{\mathrm{Ja}}{\mathrm{Pr}_v}\right)\right](R^2 - r^2) \tag{8.33}$$

The weight of the droplet must be supported by the pressure field, which requires that

$$(\rho_l - \rho_v) g H \pi R^2 = 2\pi \int_0^R (P - P_s) r \, dr \tag{8.34}$$

Substituting Eq. (8.33) into Eq. (8.34) and evaluating the integral yields the following relation for the film thickness δ:

$$\delta = \left\{ \frac{3\rho_v \nu_v^2 R^2}{2(\rho_l - \rho_v) g H} \left(\frac{\mathrm{Ja}}{\mathrm{Pr}_v} \right) \left[1 + \frac{3}{4} \left(\frac{\mathrm{Ja}}{\mathrm{Pr}_v} \right) \right] \right\}^{1/4} \tag{8.35}$$

For conduction alone, the heat transfer coefficient is given by $h = k_v/\delta$. It follows from this fact and Eq. (8.35) that

$$h = \left(\frac{k_v}{R} \right) \left\{ \frac{2(\rho_l - \rho_v) g H R}{3\rho_v \nu_v^2} \left(\frac{\mathrm{Pr}_v}{\mathrm{Ja}} \right) \left[1 + \frac{3}{4} \left(\frac{\mathrm{Ja}}{\mathrm{Pr}_v} \right) \right]^{-1} \right\}^{1/4} \tag{8.36}$$

which can be written in the form

$$\mathrm{Nu}_R = \left(\frac{hR}{k_v} \right) = \left(\frac{2}{3} \right)^{1/4} \left(\frac{H}{R} \right)^{1/4} \left(\frac{\mathrm{Ra}_R}{\mathrm{Ja}} \right)^{1/4} \left[1 + \frac{3}{4} \left(\frac{\mathrm{Ja}}{\mathrm{Pr}_v} \right) \right]^{-1/4} \tag{8.37}$$

where

$$\mathrm{Ra}_R = \frac{(\rho_l - \rho_v) g R^3 \mathrm{Pr}_v}{\rho_v \nu_v^2} \tag{8.38}$$

In a number of experimental studies of the Leidenfrost phenomenon, the time for complete vaporization of the liquid droplets as it levitates above a heated surface has been measured. A prediction of the time for complete vaporization of such a droplet can easily be developed from the analysis described above. Assuming that all heat transferred to the droplet goes into the latent heat of vaporization, an energy balance requires that

$$\rho_l h_{lv} \frac{dV}{dt} = h A_b (T_w - T_b) \tag{8.39}$$

where V and A_b are the volume and base area of the droplet, respectively. Using the fact that, for a cylindrical droplet,

$$V = \pi R^2 H \quad \text{and} \quad A_b = \pi R^2 \tag{8.40}$$

and using Eq. (8.36) to evaluate h, Eq. (8.39) can be written as

$$\frac{dV}{dt} = -C_1 V^{7/12} \tag{8.41}$$

where

$$C_1 = -\frac{C_0 k_v \pi^{2/3}}{\rho_l h_{lv}} \left(\frac{R}{H} \right)^{1/3} (T_w - T_b) \tag{8.42}$$

$$C_0 = \left\{ \frac{2(\rho_l - \rho_v)g}{3\pi\rho_v\nu_v^2} \left(\frac{\text{Pr}_v}{\text{Ja}} \right) \left[1 + \frac{3}{4} \left(\frac{\text{Ja}}{\text{Pr}_v} \right) \right]^{-1} \right\}^{1/4} \tag{8.43}$$

Integrating Eq. (8.41) from an initial droplet volume V_0 at $t = 0$ to $V = 0$ at $t = t_v$ yields

$$t_v = \frac{12V_0^{5/12}}{5C_1} \tag{8.44}$$

For small droplets, the droplet may be closer to spherical in shape than cylindrical, making the meaning of R/H for such conditions unclear. However, for the cylinder shown in Fig. 8.6, H represents the ratio of the cylinder volume to its base area. If we adopt this same interpretation of H for the spherical droplet case, it follows simply from the spherical geometry that

$$\frac{R}{H} = \frac{3}{4} \tag{8.45}$$

Taking $R/H = 3/4$ for small droplets and substituting Eqs. (8.42) and (8.43), Eq. (8.44) can be written as

$$t^* = 1.81(V_0^*)^{5/12} \tag{8.46}$$

where

$$t^* = t_v \left[\frac{\rho_l^{1/2} \mu_v h_{lv}^3 \sigma^{5/2}}{k_v^3 g^{7/2} \rho_v (T_w - T_b)^3} \right]^{-1/4} \tag{8.47}$$

$$V_0^* = V_0 \left(\frac{\sigma}{\rho_l g} \right)^{-3/2} \tag{8.48}$$

This result differs only slightly from that obtained from a more detailed model analysis developed by Baumeister et al. [8.41]. Their relation is

$$t_B^* = 1.21(V_0^*)^{5/12} \tag{8.49}$$

where V_0^* is given by Eq. (8.48), t_B^* is given by

$$t_B^* = \frac{t_v}{f} \left[\frac{\rho_l^{1/2} \mu_v h_{lv}^4 \sigma^{5/2}}{k_v^3 g^{7/2} \rho_v h_{lv}^* (T_w - T_b)^3} \right]^{-1/4} \tag{8.50}$$

$$h_{lv}^* = h_{lv} + \left(\frac{7}{20} \right) c_{pv} (T_w - T_b) \tag{8.51}$$

and f is a correction factor for radiation effects given by

$$f = \left\{ 1 + \frac{h_r/4}{\bar{h}_c [1 + (7/20) c_{pv} (T_w - T_b)/h_{lv}]} \right\}^{-3} \tag{8.52}$$

$$h_r = \epsilon_l \sigma_{SB}(T_w^2 + T_{sat}^2)(T_w + T_{sat}) \tag{8.53}$$

In the above expressions, ϵ_l is the emissivity of the liquid at the droplet surface, and $\bar{h}_c$ is the mean heat transfer coefficient for convective effects alone, given by

$$\bar{h}_c = 1.1\left[\frac{k_v^3 h_{lv}^* g \rho_v \rho_l}{(V_0 12)^{1/3}(T_w - T_b)}\right]^{1/4} \tag{8.54}$$

Baumeister et al. [8.41] found fairly good agreement between measured droplet evaporation times and times predicted using Eqs. (8.49) through (8.54) for the small-droplet regime. These investigators also found that data for larger liquid masses for a wide variety of fluids also correlated well in terms of t_B^* and V^*, although different correlations in terms of these parameters were proposed as best fits to data in the large-drop and extended-drop regimes.

In the limit of the droplet radius approaching infinity, the Leidenfrost phenomenon becomes equivalent to stable film boiling. Because of this link, for either large or small liquid masses, the minimum temperature that supports stable film boiling is often referred to as the *Leidenfrost temperature*. As noted in Chapter Seven, the minimum heat flux for pool boiling is often assumed to correspond to the minimum heat flux q''_{min} that will sustain the Taylor wave action that facilitates the escape of vapor. If a correlation for the film boiling heat transfer coefficient h_{fb} in terms of the wall superheat is available, the Leidenfrost temperature can be estimated by solving the following relation for the wall temperature:

$$q''_{min} = h_{fb}(T_{w,L} - T_{sat}) \tag{8.55}$$

Using such arguments, Berenson [8.42] obtained the following relation for the minimum film boiling temperature $T_{w,m}$:

$$T_{w,m} - T_{sat} = 0.127\frac{\rho_v h_{lv}}{k_v}\left[\frac{g(\rho_l - \rho_v)}{(\rho_l + \rho_v)}\right]^{2/3}\left[\frac{\sigma}{g(\rho_l - \rho_v)}\right]^{1/2}\left[\frac{\mu_v}{g(\rho_l - \rho_v)}\right]^{1/3} \tag{8.56}$$

As noted by Yao and Henry [8.43], the minimum wall superheat for film boiling is actually not necessarily dictated by the need to sustain the Taylor wave action. It is possible for the limiting condition to be dictated by the homogeneous nucleation of liquid brought into proximity of the surface by the collapse of the film under a detached bubble. The mechanism that is stable at the lowest wall temperature will be the one that dictates the Leidenfrost temperature. Cryogenic liquids generally require a relatively small level of superheat above the normal boiling point to initiate homogeneous nucleation. As a result, the Leidenfrost point for such fluids is generally dictated by the homogeneous nucleation condition. For most other liquids, such as water, hydrocarbons, and liquid metals, the stability of the Taylor wave action is the limiting mechanism.

It should be noted, however, that the initial collapse of the Taylor wave action may not correspond exactly to the q''_{min} condition. As explained in Section 7.6, as the heat flux is decreased in the film boiling regime, some partial liquid–surface contact may initially result in only a small departure from the film boiling conditions. Consequently, the q''_{min} condition may depend somewhat on the heated surface characteristics.

As a result of the above considerations, there is some inaccuracy in assuming that the minimum film boiling (Leidenfrost) temperature is dictated simply by the value of q'' associated with the collapse of the Taylor wave action and the heat transfer coefficient variation for film boiling. Henry [8.44] subsequently developed a model of the Leidenfrost phenomenon that included the effects of transient wetting and subsequent liquid microlayer evaporation on the Leidenfrost temperature. Based on arguments regarding the role of this mechanism, he proposed the following correlation for the Leidenfrost temperature $T_{w,m}$:

$$\frac{T_{w,m} - T_{wmB}}{T_{wmB} - T_l} = 0.42\left[\sqrt{\frac{k_l \rho_l c_{pl}}{k_w \rho_w c_{pw}}}\left(\frac{h_{lv}}{c_{pw}(T_{wmB} - T_{sat})}\right)\right]^{0.6} \tag{8.57}$$

where T_{wmB} is the value of T_{wm} given by Berenson's [8.42] correlation, Eq. (8.56). Equation (8.57) was found to provide good agreement with available film boiling data over a wide range of conditions. It was not recommended for small liquid drops, however, because such droplets tend to move about the surface, with the result that the effect of local surface cooling is much different. Discussion of other features of the Leidenfrost phenomenon can be found in references [8.45–8.51].

Example 8.5 Use the relations proposed by Berensen [8.42] and Henry [8.44] to predict the Leidenfrost temperature for saturated water on a copper surface at atmospheric pressure. Compare these predictions to the kinetic limit of superheat for water at this pressure.

For saturated water at atmospheric pressure, $T_{sat} = 100°C$, $\rho_l = 958$ kg/m^3, $\rho_v = 0.597$ kg/m^3, $c_{pl} = 4.22$ kJ/kg K, $h_{lv} = 2257$ kJ/kg, $k_v = 0.025$ W/m K, $k_l = 0.679$ W/m K, $\mu_v = 1.255 \times 10^{-5}$ Ns/m, and $\sigma = 0.0589$ W/m. Substituting in the equation (8.56) proposed by Berensen [8.42], we obtain

$$T_{wmB} = T_{sat} + 0.127\frac{\rho_v h_{lv}}{k_v}\left[\frac{g(\rho_l - \rho_v)}{\rho_l + \rho_v}\right]^{2/3}\left[\frac{\sigma}{g(\rho_l - \rho_v)}\right]^{1/2}\left[\frac{\mu_v}{g(\rho_l - \rho_v)}\right]^{1/3}$$

$$= 100 + 0.127\frac{(0.597)(2257000)}{0.025}\left[\frac{9.8(957)}{959}\right]^{2/3}\left[\frac{0.0589}{9.8(957)}\right]^{1/2}$$

$$\times\left[\frac{1.255 \times 10^{-5}}{9.8(957)}\right]^{1/3}$$

$$= 186°C$$

Henry's correlation (8.57) specifies that

$$T_{wm} = T_{wmB} + 0.42\left\{\sqrt{\frac{k_l \rho_l c_{pl}}{k_w \rho_w c_{pw}}}\left[\frac{h_{lv}}{c_{pw}(T_{wmB} - T_{sat})}\right]\right\}^{0.6}(T_{wmB} - T_l)$$

For saturated liquid water, $T_l = T_{sat}$; and for copper, k_w = 390 W/m K, ρ_w = 8954 kg/m^3, and c_{pw} = 384 J/kg K. Substituting yields

$$T_{wm} = 186 + 0.42\left[\sqrt{\frac{0.679(958)(4{,}220)}{390(8{,}954)(384)}}\left(\frac{2{,}257{,}000}{384(186-100)}\right)\right]^{0.6}(186-100)$$

$$= 257°\text{C}$$

As described in Section 5.4, the kinetic limit of superheat has been experimentally observed to be between 250°C and 280°C, which is about the same as the value of T_{wm} predicted above.

8.3 EVAPORATION OF THIN LIQUID FILMS

Evaporation of a thin liquid film is an important element in a number of technological processes, and in nature. Perhaps the most conspicuous example of such a process is the evaporation of the thin layer of perspiration on the skin, which is a key element in the mechanism that regulates human body temperature.

Convective Vaporization of a Thin Liquid Film

Considerable insight into vaporization processes of this type can be obtained by considering the specific example indicated in Fig. 8.7. A very thin film of liquid exists on an adiabatic, solid, flat surface over which a laminar boundary-layer flow of gas occurs. In the far ambient, the concentration of the evaporating species is C_∞. The concentration of the evaporating species is higher at the surface of the film where the liquid evaporates, and the vapor is transported into the ambient by diffusion and convection. At steady state, the heat of vaporization is supplied by the air. The interface temperature of the liquid is less than the ambient and equal to the saturation temperature of the evaporating species at the ambient pressure and concentration existing at the interface.

Although this analysis could apply to any species evaporating into any gas or gas mixture, to make the example more concrete, we will specifically consider

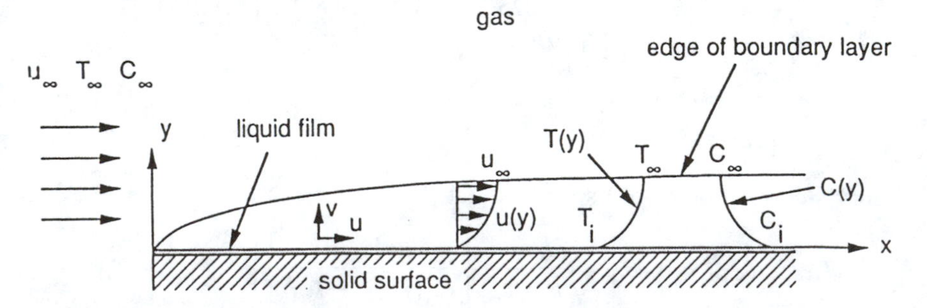

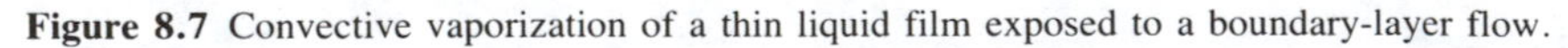
Figure 8.7 Convective vaporization of a thin liquid film exposed to a boundary-layer flow.

water evaporating into moist air. The concentration C is therefore interpreted as being the ratio of the mass of water vapor to the total mass of a given sample of air. For the laminar forced-convection boundary-layer flow shown in Fig. 8.7, the governing transport equations and boundary conditions are

$$\frac{\partial u}{\partial x} + \frac{\partial v}{\partial y} = 0 \tag{8.58}$$

$$u\frac{\partial u}{\partial x} + v\frac{\partial u}{\partial y} = \nu\left(\frac{\partial^2 u}{\partial y^2}\right) \tag{8.59}$$

$$u\frac{\partial T}{\partial x} + v\frac{\partial T}{\partial y} = \alpha_T\left(\frac{\partial^2 T}{\partial y^2}\right) \tag{8.60}$$

$$u\frac{\partial C}{\partial x} + v\frac{\partial C}{\partial y} = D\left(\frac{\partial^2 C}{\partial y^2}\right) \tag{8.61}$$

At $y = 0$:

$$u = 0 \qquad v = \frac{m_e''}{\rho} \qquad C = C_i \qquad T_i = T_{\text{sat}}(C_i, P_\infty) \tag{8.62}$$

At $y \to \infty$:

$$u \to u_\infty \qquad C \to C_\infty \qquad T \to T_\infty \tag{8.63}$$

In the above expressions, m_e'' is the mass flux from the liquid surface, ν is the kinematic viscosity, α_T is the thermal diffusivity, and D is the binary mass diffusion coefficient for the vaporizing species. This formulation implicitly assumes that the thickness of the film is negligible compared to the thickness of the boundary layer and that the viscosity of the liquid is much larger than that of the gas, so that any induced u-velocity in the liquid film will be negligible compared to u-velocities in the gas.

The above system of equations and boundary conditions can be cast into a similarity formulation by defining the following stream function ψ and similarity variables:

$$u = \frac{\partial \psi}{\partial y} \qquad v = -\frac{\partial \psi}{\partial x} \qquad \psi = \sqrt{\nu u_\infty x}\, F \tag{8.64}$$

$$\eta = y\sqrt{\frac{u_\infty}{\nu x}} \qquad \theta = \frac{T - T_i}{T_\infty - T_i} \qquad \phi = \frac{C - C_i}{C_\infty - C_i} \tag{8.65}$$

Note that in terms of these variables, u and v are given by

$$u = u_\infty F(\eta) \qquad v = -\frac{1}{2}\sqrt{\frac{\nu u_\infty}{x}}\,(F - \eta F') \tag{8.66}$$

Assuming that heat transferred to the interface goes completely into the latent heat of vaporization, an energy balance at the interface requires

$$m_e''h_{lv} = -(q'')_{y=0} = k\left(\frac{\partial T}{\partial y}\right)_{y=0} \tag{8.67}$$

Using the relation for m_e'' in terms of v given in Eq. (8.62), and expressing v, T, and y in terms of the similarity variables defined above, Eq. (8.67) can be written as

$$F(0) = -2\left(\frac{\text{Ja}}{\text{Pr}}\right)\theta'(0) \tag{8.68}$$

where

$$\text{Ja} = \frac{c_p(T_\infty - T_w)}{h_{lv}} \tag{8.69}$$

In the equations above and throughout this analysis, the prime will denote a derivative with respect to η.

The concentration at the interface is taken to be constant, but its value cannot be specified a priori. In this analysis, as in the real system, the interface concentration will be determined by the transport and thermodynamics of the system. To incorporate this behavior into the analysis, a mass balance at the interface must be considered. As indicated in Fig. 8.8, the mass flux of the evaporating species m_e'' must equal the sum of the rate of diffusion and the rate of convection of this species away from the interface:

$$m_e'' = m_{\text{diff}}'' + m_{\text{conv}}'' \tag{8.70}$$

where

$$m_{\text{diff}}'' = -\rho D\left(\frac{\partial C}{\partial y}\right)_{y=0} = -\rho D(C_\infty - C_i)\phi'(0)\sqrt{\frac{u_\infty}{\nu x}} \tag{8.71}$$

$$m_{\text{conv}}'' = \rho C_i v_{y=0} \tag{8.72}$$

Using Eq. (8.67) to evaluate m_e'' and Eqs. (8.71) and (8.72) for the other mass flux terms, Eq. (8.70) can be written in terms of similarity variables as

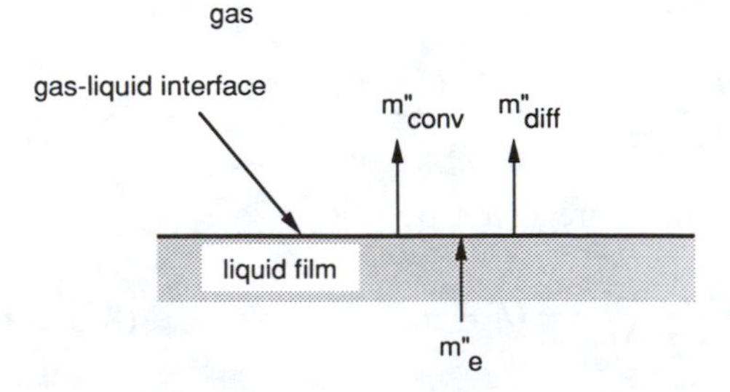

Figure 8.8 Mass flux balance at the gas–liquid interface.

$$\frac{(C_i - C_\infty)}{(1 - C_i)} \phi'(0) = \frac{\text{Sc}}{\text{Pr}} \text{Ja}\, \theta'(0) \tag{8.73}$$

where Sc is the Schmidt number, defined as $\text{Sc} = \nu/D$. Upon converting the remaining boundary conditions to similarity form, the entire mathematical formulation can be written as

$$F''' + \left(\frac{1}{2}\right) FF'' = 0 \tag{8.74}$$

$$\theta'' + \left(\frac{1}{2}\right) \text{Pr}\, F\theta' = 0 \tag{8.75}$$

$$\phi'' + \left(\frac{1}{2}\right) \text{Sc}\, F\phi' = 0 \tag{8.76}$$

$$F'(0) = 0 \qquad \theta(0) = 0 \qquad \phi(0) = 0 \tag{8.77}$$

$$F'(\infty) = 1 \qquad \theta(\infty) = 1 \qquad \phi(\infty) = 1 \tag{8.78}$$

$$\frac{(C_i - C_\infty)}{(1 - C_i)} \phi'(0) = \frac{\text{Sc}}{\text{Pr}} \frac{c_p[T_\infty - T_{\text{sat}}(C_i)]}{h_{lv}} \theta'(0) \tag{8.79}$$

$$F(0) = -\frac{2}{\text{Pr}} \left(\frac{c_p[T_\infty - T_{\text{sat}}(C_i)]}{h_{lv}}\right) \theta'(0) \tag{8.80}$$

Examination of the above system reveals that it is a seventh-order system of nonlinear ordinary differential equations, for which we have eight boundary conditions. Seven of these boundary conditions are sufficient to close the system of equations mathematically. The concentration at the liquid–vapor interface C_i, which is unknown a priori, must be chosen to satisfy the additional eighth relation.

For given values of T_∞, P_∞, C_∞, Sc, and Pr, solution of the above system can be achieved using the following procedure:

1. Guess C_i. From the guessed value of C_i, determine the corresponding $T_i = T_{\text{sat}}(C_i, P_\infty)$ from thermodynamic relations or data.
2. For the guessed value of C_i, guess $F''(0)$, $\theta'(0)$, and $\phi'(0)$. From Eq. (8.80), $F(0)$ can be determined.
3. Integrate Eqs. (8.74) through (8.76) from $\eta = 0$ to $\eta = \eta_{\text{edge}}$, where η_{edge} is large enough to approximate $\eta \to \infty$. The guessed values of $F''(0)$, $\theta'(0)$, and $\phi'(0)$ are iteratively corrected until the conditions $F'(\infty) = 1$, $\theta(\infty) = 1$, $\phi(\infty) = 1$ are satisfied.
4. Check to see if Eq. (8.79) is satisfied. If it is satisfied, the solution is complete. If it is not, the guessed value of C_i is corrected, and the solution process, beginning with step 2, is repeated.

Solution of the above system of equations and boundary conditions thus predicts the heat transfer, evaporation rate, T_i, and C_i for a specific combination of evaporating fluid and gas. It can be seen that even for the relatively simple evaporation process considered here, analytical prediction of the transport is computationally quite demanding. However, one useful aspect of the above analysis is that it indicates the nondimensional parameters on which the transport depends. Specifically, it can be easily shown that the Nusselt number, Nu_x, is given by

$$\text{Nu}_x = \frac{hx}{k} = \theta'(0)\,\text{Re}_x^{1/2} \tag{8.81}$$

where

$$\text{Re}_x = \frac{u_\infty x}{\nu} \quad \text{and} \quad h = \frac{q''_{y=0}}{T_\infty - T_i} \tag{8.82}$$

Because $\theta'(0)$ can be determined by solving the above system of equations, it is clear that it is a function of Pr, Sc, T_∞, C_∞, and the thermodynamic saturation relation for the vaporizing substance. Any empirical correlation technique for predicting the evaporation rate must therefore account for all these parametric effects.

Ultrathin Films

It should be noted that the above treatment of film evaporation implicitly assumes that the liquid film is thick enough that disjoining pressure effects on the thermodynamics are negligible. As described in Chapter Three, for very thin films, attractive forces between the liquid molecules and molecules of the solid surface effectively produces a pressure difference across the liquid–vapor interface referred to as the disjoining pressure difference. As in the case of small bubbles or droplets, a finite pressure difference across the interface will alter the equilibrium conditions from those observed for a flat interface above an extensive pool of liquid.

The equilibrium conditions for very thin films can be examined more concretely by considering the circumstances shown in Fig. 8.9. At a specified vapor ambient pressure P_{ve}, the system is at equilibrium at a uniform temperature T_e. Two necessary conditions for equilibrium are that the temperature and chemical potential in the liquid and vapor phases be equal (see Chapter One). As shown in Chapter Five, integration of the Gibbs-Duhem equation along a line of constant temperature from saturation conditions to an arbitrary point on the isotherm in the vapor and liquid regions yields the following two relations:

$$\mu_v = \mu_{\text{sat},v} + RT_e \ln\left[\frac{P_{ve}}{P_{\text{sat}}(T_e)}\right] \tag{8.83}$$

$$\mu_l = \mu_{\text{sat},l} + v_l[P_l - P_{\text{sat}}(T_e)] \tag{8.84}$$

Setting $\mu_v = \mu_l$ and solving for P_{ve} yields

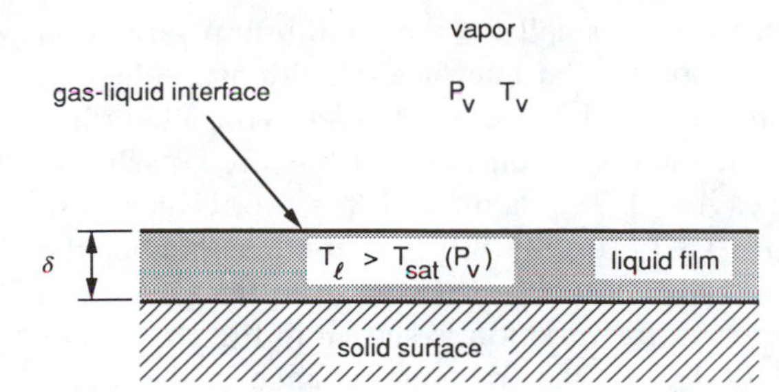

Figure 8.9 An ultrathin liquid film on a solid surface.

$$P_{ve} = P_{sat}(T_e) \exp\left\{\frac{v_l[P_l - P_{sat}(T_e)]}{RT_e}\right\} \tag{8.85}$$

As described in Chapter Three, the disjoining pressure $P_d = -(P_{ve} - P_l)$ has been empirically linked to the film thickness δ by a relation of the form

$$P_d = -A\delta^{-B} \tag{8.86}$$

from which it follows that

$$P_l = P_{ve} - A\delta^{-B} \tag{8.87}$$

Substituting this expression into Eq. (8.85), we obtain:

$$P_{ve} = P_{sat}(T_e) \exp\left\{\frac{v_l[P_{ve} - P_{sat}(T_e) - A\delta^{-B}]}{RT_e}\right\} \tag{8.88}$$

For a specified film thickness, this relation can be used to determine the equilibrium vapor pressure for a given system temperature. Alternatively, for a specified system vapor pressure and film thickness, the equilibrium temperature can be determined. Rearranging this relation, an expression for the equilibrium film thickness for a specified set of vapor pressure and temperature conditions can also be obtained:

$$\delta = \left[\frac{A}{P_{sat}(T_e)}\right]^{1/B} \left\{\frac{P_{ve}}{P_{sat}(T_e)} - 1 - \left[\frac{RT_e}{v_l P_{sat}(T_e)}\right] \ln\left[\frac{P_{ve}}{P_{sat}(T_e)}\right]\right\}^{-1/B} \tag{8.89}$$

At a given vapor pressure, this equation indicates that if the film is thin enough, it can be superheated $[T_e > T_{sat}(P_e)]$ and be in equilibrium with the vapor. This means that heating the solid surface, and hence the liquid film, to a temperature above its normal saturation temperature will not necessarily evaporate the liquid film completely. Instead, an equilibrium thin film may remain on the surface whose thickness satisfies Eq. (8.89).

Equation (8.89) can be interpreted as meaning that a thin film can exist on a solid surface even when its vapor pressure in the surrounding gas is below the normal saturation pressure for the system temperature. This is consistent with the adsorption of thin liquid films onto high-energy surfaces (such as metals) discussed in Chapter Three.

The film thickness must be exceedingly small for the equilibrium saturation conditions to differ significantly from normal flat-interface conditions. Values of A and B in the disjoining pressure relation (8.87) are not available for all surface and fluid combinations. For carbon tetrachloride on glass, however, Potash and Wayner [8.52] recommend the values $A = 1.782$ Pa m^B and $B = 0.6$. Using these values with thermodynamic data for CCl_4, Eq. (8.89) was used to predict the fractional change in the equilibrium saturation pressure with film thickness at a system temperature of 76.7°C. The resulting variation is shown in Fig. 8.10. For a liquid film thickness of 1 mm (1000 μm), the fractional change in the vapor pressure is less than 10^{-4}, or less than 0.01%. A 1% change in the vapor pressure would require a film thickness of about 0.1 μm. For films thicker than 1 μm, the deviation from normal equilibrium conditions is insignificant.

The above considerations also affect vaporization at the location where a pool of liquid comes into contact with the wall of a container. Such a circumstance is shown in Fig. 8.11. If the wall temperature is above the normal saturation temperature, heat may be conducted across the liquid film in this region, resulting in evaporation at the interface. However, because of the curved interface in the intrinsic meniscus region, the wall superheat will have to exceed that required for equilibrium with the interface curvature present, as indicated by Eq. (5.36). In addition, because of the considerations described above, beyond the intrinsic meniscus a thin film can exist on the wall at equilibrium. This thin film can exist without evaporating, provided its thickness is equal to that given by Eq. (8.89).

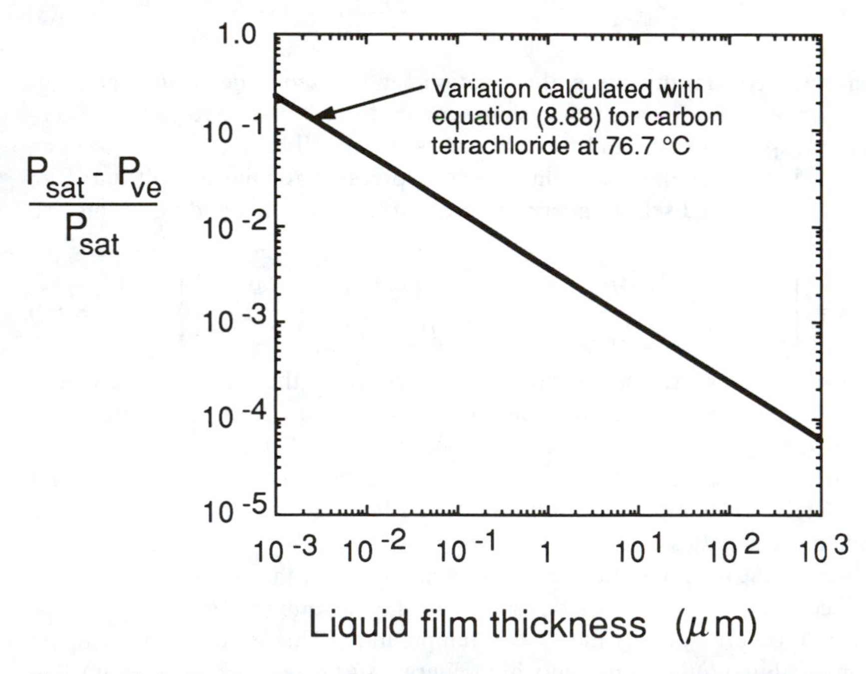

Figure 8.10 Predicted variation of the equilibrium vapor pressure with liquid film thickness due to disjoining pressure effects.

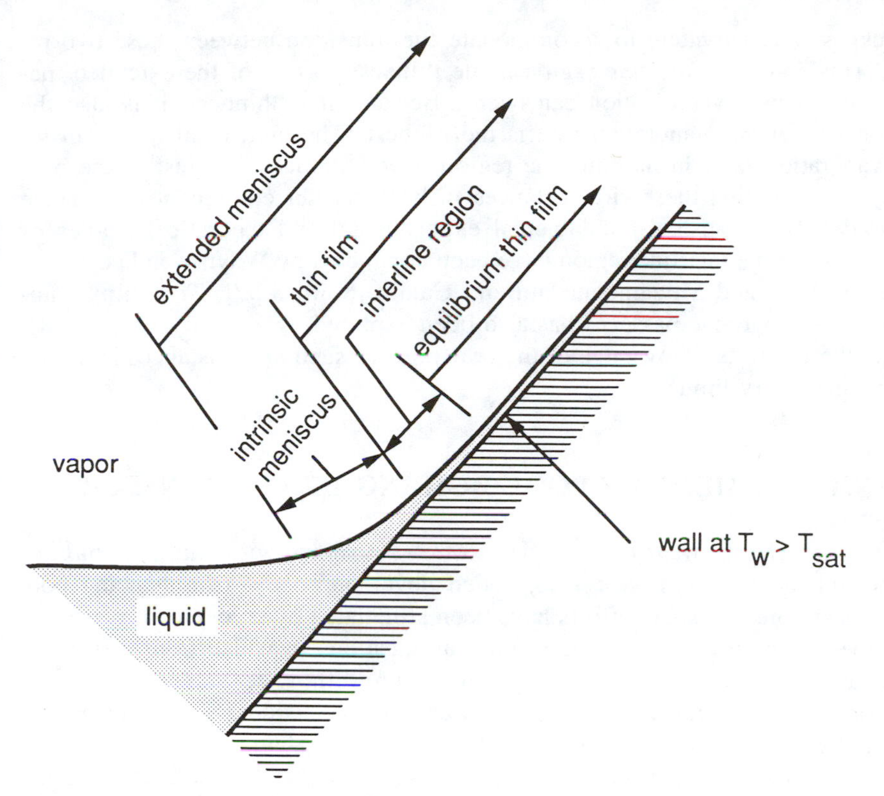

Figure 8.11 The extended meniscus region near a heated wall.

Example 8.6 Determine the thickness of an equilibrium thin film of carbon tetrachloride on glass at atmospheric pressure for a superheat of 10°C.

For these circumstances, Eq. (8.89) can be used to estimate the equilibrium film thickness:

$$\delta = \left[\frac{A}{P_{\text{sat}}(T_e)}\right]^{1/B} \left\{\frac{P_{ve}}{P_{\text{sat}}(T_e)} - 1 - \left[\frac{RT_e}{v_l P_{\text{sat}}(T_e)}\right] \ln\left[\frac{P_{ve}}{P_{\text{sat}}(T_e)}\right]\right\}^{-1/B}$$

For CCl_4 under these conditions, $T_e = 360$ K, $P_{\text{sat}}(T_e) = 142.5$ kPa, $v_l = 6.84 \times 10^{-4}$ m^3/kg, and $R = 0.0541$ kJ/kg K. Taking $A = 1.782$ Pa m^B and $B = 0.6$ as indicated by Potash and Wayner [8.52], substitution in the above equation yields

$$\delta = \left(\frac{1.782}{142{,}500}\right)^{1/0.6} \left\{\frac{101}{142.5} - 1 - \left[\frac{(0.0541)(360)}{(6.84 \times 10^{-3})(142.5)}\right] \ln\left(\frac{101}{142.5}\right)\right\}^{-1/0.6}$$

$$= 8.28 \times 10^{-10} \text{ m}$$

For the circumstances shown in Fig. 8.11, at equilibrium there is a region between the intrinsic meniscus and the equilibrium film over which the film varies in thickness and curvature to accommodate the transition between these two regions. This so-called *interline region* is the thinnest portion of the extended meniscus over which vaporization can occur. Because it is thinnest, it is also the location where the evaporation rate is the highest. The high local heat transfer and evaporation rates in the interline region have stimulated interest in the possibility of promoting these circumstances in heat transfer equipment to enhance evaporation heat transfer. Fundamental experimental and theoretical studies of heat transfer in the interline region have been conducted by Wayner and co-workers [8.53–8.55] and Mirzamoghadam and Catton [8.56, 8.57]. Thin film evaporation in microgroove passages has also been explored by Xu and Carey [8.58]. Despite these efforts, knowledge of the transport for such circumstances is, at the present time, very limited.

8.4 ENHANCEMENT OF POOL BOILING HEAT TRANSFER

Despite the high heat transfer coefficients often associated with normal pool boiling, considerable effort has been expended developing ways of enhancing pool boiling heat transfer. These efforts have been stimulated by cryogenic processing, heat pipes, electronics cooling, and other applications in which transmission of high heat flux levels with low driving temperature differences is desired.

From the discussion of the basic mechanisms in Section 7.2, it is clear that nucleate boiling is most effective when there exists an abundance of cavities on the surface to act as nucleation sites. It is also clear that the cavities are most likely to become active nucleation sites if they effectively trap vapor and gas in them when the system is filled and during periods when the system is inactive. Given the above two conclusions, it is not surprising that two of the most widely studied ways of enhancing nucleate boiling are (1) increasing the surface roughness, and (2) creation of special surfaces having artificially formed cavities designed to efficiently trap vapor.

Evidence of the effect of increasing the surface roughness on nucleate boiling is abundant in the literature. As demonstrated by the data in Fig. 8.12 from the study by Young and Hummel [8.59], the nucleate pool boiling curve generally shifts to the left as the surface roughness increases. The same trend is reflected in the fact that as the surface roughness increases, the C_{sf} constant in Rohsenow's correlation (7.43) decreases.

Many of the more recent attempts to enhance nucleate boiling have considered more sophisticated methods. For nucleate boiling in water, Young and Hummel [8.59] enhanced the heat transfer by adding spots of Teflon either on the heated surface or in pits on the surface. As indicated in Fig. 8.12, this strongly shifted the boiling curve to the left, reflecting a significant increase in the heat transfer coefficient.

The enhancement observed by Young and Hummel [8.59] is associated with the fact that water poorly wets the Teflon spots. Gas or vapor is therefore easily trapped and held in crevices of the Teflon coating. Nucleation can then be initiated

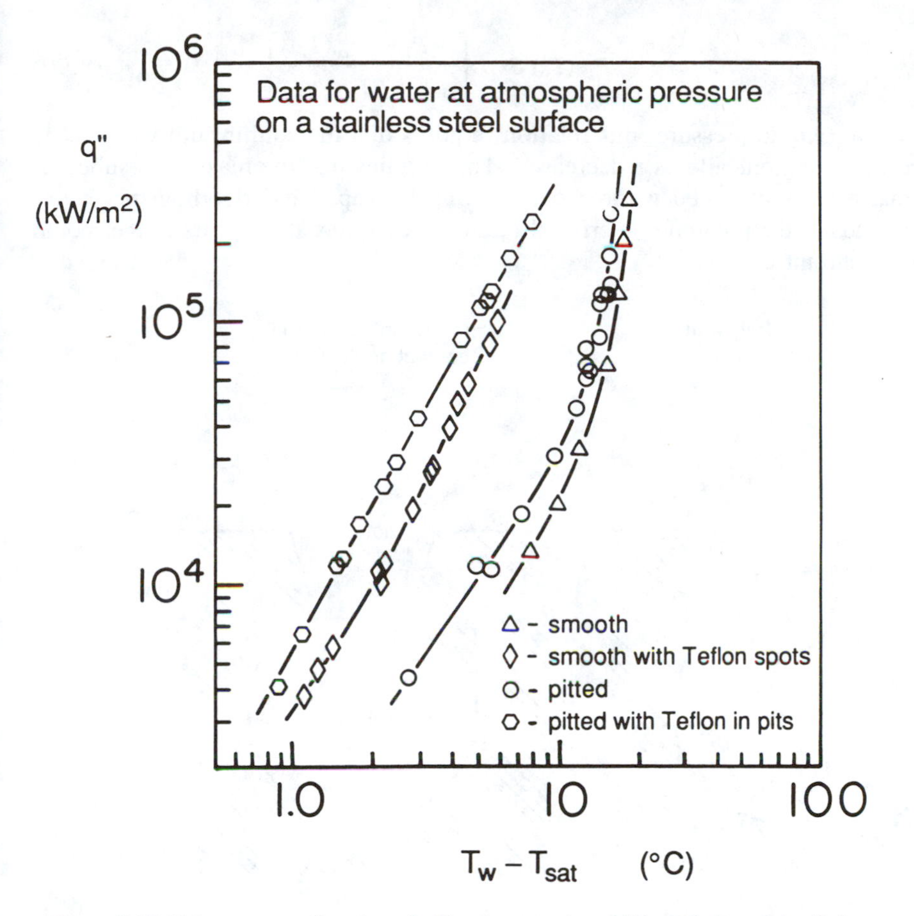

Figure 8.12 Enhancement of nucleate boiling in water by addition of pits and/or Teflon patches to the surface. (*Adapted from Young and Hummel [8.59] with permission, copyright © 1964, American Institute of Chemical Engineers.*)

at very low superheat levels, and since there is an abundance of available sites, the number of active sites increases rapidly as the superheat increases. The overall result is an increase in the effectiveness of the nucleate boiling process. This technique is not effective with Freon-type refrigerants because they wet virtually all substances, including Teflon, to a much greater degree.

Many of the more recent efforts to enhance nucleate boiling have employed surface structures designed to produce reentrant or doubly reentrant cavities. The main objective in designing the cavities in this manner is to produce a convex liquid–vapor interface (as seen from the liquid side) when liquid penetrates into the cavity. This assures that the vapor trapped in the cavity will withstand at least some subcooling of the surrounding solid surface without condensing. Results of the thermodynamic analysis in Chapter Five indicated that the saturation temperature of the vapor in the cavity T_v for a given system (liquid) pressure P_l is related to the interface radius of curvature by the relation

$$P_l - \frac{2\sigma}{r} = P_{\text{sat}}(T_v) \exp\left\{\frac{v_l[P_l - P_{\text{sat}}(T_v)]}{RT_v}\right\} \tag{8.90}$$

For a given liquid pressure, this relation requires that the equilibrium value of T_v decreases monotonically as r decreases. This implies that the maximum subcooling that can be sustained without condensing the vapor and flooding the cavity corresponds to the minimum interfacial radius of curvature that occurs at the mouth of the reentrant cavity.

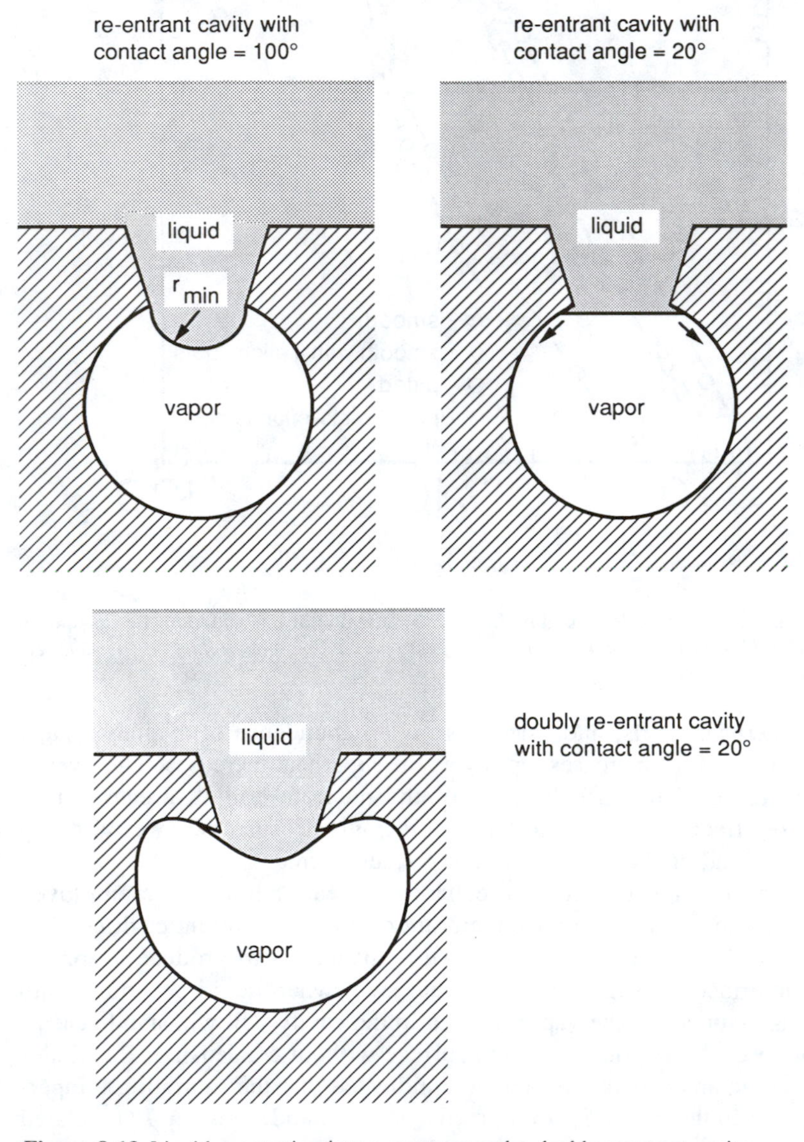

Figure 8.13 Liquid penetration into a reentrant and a doubly reentrant cavity.

As illustrated in Fig. 8.13, an ordinary reentrant cavity could fill with liquid if the contact angle is near zero. Establishing a low contact angle as the contact line goes around the corner at the mouth of the reentrant chamber forms a concave interface and allows a liquid film to flow down along the wall of the chamber, displacing the vapor and flooding the cavity. However, in the doubly reentrant cavity, the low contact angle can be established just beyond the mouth of the chamber while maintaining a convex interface. Surface tension then acts to resist further penetration into the cavity. The convex interface also depresses the effective saturation temperature of the vapor inside the cavity, preventing condensation of vapor in the cavity if the wall is subcooled slightly.

Artificially produced enhanced surfaces for boiling are generally fabricated in one of two ways. One approach is to fuse metal particles or small drops of molten metal to a flat metal surface to create a thin layer that provides an irregular matrix of reentrant cavities. The second method first produces a tube or surface having finlike protrusions, which are subsequently flattened, upset, or bent over to form reentrant cavity structures.

Surfaces created by methods falling into the first category described above have been studied by Marto and Rohsenow [8.25], O'Neil et al. [8.60], Danilova et al. [8.61], Oktay and Schmeckenbecher [8.62], and Fujii et al. [8.63]. Examples of surfaces produced by methods in the second category include those described by Kun and Czikk [8.64], Webb [8.65], Zatell [8.66], and Nakayama et al. [8.67].

Yilmaz et al. [8.68] systematically compared the performance of three enhanced surfaces of these types. The surface constructions and experimentally determined pool boiling curves for the three surfaces considered by these investigators are shown in Fig. 8.14. It can be seen that the boiling curves for all three of the enhanced surfaces are shifted well to the left of that for the plain tube, implying a substantial enhancement.

Superheat reductions of up to a factor of 10 have been observed for enhanced boiling surfaces like those described above, relative to corresponding results for a plain heater surface. Available evidence implies that the mechanism of vaporization for these surfaces is different from that for naturally occurring cavities (which are generally much smaller than the artificial cavities). Apparently, liquid flows partially down into the porous layer or reentrant cavity structure, where thin film evaporation takes place on the large interior surface provided by these structures. The vapor generated in this manner leaves as a bubble, which emerges from an opening in the structure or cavity.

Although there is little information regarding the critical heat flux for the enhanced surfaces described above, the data of O'Neill et al. [8.60] suggest that it is usually as high or higher than that for plain metal surfaces. While surfaces of this type offer clear advantages for nucleate boiling heat transfer applications, their suitability for a given application must be weighed against the added cost of producing the surface. Fouling of the surface, and subsequent degradation of its performance, are also important concerns.

Other methods of enhancing boiling heat transfer are also possible. The ex-

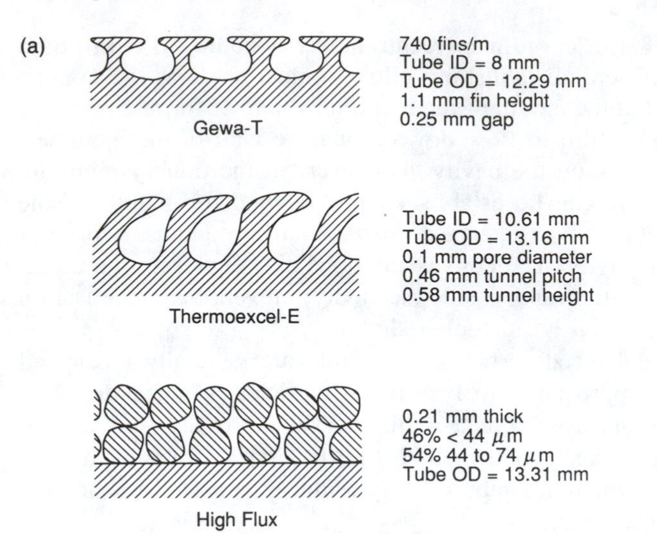

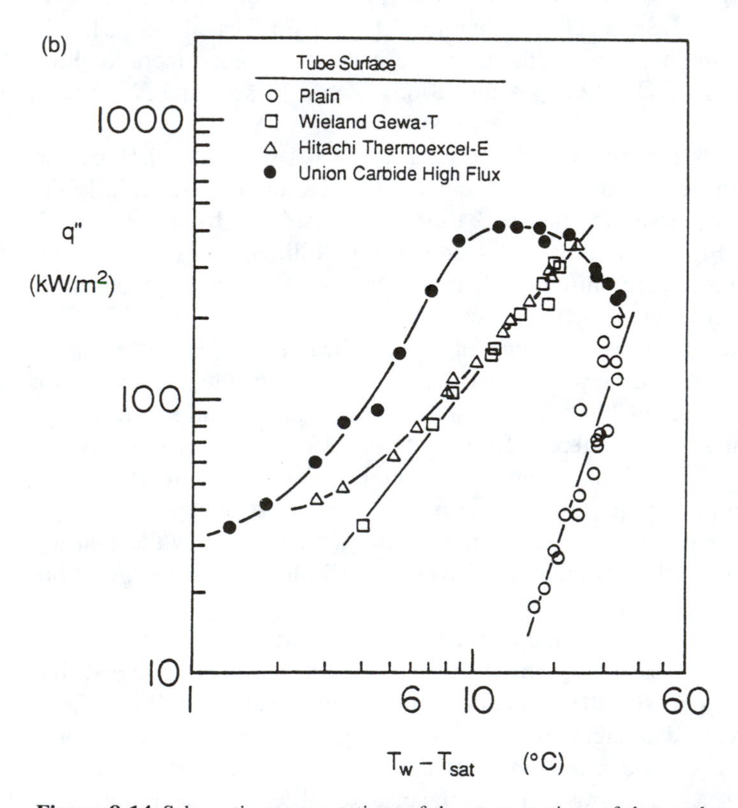

Figure 8.14 Schematic representations of the cross sections of three enhanced surfaces for nucleate boiling (*a*) and comparison of their pool boiling curves with that for a plain tube (*b*). (*Adapted from Yilmaz et al. [8.69] with permission, copyright © 1980, American Society of Mechanical Engineers.*)

periments of Chuah and Carey [8.30] demonstrated that an unconfined layer of metal beads on an upward-facing horizontal surface can enhance nucleate boiling heat transfer. Raiff and Wayner [8.69] have shown that film boiling heat transfer can be enhanced by suction of the vapor through a porous heated wall. Experiments described in the literature further suggest that liquid additives, mechanical aids, surface or fluid vibration, and electrostatic fields can enhance pool boiling heat transfer. For further discussion of these methods, the interested reader is referred to the review articles by Bergles [8.70, 8.71] or the recent book by Thome [8.72].

REFERENCES

8.1 Kutateladze, S. S., Heat transfer during condensation and boiling, translated from a publication of the State Scientific and Technical Publishers of Literature and Machinery, Moscow-Leningrad, as AEC-tr-3770, 1952.

8.2 Ivey, H. J., and Morris, D. J., On the relevance of the vapor–liquid exchange mechanism for subcooled boiling heat transfer at high pressure, British Rep. AEEW-R-137, Atomic Energy Establishment, Winfrith, 1962.

8.3 Zuber, N., Tribus, M., and Westwater, J. W., The hydrodynamic crisis in pool boiling of saturated and subcooled liquid, *Proc. Int. Heat Transfer Mtg.*, Boulder, Colorado, paper no. 27, p. 230, 1961.

8.4 Sparrow, E. M., and Cess, R. D., The effect of subcooled liquid on laminar film boiling, *J. Heat Transfer*, vol. 84, pp. 149–156, 1962.

8.5 Frederking, T. H. K., and Hopenfeld, J., Laminar two-phase boundary layers in natural convection film boiling of subcooled liquid, *Z. Angew. Math. Phys.*, vol. 15, pp. 388–399, 1964.

8.6 Bergles, A. E., and Rohsenow, W. M., The determination of forced-convection surface-boiling heat transfer, *J. Heat Transfer*, vol. 86, pp. 365–372, 1964.

8.7 Haramura, Y., and Katto, Y., A new hydrodynamic model of the critical heat flux, applicable widely to both pool and forced convective boiling on submerged bodies in saturated liquids, *Int. J. Heat Mass Transfer*, vol. 26, pp. 389–399, 1983.

8.8 Lienhard, J. H., and Eichhorn, R., Peak boiling heat flux on cylinders in a cross flow, *Int. J. Heat Mass Transfer*, vol. 19, pp. 1135–1142, 1976.

8.9 Sharan, A., and Lienhard, J. H., On predicting burnout in the jet-disk configuration, *J. Heat Transfer*, vol. 107, pp. 398–401, 1985.

8.10 Katto, Y., and Kurata, C., Critical heat flux of saturated convective boiling on uniformly heated plates, *Int. J. Multiphase Flow*, vol. 6, pp. 575–582, 1980.

8.11 Yagov, V. V., and Puszin, V. A., Critical heat fluxes in forced convection boiling of refrigerant-12 under conditions of local heat sources, *Heat Transfer—Sov. Res.*, vol. 16, pp. 47–55, 1984.

8.12 Mudawwar, I., and Maddox, D. E., Critical heat flux in subcooled flow boiling of fluorocarbon liquid on a simulated electronic chip in a vertical rectangular channel, *Int. J. Heat Mass Transfer*, vol. 32, pp. 379–394, 1989.

8.13 Bromley, L. A., LeRoy, N. R., and Robbers, J. A., Heat transfer in forced convection film boiling, *Ind. Eng. Chem.*, vol. 45, pp. 2639–2646, 1953.

8.14 Yilmaz, S., and Westwater, J. W., Effect of velocity on heat transfer to boiling freon 113, *J. Heat Transfer*, vol. 102, pp. 26–32, 1980.

8.15 Witte, L. C., Film boiling from a sphere, *Ind. Eng. Chem. Fundamentals*, vol. 7, pp. 517–518, 1965.

8.16 Orozco, J. A., and Witte, L. C., Film boiling from a sphere to subcooled freon 11, in *Fun-*

damentals of Phase Change: Boiling and Condensation, C. T. Avedisian and T. M. Rudy, Eds., ASME HTD vol. 38, pp. 35–42, 1984.

8.17 Motte, E. I., and Bromley, L. A., Film boiling of subcooled liquids, *Ind. Eng. Chem.*, vol. 49, pp. 1921–1928, 1957.

8.18 Cess, R. D., and Sparrow, E. M., Film boiling in forced convection boundary layer flow, *J. Heat Transfer,* vol. 83, pp. 370–376, 1961.

8.19 Cess, R. D., and Sparrow, E. M., Subcooled forced-convection film boiling on a flat plates, *J. Heat Transfer,* vol. 83, pp. 377–379, 1961.

8.20 Cess, R. D., Forced-convection film boiling on a flat plates with uniform heat flux, *J. Heat Transfer,* vol. 84, p. 395, 1962.

8.21 Bakhru, N., and Lienhard, J. H., Boiling from small cylinders, *Int. J. Heat Mass Transfer,* vol. 15, pp. 2011–2025, 1972.

8.22 Corty, C., and Foust, A. S., Surface variables in nucleate boiling, *Chem. Eng. Prog. Symp. Ser.,* vol. 17, p. 51, 1955.

8.23 Berenson, P., Transition boiling heat transfer from a horizontal surface, *MIT Heat Transfer Lab.,* Rep. 17, March 1960.

8.24 Chichelli, M. T., and Bonilla, C. F., Heat transfer to liquids boiling under pressure, *Trans. AIChE,* vol. 41, pp. 755–787, 1945.

8.25 Marto, P. J., and Rohsenow, W. M., Effects of surface conditions on nucleate pool boiling of sodium, *J. Heat Transfer,* vol. 88, pp. 196–204, 1966.

8.26 Joudi, K. A., and James, D. D., Surface contamination, rejuvenation and the reproducibility of results in nucleate pool boiling, *J. Heat Transfer,* vol. 103, pp. 453–458, 1981.

8.27 Costello, C. P., and Frea, W. J., A salient non-hydrodynamic effect on pool boiling burnout of small semi-cylindrical heaters, Sixth Nat. Heat Transfer Conf., Boston, AIChE Preprint 15, 1963.

8.28 Farber, E. A., and Scorah, R. L., Heat transfer to water boiling under pressure, *Trans. ASME,* vol. 70, p. 369, 1948.

8.29 Carey, V. P., Markovitz, E., and Chuah, Y. K., The effect of turbulence-suspended light particles on dryout for pool boiling from a horizontal surface, *J. Heat Transfer,* vol. 108, pp. 109–116, 1986.

8.30 Chuah, Y. K., and Carey, V. P., Boiling heat transfer in a shallow fluidized particulate bed, *J. Heat Transfer,* vol. 109, pp. 196–203, 1987.

8.31 McAdams, W. H., Kennel, W. E., Minden, C. S., Rudolf, C., and Dow, J. E., Heat transfer at high rates to water with surface boiling, *Ind. Eng. Chem.,* vol. 41, p. 1945, 1959.

8.32 You, S.M., Simon, T.W., Bar-Cohen, A. and Tong, W., Experimental investigation of nucleate boiling incipience with a highly-wetting dielectric fluid (R-113), *Int. J. Heat Mass Transfer,* vol. 33, pp. 105–117, 1990.

8.33 Pramuk, F. S., and Westwater, J. W., Effect of agitation on the critical temperature difference for boiling liquids, *Chem. Eng. Prog. Symp. Ser.,* vol. 52, no. 18, pp. 79–84, 1956.

8.34 Siegel, R., and Usiskin, C., Photographic study of boiling in absence of gravity, *J. Heat Transfer,* vol. 81, p. 3, 1959.

8.35 Merte, J., Jr., and Clarke, J. A., Study of pool boiling in an accelerating system, Univ. of Michigan Rept. 2646-21-T, Tech. Rept. 3, 1959.

8.36 Chongrungreong, S., and Sauer, H. J., Jr., Nucleate boiling performance of refrigerants and refrigerant–oil mixtures, *J. Heat Transfer,* vol. 102, pp. 701–705, 1980.

8.37 Jackman, D. L., and Jensen, M. K., Nucleate pool boiling of refrigerant–oil mixtures, ASME Paper No. 82-WA/HT-45, 1982.

8.38 Jensen, M. K., and Jackman, D. L., Prediction of nucleate pool boiling heat transfer coefficients of refrigerant–oil mixtures, *J. Heat Transfer,* vol. 106, pp. 184–190, 1984.

8.39 Jensen, M. K., Enhancement of the critical heat flux in pool boiling of refrigerant–oil mixtures, *J. Heat Transfer,* vol. 106, pp. 477–479, 1984.

8.40 Leidenfrost, J. G., *De Aquae Communis Nonnullis Qualitatibus Tractatus,* Duisburg on Rhine, 1756. (The relevant portion has been translated and published in *Int. J. Heat Mass Transfer,* vol. 9, pp. 1153–1166, 1966.)

8.41 Baumeister, K. J., Hamill, T. D., and Schoessow, G. J., A generalized correlation of vaporization times of drops in film boiling on a flat plate, *Proc. Third Int. Heat Transfer Conf.*, vol. 4, pp. 66–73, 1966.

8.42 Berenson, P. J., Film boiling heat transfer from a horizontal surface, *J. Heat Transfer,* vol. 83, pp. 351–362, 1961.

8.43 Yao, S., and Henry, R. E., An investigation of the minimum film boiling temperature on horizontal surfaces, *J. Heat Transfer,* vol. 100, pp. 260–267, 1978.

8.44 Henry, R. E., A correlation for the minimum film boiling temperature, *Chem. Eng. Prog. Symp. Ser.*, vol. 70, no. 138, pp. 81–90, 1974.

8.45 Baumeister, K. J., Hamill, T. D., Schwartz, F. L., and Schoessow, G. J., Film boiling heat transfer to water drops on a flat plate, NASA TM-X-52103, 1965.

8.46 Gottfried, B. S., Lee, C. J., and Bell, K. J., The Leidenfrost phenomenon: Film boiling of liquid droplets on a flat plate, *Int. J. Heat Mass Transfer,* vol. 9, pp. 1167–1187, 1966.

8.47 Gottfried, B. S., and Bell, K. J., Film boiling of spheroidal droplets, *Ind. Eng. Chem. Fundamentals,* vol. 5, pp. 561–568, 1966.

8.48 Patel, B. M., and Bell, K. J., The Leidenfrost phenomenon for extended liquid masses, *AIChE Chem. Eng. Prog. Symp. Ser.*, vol. 62, no. 64, pp. 62–71, 1966.

8.49 Wachters, L. H. J., Bonne, H., and van Nouhius, H. J., The heat transfer from a hot horizontal plate to sessile water drops in the spheroidal state, *Chem. Eng. Sci.*, vol. 21, pp. 923–936, 1966.

8.50 Goleski, E. S., and Bell, K. J., The Leidenfrost phenomenon for binary liquid solutions, *Proc., Third Int. Heat Transfer Conf.*, vol. 4, pp. 51–58, 1966.

8.51 Bell, K. J., The Leidenfrost phenomenon: A survey, *Chem. Eng. Prog. Symp. Ser.*, vol. 63, no. 79, pp. 73–82, 1967.

8.52 Potash, M. L., Jr., and Wayner, P. C., Jr., Evaporation from a two-dimensional extended meniscus, *Int. J. Heat Mass Transfer,* vol. 15, pp. 1851–1863, 1972.

8.53 Wayner, P. C., Jr. The effect of the London-van der Waals dispersion forces on interline heat transfer, *J. Heat Transfer,* vol. 100, pp. 155–159, 1978.

8.54 Wayner, P. C., Jr., Kao, Y. K., and Lacroix, L. V., The interline heat transfer coefficient of an evaporating wetting film, *Int. J. Heat Mass Transfer,* vol. 19, pp. 487–492, 1976.

8.55 Wayner, P. C., Jr., A constant heat flux model of the evaporating interline region, *Int. J. Heat Mass Transfer,* vol. 21, pp. 362–364, 1978.

8.56 Mirzamoghadam, A., and Catton, I. A physical model of the evaporating meniscus, *J. Heat Transfer,* vol. 110, pp. 201–207, 1988.

8.57 Mirzamoghadam, A., and Catton, I., Holographic interferometry investigation of enhanced tube meniscus behavior, *J. Heat Transfer,* vol. 110, pp. 208–213, 1988.

8.58 Xu, X., and Carey, V. P., Film evaporation from a micro-grooved surface—An approximate heat transfer model and its comparison with experimental data, *AIAA J. Thermophys. Heat Transfer,* vol. 4, pp. 512–520, 1991.

8.59 Young, R. K., and Hummel, R. L., Improved nucleate boiling heat transfer, *Chem. Eng. Prog.*, vol. 60, no. 7, pp. 53–58, 1964.

8.60 O'Neill, P. S., Gottzmann, C. F., and Terbog, C. F., Novel heat exchanger increases cascade cycle efficiency for natural gas liquification, *Adv. Cryogen. Eng.*, vol. 17, pp. 421–437, 1972.

8.61 Danilova, G. N., et al., Enhancement of heat transfer during boiling of liquid refrigerants at low heat fluxes, *Heat Transfer—Sov. Res.*, vol. 8, no. 4, pp. 1–8, 1976.

8.62 Oktay, S., and Schmeckenbecher, A. F., Preparation and performance of dendritic heat sinks, *J. Electrochem. Soc.*, vol. 21, pp. 912–918, 1974.

8.63 Fujii, M., Nishiyama, E., and Yamanaka, G., Nucleate pool boiling heat transfer from a micro-porous heating surface, in *Advances in Enhanced Heat Transfer,* ASME, New York, pp. 45–51, 1979.

8.64 Kun, L. C., and Czikk, A. M., Surface for boiling liquids, U.S. Patent 3,454,081, July 8, 1969.

8.65 Webb, R. L., Heat transfer surface having a high boiling heat transfer coefficient, U.S. Patent 3,696,861, October 10, 1972.

8.66 Zatell, V. A., Methods of modifying a finned tube for boiling enhancement, U.S. Patent 3,768,290, October 30, 1973.

8.67 Nakayama, W., et al., High-flux heat transfer surface "Thermoexcel," *Hitachi Rev.*, vol. 24, no. 8, pp. 329–333, 1975.
8.68 Yilmaz, S., Hwalek, J. J., and Westwater, J. W., Pool boiling heat transfer performance for commercially enhanced tube surfaces, ASME Paper No. 80-HT-41, presented at the 19th Nat. Heat Transfer Conf., July 1980.
8.69 Raiff, R. J., and Wayner, P. C., Jr., Evaporation from a porous flow control element on a porous heat source, *Int. J. Heat Mass Transfer,* vol. 16, pp. 1919–1930, 1973.
8.70 Bergles, A. E., Survey and evaluation of techniques to augment convective heat and mass transfer, in *Progress in Heat and Mass Transfer,* U. Grigull and E. Hahne, Eds., Pergamon Press, Oxford, vol. 1, pp. 331–334, 1969.
8.71 Bergles, A. E., Enhancement of boiling and condensation, in *Two-Phase Flow and Heat Transfer, China-U.S. Progress,* X.-J. Chen and T. N. Veziroglu, Eds., Hemisphere, New York, 1985.
8.72 Thome, J. R., *Enhanced Boiling Heat Transfer,* Hemisphere, New York, 1990.

PROBLEMS

8.1 Compute the critical heat flux using Kutateladze's [8.1] correlation and the correlation of Zuber et al. [8.3] for liquid nitrogen boiling on an infinite horizontal surface (facing upward) at atmospheric pressure for ambient pool subcoolings ranging from 5°C to 40°C and plot the variation of q''_{max} with subcooling over this range. Where do these two correlations differ the most?

8.2 Determine the critical heat flux for subcooled boiling of R-12 on a large upward-facing flat surface at pressures of 101 kPa, 793 kPa, and 2901 kPa for a pool subcooling of 20°C. Where does the subcooling have the greatest effect on q''_{max}?

8.3 For bulk flow velocities of 0.1, 1.0, and 10 m/s, determine the critical heat flux for saturated water at atmospheric pressure flowing over a flat plate 3.0 cm long in the direction of flow. Repeat the calculations for the same flow rates at a pressure of 6124 kPa. At which pressure does increasing the flow velocity have the greatest effect?

8.4 Equation (8.4) was proposed by Haramura and Katto [8.7] for predicting the critical heat flux for flow of a saturated liquid over a finite-sized flat plate. This relation predicts that q''_{max} goes to zero as the flow velocity approaches zero. However, for zero velocity, q''_{max} will not equal zero, but instead will equal the pool boiling value, as predicted, for example, by the appropriate equation from Table 7.3. Assume that the lateral width of the plate is essentially infinite and its length L in the flow direction is 2.0 cm. (*a*) Using an appropriate pool boiling equation from Table 7.3, find the flow velocity at which Eq. (8.4) predicts a value of q''_{max} that equals the pool boiling value for water and liquid nitrogen at atmospheric pressure. (*b*) What is your interpretation of the values of q''_{max} predicted by Eq. (8.4) for flow velocities below the values determined in part (*a*)?

8.5 A small heated cylinder with an outside diameter of 5 mm is immersed in saturated liquid nitrogen in a system at a pressure of 1083 kPa. The liquid nitrogen flows normal to the cylinder axis with a bulk velocity of 50 cm/s. Determine the critical heat flux for the cylinder. What happens to the boiling process if the pressure in the system suddenly drops to atmospheric pressure (101 kPa)?

8.6 Determine and plot the variation of the critical heat flux with jet velocity for a cylindrical jet of saturated liquid R-22 impinging on a heated disk with a diameter of 1.0 cm flush-mounted on an otherwise adiabatic surface. Specifically consider jet velocities between 0.1 and 10 m/s. The pressure in the system is 376 kPa, and the jet diameter is 5 mm.

8.7 It is proposed to cool a flush-mounted electronic chip containing high-temperature superconducting elements with an impinging jet of liquid nitrogen at atmospheric pressure. The chip is modeled as a circular element with a diameter of 1.0 cm dissipating heat uniformly over its surface. To operate properly, this chip must be maintained at a temperature near 80 K while dissipating a heat flux of 80 W/cm^2. (*a*) Determine the required jet velocity to meet the heat flux requirement if the jet diameter is 2 mm and the impinging liquid is saturated. (*b*) Using an appropriate nucleate boiling correlation from Chapter 7, estimate the surface temperature of the element for the specific heat flux condition. (*c*) If the wall temperature exceeds the design specification of 80 K, propose a modification to the cooling scheme that will make it possible to meet this condition without exceeding the maximum heat flux.

8.8 Estimate the film boiling heat transfer coefficient for film boiling of saturated liquid nitrogen flowing normal to a cylinder with a velocity of 20 cm/s. The cylinder wall is held at a temperature of 110 K and the system is at atmospheric pressure.

8.9 Determine the film boiling heat transfer coefficient for film boiling of liquid nitrogen on a sphere over which liquid nitrogen flows with a bulk velocity of 40 cm/s. The system pressure is 778 kPa, the sphere diameter is 3 cm, and the surface of the sphere is held at 130 K. What happens if the system pressure suddenly drops to atmospheric pressure?

8.10 Estimate the time to completely vaporize a droplet of saturated liquid nitrogen undergoing Leidenfrost vaporization over a solid surface at a temperature of 20°C. The emissivity of the surface is 0.8, and the emissivity of the liquid interface is 0.95. Does radiation have a significant effect? Justify your answer.

8.11 Droplets of liquid R-12 sprayed from a leaking refrigeration system at atmospheric pressure fall 4 cm to a copper plate at 20°C. Do you expect the droplets to strike and wet the surface? Justify your answer quantitatively.

8.12 As part of a metallurgical process, it is proposed that a 2-cm-thick stainless steel plate be cooled with a spray of water droplets. The plate, which is heated to 1200 K, is oriented vertically with the droplets sprayed horizontally at the plate. How effective do you expect this spray to be in cooling the plate? Does radiation play a significant role in the process? Justify your answers quantitatively where possible.

8.13 For liquid nitrogen at atmospheric pressure, estimate the Leidenfrost temperature for a solid copper surface. Compare your result to the value of the kinetic limit of superheat predicted by Eq. (5.74). Which is larger? Based on these results, what do expect to observe if liquid nitrogen droplets fall into a container with liquid water in it at 20°C? Explain briefly.

8.14 Use the relation proposed by Henry [8.44] to estimate the Leidenfrost temperatures for acetone, methanol, and R-12 at atmospheric pressure on a solid copper surface.

8.15 Develop an integral analysis of forced-convection-dominated laminar film boiling over a vertical surface exposed to a saturated liquid flowing upward over the surface at a uniform free-stream velocity u_∞. Assume that a constant heat flux q'' is dissipated to the flow at the surface and adopt the following idealizations:

Neglect downstream convection of momentum and energy compared to transport across the film.
Neglect radiation effects.
Assume that the interface is smooth.
Assume that the upward velocity u at the liquid–vapor interface is equal to the free-stream velocity u_∞.

From the analysis, derive relations for the temperature and velocity profiles in the film, the variation of the film thickness with downstream location x, and show that the local heat transfer coefficient h is given by

$$h = \frac{\rho_v k_v u_\infty h_{lv}}{2\,x\,q''}$$

8.16 Consider the problem of cooling a circular flush-mounted heated element on a solid surface by boiling of an impinging jet of saturated liquid nitrogen at atmospheric pressure. Use the correlation of Sharan and Lienhard [8.9] to predict the value of the jet velocity that would yield a value of

q''_{max} equal to the maximum possible flux as specified by the curve in Fig. 4.13. What is your opinion of using such a jet velocity in a real system for cooling electronic components?

8.17 Assuming that the A and B values in Eq. (8.89) recommended by Potash and Wayner [8.52] for CCl_4 also apply to acetone, estimate the thickness of an equilibrium film of acetone on glass at atmospheric pressure and a superheat of 10°C.

CHAPTER
NINE

EXTERNAL CONDENSATION

9.1 HETEROGENEOUS NUCLEATION IN VAPORS

In most applications involving condensation, the process is initiated by removing heat through the walls of the structure containing the vapor to be condensed. If enough heat is removed, the vapor near the wall may be cooled below its equilibrium saturation temperature for the specified system pressure. Since the heat removal process will establish a temperature field in which the temperature is lowest right at the wall of the containment, the formation of a liquid droplet embryo is most likely to occur right on the solid–vapor surface. The formation of a liquid embryo at the interface between a metastable supersaturated vapor and another solid phase is one type of heterogeneous nucleation. (See Chapter Five for further description of the differences between heterogeneous and homogeneous nucleation.)

As in the case of heterogeneous nucleation of vapor bubbles, the analysis of homogeneous nucleation of liquid droplets can be extended to heterogeneous nucleation at a solid–vapor interface. Because the analysis of the kinetics of the heterogeneous nucleation process is very similar to that described in Chapter Five for the comparable homogeneous nucleation process, the analysis for heterogeneous nucleation will be only briefly summarized here. If the solid surface is idealized as being perfectly smooth, in general, the shape of a droplet at the surface will be dictated by the shape of the surface itself, the interfacial tension σ, and the contact angle θ. For a flat, solid surface, the embryo liquid droplet will have a profile like that shown in Fig. 9.1.

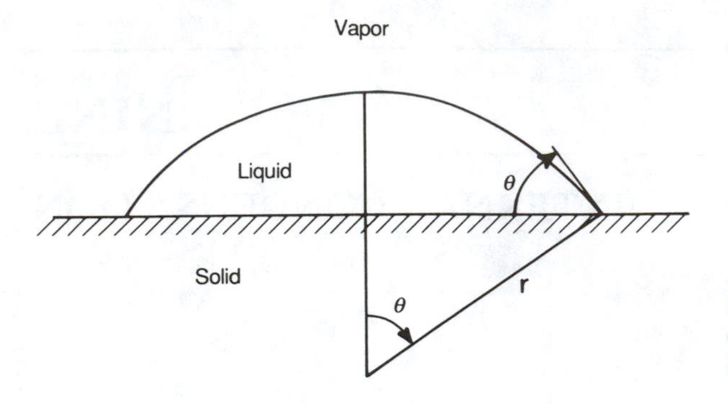

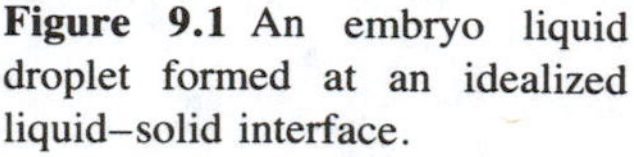

Figure 9.1 An embryo liquid droplet formed at an idealized liquid–solid interface.

We will specifically consider the heterogeneous nucleation process in which formation of a droplet embryo occurs in a system held at constant temperature T_v and pressure P_v, as shown schematically in Fig. 9.2. If the embryo shape is idealized as being a portion of a sphere, it follows directly from its geometry that the embryo volume V_l and the areas of the liquid–vapor (A_{lv}) and the solid–liquid interfaces (A_{sl}) are given by

$$V_l = \left(\frac{\pi r^3}{3}\right)(2 - 3\cos\theta + \cos^3\theta) \tag{9.1}$$

$$A_{lv} = 2\pi r^2(1 - \cos\theta) \tag{9.2}$$

$$A_{sl} = \pi r^2(1 + \cos\theta) \tag{9.3}$$

In the above relations, θ is the liquid contact angle and r is the spherical cap radius indicated in Fig. 9.1.

The system shown in Fig. 9.2 initially contains only supersaturated vapor. The initial availability Ψ_0, referenced to T_v and P_v, is therefore given by

$$\Psi_0 = m_T g_v(T_v, P_v) + (A_{sv})_i \sigma_{sv} \tag{9.4}$$

where, by definition, $g_v(T_v, P_v) = u_v - T_v s_v + P_v v_v$. After the formation of the embryo, the total availability for the system is the sum of contributions associated with the liquid Ψ_l, the vapor Ψ_v, and the interfaces Ψ_i:

$$\Psi = \Psi_l + \Psi_v + \Psi_i \tag{9.5}$$

The availability of each of these system components is given by

$$\Psi_l = m_l[g_l(T_v, P_l) + (P_v - P_l)v_l] \tag{9.6}$$

$$\Psi_v = (m_T - m_l)g_v(T_v, P_v) \tag{9.7}$$

$$\Psi_i = A_{lv}\sigma_{lv} + (A_{sv})_f \sigma_{sv} + A_{sl}\sigma_{sl} \tag{9.8}$$

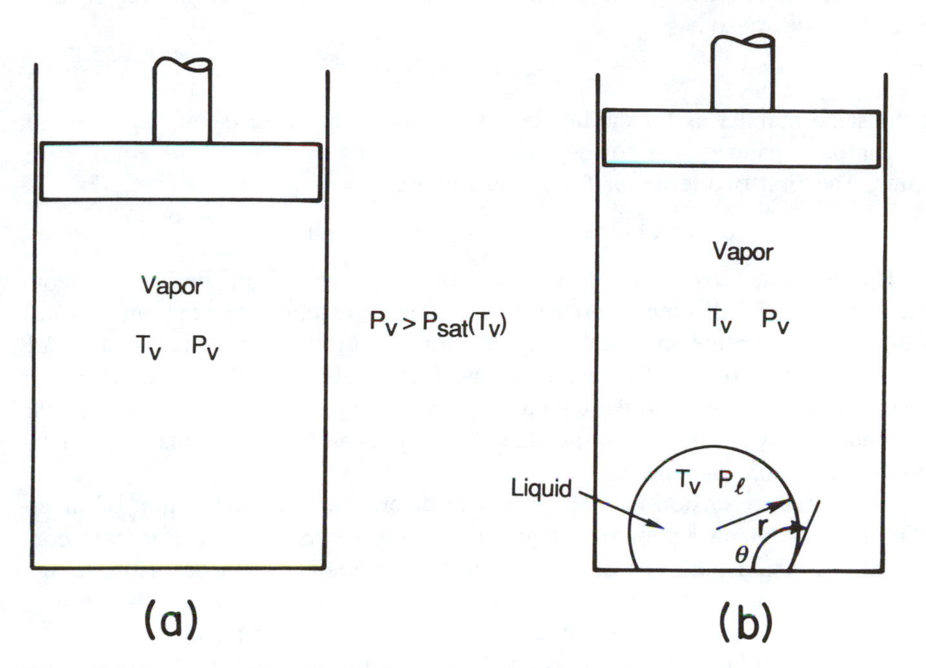

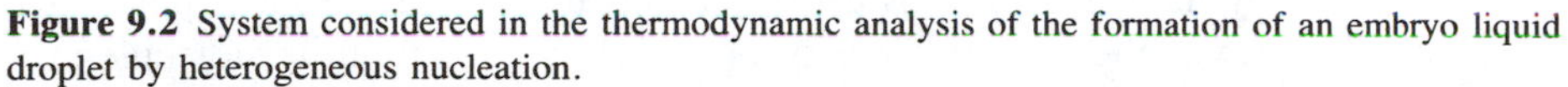

Figure 9.2 System considered in the thermodynamic analysis of the formation of an embryo liquid droplet by heterogeneous nucleation.

Using the fact that

$$(A_{sv})_f = (A_{sv})_i - A_{sl} \tag{9.9}$$

and the requirement of a tangential force balance along the interline (see Chapter Three),

$$\sigma_{lv} \cos \theta = \sigma_{sl} - \sigma_{sv} \tag{9.10}$$

Eqs. (9.1) through (9.8) can be combined to obtain the following expression for the change in the availability function $\Delta\Psi$ associated with formation of the embryo:

$$\Delta\Psi = \Psi - \Psi_0 = m_l[g_l(T_v, P_l) - g_v(T_v, P_v)] + V_l(P_v - P_l) + 4\pi r^2 \sigma_{lv} F \tag{9.11}$$

where

$$F = \frac{2 - 3 \cos \theta + \cos^3 \theta}{4} \tag{9.12}$$

Equation (9.11) is identical to the expression obtained for $\Delta\Psi$ in Chapter Five for homogeneous nucleation except that σ_{lv} has been replaced with $\sigma_{lv}F$. If the embryo has exactly the right radius $r = r_e$ to be in thermodynamic equilibrium with the surrounding vapor, the g_l and g_v terms in Eq. (9.11) are equal and the relation for $\Delta\Psi = \Delta\Psi_e$ becomes

$$\Delta\Psi_e = (4/3)\pi r_e^2 \sigma_{lv} F \tag{9.13}$$

In the same manner as for the homogeneous nucleation case (see Chapter Five), $\Delta\Psi$ can be expanded in a power series in terms of $r - r_e$ about the equilibrium radius. The first two terms of the expansion are

$$\Delta\Psi = (4/3)\pi r_e^2 \sigma_{lv} F - 4\pi\sigma_{lv} F(r - r_e)^2 + \ldots \tag{9.14}$$

It follows directly from the same arguments presented for the homogeneous case in Section 5.5 that the equilibrium condition corresponds to a maximum value of $\Delta\Psi$ and is therefore an unstable equilibrium. As in the homogeneous case, $\Delta\Psi$ is expected to increase to a maximum and then decrease with increasing radius r. This once again leads to the conclusion that embryos having a radius less than r_e spontaneously disappear, while those having a radius greater than r_e spontaneously grow (see Section 5.5).

The above expansion for $\Delta\Psi$ is used to determine the kinetic limit of supersaturation in a manner similar to that for the homogeneous nucleation case considered in Section 5.6. The details are virtually identical to those of the homogeneous nucleation analysis presented in Section 5.6, and hence they will not be presented here. There are, however, two important differences in the heterogeneous nucleation analysis. First, as an initial step in the analysis, it is postulated that, at equilibrium, the number density of embryos containing n molecules per unit of interface area N_n is given by

$$N_n = N_v^{2/3} \exp\left[\frac{-\Delta\Psi(r)}{k_B T_v}\right] \tag{9.15}$$

where N_v is the number density of vapor molecules per unit volume and $\Delta\Psi(r)$ is the availability function previously defined. For the heterogeneous nucleation process considered here, only vapor molecules near the solid surface can participate in embryo droplet formation. To account for this condition, the factor multiplying the exponential term in Eq. (9.15) is taken to be $N_v^{2/3}$, which is representative of the number of vapor molecules immediately adjacent to the solid surface per unit of surface area.

The second different aspect of the heterogeneous analysis is the relationship between the number of molecules n in the embryo and its radius:

$$n = \frac{N_A \pi r^3}{3\overline{M} v_l}(2 - 3\cos\theta + \cos^3\theta) \tag{9.16}$$

This relation differs from that used in the analysis of homogeneous nucleation because the embryo geometries are different.

Analysis of the kinetics of the heterogeneous nucleation process incorporates these two changes and makes use of the expansion for $\Delta\Psi$ developed for this case.

Otherwise the analysis is identical to that presented in Section 5.6 for homogeneous nucleation. Carrying the analysis to completion yields the following relation between the rate of embryo formation J (m^{-2} s^{-1}) and the system conditions and properties:

$$J = \left(\frac{2\sigma_{lv}FN_A}{\pi\overline{M}}\right)^{1/2}\left(\frac{P_v}{RT_v}\right)^{5/3}\left(\frac{N_A}{\overline{M}}\right)^{2/3} v_l F\left(\frac{1-\cos\theta}{2}\right)\exp\left(\frac{-16\pi(\sigma_{lv}F/RT_v)^3 v_l^2 N_A}{3\overline{M}\{\ln[P_v/P_{\text{sat}}(T_v)]\}^2}\right) \tag{9.17}$$

where F is defined by Eq. (9.12). If θ is taken to be 180° and $N_v^{2/3}$ is replaced by N_v, Eq. (9.17) becomes identical to the expression (5.100) obtained in Section 5.6 for homogeneous droplet nucleation.

As in the homogeneous nucleation case, J is interpreted as the rate at which embryos of critical size are generated. As J increases, the probability that a bubble will exceed critical size and grow spontaneously becomes greater. If a threshold value of J is specified as corresponding to the onset of nucleation, the corresponding vapor temperature $T_v = T_{\text{SSL}}$ for the specified system pressure can be determined from Eq. (9.17). Alternatively, for the specified threshold J value, the limiting supersaturation pressure can be determined for a given system temperature.

For water vapor at 100°C, the variation of J with vapor pressure as predicted by Eq. (9.17) is shown in Fig. 9.3 for several values of liquid contact angle.

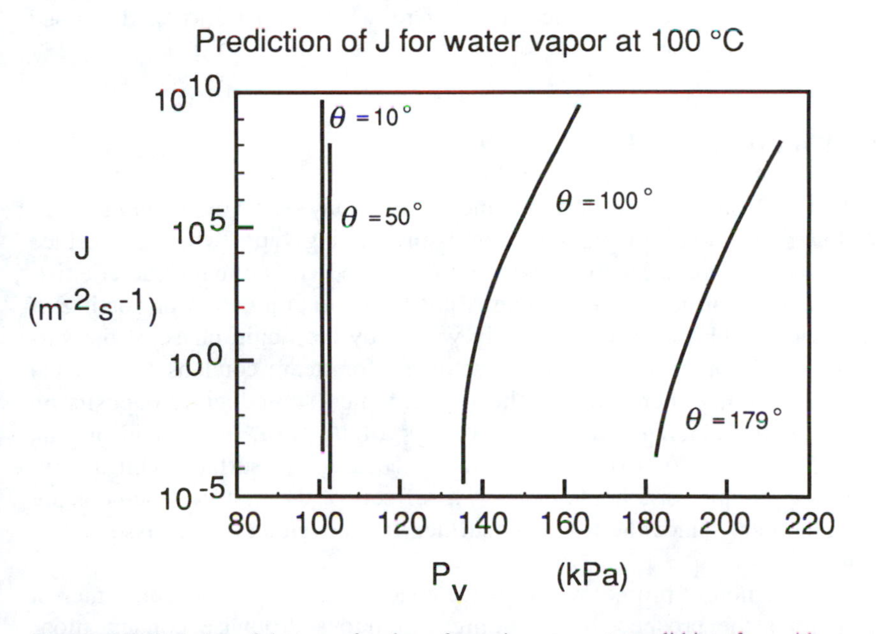

Figure 9.3 Variation of the rate of embryo formation at a water–solid interface with vapor pressure as predicted for different contact angles by analysis of the kinetics of embryo droplet formation.

Assuming that a fixed threshold value of J would apply for all contact angles, it is clear that the predicted value of $(P_v)_{SSL}$ decreases with decreasing contact angle toward the normal saturation vapor pressure. At a liquid contact angle of 10°, the difference between the predicted $(P_v)_{SSL}$ value and $P_{sat}(T_v)$ is negligible for virtually any threshold value of J between 10^{-5} and 10^{10}.

Contact angles for virtually all real systems lie between zero and about 110°, and for metal surfaces with nonmetallic liquids, the contact angle is often below 50° (see Chapter Three). The results of the above analysis therefore suggest that condensation can be initiated at a solid surface in contact with the vapor at supersaturation levels significantly below those required for homogeneous nucleation, if the liquid phase of the vapor wets the surface reasonably well.

As discussed in Chapters Three and Eight, it is quite possible for a thin microfilm of liquid to be absorbed on all or part of a solid surface. This is particularly true for high-energy surfaces such as metals. In addition, when water is the fluid, its polar nature can enhance the tendency of water molecules to attach to portions of the solid surface. (Many oxides and corrosion-produced compounds on metals surfaces are *hydrophilic*.) Patches of adsorbed liquid molecules on the solid surface can thus serve as nuclei for condensation of the liquid phase when the vapor is supersaturated. Condensation on the surface can begin as the formation of very small droplets on the surface at the sites of these nuclei. This so-called dropwise condensation process is, in fact, commonly observed when water vapor in air condenses on a cold beverage glass. This is usually interpreted as being a direct consequence of the fact that the liquid poorly wets the glass, except at nuclei locations where water molecules have adsorbed to crevices (scratches) or foreign matter (such as dust particles) on the surface. Dropwise condensation is discussed further in the next section.

9.2 DROPWISE CONDENSATION

As described in Section 9.1, dropwise condensation may occur on a solid surface cooled below the saturation temperature of a surrounding vapor when the surface is poorly wetted except at locations where well-wetted contaminant nuclei exist. The poorly wetted surface condition can result from contamination or coating of the surface with a substance that is poorly wetted by the liquid phase of the surrounding vapor. In practice, this can be achieved for steam condensation by (1) injecting a nonwetting chemical into the vapor, which subsequently deposits on the surface; (2) introducing a substance such as a fatty (i.e., oleic) acid or wax onto the solid surface; or (3) by permanently coating the surface with a low-surface-energy polymer or a noble metal. The effects of the first two methods are generally temporary, since the resulting surface films eventually are dissolved or eroded away.

The third method of promoting dropwise condensation is of particular interest because it holds the prospect of providing continuous dropwise condensation. Dropwise condensation is generally the preferred mode of condensation because the resulting heat transfer coefficient may be as much as an order of magnitude higher than that for film condensation under comparable circumstances. Recent studies by Westwater and co-workers [9.1, 9.2] have demonstrated that dropwise condensation of steam can be consistently obtained on gold and silver surfaces.

The occurrence of dropwise condensation on gold and silver surfaces would appear to contradict the reasoning that high-surface-energy metal surfaces should be well wetted by the liquid phase, producing film condensation instead of dropwise condensation. It has been demonstrated in recent experiments [9.3], however, that a gold surface applied under highly controlled, ultraclean conditions will spontaneously wet with liquid water, as suggested by the above arguments. Apparently because of its high surface energy, the gold tends to attract and retain organics, which render the surface hydrophobic and thereby produce dropwise condensation. The studies of Woodruff and Westwater [9.1] also indicate that the promotion of dropwise condensation on gold-plated surfaces is somewhat affected by the presence of very small amounts of carbon, copper, aluminum, and oxygen in the coating.

During dropwise condensation, the condensate is usually observed to appear in the form of droplets, which grow on the surface and coalesce with adjacent droplets. When droplets become large enough, they are generally removed from the surface by the action of gravity or drag forces resulting from the motion of the surrounding gas. As the drops roll or fall from the surface, they merge with droplets in their path, effectively sweeping the surface clean of droplets. Droplets then begin to grow anew on the freshly exposed solid surface. This sweeping and renewal of the droplet growth process is responsible for the high heat transfer coefficients associated with dropwise condensation. The visual appearance of this process is indicated in Fig. 9.4.

Despite numerous studies of dropwise condensation over the years, its mechanism remains the subject of debate. Two different types of models have been proposed. The first model type is based on the premise that droplet formation is a heterogeneous nucleation process like that described in Section 9.1. Droplet embryos are postulated to form and grow at nucleation sites, while portions of the surface between the growing droplets remain dry. This type of model apparently was first proposed by Eucken [9.4] in 1937. Experimental evidence supporting this physical model of the condensation process has emerged from several experimental investigations.

In an early study, McCormick and Baer [9.5] presented experimental results supporting the contention that microscopic droplets are nucleated at active nucleation sites on the cooled surface. These active sites were identified as wetted crevices and grooves in the surface that were repeatedly reexposed to supersaturated vapor as a result of the coalescence of droplets and their removal from the surface by drag or gravity body forces. Using an optical technique to indicate changes in the thickness of very thin liquid films, Umur and Griffith [9.6] found that, at least for low temperature differences, the area between growing droplets on the surface was, in fact, dry. Their results indicate that no film greater than a monolayer existed between the droplets, and that no condensation took place in

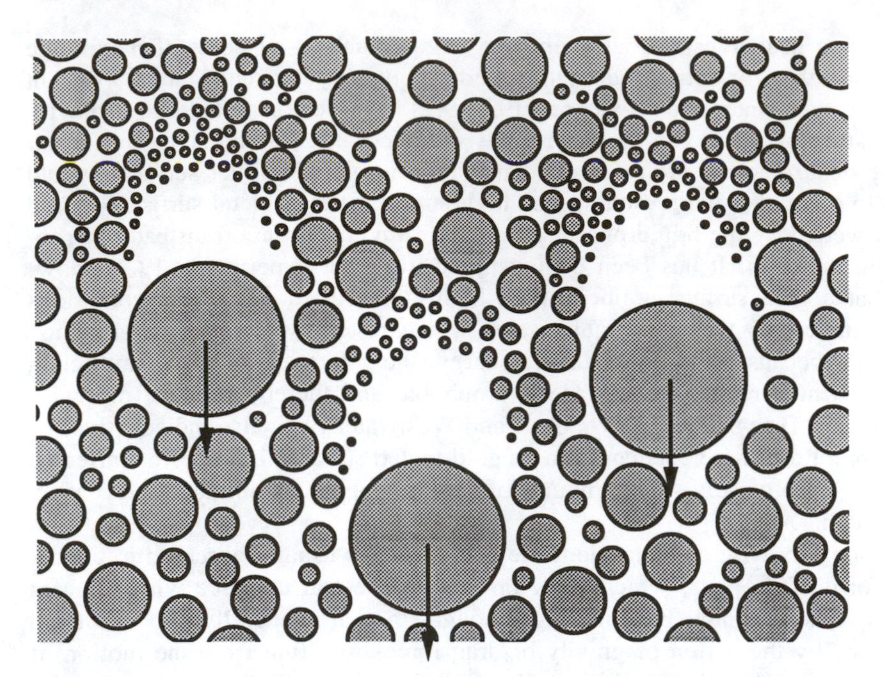

Figure 9.4 Droplet array appearance for dropwise condensation.

those areas. A model of the dropwise condensation process that includes droplet nucleation, growth, removal, and renucleation on reexposed sites was developed by Gose et al. [9.7].

In the second type of dropwise condensation model, it is postulated that condensation occurs initially in a filmwise manner, forming an extremely thin film on the solid surface. As condensation continues, this film eventually reaches a critical thickness, estimated to be about 1 μm, at which point it ruptures and droplets form. Condensation then continues on the surface between the droplets that form when the film ruptures. Condensate produced in these regions is drawn to adjacent drops by surface-tension effects. Droplets also grow by direct condensation on the droplet surfaces themselves.

This second model of the dropwise condensation process apparently was proposed by Jakob [9.8] as early as 1936. Modified versions of this model have also been proposed by Kast [9.9] and Silver [9.10]. The results of several investigations seem to support this type of interpretation of the condensation process. Results presented by Welch and Westwater [9.11] and Sugawara and Katsuda [9.12] indicate that condensation occurs entirely between droplets on a very thin liquid film. Welch and Westwater [9.11] observed the process by taking high-speed movies through a microscope. It appeared to them that droplets large enough to be visible grew mainly by coalescence, leaving behind a "lustrous bare area," which quickly took on a faded appearance. They interpreted the lustrous appearance as corresponding to thin film, which, upon rupturing at a thickness of around 1 μm, took on a more dull appearance.

In contrast, it is postulated in the first model described above that condensation occurs only on the droplets, and not on the surface between them. The rate of condensation on the larger droplets is less than on the smaller ones because of the higher resistance to heat conduction through larger drops. The large drops therefore grow primarily through coalescence. This model implies that most of the heat transfer during dropwise condensation is transferred to that portion of the surface covered with the smallest droplets.

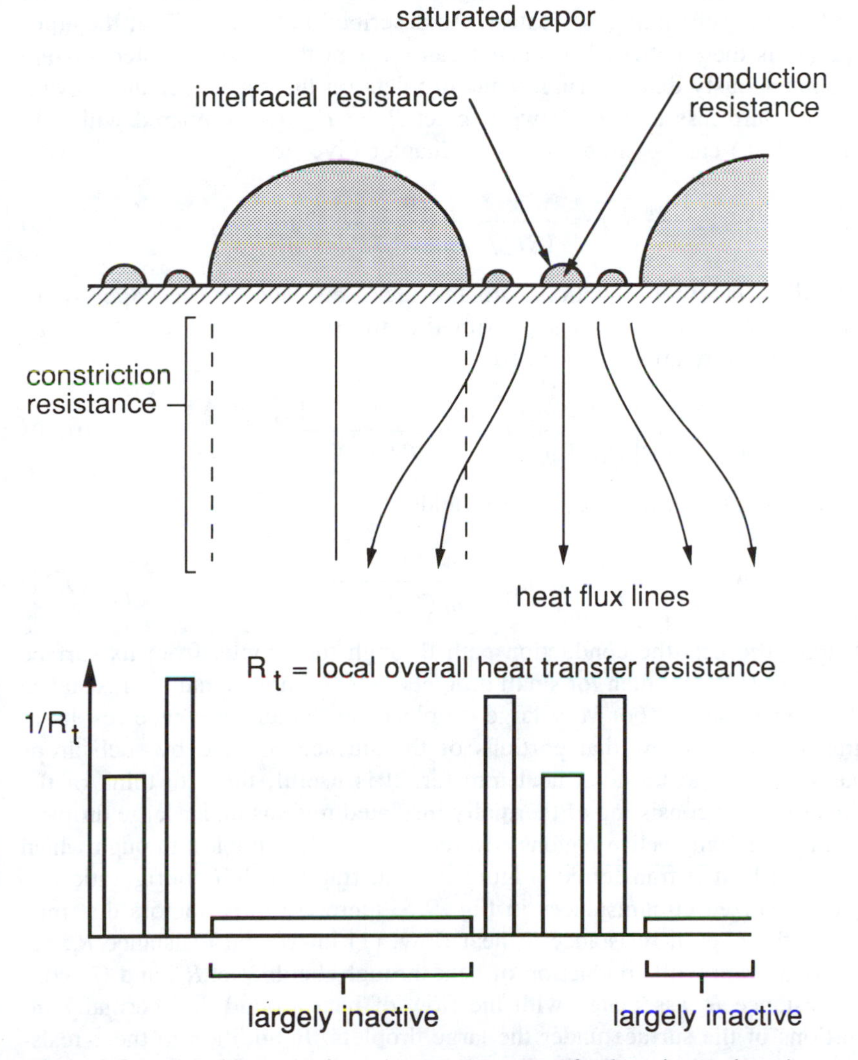

Figure 9.5 Model of heat transfer resistance components for dropwise condensation.

Detailed modeling of dropwise condensation heat transfer based on the first model hypothesis has, in fact, been attempted by several investigators. These models generally idealize the heat transfer process as indicated in Fig. 9.5. Droplets in a wide range of sizes are expected to exist on the surface. The sizes are designated by the diameter of the droplets D, all of which are taken to be hemispheres. The smallest droplet size possible corresponds to the equilibrium radius of curvature r_e for the specified wall subcooling $T_{sat} - T_w$:

$$\frac{D_{min}}{2} = r_e = \frac{2\sigma}{(RT_w/v_l) \ln [P_v/P_{sat}(T_w)] - P_v + P_{sat}(T_w)} \tag{9.18}$$

where T_w is the wall temperature. The highest supersaturation exists at the cold surface, where the subcooling is greatest. As described in Chapter Five, the equilibrium radius is the smallest droplet that can exist in the supersaturated system without spontaneously disappearing. Thus droplets on the surface cannot have a radius of curvature less than r_e. If we neglect $P_v - P_{sat}(T_v)$ compared with $2\sigma/r_e$, then Eq. (9.18) can be simplified (see Chapter Five) to

$$\frac{D_{min}}{2} = r_e = \frac{2\sigma}{(RT_w/v_l) \ln [P_v/P_{sat}(T_w)]} \tag{9.19}$$

Combining the Clapeyron equation with the ideal gas law for the vapor, integrating between P_v and P_{sat}, and approximating the product $T_w T_{sat}$ as T_w^2 and v_{lv} as v_v, the following relation is obtained:

$$\ln \left[\frac{P_v}{P_{sat}(T_w)} \right] = \frac{h_{lv}[T_{sat}(P_v) - T_w]}{RT_w^2} \tag{9.20}$$

Substituting this relation into Eq. (9.19) yields

$$\frac{D_{min}}{2} = (r_e)_{min} = \frac{2v_l \sigma T_w}{h_{lv}(T_{sat} - T_w)} \tag{9.21}$$

For larger droplets the conduction path through the droplet from its surface to the solid wall is longer than for small droplets, which implies that the resistance to heat transfer is larger. For very large droplets, the larger resistance results in a heat transfer rate so low that portions of the surface covered by such drops contribute very little to the total heat transfer. It is useful, then, to think of the cold solid surface as consisting of thermally insulated regions under large droplets surrounded by thermally active regions covered with smaller droplets through which virtually all the heat is transferred. Figure 9.5 illustrates such a configuration.

For the idealized circumstances in Fig. 9.5, there are three factors that may contribute to the overall resistance to heat flow: (1) interfacial resistance R_i, (2) resistance associated with conduction of heat through the droplet R_c, and (3) constriction resistance R_s associated with the flow of heat around the thermally inactive portions of the surface under the large droplets. In addition to these resistances, there is also a loss of driving temperature potential due to droplet interface curvature, which can be thought of as an additional resistance. If the vapor is one component in a mixture, there may also be resistance associated with the mass transport of the condensing vapor to the surface. For the purposes of further discussion here, however, condensation of pure vapors will be considered, and this effect will not arise.

Initially, we will neglect the constriction resistance, but we will return to discuss it further after considering the other two effects. If the solid surface is assumed to be held at a constant mean temperature T_w, the overall temperature difference $\Delta T_t = T_{sat} - T_w$ between the vapor and the wall must equal the sum of temperature differences associated with the resistances to heat flow described above, plus the capillary depression of the saturation temperature. The overall temperature difference can thus be written as

$$\Delta T_t = \Delta T_i + \Delta T_{cap} + \Delta T_{con} \tag{9.22}$$

where ΔT_i, ΔT_{cap}, and ΔT_{con} are the temperature differences associated with interfacial resistance, capillary depression of the equilibrium saturation temperature, and conduction resistance through the droplet, respectively.

In Section 4.6 the following relation for the interfacial heat transfer coefficient h_i was derived for small values of the parameter $a = q_i''(\overline{M}/2\overline{R}T_v)^{1/2}/\rho_v h_{lv}$:

$$h_i = \left(\frac{2\hat{\sigma}}{2-\hat{\sigma}}\right)\frac{h_{lv}^2}{T_v v_l}\left(\frac{\overline{M}}{2\pi\overline{R}T_v}\right)^{1/2}\left[1 - \frac{P_v v_{lv}}{2h_{lv}}\right] \tag{9.23}$$

In this relation, $\hat{\sigma}$ is the accommodation coefficient (see Chapter Four). The second term in the square brackets is very small compared to 1 for most systems and can be neglected, reducing the relation to

$$h_i = \left(\frac{2\hat{\sigma}}{2-\hat{\sigma}}\right)\frac{h_{lv}^2}{T_v v_l}\left(\frac{\overline{M}}{2\pi\overline{R}T_v}\right)^{1/2} \tag{9.24}$$

From the definition of this coefficient, it follows that the temperature drop associated with the interface resistance ΔT_i is given by

$$\Delta T_i = \frac{q_d}{h_i \pi D^2/2} \tag{9.25}$$

where q_d is the heat transfer rate from the droplet.

Following the analysis of Graham and Griffith [9.13], we estimate the depression of the equilibrium interface temperature below the normal saturation temperature for a droplet of radius $r = D/2$ by replacing $T_{sat} - T_w$ by ΔT_{cap} and D_{min} by the droplet diameter D in Eq. (9.21). The resulting relation for ΔT_{cap} is

$$\Delta T_{cap} = \frac{4 v_l \sigma T_w}{h_{lv} D} \tag{9.26}$$

Combining this relation with Eq. (9.21), the relation for ΔT_{cap} can be written as

$$\Delta T_{cap} = \frac{(T_{sat} - T_w) D_{min}}{D} \tag{9.27}$$

where D_{min} is given by Eq. (9.21).

Graham and Griffith [9.13] also argued that the conduction resistance through a liquid droplet from the wall to the liquid–vapor interface is such that the effective temperature drop associated with this resistance is given by

$$\Delta T_{\text{con}} = \frac{q_d(D/2)}{4\pi k_l(D/2)^2} \tag{9.28}$$

Substituting Eqs. (9.25), (9.27), and (9.28) into Eq. (9.22) yields

$$\Delta T_t = T_{\text{sat}} - T_w = \frac{2q_d}{h_i \pi D^2} + \frac{(T_{\text{sat}} - T_w)D}{D_{\min}} + \frac{q_d}{2k_l \pi D} \tag{9.29}$$

where h_i and $D_{\min}$ are given by Eqs. (9.24) and (9.21), respectively. Equation (9.29) can be rearranged to obtain

$$q_d = \left(\frac{\pi D^2}{2}\right) \Delta T_t \frac{(1 - D_{\min}/D)}{(1/h_i + D/4k_l)} \tag{9.30}$$

This equation gives the heat transfer rate for one droplet of size D. To get the total heat flux from the surface, the contributions for all droplets must be included. We can do so if we know the droplet size distribution n''_D equal to the number of droplets per unit area of surface having a diameter between D and $D + dD$. We obtain an expression for the total heat flux by multiplying q_d by the number density $n''_D \, dD$ of droplets of size D and integrating over the entire range of possible D sizes:

$$q'' = \left(\frac{\pi \Delta T_t}{2}\right) \int_{D_{\min}}^{D_{\max}} n''_D D^2 \frac{(1 - D_{\min}/D)}{(1/h_i + D/4k_l)} \, dD \tag{9.31}$$

Equivalently, we can write the above expression in terms of the heat transfer coefficient $h_{dci} = q''/\Delta T_t$:

$$h_{dci} = \left(\frac{\pi}{2}\right) \int_{D_{\min}}^{D_{\max}} n''_D D^2 \frac{(1 - D_{\min}/D)}{(1/h_i + D/4k_l)} \, dD \tag{9.32}$$

It is clear from this expression that the droplet size distribution n''_D must be known over the entire size range D in order to predict the heat transfer coefficient using this theoretical model. Unfortunately, this type of information is not readily available for most systems of interest. From visual observations, Graham and Griffith [9.13] found that for water condensing on a mirror-smooth copper surface, n''_D varied about proportional to $D^{-3.5}$ for sizes between 10 and 1000 μm. However, they were unable to observe and count droplets smaller than 10 μm. They also demonstrated that a significant fraction of the heat is transferred by droplets smaller than 10 μm. Hence the number distribution could not be determined over an important portion of the size range. The only hope for using Eq. (9.32) would be to extrapolate the observed number distribution to smaller sizes in some reasonable manner.

In deriving Eq. (9.32) for the heat transfer coefficient, the effects of constriction resistance in the solid wall (due to the flow of heat around large droplets) were neglected. Based on analysis of the associated conduction problem, Mikic [9.14] argued that the constriction resistance to heat flow is given approximately by

$$R_s = \left(\frac{1}{3\pi k_l}\right) \int_{D_{\min}}^{D_{\max}} \frac{n''_D D^2 \, dD}{[1 - f(D)]} \tag{9.33}$$

where $f(D)$ is the fraction of the surface area covered by droplets having a diameter greater than D. If this resistance is considered to be in series with the additional resistance mechanisms considered above, the heat transfer coefficient is then given by

$$h_{dc} = \left(\frac{1}{h_{dci}} + R_s\right)^{-1} \tag{9.34}$$

where h_{dci} and R_s are given by Eqs. (9.32) and (9.33), respectively. Mikic [9.14] found that the constriction resistance for stainless steel as the condensing surface was about 84% of the total resistance associated with the dropwise condensation process. For a copper surface his results implied that the contribution of the constriction resistance was about 20% of the total.

While the analytical models described above provide a useful framework for exploring the mechanisms of dropwise condensation, the droplet size distribution must be known for the conditions of interest before computation of the heat transfer coefficient is possible. These modeling efforts do not address the question of whether condensation occurs in a film on the surface between droplets.

Further investigation of dropwise condensation is clearly needed to resolve the differing interpretations of this process. As suggested by Collier [9.15], it may be that droplet nucleation dominates at low condensation rates, with the film disruption mechanism taking over at higher condensation rates. Because, as indicated in Fig. 9.6 [9.16], a transition to dropwise condensation generally enhances the heat transfer coefficient by a significant amount, efforts to better understand the mechanisms of dropwise condensation will, no doubt, continue.

The use of organic coatings to promote dropwise condensation of steam is discussed in some detail by Holden et al. [9.17]. The effects of varying the system pressure on dropwise condensation are described by O'Bara et al. [9.18]. A detailed discussion of theoretical aspects of dropwise condensation can also be found in a series of recent publications by Tanaka [9.19–9.21]. The droplet size distribution and the effects of surface characteristics on dropwise condensation are also discussed in references [9.22–9.25].

Correlations for the heat transfer coefficient associated with dropwise condensation have been proposed by a number of investigators [9.11, 9.26–9.29]. One example is the following correlation, proposed by Peterson and Westwater [9.26] for dropwise condensation of steam and ethylene glycol:

$$\mathrm{Nu} = 1.46 \times 10^{-6} (\mathrm{Re}^*)^{-1.63} \, \Pi_k^{1.16} \, \mathrm{Pr}_l^{0.5} \tag{9.35}$$

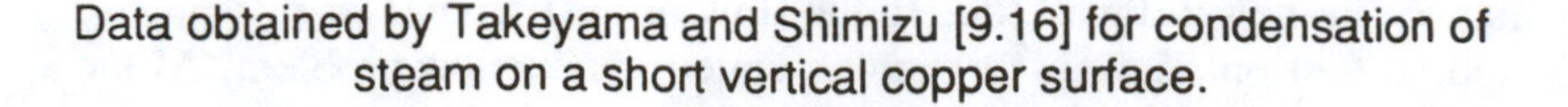

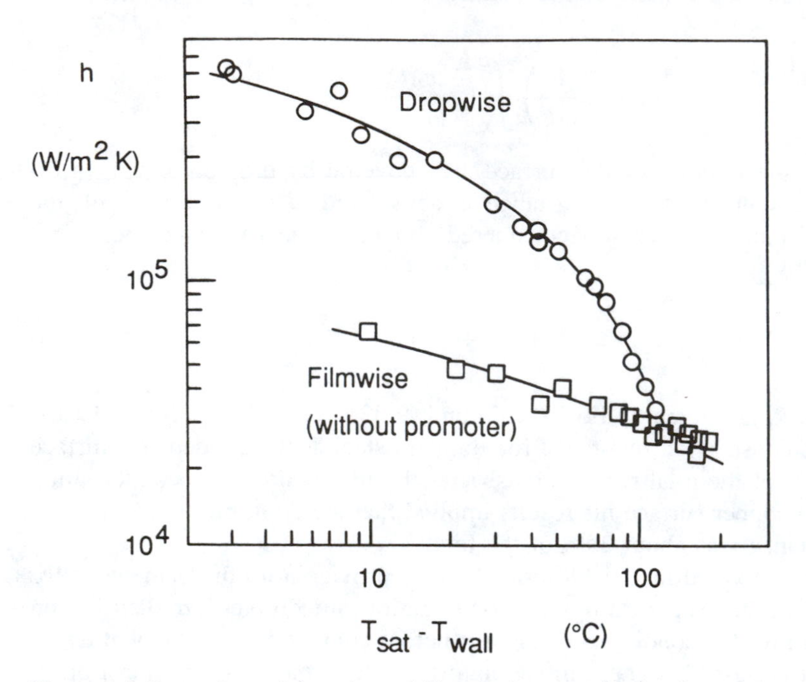

Figure 9.6 Comparison of dropwise and filmwise heat transfer data for steam at atmospheric pressure. The lines are approximate best fits to the data.

where

$$\mathrm{Nu} = \frac{2h\sigma T_{\mathrm{sat}}}{\rho_l h_{lv} k_l (T_{\mathrm{sat}} - T_w)} \tag{9.36}$$

$$\mathrm{Re}^* = \frac{k_l (T_{\mathrm{sat}} - T_w)}{\mu_l h_{lv}} \tag{9.37}$$

$$\Pi_k = -\frac{2\sigma T_{\mathrm{sat}}}{\mu_l^2 h_{lv}} \left(\frac{d\sigma}{dT} \right) \tag{9.38}$$

and Pr_l is the liquid Prandtl number. This modified version of the correlation proposed by Isachenko [9.29] was recommended for $1.75 \leq \mathrm{Pr}_l \leq 23.6$, $7.8 \times 10^{-4} \leq \Pi_k \leq 2.65 \times 10^{-2}$, and $2 \times 10^{-4} \leq \mathrm{Re}^* \leq 3 \times 10^{-2}$. While correlations of this type can be made to agree quite well with data for a specific surface and fluid combination, their general applicability has not been demonstrated. The use of such a correlation for circumstances other than those for which it was developed is questionable at best. Further discussion of correlations for dropwise condensation is provided in the review article by Merte [9.30].

Example 9.1 Use the correlation proposed by Peterson and Westwater [9.26] to predict the heat transfer coefficient for dropwise condensation of steam at atmospheric pressure on a surface held at 80°C.

For saturated water at atmospheric pressure, $T_{sat} = 100°C$, $\sigma = 0.0589$ N/m, $\rho_l = 958$ kg/m^3, $h_{lv} = 2257$ kJ/kg, $k_l = 0.679$ W/m K, $Pr_l = 1.72$, and $\mu_l = 2.78 \times 10^{-4}$ N s/m^2. Estimating $T_{sat}(d\sigma/dT)$ using the values of σ at 373 K and 400 K yields

$$T_{sat}\left(\frac{d\sigma}{dT}\right) = (373)\left(\frac{0.0535 - 0.0589}{400 - 373}\right) = -0.0746 \text{ N/m}$$

Substituting to determine Π_k and Re^* yields

$$\Pi_k = \frac{2\sigma}{\mu_l^2 h_{lv}}(T_{sat})\left(-\frac{d\sigma}{dT}\right) = \frac{2(0.0589)}{(2.78 \times 10^{-4})^2(2{,}257{,}000)}(0.0746)$$

$$= 0.0506$$

$$Re^* = \frac{k_l(T_{sat} - T_w)}{\mu_l h_{lv}} = \frac{0.679(20)}{(2.78 \times 10^{-4})(2{,}257{,}000)}$$

$$= 0.0217$$

The Nusselt number is then determined as

$$Nu = 1.46 \times 10^{-6}(Re^*)^{-1.67}\,\Pi_k^{1.16}\,Pr_l^{0.5}$$

$$= 1.46 \times 10^{-6}(0.0217)^{-1.63}(0.0506)^{1.16}(1.72)^{0.5}$$

$$= 3.10 \times 10^{-5}$$

Solving for h using the definition (9.36) of Nu, we obtain

$$h = \frac{\rho_l h_{lv} k_l (T_{sat} - T_w)\,Nu}{2\sigma T_{sat}}$$

$$= \frac{(958)(2{,}257{,}000)(0.679)(100 - 80)(3.10 \times 10^{-5})}{2(0.0589)(373)}$$

$$= 20{,}700 \text{ W/m}^2\text{ K} = 2.07 \times 10^4 \text{ W/m}^2\text{ K}$$

Note that this predicted value of h is an order of magnitude lower than the data obtained by Takeyama and Shimizu [9.16] for dropwise condensation of steam at atmospheric pressure for a wall subcooling of about 20°C, as indicated in Fig. 9.6. The data to which the correlation was fitted were for a different surface size and promoter, although in both cases vertical copper surfaces were tested.

9.3 FILM CONDENSATION ON A FLAT, VERTICAL SURFACE

If the liquid phase fully wets a cold surface in contact with a vapor near saturation conditions, the conversion of vapor to liquid will take the form of film condensation. As the name implies, the condensation takes place at the interface of a liquid film covering the solid surface. Because the latent heat of vaporization must be removed at the interface to sustain the process, the rate of condensation is directly linked to the rate at which heat is transported across the liquid film from the interface to the surface.

Integral Analysis of Laminar Film Condensation

The classic Nusselt integral analysis [9.31] of laminar film condensation on a vertical surface considers the physical circumstances shown in Fig. 9.7. The surface exposed to a motionless ambient of saturated vapor is taken to be isothermal with a temperature below the saturation temperature. Note that although a vertical surface is considered here, the analysis is identical for an inclined surface, except that the gravitational acceleration g is replaced by $g \sin \Omega$, where Ω is the angle between the surface and the horizontal. Because the liquid film flows down the surface due to gravity, this situation is sometimes referred to as *falling-film condensation*.

A simple force balance on the shaded film element in Fig. 9.7 requires that

$$(\delta - y)\, dx(\rho_l - \rho_v)g = \mu_l\left(\frac{du}{dy}\right) dx \tag{9.39}$$

Note that this force balance includes the effects of gravity body forces and vis-

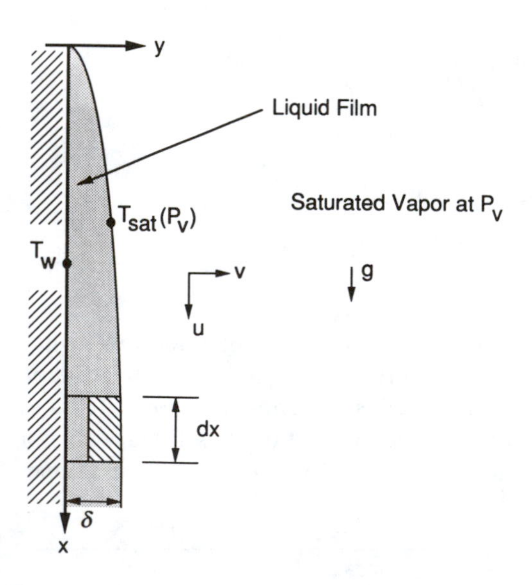

Figure 9.7 Model system for analysis of falling-film condensation on a vertical surface.

cosity, but neglects inertia effects. Integrating this equation from $y = 0$ to $y = \delta$ and using the fact that $u = 0$ at $y = 0$, we obtain the following relation for the velocity profile:

$$u = \frac{(\rho_l - \rho_v)g}{\mu_l}\left(y\delta - \frac{y^2}{2}\right) \tag{9.40}$$

Integrating this velocity across the film yields the following relation for the mass flow rate per unit width of surface $\dot{m}'$:

$$\dot{m}' = \rho_l \int_0^\delta u\, dy = \frac{\rho_l(\rho_l - \rho_v)g\delta^3}{3\mu_l} \tag{9.41}$$

Differentiating the above relation with respect to δ yields

$$\frac{d\dot{m}'}{d\delta} = \frac{\rho_l(\rho_l - \rho_v)g\delta^2}{\mu_l} \tag{9.42}$$

Assuming that heat transfer across the film is mainly due to conduction, convective effects are neglected, whereupon the heat flux to the interface is given by

$$q'' = \frac{k_l\, \Delta T}{\delta}, \qquad \Delta T = T_{\text{sat}} - T_w \tag{9.43}$$

This implies that the heat flow dq across the differential element shown in Fig. 9.7 is given by

$$dq = \frac{k_l\, \Delta T}{\delta}\, dx \tag{9.44}$$

Ignoring sensible subcooling of the liquid film compared to latent heat effects, conservation of mass and energy dictates that

$$dq = h_{lv}\, d\dot{m}' \tag{9.45}$$

Combining Eqs. (9.42), (9.44), and (9.45), the following differential equation is obtained for δ:

$$\frac{d\delta}{dx} = \frac{k_l\mu_l\, \Delta T}{\rho_l(\rho_l - \rho_v)gh_{lv}\delta^3} \tag{9.46}$$

Integrating Eq. (9.46) using the condition that $\delta = 0$ at $x = 0$ yields

$$\delta = \left[\frac{4k_l\mu_l x\, \Delta T}{\rho_l(\rho_l - \rho_v)gh_{lv}}\right]^{1/4} \tag{9.47}$$

Because heat transfer across the film is by conduction alone, the local heat transfer coefficient is given by $h_l = k_l/\delta$. It follows from Eq. (9.47) that the local Nusselt number is given by

$$\text{Nu}_x = \frac{h_l x}{k_l} = \left[\frac{\rho_l(\rho_l - \rho_v)gh_{lv}x^3}{4k_l\mu_l(T_{\text{sat}} - T_w)}\right]^{1/4} \tag{9.48}$$

We define a mean heat transfer coefficient as

$$\overline{h}_l = \frac{1}{x_e} \int_0^{x_e} h_l(x)\, dx \tag{9.49}$$

Using Eq. (9.48) to execute the integration reveals that $\overline{h}_l$ is related to the local liquid flow rate in the film per unit width of surface $\dot{m}'$ as

$$\frac{\overline{h}_l}{k_l} \left[\frac{\mu_l^2}{\rho_l(\rho_l - \rho_v)g} \right]^{1/3} = 1.47\, \mathrm{Re}_L^{-1/3}, \qquad \mathrm{Re}_L = \frac{4\dot{m}'}{\mu_l} \tag{9.50}$$

Using this result, the following relation for the mean Nusselt number can be obtained:

$$\overline{\mathrm{Nu}}_x = \frac{\overline{h}_l x}{k_l} = 0.943 \left[\frac{\rho_l(\rho_l - \rho_v) g h_{lv} x^3}{k_l \mu_l (T_{\mathrm{sat}} - T_w)} \right]^{1/4} \tag{9.51}$$

The classic Nusselt analysis described above incorporates the following idealizations:

1. Laminar flow.
2. Constant properties.
3. Subcooling of liquid is negligible in the energy balance.
4. Inertia effects are negligible in the momentum balance.
5. The vapor is stationary and exerts no drag.
6. The liquid–vapor interface is smooth.
7. Heat transfer across film is only by conduction (convection is neglected).

Modified versions of this analysis have been developed subsequently that relax many of these assumptions. For example, idealization 3 can be removed by including the effect of liquid subcooling in the energy balance:

$$\frac{dq}{dx} = \frac{k_l\, \Delta T}{\delta} = h_{lv} \frac{d\dot{m}'}{dx} + \frac{d}{dx} \int_0^{\delta} \rho_l c_{pl} u (T_{\mathrm{sat}} - T_w)\, dy \tag{9.52}$$

If we use the velocity profile given by Eq. (9.40) and the linear temperature profile,

$$\frac{T_{\mathrm{sat}} - T}{T_{\mathrm{sat}} - T_w} = 1 - \frac{y}{\delta} \tag{9.53}$$

to evaluate the integral in Eq. (9.52), the energy balance can be written as

$$\frac{k_l\, \Delta T}{\delta} = h_{lv}^* \frac{d\dot{m}'}{dx} \tag{9.54}$$

where

$$h_{lv}^* = h_{lv} \left\{ 1 + \frac{3}{8} \left[\frac{c_{pl}(T_{\mathrm{sat}} - T_w)}{h_{lv}} \right] \right\} \tag{9.55}$$

Equation (9.54) is identical to the energy balance given by Eqs. (9.44) and (9.45) in the original analysis, except that h_{lv} has been replaced by h_{lv}^*. Consequently, the effect of liquid subcooling can be incorporated by simply replacing h_{lv} with h_{lv}^* in Eqs. (9.48) and (9.51) for the heat transfer coefficient. More detailed integral analyses of laminar film condensation on a vertical surface have been developed by Koh [9.32] and by Rohsenow [9.33]. The analysis of Rohsenow [9.33] indicated that the effect of subcooling and energy convection can be incorporated by replacing h_{lv} in the above relations by h'_{lv}, given by

$$h'_{lv} = h_{lv}\left\{1 + 0.68\left[\frac{c_{pl}(T_{sat} - T_w)}{h_{lv}}\right]\right\} \tag{9.56}$$

Example 9.2 Use Eq. (9.51) to predict the mean heat transfer coefficient for film condensation of steam at atmospheric pressure on a vertical flat plate. The plate, which is 10 cm high, is held at a uniform temperature of 80°C. Compare the resulting $\bar{h}_l$ with the dropwise condensation result from Example 9.1.

Equation (9.51) can be written as

$$\bar{h}_l = 0.943\left(\frac{k_l}{x}\right)\left[\frac{\rho_l(\rho_l - \rho_v)gh_{lv}x^3}{k_l\mu_l(T_{sat} - T_w)}\right]^{1/4}$$

For water at atmospheric pressure, $\rho_l = 958$ kg/m^3, $\rho_v = 0.597$ kg/m^3, $h_{lv} = 2257$ kJ/kg, $k_l = 0.679$ W/m K, and $\mu_l = 2.78 \times 10^{-4}$ N s/m^2. Substituting in the above equation,

$$\bar{h}_l = 0.943\left(\frac{0.679}{0.1}\right)\left[\frac{958(958 - 0.597)(9.8)(2{,}257{,}000)(0.1)^3}{0.679(2.78 \times 10^{-4})(1{,}000 - 80)}\right]^{1/4}$$

$$= 9.75 \times 10^3 \text{ W/m}^2\text{ K}$$

This value for the film condensation heat transfer coefficient is about a factor of 2 smaller than the dropwise heat transfer coefficient.

Boundary-Layer Analysis of Film Condensation

Laminar film condensation on a vertical surface can also be analyzed with a full boundary-layer formulation. In terms of the Cartesian coordinates and the corresponding velocities shown in Fig. 9.8, the governing two-dimensional boundary-layer equations are

$$\frac{\partial u}{\partial x} + \frac{\partial v}{\partial y} = 0 \tag{9.57}$$

$$u\frac{\partial u}{\partial x} + v\frac{\partial u}{\partial y} = \nu_l\frac{\partial^2 u}{\partial y^2} + \frac{g(\rho_l - \rho_v)}{\rho_l} \tag{9.58}$$

$$u\frac{\partial T}{\partial x} + v\frac{\partial T}{\partial y} = \alpha_{T,l}\left(\frac{\partial^2 T}{\partial y^2}\right) \tag{9.59}$$

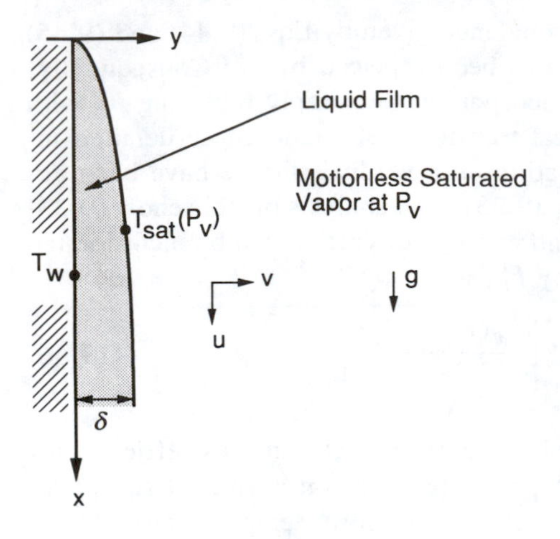

Figure 9.8 System model for similarity analysis of film condensation.

where ν_l and $\alpha_{T,l}$ are the kinematic viscosity and thermal diffusivity of the liquid, respectively. In this analysis, shear exerted by the vapor on the falling liquid film will be neglected. The temperature in the vapor is assumed to be equal to $T_{\text{sat}}(P_v)$ everywhere, since the vapor in the ambient is at the saturation temperature.

The boundary conditions at the cooled wall ($y = 0$), and at the liquid–vapor interface ($y = \delta$) are

$$\text{at } y = 0: \qquad u = v = 0, \qquad T = T_w \tag{9.60}$$

$$\text{at } y = \delta: \qquad \frac{\partial u}{\partial y} = 0, \qquad T = T_{\text{sat}}(P_v) \tag{9.61}$$

At the wall, the boundary conditions are a consequence of the no-slip condition and the isothermal wall specification. The relations (9.61), which apply for $y = \delta$, specify continuity of the temperature profile across the interface and a negligible shear stress exerted by the surrounding vapor on the liquid film.

Following the analysis presented by Sparrow and Gregg [9.34], these equations and boundary conditions can be cast into a similarity formulation. Defining a stream function in the liquid film so that

$$u = \frac{\partial \psi}{\partial y}, \qquad v = -\frac{\partial \psi}{\partial x} \tag{9.62}$$

it follows that the continuity relation (9.57) is automatically satisfied. The following similarity transformation is then imposed:

$$\eta = C_l y x^{-1/4} \tag{9.63}$$

$$\psi = 4\alpha_{T,l} C_l x^{3/4} f(\eta) \tag{9.64}$$

$$\theta(\eta) = \frac{T_{\text{sat}} - T}{T_{\text{sat}} - T_w} \tag{9.65}$$

where

$$C_l = \left[\frac{g c_{pl}(\rho_l - \rho_v)}{4 \nu_l k_l} \right]^{1/4} \tag{9.66}$$

The velocity components are related to the similarity variables as

$$u = 4C_l^2 \alpha_{T,l} x^{1/2} f'(\eta) \tag{9.67}$$

$$v = C_l \alpha_{T,l} x^{-1/4} [\eta f'(\eta) - 3f(\eta)] \tag{9.68}$$

where the primes denote differentiation with respect to η.

In terms of the similarity variables, the governing momentum and energy equations and associated boundary conditions become

$$f''' + \frac{1}{\Pr_l} [3ff'' - 2(f')^2] + 1 = 0 \tag{9.69}$$

$$\theta'' + 3f\theta' = 0 \tag{9.70}$$

at $\eta = 0$:

$$f(0) = f'(0) = 0, \qquad \theta(0) = 1 \tag{9.71}$$

at $\eta = \eta_\delta$:

$$f''(\eta_\delta) = \theta(\eta_\delta) = 0 \tag{9.72}$$

The mathematical problem posed by Eqs. (9.69) and (9.70) with boundary conditions (9.71) and (9.72) is a fifth-order system of nonlinear ordinary differential equations with five boundary conditions. The system is therefore closed if we can specify the location of the interface $y = \delta$. However, the location of the interface is not known a priori. The interface location is dictated by the transport, and it therefore must be determined as part of the solution process.

The additional relation needed to allow determination of the interface location is obtained from an energy balance over a segment of the film. If we neglect sensible cooling of the condensate, this balance can be written as

$$\int_0^x \left[k_l \left(\frac{\partial T}{\partial y} \right)_{y=\delta} \right] dx = \int_0^\delta \rho_l u_l h_{lv} \, dy \tag{9.73}$$

In terms of the similarity variables defined above, Eq. (9.73) can be written as

$$-\frac{3f(\eta_\delta)}{\theta'(\eta_\delta)} = \mathrm{Ja}_l = \frac{c_{pl}(T_{\text{sat}} - T_w)}{h_{lv}} \tag{9.74}$$

where, as before, η_δ is η evaluated at $y = \delta$. The problem is now completely closed mathematically. If a value of η_δ is assumed, Eqs. (9.69) and (9.70) can be solved numerically using boundary conditions (9.71) and (9.72). Using the results of the computed solution and Eq. (9.74), the corresponding value of $\mathrm{Ja}_l = c_{pl}(T_{\text{sat}} - T_w)/h_{lv}$ can be determined. Thus η_δ is a function of Ja_l. Also, if we begin with the relation

$$h = \frac{q''}{T_{\text{sat}} - T_w} = \frac{k_l}{T_{\text{sat}} - T_w} \left(\frac{\partial T}{\partial y} \right)_{y=0} \tag{9.75}$$

then applying the similarity transfer to the temperature gradient term yields the relation

$$\mathrm{Nu}_x = \frac{hx}{k_l} = [-\theta'(0)]\,\mathrm{Ja}_l^{1/4}\left[\frac{g\rho_l(\rho_l-\rho_v)x^3h_{lv}}{4\mu_l k_l(T_{\text{sat}}-T_w)}\right]^{1/4} \tag{9.76}$$

Thus, for a given set of physical conditions, the value of $\theta'(0)$ can be computed using the scheme described above and then the variation of the Nusselt number or heat transfer coefficient along the surface can be determined from Eq. (9.76). Note that this relation is the same form as that obtained from the Nusselt analysis except that the term in the large square brackets now has the prefactor $[-\theta'(0)]\,\mathrm{Ja}_l^{1/4}$. The variation of the Nusselt number with the system parameters, as determined by the similarity solution calculations of Sparrow and Gregg [9.34], is indicated in Fig. 9.9. These results indicate that for $\mathrm{Pr}_l \geq 100$, the variation of the Nusselt number is identical to that predicted by an integral analysis neglecting the inertia terms. For such conditions, the variation is closely approximated by the relation

$$-\theta'(0) = (0.68 + \mathrm{Ja}_l^{-1})^{1/4}$$

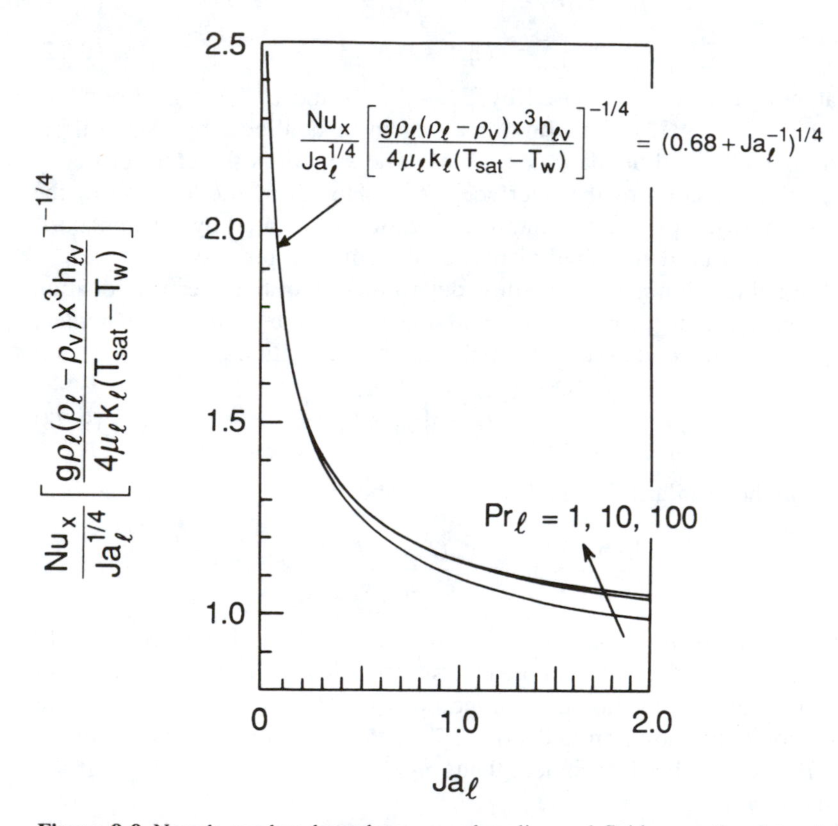

Figure 9.9 Nusselt number dependence on subcooling and fluid properties determined from similarity solution calculations. *(Adapted from reference* [9.34] *with permission, copyright © 1959, American Society of Mechanical Engineers.)*

Substituting this result into Eq. (9.76), the resulting relation for the local Nusselt number can be written as

$$\mathrm{Nu}_x = \left[\frac{g\rho_l(\rho_l - \rho_v)x^3 h'_{lv}}{4\mu_l k_l (T_{\mathrm{sat}} - T_w)}\right]^{1/4} \quad \text{where } h'_{lv} = (1 + 0.68Ja_l)h_{lv} \tag{9.77}$$

In this form it is clear that the similarity solution prediction for $\mathrm{Pr}_l \to \infty$ is identical to the correlation obtained by Rohsenow [9.33] based on an integral analysis that included subcooling and energy convection but neglected inertia terms. Sadasivan and Lienhard [9.35] showed that the similarity solution values for Nu_x could be accurately predicted for all likely values of Pr_l and Ja_l using Eq. (9.77) with

$$h'_{lv} = h_{lv}(1 + C_c\,\mathrm{Ja}_l) \tag{9.78}$$

$$C_c = 0.683 - \frac{0.228}{\mathrm{Pr}_l} \tag{9.79}$$

Equation (9.79) for C_c was developed as a best fit to the variation of C_c with Pr_l inferred from results computed numerically using the similarity analysis of Sparrow and Gregg [9.34].

The analyses described above predict the main features of laminar film condensation on a vertical or inclined surface reasonably well. There are, however, two physical mechanisms not included in these analyses that can significantly affect the transport: (1) the effects of waves on the liquid–vapor interface and (2) interfacial vapor drag on the interface. The effects of interfacial vapor drag were examined analytically by Koh, Sparrow, and Hartnett [9.36]. These investigators presented a similarity solution of the governing boundary-layer transport equations in the liquid film and surrounding vapor that was basically an extension of the similarity solution of Sparrow and Gregg [9.34] described above.

The results of Koh et al. [9.36] indicate that interfacial shear has very little effect on heat transfer for large liquid Prandtl numbers (≥ 10). For $\mathrm{Pr}_l = 1.0$, the effect is less than 10% for most practical circumstances. On the other hand, for very small Pr_l values characteristic of liquid metals, the vapor drag has a strong effect, significantly reducing the heat transfer below that predicted when the interfacial shear is neglected. The reduction of the heat transfer data for film condensation of liquid metals below that predicted by the classical Nusselt theory may also be due, at least in part, to interfacial resistance [9.37] (see Section 4.6).

The effects of surface waves on laminar film condensation are more difficult to incorporate into theoretical analyses. It was shown in Chapter Four that a falling film on an inclined (or, in the limit, vertical) wall is unstable at any finite film Reynolds number. However, the disturbance amplification rate increases rapidly as the film Reynolds number increases, with the result that there typically exists a quasi-critical Reynolds number beyond which waves of finite amplitude are so highly probable that they invariably are found. Based on experimental observation, surface waves are expected when $\mathrm{Re}_L = 4\dot{m}'/\mu_l$ is greater than 33 [9.38].

In general, interfacial waves are expected to enhance convective heat transport in the film since it intermittently thins the film, increases the interfacial area, and induces mixing. Because of these effects, laminar film condensation heat transfer data are often significantly higher than the values predicted by Eq. (9.50) or (9.51). In some cases, however, the data have also been found to be significantly lower than the values predicted by these relations, as indicated in Fig. 9.10, which shows data from a study by Spencer and Ibele [9.39]. Deviations of ±50% from the prediction of the Nusselt relation (9.50) or (9.51) for data reported in the literature are not uncommon.

The results of the above laminar analyses for falling-film condensation on a flat, vertical surface can also be applied to condensation on the exterior of a vertical tube if the liquid film thickness is small relative to the tube diameter. However, for such circumstances, McAdams [9.40] recommends that the coefficient on the right side of Eqs. (9.50) and (9.51) be changed to 1.88 and 1.13, respectively, to better match available data.

Turbulent Film Condensation

As for any boundary-layer flow, when the film Reynolds number $Re_L = 4\dot{m}'/\mu_l$ becomes large enough, it is expected that a transition to turbulent flow will occur. In general, there may exist a regime of laminar flow over a portion of the surface $0 < x \leq x_L$ near the leading edge of the film flow, with turbulent flow for $x > x_L$, as indicated in Fig. 9.11. The length of the laminar regime

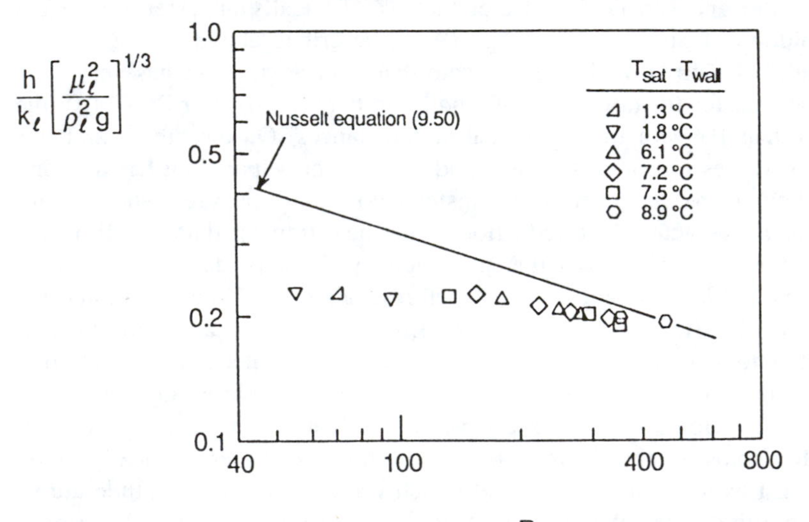

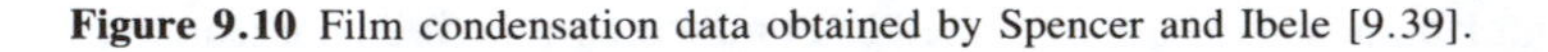

Figure 9.10 Film condensation data obtained by Spencer and Ibele [9.39].

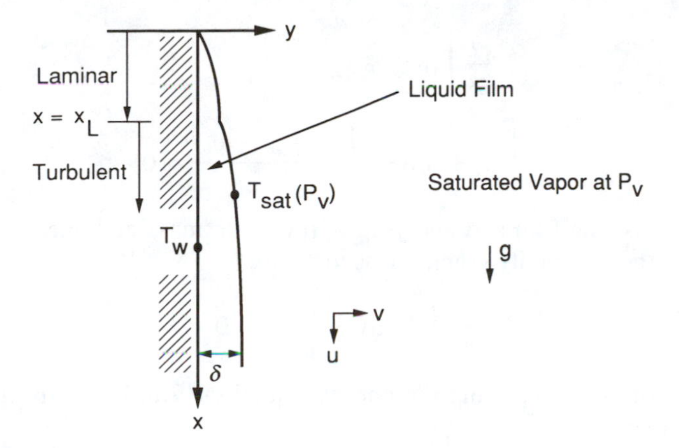

Figure 9.11 System model for analysis of turbulent film condensation.

compared to the overall surface length dictates whether the flow is all laminar, partially laminar and partially turbulent, or mostly turbulent.

In the turbulent regime, the governing two-dimensional boundary-layer forms of the mass, momentum, and energy balance equations for the film can be written approximately as

$$\frac{\partial u}{\partial x} + \frac{\partial v}{\partial y} = 0 \tag{9.80}$$

$$u\frac{\partial u}{\partial x} + v\frac{\partial u}{\partial y} = \frac{\partial}{\partial y}\left[(\nu_l + \epsilon_M)\frac{\partial u}{\partial y}\right] + \frac{g(\rho_l - \rho_v)}{\rho_l} \tag{9.81}$$

$$u\frac{\partial T}{\partial x} + v\frac{\partial T}{\partial y} = \frac{\partial}{\partial y}\left[(\alpha_{T,l} + \epsilon_H)\frac{\partial T}{\partial y}\right] \tag{9.82}$$

where ϵ_M and ϵ_H are the eddy diffusivities of momentum and heat, respectively. The other variables are identical to those for the laminar case described above. Neglecting interfacial shear effects and assuming that the surrounding vapor is saturated, the corresponding boundary conditions are

$$\text{at } y = 0: \qquad u = v = 0 \qquad T = T_w \tag{9.83}$$

$$\text{at } y = \delta: \qquad \frac{\partial u}{\partial y} = 0 \qquad T = T_{\text{sat}}(P_v) \tag{9.84}$$

The convection terms in the energy and momentum equations are often neglected because the transport rate across the film is much greater than downstream convection. Ignoring the downstream (x) variation in the u and temperature fields compared to the cross-stream variation, the energy and momentum equations simplify to:

$$\frac{\partial}{\partial y}\left[(\alpha_{T,l} + \epsilon_H)\frac{\partial T}{\partial y}\right] = 0 \tag{9.85}$$

$$\frac{\partial}{\partial y}\left[(\nu_l + \epsilon_M)\frac{\partial u}{\partial y}\right] + \frac{g(\rho_l - \rho_v)}{\rho_l} = 0 \tag{9.86}$$

Integrating across the film and applying appropriate mass and energy balances at the interface, the continuity equation becomes

$$\frac{d}{dx}\int_0^\delta u\,dy - \frac{q''_{y=\delta}}{\rho_l h_{lv}} = 0 \tag{9.87}$$

In a similar fashion, integrating the energy equation (9.85) first from 0 to y and then from 0 to δ yields

$$\frac{T_{\text{sat}} - T_w}{q''_w/k_l} = \frac{k_l}{h} = \int_0^\delta \frac{dy}{1 + (\text{Pr}_l/\text{Pr}_t)(\epsilon_M/\nu_l)} \tag{9.88}$$

where h is the local heat transfer coefficient and Pr_t is the turbulent Prandtl number, defined as $\text{Pr}_t = \epsilon_M/\epsilon_H$.

Before the above system of equations (9.86) through (9.88) can be analyzed further, some means of determining eddy diffusivities must be specified. If the variations of the turbulent Prandtl number and ϵ_M with y are known, the heat transfer coefficient h at a given x location can be determined by the following sequence of steps:

1. The velocity profile $u(y)$ can be determined by integration of Eq. (9.86).
2. Using the velocity profile determined in step 1, the mass flow rate per unit width of surface is determined as

$$\dot{m}' = \rho_l \int_0^\delta u\,dy \tag{9.89}$$

3. Differentiating the expression for $\dot{m}'$ obtained in step 2 with respect to δ, a relation for $d\dot{m}'/d\delta$ is obtained.
4. An energy balance at the interface requires that

$$h_{lv}\left(\frac{d\dot{m}'}{dx}\right) = h(T_{\text{sat}} - T_w) \tag{9.90}$$

   This equation is solved for $d\dot{m}'/dx$.
5. Executing the integration in Eq. (9.88) yields a relation for the heat transfer coefficient h as a function of δ.
6. The relations for $d\dot{m}'/d\delta$, $d\dot{m}'/dx$, and $h(x)$ obtained in steps 3 through 5 are combined to obtain a relation for $d\delta/dx$. The value of δ for a given x location is obtained by integrating the resulting relation for $d\delta/dx$.
7. Substituting the value of δ obtained in step 6 into the relation for $h(x)$ obtained in step 5, the value of the local heat transfer coefficient h for a given x location is obtained.

Seban [9.41] was apparently the first to propose using known information about fully developed turbulent pipe flows to evaluate the eddy diffusivities in the falling-film governing equations. He postulated that turbulent flow in the film consisted of a viscous sublayer near the wall and a fully turbulent region farther away, as indicated in Fig. 9.12, with a buffer layer between. Using the reduced momentum equation (9.86), it can easily be shown that the shear stress must vary linearly from a finite value of $\tau_w = \rho_l g \delta$ at the wall to zero at the interface (see Fig. 9.12):

$$\tau = \tau_w \left(\frac{1 - y}{\delta} \right) \tag{9.91}$$

Using Eq. (9.91) for the shear stress together with the universal velocity profiles for turbulent flow,

$$0 \le y^+ \le 5: \qquad u^+ = y^+ \tag{9.92a}$$

$$5 \le y^+ \le 30: \qquad u^+ = -3.05 + 5 \ln y^+ \tag{9.92b}$$

$$30 < y^+: \qquad u^+ = 5.5 + 2.5 \ln y^+ \tag{9.92c}$$

$$y^+ = \frac{y}{\nu_l} \sqrt{\frac{\tau_w}{\rho_l}} \qquad u^+ = u \sqrt{\frac{\rho_l}{\tau_w}} \tag{9.92d}$$

and the definition of the eddy diffusivity,

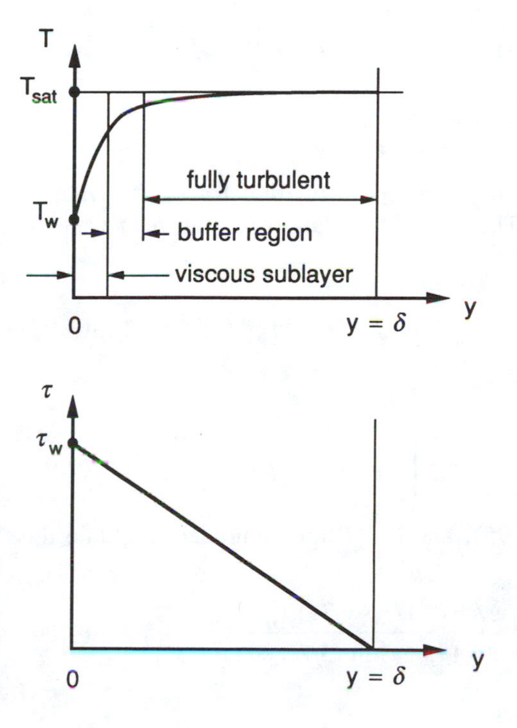

Figure 9.12 Temperature and shear stress distribution in a falling liquid film.

$$\epsilon_M = \frac{\tau/\rho_l}{\partial u/\partial y} - \nu_l \tag{9.93}$$

Seban [9.41] obtained the following relations for the eddy diffusivity:

$$0 \leq y^+ \leq 5: \qquad \epsilon_M = 0 \tag{9.94a}$$

$$5 \leq y^+ \leq 30: \qquad \epsilon_M = \frac{\sqrt{g\delta}}{5}\left(1 - \frac{y}{\delta}\right)y - \nu_l \tag{9.94b}$$

$$30 < y^+: \qquad \epsilon_M = \frac{\sqrt{g\delta}}{2.5}\left(1 - \frac{y}{\delta}\right)y \tag{9.94c}$$

In the fully turbulent regime $30 < y^+$, the molecular diffusivity has been neglected compared to the turbulent eddy diffusivity.

Assuming that the viscous sublayer and buffer regions are thin, the total mass flow per unit length can be computed using the velocity profile in the fully turbulent range. Following steps 1 through 4, Seban [9.41] obtained the following relation for $d\dot{m}'/d\delta^+$:

$$\frac{d\dot{m}'}{d\delta^+} = \mu_l(5.5 + 2.5 \ln \delta^+) \tag{9.95}$$

The integral relation (9.88) for the heat transfer coefficient can then be written in terms of y^+ and broken down into three parts:

$$\frac{\rho_l c_{pl}\sqrt{g\delta}}{h} = \int_0^5 \frac{dy^+}{(1/\text{Pr}_l) + (\epsilon_M/\text{Pr}_t\,\nu_l)} + \int_5^{30} \frac{dy^+}{(1/\text{Pr}_l) + (\epsilon_M/\text{Pr}_t\,\nu_l)} + \int_{30}^{\delta^+} \frac{dy^+}{(1/\text{Pr}_l) + (\epsilon_M/\text{Pr}_t\,\nu_l)} \tag{9.96}$$

Seban [9.41] took $\text{Pr}_t = 1$ and used the appropriate relation for ϵ_M to evaluate each of the integrals in Eq. (9.96). The resulting relation for the heat transfer coefficient is

$$\frac{h}{k_l}\left(\frac{\nu^2}{g}\right)^{1/3} = \text{Pr}_l(\delta^+)^{1/3}[5\,\text{Pr}_l + 5 \ln(5\,\text{Pr}_l + 1) + I^*]^{-1} \tag{9.97}$$

where

$$I^* = \frac{2.5}{S}\left[\frac{(1+S)(60/\delta^+) - 1 - S}{(1-S)(60/\delta^+) - 1 + S}\right] \qquad S = \sqrt{1 + \frac{10}{\text{Pr}_l\,\delta^+}} \tag{9.98}$$

Following step 6, Eqs. (9.90), (9.95), and (9.97) are combined to obtain the differential equation

$$\frac{d\delta^+}{dx} = \frac{k_l\,\text{Pr}_l(\delta^+ g/\nu_l^2)^{1/3}(T_{\text{sat}} - T_w)}{\mu_l h_{lv}(5.5 + 2.5 \ln \delta^+)[5\,\text{Pr}_l + 5 \ln(5\,\text{Pr}_l + 1) + I^*]} \tag{9.99}$$

Integrating from the point of transition to turbulence (x_L and δ_L^+ at $Re_L = 4\dot{m}'/\mu_l = 1600$) to any location of interest (x and δ^+), the following implicit relation for δ^+ is obtained:

$$\int_{\delta_L^+}^{\delta^+} \frac{\mu_l h_{lv}(5.5 + 2.5 \ln \delta^+)[5 \, Pr_l + 5 \ln (5 \, Pr_l + 1) + I^*]}{(\delta^+)^{1/3}} \, d\delta^+ = \left(\frac{g}{\nu_l^2}\right)^{1/3} \left[\frac{c_{pl}(T_{sat} - T_w)}{h_{lv}}\right](x - x_L) \tag{9.100}$$

Finally, in step 7, upon integrating Eq. (9.100), the resulting value of δ^+ can be substituted into Eq. (9.97) to determine the local value of h at the specified x location. Note that, to do the integration indicated in Eq. (9.100), we must first use the laminar solution to get x_L and δ_L^+ for $Re_L = 1600$.

Although subsequent investigations by Dukler [9.42] and Lee [9.43] refined this method of predicting the heat transfer during falling-film condensation on a flat, vertical surface, the structure of the methods, as indicated in steps 1 through 7 above, was basically unchanged from that in Seban's [9.41] original study. This methodology was also subsequently extended to evaporation of a falling liquid film by Dukler [9.42], Kunz and Yerazunis [9.44], Chun and Seban [9.45], Mills and Chung [9.46], and Mostofizadeh and Stephan [9.47].

More recent studies (see, e.g., Mills and Chung [9.46]) have suggested that the presence of the interface tends to damp larger turbulent eddies near the interface in the liquid film. This implies that a viscous sublayer exists at the interface as well as at the wall. Recent efforts to model falling-film evaporation and condensation processes by Kutateladze [9.48] and by Sandall et al. [9.49] have therefore included a variation of the eddy viscosity in which it goes to zero at both the wall and the interface. Kutateladze [9.48] used the following variation of ϵ_M with y^+:

$$0 < y^+ < 6.8: \qquad \epsilon_M = 0 \tag{9.101a}$$

$$6.8 < y^+ < 6.8 + 0.2(\delta^+ - 6.8): \qquad \epsilon_M = 0.4(y^+ - 6.8)\sqrt{1 - \frac{y^+}{\delta^+}} \tag{9.101b}$$

$$6.8 + 0.2(\delta^+ - 6.8) < y^+ < \delta^+: \qquad \epsilon_M = 0.08(y^+ - 6.8)\sqrt{1 - \frac{y^+}{\delta^+}} \tag{9.101c}$$

The variation of the mean Nusselt number with Re_L and Pr_l predicted using these equations in the analysis described above is shown in Fig. 9.13. The variation for fully turbulent flow is qualitatively similar to that predicted by the following empirical correlation recommended by Colburn [9.50]:

$$\frac{\bar{h}}{k_l}\left(\frac{\nu_l^2}{g}\right)^{1/3} = 0.056 \, Pr_l^{1/3} \, Re_L^{0.2} \tag{9.102}$$

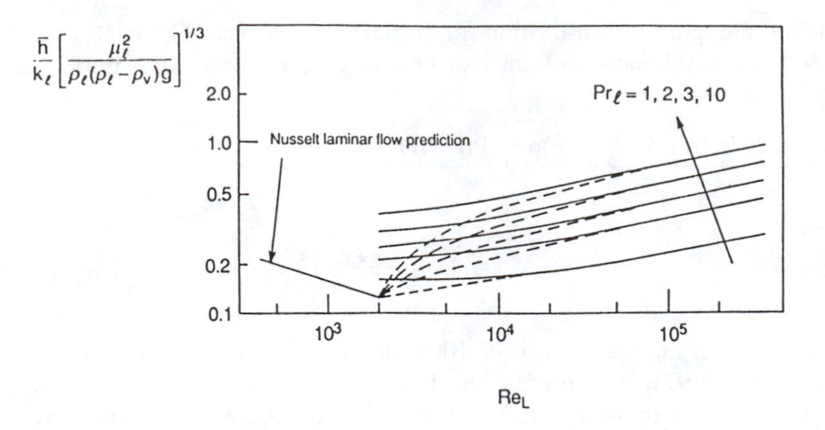

Figure 9.13 Variation of the Nusselt number with Reynolds and Prandtl numbers as predicted by the analysis of Kutateladze [9.48] for turbulent film condensation. *(Adapted with permission from reference [9.48], copyright © 1982, Pergamon Press.)*

The dotted lines in Fig. 9.13 indicate the interpolated variation predicted by the relation

$$\overline{Nu} = \frac{\bar{h}}{k_l}\left(\frac{\nu_l^2}{g}\right)^{1/3} = (\overline{Nu})_{lam}\left(\frac{Re_{L,trans}}{Re_L}\right) + (\overline{Nu})_{turb}\left(\frac{Re_L - Re_{L,trans}}{Re_L}\right) \tag{9.103}$$

where the transition Reynolds number $Re_{L,trans}$ is taken to be 2000 and $(\overline{Nu})_{lam}$ is the value predicted by the laminar Nusselt correlation (9.50) at $Re_{L,trans}$.

Based on experimental data for liquid Prandtl numbers between 1 and 5, Grober et al. [9.51] proposed a film condensation heat transfer correlation that can be cast in the form

$$\frac{\bar{h}}{k_l}\left(\frac{\nu_l^2}{g}\right)^{1/3} = 0.0131\ Re_L^{1/3} \tag{9.104}$$

This relation was recommended at Re_L values above the transition value, which was taken to be Re_L = 1400.

Example 9.3 Estimate the mean heat transfer coefficient using Eqs. (9.50), (9.102), and (9.103) for condensation of steam at atmospheric pressure on a 1.0- by 1.0-m vertical plate with a heat removal rate of 500 kW/m^2.

For saturated water at atmospheric pressure, T_{sat} = 100°C, ρ_l = 958 kg/m^3, ρ_v = 0.597 kg/m^3, h_{lv} = 2257 kJ/kg, k_l = 0.679 W/m K, Pr_l = 1.72, and $\mu_l = 2.78 \times 10^{-4}$ N s/m^2. It follows that

$$\dot{m}_L' = \frac{q''L}{h_{lv}} = \frac{(5 \times 10^5)(1.0)}{2{,}257{,}000} = 0.222 \text{ kg/ms}$$

$$Re_L = \frac{4\dot{m}_L'}{\mu_l} = \frac{4(0.222)}{2.78 \times 10^{-4}} = 3{,}187$$

Equation (9.50) can be written approximately as

$$\overline{\mathrm{Nu}}_{\mathrm{lam}} = \frac{\bar{h}_l(\mu_l^2/g\rho_l^2)^{1/3}}{k_l} = 1.47\ \mathrm{Re}_L^{-1/3}$$

Substituting for $\mathrm{Re}_{L,\mathrm{trans}} = 2{,}000$,

$$\overline{\mathrm{Nu}}_{\mathrm{lam}} = 1.47(2{,}000)^{-1/3} = 0.117$$

Using Eq. (9.102) for turbulent flow,

$$\overline{\mathrm{Nu}}_{\mathrm{turb}} = 0.056\ \mathrm{Pr}_l^{1/3}\ \mathrm{Re}_L^{0.2}$$

$$= 0.056(1.72)^{1/3}(3187)^{0.2} = 0.337$$

The overall Nusselt number is given by Eq. (9.103):

$$\overline{\mathrm{Nu}} = \overline{\mathrm{Nu}}_{\mathrm{lam}}\left(\frac{2{,}000}{\mathrm{Re}_L}\right) + \overline{\mathrm{Nu}}_{\mathrm{turb}}\left(\frac{\mathrm{Re}_L - 2{,}000}{\mathrm{Re}_L}\right)$$

$$= (0.117)\left(\frac{2{,}000}{3{,}187}\right) + (0.337)\left(\frac{3{,}187 - 2{,}000}{3{,}187}\right)$$

$$= 0.199$$

It follows that the mean heat transfer coefficient is given by

$$\bar{h} = k_l\left(\frac{g\rho_l^2}{\mu_l^2}\right)^{1/3}\overline{\mathrm{Nu}}$$

$$= 0.679\left[9.8\left(\frac{958}{2.78 \times 10^{-4}}\right)^2\right]^{1/3}(0.199)$$

$$= 6{,}596\ \mathrm{W/m^2\ K} = 6.596\ \mathrm{kW/m^2\ K}$$

The analysis tools and correlations described above work reasonably well for Pr_l values above 1. However, deviation of the predictions using these methods from heat transfer data for liquid metals can be quite significant. Generally, Dukler's [9.42] somewhat idealized treatment for film condensation has been found to agree reasonably well with falling-film condensation heat transfer data for liquid metals. On the other hand, computed results obtained using Lee's [9.43] more detailed analysis were found to deviate more from available liquid metal data. Based on the currently available data, it appears that no model or correlation presently exists that will accurately predict falling-film condensation heat transfer coefficients for both liquid metals and liquids with Pr_l values greater than 1.0. Further experimental investigation appears to be necessary to resolve the differences between trends predicted by analytical models of turbulent falling-film condensation heat transfer and the current body of experimental data for liquid metals.

9.4 FILM CONDENSATION ON CYLINDERS AND AXISYMMETRIC BODIES

Because of its importance to the design of tube-and-shell condensers, condensation on the outside of horizontal tubes has been the subject of numerous studies. The length of the tube perimeter over which the condensate flows is usually small for commonly used tubes. Consequently, the film Reynolds number is usually low, and the flow in the liquid film is laminar.

With slight modification, the Nusselt [9.31] analysis of laminar falling-film condensation over a flat plate can be adapted to film condensation on an isothermal horizontal cylinder, as shown in Fig. 9.14. The analysis is, in fact, very similar to that for laminar film boiling on a horizontal cylinder discussed in Section 7.6. (Historically, however, Nusselt's [9.31] analysis of film condensation preceded the comparable analysis for film boiling.)

As for vertical surfaces, if convective terms are neglected, an energy balance on a differential element of the film requires that

$$h_{lv}\frac{d\dot{m}'}{dx} = \frac{k_l[T_{\text{sat}}(P_v) - T_w]}{\delta} \tag{9.105}$$

The component of the gravitational body force acting tangential to the surface of the cylinder is $g \sin \Omega$. Assuming that the momentum balance is dominated by viscous and gravity body forces, Eq. (9.39) again applies, except that g is replaced

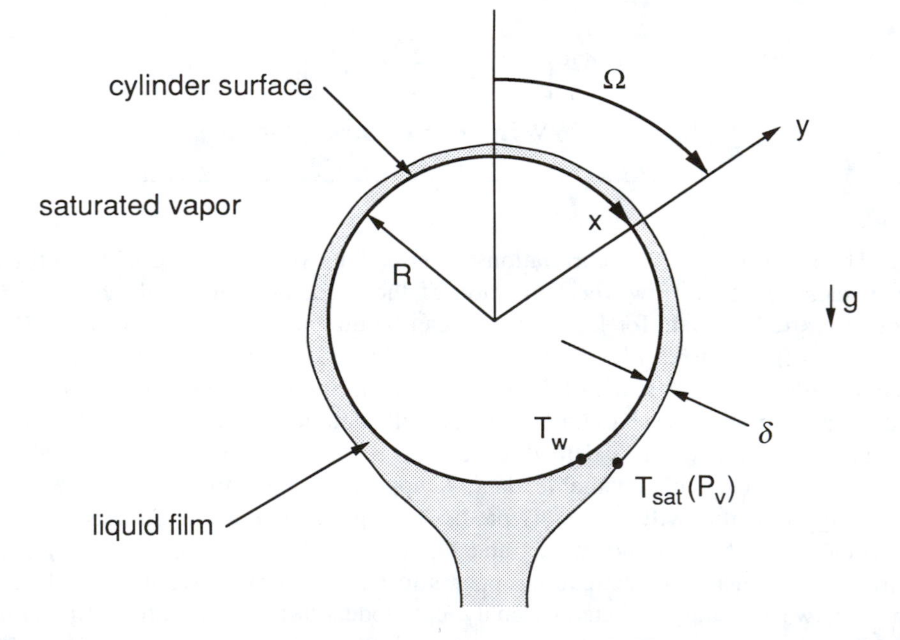

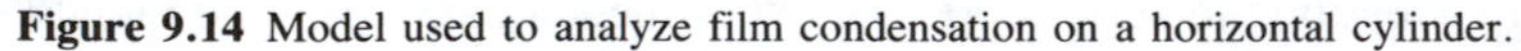

Figure 9.14 Model used to analyze film condensation on a horizontal cylinder.

by $g \sin \Omega$. Integrating this equation and using the boundary condition $u = 0$ at $y = 0$ yields

$$u = \frac{(\rho_l - \rho_v) g \sin \Omega}{\mu_l} \left(y\delta - \frac{y^2}{2} \right) \tag{9.106}$$

The condensate mass flow rate per unit of tube length $\dot{m}'$ is obtained, as before, by integrating the above velocity profile:

$$\dot{m}' = \rho_l \int_0^\delta u \, dy = \frac{\rho_l (\rho_l - \rho_v) \delta^3 g \sin \Omega}{3\mu_l} \tag{9.107}$$

Using the fact that $\Omega = x/R$ and substituting Eq. (9.107) into Eq. (9.105) yields, after some manipulation,

$$(\dot{m}')^{1/3} \, d\dot{m}' = \frac{Rk_l (T_{\text{sat}} - T_w)}{h_{lv}} \left[\frac{(\rho_l - \rho_v) g}{3\nu_l} \right]^{1/3} \sin^{1/3} \Omega \, d\Omega \tag{9.108}$$

This relation can be integrated from $\Omega = 0$ to $\Omega = \pi$ to obtain the liquid condensed over half the cylinder per unit length (in the axial direction):

$$\dot{m}' = 1.924 \left[\frac{R^3 k_l^3 (T_{\text{sat}} - T_w)^3 (\rho_l - \rho_v) g}{h_{lv}^3 \nu_l} \right]^{1/4} \tag{9.109}$$

An overall energy balance over a control volume surrounding the cylinder requires that

$$2h_{lv} \dot{m}' = 2\pi R \overline{h} (T_{\text{sat}} - T_w) \tag{9.110}$$

Substituting Eq. (9.109) for $\dot{m}'$ and solving the above equation for $\overline{h}$, we obtain the following relation for the Nusselt number:

$$\overline{\text{Nu}_D} = \frac{\overline{h} D}{k_l} = 0.728 \left(\frac{\text{Ra}}{\text{Ja}} \right)^{1/4} \tag{9.111a}$$

where

$$\text{Ra} = \frac{g(\rho_l - \rho_v) \, \text{Pr}_l \, D^3}{\rho_l \nu_l^2} \tag{9.111b}$$

$$\text{Ja} = \frac{c_{pl} (T_{\text{sat}} - T_w)}{h_{lv}} \tag{9.111c}$$

and $D = 2R$. Alternatively, Eqs. (9.109) and (9.111) can be combined to eliminate $(T_{\text{sat}} - T_w)$, which allows the relation for the mean heat transfer coefficient to be written in the form

$$\frac{\overline{h}}{k_l} \left[\frac{\mu_l^2}{\rho_l (\rho_l - \rho_v) g} \right]^{1/3} = 1.92 \, \text{Re}_L^{-1/3}, \qquad \text{Re}_L = \frac{4\dot{m}'}{\mu_l} \tag{9.112}$$

Selin [9.52] found that better agreement with film condensation data for horizontal tubes was obtained by replacing the constants in Eqs. (9.111*a*) and (9.112) by 0.61 and 1.27, respectively.

If the film flow around each half of the tube is considered to be equivalent to film flow over a flat plate of equal length, the flow is expected to become turbulent when Re_L is greater than 1400. For tubes in most condensers, Reynolds numbers this high are usually not reached, and the condensate film flow is rarely turbulent. As in the case of vertical plates, film condensation heat transfer coefficients for liquid metals on horizontal tubes are typically 10 to 70% below the values predicted by these correlations [9.53].

By replacing g with $g \cos \phi$, where ϕ is the angle of inclination, Selin [9.53] found that the correlation (9.111) for horizontal round tubes (with his recommended coefficient) matched inclined tube data within 15% for $0° \leq \phi \leq 60°$. This simple modification works well for predicting the mean heat transfer coefficient, apparently because most of the condensation occurs near the top of the tube where the film is thin and the effects of flow axially along the tube are small compared to transport across the film. At the bottom of the tube, where the film is thicker, although axial effects may be significant, this region contributes little to the overall heat transfer. Hence the inaccurate treatment of this region does not strongly affect the accuracy of the overall prediction of the mean heat transfer coefficient. A more detailed model for film condensation on inclined tubes has been proposed by Sheynkman and Linetskiy [9.54].

As noted by Nusselt [9.31] in his pioneering study of laminar film condensation, the film-flow analysis for horizontal tubes can be extended in a straightforward manner to an in-line bank of tubes (see Fig. 9.15). The analysis of each individual tube in the bank is identical to that for a single tube, as described above, up to the derivation of Eq. (9.108). Integration of that equation then requires the use of a different boundary condition at $\Omega = 0$ for each tube, because condensate from the tube immediately above is assumed to attach to the tube below at that location. It can easily be shown by integrating Eq. (9.108) that, for tube i in the bank,

$$(\dot{m}')^{4/3}_{i,\mathrm{bottom}} = (\dot{m}')^{4/3}_{i,\mathrm{top}} + M \tag{9.113}$$

where

$$M = 2.393 \left[\frac{R^3 k_l^3 (T_{\mathrm{sat}} - T_w)^3 (\rho_l - \rho_v) g}{h_{lv}^3 \nu_l} \right]^{1/3} \tag{9.114}$$

Conservation of liquid also requires that

$$(\dot{m}')^{4/3}_{i,\mathrm{top}} = (\dot{m}')^{4/3}_{i-1,\mathrm{bottom}} \tag{9.115}$$

Combining these relations yields

$$(\dot{m}')^{4/3}_{i,\mathrm{bottom}} - (\dot{m}')^{4/3}_{i-1,\mathrm{bottom}} = M \tag{9.116}$$

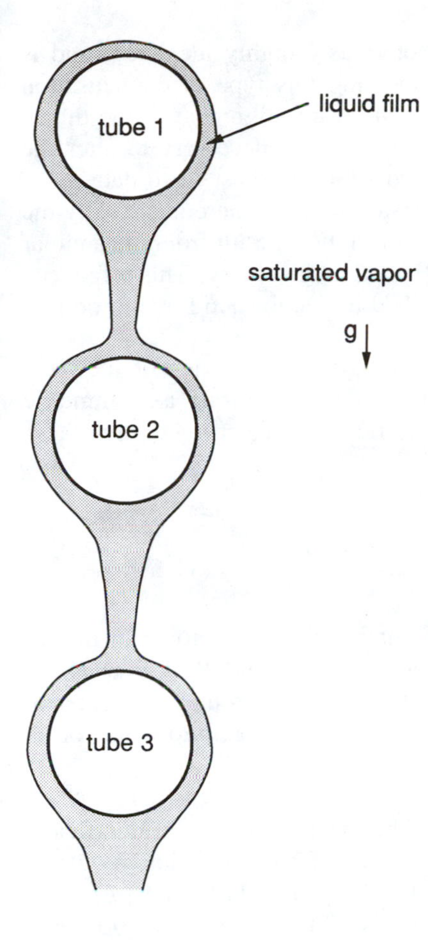

Figure 9.15 Idealized model of film condensation on an in-line bank of round tubes.

If i is interpreted as a continuous variable instead of an integer, the above expression can be considered to be equivalent to

$$\frac{d}{di}[(\dot{m}')^{4/3}_{\text{bottom}}] = M \tag{9.117}$$

Integrating this relation, using the fact that $(\dot{m}')^{4/3}_{\text{bottom}} = M$ at $i = 1$ yields

$$(\dot{m}')^{4/3}_{i,\text{bottom}} = iM \tag{9.118}$$

Neglecting sensible cooling of the film, the mean heat transfer coefficient for the bank of tubes $\bar{h}$ is then found from conservation of mass and energy as

$$2\pi Ri\bar{h}(T_{\text{sat}} - T_w) = 2h_{lv}(\dot{m}')_{i,\text{bottom}} \tag{9.119}$$

Combining the above relations, the following relation for $\bar{h}$ can be obtained:

$$\frac{\bar{h}(nD)}{k_l} = 0.728\left[\frac{g(\rho_l - \rho_v)(nD)^3 h_{lv}}{k_l \nu_l (T_{\text{sat}} - T_w)}\right]^{1/4} \tag{9.120}$$

Nusselt's [5.31] treatment of this problem is obviously highly idealized, and is perhaps best viewed as a crude first cut at modeling this type of condensation process. Effects such as liquid subcooling, splashing, misalignment of the tubes, vibration of the tubes, and film waviness usually make it necessary to alter the coefficient on the right side of this correlation to achieve a best fit to data.

Analytical treatment of laminar film condensation on a sphere is virtually the same as that for a horizontal cylinder. The only differences result from the angular variation of the body perimeter because of the spherical geometry. This effect can be included in the analysis in the manner described in Section 7.6 for film boiling from a sphere.

Dhir and Lienhard [9.55] have shown that analytical prediction of the local heat transfer coefficient for laminar film condensation on arbitrary axisymmetric bodies can be generated by replacing g in the vertical surface correlation with an effective value g_{eff}, given by

$$g_{\text{eff}} = \frac{x(gR)^{4/3}}{\int_0^x g^{1/3} R^{4/3}\, dx} \tag{9.121}$$

In this formula, $g = g(x)$ is the local component of the gravity force in the direction of the streamwise coordinate along the body (x), and $R = R(x)$ is the body's local radius of curvature in a vertical plane cutting through the center of the body. Based on the above discussion, this method is expected to work for all fluids except liquid metals.

Example 9.4 Use Eq. (9.120) to estimate the mean heat transfer coefficient for condensation of R-12 on the outside of a row of 10 tubes with an outside diameter of 2.0 cm. Motionless saturated R-12 vapor at 1145 kPa surrounds the tubes. The wall temperature of the tubes is 27°C.

For saturated R-12 at 1145 kPa, $T_{\text{sat}} = 320$ K, $\rho_l = 1{,}225$ kg/m^3, $\rho_v = 65.4$ kg/m^3, $h_{lv} = 127.2$ kJ/kg, $\mu_l = 187 \times 10^{-4}$ N s/m^2, and $k_l = 0.0598$ W/m K. Equation (9.120) can be written as

$$\bar{h} = 0.728\left(\frac{k_l}{nD}\right)\left[\frac{g\rho_l(\rho_l - \rho_v)(nD)^3 h_{lv}}{k_l \mu_l (T_{\text{sat}} - T_w)}\right]^{1/4}$$

Substituting yields

$$\bar{h} = 0.728\left[\frac{0.0598}{10(0.02)}\right]\left[\frac{9.8(1{,}225)(1{,}225 - 65.4)[10(0.02)]^3 127{,}200}{(0.0598)(1.87 \times 10^{-4})(320 - 300)}\right]^{1/4}$$

$$= 1{,}092 \text{ W/m}^2\text{ K}$$

Note that $\bar{h}_1 = \bar{h}n^{1/4} = 1{,}092(10)^{1/4} = 1{,}942$ W/m^2 K.

It can be seen that the mean heat transfer coefficient for the first tube is substantially higher than the mean for the entire bank of tubes.

9.5 EFFECTS OF VAPOR MOTION AND INTERFACIAL WAVES

The effects of downward motion of the surrounding vapor on falling-film condensation on vertical plates or tubes were examined analytically in an early study by Rohsenow, Webber, and Ling [9.56]. For laminar flow under such circumstances, the assumption of a linear variation of the shear stress across the film is still justifiable. At the interface, however, the shear stress in the liquid film approaches a constant value, dictated mainly by the vapor flow field near the interface (see Fig. 9.16).

The integral analysis of falling-film condensation with downward interfacial shear basically follows the Nusselt-type analysis for zero interfacial shear described in Section 9.4. Considering the shaded differential element in Fig. 9.7 with interfacial shear τ_i present, the momentum balance (neglecting the inertia and downstream diffusion terms) requires that

$$(\delta - y)\, dx\left(\rho_l g - \frac{dP}{dx}\right) + \tau_i\, dx = \mu_l\left(\frac{du}{dy}\right) dx \tag{9.122}$$

In this momentum balance the pressure gradient along the surface dP/dx is equal to the hydrostatic gradient plus an imposed gradient that drives the vapor motion:

$$\frac{dP}{dx} = \left(\frac{dP}{dx}\right)_{\text{hyd}} + \left(\frac{dP}{dx}\right)_m = \rho_v g + \left(\frac{dP}{dx}\right)_m \tag{9.123}$$

For convenience, we define a fictitious vapor density ρ_v^* such that

$$\rho_v^* g = \rho_v g + \left(\frac{dP}{dx}\right)_m \tag{9.124}$$

The momentum balance (9.122) can be integrated using $u = 0$ at $y = 0$ to obtain

$$u = \frac{(\rho_l - \rho_v^*)g}{\mu_l}\left(y\delta - \frac{y^2}{2}\right) + \frac{\tau_i y}{\mu_l} \tag{9.125}$$

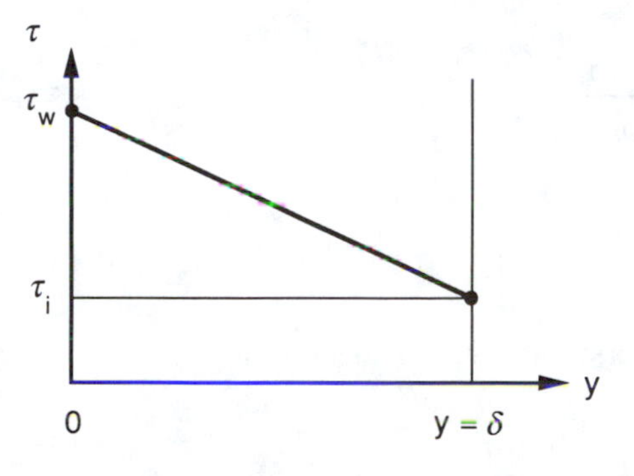

Figure 9.16 Model shear stress variation in a liquid film subjected to interfacial shear.

Using this velocity profile to evaluate the integral relation (see Eq. [9.41]) for $\dot{m}'$ yields

$$\dot{m}' = \frac{\rho_l(\rho_l - \rho_v^*)g\delta^3}{3\mu_l} + \frac{\tau_i \rho_l \delta^2}{2\mu_l} \tag{9.126}$$

Differentiating with respect to δ, we obtain

$$\frac{d\dot{m}'}{d\delta} = \frac{\rho_l(\rho_l - \rho_v^*)g\delta^2}{\mu_l} + \frac{\tau_i \rho_l \delta}{\mu_l} \tag{9.127}$$

The energy and mass balance relations (9.44) and (9.45) are still valid for these circumstances. Combining these with Eq. (9.127) yields

$$\frac{d\delta}{dx} = \frac{k_l \mu_l (T_{\text{sat}} - T_w)}{\rho_l(\rho_l - \rho_v^*)g h_{lv} \delta^3 + \tau_i \rho_l h_{lv} \delta^2} \tag{9.128}$$

This equation is integrated using the condition that $\delta = 0$ at $x = 0$ to obtain

$$\frac{4xk_l \mu_l (T_{\text{sat}} - T_w)}{\rho_l(\rho_l - \rho_v^*)g h_{lv}} = \delta^4 + \frac{4\tau_i \delta^3}{3(\rho_l - \rho_v^*)g} \tag{9.129}$$

If sensible subcooling in the liquid film is included in the energy balance, the above analysis results in the same equation, except that h_{lv} is replaced by h'_{lv} given by Eq. (9.55):

$$\frac{4xk_l \mu_l (T_{\text{sat}} - T_w)}{\rho_l(\rho_l - \rho_v^*)g h'_{lv}} = \delta^4 + \frac{4\tau_i \delta^3}{3(\rho_l - \rho_v^*)g} \tag{9.130}$$

In analyzing this process, Rohsenow et al. [9.56] made use of the following dimensionless parameters:

$$\delta^* = \frac{\delta}{L_F} \tag{9.131}$$

$$x^* = \left(\frac{x}{L_F}\right) \frac{4c_{pl}(T_{\text{sat}} - T_w)}{\text{Pr}_l \, h'_{lv}} \tag{9.132}$$

$$\tau_i^* = \frac{\tau_i}{L_F(\rho_l - \rho_v^*)g} \tag{9.133}$$

where

$$L_F = \left[\frac{\mu_l^2}{\rho_l(\rho_l - \rho_v^*)g}\right]^{1/3} \tag{9.134}$$

In terms of these dimensionless parameters, Eq. (9.130) becomes

$$x^* = (\delta^*)^4 + \frac{4}{3}(\delta^*)^3 \tau_i^* \tag{9.135}$$

and it can be shown that

$$\overline{\mathrm{Nu}_F} = \frac{\bar{h}_L L_F}{k_l} = \frac{4}{3}\frac{(\delta^*)^3}{x^*} + \frac{2(\delta^*)^2\tau_i^*}{x^*} \tag{9.136}$$

$$\mathrm{Re}_L = \frac{4\dot{m}'}{\mu_l} = \frac{4}{3}(\delta^*)^3 + 2(\delta^*)^2\tau_i^* \tag{9.137}$$

If the length of the surface x, its temperature, the physical properties of the fluids, and the interfacial shear are specified, then x^* and τ_i^* can be computed from Eqs. (9.132) through (9.134). For the x^* and τ_i^* values thus determined, the three nonlinear equations (9.135) through (9.137) can be solved for δ^*, $\overline{\mathrm{Nu}_F}$, and Re_L. Alternatively, if τ_i^* and Re_L are specified, the equations can be solved for x^*, $\overline{\mathrm{Nu}_F}$, and δ^*. Rohsenow et al. [9.56] presented plots of $\overline{\mathrm{Nu}_F}$ as a function of τ_i^* and Re_L computed in this manner.

For low values of interfacial shear stress, Rohsenow et al. [9.56] also presented theoretical arguments from which they concluded that transition to turbulent film flow occurs at a transition Reynolds number $\mathrm{Re}_{L,\mathrm{tr}}$, given by

$$\mathrm{Re}_{L,\mathrm{tr}} = 1{,}800 - 246\left(1 - \frac{\rho_v}{\rho_l}\right)^{1/3}\tau_i^* + 0.667\left(1 - \frac{\rho v}{\rho l}\right)(\tau_i^*)^3 \tag{9.138}$$

For condensation at film Reynolds numbers beyond this transition point, Rohsenow et al. [9.56] extended Seban's [9.41] falling-film condensation analysis for turbulent flow with no vapor shear to include the effect of vapor shear at the interface. The results of this extended analysis were combined with those for the above laminar analysis to predict the variation of $\bar{h}_L$ over the a wide range of Re_L values. Specifically, the laminar analysis was used to predict δ and $\dot{m}'$ at the transition point. These values were used as boundary conditions in the integration of the equation governing the turbulent transport. Computed results for $\mathrm{Pr}_l = 10$ are shown in Fig. 9.17. For $2 \leq \mathrm{Pr}_l \leq 3$ and $5 \leq \tau_i^* \leq 50$, Rohsenow et al. [9.56] found that their computed results agreed well with the relation obtained by Carpenter and Colburn [9.57], which can be cast in the form

$$\frac{\bar{h}_L}{k_l}\left(\frac{\nu_l^2}{g}\right)^{1/3} = 0.065\,\mathrm{Pr}_l^{1/2}(\tau_i^*)^{1/2} \tag{9.139}$$

The later analysis by Dukler [9.42] similarly treated falling-film condensation with interfacial shear present. However, for fully turbulent flow conditions, Dukler used Deissler's [9.58] equation for the eddy viscosity variation near the solid wall ($y^+ < 20$) and von Karman's relation farther away.

for $y^+ \leq 20$:
$$\epsilon_M = \epsilon_H = n^2uy\left[1 - \exp\left(\frac{-n^2uy}{\nu_l}\right)\right] \tag{9.140a}$$

for $y^+ > 20$:
$$\epsilon_M = \epsilon_H = 0.4\left(\frac{du}{dy}\right)^3\left(\frac{d^2u}{dy^2}\right)^2 \tag{9.140b}$$

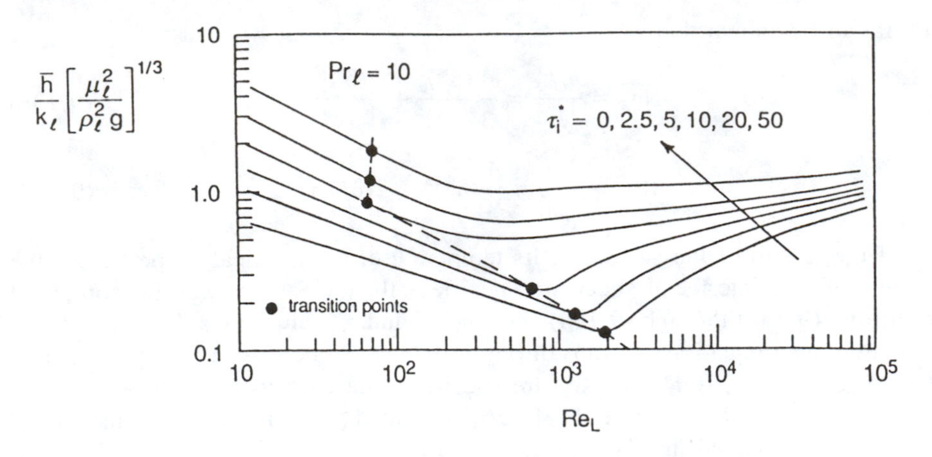

Figure 9.17 Variation of the mean film condensation heat transfer coefficient with Reynolds number and τ_i^* as predicted by the analytical model of Rohsenow et al. [9.56]. *(Adapted from [9.56] with permission, copyright © 1956, American Society of Mechanical Engineers.)*

where n is an experimentally determined constant equal to 0.124. Dukler's [9.42] analysis of laminar film flow was essentially equivalent to the Nusselt-type analysis used by Rohsenow et al. [9.56]. The governing equations were solved numerically to determine the local heat transfer coefficient as a function of downstream position and film Reynolds number. Dukler's computed results for Pr_l = 1.0 are plotted in nondimensional form in Fig. 9.18.

One questionable aspect of Dukler's analysis is that it neglects the molecular diffusion relative to the turbulent diffusivity in the fully turbulent region away from the wall. This is generally appropriate for high-Prandtl-number fluids, but could result in significant error for low-Prandtl-number, high-conductivity fluids. The analyses of Dukler [9.42] and Rohsenow et al. [9.56] both rely heavily on the assumption that the eddy diffusivity variation across the film is essentially the same as that near the wall in a turbulent pipe flow. In particular, they do not accoount explicitly for the damping effect that the interface has on turbulent transport. These analyses nevertheless predict film condensation heat transfer coefficients that are reasonably close to measured data, at least for Pr_l values greater than 1.

Forced-convection laminar film condensation on a horizontal tube can also be analyzed in a manner similar to that for condensation on a vertical surface described above. Such an analysis has been developed by Denny and Mills [9.59]. The effects of interfacial shear on film condensation in internal flow circumstances will be discussed in detail in Chapter Eleven.

The effects of interfacial waves on condensation heat transfer are more difficult to analyze exactly. However, consideration of such effects by Brauer [9.60] and Kutateladze [9.48] has shed some light on these matters. Data presented by Brauer [9.60] imply that waves begin to affect laminar film condensation for

$$Re_L > 9.3\ Ar^{1/5} \tag{9.141}$$

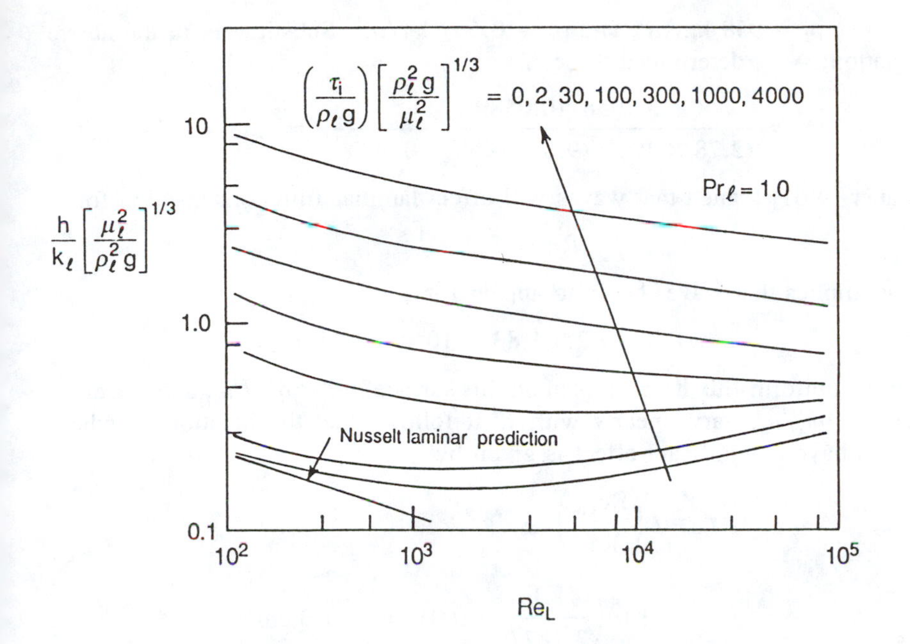

Figure 9.18 Variation of the local film condensation heat transfer coefficient with Reynolds number and τ_i as predicted by the analytical model of Dukler [9.42]. *(Adapted from [9.42] with permission, copyright © 1960, American Institute of Chemical Engineers.)*

where Ar is the Archimedes number, defined as

$$\text{Ar} = \frac{\sigma}{\mu_l g^{1/2}(\rho_l - \rho_v)^{3/2}} \tag{9.142}$$

For $0 < \text{Re}_L < 400$ with wave effects present, Kutateladze [9.48] recommends the following correlation for the mean heat transfer coefficient for falling-film condensation as a best fit to data:

$$\overline{\text{Nu}_L} = \frac{\bar{h}_L}{k_l}\left(\frac{\rho_v \nu_l^2}{g(\rho_l - \rho_v)}\right)^{1/3} = 1.23\ \text{Re}_L^{-1/4} \tag{9.143}$$

While this relation appears to agree well with data for liquid Prandtl numbers greater than 1, it has not been tested against data for liquid metals.

Example 9.5 For the film condensation process considered in Example 9.3, estimate where waves on the interface are likely to be observed.

The Archimedes number Ar is given by Eq. (9.142):

$$\text{Ar} = \frac{\rho_l^2 \sigma^{3/2}}{\mu_l^2 g^{1/2}(\rho_l - \rho_v)^{3/2}}$$

For steam at atmospheric pressure, $\sigma = 0.0589$ N/m, $\mu_l = 2.78 \times 10^{-4}$

N s/m^2, ρ_l = 958 kg/m^3, and ρ_v = 0.597 kg/m^3. Substituting in the above equation, Ar is determined to be

$$\text{Ar} = \frac{(958)^2(0.0589)^{3/2}}{(2.78 \times 10^{-4})^2(9.8)^{1/2}(958 - 0.597)^{3/2}} = 1.83 \times 10^6$$

Brauer [9.61] argued that waves will affect laminar film condensation for

$$\text{Re}_L > 2.3\ \text{Ar}^{1/5}$$

This implies that waves begin to appear for

$$\text{Re}_{L,w} \simeq 2.3(1.83 \times 10^6)^{1/5} = 41.1$$

For the uniform-flux heat removal on this surface, $\dot{m}'_x = q''x/h_{lv}$, and therefore $\text{Re}_L = 4\dot{m}'_x/\mu_l$, vary linearly with x. It follows that the location x_w where waves have a significant effect is given by

$$x_w = L\left(\frac{\text{Re}_{L,w}}{\text{Re}_L}\right)$$

$$= (1.0)\left(\frac{41.1}{3{,}187}\right) = 0.013 \text{ m} = 1.3 \text{ cm}$$

Thus waves are expected to be observable and to have a significant effect on film condensation heat transfer at only a short distance downstream of the leading (upper) edge of the plate.

9.6 CONDENSATION IN THE PRESENCE OF A NONCONDENSABLE GAS

In nature and a number of technological applications, condensation of one component vapor in a mixture may occur in the presence of other noncondensable components. The most common example is the condensation of water vapor in the air on a cold, solid surface. If the component gases are considered to be a mixture of independent substances, condensation of one component vapor will occur if the temperature of the surface is below the saturation temperature of the pure vapor at its partial pressure in the mixture. This temperature threshold is referred to as the *dew point* of the mixture.

The general effects of noncondensable gas on a film condensation process can be envisioned by considering Fig. 9.19. Once steady state is achieved, condensation occurs at the interface of a liquid film on the wall. Due to the condensation process at the interface, there is a bulk velocity of the gas toward the wall, as if there were suction at the interface. Because only the vapor is condensed, the concentration of the noncondensable gas at the interface W_i is higher than its value W_∞ in the far ambient. This, in turn, decreases the partial pressure of the vapor

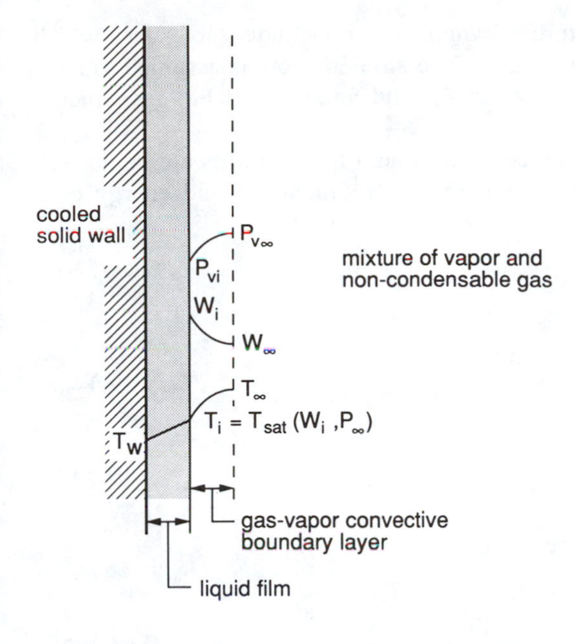

Figure 9.19 System model for analysis of film condensation in the presence of a noncondensable gas.

at the interface below its ambient value. The corresponding saturation temperature at the interface is therefore lower than the bulk temperature T_∞. At equilibrium, the interface concentration of the noncondensable gas is high enough so that the resulting diffusion and/or convection of this component away from the interface into the ambient just balances the rate at which its concentration increases due to the condensation process.

In systems where the noncondensable gas is a low-concentration contaminant, the phenomena described above can lead to a high concentration of the noncondensable contaminant at the interface. The resulting depression of the interface temperature generally reduces the condensation heat transfer rate below that which would result for pure vapor alone under the same conditions.

For concreteness, we will first consider the forced-convection film condensation process shown in Fig. 9.20. The flat wall, taken to be isothermal, is ex-

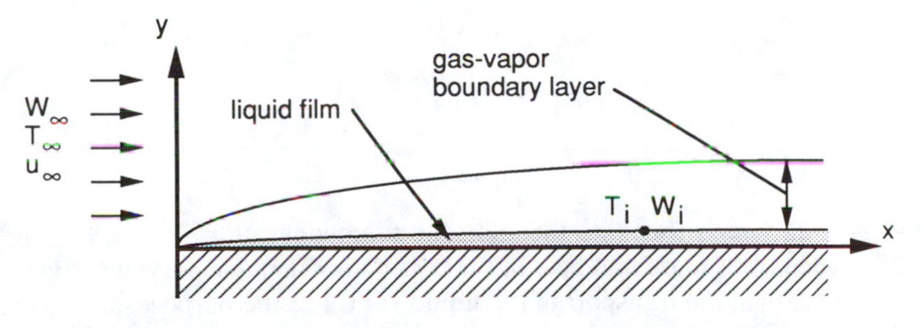

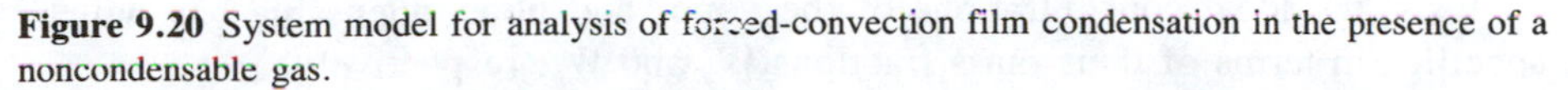

Figure 9.20 System model for analysis of forced-convection film condensation in the presence of a noncondensable gas.

posed to a forced flow of a mixture of vapor and noncondensable gas, which results in film condensation on the surface. The ambient flow is assumed to have uniform values of velocity u_∞, temperature T_∞, and noncondensable gas concentration W_∞.

For the liquid film, if the energy convection and inertia terms are neglected, the boundary-layer forms of the governing mass, momentum, and energy conservation equations become

$$\frac{\partial u}{\partial x} + \frac{\partial v}{\partial y} = 0 \tag{9.144a}$$

$$\frac{\partial^2 u}{\partial y^2} = 0 \tag{9.144b}$$

$$\frac{\partial^2 T}{\partial y^2} = 0 \tag{9.144c}$$

Appropriate boundary conditions are

at $y = 0$: $$u = v = 0, \qquad T = T_w \tag{9.145a}$$

at $y = \delta$: $$T = T_i \tag{9.145b}$$

For the gas–vapor boundary layer adjacent to the film, the continuity and momentum equations are

$$\frac{\partial u}{\partial x} + \frac{\partial v}{\partial y} = 0 \tag{9.146a}$$

$$u\frac{\partial u}{\partial x} + v\frac{\partial u}{\partial y} = \nu_m \frac{\partial^2 u}{\partial y^2} \tag{9.146b}$$

At the interface, the streamwise velocity u must be continuous. However, the liquid velocity at the interface is generally very much smaller than the free-stream velocity u_∞ of the vapor. Consequently, for the vapor, the u velocity can be taken to be zero at the interface with very little loss in accuracy. With this approximation, the boundary conditions on the gas velocity fields are

at $y = \delta$: $$u = 0 \tag{9.147a}$$

$$\dot{m}_l'' = \rho_m \left(u \frac{d\delta}{dx} - v \right) \tag{9.147b}$$

at $y \to \infty$: $$u \to u_\infty \tag{9.147c}$$

where $\dot{m}_l''$ is the mass flux into the film due to condensation.

The transport of mass and heat in the two-component gas mixture is generally more complex than the transport in the liquid film. For the purposes of the present analysis, the local concentrations of the vapor and noncondensable gas will be specified in terms of their mass fractions W_v and W_g, respectively,

$$W_v = \frac{\rho_v}{\rho_m} \qquad W_g = \frac{\rho_g}{\rho_m} \tag{9.148}$$

where ρ_m is the local density of the mixture and ρ_g and ρ_v are the local densities of the gas and vapor, respectively. It follows directly from these definitions that $W_g + W_v = 1$. Because these concentrations are not independent, one can be eliminated from the problem. For this analysis, we will therefore deal strictly with the concentration of the noncondensable gas W_g.

In a mixture of the type considered here, the diffusive mass flux of the noncondensable gas j_g is given by

$$j_g = -\rho_m D \frac{\partial W_g}{\partial y} - \rho_m D \frac{\alpha_D^* W_g(1 - W_g)}{T}\left(\frac{\partial T}{\partial y}\right) \tag{9.149}$$

The first term on the right side of the above expression is the well-known Fickian diffusion term, with D being the binary diffusion coefficient. The second term represents mass diffusion induced by the temperature gradient. This latter mechanism is the so-called *thermal diffusion* or *Soret effect*. The parameter α_D^* is a dimensionless property of the system sometimes referred to as the *thermal diffusion factor*.

A relation similar to Eq. (9.149) can also be written for the diffusion flux of the vapor species:

$$j_v = -\rho_m D \frac{\partial W_v}{\partial y} - \rho_m D \frac{\alpha_D^* W_v(1 - W_v)}{T}\left(\frac{\partial T}{\partial y}\right) \tag{9.150}$$

from which it follows directly that $j_v = -j_g$.

At any location in the flow, the heat flux in the direction normal to the surface is given by

$$q'' = -k_g\left(\frac{\partial T}{\partial y}\right) + \alpha_D^* R_m T \frac{M_m^2}{M_g M_v} j_g \tag{9.151}$$

where R_m and M_m are the ideal gas constant and the molecular weight of the mixture at the local concentration, respectively, and M_g and M_v are the molecular weights of the noncondensable gas and the vapor species, respectively. The first term on the right side of Eq. (9.151) represents Fourier conduction, while the second term accounts for additional transport of thermal energy associated with species diffusion. This latter effect is termed the *diffusion thermo* effect or the *Dufour effect*.

The boundary-layer equations governing the species and heat transport in the gas mixture near the interface are

$$\rho_m\left(u \frac{\partial W_g}{\partial x} + v \frac{\partial W_g}{\partial y}\right) = -\frac{\partial j_g}{\partial y} \tag{9.152}$$

$$\rho_m c_{pm}\left(u \frac{\partial T}{\partial x} + v \frac{\partial T}{\partial y}\right) + (c_{pg} - c_{pv}) j_g \frac{\partial T}{\partial y} = -\frac{\partial q''}{\partial y} \tag{9.153}$$

where c_{pm}, c_{pg}, and c_{pv} are the specific heat of the mixture, the gas, and the vapor, respectively, and j_g and q'' are given by Eqs. (9.149) and (9.151). The second term on the left side of Eq. (9.153) accounts for the net enthalpy flux associated with the diffusive flux of the individual components.

Because the thermal diffusion and diffusion thermo effects are small in many circumstances of practical interest, we will neglect these effects in our subsequent analytical developments here. Furthermore, for low to moderate concentrations of the noncondensable gas, the sensible heat transfer to the interface is usually negligible compared to the heat transferred as a result of the transport and condensation of the vapor at the interface. Restricting our attention to circumstances of this type in which the transport of heat is mass transfer-dominated, the equation governing sensible transport of thermal energy will be neglected. Of primary concern is the resulting reduced form of the equation for mass transport,

$$\left(u\frac{\partial W_g}{\partial x} + v\frac{\partial W_g}{\partial y}\right) = D\frac{\partial^2 W_g}{\partial y^2} \tag{9.154}$$

Boundary conditions on the concentration field are

at $y = \delta$:
$$j_g = -\rho_m D\frac{dW_g}{dy} = \rho_g\left(u\frac{d\delta}{dx} - v\right) = -\rho_g v \tag{9.155a}$$

at $y \to \infty$:
$$W_g \to W_{g,\infty} \tag{9.155b}$$

The relations (9.144) through (9.147) and (9.154) through (9.155) form a mathematically complete system of equations and boundary conditions. However, it is presumed in this formulation that the location $y = \delta$ is known. The additional relation that makes it possible to determine δ comes from the requirement of an energy balance at the liquid–gas interface:

$$\dot{m}_l'' h_{lv} + k_m\left(\frac{\partial T_m}{\partial y}\right)_{y=\delta} = k_l\left(\frac{\partial T_l}{\partial y}\right)_{y=\delta} \tag{9.156}$$

Consistent with the arguments given above, the contribution of conduction from the gas mixture, represented by the second term on the left side of Eq. (9.156), is taken to be negligibly small. The relation can therefore be simplified to

$$\dot{m}_l'' h_{lv} = k_l\left(\frac{\partial T_l}{\partial y}\right)_{y=\delta} \tag{9.157}$$

With the addition of Eq. (9.157), the mathematical problem is completely closed. To facilitate solution of this problem, Sparrow, Minkowycz, and Saddy [9.61] introduced the following similarity variables:

In the liquid film:

$$\eta = y\sqrt{\frac{u_\infty}{\nu_l x}} \tag{9.158a}$$

$$\psi = f(\eta)\sqrt{u_\infty \nu_l x} \qquad \theta = \frac{T_l - T_w}{T_{li} - T_w} \tag{9.158b}$$

$$u = \frac{\partial \psi}{\partial y} = u_\infty f(\eta) \qquad v = -\frac{\partial \psi}{\partial x} = \frac{1}{2}\sqrt{\frac{u_\infty}{\nu_l x}}\,(\eta f' - f) \tag{9.158c}$$

In the vapor:

$$\xi = (y - \delta)\sqrt{\frac{u_\infty}{\nu_m x}} \tag{9.159a}$$

$$\psi = F(\xi)\sqrt{u_\infty \nu_m x} \qquad \phi = \frac{W_g - W_{g,\infty}}{W_{g,i} - W_{g,\infty}} \tag{9.159b}$$

In terms of these variables, the governing equations and boundary conditions become

In the liquid film:
$$f''' = 0 \qquad \theta'' = 0 \tag{9.160a}$$

$$f(0) = 0 \qquad f'(0) = 0 \qquad \theta(0) = 0 \qquad \theta(\eta_\delta) = 1 \tag{9.160b}$$

In the gas boundary layer:
$$F''' + \frac{1}{2}FF'' = 0 \tag{9.161a}$$

$$\phi'' + \frac{1}{2}\,\mathrm{Sc}\,F\phi' = 0 \tag{9.161b}$$

$$F'(0) = 0 \qquad \phi(0) = 1 \qquad F'(\infty) = 1 \qquad \phi(\infty) = 0 \tag{9.161c}$$

The required conditions at the interface are

$$F(0) = Rf(\eta_\delta) \qquad F''(0) = Rf''(\eta_\delta) \tag{9.162a}$$

$$\frac{f(\eta_\delta)}{2\theta'(\eta_\delta)} = \frac{\mathrm{Ja}_l}{\mathrm{Pr}_l} \tag{9.162b}$$

$$-\frac{\mathrm{Sc}\,F(0)}{2\phi'(0)} = 1 - \frac{W_{g,\infty}}{W_{g,i}} \tag{9.162c}$$

where

$$R = \left[\frac{\rho_l \mu_l}{\rho_m \mu_m}\right]^{1/2} \qquad \mathrm{Ja} = \frac{c_{pl}(T_i - T_w)}{h_{lv}} \tag{9.162d}$$

and

$$\eta_\delta = \delta\sqrt{\frac{u_\infty}{\nu_l x}} \tag{9.162e}$$

It can be easily shown that solutions of the equations (9.160*a*) that satisfy the boundary conditions (9.160*b*) are

$$f = \frac{1}{2} f''(0)\eta^2 \qquad \theta = \frac{\eta}{\eta_\delta} \tag{9.163}$$

Substituting these relations, the interface conditions (9.162*a*) through (9.162*b*) can be combined to obtain

$$\frac{R\,\text{Ja}_l}{\text{Pr}_l} = \left\{ \frac{[F(0)]^3}{2F''(0)} \right\}^{1/2} \tag{9.164a}$$

$$\eta_\delta = \left[\frac{2F(0)}{F''(0)} \right]^{1/2} \tag{9.164b}$$

With these relations, the complete solution to the problem can be obtained as follows. For a chosen value of $F(0)$, Eq. (9.161*a*) with appropriate boundary conditions (9.161*c*) can be solved numerically. Then, using the chosen $F(0)$ value and the numerically determined value of $F''(0)$, Eqs. (9.164*a*) and (9.164*b*) are used to determine the corresponding values of $R\,\text{Ja}_l/\text{Pr}_l$ and η_δ. This implies that η_δ is a unique function of $R\,\text{Ja}_l/\text{Pr}_l$. By doing such calculations for a sequence of $F(0)$ values, the relation between η_δ and $R\,\text{Ja}_l/\text{Pr}_l$ can be determined for a wide range of conditions. Figure 9.21 shows the predicted variation of η_δ with $R\,\text{Ja}_l/\text{Pr}_l$ as determined numerically by Sparrow et al. [9.61]. Once this is accomplished, for any real circumstances of interest, η_δ can be determined from the established η_δ variation with $R\,\text{Ja}_l/\text{Pr}_l$. $F(0)$ and $F''(0)$ can then be determined from Eqs. (9.164*a*) and (9.164*b*), and Eq. (9.161*a*) can be integrated to determine the $F(\xi)$ and $F'(\xi)$ variations.

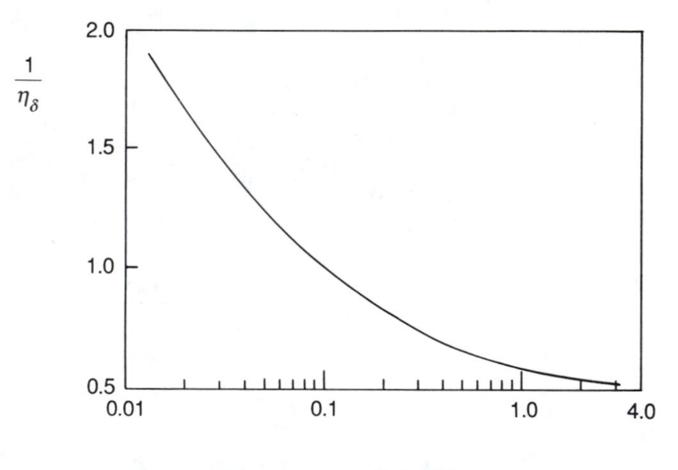

Figure 9.21 Variation of η_δ with R Ja$_l$ Pr$_l$ as determined numerically by Sparrow et al. [9.61]. *(Adapted from [9.61] with permission, copyright © 1967, Pergamon Press.)*

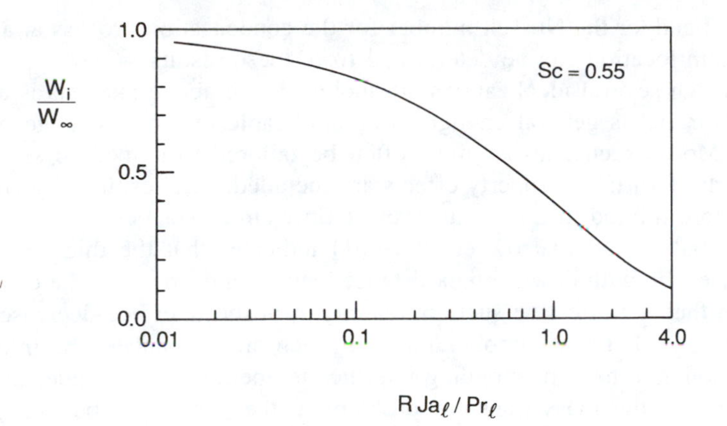

Figure 9.22 Variation of W_i/W_∞ with R Ja_l Pr_l for Sc = 0.55 as determined numerically by Sparrow et al. [9.61]. *(Adapted from [9.61] with permission, copyright © 1967, Pergamon Press.)*

With the solution for $F(\xi)$ and the value of η_δ determined, Eq. (9.161*b*) can be solved numerically for a specific value of the Schmidt number $\mathrm{Sc} = \nu_m/D$, subject to the boundary conditions on ϕ given in Eq. (9.161*c*). The numerically determined value of $\phi'(0)$ with the corresponding value of $F(0)$ can then be used in Eq. (9.162*c*) to determine the interface condition $W_{g,i}$. This solution scheme implies that $W_{g,i}/W_{g,\infty}$ is a function only of Sc and $R\,\mathrm{Ja}_l/\mathrm{Pr}_l$. The computed variation of $W_{g,i}/W_{g,\infty}$ for Sc = 0.55 determined numerically by Sparrow et al. [9.61] is shown in Fig. 9.22.

Treating the vapor and noncondensable gas as a mixture of ideal gases, the partial pressure of the vapor at the interface $P_{v,i}$ can be computed as

$$\frac{P_{v,i}}{P_\infty} = \frac{1 - W_{g,i}}{1 - W_{g,i}(1 - M_v/M_g)} \tag{9.165}$$

The interface temperature is the saturation temperature at the vapor partial pressure there:

$$T_i = T_{\mathrm{sat}}(P_{v,i}) \tag{9.166}$$

It also follows directly from the above formulation that

$$\mathrm{Nu}_x = \frac{hx}{k_l} = \left(\frac{T_i - T_w}{T_\infty - T_w}\right)\frac{\mathrm{Re}_x^{1/2}}{\eta_\delta} \tag{9.167}$$

where

$$\mathrm{Re}_x = \frac{u_\infty x}{\nu_l}, \qquad h = \frac{q''}{T_\infty - T_w} \tag{9.168}$$

Thus, once T_i and η_δ are determined for a given flow condition, the heat transfer coefficient and/or the Nusselt number for the condensation process at a specified downstream location can be determined from these results.

Although several idealizations are included, the above analysis is a compact formulation that is general enough to be applicable to a broad range of circumstances. More exact analyses must often be tailored to a specific set of fluids, particularly if variable property effects are included. The results of such analyses are therefore limited to a particular set of flow circumstances.

The analysis of Sparrow et al. [9.61] indicates that the thicknening of the condensate film with downstream distance leads to a decrease in the condensation rate such that the effective suction velocity v_i at the interface decreases proportional to $x^{-1/2}$. It further implies that, for such circumstances, the interface gas composition and the corresponding interface temperature T_i are independent of x. Consequently, the mass transport problem in the vapor–gas boundary layer is identical to the case of heat transfer for flow over an isothermal plate with a surface suction velocity varying proportional to $x^{-1/2}$.

Rose [9.62, 9.63] showed that this similarity makes it possible to apply an approximate relation developed for the heat transfer problem to the mass transfer for the condensation process considered above. The resulting relation for the mass transfer coefficient is

$$\frac{h_m x}{\rho_m D} = [\zeta(1 + 0.941\beta_x^{1.14}\,\mathrm{Sc}^{0.93})^{-1} + \beta_x\,\mathrm{Sc}]\,\mathrm{Re}_x^{1/2} \tag{9.169a}$$

where

$$\zeta = \mathrm{Sc}^{1/2}(27.8 + 75.9\,\mathrm{Sc}^{0.306} + 657\,\mathrm{Sc})^{-1/6} \tag{9.169b}$$

$$\beta_x = -\left(\frac{v_i}{u_\infty}\right)\mathrm{Re}_x^{1/2} \tag{9.169c}$$

$$h_m = \frac{-\rho_m D}{(W_{g,i} - W_{g,\infty})}\left(\frac{\partial W_g}{\partial y}\right)_{y=\delta} \tag{9.169d}$$

A mass balance for the gas at the interface also requires that

$$h_m = -\frac{\rho_m v_i W_{g,i}}{(W_{g,i} - W_{g,\infty})} = -\frac{\rho_m v_i}{(1 - W_{g,\infty}/W_{g,i})} \tag{9.170}$$

Substituting this relation into the equation (9.169*a*) for h_m and rearranging yields the following explicit relation for the interface condition:

$$\frac{W_{g,i}}{W_{g,\infty}} = 1 + \frac{\beta_x\,\mathrm{Sc}(1 + 0.941\,\mathrm{Sc}^{0.93})}{\zeta} \tag{9.171}$$

For a given set of free-stream and wall temperature conditions, this equation relates the interface velocity v_i and $W_{g,i}$ directly. Rose [9.63] found that this relation predicts values of $W_{g,i}/W_{g,\infty}$ that agree well with the numerical solutions of Sparrow et al. [9.61] for Sc = 0.55. If it is presumed that the heat transfer at the interface is due entirely to condensation of vapor, energy balance at the interface also requires that

$$v_i = -\frac{h(T_\infty - T_w)}{\rho_m h_{lv}(1 - W_{g,i})} \tag{9.172}$$

Equations (9.171) and (9.172) can be used to greatly simplify the computation of the condensation heat transfer coefficient for the flat-plate forced-convection condensation process considered here. As noted above, Eq. (9.161*a*) can be solved numerically subject to the boundary conditions $F'(0) = 0$ and $F'(\infty) = 1$ for various values of $F(0)$. The resulting numerically computed variation of $F''(0)$ with $F(0)$ is shown in Fig. 9.23.

Using the computed results represented in Fig. 9.23, the heat transfer coefficient for a given set of flow circumstances can be iteratively computed as follows. (It is assumed that T_w, T_∞, u_∞, $W_{g,\infty}$, P_∞, and x have been specified and that the fluid properties are known.)

1. A value of v_i is guessed and the corresponding value of $F(0) = -2v_i(xu_\infty/\nu_m)^{1/2}$ is calculated.
2. $F''(0)$ can then be determined from interpolation of tabulated numerical results or graphically from Fig. 9.23.
3. Using the $F(0)$ and $F''(0)$ values, Eqs. (9.164*a*) and (9.164*b*) can be used to compute $R\ \text{Ja}_l/\text{Pr}_l$ and η_δ.
4. Equation (9.171) can then be used to determine $W_{g,i}$, the value of which can be substituted into Eq. (9.165) to determine $P_{v,i}$.
5. T_i is calculated as $T_i = T_{\text{sat}}(P_{v,i})$ from thermodynamic data for the pure vapor.
6. The heat transfer coefficient h is calculated using Eq. (9.167).
7. Finally, Eq. (9.172) is used to recalculate v_i using the value of h computed in

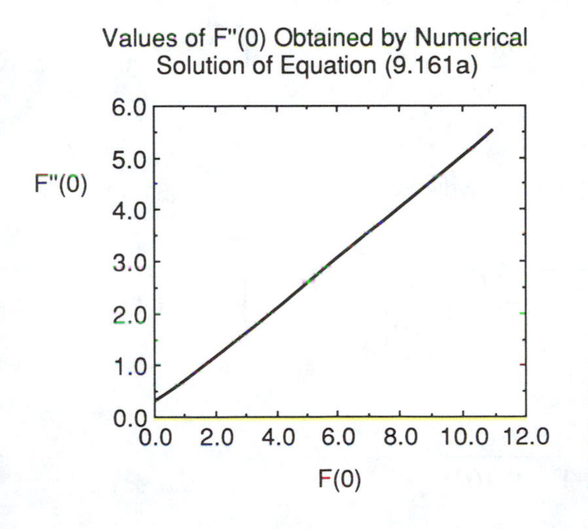

Figure 9.23 Variation of $F''(0)$ with $F(0)$ as determined from numerical solution of Eq. (9.161*a*) with boundary conditions $F'(0) = 0$ and $F'(\infty) = 1$.

step 6. If this value of v_i agrees with that initially guessed, the solution is complete. If it does not agree to an acceptable degree, an improved guess for v_i is generated and the process is repeated, beginning with step 2, until convergence is achieved.

While the above computational scheme to determine h is iterative, it does not require integration of the differential equations governing the transport, if the variation of $F''(0)$ with $F(0)$ is known. It is therefore somewhat easier to use than the full similarity solution analysis of Sparrow et al. [9.61], and yet has comparable accuracy.

Example 9.6 Forced-convection film condensation of water occurs on a 50-cm-long flat plate exposed to a flow of a mixture of steam and air. The ambient conditions are $T_\infty = 95°\text{C}$, $u_\infty = 5$ m/s, $P_\infty = 101$ kPa, and $W_{g,\infty} = 0.02$. The plate is held at a constant temperature of 60°C. Experimentally, the heat transfer coefficient at a distance $x = 15$ cm from the leading edge of the plate is determined to be 1.5 kW/m^2 K. Use Eqs. (9.171) and (9.172) to determine the interface velocity v_i and interface concentration $W_{g,i}$.

For air and water vapor at the specified conditions, $\rho_m = 0.60$ kg/m^3, Sc = 0.9, $\nu_m = 2.10 \times 10^{-5}$ m^2/s, and at $T_w = 60°\text{C}$, h_{lv} for the condensate water is 2359 kJ/kg. Equations (9.171) and (9.172) must be satisfied simultaneously to determine v_i and $W_{g,i}$. Initially taking $W_{g,i} = W_{g,\infty}$, v_i is determined from Eq. (9.172) to be

$$v_i = \frac{-h(T_\infty - T_w)}{\rho_m h_{lv}(1 - W_{g,i})} = \frac{-1.5(95 - 60)}{0.60(2359)(1 - 0.02)} = -0.0378 \text{ m/s}$$

Using Eqs. (9.169b) and (9.169c),

$$\zeta = \text{Sc}^{1/2}(27.8 + 75.9\,\text{Sc}^{0.306} + 657\,\text{Sc})^{-1/6}$$

$$= (0.9)^{1/2}[27.8 + 75.9(0.9)^{0.306} + 657(0.9)]^{-1/6} = 0.336$$

$$\beta_x = -\left(\frac{v_i}{u_\infty}\right)\left(\frac{u_\infty x}{\nu_m}\right)^{1/2} = \left(\frac{0.0378}{5.0}\right)\left[\frac{5(0.15)}{2.10 \times 10^{-5}}\right]^{1/2} = 1.43$$

Substituting into Eq. (9.171),

$$W_{g,i} = W_{g,\infty}\left[1 + \frac{\beta_x\,\text{Sc}(1 + 0.941\,\text{Sc}^{0.93})}{\zeta}\right]$$

$$= (0.02)\left[1 + \frac{1.43(0.9)(1 + 0.941(0.9)^{0.93})}{0.336}\right] = 0.162$$

Substituting back into Eq. (9.172),

$$v_i = \frac{-1.5(95 - 60)}{0.60(2359)(1 - 0.162)} = -0.0443 \text{ m/s}$$

Repeating the above calculations of β_x and $W_{g,i}$ yields $\beta_x = 1.67$, $W_{g,i} = 0.186$. After a few iterations, the following converged values are obtained:

$$v_i = -0.046 \text{ m/s} \qquad W_{g,i} = 0.19$$

Thus there is a "suction velocity" of 4.6 cm/s at the interface and the concentration of noncondensable gas is almost 10 times the ambient concentration.

Analyses similar to those described above have also been developed for falling-film condensation on a vertical plate in the presence of a noncondensable gas [9.64–9.67]. Combined forced- and free-convection film condensation on a vertical plate immersed in a steam–air mixture has also been analyzed by Denny, Mills, and Jusionis [9.68]. Film condensation of liquid metals from a metal vapor and gas mixture on a vertical surface for forced- and free-convection conditions has also been treated theoretically by Turner, Mills, and Denny [9.69]. Where necessary, their analysis also included the effects of interfacial resistance. Experimentally determined heat transfer coefficients for film condensation on vertical flat surfaces in the presence of a noncondensable gas reported by Al-Diwany and Rose [9.70] and by Slegers and Seban [9.71] generally are consistent with the predictions of the boundary-layer analyses proposed in the studies noted above.

Film condensation on a horizontal round tube in the presence of a noncondensable gas may occur in tube-and-shell condensers when a contaminant such as air is introduced into the system. Berman [9.72] has proposed a correlation to predict the mass transfer coefficient for such circumstances. A more complete analysis of such processes has been presented by Al-Diwany and Rose [9.63]. Their model analysis was found to agree well with experimental data reported by Mills, Tan, and Chung [9.73].

9.7 ENHANCEMENT OF CONDENSATION HEAT TRANSFER

In many circumstances of practical interest, most, if not all of the resistance to heat flow during the condensation process is due to conduction and/or convection through the liquid condensation on the cooled surface. Consequently, most techniques for enhancing condensation heat transfer involve controlling the condensate on the surface so as to reduce the heat transfer resistance associated with its presence on the surface.

One means of reducing the average heat transfer resistance of the liquid on the surface is to promote dropwise condensation. Note that, in doing so, a uniformly thick film is replaced by droplets with no film, or a microfilm between them. The usual result is that the mean heat transfer coefficient is significantly higher for dropwise condensation than for filmwise condensation under the same conditions.

As described in Section 9.2, dropwise condensation of steam can be promoted by introducing a fatty acid, wax, or other substance onto the cooled solid surface, or by permanently coating the surface with a low-surface-energy polymer or noble metal. Each of these methods has its drawbacks and/or limitations. Fatty acids and other injected promoters are washed off of condenser surfaces under common industrial operating conditions. To sustain dropwise condensation, it is therefore necessary to periodically reinject the promoter into the system. Using permanent coatings of the condenser surface avoids this difficulty, but generally add considerable cost to the fabrication of the condenser.

In practice, the above techniques for promoting dropwise condensation have been successful primarily for condensation of steam. There are a number of substances that are poorly wetted by liquid water, as a consequence of its relatively high interfacial tension. For liquids having lower interfacial tension, such as the fluorocarbon refrigerants R-11, R-12, R-22, R-113, and many hydrocarbons, no dropwise condensation promoters have been reported in the literature (see the survey by Iltscheff [9.74]). Hence, promotion of dropwise condensers for refrigeration systems using Freon-like working fluids awaits the discovery of substances that are poorly wetted by these liquids. Further information on techniques for promoting dropwise condensation of steam can be found in the survey paper by Tanasawa [9.75].

For film condensation, the use of formed or extended surfaces as a means of enhancing the heat transfer coefficient has received considerable attention. Gregorig [9.76] pioneered the use of a wavy surface to enhance film-condensation heat transfer. The effect of such a surface profile on the condensate can be seen by considering Fig. 9.24. As indicated in this figure, the curvature of the liquid–vapor interface will reverse at the "peak" and "valley" locations to allow the film to conform to the wavy surface. For a two-dimensional surface of this type, it follows from Young's equation (see Chapter Two) that

$$P_1 = P_v + \frac{\sigma}{r_1} \tag{9.173a}$$

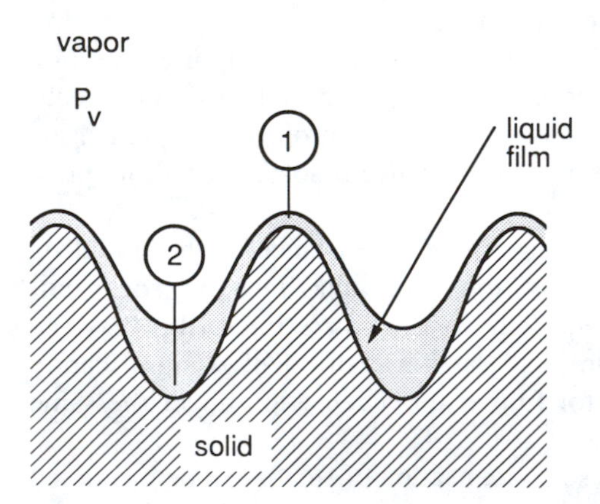

Figure 9.24 Schematic representation of a wavy surface used to enhance film condensation heat transfer.

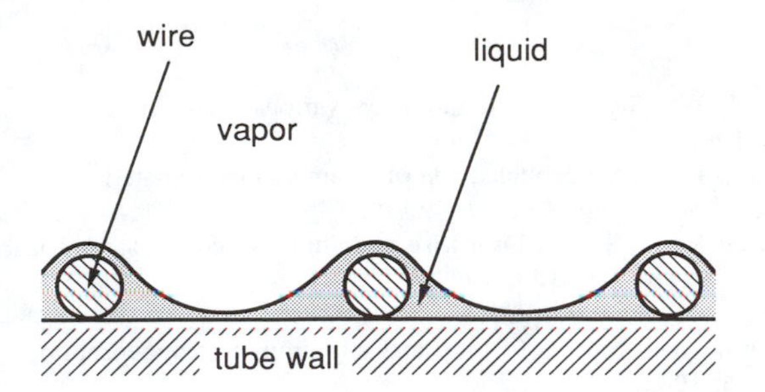

Figure 9.25 Schematic representation of wires used to enhance film condensation on vertical tubes.

$$P_v = P_2 + \frac{\sigma}{r_2} \tag{9.173b}$$

where r_1 and r_2 are the radii of curvature of the interface at 1 and 2, respectively. These equations are easily combined to show that

$$P_1 - P_2 = \sigma\left(\frac{1}{r_1} + \frac{1}{r_2}\right) \tag{9.174}$$

Thus interfacial tension induces a pressure difference in the liquid film, which causes liquid to flow from the peak regions to the valley regions of the surface. This thins the film in the peak regions, allowing more rapid condensation there, at the expense of the valley regions where the film is thicker. The overall mean heat transfer coefficient for the surface can be substantially greater than if the surface were covered with a film of uniform thickness. This effect of surface tension on film condensation on a wavy surface is sometimes referred to as the *Gregorig effect*.

A number of variations of the above theme have been proposed as means of enhancing film condensation heat transfer. Thomas [9.77], for example, proposed attaching vertical wires onto vertical condenser tubes to take advantage of this effect. Liquid condensate is pulled into the crevice between the round tube wall and the wire, thinning the liquid film between the wires, as shown in Fig. 9.25. The tendency for circumferentially finned condenser tubes to modify the condensate distribution similarly has also been studied by a number of investigators (see, for example, Winniarachchi et al. [9.78], Zozulya et al. [9.79], Rudy and Webb [9.80], and Hirasawa et al. [9.81]). Excellent summaries of the works in this area can be found in the review articles by Marto and Nunn [9.82], Webb [9.83], Cooper and Rose [9.84], and Marto [9.85].

Enhancement of external film condensation using mechanical aids such as rotating cylinders, disks, and wiper blades has also been examined [9.86–9.89]. Other enhancement methods that have been considered include surface vibration [9.90], electrostatic fields [9.91], and suction [9.92]. Obviously, combinations of these different methods can also be employed. Further discussion of such methods is provided in the review article by Bergles [9.93].

REFERENCES

9.1 Woodruff, D. M., and Westwater, J. W., Steam condensation on various gold surfaces, *J. Heat Transfer,* vol. 103, 685–692, 1981.

9.2 O'Neill, G. A., and Westwater, J. W., Dropwise condensation of steam on electroplated silver surfaces, *Int. J. Heat Mass Transfer,* vol. 27, pp. 1539–1549, 1984.

9.3 Wilkins, D. G., Bromley, L. A., and Read, S. M., Dropwise and filmwise condensation of water vapor on gold, *AIChE J.,* vol. 19, pp. 119–123, 1973.

9.4 Eucken, A., *Naturwissenschaften,* vol. 25, p. 209, 1937.

9.5 McCormick, J. L., and Baer, E., On the mechanism of heat transfer in dropwise condensation, *J. Colloid Sci.,* vol. 18, pp. 208–216, 1963.

9.6 Umur, A., and Griffith, P., Mechanism of dropwise condensation, *J. Heat Transfer,* vol. 87, pp. 275–282, 1965.

9.7 Gose, E., Mucciordi, A. N., and Baer, E., Model for dropwise condensation on randomly distributed sites, *Int. J. Heat Mass Transfer,* vol. 10, pp. 15–22, 1967.

9.8 Jakob, M., *Mech. Eng.,* vol. 58, p. 729, 1936.

9.9 Kast, W., *Chem.-Ing. Tech.,* Heat transfer during dropwise condensation (in German), vol. 35, pp. 163–168, 1963.

9.10 Silver, R. S., An approach to a general theory of surface condensers, *Proc. Inst. Mech. Eng.,* vol. 178, pp. 339–376, 1964.

9.11 Welch, J. F., and Westwater, J. W., Microscopic study of dropwise condensation, *International Developments in Heat Transfer, Proc. Int. Heat Transfer Conf.,* ASME, Part II, pp. 302–309, 1961.

9.12 Sugawara, S., and Katusuta, K., Fundamental study on dropwise condensation, *Proc. 3rd Int. Heat Transfer Conf.,* vol. 2, pp. 354–361, 1966.

9.13 Graham, C., and Griffith, P., Drop size distribution and heat transfer in dropwise condensation, *Int. J. Heat Mass Transfer,* vol. 16, pp. 337–346, 1973.

9.14 Mikic, B. B., On mechanism of dropwise condensation, *Int. J. Heat Mass Transfer,* vol. 12, pp. 1311–1323, 1969.

9.15 Collier, J. G., *Convective Boiling and Condensation,* 2nd ed., McGraw-Hill, New York, 1981.

9.16 Takeyama, T., and Shimizu, S., On the transition of dropwise condensation, *Heat Transfer, 1974—Proc. 5th Int. Heat Transfer Conf.,* vol. 3, p. 274, 1974.

9.17 Holden, K. M., Wanniarachchi, A. S., Marto, P. J., Boone, D. H., and Rose, J. W., The use of organic coatings to promote dropwise condensation of steam, *J. Heat Transfer,* vol. 109, pp. 768–774, 1987.

9.18 O'Bara, J. T., Killion, E. S., and Roblee, L. H. S., Dropwise condensation of steam at atmospheric and above atmospheric pressure, *Chem. Eng. Sci.,* vol. 22, pp. 1305–1314, 1967.

9.19 Tanaka, H., A theoretical study of dropwise condensation, *J. Heat Transfer,* vol. 97, pp. 72–78, 1975.

9.20 Tanaka, H., Further developments of dropwise condensation theory, *J. Heat Transfer,* vol. 101, pp. 603–611, 1979.

9.21 Tanaka, H., Effect of Knudsen number on dropwise condensation. *J. Heat Transfer,* vol. 103, pp. 606–607, 1981.

9.22 Tanaka, H., Measurements of drop-size distributions during transient dropwise condensation, *J. Heat Transfer,* vol. 97, pp. 341–346, 1975.

9.23 Griffith, P., and Lee, M. S., The effect of the surface thermal properties and finish on dropwise condensation, *Int. J. Heat Mass Transfer,* vol. 10, pp. 697–707, 1967.

9.24 Hannemann, R. J., and Mikic, B. B., An analysis of the effect of surface thermal conductivity on the rate of heat transfer in dropwise condensation, *Int. J. Heat Mass Transfer,* vol. 19, pp. 1299–1307, 1976.

9.25 Hannemann, R. J., Condensing surface thickness effects in dropwise condensation, *Int. J. Heat Mass Transfer,* vol. 21, pp. 65–66, 1978.

9.26 Peterson, A. C., and Westwater, J. W., Dropwise condensation of ethylene glycol, *Chem. Eng. Prog. Symp. Ser.,* vol. 62, no. 64, pp. 135–142, 1966.

9.27 Fatica, N., and Katz, D. L., Dropwise condensation, *Chem. Eng. Prog.*, vol. 45, pp. 661–674, 1949.
9.28 Sugawara, S., and Michiyoshi, I., *Mem. Fac. Eng.*, Kyoto Univ., vol. 18, p. 11, 1956.
9.29 Isachenko, V. P., *Teploenerg*, vol. 9, p. 81, 1962.
9.30 Merte, H., Condensation heat transfer, *Adv. Heat Transfer*, vol. 9, pp. 181–272, 1973.
9.31 Nusselt, W., Die Oberflachenkondensation des Wasser dampfes, *Z. Vereins deutscher Ininuere*, vol. 60, pp. 541–575, 1916.
9.32 Koh, J. C. Y., An integral treatment of two-phase boundary layer in film condensation, *J. Heat Transfer*, vol. 83, pp. 359–362, 1961.
9.33 Rohsenow, W. M., Heat transfer and temperature distribution in laminar film condensation, *Trans. ASME*, vol. 78, pp. 1645–1648, 1956.
9.34 Sparrow, E. M., and Gregg, J. L., A boundary-layer treatment of laminar film condensation, *J. Heat Transfer*, vol. 81, pp. 13–23, 1959.
9.35 Sadasivan, P., and Lienhard, J. H., Sensible heat correction in laminar film boiling and condensation, *J. Heat Transfer*, vol. 109, pp. 545–546, 1987.
9.36 Koh, J. C. Y., Sparrow, E. M., and Hartnett, J. P., The two-phase boundary layer in laminar film condensation, *Int. J. Heat Mass Transfer*, vol. 2, pp. 69–82, 1961.
9.37 Sukhatme, S. P., and Rohsenow, W. M., Heat transfer during film condensation of a liquid metal vapor, *J. Heat Transfer*, vol. 88, pp. 19–28, 1966.
9.38 Kapitsa, P. L., Wave flow of thin layers of a viscous fluid, *Zh. Eksp. Teoret. Fiz.*, vol. 18, p. 1, 1948.
9.39 Spencer, D. L., and Ibele, W. E., Laminar film condensation of a saturated and superheated vapor on a surface with a controlled temperature distribution, *Proc. Third Int. Heat Transfer Conf.*, Chicago, vol. 2, pp. 337–347, 1966.
9.40 McAdams, W. H., *Heat Transmission*, 3rd ed., ch. 9, McGraw-Hill, New York, 1954.
9.41 Seban, R., Remarks on film condensation with turbulent flow, *Trans. ASME*, vol. 76, pp. 299–303, 1954.
9.42 Dukler, A. E., Fluid mechanics and heat transfer in vertical falling film systems, *Chem. Eng. Prog. Symp. Ser.*, vol. 56, no. 30, pp. 1–10, 1960.
9.43 Lee, J., Turbulent condensation, *AIChE J.*, vol. 10, pp. 540–544, 1964.
9.44 Kunz, H. R., and Yerazunis, S., An analysis of film condensation film evaporation and single-phase heat transfer for liquid Prandtl numbers from 10^{-3} to 10^4, *J. Heat Transfer*, vol. 91, pp. 413–420, 1969.
9.45 Chun, K. R., and Seban, R. A., Heat transfer to evaporating liquid films, *J. Heat Transfer*, vol. 93, pp. 391–396, 1971.
9.46 Mills, A. F., and Chung, D. K., Heat transfer across turbulent falling films, *Int. J. Heat Mass Transfer*, vol. 16, pp. 694–696, 1973.
9.47 Mostofizadeh, C., and Stephan, K., Stomung und Warmeubergang beider Oberflachenver dampfung und Filmkondesation, *Warme und Stoffubertragung*, vol. 15, pp. 93–115, 1981.
9.48 Kutateladze, S. S., Semi-empirical theory of film condensation of pure vapors, *Int. J. Heat Mass Transfer*, vol. 25, pp. 653–660, 1982.
9.49 Sandall, O. C., Hanna, O. T., and Wilson, C. L., III, Heat transfer across falling liquid films, *AIChE Symp. Ser.*, vol. 80, no. 263, pp. 3–9, 1984.
9.50 Colburn, A. P., The calculation of condensation where a portion of the condensate layer is in turbulent flow, *Trans. AIChE*, vol. 30, pp. 187–193, 1933.
9.51 Grober, H., Erk, S., and Grigull, U., *Fundamentals of Heat Transfer*, McGraw-Hill, New York, 1961.
9.52 Selin, G., Heat transfer by condensating pure vapors outside inclined tubes, *Proc. Int. Heat Transfer Conf.*, Univ. of Colorado, Boulder, Colo., part II, pp. 279–289, 1961.
9.53 Chen, M. M., An analytical study of laminar film condensation, part I—flat plates, *J. Heat Transfer*, vol. 83, pp. 48–54, 1961.
9.54 Sheynkman, A. G., and Linetskiy, V. N., Hydrodynamics and heat transfer by film condensation of stationary steam on an inclined tube, *Heat Transfer—Sov. Res.*, vol. 1, pp. 90–97, 1969.

9.55 Dhir, V., and Lienhard, J., Laminar film condensation on plane and axisymmetric bodies in nonuniform gravity, *J. Heat Transfer,* vol. 93, pp. 97–100, 1971.

9.56 Rohsenow, W. M., Webber, J. H., and Ling, A. T., Effect of vapor velocity on laminar and turbulent-film condensation, *Trans. ASME,* vol. 78, pp. 1637–1643, 1956.

9.57 Carpenter, F. S., and Colburn, A. P., The effect of vapor velocity on condensation inside tubes, *Proc. General Discussion of Heat Transfer,* Institute of Mechanical Engineers and American Society of Mechanical Engineers, pp. 20–26, 1951.

9.58 Deissler, R. G., Analysis of turbulent heat transfer, mass transfer and friction in smooth tubes at high Prandtl and Schmidt numbers, NACA Technical Note 3145, 1954.

9.59 Denny, V. E., and Mills, A. F., Laminar film condensation of a flowing vapor on a horizontal cylinder at normal gravity, *J. Heat Transfer,* vol. 91, pp. 495–510, 1969.

9.60 Brauer, H., Stromung und Warmeubergang bei Reiselfilmen, *VDI Forschung,* vol. 22, pp. 1–40, 1956.

9.61 Sparrow, E. M., Minkowycz, W. J., and Saddy, M., Forced convection condensation in the presence of noncondensables and interfacial resistance, *Int. J. Heat Mass Transfer,* vol. 10, pp. 1829–1845, 1967.

9.62 Rose, J. W., Boundary-layer flow with transpiration on an isothermal flat plate, *Int. J. Heat Mass Transfer,* vol. 22, pp. 1243–1244, 1979.

9.63 Rose, J. W., Approximate equations for forced convection condensation in the presence of a non-condensing gas on a flat plate and horizontal tube, *Int. J. Heat Mass Transfer,* vol. 23, pp. 539–546, 1980.

9.64 Sparrow, E. M., and Eckert, E. R. G., Effects of superheated vapor and noncondensable gases on laminar film condensation, *AIChE J.,* vol. 7, p. 473, 1967.

9.65 Sparrow, E. M., and Lin, S. H., Condensation heat transfer in the presence of a noncondensable gas, *J. Heat Transfer,* vol. 86, p. 430, 1964.

9.66 Minkowycz, W. J., and Sparrow, E. M., Condensation heat transfer in the presence of noncondensables, interfacial resistance, superheating, variable properties and diffusion, *Int. J. Heat Mass Transfer,* vol. 9, p. 1125–1144, 1966.

9.67 Rose, J. W., Condensation of a pure vapor in the presence of a noncondensable gas, *Int. J. Heat Mass Transfer,* vol. 12, p. 233, 1969.

9.68 Denny, V. E., Mills, A. F., and Jusionis, V. J., Laminar film condensation from a steam–air mixture undergoing forced flow down a vertical surface, *J. Heat Transfer,* vol. 10, p. 1829, 1967.

9.69 Turner, R. H., Mills, A. F., and Denny, V. E., The effect of noncondensable gas on laminar film condensation of liquid metals, *J. Heat Transfer,* vol. 95, pp. 6–11, 1973.

9.70 Al-Diwany, H. K., and Rose, J. W., Free convection film condensation of steam in the presence of non-condensing gases, *Int. J. Heat Mass Transfer,* vol. 16, pp. 1359–1369, 1973.

9.71 Slegers, L., and Seban, R. A., Laminar film condensation of steam containing small concentrations of air, *Int. J. Heat Mass Transfer,* vol. 13, p. 1941, 1970.

9.72 Berman, L. D., Determining the mass transfer coefficient in calculations on condensation of steam containing air, *Teploenergetika,* vol. 16, 68–71, 1969.

9.73 Mills, A. F., Tan, C., and Chung, D. K., Experimental study of condensation from steam–air mixtures flowing over a horizontal tube: Overall condensation rates, *Proc. 5th Int. Heat Transfer Conf.,* Tokyo, paper CT 1.5, vol. 5, pp. 20–23, 1974.

9.74 Iltscheff, S., Uber einege Versuche zur Erzeilung von Tropfkondensation mit Fluorierten Kaltenmitteln, *Kaltetechnik-Klimatiserung,* vol. 23, pp. 237–241, 1971.

9.75 Tanasawa, I., Dropwise condensation—The way to practical applications, *Heat Transfer 1978,* Proc. Sixth Int. Heat Transfer Conf., Toronto, vol. 6, pp. 393–405, 1978.

9.76 Gregorig, R., Film condensation on finely rippled surfaces with consideration of surface tension, *Z. Angewa. Mathe. Phys.,* vol. 5, pp. 36–49, 1954.

9.77 Thomas, D. G., Enhancement of film condensation rates on vertical tubes by vertical wires, *Ind. Eng. Chem. Fund.,* vol. 6, pp. 97–102, 1967.

9.78 Winniarachchi, A. S., Marto, P. J., and Rose, J. W., Film condensation of steam on horizontal finned tubes: Effects of fin spacing, *J. Heat Transfer,* vol. 108, pp. 960–966, 1986.

9.79 Zozulya, N. V., Karkhu, V. A., and Borovkov, V. P., An analytical and experimental study of heat transfer in condensation of vapor on finned surfaces, *Heat Transfer—Sov. Res.*, vol. 9, no. 2, pp. 18–22, 1977.

9.80 Rudy, T. M., and Webb, R. L., An analytical model to predict condensate retention on horizontal integral-fin tubes, *J. Heat Transfer,* vol. 107, pp. 361–368, 1985.

9.81 Hirasawa, S., Hijikata, K., Mori, Y., and Nakayama, W., Effect of surface tension on condensate motion in laminar film condensation, *Int. J. Heat Mass Transfer,* vol. 23, pp. 1471–1478, 1980.

9.82 Marto, P. J., and Nunn, R. H., The potential for heat transfer enhancement in surface condensers, *AIChE Symp. Ser.*, no. 75, pp. 23–39, 1982.

9.83 Webb, R. L., The use of enhanced surface geometries in condensers: An overview, in *Power Condenser Heat Transfer Technology,* P. J. Marto and R. H. Nunn, Eds., Hemisphere, New York, pp. 287–324, 1981.

9.84 Cooper, J. R., and Rose, J. W., Condensation heat transfer enhancement by vapour-side surface geometry modification, *Proc. 1980 HTFS Res. Symp.*, Oxford, paper RS402, pp. 642–672, 1981.

9.85 Marto, P. J., Recent progress in enhancing film condensation heat transfer on horizontal tubes, *Heat Transfer 1986,* Proc. Eighth Int. Heat Transfer Conf., vol. 1, pp. 161–170, 1986.

9.86 Nicol, A. A., and Gacesa, M., Condensation of steam on a rotating vertical cylinder, *J. Heat Transfer,* vol. 92, pp. 144–152, 1970.

9.87 Sparrow, E. M., and Gregg, J. L., A theory of rotating condensation, *J. Heat Transfer,* vol. 81, p. 113, 1959.

9.88 Tleimat, B. W., Performance of a rotating flat-disk wiped-film evaporator, Paper no. 71-HT-37, ASME-AIChE Heat Transfer Conference, Tulsa, OK, August 15–18, 1971.

9.89 Lustenader, E. L., Richter, R., and Neugebauer, F. J., The use of thin films for increasing evaporation and condensation rates in process equipment, *J. Heat Transfer,* vol. 81, pp. 297–307, 1959.

9.90 Brodov, Y. M., Savel'yev, R. Z., Permyakov, V. A., Kuptsov, V. K., and Gal'perin, A. G., The effect of vibration on heat transfer and flow of condensing steam on a single tube, *Heat Transfer—Sov. Res.*, vol. 9, no. 1, pp. 152–156, 1977.

9.91 Seth, A. K., and Lee, L., The effect of an electric field on the presence of noncondensable gas on film condensation heat transfer, *J. Heat Transfer,* vol. 96, pp. 257–258, 1974.

9.92 Lienhard, J., and Dhir, V., A simple analysis of laminar film condensation with suction, *J. Heat Transfer,* vol. 94, pp. 334–336, 1977.

9.93 Bergles, A. E., Enhancement of boiling and condensation, in *Two-Phase Flow and Heat Transfer, China–U.S. Progress,* X.-J. Chen and T. N. Veziroglu, Eds., Hemisphere, New York, 1985.

PROBLEMS

9.1 Use the correlation of Peterson and Westwater [9.26] to predict the heat transfer coefficient for dropwise condensation of steam at a surface subcooling of 5°C for pressures ranging between atmospheric pressure and 9460 kPa. Plot the variation of the heat transfer coefficient over this range of pressures. What properties affect this predicted variation the most?

9.2 Using the correlation of Peterson and Westwater [9.26], predict the heat transfer coefficient for dropwise condensation of steam at atmospheric pressure for wall subcooling values of 2, 5, 10, 20, and 50°C. Plot the values of the heat transfer coefficient as a function of T_{sat}-T_{wall} and compare the resulting variation to the data shown in Fig. 9.6. Discuss possible reasons for the differences.

9.3 Film condensation of acetone occurs on a vertical flat plate at atmospheric pressure. The plate is 30 cm long by 100 cm wide and is held at a constant temperature of 40°C. Determine the mean heat transfer coefficient for the plate.

9.4 Film condensation of ethanol occurs at atmospheric pressure on a 50-cm-long, vertical plate held at 20°C. Determine and plot the estimated variation of the local heat transfer coefficient with local position along the plate. Assuming that transition to turbulent flow occurs at $\mathrm{Re}_l = 2000$, be sure to account for laminar or turbulent conditions at each location.

9.5 Derive a modified form of Eq. (9.74) that also accounts for sensible cooling of the liquid in the overall energy balance. Describe how you would organize the overall computational scheme to use this improved relation to solve the system of similarity equations that govern the flow and temperature fields.

9.6 Laminar film condensation of R-12 occurs at a pressure of 1602 kPa over a flat plate that is 20 cm long. The plate is held at a constant temperature of 40°C. For this process, determine the mean heat transfer coefficient over the surface of the plate using (*a*) the Nusselt correlation (Eq. [9.51]), and (*b*) Equation (9.77) from the similarity analysis of Sparrow and Gregg [9.34]. Discuss the reasons for the difference in the results.

9.7 Film condensation of ammonia occurs on a flat, upward-facing surface, 50 cm long and 50 cm wide, inclined at 45° to the horizontal. The surface is exposed to a motionless ambient of ammonia gas at a pressure of 775 kPa. If heat is removed uniformly over the surface (i.e., a uniform heat flux), estimate the highest heat flux (into the surface) for which the condensate flow will be laminar over the entire surface.

9.8 Estimate the mean heat transfer coefficient for laminar film condensation of R-22 on a horizontal round tube with an outside diameter of 2 cm. The tube wall is held at 20°C. The tube is surrounded by motionless, saturated R-22 vapor at 1420 kPa. For the same wall subcooling, how does the heat transfer coefficient change as the pressure increases toward the critical point?

9.9 Equation (9.111*a*), which predicts the mean heat transfer coefficient for laminar film condensation on a horizontal cylinder, was derived using an integral boundary-layer analysis of the transport in the liquid film. The boundary-layer approximations in this analysis are generally expected to be valid if the film thickness is small compared to the diameter of the cylinder. Film condensation of steam occurs at atmospheric pressure on a horizontal cylinder with a wall temperature of 80°C. Use the results of the analysis presented in Section 9.4 to estimate the cylinder diameter below which the accuracy of the boundary-layer result is questionable.

9.10 Estimate the mean heat transfer coefficient for laminar film condensation of oxygen on a vertical row of 10 horizontal tubes. The tubes are exposed to saturated oxygen vapor at atmospheric pressure. The walls of the tubes are held at 77 K. The diameter and length of the tubes are 1.0 cm and 1.5 m, respectively. For the specified conditions, also estimate the fraction of the total heat duty for the row done by each tube.

9.11 In Example 9.5 it was estimated that for steam condensing on a vertical plate at atmospheric pressure, waves were expected to affect the transport at distances greater than 1.3 cm downstream of the leading edge. As the pressure of the system is increased, how does this location for the onset of wave-affected transport change? What happens to this location if a surfactant is added to the water? Explain briefly.

9.12 In Example 9.5 it was estimated that for steam condensing on a vertical plate at atmospheric pressure, waves were expected to affect the transport at distances greater than 1.3 cm downstream of the leading edge. Use the equation (9.143) recommended by Kutateladze [9.48] to estimate the mean heat transfer coefficient for film condensation of water on a vertical surface 10 cm long for a uniform heat flux into the surface of 500 kW/m^2. Also compute the mean heat transfer coefficient for these conditions using the classical Nusselt relation for film condensation and compare the results for the two methods.

9.13 Forced-convection film condensation of water occurs on a 30-cm-long, flat plate exposed to a flow of steam containing a small amount of air. The ambient conditions are T_∞ = 95°C, u_∞ = 3 m/s, P_∞ = 101 kPa, and $W_{g,\infty}$ = 0.03. The plate is held at a constant temperature of T_w = 55°C. Experimentally, the interface concentration $W_{g,i}$ at a distance 18 cm from the leading edge has been determined to be 0.11. Use results from the analysis presented in Section 9.6 to determine the "suction" velocity normal to the interface and the mean overall heat transfer coefficient for the process at this x location.

9.14 Laminar film condensation occurs on a semiinfinite, vertical, isothermal surface as shown in Fig. P9.1. The cold surface at temperature T_w is surrounded by saturated steam at atmospheric pressure. An infinite hot isothermal surface at temperature T_h is positioned parallel to the cold surface.

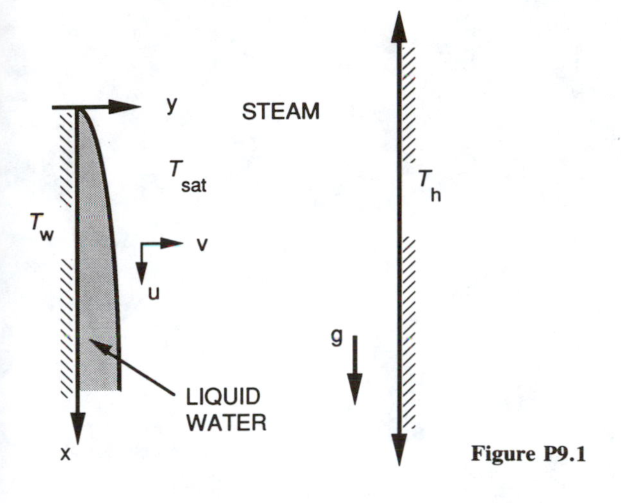

Figure P9.1

The liquid–vapor interface and the hot surface radiate like blackbodies so that the net radiative heat flux to the interface at any location is given by

$$q_R'' = \sigma_{\mathrm{SB}}(T_h^4 - T_{\mathrm{sat}}^4)$$

where

$$\sigma_{\mathrm{SB}} = 5.67 \times 10^{-8}\ \mathrm{W/m^2\ K^4}$$

(*a*) Write down an appropriate set of governing equations and boundary conditions that could be solved to predict the variation of the film thickness and heat transfer coefficient with x for the condensation process on the cooled surface. Be sure to include all necessary relations to close the system. (*b*) Assuming that the film remains laminar and that heat transfer across the liquid film is by conduction only, determine the asymptotic variation of the liquid film thickness far from the leading edge (large x) for $T_w = 50°\mathrm{C}$ and $T_h = 400°\mathrm{C}$.

9.15 Laminar film condensation occurs on a horizontal isothermal cylinder with a diameter of D surrounded by motionless saturated nitrogen vapor at atmospheric pressure. The cold surface of the cylinder is at a temperature T_w. The surrounding walls, which contain the vapor, are far from the surface, but they are at a temperature T_h, which is much higher than the saturation temperature of the nitrogen. Consequently, these walls radiate a net heat flux to the liquid vapor interface that is given by the blackbody relation

$$q_R'' = \sigma_{\mathrm{SB}}(T_h^4 - T_{\mathrm{sat}}^4)$$

where σ_{SB} is the Stephan-Boltzmann constant. Extend the approximate integral analysis of laminar film condensation on a horizontal tube presented in Section 9.4 to include the radiation heat flux to the interface. Show that the mean heat transfer coefficient is given by $\bar{h} = \dot{m}h_{lv}/\pi R(T_{\mathrm{sat}} - T_w)$, where $R = D/2$ and $\dot{m}$ is obtained by solving the following equation with the initial condition $\dot{m} = 0$ at $\Omega = 0$:

$$\frac{d\dot{m}}{d\Omega} = \frac{k_l R(T_{\mathrm{sat}} - T_w)\sin^{1/3}\Omega}{h_{lv}\dot{m}^{1/3}[3\mu_l/\rho_l g(\rho_l - \rho_v)]^{1/3}} - \frac{Rq_R''}{h_{lv}}$$

PART THREE

INTERNAL FLOW CONVECTIVE BOILING AND CONDENSATION

CHAPTER
TEN

INTRODUCTION TO TWO-PHASE FLOW

10.1 TWO-PHASE FLOW REGIMES

In internal convective vaporization and condensation processes, the vapor and liquid are in simultaneous motion inside the channel or pipe. The resulting two-phase flow is generally more complicated physically than single-phase flow. In addition to the usual inertia, viscous, and pressure forces present in single-phase flow, two-phase flows are also affected by interfacial tension forces, the wetting characteristics of the liquid on the tube wall, and the exchange of momentum between the liquid and vapor phases in the flow. Before describing convective condensation and boiling process, it is useful and necessary to develop a framework for treating the associated two-phase flow. As will be seen in later chapters, the morphology of the two-phase flow very often plays a critical role in the determination of the heat and mass transfer during vaporization and condensation processes.

For concreteness, we will begin by considering the very simple two-phase flow shown in Fig. 10.1. Although we will refer specifically to this flow configuration, the basic definitions and terminology developed here will be applicable to any gas–liquid flow circumstance. The total mass flow rate through the tube $\dot{m}$ is equal to the sum of the mass flow rates of gas $\dot{m}_v$ and liquid $\dot{m}_l$,

$$\dot{m} = \dot{m}_v + \dot{m}_l \tag{10.1}$$

The ratio of vapor flow to total flow x,

$$x = \frac{\dot{m}_v}{\dot{m}} \tag{10.2}$$

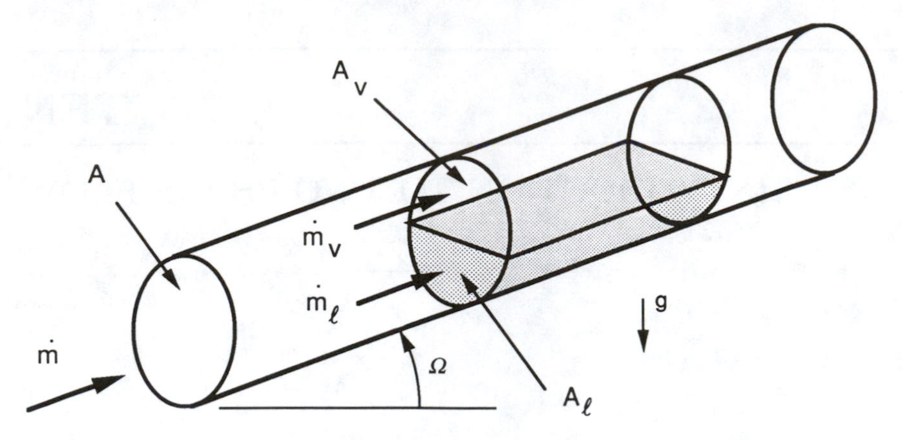

Figure 10.1 Idealized model of two-phase liquid–vapor flow in an inclined tube.

is sometimes called the *dryness fraction* or the *quality,* since it is often taken to be equal to the thermodynamic quality during convective vaporization and condensation processes. In similar fashion, the value of $1 - x = \dot{m}_l/\dot{m}$ is sometimes referred to as the *wetness fraction*.

For a channel with cross-sectional area A, the *mass flux* or *mass velocity* G is defined as

$$G = \frac{\dot{m}}{A} \tag{10.3}$$

The *void fraction* α is defined as the ratio of the gas-flow cross-sectional area A_v to the total cross-sectional area A,

$$\alpha = \frac{A_v}{A} \tag{10.4}$$

where A must equal the sum of the cross-sectional areas occupied by the two phases:

$$A = A_v + A_l \tag{10.5}$$

It follows directly that the liquid volume fraction α_l is given by

$$\alpha_l = 1 - \alpha = \frac{A_l}{A} \tag{10.6}$$

It is also useful to define superficial gas and liquid fluxes, j_v and j_l, respectively, as

$$j_v = \frac{Gx}{\rho_v} \tag{10.7a}$$

$$j_l = \frac{G(1-x)}{\rho_l} \tag{10.7b}$$

Although these parameters have units of velocity, they can also be thought of as the volume flux of each phase through the channel. Numerically they are equal to the velocity that each phase would have if it flowed at its specified mass flow rate through the channel alone.

In the two-phase flow shown in Fig. 10.1, the arrangement of the phases is relatively simple. In general, however, the morphology of the two phases can be quite complex, and it can vary depending on the fluid properties and flow conditions.

Upward Vertical Flow

For co-current upward flow in a vertical round tube, the possible observed flow regimes are indicated in Fig. 10.2. At very low quality, the flow is usually found to be in the *bubbly flow* regime, which is characterized by discrete bubbles of vapor dispersed in a continuous liquid phase. In bubbly flow, the mean size of the bubbles is generally small compared to the diameter of the tube. At slightly higher qualities, smaller bubbles may coalesce into slugs that span almost the entire cross section of the channel. The resulting flow regime is usually referred to as *slug flow*.

At much higher quality levels, the two-phase flow generally assumes an annular configuration, with most of the liquid flowing along the wall of the tube and the gas flowing in the central core. For obvious reasons, this regime is termed the *annular flow* regime. When the vapor flow velocity is high, the interface of the liquid film may become Helmholtz-unstable, leading to the formation of waves at the interface. Liquid droplets formed by breaking waves may then be entrained in the vapor core flow.

At intermediate qualities, one of two addition regimes may be observed. If both the liquid and vapor flow rates are high, an annular-type flow is observed with heavy "wisps" of entrained liquid flowing in the vapor core. Although this

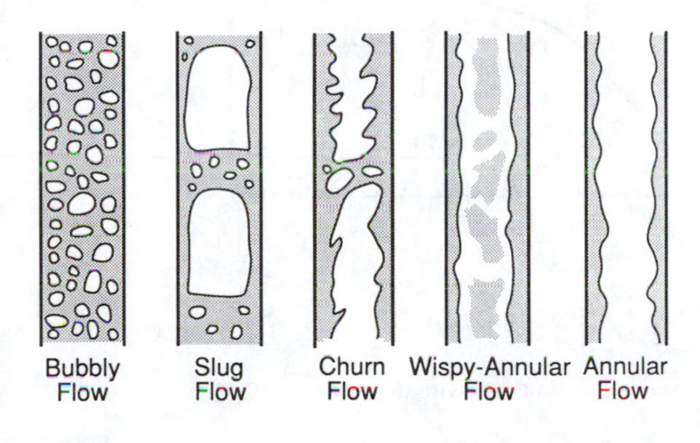

Figure 10.2 Schematic representations of flow regimes observed in vertical upward co-current gas–liquid flow.

is a form of annular flow, it is sometimes designated as a separate regime, referred to as *wispy annular flow*.

For intermediate qualities and lower flow rates, the vapor shear on the liquid–vapor interface may be near the value where it just balances the combined effects of the imposed pressure gradient and the downward gravitational body force on the liquid film. As a result, the liquid flow tends to be unstable and oscillatory. The vapor flow in the center of the tube flows continuously upward. Although the mean velocity of the liquid film is upward, the liquid experiences intermittent upward and downward motion. The flow for these conditions is highly agitated, resulting in a highly irregular interface. This oscillatory flow is referred to as *churn flow*.

The conditions corresponding to the flow regimes described above can be represented on the flow regime map shown in Fig. 10.3. The form of this map was proposed by Hewitt and Roberts [10.1]. The vertical coordinate is equal to the superficial momentum flux of the vapor, and the horizontal coordinate is the superficial momentum flux of liquid through the tube. The boundaries between the flow regimes have been established from visual observation of the two-phase flow in a series of experiments (using a transparent tube) that spanned the entire flow regime map. Since the flow regime for a given set of conditions is a matter of judgment regarding the appearance of the flow, the boundaries should be interpreted as specifying the middle of a transition between two regimes.

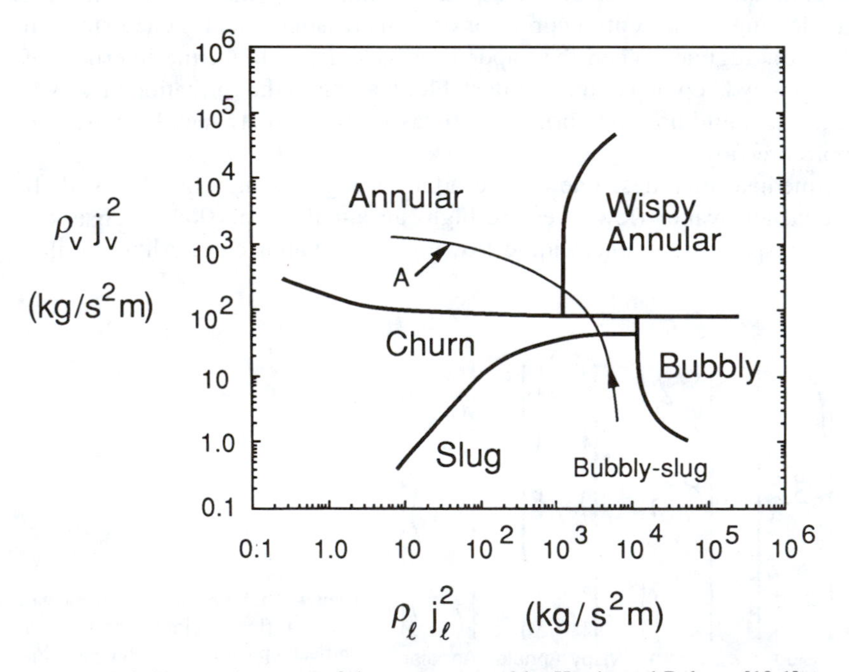

Figure 10.3 Flow regime map of the type proposed by Hewitt and Roberts [10.1].

From theoretical considerations, analytical expressions for the transition conditions between the two-phase flow regimes have also been obtained. Radovcich and Moissis [10.2] presented arguments about the frequency of bubble collisions, which suggest that the transition from bubbly to slug flow is highly probable at void fractions above $\alpha = 0.3$. Based on a more detailed analysis, Taitel and Dukler [10.3] proposed the relation

$$\frac{j_l}{j_v} = 2.34 - 1.07 \frac{[g(\rho_l - \rho_v)\sigma]^{1/4}}{j_v \rho_l^{1/2}} \tag{10.8}$$

as defining the incipient conditions for the transition from bubbly to slug flow.

As noted above, increasing quality can lead to a transition from slug flow to churn flow. This breakdown of slug flow is a consequence of the interaction between the rising slug bubble and the liquid film between the slug and the wall. In a flow of this type, this liquid film actually moves downward as the slug moves upward at a velocity higher than the mean velocity of the two-phase flow, due to its buoyancy. As the quality and void fraction increase, this countercurrent flow becomes unstable in a manner similar to the Helmholtz instability described in Chapter Four. This instability eventually leads to the breakup of the large bubbles characteristic of slug flow, initiating a transition to churn flow. Porteus [10.4] presented theoretical arguments that suggest that this transition corresponds to conditions defined by the relation

$$\frac{j_l}{j_v} = 0.105 \frac{[gD(\rho_l - \rho_v)]^{1/2}}{j_v \rho_v^{1/2}} - 1 \tag{10.9}$$

where D is the tube diameter. Taitel and Dukler [10.3], on the other hand, argued that for $(j_l + j_v)/(gD)^{1/2}$ greater than 50, the slug-to-churn transition occurs at conditions that correspond to $j_l/j_v = 0.16$.

The transition from churn flow to annular flow occurs at conditions where the upward shear stress of the vapor core flow plus the imposed pressure gradient just balances the downward gravitational force on the liquid film. These conditions correspond to the lower vapor velocity limit for which steady upward annular flow can be sustained. Based on theoretical arguments, Wallis [10.5] concluded that this transition occurred approximately at conditions specified by the relation

$$\left[\frac{j_v^2 \rho_v}{gD(\rho_l - \rho_v)}\right]^{0.5} = 0.9 \tag{10.10}$$

Taitel and Dukler [10.3] proposed the following relation as a means of predicting the transition from churn to annular flow:

$$\frac{j_v \rho_v^{0.5}}{[g(\rho_l - \rho_v)\sigma]^{0.25}} = 3.09 \frac{(1 + 20X + X^2)^{0.5} - X}{(1 + 20X + X^2)^{0.5}} \tag{10.11a}$$

where X is the Martinelli parameter, defined as

$$X = \left[\frac{(dP/dz)_l}{(dP/dz)_v}\right]^{1/2} \tag{10.11b}$$

In Eq. (10.11), $(dP/dz)_l$ and $(dP/dz)_v$ are the frictional pressure gradients for the liquid and vapor phases flowing alone in the pipe, respectively. These frictional gradients can be computed as

$$\left(\frac{dP}{dz}\right)_l = -\frac{2f_l G^2(1-x)^2}{\rho_l D} \tag{10.12a}$$

$$\left(\frac{dP}{dz}\right)_v = -\frac{2f_v G^2 x^2}{\rho_v D} \tag{10.12b}$$

$$f_l = B\,\mathrm{Re}_l^{-n}, \qquad \mathrm{Re}_l = \frac{G(1-x)D}{\mu_l} \tag{10.13a}$$

$$f_v = B\,\mathrm{Re}_v^{-n}, \qquad \mathrm{Re}_v = \frac{GxD}{\mu_v} \tag{10.13b}$$

In the above friction-factor relations, for round tubes the constants can be taken to be $B = 16$ and $n = 1$, respectively, for laminar flow (Re_l or $\mathrm{Re}_v < 2000$), or $B = 0.079$ and $n = 0.25$ for turbulent flow (Re_l or $\mathrm{Re}_v \geq 2000$).

The transition between wispy annular flow and annular flow is difficult to distinguish precisely because the regimes are so similar. Based on experiments that used a probe to detect wispy filaments in the core flow, Wallis [10.5] proposed the following correlation for the transition condition:

$$\frac{j_v}{j_l} = \left(7 + 0.06\frac{\rho_l}{\rho_v}\right) \tag{10.14}$$

This relation is recommended for $j_l \rho_l^{0.5}[gD(\rho_l - \rho_v)]^{-0.5} > 1.5$.

Horizontal Flow

For two-phase flow in horizontal round tubes, the flow regimes that may be encountered are shown in Fig. 10.4. One of the main differences between the regimes observed for horizontal flow and those for vertical flow is that there is often a tendency for stratification of the flow. Regardless of the flow regime, the vapor tends to migrate toward the top of the tube while the lower portion of the channel carries more of the liquid. At very low quality, *bubbly flow* is often observed for horizontal flow. However, as indicated in Fig. 10.4, the bubbles, because of their buoyancy, flow mainly in the upper portion of the tube.

As the quality is increased in the bubbly regime, coalescence of small bubbles produces larger plug-type bubbles, which flow in the upper portion of the tube (see Fig. 10.4). This is referred to as the *plug flow* regime. At low flow rates and somewhat higher qualities, *stratified flow* may be observed in which liquid flowing in the bottom of the pipe is separated from vapor in the upper portion of the pipe by a relatively smooth interface.

If the flow rate and/or the quality is increased in the stratified flow regime,

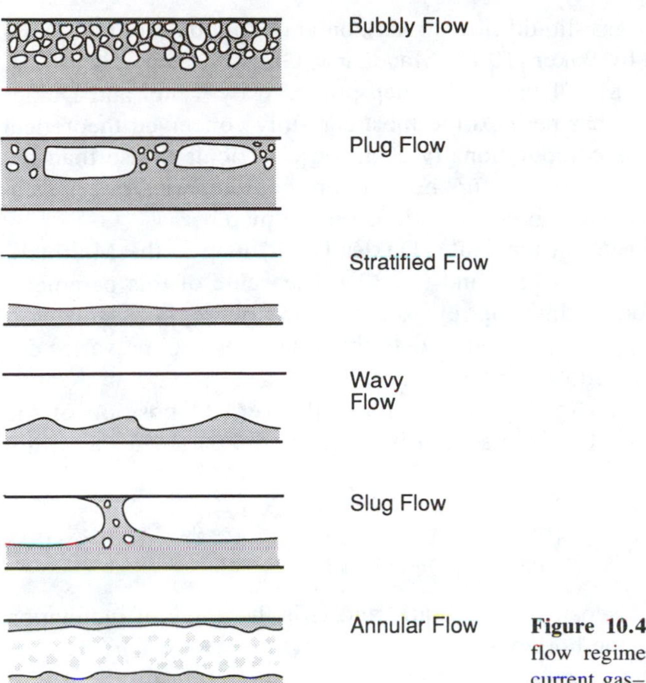

Figure 10.4 Schematic representations of flow regimes observed in horizontal, cocurrent gas–liquid flow.

eventually the interface becomes Helmholtz-unstable, whereupon the interface becomes wavy. This type of flow is categorized as *wavy flow*. The strong vapor shear on the interface for these circumstances, together with the formation and breaking of waves on the interface, may lead to significant entrainment of liquid droplets in the vapor core flow. At high liquid flow rates, the amplitude of the waves may grow so that the crests span almost the entire width of the tube, effectively forming large slug-type bubbles. Because of their buoyancy, the slugs of vapor flowing along the tube tend to skew toward the upper portion of the tube. In other respects it is identical to slug flow in vertical tubes, and hence it too is referred to as *slug flow*.

At high vapor velocities and moderate liquid flow rates, *annular flow* is observed for horizontal gas–liquid flow. For such conditions, buoyancy effects may tend to thin the liquid film on the top portion of the tube wall and thicken it at the bottom. However, at sufficiently high vapor flow rates, the vapor flow is invariably turbulent, and strong lateral Reynolds stresses and the shear resulting from secondary flows may served to distribute liquid more evenly around the tube perimeter against the tendency of gravity to stratify the flow. The strong vapor shear may also result in significant entrainment of liquid in the vapor core. Because gravitational body forces are often small compared to inertia effects and turbulent transport of momentum, the resulting flow for these circumstances is generally expected to differ little from annular flow in a vertical tube under similar flow conditions.

Flow regime maps for gas–liquid flow in horizontal or slightly inclined round tubes have been proposed by Baker [10.6], Mandhane, Gregory, and Aziz [10.7], and Taitel and Dukler [10.8]. Of these, the map proposed by Taitel and Dukler [10.8], shown in Fig. 10.5, has perhaps the most carefully conceived theoretical basis. Although this map is computationally a bit more difficult to use than the others, it at least attempts to account for the different combinations of physical parameters that affect different regime transitions on the map.

The horizontal coordinate on the Taitel-Dukler [10.8] map is the Martinelli parameter X defined by Eqs. (10.11b) and (10.12). The value of this parameter fixes the horizontal position on this map regardless of the flow regime. However, the second dimensionless parameter used to determine the flow regime varies depending on the specific transition being considered.

For the stratified flow-to-wavy flow transition, the vertical position of the corresponding point in Fig. 10.5 is specified in terms of the parameter K_{TD}, defined as

$$K_{\text{TD}} = \left[\frac{\rho_v j_v^2 j_l}{\nu_l(\rho_l - \rho_v) g \cos \Omega} \right]^{0.5} \tag{10.15a}$$

where ν_l is the kinematic viscosity of the liquid and Ω is the angle of inclination between the tube axis and the horizontal.

The wavy–annular and wavy–intermittent (plug or slug) transitions in Fig. 10.5 are evaluated in terms of X and the parameter F_{TD}, defined as

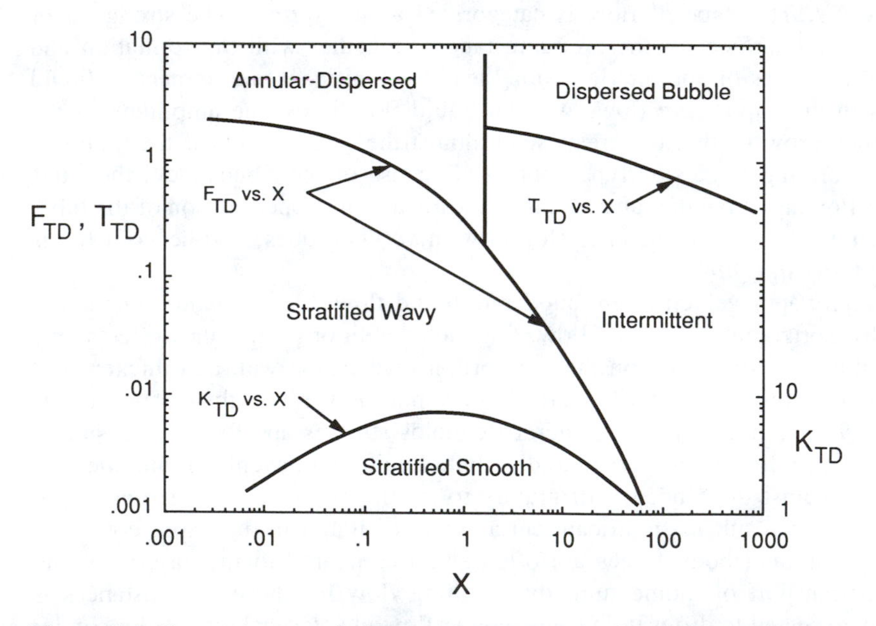

Figure 10.5 Flow regime map for horizontal co-current gas–liquid flow of the type proposed by Taitel and Dukler [10.8].

$$F_{\mathrm{TD}} = \left[\frac{\rho_v j_v^2}{(\rho_l - \rho_v) D g \cos \Omega} \right]^{0.5} \tag{10.15b}$$

The transition from bubbly flow to intermittent flow is specified in terms of X and yet a third parameter T_{TD}:

$$T_{\mathrm{TD}} = \left[\frac{-(dP/dz)_l}{(\rho_l - \rho_v) g \cos \Omega} \right]^{0.5} \tag{10.15c}$$

where $(dP/dz)_l$ is given by Eq. (10.12*a*). The transition between intermittent and annular flow or between bubbly and annular flow corresponds simply to $X = 1.6$ on this map.

General Observations

Co-current upward and co-current horizontal two-phase flows are by far the most commonly encountered configurations, and a round tube is the most common geometry used in technological applications. However, a few other flow circumstances and channel geometries, which arise in other applications, have also been investigated.

Studies of vertical upward and horizontal co-current gas–liquid flow in plain rectangular channels [10.9], in channels with offset strip fins [10.10], and in cross-ribbed channel geometries [10.11] have indicated that the flow regimes and transitions in these geometries are similar to those in round tubes. The two-phase flow regimes for upward or horizontal co-current flow in tubes with internal grooves [10.12], expansions and contractions [10.13] and helical inserts [10.14], and in bends and helically wrapped coils [10.15, 10.16, 10.17] have also been explored. Although flow regimes similar to those observed in corresponding round tube flows are often observed, in some cases these geometry variations can produce significant differences in the flow behavior.

Based on the results of experimental studies, Oshinowo and Charles [10.18] have proposed flow regime maps for vertical downward co-current gas–liquid flow in a round tube. For such circumstances, regimes of bubbly, slug, churn, and annular flow similar to the regimes observed in upward co-current flow may be observed.

Additional flow regimes can arise in upward or horizontal co-current flow when the tube is very small or when the liquid poorly wets the tube wall. For very small tubes, the *capillary bubble flow* shown in Fig. 10.6 can result. In this regime the vapor bubble completely fills the tube cross section, leaving the wall at that location completely dry. In larger tubes, if the liquid poorly wets the wall, so-called *capillary flow* may result, in which the liquid flows in rivulets on the wall, driven by vapor shear forces. If the liquid on the wall breaks up into droplets, the drop flow regime may be encountered, in which the vapor flow drags individual droplets along the wall (see Fig. 10.6).

When a phase change occurs as the two-phase mixture flows along the chan-

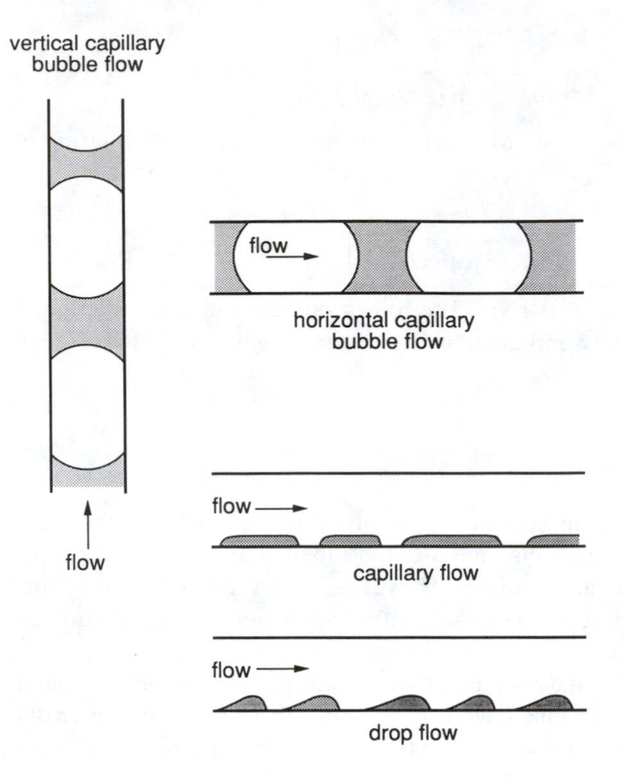

Figure 10.6 Additional flow regimes observed under special conditions.

nel, different flow regimes are generally observed at different positions along its length. The sequence of flow regimes observed will depend primarily on the flow rate, channel orientation, fluid properties, and the distribution and magnitude of the heat flux into or out of the flow at the channel wall. The sequence of flow regimes for upward flow boiling in a vertical heated tube at low to moderate heat flux levels is shown in Fig. 10.7. Boiling may be initiated before the bulk liquid reaches the saturation temperature. At this initial stage of the boiling process, the void fraction is low and bubbly flow results.

As the vaporization process continues, and liquid is converted to vapor, the void fraction increases, j_v increases, and j_l decreases. Consequently, the flow regime progressively changes from bubbly flow to slug flow, from slug flow to churn flow, and from churn flow to annular flow. On the flow regime map in Fig. 10.3, the sequence of system state points would trace a curve similar to curve *A*. Note that since the total mass flux is the same everywhere along the tube, the path of the system on this map is defined by the relation

$$\rho_v j_v + \rho_l j_l = G = \text{constant}$$

In a similar fashion, horizontal co-current flow with vaporization or condensation results in the sequences of flow regimes indicated in Figs. 10.8 and 10.9,

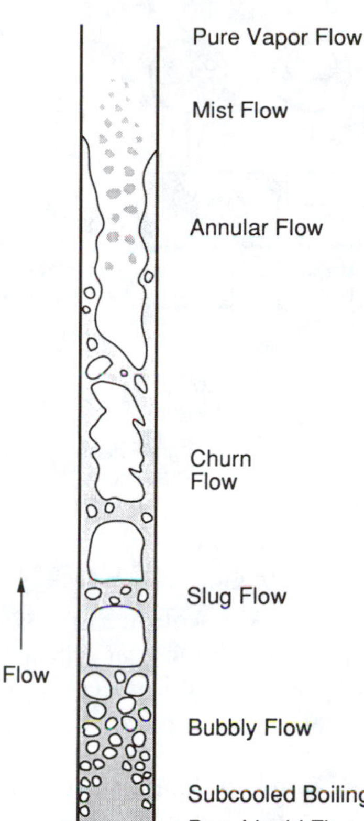

Figure 10.7 Sequence of flow regimes observed during upward flow boiling in a vertical tube.

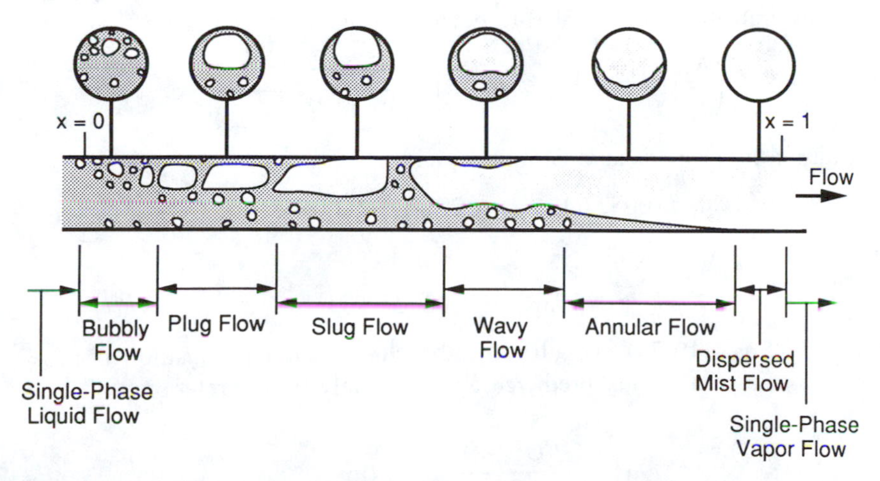

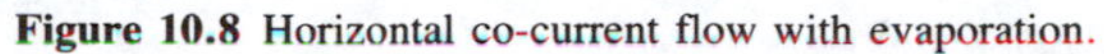

Figure 10.8 Horizontal co-current flow with evaporation.

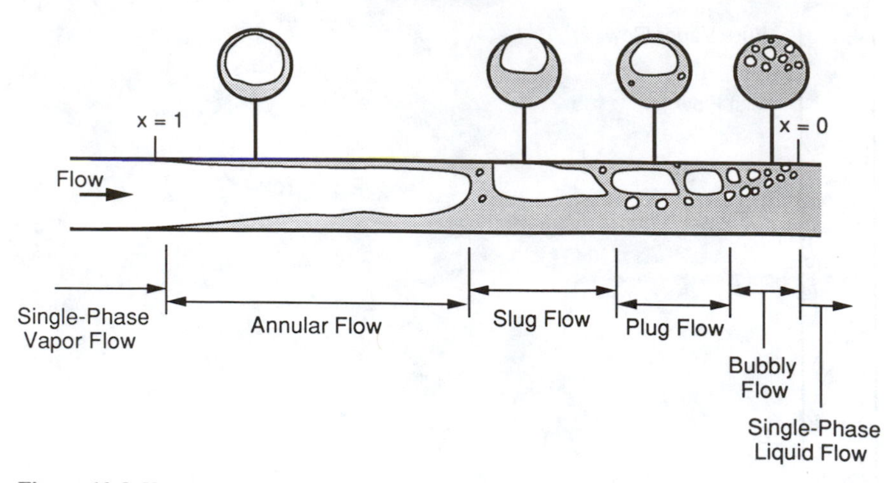

Figure 10.9 Horizontal co-current flow with condensation.

respectively. As discussed in Chapters Eleven and Twelve, the sequence of flow regimes that occurs along the tube has a very strong impact on the heat transfer characteristics associated with convective boiling and condensation processes in tubes and ducts.

Example 10.1 A two-phase mixture of saturated nitrogen liquid and vapor at atmospheric pressure flows upward in a vertical tube. The tube inside diameter is 1 cm, and the mass flux G is 300 kg/m^2 s. Estimate the values of quality at which the transitions from bubbly to slug and churn to annular flow will occur.

For saturated nitrogen at atmospheric pressure, $\rho_l = 807$ kg/m^3, $\rho_v = 4.62$ kg/m^3, and $\sigma = 0.00885$ N/m. Combining Eqs. (10.7) and (10.8), the bubbly-to-slug transition is estimated to occur at a quality that satisfies

$$\left(\frac{\rho_v}{\rho_l}\right)\left(\frac{1-x}{x}\right) = 2.34 - 1.07\,\frac{[g(\rho_l - \rho_v)\sigma]^{1/4}}{\rho_l^{1/2}Gx/\rho_v}$$

Substituting the property and G values yields

$$0.00572(1 - x) = 2.34x - 0.00168$$

Solving for $x = x_{b-s}$ yields

$$x_{b-s} = 0.0032$$

Combining Eqs. (10.7*a*) and (10.10), the churn–annular transition is predicted to occur at a quality predicted approximately by the relation

$$x\left[\frac{G^2}{gD\rho_v(\rho_l - \rho_v)}\right]^{0.5} = 0.9$$

Substituting the appropriate values of G, D, and the densities and solving for $x = x_{c-a}$ yields

$$x_{c-a} = 0.057$$

Thus the flow is predicted to be in the annular flow regime at qualities as low as 0.06.

10.2 BASIC MODELS AND GOVERNING EQUATIONS FOR ONE-DIMENSIONAL TWO-PHASE FLOW

General Considerations

Generalized analytical treatment of liquid–vapor two-phase flow is an extremely difficult task. Despite the dramatic increase in the speed of computers over the past 30 years, computational prediction of two-phase flow has lagged far behind the advances made in computational techniques for single-phase flows. The more primitive state of currently used computational techniques for two-phase flow is a direct consequence of the higher level of difficulty associated with such computations.

Developments of the full governing equations for three-dimensional, time-varying two-phase flow can be found in works by Ishii [10.19] and Boure [10.20]. The form of the governing equations can be simplified by invoking time and/or space averaging. In doing so, however, information regarding the instantaneous localized behavior of the flow is lost. In the treatment of internal gas–liquid flows considered here, the flow is considered to be steady and one-dimensional in the sense that all dependent variables are idealized as being constant over any cross section of the tube or duct, varying only in the axial direction.

To facilitate development of a one-dimensional analysis of gas–liquid flow, we will consider the system shown in Fig. 10.10. Although a stratified flow is shown for concreteness, the relations derived for conservation of mass and momentum will be generally applicable to gas–liquid flows in any of the regimes described in the previous section, within the one-dimensional approximations adopted here. The individual mean phase velocities indicated in this diagram are related to the void fraction as

$$u_v = \frac{Gx}{\rho_v \alpha} \tag{10.16a}$$

$$u_l = \frac{G(1 - x)}{\rho_l (1 - \alpha)} \tag{10.16b}$$

Conservation of mass requires that

$$\dot{m} = \dot{m}_v + \dot{m}_l \tag{10.17a}$$

$$\dot{m} = \rho_v G x A + \rho_l G(1 - x)A \tag{10.17b}$$

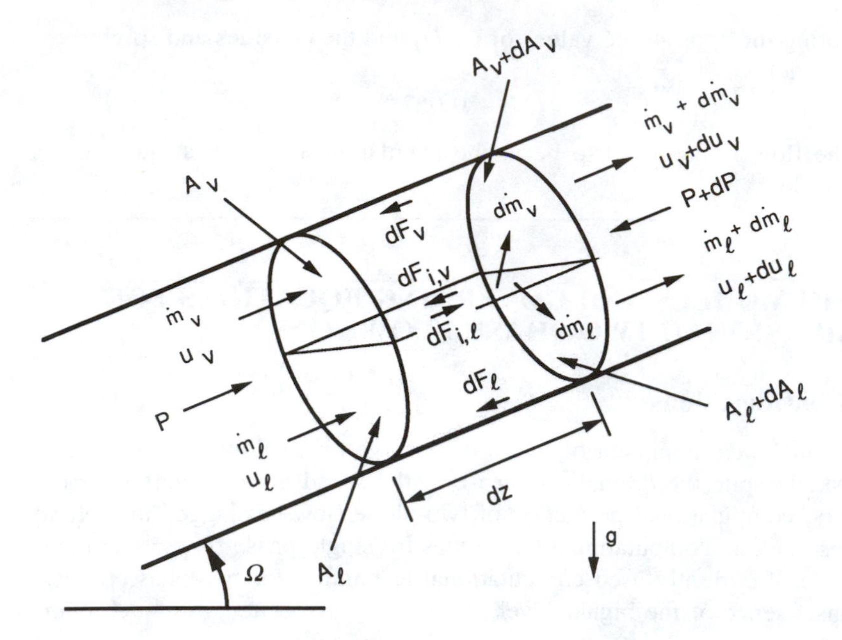

Figure 10.10 Idealized model of momentum transport during gas–liquid two-phase flow in an inclined tube.

Differentiating these relations, the conservation of mass requirement can also be expressed in the forms

$$d\dot{m}_v = -\, d\dot{m}_l \tag{10.18}$$

or

$$\frac{d}{dz}(\rho_v GxA) + \frac{d}{dz}(\rho_l G(1-x)A) = 0 \tag{10.19}$$

Alternative forms of the continuity relation can also be derived by combining one of the above equations with the variable definitions (10.2) through (10.7) or (10.16).

The second governing equation for the flow in Fig. 10.10 is derived from a force–momentum balance in the axial direction. A balance of this type on the vapor phase only gives

$$PA_v - (P + dP)(A_v + dA_v) - dF_v - dF_{i,v} - A_v\, dz\, \rho_v g \sin\Omega$$
$$= (\dot{m}_v + d\dot{m}_v)(u_v + du_v) - \dot{m}_v u_v - d\dot{m}_v u_l \tag{10.20}$$

The first two terms on the left side of this relation represent pressure forces on the element. The terms of dF_v and $dF_{i,v}$ represent the frictional effect of the vapor on the channel wall and the interfacial shear force, respectively. The first two terms on the right side of the equation account for the momentum change due to acceleration (or deceleration) of the vapor. The last term on the right side represents the momentum exchange between the liquid and vapor due to a phase change at the interface, which transfers liquid molecules into the vapor. (An equivalent term could be included for transfer in the opposite direction instead, if necessary. Terms of this type cancel later anyway.) In a completely analogous way, a momentum balance for the liquid can be written as

$$PA_l - (P + dP)(A_l + dA_l) - dF_l + dF_{i,l} - A_l\,dz\,\rho_l g \sin\Omega$$
$$= (\dot{m}_l + d\dot{m}_l)(u_l + du_l) - d\dot{m}_l u_l - \dot{m}_l u_l \qquad (10.21)$$

For steady flow, the interfacial shear forces must balance so that

$$dF_{i,l} = -dF_{i,v} \qquad (10.22)$$

Adding Eqs. (10.20) and (10.21) and using Eqs. (10.5), (10.18), and (10.22), the following relation for the overall momentum balance is obtained:

$$-A\,dP - P\,dA - dF_l - dF_v - (A_l\rho_l + A_v\rho_v)g\,dz\sin\Omega = d(\dot{m}_v u_v + \dot{m}_l u_l) \qquad (10.23)$$

Defining a fictitious pressure gradient $(dP/dz)_{fr}$ to account for the combined frictional effect of the two phases on the channel wall,

$$dF_l + dF_v = -\left(\frac{dP}{dz}\right)_{fr} A\,dz \qquad (10.24)$$

and using Eqs. (10.4) and (10.16), the momentum balance relation (10.23) can be written as

$$-\left(\frac{dP}{dz}\right) = -\frac{P}{A}\left(\frac{dA}{dz}\right) - \left(\frac{dP}{dz}\right)_{fr} + [(1-\alpha)\rho_l + \alpha\rho_v]g\sin\Omega$$
$$+ \left(\frac{1}{A}\right)\frac{d}{dz}\left[\frac{G^2x^2A}{\rho_v\alpha} + \frac{G^2(1-x)^2A}{\rho_l(1-\alpha)}\right] \qquad (10.25)$$

There are four contributions to the overall pressure gradient for the two-phase flow in the channel represented on the right side of the above equation. The first term represents pressure variations due to changes in the cross-sectional area of the duct. The second and third terms account for frictional and gravitational head effects, respectively, and the last term represents acceleration (or deceleration) of the flow. If no phase change is occuring, x and α will be constant and if A is constant, the last term will be zero.

To further develop this one-dimensional model of momentum transport we will invoke the following additional idealizations:

1. At a given axial location in the channel, the velocities of the liquid (u_l) and vapor (u_v) phases are uniform over the portion of the channel occupied by each, but are not necessarily equal.
2. The two phases are in local thermodynamic equilibrium.
3. Empirical correlations or relations derived from simplified theories are avail-

able to predict the void fraction α and one of the two-phase multipliers ϕ_l, ϕ_{lo}, or ϕ_v from parameters that quantify the local flow conditions.

The two-phase multipliers mentioned above are defined as

$$\phi_l = \left[\frac{(dP/dz)_{fr}}{(dP/dz)_l}\right]^{1/2} \tag{10.26}$$

$$\phi_{lo} = \left[\frac{(dP/dz)_{fr}}{(dP/dz)_{lo}}\right]^{1/2} \tag{10.27}$$

$$\phi_v = \left[\frac{(dP/dz)_{fr}}{(dP/dz)_v}\right]^{1/2} \tag{10.28}$$

In the above expressions the subscript designations are as follows:

fr denotes the frictional component of the two-phase pressure drop in Eq. (10.25).
l denotes the frictional pressure gradient that would result if the liquid flowed alone through the channel [at a mass flow rate equal to $G(1 - x)A$].
lo denotes the frictional pressure gradient that would result if liquid only flowed through the channel at the same total mass flow rate (GA).
v denotes the frictional pressure gradient that would result if the vapor flowed alone through the channel (at a mass flow rate equal to GxA).

In terms of the two-phase multiplier ϕ_l, the momentum equation (10.25) can be written as

$$-\left(\frac{dP}{dz}\right) = -\frac{P}{A}\left(\frac{dA}{dz}\right) - \phi_l^2\left(\frac{dP}{dz}\right)_l + [(1-\alpha)\rho_l + \alpha\rho_v]g \sin \Omega$$
$$+ \left(\frac{1}{A}\right)\frac{d}{dz}\left[\frac{G^2x^2A}{\rho_v\alpha} + \frac{G^2(1-x)^2A}{\rho_l(1-\alpha)}\right] \tag{10.29}$$

This relation can also be written in terms of ϕ_{lo} as

$$-\left(\frac{dP}{dz}\right) = -\frac{P}{A}\left(\frac{dA}{dz}\right) - \phi_{lo}^2\left(\frac{dP}{dz}\right)_{lo} + [(1-\alpha)\rho_l + \alpha\rho_v]g \sin \Omega$$
$$+ \left(\frac{1}{A}\right)\frac{d}{dz}\left[\frac{G^2x^2A}{\rho_v\alpha} + \frac{G^2(1-x)^2A}{\rho_l(1-\alpha)}\right] \tag{10.30}$$

If the cross-sectional area of the channel A is constant, these relations reduce to

$$-\left(\frac{dP}{dz}\right) = -\phi_l^2\left(\frac{dP}{dz}\right)_l + [(1-\alpha)\rho_l + \alpha\rho_v]g \sin \Omega$$
$$+ \frac{d}{dz}\left[\frac{G^2x^2}{\rho_v\alpha} + \frac{G^2(1-x)^2}{\rho_l(1-\alpha)}\right] \tag{10.31}$$

and

$$-\left(\frac{dP}{dz}\right) = -\phi_{lo}^2\left(\frac{dP}{dz}\right)_{lo} - [(1-\alpha)\rho_l + \alpha\rho_v]g\sin\Omega$$
$$+\frac{d}{dz}\left[\frac{G^2x^2}{\rho_v\alpha} + \frac{G^2(1-x)^2}{\rho_l(1-\alpha)}\right] \tag{10.32}$$

Homogeneous Flow

The so-called *homogeneous model* of momentum transport in gas–liquid flows can be considered to be a special case of the separated flow analysis described thus far. For reasons that will become obvious, this model is also sometimes called the *friction-factor model* or *fog flow model*. The idealizations adopted under this model, in addition to those noted above for separated flow, are:

1. The vapor and liquid velocities are equal.
2. The two-phase flow behaves like a single phase having fluid properties whose values are, in some sense, mean values for the flow.

Equating the expressions (10.16*a*) and (10.16*b*) for u_v and u_l, the following relation between the void fraction α and the quality is obtained:

$$\alpha = \frac{x/\rho_v}{[(1-x)/\rho_l] + (x/\rho_v)} = \frac{xv_v}{(1-x)v_l + xv_v} \tag{10.33}$$

where v_v and v_l denote the specific volumes of the vapor and the liquid, respectively.

Treating the two-phase flow as an equivalent single-phase flow, the frictional contribution to the overall pressure gradient can be determined using a conventional friction factor:

$$-\left(\frac{dP}{dz}\right)_{fr} = \frac{2f_{tp}G^2}{\bar{\rho}d_h} \tag{10.34}$$

In this expression, f_{tp} is an effective (Fanning) friction factor for the two-phase flow and $\bar{\rho}$ is the mean density of the flow given by the usual thermodynamic definition:

$$\frac{1}{\bar{\rho}} = \bar{v} = v_v x + (1-x)v_l = \frac{x}{\rho_v} + \frac{1-x}{\rho_l} \tag{10.35}$$

The fictitious single-phase pressure gradient for the entire flow as liquid can be similarly evaluated in terms of a friction factor:

$$-\left(\frac{dP}{dz}\right)_{lo} = \frac{2f_{lo}G^2}{\rho_l d_h} \tag{10.36}$$

Substituting Eqs. (10.34) and (10.36) into Eq. (10.27), the following relation can be obtained for ϕ_{lo}^2:

$$\phi_{lo}^2 = \frac{f_{tp}}{f_{lo}}\left[1 + \left(\frac{\rho_l}{\rho_v} - 1\right)x\right] \tag{10.37}$$

The friction factors can usually be expressed as power-law functions of Reynolds number:

$$f_{lo} = M\,\mathrm{Re}_{lo}^{-m} = M\left(\frac{Gd_h}{\mu_l}\right)^{-m} \tag{10.38a}$$

$$f_{tp} = N\,\mathrm{Re}_{lo}^{-n} = N\left(\frac{Gd_h}{\overline{\mu}}\right)^{-n} \tag{10.38b}$$

where d_h is the hydraulic diameter and $\overline{\mu}$ is the effective mean viscosity for the two-phase flow. The ratio of the friction factors is then given by

$$\frac{f_{tp}}{f_{lo}} = \left(\frac{N}{M}\right)\left(\frac{Gd_h}{\mu_l}\right)^{m-n}\left(\frac{\overline{\mu}}{\mu_l}\right)^{n} \tag{10.39}$$

Substituting this result into Eq. (10.37), the relation for ϕ_{lo}^2 becomes

$$\phi_{lo}^2 = \left(\frac{N}{M}\right)\left(\frac{Gd_h}{\mu_l}\right)^{m-n}\left(\frac{\overline{\mu}}{\mu_l}\right)^{n}\left[1 + \left(\frac{\rho_l}{\rho_v} - 1\right)x\right] \tag{10.40}$$

In the above treatment, the friction factor relations for the two-phase flow and the entire flow as liquid have been assumed to be different. This is appropriate if the flow of the two-phase mixture is turbulent whereas the entire flow as liquid would be laminar. If both flows are in the same regime, the relation should be the same (i.e., $M = N$, $n = m$) and the relation for ϕ_{lo}^2 simplifies to

$$\phi_{lo}^2 = \left(\frac{\overline{\mu}}{\mu_l}\right)^{n}\left[1 + \left(\frac{\rho_l}{\rho_v} - 1\right)x\right] \tag{10.41}$$

This simplification removes the dependence of ϕ_{lo}^2 on the mass flux G.

The only remaining detail to be completed at this point is the evaluation of the mean viscosity $\overline{\mu}$. In general, a weighted average with respect to the quality of the flow is perhaps the most logical means of defining $\overline{\mu}$. A mean value defined in this manner should approach the liquid viscosity and vapor viscosity as $x \to 0$ and $x \to 1$, respectively. Three proposed relations that satisfy these criteria are

$$\frac{1}{\overline{\mu}} = \frac{x}{\mu_v} + \frac{1 - x}{\mu_l} \tag{10.42}$$

$$\overline{\mu} = x\mu_v + (1 - x)\mu_l \tag{10.43}$$

$$\frac{\overline{\mu}}{\overline{\rho}} = \frac{x\mu_v}{\rho_v} + \frac{(1 - x)\mu_l}{\rho_l} \tag{10.44}$$

Equations (10.42), (10.43), and (10.44) were proposed by McAdams et al. [10.21], Cicchitti et al. [10.22], and Dukler et al. [10.23], respectively. Equation (10.42) is probably the most commonly used definition of $\overline{\mu}$. For turbulent single-phase flow in round tubes, the Fanning friction factor can be determined from the well-known Blasius correlation,

$$f = 0.079\left(\frac{Gd_h}{\mu}\right)^{-0.25} \tag{10.45}$$

If both the two-phase flow and the entire flow as liquid are turbulent, then $M = N = 0.079$ and $n = m = 0.25$. Combining Eq. (10.41) with Eq. (10.42) for such circumstances yields

$$\phi_{lo}^2 = \left[1 + \left(\frac{\mu_l}{\mu_v} - 1\right)x\right]^{-0.25}\left[1 + \left(\frac{\rho_l}{\rho_v} - 1\right)x\right] \tag{10.46}$$

Substituting the void-fraction relation (10.33), the acceleration term on the right side of Eq. (10.30) simplifies to

$$\left(\frac{1}{A}\right)\frac{d}{dz}\left[\frac{G^2x^2A}{\rho_v\alpha} + \frac{G^2(1-x)^2A}{\rho_l(1-\alpha)}\right] = \left(\frac{1}{A}\right)\frac{d}{dz}(G^2A\overline{v}) \tag{10.47}$$

where, from Eq. (10.35),

$$\bar{v} = v_l + (v_v - v_l)x \tag{10.48}$$

Applying the derivative to the product of terms in the parentheses on the right side of Eq. (10.47), and assuming that the variation of the liquid density due to changes in temperature and pressure along the channel is negligible, the following relation is obtained:

$$\left(\frac{1}{A}\right)\frac{d}{dz}\left[\frac{G^2x^2A}{\rho_v\alpha} + \frac{G^2(1-x)^2A}{\rho_l(1-\alpha)}\right] = \left(\frac{\overline{v}}{A}\right)\frac{d}{dz}(G^2A) + G^2\left[v_{lv}\left(\frac{dx}{dz}\right) + x\left(\frac{dv_v}{dz}\right)\right] \tag{10.49}$$

where v_{lv} has been used to designate $v_v - v_l$. Using the chain rule to evaluate dv_v/dz as

$$\frac{dv_v}{dz} = \frac{dv_v}{dP}\left(\frac{dP}{dz}\right) \tag{10.50}$$

Eq. (10.49) can be written as

$$\left(\frac{1}{A}\right)\frac{d}{dz}\left[\frac{G^2x^2A}{\rho_v\alpha} + \frac{G^2(1-x)^2A}{\rho_l(1-\alpha)}\right] = \left(\frac{\overline{v}}{A}\right)\frac{d}{dz}(G^2A) + G^2\left[v_{lv}\left(\frac{dx}{dz}\right) + x\frac{dv_v}{dP}\left(\frac{dP}{dz}\right)\right] \tag{10.51}$$

This decomposes the acceleration term into three parts. The first term accounts for acceleration effects associated with changes in the duct cross section. The second and third terms account for acceleration or deceleration of the flow due to changes in the flow density, either due to changes in quality or due to expansion of the vapor as the pressure varies along the duct.

Substituting Eq. (10.51) into Eq. (10.30), using Eq. (10.33) to evaluate α in the gravitational head terms, and solving for $-\,dP/dz$ yields

$$-\left(\frac{dP}{dz}\right) = \left[-\frac{P}{A}\left(\frac{dA}{dz}\right) + \phi_{lo}^2\left(\frac{2f_{lo}G^2v_l}{d_h}\right) + \left(\frac{v_v + xv_{lv}}{A}\right)\frac{d}{dz}(G^2A) + G^2v_{lv}\left(\frac{dx}{dz}\right) + \frac{g\sin\Omega}{v_l + v_{lv}x}\right] \Big/ [1 + G^2x(dv_v/dP)] \tag{10.52}$$

If the cross-sectional area A does not vary along the duct, this equation reduces to

$$-\left(\frac{dP}{dz}\right) = \left[\phi_{lo}^2\left(\frac{2f_{lo}G^2v_l}{d_h}\right) + G^2v_{lv}\left(\frac{dx}{dz}\right) + \frac{g\sin\Omega}{v_l + v_{lv}x}\right] \Big/ [1 + G^2x(dv_v/dP)] \tag{10.53}$$

Given appropriate information about the flow conditions along the channel, either of these equations can be integrated to determine the pressure variation along the duct, as predicted by the homogeneous model.

Separated Flow

Returning to the more general separated flow case, the acceleration term in Eq. (10.29) can be expanded by applying the chain rule and invoking conservation of mass as $d(GA)/dz = 0$.

$$\begin{aligned}\left(\frac{1}{A}\right)\frac{d}{dz}\left[\frac{G^2x^2A}{\rho_v\alpha} + \frac{G^2(1-x)^2A}{\rho_l(1-\alpha)}\right] = {} & -G^2\left(\frac{1}{A}\right)\frac{dA}{dz}\left[\frac{x^2v_v}{\alpha} + \frac{(1-x)^2v_l}{(1-\alpha)}\right] \\ & + G^2\frac{dx}{dz}\left\{\left[\frac{2xv_v}{\alpha} - \frac{2(1-x)v_l}{(1-\alpha)}\right]\right. \\ & \left. + \frac{d\alpha}{dx}\left[\frac{(1-x)^2v_l}{(1-\alpha)^2} - \frac{x^2v_v}{\alpha^2}\right]\right\} + G^2\frac{dP}{dz}\left\{\frac{x^2}{\alpha}\left(\frac{dv_v}{dP}\right)\right. \\ & \left. + \frac{d\alpha}{dP}\left[\frac{(1-x)^2v_l}{(1-\alpha)^2} - \frac{x^2v_v}{\alpha^2}\right]\right\}\end{aligned} \tag{10.54}$$

As for the homogeneous model above, the frictional pressure drop for the liquid phase alone in Eq. (10.29) can be evaluated in terms of friction factor f_l:

$$-\left(\frac{dP}{dz}\right)_l = \frac{2f_l G^2(1-x)^2}{\rho_l d_h} \tag{10.55}$$

Substituting Eqs. (10.54) and (10.55) into Eq. (10.29) and solving for $-dP/dz$ yields

$$\begin{aligned}-\left(\frac{dP}{dz}\right) = \left(\frac{1}{\Lambda}\right)\Bigg(&\phi_l^2\left[\frac{2f_l G^2(1-x)^2}{\rho_l d_h}\right] + [(1-\alpha)\rho_l + \alpha\rho_v]g \sin\Omega \\ &- \left(\frac{1}{A}\right)\frac{dA}{dz}\left\{P + G^2\left[\frac{x^2 v_v}{\alpha} + \frac{(1-x)^2 v_l}{(1-\alpha)}\right]\right\} \\ &+ G^2\frac{dx}{dz}\left\{\left[\frac{2xv_v}{\alpha} - \frac{2(1-x)v_l}{(1-\alpha)}\right]\right. \\ &\left.+ \frac{d\alpha}{dx}\left[\frac{(1-x)^2 v_l}{(1-\alpha)^2} - \frac{x^2 v_v}{\alpha^2}\right]\right\}\Bigg)\end{aligned} \tag{10.56a}$$

where

$$\Lambda = 1 + G^2\left\{\frac{x^2}{\alpha}\left(\frac{dv_v}{dP}\right) + \frac{d\alpha}{dP}\left[\frac{(1-x)^2 v_l}{(1-\alpha)^2} - \frac{x^2 v_v}{\alpha^2}\right]\right\} \tag{10.56b}$$

For a duct with a constant cross-sectional area, Eq. (10.56*a*) reduces to

$$\begin{aligned}-\left(\frac{dP}{dz}\right) = \left(\frac{1}{\Lambda}\right)\Bigg(&\phi_l^2\left[\frac{2f_l G^2(1-x)^2}{\rho_l d_h}\right] + [(1-\alpha)\rho_l + \alpha\rho_v]g \sin\Omega \\ &+ G^2\frac{dx}{dz}\left\{\left[\frac{2xv_v}{\alpha} - \frac{2(1-x)v_l}{(1-\alpha)}\right]\right. \\ &\left.+ \frac{d\alpha}{dx}\left[\frac{(1-x)^2 v_l}{(1-\alpha)^2} - \frac{x^2 v_v}{\alpha^2}\right]\right\}\Bigg)\end{aligned} \tag{10.57a}$$

Note that Eq. (10.57*a*) can be written equivalently in terms of f_{lo} and ϕ_{lo}^2 as

$$\begin{aligned}-\left(\frac{dP}{dz}\right) = \left(\frac{1}{\Lambda}\right)\Bigg(&\phi_{lo}^2\left(\frac{2f_{lo} G^2}{\rho_l d_h}\right) + [(1-\alpha)\rho_l + \alpha\rho_v]g \sin\Omega \\ &+ G^2\frac{dx}{dz}\left\{\left[\frac{2xv_v}{\alpha} - \frac{2(1-x)v_l}{(1-\alpha)}\right]\right. \\ &\left.+ \frac{d\alpha}{dx}\left[\frac{(1-x)^2 v_l}{(1-\alpha)^2} - \frac{x^2 v_v}{\alpha^2}\right]\right\}\Bigg)\end{aligned} \tag{10.57b}$$

Integration of Eq. (10.56*a*), (10.57*a*), or (10.57*b*) to determine the pressure variation along the duct can be accomplished only if the two-phase multiplier and the void fraction can be determined from the local conditions all along the duct. Methods of determining these parameters are discussed in detail in the next section.

10.3 DETERMINATION OF THE TWO-PHASE MULTIPLIER AND VOID FRACTION

The separated flow analysis described in Section 10.2 can be used to predict the pressure gradient for a gas–liquid flow only if some means of determining the two-phase multiplier and void fraction are available. The methodologies developed to predict these quantities have typically been based on a mixture of semi-theoretical arguments and empirical evidence. The pioneering work of Martinelli and his co-workers in the 1940s provided the first widely successful methods for predicting ϕ_l and α, and established an analytical foundation on which much of the subsequent work on two-phase flow has been built.

Lockhart and Martinelli [10.24] proposed a generalized correlation method for determining the two-phase multiplier ϕ_l or ϕ_v, from which the frictional pressure gradient can be predicted for adiabatic gas–liquid flow in a round tube. This correlation was based on data from a series of studies of adiabatic two-phase flow in horizontal tubes. The relations used to predict ϕ_l and ϕ_v are

$$\phi_l = \left(1 + \frac{C}{X} + \frac{1}{X^2}\right)^{1/2} \tag{10.58a}$$

$$\phi_v = (1 + CX + X^2)^{1/2} \tag{10.58b}$$

where X is the Martinelli parameter previously defined by Eq. (10.11*b*).

The recommended value of the constant C differs, depending on the flow regimes associated with the flow of the vapor and the liquid alone in the duct. The constants associated with each of the four possible combinations are indicated in Table 10.1. For round tubes, the liquid flow is generally regarded as being turbulent for $Re_l > 2000$, and laminar for $Re_l \leq 2000$. For the gas flow, the transition from laminar to turbulent may similarly be taken to occur at $Re_v = 2000$.

At first glance, it may not be apparent that ϕ_l and ϕ_v should depend on the Martinelli parameter X. However, by considering the variation of these parameters with quality, the link can be made more obvious. First, as the quality increases, for a given total flow rate, $(dP/dz)_l$ decreases and $(dP/dz)_v$ increases. As a result, X monotonically decreases as the quality increases.

It also follows directly from the definitions of ϕ_l and ϕ_v that as $x \to 0$ ($X \to \infty$), $\phi_l \to 1$, $\phi_v \to X$, and as $x \to 1$ ($X \to 0$), $\phi_l \to 1/X$, $\phi_v \to 1$. The asymptotic behavior noted above requires that ϕ_l vary from 1 at low quality to $1/X$ at high

Table 10.1

Liquid	Gas	Subscript designation	C
Turbulent	Turbulent	*tt*	20
Viscous	Turbulent	*vt*	12
Turbulent	Viscous	*tv*	10
Viscous	Viscous	*vv*	5

quality. It is not unreasonable, as a first guess, to presume that ϕ_l varies solely as a function of X between these limits. Similar arguments can be applied to the variation of ϕ_v.

The dependence of ϕ_l and ϕ_v on X can be made more concrete by considering the separate cylinders model of gas–liquid flow described by Wallis [10.5]. In this model, an actual two-phase flow like that shown in Fig. 10.11*a* is taken to be equivalent to the model flow cirumstances shown in Fig. 10.11*b*, in which the vapor and liquid phases flow at the same flow rates through separate cylinders. The radii of the vapor (r_{ve}) and liquid (r_{le}) cylinders are required to satisfy the relations

$$\frac{r_{ve}^2}{r_0^2} = \alpha \tag{10.59a}$$

$$\frac{r_{le}^2}{r_0^2} = 1 - \alpha \tag{10.59b}$$

This ensures that the combined total flow area of the two cylinders is the same as that for the actual pipe, and that the cross-sectional area associated with each phase is the same. The pressure gradients in each of the model cylinders are assumed to be equal to each other, and their value is taken to be equal to the two-phase frictional pressure gradient in the actual flow.

With the idealizations described above, the pressure gradient in the separate cylinder carrying the vapor flow is given by

$$-\left(\frac{dP}{dz}\right) = \frac{2f_{ve}}{2r_{ve}}\left[\frac{(\pi r_0^2 G)^2 x^2/(\pi r_{ve}^2)^2}{\rho_v}\right] = \frac{f_{ve}G^2x^2}{\rho_v r_{ve}\alpha^2} \tag{10.60}$$

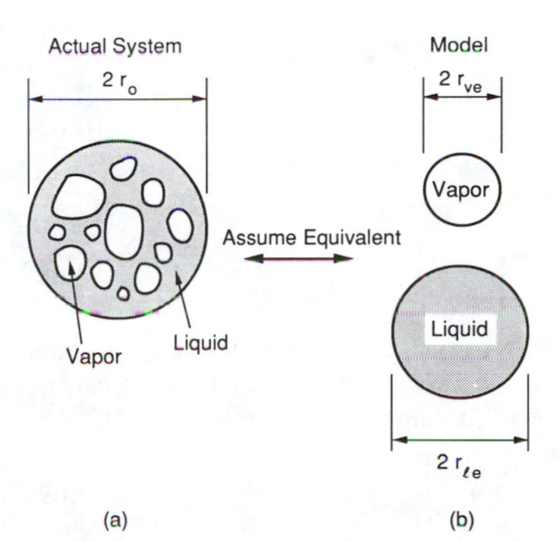

Figure 10.11 System model for separate cylinders analysis of two-phase flow.

where f is presumed to be a function of the Reynolds number Re_{ve}, defined as

$$\mathrm{Re}_{ve} = \frac{(\pi r_0^2 G)x(2r_{ve})/(\pi r_{ve}^2)^2}{\mu_v} = \frac{2r_{ve}Gx}{\mu_v \alpha} \tag{10.61}$$

For the vapor flowing alone through the actual pipe, the pressure gradient is given by

$$-\left(\frac{dP}{dz}\right)_v = \frac{f_v G^2 x^2}{\rho_v r_0} \tag{10.62}$$

where the friction factor f_v is a function of Re_v, defined as

$$\mathrm{Re}_v = \frac{2r_0 Gx}{\mu_v} \tag{10.63}$$

Assuming that f_v and f_{ve} both obey the same power-law dependence on Reynolds number $f = B\,\mathrm{Re}^{-n}$, and making use of the assumption that dP/dz in the separate cylinders equals $(dP/dz)_{fr}$ in the actual cylinder, Eqs. (10.28) and (10.60) through (10.63) can be combined to obtain the following expressions for ϕ_v^2:

$$\phi_v^2 = \left(\frac{r_0}{r_{ve}}\right)^{5-n} = \frac{1}{\alpha^{(5-n)/2}} \tag{10.64}$$

Based on similar arguments, it can be shown that for the liquid flowing in the separate model cylinder,

$$-\left(\frac{dP}{dz}\right) = \frac{2f_{le}}{2r_{le}}\left[\frac{(\pi r_0^2 G)^2(1-x)^2/(\pi r_{le}^2)^2}{\rho_l}\right] = \frac{f_{le}G^2(1-x)^2}{\rho_l r_{le}(1-\alpha)^2} \tag{10.65}$$

where f_{le} is a function of

$$\mathrm{Re}_{le} = \frac{(\pi r_0^2 G)(1-x)(2r_{le})/(\pi r_{le}^2)^2}{\mu_l} = \frac{2r_{le}G(1-x)}{\mu_l(1-\alpha)} \tag{10.66}$$

For the liquid flowing alone in the actual pipe,

$$-\left(\frac{dP}{dz}\right)_l = \frac{f_l G^2(1-x)^2}{\rho_l r_0} \tag{10.67}$$

where f_l depends on Re_l, given by

$$\mathrm{Re}_l = \frac{2r_0 G(1-x)}{\mu_l} \tag{10.68}$$

Assuming that both liquid model flows obey the same power-law variation of f with Reynolds number as for the vapor flows, the above equations can be combined with Eq. (10.26) to obtain the following relation for ϕ_l^2:

$$\phi_l^2 = \left(\frac{r_0}{r_{le}}\right)^{5-n} = \frac{1}{(1-\alpha)^{(5-n)/2}} \tag{10.69}$$

Equations (10.64) and (10.69) can be combined to eliminate α, which yields the relation

$$\frac{1}{\phi_l^{4/(5-n)}} + \frac{1}{\phi_v^{4/(5-n)}} = 1 \tag{10.70}$$

Multiplying through by $\phi_l^{4/(5-n)}$ and using the fact that

$$X^2 = \frac{(dP/dz)_l}{(dP/dz)_v} = \frac{\phi_v^2}{\phi_l^2} \tag{10.71}$$

the following relation is obtained for ϕ_l^2:

$$\phi_l^2 = \left[1 + \left(\frac{1}{X}\right)^{4/(5-n)}\right]^{(5-n)/2} \tag{10.72}$$

It can similary be shown that ϕ_v^2 is given by

$$\phi_v^2 = [1 + X^{4/(5-n)}]^{(5-n)/2} \tag{10.73}$$

If both the liquid and vapor flows are laminar, $n = 1$ for a round tube, and the relation for ϕ_l^2 becomes

$$\phi_l^2 = \left[1 + \left(\frac{1}{X_{vv}}\right)\right]^2 \tag{10.74}$$

If the flows associated with the two phases are both turbulent, $n = 0.25$ for a round tube, which reduces Eq. (10.72) to

$$\phi_l^2 = \left[1 + \left(\frac{1}{X_{tt}}\right)^{16/19}\right]^{19/8} \tag{10.75}$$

The above relations for ϕ_l differ only slightly from the corresponding relations in the Lockhart-Martinelli correlation. Perhaps of greater significance is the fact that this analysis directly demonstrates that different $\phi(X)$ relations are expected to apply, depending on the flow regimes associated with each phase.

Lockhart and Martinelli [10.24] also correlated the vapor void fraction as a function of the Martinelli parameter X. As noted by Butterworth [10.25], their correlation is well represented by the relation

$$\alpha = [1 + 0.28X^{0.71}]^{-1} \tag{10.76}$$

The Lockhart-Martinelli correlations for the two-phase multipliers and the void fraction are plotted in Fig. 10.12.

Martinelli and Nelson [10.26] later developed a correlation technique for predicting the pressure drop during flow boiling inside tubes. Their correlation is based on the Lockhart-Martinelli methodology described above, with the following additional idealizations:

1. Thermodynamic equilibrium exists at each location along the tube in which flow boiling is taking place.

2. For forced-flow boiling, the two-phase flow corresponds to the turbulent-turbulent case.
3. The frictional contribution to the overall pressure gradient for flow boiling is equal to that predicted by a correlation for horizontal adiabatic gas–vapor flow under comparable conditions.

Based on these idealizations, it should be possible to predict the frictional contribution to the flow-boiling pressure gradient using the Lockhart-Martinelli correlation. For such circumstances, the two-phase multiplier ϕ_{lo} should approach 1 as the system saturation pressure approaches the critical point. However, Martinelli and Nelson [10.26] found that the values of ϕ_{lo} predicted by the Lockhart-Martinelli correlation near the critical point were about 5. From these results, they surmised that the two-phase multiplier must vary with pressure as well as with the Martinelli parameter X.

Assuming that the predictions of the Lockhart-Martinelli correlation were correct at atmospheric pressure and adopting a variation of the two-phase multiplier that was consistent with the critical-point limit and available data at intermediate pressures, Martinelli and Nelson [10.26] developed the correlations for $(\phi_l)_{tt}$ and $1 - \alpha$ as functions of pressure and a parameter χ_{tt}, defined as

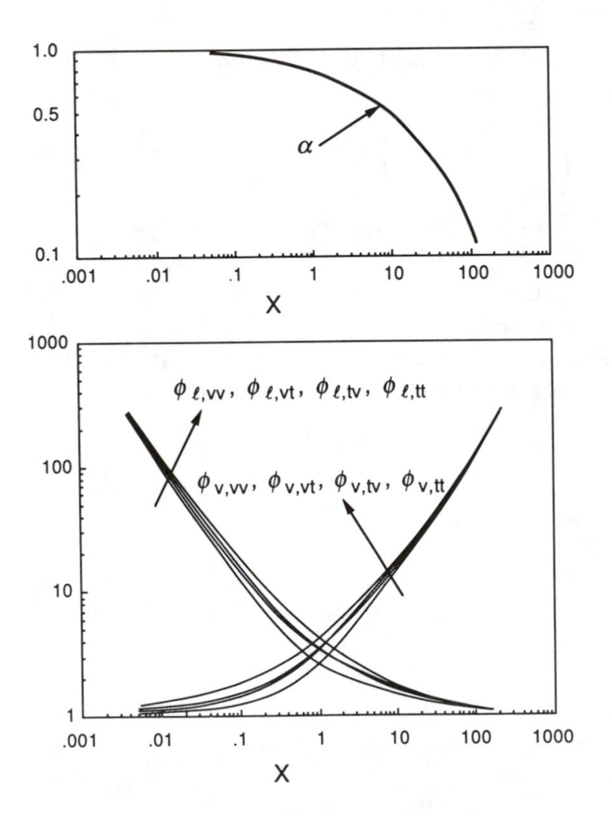

Figure 10.12 Graphical representation of the Lockhart-Martinelli correlation.

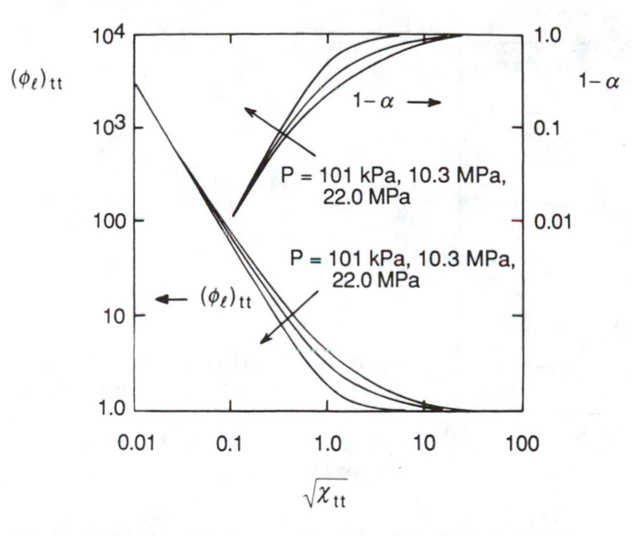

Figure 10.13 Graphical representation of the Martinelli-Nelson correlation. (*Adapted from* [*10.26*] *with permission, copyright © 1948, American Society of Mechanical Engineers.*)

$$\chi_{tt} = \left(\frac{\rho_v}{\rho_l}\right)^{0.571}\left(\frac{\mu_l}{\mu_v}\right)^{0.143}\left(\frac{1-x}{x}\right) \tag{10.77}$$

These correlations for flow boiling in a round tube are shown in Fig. 10.13. The definition of χ_{tt} differs only slightly from that of the Martinelli parameter for turbulent-turbulent flow X_{tt} as given by Eqs. (10.11) through (10.13) with $n = 0.25$:

$$X_{tt} = \left(\frac{\rho_v}{\rho_l}\right)^{0.5}\left(\frac{\mu_l}{\mu_v}\right)^{0.125}\left(\frac{1-x}{x}\right)^{0.875} \tag{10.78}$$

It can be seen in Fig. 10.13 that increasing pressure at a given χ_{tt} value tends to decrease α and $(\phi_l)_{tt}$.

From Eqs. (10.26), (10.27), (10.36), and (10.55), it can easily be shown that

$$\frac{\phi_{lo}^2}{\phi_l^2} = \frac{f_l}{f_{lo}}(1-x)^2 \tag{10.79}$$

Using Eqs. (10.45) to evaluate f_l and f_{lo} for round tubes, the relation between ϕ_{lo}^2 and ϕ_l^2 becomes

$$\phi_{lo}^2 = \phi_l^2(1-x)^{1.75} \tag{10.80}$$

Martinelli and Nelson [10.26] used this relation together with the correlation for $(\phi_l)_{tt}$ represented in Fig. 10.13 to obtain a prediction of ϕ_{lo}^2 for flow boiling of water in round pipes as a function of quality and pressure. Their predicted variation of ϕ_{lo}^2 with quality and pressure is shown graphically in Fig. 10.14.

In general, integration of Eq. (10.52) or (10.56*a*) to determine the pressure

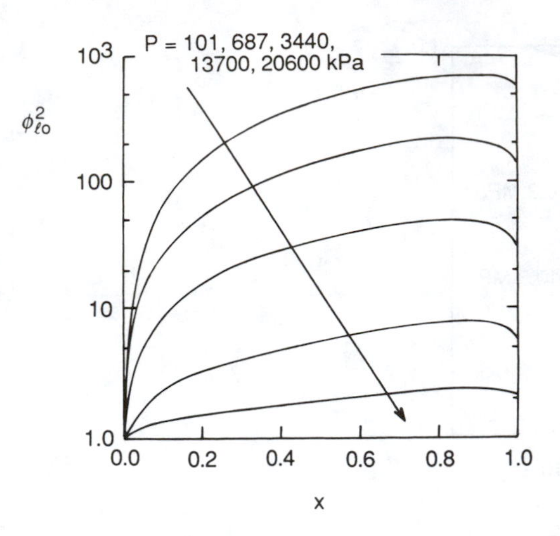

Figure 10.14 Variation of ϕ_{lo}^2 with quality and pressure as predicted by the Martinelli-Nelson correlation. (*Adapted from [10.26] with permission, copyright © 1948, American Society of Mechanical Engineers.*)

variation along the pipe is so complex that closed-form integration is impossible. In the most general circumstances, numerical integration is necessary to predict the pressure variation. However, under some special circumstances, calculation of the total pressure drop can be greatly simplified. In particular, for separated flow, simplified calculation of the total pressure drop is possible if the following conditions are satisfied:

1. The cross-sectional area of the duct is constant.
2. $\Lambda \approx 1$, which implies that the compressibility of the gas phase is negligible.
3. The friction factor f_{lo} and the liquid and vapor densities ρ_l and ρ_v are constant along the entire flow passage.
4. Liquid is evaporated from saturated liquid at the inlet ($x = 0$ at $z = 0$) to some arbitrary quality x_e at $z = z_e$, and the quality varies linearly with z along the tube (dx/dz = constant).

Invoking these conditions and integrating Eq. (10.57*b*) from $z = 0$ to z_e, the following relation for the overall pressure drop ΔP is obtained:

$$-\Delta P = \frac{2f_{lo}G^2z_e}{d_h\rho_l}\left(\frac{1}{x_e}\right)\int_0^{x_e}\phi_{lo}^2\,dx + \frac{G^2}{\rho_l}\left[\frac{x^2\rho_l}{\alpha\rho_v} + \frac{(1-x)^2}{(1-\alpha)} - 1\right]_{x=x_e}$$
$$+ \frac{z_e g\sin\Omega}{x_e}\int_0^{x_e}[\rho_v\alpha + \rho_l(1-\alpha)]\,dx \tag{10.81}$$

The correlations for the two-phase multiplier and void fraction described above can be used to evaluate the terms

$$I_\phi = \left(\frac{1}{x_e}\right)\int_0^{x_e}\phi_{lo}^2\,dx \tag{10.82a}$$

$$a_\alpha = \left[\frac{x^2\rho_l}{\alpha\rho_v} + \frac{(1-x)^2}{(1-\alpha)} - 1\right]_{x=x_e} \tag{10.82b}$$

$$I_\rho = \frac{1}{x_e}\int_0^{x_e} [\rho_v\alpha + \rho_l(1-\alpha)]\,dx \tag{10.82c}$$

as functions of exit quality and pressure alone. The variations of I_ϕ and a_α as functions of x_e and pressure were determined by Martinelli and Nelson [10.26] using their correlations for ϕ_{lo} and α for water and steam. For flow boiling of water under the conditions described above, these two computed variations are sufficient to compute the overall pressure drop if the gravitational head term is small. Thom [10.27] later published alternative variations of α, ϕ_{lo}, I_ϕ, a_α, and I_ρ developed from a more extensive body of two-phase pressure drop data for steam–water flow in horizontal and vertical tubes. The resulting variations of I_ϕ, a_α, and I_ρ are plotted in Figs. 10.15 through 10.17.

The simplistic Lockhart-Martinelli-Nelson correlations for void fraction and two-phase multiplier clearly do not account for effects other than the fluid properties and quality. Comparison of the predictions of these correlations with data suggests that there is a systematic variation of the two-phase multiplier with mass flux that these correlations do not take into account.

Several attempts have been made to develop improved correlations for the two-phase multiplier and void fraction. However, it has been generally found that

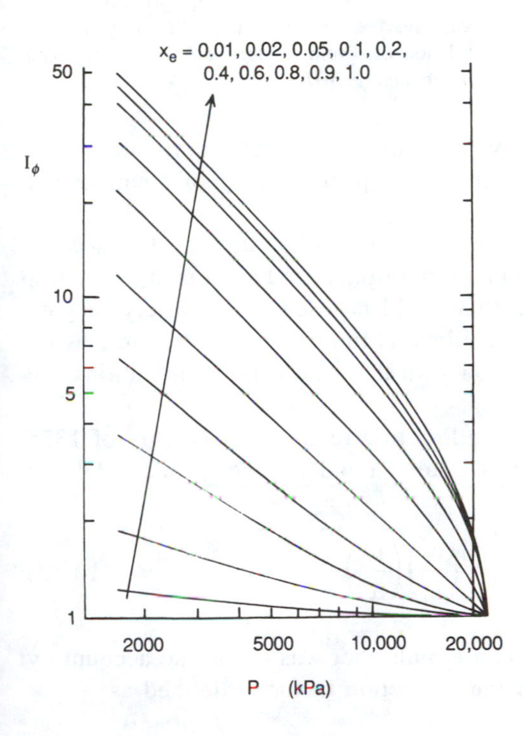

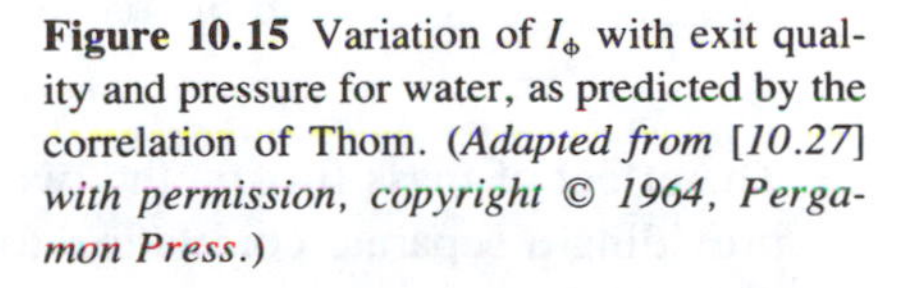
Figure 10.15 Variation of I_ϕ with exit quality and pressure for water, as predicted by the correlation of Thom. (*Adapted from [10.27] with permission, copyright © 1964, Pergamon Press.*)

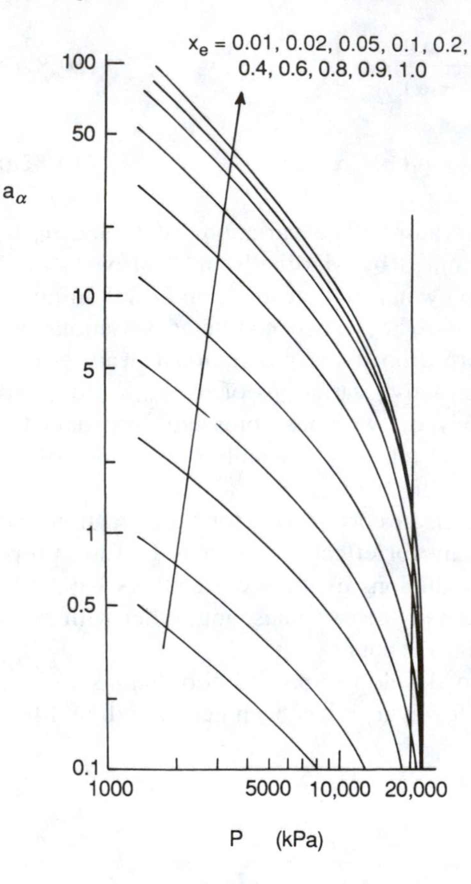

Figure 10.16 Variation of a_α with exit quality and pressure for water, as predicted by the correlation of Thom. (*Adapted from [10.27] with permission, copyright © 1964, Pergamon Press.*)

improved accuracy can be obtained only with more complicated correlation techniques or by developing correlations only for a specific gas–liquid pair over a limited range of conditions.

A subsequent correlation proposed by Baroczy [10.28] attempted to account for the effects of mass flux on the two-phase multiplier, and attempted to develop a broadly based correlation technique that could be used for a variety of gas–liquid combinations. This correlation technique is fundamentally the same as the Martinelli-Nelson (separated flow) method with the following modifications:

1. The two-phase multiplier ϕ_{lo}^2 corresponding to a reference mass flux of 1355 kg/m^2 s was correlated as a function of quality x and a property index, defined as

$$\text{property index} = \left(\frac{\rho_v}{\rho_l}\right)\left(\frac{\mu_l}{\mu_v}\right)^{0.2} \tag{10.83}$$

2. The effect of mass flux on the two-phase multiplier was taken into account by providing a separate correlation for the correction factor γ, defined as

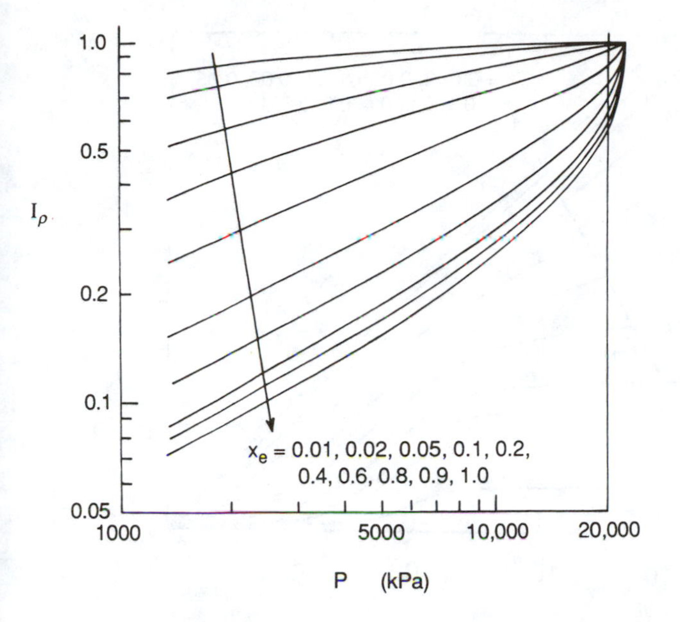

Figure 10.17 Variation of I_ρ with exit quality and pressure for water, as predicted by the correlation of Thom. (*Adapted from [10.27] with permission, copyright © 1964, Pergamon Press.*)

$$\gamma = \frac{\phi_{lo}^2}{(\phi_{lo}^2)_{G=1355\ \mathrm{kg/m^2\ s}}} \tag{10.84}$$

This correlation technique is represented graphically in Figs. 10.18 and 10.19. At a given flow condition, the property index can be evaluated, and it and the quality can be used with Fig. 10.18 to determine ϕ_{lo}^2 at the reference mass flux. Figure 10.19 is used to determine the correction factor that, when multiplied by the value of ϕ_{lo}^2 for the reference mass flux, yields the predicted value of ϕ_{lo}^2 for the conditions of interest. This value can then be used to evaluate the frictional contribution to the overall pressure gradient in Eq. (10.57*b*). Baroczy [10.29] also proposed a void-fraction correlation in which the void fraction was correlated in terms of the turbulent-turbulent Martinelli parameter X_{tt} and the property index used in his two-phase multiplier correlation. This correlation is shown graphically in Fig. 10.20.

The complex behavior of the multiplier correction (γ) curves has been perhaps the most controversial aspect of this correlation method. No interpretation of the complex tangle of curves in Fig. 10.19 has been offered. However, since the conditions represented in this figure span several quite different flow regimes, it is possible that some of the complex changes in the correction factor variations may be associated with flow regime transitions.

More recently, Friedel [10.30] used a data base of 25,000 points to develop

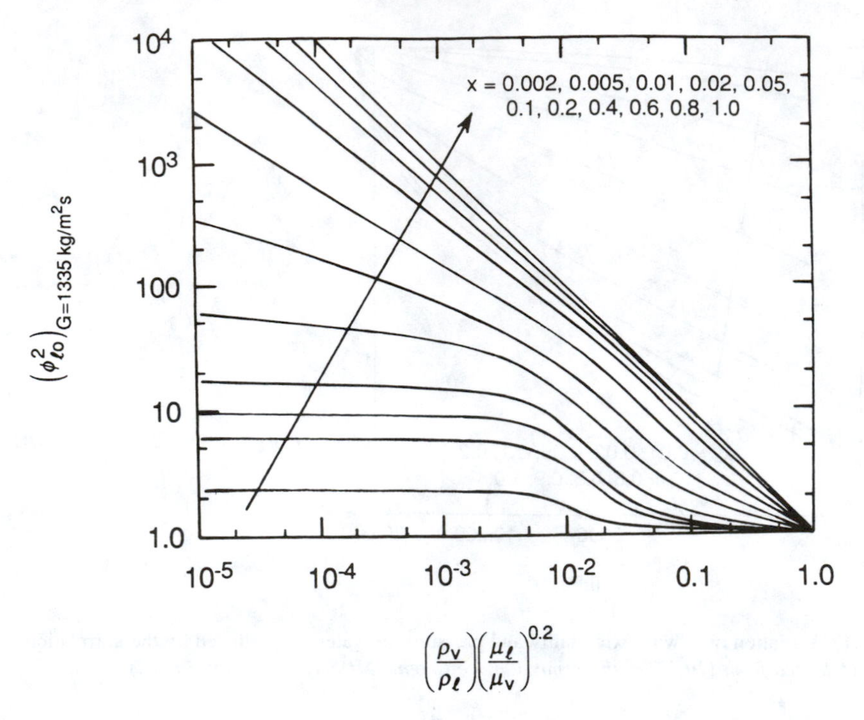

Figure 10.18 Correlation for ϕ^2_{lo} at $G = 1335$ kg/m^2 s proposed by Baroczy. (*Adapted from [10.28] with permission, copyright © 1965, American Institute of Chemical Engineers.*)

the following correlation for predicting the two-phase multiplier ϕ^2_{lo} for vertical upward and horizontal flow in round tubes.

$$\phi^2_{lo} = C_{F1} + \frac{3.24\ C_{F2}}{\mathrm{Fr}^{0.045}\mathrm{We}^{0.035}} \tag{10.85}$$

where

$$C_{F1} = (1 - x)^2 + X^2 \left(\frac{\rho_l}{\rho_v}\right)\left(\frac{f_{vo}}{f_{lo}}\right) \tag{10.86}$$

$$C_{F2} = x^{0.78}(1 - x)^{0.24}\left(\frac{\rho_l}{\rho_v}\right)^{0.91}\left(\frac{\mu_v}{\mu_l}\right)^{0.19}\left(1 - \frac{\mu_v}{\mu_l}\right)^{0.7} \tag{10.87}$$

$$\mathrm{Fr} = \frac{G^2}{g d_n \rho^2_{tp}} \tag{10.88}$$

$$\mathrm{We} = \frac{G^2 d_n}{\rho_{tp}\sigma} \tag{10.89}$$

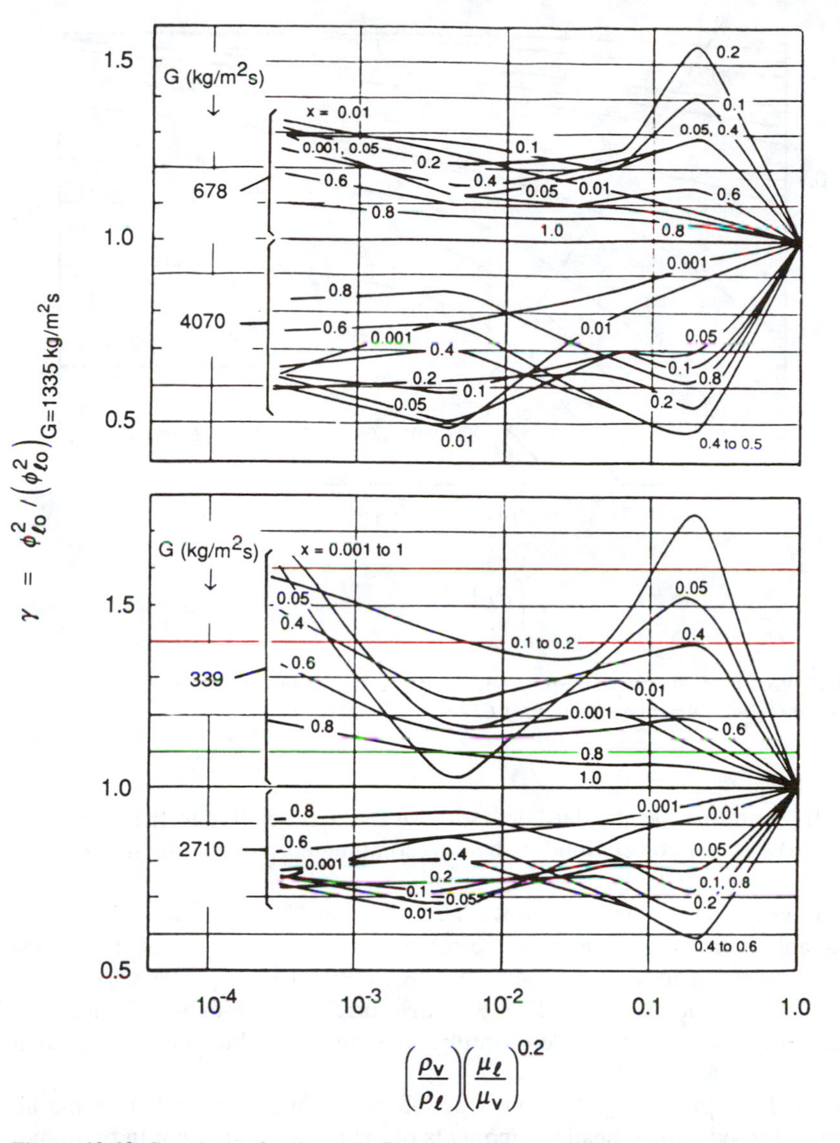

Figure 10.19 Correlation for the mass flux and property effects correction factor proposed by Baroczy. (*Adapted from* [*10.28*] *with permission, copyright © 1965, American Institute of Chemical Engineers*.)

$$\rho_{tp} = \left(\frac{x}{\rho_v} + \frac{1-x}{\rho_l} \right)^{-1} \tag{10.90}$$

In the above correlation, f_{vo} and f_{lo} are friction factors for the total mass flowing as vapor and liquid, respectively. For single component two-phase flows, Frie-

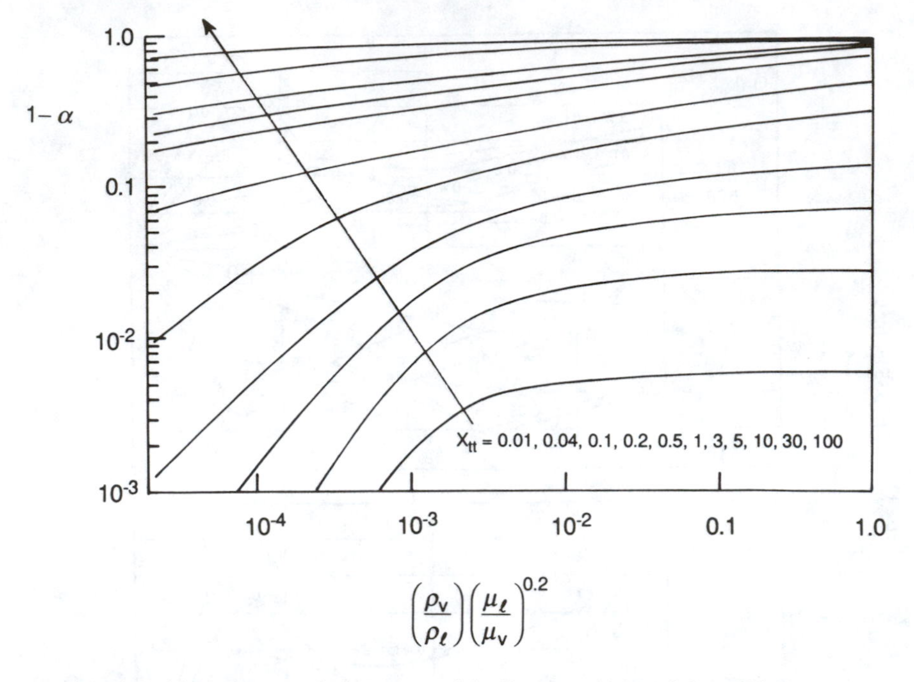

Figure 10.20 Void fraction correlation proposed by Baroczy. (*Adapted from* [*10.29*] *with permission, copyright © 1965, American Institute of Chemical Engineers.*)

del [10.30] found that the standard deviation of the data relative to the correlation was about 30%. This correlation has been recommented for use when $(\mu_l/\mu_v) < 1000$ (see Hetsroni [10.31]).

For large-diameter tubes, Chenoweth and Martin [10.32] found that a good fit to available data was provided by correlating ϕ_{lo}^2 as a function of the inverse of the property index used by Baroczy [10.28] (defined by Eq. [10.83]) and the liquid volume fraction, equal to $1 - \beta$. This correlation, shown graphically in Fig. 10.21, is usually recommended for tubes having inside diameters greater than 5 cm.

A method for predicting the void fraction is essential for predicting the acceleration and gravitational head components of the pressure gradient in two-phase flows. Butterworth [10.25] has shown that several of the available void-fraction correlations can be cast in the general form

$$\alpha = \left[1 + B_B\left(\frac{1-x}{x}\right)^{n_1}\left(\frac{\rho_v}{\rho_l}\right)^{n_2}\left(\frac{\mu_l}{\mu_v}\right)^{n_3}\right]^{-1} \tag{10.91}$$

where the values of the unspecified constants in this relation corresponding to different correlations are listed in Table 10.2. This formulation facilitates comparison of the different models and makes using them somewhat easier. It also illustrates some inconsistencies. When $\rho_l = \rho_v$ and $\mu_l = \mu_v$, the void fraction

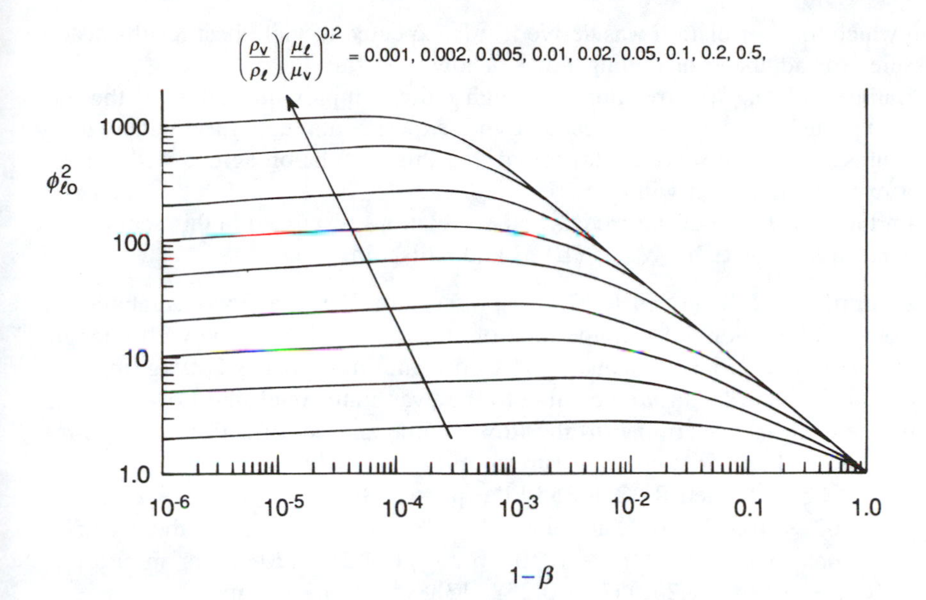

Figure 10.21 Correlation for ϕ^2_{lo} proposed by Chenoweth and Martin [10.32] for large-diameter tubes.

should equal the quality, implying that B_B and n_1 should equal 1 in all the correlations. In those correlations where these constants are not 1, some inaccuracy may be expected when the densities and the viscosities of the two phases are nearly the same (i.e., near the critical point).

For engineering design calculations, selection of a two-phase flow model should be done carefully. The homogeneous model yields reasonably accurate results only for limited cirumstances. The best agreement is expected for bubbly or dispersed droplet (mist) flows, where the slip velocity between the phases is small.

Martinelli and Nelson's [10.26] correlation has been shown to yield reasonably accurate results for a wide variety of two-phase flow circumstances in round tubes and simple channel geometries. This type of correlation has also been adapted to fit data for complex finned channel geometries [10.34]. Given the data base

Table 10.2

Correlation or model	B_B	n_1	n_2	n_3
Homogeneous model	1	1	1	0
Zivi [10.33] model	1	1	0.67	0
Wallis [10.5] separate-cylinder model	1	0.72	0.40	0.08
Lockhart and Martinelli [10.24]	0.28	0.64	0.36	0.07
Thom correlation [10.27]	1	1	0.89	0.18
Baroczy correlation [10.29]	1	0.74	0.65	0.13

from which this correlation was derived, it is expected to yield best results at low pressures for adiabatic or boiling flows at low heat flux levels.

Baroczy's [10.28] correlation, although more complex than some of the others, is expected to give good agreement with experimental data for a wide variety of circumstances. Baroczy [10.28] found that this correlation agreed with a large set of experimental data within ±20%.

Further discussion of the models and correlations discussed in this section can be found in the books by Wallis [10.5] and Chisholm [10.35].

Example 10.2 Saturated R-12 at a pressure of 333 kPa flows adiabatically from an expansion valve to the inlet of an evaporator in a tube with a length of 0.7 m and an inside diameter of 1 cm. The mass flux is 250 kg/m^2, and the quality is 0.25. From the valve to the evaporator inlet, the flow rises 0.5 m in elevation. Determine (*a*) the flow regime and void fraction, and (*b*) the gravitational and frictional pressure drop.

(*a*) For saturated R-12 at 333 kPa, $\rho_l = 1388$ kg/m^3, $\rho_v = 19.2$ kg/m^3, $\mu_l = 262 \times 10^{-6}$ N s/m^2, and $\mu_v = 11.7 \times 10^{-6}$ N s/m^2. For the specified conditions, $\rho_v j_v^2 = G^2 x^2/\rho_v = (250)^2(0.25)^2/(19.2) = 203$ kg/s^2 m and $\rho_l j_l^2 = G^2(1 - x)^2/\rho_l = (250)^2(1 - 0.25)^2/1388 = 25.3$ kg/s^2 m.

From Fig. 10.3, and the above values of $\rho_v j_v^2$ and $\rho_l j_l^2$, the flow is predicted to be in the annular regime. Using Eq. (10.91) with the constants from Table 10.2 for the Lockhart and Martinelli model,

$$\alpha = \left[1 + 0.28\left(\frac{1-x}{x}\right)^{0.64}\left(\frac{\rho_v}{\rho_l}\right)^{0.36}\left(\frac{\mu_l}{\mu_v}\right)^{0.07}\right]^{-1}$$

Substituting and solving for α yields

$$\alpha = \left[1 + 0.28\left(\frac{0.75}{0.25}\right)^{0.64}\left(\frac{19.2}{1388}\right)^{0.36}\left(\frac{262}{11.7}\right)^{0.07}\right]^{-1} = 0.869$$

(*b*) Using Eq. (10.57*b*), neglecting compressibility effects ($\Lambda = 1$), and setting $dx/dz = 0$, the gravitational contribution $-(dP/dz)_G$ is given by

$$-(dP/dz)_G = [(1-\alpha)\rho_l + \alpha\rho_v]g \sin\Omega$$

Since α is constant, it follows that

$$-\Delta P_G = [(1-\alpha)\rho_l + \alpha\rho_v]g\,\Delta z \sin\Omega$$

Substituting,

$$-\Delta P_G = [(1 - 0.869)(1388) + 0.869(19.2)](9.8)(0.7)(.5/.7)$$

$$= 972\text{Pa} = 0.972 \text{ kPa}$$

For the respective phases flowing alone,

$$\text{Re}_v = \frac{GxD}{\mu_v} = \frac{250(0.25)(0.01)}{11.7 \times 10^{-6}} = 53{,}400$$

$$\text{Re}_l = \frac{G(1-x)D}{\mu_l} = \frac{250(0.75)(0.01)}{262 \times 10^{-6}} = 7160$$

Thus the flow is turbulent-turbulent, and the Martinelli parameter is given by Eq. (10.78):

$$X_{tt} = \left(\frac{\rho_v}{\rho_l}\right)^{0.5}\left(\frac{\mu_l}{\mu_v}\right)^{0.125}\left(\frac{1-x}{x}\right)^{0.875}$$

$$= \left(\frac{19.2}{1{,}388}\right)^{0.5}\left(\frac{262}{11.7}\right)^{0.125}\left(\frac{0.75}{0.25}\right)^{0.875} = 0.454$$

From Eq. (10.58*a*) with $C = 20$,

$$\phi_l = \left[1 + \frac{C}{X} + \frac{1}{X^2}\right]^{1/2} = \left[1 + \frac{20}{0.454} + \frac{1}{(0.454)^2}\right]^{1/2} = 7.06$$

Using Eqs. (10.45) and (10.80),

$$f_{lo} = 0.079\left(\frac{GD}{\mu_l}\right)^{-0.25} = 0.079\left(\frac{250 \times 0.01}{262 \times 10^{-6}}\right)^{-0.25} = 0.00799$$

$$\phi_{lo}^2 = \phi_l^2(1-x)^{1.75} = (7.06)^2(0.75)^{1.75} = 30.1$$

From Eq. (10.57*b*) it follows that the frictional component of the pressure drop $-\Delta P_F$ is given by

$$-\Delta P_F = -\left(\frac{dP}{dz}\right)_F \Delta z = \phi_{lo}^2\left(\frac{2f_{lo}G^2}{\rho_l D}\right)\Delta z$$

$$= (30.1)\left[\frac{2(0.00799)(250)^2}{1388(0.01)}\right](0.7)$$

$$= 1{,}517 \text{ Pa} = 1.52 \text{ kPa}$$

Example 10.3 Water leaves an evaporator at 90% quality and flows along a horizontal tube 3 m long and 2 cm in inside diameter. Heat is added uniformly along the tube so that the quality at the exit is exactly 100%. The flow conditions are such that $G = 600$ kg/m^2 s and the pressure at the tube inlet is 3773 kPa. Determine the pressure drop along the tube and compare it to that if pure saturated vapor alone flowed through it.

For saturated water at 3773 kPa, $\rho_l = 804$ kg/m^3, $\rho_v = 18.9$ kg/m^3, $\mu_l = 127 \times 10^{-6}$ N s/m^2, and $\mu_v = 17.0 \times 10^{-6}$ N s/m^2. Writing the frictional component of the pressure gradient in terms of ϕ_v, if $\Lambda = 1$, the separated-flow pressure gradient relation for a horizontal tube ($\Omega = 0$) can be written as

$$-\left(\frac{dP}{dz}\right) = \phi_v^2\left(\frac{2f_v G^2 x^2}{\rho_v D}\right) + G^2\frac{d}{dz}\left[\frac{x^2 v_v}{\alpha} + \frac{(1-x)^2 v_l}{1-\alpha}\right]$$

Integrating this equation from $z = z_1$ to $z = z_2$ yields the following relation for the pressure drop $-\Delta P$:

$$-\Delta P = \frac{2G^2}{\rho_v D} I_z + G^2 \left[\frac{x^2 v_v}{\alpha} + \frac{(1-x)^2 v_l}{1-\alpha} \right]_{z=z_2} - G^2 \left[\frac{x^2 v_v}{\alpha} + \frac{(1-x)^2 v_l}{1-\alpha} \right]_{z=z_1}$$

where

$$I_z = \int_{z=z_1}^{z=z_2} f_v x^2 \phi_v^2 \, dz$$

Note that in obtaining the above relations, the densities of the vapor and liquid have been assumed to be constant along the tube. At $x = 0.95$,

$$\text{Re}_l = \frac{G(1-x)D}{\mu_l} = \frac{600(0.05)(0.02)}{127 \times 10^{-6}} = 4724$$

$$\text{Re}_v = \frac{GxD}{\mu_v} = \frac{600(0.95)(0.02)}{17 \times 10^{-6}} = 6.71 \times 10^5$$

$$f_v = 0.079 \, \text{Re}_v^{-0.25} = 0.079(6.71 \times 10^5)^{-0.25} = 0.00276$$

$$X_{tt} = \left(\frac{\rho_v}{\rho_l} \right)^{0.5} \left(\frac{\mu_l}{\mu_v} \right)^{0.125} \left(\frac{1-x}{x} \right)^{0.875}$$

$$= \left(\frac{18.9}{804} \right)^{0.5} \left(\frac{127}{17} \right)^{0.125} \left(\frac{0.05}{0.95} \right)^{0.875} = 0.0150$$

$$\phi_v^2 = 1 + CX + X^2 = 1 + 20X_{tt} + X_{tt}^2 = 1 + 20(0.015) + (0.0150)^2 = 1.300$$

Since both Reynolds numbers are above 2000, turbulent relations for f_v and X have been used, and the turbulent-turbulent Martinelli correlation was used to determine ϕ_v. It can similarly be shown that

$$f_v = 0.00280, \ \phi_v^2 = 1.578 \qquad \text{at } x = 0.90$$

$$f_v = 0.00273, \ \phi_v^2 = 1 \qquad \text{at } x = 1.00$$

Using the trapezoidal rule and the results of the above calculations, I_z is approximated numerically as

$$I_z = \left[\frac{1}{2} (f_v x^2 \phi_v^2)_{z=0} + (f_v x^2 \phi_v^2)_{z=1.5} + \frac{1}{2} (f_v x^2 \phi_v^2)_{z=3.0} \right] \Delta z$$

Since heat is applied uniformly along the tube, $z = 0, 1.5, 3.0$ correspond to $x = 0.9, 0.95, 1.0$, respectively. It follows that

$$I_z = \left[\frac{1}{2} (0.00280)(0.9)^2(1.578) + (0.00276)(0.95)^2(1.300) + \frac{1}{2} (0.00273)(1)^2(1)^2 \right] (1.5) = 0.00959$$

At $x = 1$, $\alpha = 1$ and the second term in the above expression for $-\Delta P$ reduces to $G^2 v_v$. At $x = 0.9$, α is obtained using Eq. (10.91) with the constants from Table 10.2 for the Lockhart and Martinelli model:

$$\alpha = \left[1 + 0.28\left(\frac{1-x}{x}\right)^{0.64}\left(\frac{\rho_v}{\rho_l}\right)^{0.36}\left(\frac{\mu_l}{\mu_v}\right)^{0.07}\right]^{-1}$$

$$= \left[1 + 0.28\left(\frac{0.1}{0.9}\right)^{0.64}\left(\frac{18.9}{804}\right)^{0.36}\left(\frac{127}{17}\right)^{0.07}\right]^{-1} = 0.980$$

Substituting into the relation for $-\Delta P$ yields

$$-\Delta P = \frac{2(600)^2(0.00959)}{18.9(0.02)} + \frac{(600)^2}{18.9} - (600)^2\left[\frac{(0.9)^2}{0.980(18.4)} - \frac{(0.1)^2}{(0.020)(804)}\right]$$

$$= 18{,}267 + 19{,}048 - 15{,}520$$

$$= 21{,}800 \text{ Pa} = 21.8 \text{ kPa}$$

The pressure drop for pure vapor flow is determined in the following sequence of steps:

$$\text{Re} = \frac{GD}{\mu_v} = \frac{600(0.02)}{17.0 \times 10^{-6}} = 7.06 \times 10^5$$

$$f = 0.079\,\text{Re}^{-0.25} = 0.079(7.06 \times 10^5)^{-0.25} = 0.00273$$

$$-\Delta P = \frac{2LfG^2}{\rho_v D} = \frac{2(3)(0.00273)(600)^2}{18.9(0.02)} = 15{,}600 \text{ Pa} = 15.6 \text{ kPa}$$

Thus the pressure drop for the vaporizing flow is more than for pure vapor flow. Note that this is due to the higher frictional contribution and the added acceleration effect.

Example 10.4 Saturated R-22 flows adiabatically through a horizontal tube with an inside diameter of 1.5 cm. At a specific location in the tube $x = 0.4$ and $P = 619$ kPa. Compute the pressure gradient at this location using the Baroczy and Martinelli correlations if (*a*) $G = 339$ kg/m^2 s and (*b*) $G = 2710$ kg/m^2 s.

For saturated R-22 at 619 kPa, $T_{sat}(P) = 280$ K $= 7°$C $= 44.6°$F, $\rho_l = 1260$ kg/m^3, $\rho_v = 263$ kg/m^3, $\mu_l = 225 \times 10^{-6}$ N s/m^2, and $\mu_v = 12.3 \times 10^{-6}$ N s/m^2.

(*a*) The property index used in the Baroczy correlation is determined as

$$\left(\frac{\mu_l}{\mu_v}\right)^{0.2}\left(\frac{\rho_v}{\rho_l}\right) = \left(\frac{225}{12.3}\right)^{0.2}\left(\frac{26.3}{1260}\right) = 0.0373$$

From Fig. 10.18, at $x = 0.4$ $(\phi_{lo}^2)_{1335} = 17$. Using Fig. 10.19, the correction factor γ for $G = 339$ is 1.30. Thus

$$\phi_{lo}^2 = (\phi_{lo}^2)_{1335}\gamma = (17)(1.30) = 22.1$$

$$\mathrm{Re}_{lo} = \frac{GD}{\mu_l} = \frac{339(0.015)}{225 \times 10^{-6}} = 2.26 \times 10^4$$

$$f_{lo} = 0.079\,\mathrm{Re}_{lo}^{-0.25} = 0.079(2.26 \times 10^4)^{-0.25} = 0.00644$$

$$\left(-\frac{dP}{dz}\right)_{\mathrm{Bar}} = \phi_{lo}^2\left[\frac{2f_{lo}G^2}{\rho_l D}\right] = 22.1\left[\frac{2(0.00644)(339)^2}{1260(0.015)}\right] = 1620\ \mathrm{Pa/m}$$

Using the Martinelli correlation,

$$\mathrm{Re}_l = \frac{G(1-x)D}{\mu_l} = \frac{339(0.6)(0.015)}{225 \times 10^{-6}} = 1.36 \times 10^4$$

$$\mathrm{Re}_v = \frac{GxD}{\mu_v} = \frac{339(0.4)(0.015)}{12.3 \times 10^{-6}} = 1.65 \times 10^5$$

(thus the flow is turbulent-turbulent)

$$X_{tt} = \left(\frac{\rho_v}{\rho_l}\right)^{0.5}\left(\frac{\mu_l}{\mu_v}\right)^{0.125}\left(\frac{1-x}{x}\right)^{0.875}$$

$$= \left(\frac{26.3}{1260}\right)^{0.5}\left(\frac{225}{12.3}\right)^{0.125}\left(\frac{0.6}{0.4}\right)^{0.875} = 0.296$$

$$\phi_l^2 = 1 + \frac{20}{X_{tt}} + \frac{1}{X_{tt}^2} = 1 + \frac{20}{0.296} + \frac{1}{(0.296)^2} = 71.9$$

$$f_l = 0.079\,\mathrm{Re}_l^{-0.25} = 0.079(1.36 \times 10^4)^{-0.25} = 0.00732$$

$$\left(-\frac{dP}{dz}\right)_{\mathrm{Mar}} = \phi_l^2\left[\frac{2f_lG^2(1-x)^2}{\rho_l D}\right]$$

$$= (71.9)\left[\frac{2(0.00732)(339)^2(0.6)^2}{1260(0.015)}\right] = 2300\ \mathrm{Pa/m}$$

(*b*) Following the same sequence of steps as above, it is easily shown that for $G = 2710\ \mathrm{kg/m^2\,s}$:

$$\left(-\frac{dP}{dz}\right)_{\mathrm{Bar}} = 37{,}500\ \mathrm{Pa/m}$$

$$\left(-\frac{dP}{dz}\right)_{\mathrm{Mar}} = 87{,}700\ \mathrm{Pa/m}$$

Thus the two correlations agree within about 40% at the lower G value, but at the higher mass flux they differ by more than a factor of 2.

10.4 ANALYTICAL MODELS OF ANNULAR FLOW

During convective vaporization and condensation processes inside tubes, the large difference between the liquid and vapor densities away from the critical point can result in high vapor void fractions even at relatively low quality. As an example, consider the evaporation of R-12 in the tubes of an air-conditioning evaporator at a saturation pressure of 384 kPa (corresponding to a saturation temperature of 7°C). For turbulent-turbulent flow in a round tube at a quality of 20%, the Martinelli parameter X_{tt} is 0.6 and the Lockhart-Martinelli correlation (10.76) predicts that the void fraction is 0.83. At the moderate wall superheat levels typical of air-conditioning evaporators, this would very likely correspond to the annular flow regime with a thin film of liquid flowing along the tube walls, while vapor and possibly some liquid droplets flow in the central core region.

It is noteworthy that the 20% quality condition considered above would correspond to the inlet condition of a typical evaporator, after the refrigerant completes the throttling process in the expansion valve. Thus, the entire vaporization process occurs at void fractions greater than 0.83. Vapor–liquid flows for such conditions will generally correspond to the annular flow regime, up to the point where the liquid film on the wall begins to dry out. Hence annular flow is the predominant flow regime over most of the length of the flow passage. Similar arguments for the convective condensation process in the condenser of a typical air-conditioning system indicate that annular flow will also exist for most of the length of the condensing-side passage.

The importance of annular flow to air-conditioning systems and other applications has stimulated efforts to develop accurate methods to predict the transport for such circumstances. In this section, analytical treatments of annular flow will be considered in detail. Although annular flow may occur in channels of any cross section, the development here will focus on annular flow in round tubes. Extensions of the treatment for round tubes are often adequate to analyze the transport for annular flow in other geometries.

Annular Flow with No Entrainment

To develop the model analysis of annular flow in a vertical round tube, we consider the configuration shown in Fig. 10.22. The model assumes (1) that the flow is steady, (2) that the downstream pressure gradients felt in the core and the liquid film are the same, and (3) that liquid flows in an annular film on the inside wall of the tube that has a uniform thickness δ and a smooth liquid–vapor interface. In most cases, the interface is expected to be wavy. However, the effects of the wavy interface on heat transfer and pressure drop are often small or can be treated by empirically modifying a smooth-interface model. As a first approximation,

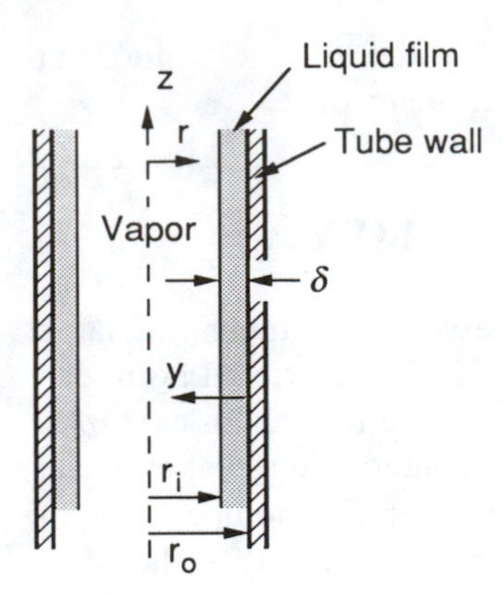

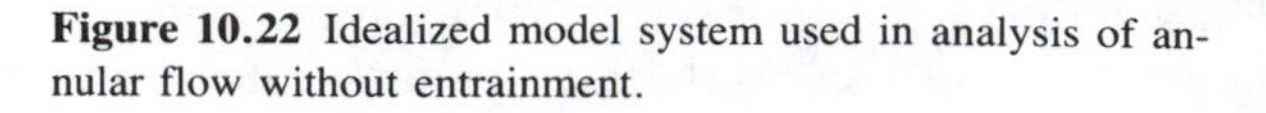

Figure 10.22 Idealized model system used in analysis of annular flow without entrainment.

these effets will be neglected here. In addition, entrainment of liquid in the vapor core will, at least initially, be assumed to be zero. Modifications of the analysis to include entrainment effects will be discussed later in this section.

As shown by Hewitt and Hall-Taylor [10.36], the streamwise force–momentum balance for the vapor core flow in Fig. 10.22 can be expressed as

$$\tau_i = -\frac{r_i}{2}\left\{\frac{dP}{dz} + \left(\frac{r_o}{r_i}\right)^2 \frac{d}{dz}\left[G^2\left(\frac{r_o}{r_i}\right)^2 \frac{x^2}{\rho_v}\right] + \rho_v g\right\} \tag{10.92}$$

where τ_i is the shear stress at the liquid–vapor interface. The terms inside the curly brackets on the right side represent, from left to right, the pressure gradient, acceleration, and body force terms. Using the fact that the core void fraction $\alpha = (r_o/r_i)^2$, and expanding the derivative with respect to z, the acceleration term can be written in the form

$$\left(\frac{1}{\alpha}\right)\frac{d}{dz}\left[\frac{G^2x^2}{\alpha\rho_v}\right] = \frac{2G^2x}{\alpha^2\rho_v}\left(\frac{dx}{dz}\right)\left[1 - \frac{x}{2\alpha}\left(\frac{d\alpha}{dx}\right)\right] \tag{10.93}$$

For most systems of practical interest, the term $(x/2\alpha)(d\alpha/dx)$ is small compared to 1. This term will therefore be neglected, whereupon Eq. (10.92) can be written as

$$\frac{dP}{dz} = -\frac{4}{d_t\sqrt{\tau_i}} - \frac{2xG^2}{\alpha^2\rho_v}\left(\frac{dx}{dz}\right) - \rho_v g \tag{10.94}$$

where $d_t = 2r_o$.

Neglecting inertia effects, the steady-state balance of shear stress, pressure, and gravitational body forces at a given r location in the liquid film requires that

$$\tau(r) = \tau_i\left(\frac{r_i}{r}\right) + \frac{1}{2}\left(\frac{dP}{dz} + \rho_l g\right)\left(\frac{r_i^2 - r^2}{r}\right) \tag{10.95}$$

The cylindrical geometry requires that

$$\frac{r_i}{r} = \frac{d_t/2 - \delta}{d_t/2 - y} \tag{10.96}$$

where y is the distance measured from the tube wall toward the centerline. Making use of the relation

$$\frac{du}{dy} = \frac{\tau}{\mu_l + \epsilon\rho_l} \tag{10.97}$$

Eq. (10.95) can be rearranged to obtain

$$\frac{du}{dy} = \frac{\tau_i}{\mu_l + \epsilon\rho_l}\left(\frac{d_t/2 - \delta}{d_t/2 - y}\right) + \frac{1}{2}\left(\frac{dP}{dz} + \rho_l g\right)\left(\frac{d_t/2 - y}{\mu_l + \epsilon\rho_l}\right)\left[\left(\frac{d_t/2 - \delta}{d_t/2 - y}\right)^2 - 1\right] \tag{10.98}$$

In the above relations ϵ is the eddy diffusivity for turbulent momentum transport in the liquid film.

An additional conservation requirement that must be satisfied is that the total mass flow rate in the film must be equal to $\pi(d_t/2)^2 G(1 - x)$. This condition is expressed mathematically as

$$\left(\frac{d_t}{4}\right) G(1 - x) = \rho_l \int_0^\delta u\, dy \tag{10.99}$$

Conservation of energy (neglecting sensible cooling of the film) also requires that dx/dz is related to the wall heat flux as

$$\frac{dx}{dz} = \frac{4q''}{G d_t h_{lv}} \tag{10.100}$$

Boundary conditions for the liquid film are

at $y = 0$: $$u = 0 \tag{10.101a}$$

at $y = \delta$: $$\frac{du}{dy} = \frac{\tau_i}{\mu_l + \epsilon\rho_l} \tag{10.101b}$$

The second condition (10.101*b*) is automatically satisfied by Eq. (10.98). Equation (10.101*a*) thus provides the needed boundary condition for the first-order differential equation for u.

For specified values of q'', G, x, d_t, and fluid properties, the differential equation (10.98) with boundary condition (10.101*a*) can be solved for the velocity profile in the film only if relations are available to determine (dP/dz), τ_i, ϵ, and δ simultaneously. There are, however, only two additional relations, Eqs. (10.94) and (10.99), available, since Eq. (10.94) requires the use of Eq. (10.100) to evaluate dx/dz. Two additional relations are therefore required to close the system of equations mathematically. Closure of the system is usually achieved by including empirical relations for determining the interfacial shear stress and the eddy diffusivity.

The interfacial shear stress is usually evaluated in terms of an interfacial friction factor f_i, defined as

$$\tau_i = \frac{f_i G^2 x^2}{2\rho_v \alpha^2} \tag{10.102}$$

In general, the interfacial friction factor f_i may be a function of the film thickness, quality, mass flux, and the tube diameter:

$$f_i = f_i(\delta, x, G, d_t) \tag{10.103}$$

Perhaps the simplest correlation of this type is that proposed by Wallis [10.5]:

$$f_i = 0.005\left(1 + 300\frac{\delta}{d_t}\right) \tag{10.104}$$

The eddy diffusivity in the liquid film for turbulent flow conditions has typically been evaluated using relations like those developed for single-phase turbulent flow in tubes. The Deissler correlation is one example of such a relation:

$$\epsilon = n^2 uy(1 - e^{-\rho_l n^2 uy/\mu_l}) \tag{10.105}$$

where n is a constant equal to 0.1. More recently, for thin films, variations of the eddy diffusivity have been used in which the eddy diffusivity goes to zero at both the wall and the interface, reflecting the damping effect of the interface on turbulence. An example of a variation of this type is the relation used by Blanghetti and Schlunder [10.37]:

$$\frac{\epsilon}{\nu_l} = -0.5 + 0.5[1 + 0.64(y^+)^2(1 - e^{-(y^+)^2/26})]^{1/2} \qquad (y \leq \delta') \tag{10.106a}$$

$$\frac{\epsilon}{\nu_l} = 0.0161\,\text{Ka}^{1/3}\,\text{Re}_l^{1.34}\left[\frac{\tau}{g(\rho_l - \rho_v)(\nu_l^2/g)^{1/3}} + \frac{\delta}{(\nu_l^2/g)^{1/3}}\left(1 - \frac{y^+}{\delta^+}\right)\right](\delta^+ - y^+) \qquad (y > \delta') \tag{10.106b}$$

where

$$\text{Ka} = \frac{\rho_l^3 g^3 (\nu_l^2/g)^2}{\sigma}, \quad y^+ = \frac{y(\tau_w/\rho_l)^{1/2}}{\nu_l}, \quad \delta^+ = \frac{\delta(\tau_w/\rho_l)^{1/2}}{\nu_l} \tag{10.106c}$$

and Re_l is the Reynolds number defined by Eq. (10.68). Note that τ_w is the wall shear stress (at $y = 0$) that can be determined from Eqs. (10.97) and (10.98). The delimiting value of $y = \delta'$ is the location where the above two expressions for ϵ/ν_l, (10.106*a*) and (10.106*b*), intersect, which is usually near $y = \delta/2$. In practice, both expressions can be evaluated and the smaller value of ϵ/ν_l taken to be correct. It can be seen that since ϵ/ν_l is a function of the local shear stress, Eq. (10.97) must be used to eliminate τ, resulting in a very complex relation for du/dy.

With the addition of relations for f_i and ϵ/ν_l, solution of the governing equations for specified fluid properties and values of G, q'', and d_t may proceed as follows:

1. A value of δ is guessed.
2. Equations (10.102) and (10.104) can be used to determine τ_i.
3. Equations (10.94) and (10.100) can be used to compute dP/dz.
4. Using a relation like Eq. (10.105) to evaluate the eddy diffusivity, Eq. (10.98) is numerically integrated using boundary condition (10.101*a*) to determine the u velocity across the liquid film. The integral of u from $y = 0$ to $y = \delta$ is also determined numerically as part of this computation.
5. If the numerically determined integral of u across the film satisfies Eq. (10.99) for the specified conditions, then the solution is complete. If this condition is not satisfied, a new value of δ must be guessed and the sequence of calculations must be repeated beginning with step 2. This process is repeated until Eq. (10.99) is satisfied.

Programming of this algorithm on a computer is relatively straightforward. The converged solution thus provides a prediction of the two-phase pressure gradient for the imposed flow conditions. In addition, since the void fraction can be computed directly from the film thickness, the computed solution also predicts the void fraction.

Annular Flow with Entrainment

The model analysis described above can be significantly in error when liquid is entrained into the vapor core flow. Perhaps the most commonly used analytical treatment of annular flow with entrainment makes use of a modified form of the separated flow analysis described in Section 10.2. The total flow is postulated to consist of three "separated" flows: (1) the liquid flow in the film on the walls, (2) the vapor flow in the core, and (3) the liquid entrained in the core flow.

In the one-dimensional force–momentum balance, friction, gravitational, and acceleration contributions to the overall pressure gradient arise, just as in the separated flow model without entrainment. The frictional and gravitational contributions to the overall pressure gradient for separated flow with entrainment are identical to those for separated flow without entrainment (see Eq. [10.31]). As formulated here, the only difference between these two cases is in the acceleration term.

Following the annular-flow model analysis of Hewitt and Hall-Taylor [10.36], we designate the volume fraction of the liquid in the film on the walls as β_f, and the mass fraction of the liquid phase entrained in the core as E. The mean streamwise velocity associated with the vapor (u_v), liquid film (u_{lf}), and entrained liquid (u_{le}) are given by

$$u_v = \frac{Gx}{\alpha \rho_v} \tag{10.107a}$$

$$u_{lf} = \frac{G(1-x)(1-E)}{\beta_f \rho_l} \tag{10.107b}$$

$$u_{le} = \frac{G(1-x)E}{(1-\alpha-\beta_f)\rho_l} \tag{10.107c}$$

The acceleration contribution to the overall pressure gradient $-(dP/dz)_{\rm acc}$ can be written in terms of these velocities as

$$-\left(\frac{dP}{dz}\right)_{\rm acc} = \frac{d}{dz}[\alpha\rho_v u_v^2 + \beta_f \rho_l u_{lf}^2 + (1-\alpha-\beta_f)\rho_l u_{le}^2] \tag{10.108}$$

Substituting the relations (10.107) yields

$$-\left(\frac{dP}{dz}\right)_{\rm acc} = G^2 \frac{d}{dz}\left[\frac{x^2}{\alpha\rho_v} + \frac{(1-E)^2(1-x)^2}{\beta_f \rho_l} + \frac{E^2(1-x)^2}{(1-\alpha-\beta_f)\rho_l}\right] \tag{10.109}$$

If it is further assumed that the droplets in the core flow at the same velocity as the gas, setting $u_{le} = u_v$ yields the following relation for β_f:

$$\beta_f = 1 - \alpha - \frac{\alpha E(1-x)\rho_v}{x\rho_l} \tag{10.110}$$

This relation can be substituted into Eq. (10.109) to eliminate β_f:

$$-\left(\frac{dP}{dz}\right)_{\rm acc} = G^2 \frac{d}{dz}\left[\frac{x^2}{\alpha\rho_v} + \frac{(1-E)^2(1-x)^2 x}{\rho_l(1-\alpha) - \rho_v \alpha E(1-x)} + \frac{E(1-x)x}{\alpha\rho_v}\right] \tag{10.111}$$

Magiros and Dukler (1961) proposed the following additional idealizations: (1) Acceleration of the liquid film is negligible compared to acceleration of the core flow; and (2) the void fraction is, to a first approximation, equal to 1.0. With these additional idealizations, Eq. (10.111) can be written approximately as

$$-\left(\frac{dP}{dz}\right)_{\rm acc} = \left(\frac{G^2}{\rho_v}\right)\frac{d}{dz}[x^2 + x(1-x)E] \tag{10.112}$$

Replacing the last term on the right side of Eq. (10.31) with the acceleration contribution specified by Eq. (10.111) yields the following separated flow expression for the total pressure gradient:

$$-\left(\frac{dP}{dz}\right) = \phi_l^2 \left(\frac{dP}{dz}\right)_l + [(1-\alpha)\rho_l + \alpha\rho_v]g \sin\Omega$$
$$+ G^2 \frac{d}{dz}\left[\frac{x^2}{\alpha\rho_v} + \frac{(1-E)^2(1-x)^2 x}{\rho_l(1-\alpha) - \rho_v \alpha E(1-x)} + \frac{E(1-x)x}{\alpha\rho_v}\right] \tag{10.113}$$

Similarly, for the core flow, the streamwise force–momentum balance obtained by modifying Eq. (10.92) to account for entrainment yields

$$\frac{dP}{dz} = -\frac{4}{d_t\sqrt{\tau_i}} - \frac{\rho_v g[x + E(1-x)]}{x + E(1-x)\rho_v/\rho_l}$$

$$- G^2 \frac{d}{dz}\left[\frac{x^2}{\alpha\rho_v} + \frac{(1-E)^2(1-x)^2 x}{\rho_l(1-\alpha) - \rho_v \alpha E(1-x)} + \frac{E(1-x)x}{\alpha\rho_v}\right] \tag{10.114}$$

In addition, for conditions under which entrainment occurs, conservation of mass in the liquid film requires that

$$\left(\frac{d_t}{4}\right)G(1-x)(1-E) = \rho_l \int_0^{\delta} u\, dy \tag{10.115}$$

If a means is available to predict the entrainment E, the pressure gradient and void fraction for annular flow with entrainment can be determined using the same iterative algorithm described above for the no-entrainment case. Methods used to predict the entrainment are generally empirical in nature. Several of the more recent efforts to develop entrainment models have been reviewed in the paper by Govan et al. [10.39]. Upon reviewing the available entrainment and deposition data, these investigators proposed an improved correlation scheme for predicting the amount of entrainment in vertical upward annular flow in round tubes. Using their methodology, the equilibrium entrainment is computed as follows. First, the minimum liquid film mass flux at which entrainment will occur, G_{lfo}, is computed from the relation

$$\frac{G_{lfo} D}{\mu_l} = \exp\left[5.8504 + 0.429\left(\frac{\mu_v}{\mu_l}\right)\left(\frac{\rho_l}{\rho_v}\right)^{0.5}\right] \tag{10.116}$$

The rate of entrainment E'' (mass of droplets entrained per second, per unit of wall area) is then computed from the relation

$$\frac{E''}{Gx} = 5.75 \times 10^{-5}\left[(G_{lf} - G_{lfo})^2\left(\frac{D\rho_l}{\sigma\rho_v^2}\right)\right]^{0.316} \qquad (\text{for } G_{lf} > G_{lfo}) \tag{10.117a}$$

where G_{lf} is the liquid film mass flux, given by

$$G_{lf} = G(1-x)(1-E) \tag{10.117b}$$

The entrainment correlation embodied in Eq. (10.117*a*) is compared to the envelope of measured annular-flow entrainment data summarized by Govan et al. [10.39] in Fig. 10.23.

Because the flow is assumed to be in equilibrium, the rate of entrainment must equal the rate of deposition, which is postulated to equal a deposition coefficient k_d multiplied by the droplet mass concentration in the core flow C_e (in kg/m^3):

$$E'' = k_d C_e \tag{10.118}$$

Based on a fit to deposition rate data, Govan et al. [10.39] proposed the following correlation for the deposition rate constant:

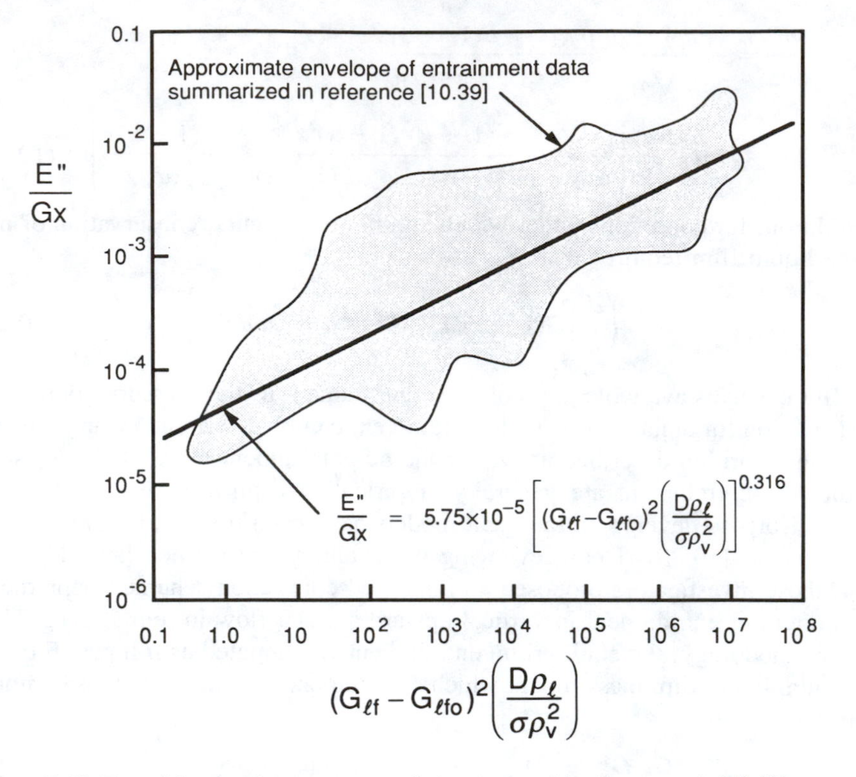

Figure 10.23 Plot of the entrainment rate correlation proposed by Govan et al. [10.39].

$$k_d\sqrt{\frac{\rho_v D}{\sigma}} = 0.18 \qquad \text{for } \left(\frac{C_e}{\rho_v}\right) < 0.3 \tag{10.119a}$$

$$k_d\sqrt{\frac{\rho_v D}{\sigma}} = 0.083\left(\frac{C_e}{\rho_v}\right)^{-0.65} \qquad \text{for } \left(\frac{C_e}{\rho_v}\right) \geqslant 0.3 \tag{10.119b}$$

The entrainment fraction is related to the mass contration of droplets as

$$E = \frac{C_e x/\rho_v}{(1-x)[1-C_e/\rho_l]} \tag{10.120}$$

Equations (10.117*a*), (10.117*b*), and (10.118) can be combined to obtain

$$\frac{k_d C_e}{Gx} = 5.75\times10^{-5}\left[[G(1-x)(1-E)-G_{lfo}]^2\left(\frac{D\rho_l}{\sigma\rho_v^2}\right)\right]^{0.316} \tag{10.121}$$

Once G_{lfo} is computed using Eq. (10.116), the nonlinear system of equations (10.119) through (10.121) must be solved simultaneously to determine k_d, E and C_e. The solution of this system thus provides a prediction of the entrainment of droplets in the core flow at the specified local conditions. While, in general, simultaneous solution of these relations requires an iterative procedure, it is possible to simplify the computation in some cases (see Example 10.5).

With the above method of determining the entrainment, computation of the pressure gradient and void fraction for specified fluid properties and values of q'', G, x, and d_t may be achieved with the following algorithm:

1. A value of δ is guessed.
2. Equations (10.102) and (10.104) can be used to determine τ_i.
3. Equations (10.119) through (10.121) can be solved simultaneously using an iterative scheme (or a similar entrainment model) to determine k_d, E, and C_e.
4. Equations (10.100) and (10.114) can be used to compute dP/dz.
5. Using a relation such as Eq. (10.105) to evaluate the eddy diffusivity, Eq. (10.98) is integrated numerically using boundary condition (10.101*a*) to determine the u velocity across the liquid film. The integral of u from $y = 0$ to $y = \delta$ is also determined numerically as part of this computation.
6. If the numerically determined integral of u across the film satisfies Eq. (10.115) for the specified conditions, then the solution is complete. If this condition is not satisfied, a new value of δ must be guessed and the sequence of calculations must be repeated beginning with step 2. This process is repeated until Eq. (10.115) is satisfied.

In form, this algorithm is quite similar to that described above for the case of no entrainment. With suitable closure models for τ_i, the eddy diffusivity, and E, the above scheme can, in principle, be used iteratively to determine the pressure drop and void fraction for the flow. In practice, recommended closure models for these quantities are available for only a limited range of circumstances. Analysis of heat transfer during annular film flow transport will be considered in detail in Chapter Twelve. For further information about annular film flow transport, the interested reader is referred to the excellent treatise by Hewitt and Hall-Taylor [10.36].

Example 10.5 A two-phase flow of saturated nitrogen at atmospheric pressure flows upward through a vertical tube with an inside diameter of 0.5 cm at $G = 200$ kg/m^2 s. The nitrogen enters the tube at a quality of 20% and slowly vaporizes (at low heat flux) as it flows along the passage. Estimate the equilibrium droplet mass concentration and the fraction of the liquid entrained for $x = 0.5$ and $x = 0.7$.

For saturated nitrogen at atmospheric pressure, $\rho_l = 807$ kg/m^3, $\rho_v = 4.62$ kg/m^3, $\mu_l = 163 \times 10^{-6}$ N s/m^2, $\mu_v = 5.41 \times 10^{-6}$ N s/m^2, and $\sigma = 0.00885$ N/m. Using Eq. (10.116), the minimum liquid film mass flux for entrainment to occur, G_{fo}, is determined as

$$G_{lfo} = \left(\frac{\mu_l}{D}\right) \exp\left[5.8504 + 0.429\left(\frac{\mu_v}{\mu_l}\right)\left(\frac{\rho_l}{\rho_v}\right)^{0.5}\right]$$

$$= \left(\frac{163 \times 10^{-6}}{0.005}\right) \exp\left[5.8504 + 0.429\left(\frac{5.41}{163}\right)\left(\frac{807}{4.62}\right)^{0.5}\right]$$

$$= 13.7 \text{ kg/m}^2 \text{ s}$$

Inverting Eq. (10.120) to solve for C_e yields

$$C_e = [(1 - x)E] \bigg/ \left[\frac{x}{\rho_v} + \frac{(1 - x)}{\rho_l} E \right]$$

Since $0 \leqslant E \leqslant 1$ and $\rho_l \gg \rho_v$, this relation for C_e is well approximated as

$$C_e = \frac{\rho_v E(1 - x)}{x}$$

which implies that

$$E = \frac{(C_e/\rho_v)x}{1 - x}$$

It is further noted that for the specified flow conditions k_d, as given by Eqs. (10.119*a*) and (10.119*b*), is a function only of C_e:

$$k_d = 0.18 \sqrt{\frac{\sigma}{\rho_v D}} \qquad \text{for } \frac{C_e}{\rho_v} < 0.3$$

$$= 0.083 \sqrt{\frac{\sigma}{\rho_v D}} \left(\frac{C_e}{\rho_v} \right)^{-0.65} \qquad \text{for } \frac{C_e}{\rho_v} \geqslant 0.3$$

Substituting the relation obtained above for E into Eq. (10.121), after some rearrangement, the following equation may be obtained:

$$0 = C_e - \frac{5.75 \times 10^{-5} Gx}{k_d} \cdot \left(\left\{ G(1 - x) \left[1 - \left(\frac{x}{1 - x} \right) \left(\frac{C_e}{\rho_v} \right) \right] - G_{lfo} \right\}^2 \left(\frac{D\rho}{\sigma \rho_v^2} \right) \right)^{0.316}$$

Substituting the specified values of G, x, D, σ, ρ_l, ρ_v, and G_{lfo}, and using the above relations for k_d, the right side of this equation can be evaluated iteratively for different values of C_e to determine the value that makes the right side equal to zero. Doing so for $x = 0.5$ and $x = 0.7$ and determining the corresponding E values yields

$$C_e = 1.90 \text{ kg/m}^3, \; E = 0.411 \qquad \text{at } x = 0.5$$

$$C_e = 1.04 \text{ kg/m}^3, \; E = 0.525 \qquad \text{at } x = 0.7$$

Note that the mass concentration of liquid droplets is lower at the higher quality, where the liquid inventory is lower. However, the fraction of liquid entrained is actually larger at higher quality. Linear extrapolation of these results further suggests that $C_e = 0$ at about $x = 0.94$, presumably because the liquid film Reynolds number becomes too small for entrainment to occur (i.e., it falls below the value specified by Eq. [10.116]).

REFERENCES

10.1 Hewitt, G. F., and Roberts, D. N., Studies of two-phase flow patterns by simultaneous X-ray and flash photography, AERE-M 2159, Her Majesty's Stationery Office, London, 1969.

10.2 Radovcich, N. A., and Moissis, R., The transition from two-phase bubble flow to slug flow, Report No. 7-7673-22, Mechanical Engineering Department, MIT, Cambridge, MA, 1962.

10.3 Taitel, Y., and Dukler, A. E., Flow regime transitions for vertical upward gas–liquid flow: A preliminary approach through physical modeling, Paper presented at Session on Fundamental Research in Fluid Mechanics at the 70th AIChE Annual Meeting, New York, 1977.

10.4 Porteus, A., Prediction of the upper limit of the slug flow regime, *Brit. Chem. Eng.*, vol. 14, no. 9, pp. 117–119, 1969.

10.5 Wallis, G. B., *One-Dimensional Two-Phase Flow,* Wiley, New York, 1965.

10.6 Baker, O., Simultaneous flow of oil and gas, *Oil and Gas J.*, vol. 53, pp. 185–195, 1954.

10.7 Mandhane, J. M., Gregory, G. A., and Aziz, K., Flow pattern map for gas–liquid flow in horizontal pipes, *Int. J. Multiphase Flow,* vol. 1, pp. 537–553, 1974.

10.8 Taitel, Y., and Dukler, A. E., A model for predicting flow regime transitions in horizontal and near horizontal gas–liquid flow, *AIChE J.*, vol. 22, pp. 47–55, 1976.

10.9 Holser, E. R., Flow patterns in high pressure two-phase (steam–water) flow with heat addition, AIChE Preprint 22, Paper presented at 9th Nat. Heat Transfer Conf., Seattle, August 1967.

10.10 Carey, V. P., and Mandrusiak, G. D., Annular film-flow boiling of liquids in a partially heated vertical channel with offset strip fins, *Int. J. Heat Mass Transfer,* vol. 29, pp. 927–939, 1986.

10.11 Xu, X., and Carey, V. P., Heat transfer and two-phase flow during convective boiling in a partially-heated cross-ribbed channel, *Int. J. Heat Mass Transfer,* vol. 30, pp. 2385–2397, 1987.

10.12 Nishikawa, K., et al., Two-phase annular flow in a smooth tube and grooved tubes, Paper presented at Int. Symp. on Research in Co-current Gas–Liquid Flow, University of Waterloo, September 1968.

10.13 Richardson, B. L., Some problems in horizontal two-phase two-component flow, Report ANL-5949, Argonne National Lab., Argonne, IL, 1958.

10.14 Zarnett, G. D., and Charles, M. E., Co-current gas–liquid flow in horizontal tubes with internal spiral ribs, Paper presented at Int. Symp. on Research in Co-current Gas–Liquid Flow, University of Waterloo, September 1968.

10.15 Zahn, W. R., A visual study of two-phase flow while evaporating in horizontal tubes, *J. Heat Transfer,* vol. 86, pp. 417–429, 1964.

10.16 Boyce, B. E., Collier, J. G., and Levy, J., Hold up and pressure drop measurements in the two-phase flow of air–water mixtures in helical coils, Paper presented at Int. Symp. on Research in Co-current Gas–Liquid Flow, University of Waterloo, September 1968.

10.17 Banerjee, S., Rhodes, E., and Scott, D. S., Film inversion of co-current two-phase flow in helical coils, *AIChE J.*, vol. 13, pp. 189–191, 1967.

10.18 Oshinowo, T., and Charles, M. E., Vertical two-phase flow. Part I. Flow pattern correlations, *Can. J. Chem. Eng.*, vol. 52, pp. 25–35, 1974.

10.19 Ishii, M., *Thermo-Fluid Dynamic Theory of Two-Phase Flow,* Eyrolles, Paris, 1975.

10.20 Boure, J. A., Constitutive equations for two-phase flows, in *Two-Phase Flows and Heat Transfer with Application to Nuclear Reactor Design Problems,* Von Karman Institute Book, Hemisphere, New York, chap. 9, 1978.

10.21 McAdams, W. H., Woods, W.K., and Heroman, L.C., Jr., Vaporization inside horizontal tubes—II—Benzene–oil mixtures, *Trans. ASME,* vol. 64, p. 193, 1942.

10.22 Cicchitti, A., et al., Two-phase cooling experiments—pressure drop, heat transfer and burnout measurements, *Energia Nucleare,* vol. 7, no. 6, pp. 407–425, 1960.

10.23 Dukler, A. E., Wicks, M. III, and Cleveland, R.G., Pressure drop and hold-up in two-phase

flow, Part A—A comparison of existing correlations; and Part B—An approach through similarity analysis, *AIChE J.*, vol. 10, pp. 38–51, 1964.
10.24 Lockhart, R. W., and Martinelli, R. C., Proposed correlation of data for isothermal two-phase, two-component flow in pipes, *Chem. Eng. Prog.*, vol. 45, no. 1, pp. 39–48, 1949.
10.25 Butterworth, D., A comparison of some void-fraction relationships for co-current gas–liquid flow, *Int. J. Multiphase Flow*, vol. 1, pp. 845–850, 1975.
10.26 Martinelli, R. C., and Nelson, D. B., Prediction of pressure drop during forced-circulation boiling of water, *Trans. ASME*, vol. 70, pp. 695–702, 1948.
10.27 Thom, J. R. S., Prediction of pressure drop during forced circulation boiling of water, *Int. J. Heat Mass Transfer*, vol. 7, pp. 709–724, 1964.
10.28 Baroczy, C. J., A systematic correlation for two-phase pressure drop, AIChE reprint 37, presented at the 8th Nat. Heat Transfer Conf., Los Angeles, August 1965.
10.29 Baroczy, C. J., Correlation of liquid fraction in two-phase flow with applications to liquid metals, *Chem. Eng. Prog. Symp. Ser.*, vol. 61, no. 57, pp. 179–191, 1965.
10.30 Friedel, L., Improved friction pressure drop correlations for horizontal and vertical two phase pipe flow, paper E2, European Two Phase Flow Group Meeting, Ispra, Italy, 1979.
10.31 Hetsroni, G. (editor), *Handbook of Multiphase Systems*, section 2.2.3.2, Hemisphere, New York, 1982.
10.32 Chenoweth, J. M., and Martin, M. W., Turbulent two-phase flow, *Petroleum Refiner*, vol. 34, no. 10, pp. 151–155, 1955.
10.33 Zivi, S. M., Estimation of steady-state steam void-fraction by means of the principle of minimum entropy production, *J. Heat Transfer*, vol. 86, pp. 247–252, 1964.
10.34 Mandrusiak, G. D., and Carey, V. P., Pressure drop characteristics of two-phase flow in a vertical channel with offset strip fins, *Exp. Thermal and Fluid Sci.*, vol. 1, pp. 41–50, 1988.
10.35 Chisholm, D., *Two-Phase Flow in Pipelines and Heat Exchangers*, George Godwin of Longman Group Limited, New York, 1983.
10.36 Hewitt, G. F., and Hall-Taylor, N. S., *Annular Two-Phase Flow*, Pergamon Press, Oxford, 1970.
10.37 Blanghetti, F., and Schlunder, E. U., Local heat transfer coefficients on condensation in vertical tube, *Proc. 6th Int. Heat Transfer Conf.*, vol. 2, pp. 437–442, 1978.
10.38 Margiros, P. G., and Dukler, A. E., Entrainment and pressure drop in concurrent gas–liquid flow: II, liquid property and momentum effects, in *Developments in Mechanics*, Plenum Press, New York, vol. 1, 1961.
10.39 Govan, A. H., Hewitt, G. F., Owen, D. G., and Bott, T. R., An improved CHF modelling code, *Proc. Second U.K. Conf. on Heat Transfer*, Institute of Mechanical Engineers, London, vol. 1, pp. 33–48, 1988.

PROBLEMS

10.1 A two-phase mixture of saturated oxygen liquid and vapor at a pressure of 196 kPa flows upward in a vertical tube. The tube inside diameter is 8 mm and the mass flux is 200 kg/m^2 s. Determine the flow regime at qualities of 0.1, 0.5, and 0.9.

10.2 Solve Problem 10.1 for a horizontal tube.

10.3 A two-phase mixture of saturated liquid and vapor water at atmospheric pressure flows through a round tube with a diameter of 1.2 cm. The flow rate is such that the mass flux is 400 kg/m^2 s. Estimate the values of quality at which transitions from bubbly to slug and churn to annular flow are expected to occur.

10.4 Saturated ammonia at 165 kPa flows from the expansion valve of a refrigeration system to the inlet of an evaporator through a horizontal tube with a diameter of 2.2 cm and a length of 0.5 m. The mass flux is 500 kg/m^2 s and the quality is 0.22. Determine (*a*) the flow regime, (*b*) the void fraction, and (*c*) the total pressure drop using the separated flow model.

10.5 Solve parts (*b*) and (*c*) of Problem 10.4 using the homogeneous flow model.

10.6 Adiabatic two-phase flow of steam and water at 6124 kPa flows upward in a vertical tube 4 m long and 2.5 cm in diameter. The mass flux is 700 kg/m^2 s and the quality is 0.7. Determine the pressure drop through the tube using (*a*) the homogeneous flow model, (*b*) the Martinelli correlation, and (*c*) the Baroczy correlation.

10.7 R-12 leaves an evaporator at 90% quality and flows through a vertical tube 2 m long and 1 cm in inside diameter. Heat is added uniformly so that the quality at the exit of the tube is 95%. The mass flux through the tube is 400 kg/m^2 s and the pressure at the tube inlet is 333 kPa. Determine the pressure drop along the tube using the separated flow model by dividing the tube into four segments and assuming that the pressure gradient is uniform over each segment.

10.8 For water at 3773 kPa, plot and compare the $\bar{\mu}/\mu_l$-versus-x variations predicted by Eqs. (10.42), (10.43), and (10.44). Based on these variations, what conclusions do you draw regarding the suitability of each of these relations for bubbly flow and droplet mist flow?

10.9 Saturated two-phase flow of nitrogen occurs adiabatically in a vertical tube with an inside diameter of 1.7 cm. At a specific location in the tube, $x = 0.3$ and $P = 778$ kPa. The mass flux is 350 kg/m^2 s. Compute the void fraction and pressure gradient at this location using (*a*) the Baroczy correlation and (*b*) the Martinelli correlation.

10.10 For two-phase flow of saturated nitrogen at atmospheric pressure, compute and plot the variation of void fraction with quality predicted by each of the void fraction models listed in Table 10.2.

10.11 A water evaporator consists of a vertical metal tube 1.5 m long with an inside diameter of 1.0 cm. A uniform heat flux of 800 kW/m^2 is applied to the tube wall. Saturated liquid water enters the tube at a pressure of 2185 kPa. Using the Martinelli-Nelson correlation, determine and plot the pressure drop in the tube for flow rates between 21 and 800 g/s. You have been asked to select a mass flow rate that (1) assures that the exit flow is either in the bubbly or slug flow regime (but not in the churn or annular regimes) and (2) minimizes the overall pressure drop within the limits imposed by constraint (1). What flow rate do you recommend?

10.12 A vaporizing flow of water flows downward in a tube having an inside diameter of 1.0 cm and a length of 1.5 m. The flow enters as saturated liquid at 1172 kPa. A heat flux of 600 kW/m^2 (based on inside wall area) is applied uniformly to the walls of the tube. Determine the pressure drop over the length of the tube for mass flux values between 50 and 300 kg/m^2 s and plot the pressure drop for the system over this range of G.

10.13 Saturated liquid R-12 enters a tube at a pressure of 333 kPa. The tube inside diameter is 1.2 cm and its length is 1.0 m. The mass flux into the tube is 400 kg/m^2 s, and a uniform heat flux of 15 W/cm^2 is delivered to the flow uniformly over the inside surface of the tube. Determine the void fraction variation and the sequence of flow regimes encountered along the tube (*a*) if the tube is vertical and (*b*) if the tube is horizontal.

10.14 Saturated ammonia vapor enters a tube at a pressure of 2422 kPa and a mass flux of 400 kg/m^2 s. The tube inside diameter is 1.5 cm and the tube length is 2.0 m. The tube wall is cooled in such a way that a heat flux of 75 W/cm^2 is removed uniformly along the inside wall of the tube. Determine the void fraction variation along the tube and the sequence of flow regimes encountered if (*a*) the tube is horizontal and (*b*) the tube is inclined at an angle of 40° to the horizontal.

10.15 Use the algorithm given in Section 10.4 for annular flow with no entrainment to write a computer program to iteratively compute the film thickness and interfacial shear stress for annular flow in vertical tube. Make use of Eqs. (10.94) through (10.105) as appropriate. Use the model to compute the film thickness and two-phase frictional pressure gradient for two-phase flow of saturated R-12 in a tube with an inside diameter of 1.0 cm at 333 kPa for $x = 0.8$ and $G = 300$ kg/m^2 s. Compare the results to those predicted by the Martinelli correlation.

10.16 A two-phase flow of R-12 at 200 kPa flows upward through a vertical tube with an inside diameter of 7 mm at $G = 300$ kg/m^2 s. The R-12 enters the tube at a quality of 20% and slowly vaporizes (at low heat flux) along the passage. Estimate the equilibrium droplet concentration and the fraction of liquid entrained for $x = 0.5$ and $x = 0.75$.

10.17 Use the algorithm given in Section 10.4 for annular flow with entrainment to write a computer program to iteratively compute the film thickness and interfacial shear stress for annular flow in vertical tube. Make use of the equations in that section and the entrainment correlation, as appropriate. Use the model to compute the film thickness and two-phase frictional pressure gradient for two-phase flow of saturated nitrogen in a tube with an inside diameter of 0.8 cm at 540 kPa for $x = 0.7$ and $G = 300 \text{ kg/m}^2 \text{ s}$. Compare the results to those predicted by the Baroczy correlation.

CHAPTER

ELEVEN

INTERNAL CONVECTIVE CONDENSATION

11.1 REGIMES OF CONVECTIVE CONDENSATION

As noted in Chapter Ten, the two-phase flow regime for gas–liquid flow in a tube generally depends on the mean velocities of the two phases and their properties. During convective condensation in a tube, the properties often vary only slightly over the flow length. However, because the phase change produces an appreciable variation in the relative flow velocities of the two phases, the flow regime can change dramatically over the length of the passage.

Perhaps the most common flow configuration in which convective condensation occurs is flow in a horizontal circular tube. This configuration is encountered in air-conditioning and refrigeration condensers as well as condensers in Rankine power cycles. Although convective condensation is also sometimes contrived to occur in co-current vertical downward flow, horizontal flow is often preferred because the flow can be repeatedly passed through the heat exchanger core in a serpentine fashion, as indicated in Fig. 11.1, without trapping liquid or vapor in the return bends. If the straight sections of the serpentine tubes are oriented vertically, intermittent trapping of liquid or vapor in the bends can cause the flow to oscillate in a manner that may reduce the system's performance. If the flow to the tubes is fed by a header or inlet manifold, as shown in Fig. 11.2, downflow condensation can also be accomplished without this problem.

The sequence of flow regimes that can be encountered during condensation in a horizontal tube is shown schematically in Fig. 11.3. A clearer picture of the regimes typically encountered, and their impact on the heat transfer process, can be obtained by considering the flow regime map shown in Fig. 11.4. This type of map was proposed by Baker [11.1] for horizontal gas–liquid two-phase flow.

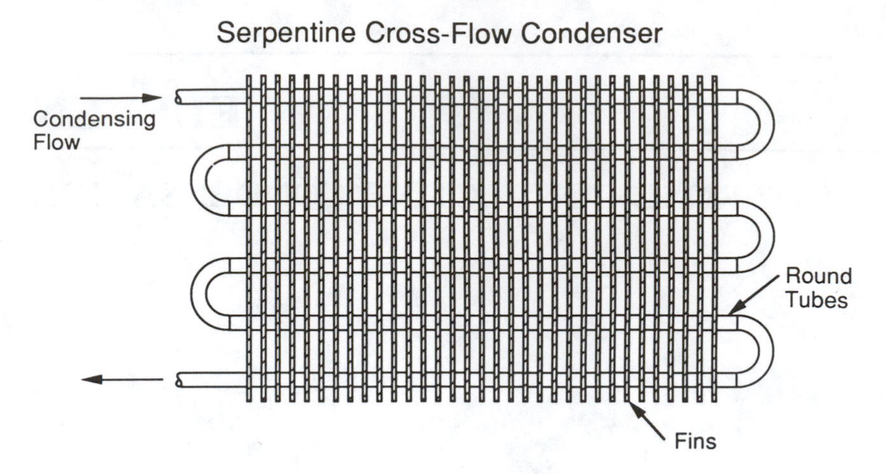

Figure 11.1 Schematic of a typical cross-flow air-cooled condenser.

The vertical and horizontal coordinates are the vapor and liquid superficial mass fluxes, respectively, scaled with parameters that account for property variations. The scaling parameters λ and ψ are defined as

$$\lambda = \left[\left(\frac{\rho_v}{\rho_a}\right)\left(\frac{\rho_l}{\rho_w}\right)\right]^{1/2} \tag{11.1}$$

$$\psi = \left(\frac{\sigma_w}{\sigma}\right)\left[\left(\frac{\mu_l}{\mu_w}\right)\left(\frac{\rho_w}{\rho_l}\right)^2\right]^{1/3} \tag{11.2}$$

In these relations, the subscripts a and w denote properties for air and liquid water, respectively, at room temperature and atmospheric pressure.

While this flow regime map is considered by some to be less general than the Taitel-Dukler [11.2] map considered in Chapter Ten, it is particularly convenient for considering the sequence of flow regimes exhibited during a convective condensation process in a horizontal tube. Shown in Fig. 11.4 is a curve representing the locus of flow regimes that exist along the tube during condensation of R-12 from $x = 0.95$ to $x = 0.05$ at a pressure of 1.60 MPa. Curves are shown for mass flux values of 100, 200, and 500 kg/m^2 s, which span the range of conditions typically encountered in air-conditioning and refrigeration applications.

Increasing the mass flux moves the process line upward and to the right on the flow regime map in Fig. 11.4. For $G = 500$ kg/m^2 s, which is typical of air-conditioning system condensers, the flow regime is annular to qualities as low as 30%. In fact, for many applications in which condensation occurs at low to moderate pressure, the void fraction is relatively high even at low qualities near the end of the condensation process.

For the process lines shown in Fig. 11.4, flow regimes other than annular flow may occur, but they typically exist during only a small portion of the heat transfer process. At low flow rates, Fig. 11.4 indicates that a small region of wavy flow may be encountered at higher qualities before annular flow is established. However, even for conditions at which Fig. 11.4 predicts stratified or wavy flow, condensation on the upper portion of the inside wall of the tube will cover it with a liquid film. Hence, the condensing flow tends to take on an annular configuration even for circumstances that would produce stratified or wavy flow in an adiabatic system. At the very end of the condensation process, small portions of the flow may be in the slug, plug, and/or bubbly flow regimes.

Because most of the heat transfer associated with the condensation process takes place under annular flow conditions, prediction of the condensation heat transfer coefficient for annular flow is obviously of primary importance to the design of condensers operating at low to moderate pressures. It is not surprising, therefore, that most efforts to develop methods for predicting internal convective condensation heat transfer have focused on annular flow. Condensation during slug, plug, or bubbly flow at the end of the condensation process has received much less attention. Annular flow is also somewhat easier to model analytically than the intermittent slug or plug flows. Analytical models and empirical prediction techniques for condensation in these regimes are discussed in detail in the next two sections.

Example 11.1 R-22 condenses at 2800 kPa in a horizontal tube with an inside diameter of 1 cm. Determine the condensation regime for $x = 0.9$, 0.4 and 0.05 if $G = 400\ \text{kg/m}^2\ \text{s}$.

For saturated R-22 at 2800 kPa, $\rho_l = 991\ \text{kg/m}^3$, $\rho_v = 134\ \text{kg/m}^3$, μ_l

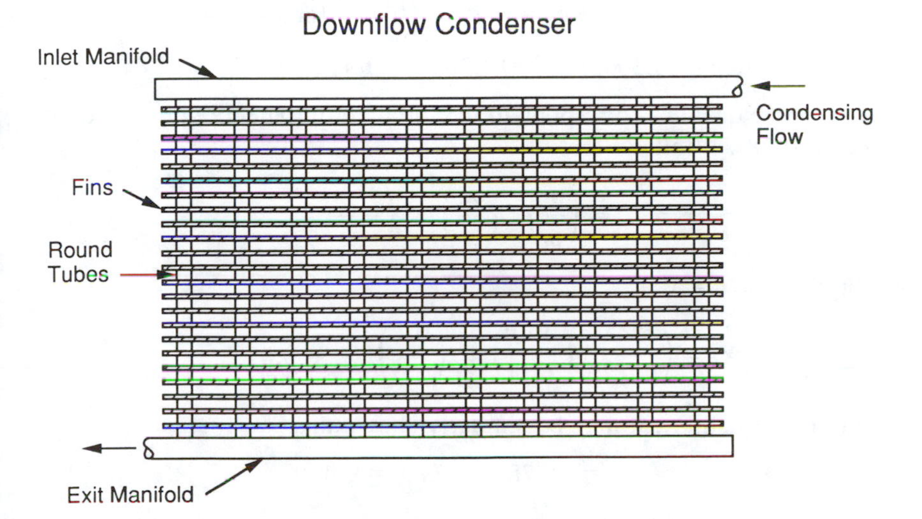

Figure 11.2 Schematic of a typical downflow air-cooled condenser.

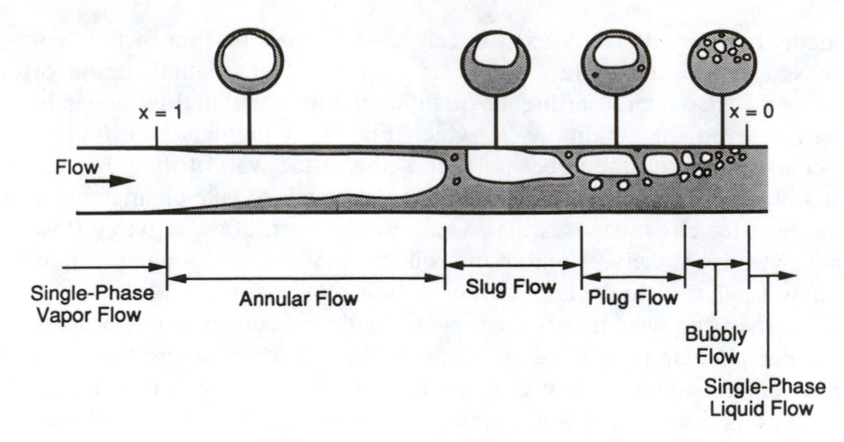

Figure 11.3 Horizontal co-current flow with condensation.

$= 150 \times 10^{-6}$ N s/m², $\mu_v = 16.4 \times 10^{-6}$ N s/m², and $\sigma = 0.0025$ N/m. Substituting for $x = 0.9$,

$$\text{Re}_l = \frac{G(1-x)D}{\mu_l} = \frac{400(0.1)(0.01)}{150 \times 10^{-6}} = 2667$$

$$\text{Re}_v = \frac{GxD}{\mu_v} = \frac{400(0.9)(0.01)}{16.4 \times 10^{-6}} = 2.20 \times 10^5$$

Similar calculations for the other qualities yield

$$\text{Re}_l = 16{,}000,\ \text{Re}_v = 9.78 \times 10^4 \qquad \text{for } x = 0.4$$

$$\text{Re}_l = 25{,}300,\ \text{Re}_v = 12{,}200 \qquad \text{for } x = 0.05$$

Thus, all three qualities correspond to the turbulent-turbulent regime, and using Eq. (10.78), it follows that

$$X = X_{tt} = \left(\frac{\rho_v}{\rho_l}\right)^{0.5} \left(\frac{\mu_l}{\mu_v}\right)^{0.125} \left(\frac{1-x}{x}\right)^{0.875}$$

Substituting for $x = 0.9$,

$$X_{tt} = \left(\frac{134}{991}\right)^{0.5} \left(\frac{150}{16.4}\right)^{0.125} \left(\frac{0.1}{0.9}\right)^{0.875} = 0.0709$$

For the other qualities, it may be similarly shown that

$$X_{tt} = 0.378 \qquad \text{for } x = 0.4$$

$$X_{tt} = 6.38 \qquad \text{for } x = 0.05$$

For a horizontal tube, $\Omega = 0$, and combining Eqs. (10.7*a*) and (10.15*b*) yields

$$F_{TD} = \left[\frac{G^2x^2}{\rho_v(\rho_l - \rho_v)Dg}\right]^{0.5}$$

Substituting for $x = 0.9$, we find

$$F_{TD} = \left[\frac{(400)^2(0.9)^2}{(134)(991 - 134)(0.01)9.8}\right]^{0.5} = 3.39$$

Using this value of F_{TD} with $X_{tt} = 0.0709$, it follows from the flow regime map in Fig. 10.5 that the flow is in the annular flow regime. Thus annular film-flow condensation is expected for $x = 0.9$. Substituting in a similar manner yields

$$F_{TD} = 1.51 \qquad \text{for } x = 0.4$$

$$F_{TD} = 0.118 \qquad \text{for } x = 0.05$$

Using $F_{TD} = 1.51$ and $X_{tt} = 0.378$, Fig. 10.5 indicates that the flow is in the

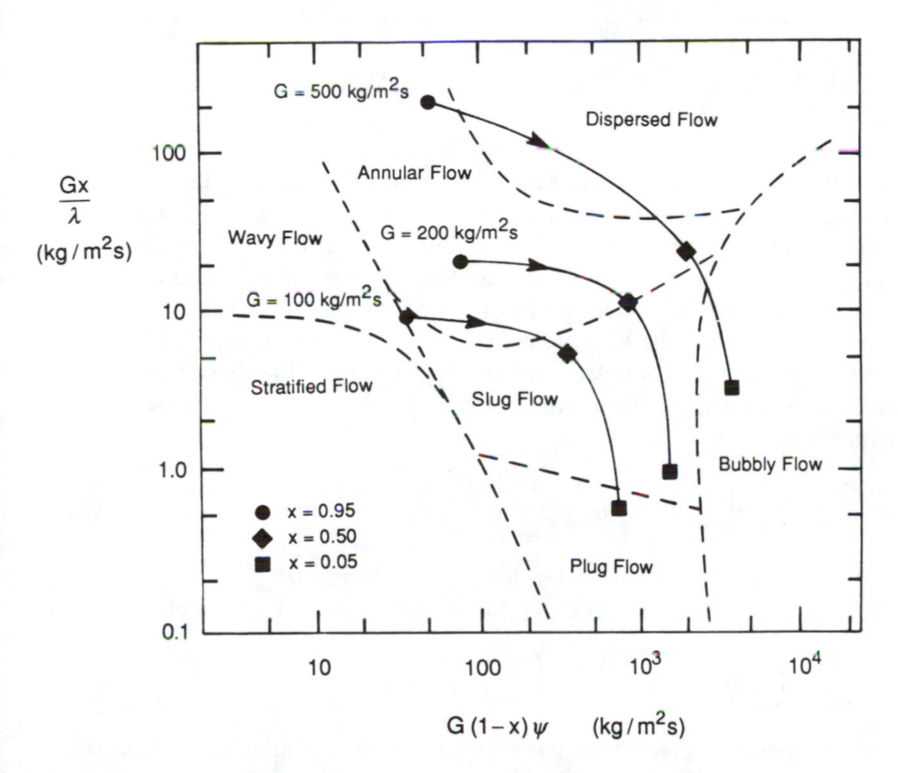

Figure 11.4 Curves indicating the locus of states that exist along the flow path of a condensing two-phase flow of R-12 in a horizontal tube.

annular flow regime. Thus annular flow film condensation is also expected at $x = 0.4$. Using $F_{TD} = 0.118$ and $X_{tt} = 6.38$, Fig. 10.5 indicates that the flow may be in the intermittent or dispersed bubble regime. Evaluating T_{TD} using Eqs. (10.12*a*) and (10.17) for $x = 0.05$ yields

$$f_l = 0.079 \, \text{Re}_l^{-0.25} = 0.079(25{,}300)^{-0.25} = 0.00626$$

$$-\left(\frac{dP}{dz}\right)_l = \frac{2f_l G^2(1-x)^2}{\rho_l D} = \frac{2(0.00626)(400)^2(0.95)^2}{991(0.01)} = 183 \text{ Pa/m}$$

$$T_{TD} = \left[\frac{-(dP/dz)_l}{(\rho_l - \rho_v)g}\right]^{0.5} = \left[\frac{183}{(991 - 134)9.8}\right]^{0.5} = 0.148$$

For $X_{tt} = 6.38$ and $T_{TD} = 0.148$, Fig. 10.5 indicates that the state point for the flow is below the intermittent/dispersed-bubble transition. The flow is therefore in the intermittent regime for these conditions. Thus, slug flow or plug flow condensation is expected for $x = 0.05$. Note that these conditions are typical of those encountered in the horizontal tubes of the condenser in a heat pump system used for heating and/or air conditioning.

11.2 ANALYTICAL MODELING OF DOWNFLOW INTERNAL CONVECTIVE CONDENSATION

As noted in Section 11.1, for many applications, convective condensation inside tubes occurs under annular flow conditions over much of the tube length. If the flow is downward and laminar with no entrainment, the Nusselt-type analysis can be adapted to internal convective condensation in a round tube. The circumstances of interest are shown schematically in Fig. 11.5. An integral analysis of the condensation heat transfer in the tube basically follows the Nusselt analysis for falling-film condensation described in Section 9.3. Neglecting the inertia and downstream diffusion contributions, the momentum balance for the differential element in Fig. 11.5 requires that

$$(\delta - y)\, dz\left[\rho_l g - \frac{dP}{dz}\right] + \tau_i \, dz = \mu_l \left(\frac{du}{dy}\right) dz \tag{11.3}$$

The total pressure gradient along the tube in the vapor (dP/dz) is equal to the hydrostatic gradient plus contributions due to frictional and deceleration effects:

$$\frac{dP}{dz} = \left(\frac{dP}{dz}\right)_{\text{hyd}} + \left(\frac{dP}{dz}\right)_{\text{fr}} + \left(\frac{dP}{dz}\right)_{\text{dec}} = \rho_v g + \left(\frac{dP}{dz}\right)_{\text{fr}} + \left(\frac{dP}{dz}\right)_{\text{dec}} \tag{11.4}$$

The frictional pressure gradient in the vapor is due to the interfacial shear stress:

$$\left(\frac{dP}{dz}\right)_{\text{fr}} = -\frac{4\tau_i}{(D - 2\delta)} \tag{11.5}$$

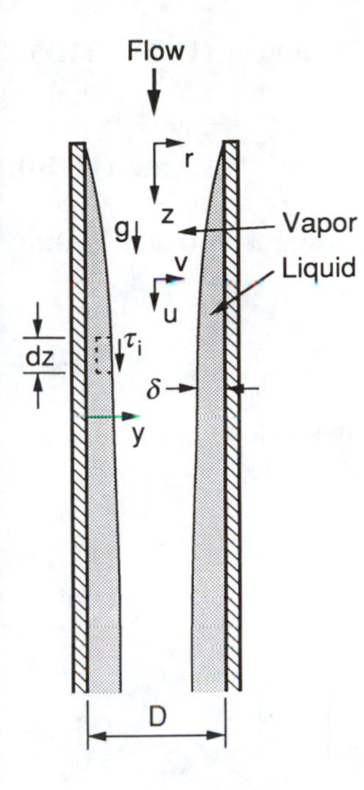

Figure 11.5 Model system for analysis of downflow internal condensation in a tube.

Based on the results of the one-dimensional two-phase separated-flow analysis presented in Chapter Ten, the deceleration pressure gradient is given by

$$\left(\frac{dP}{dz}\right)_{\text{dec}} = -G^2 \frac{d}{dz}\left[\frac{x^2}{\rho_v \alpha} + \frac{(1-x)^2}{\rho_l(1-\alpha)}\right] \tag{11.6}$$

If the vapor density is low compared to the liquid density and the film is thin, the second term in the square brackets is small compared to the first. Assuming that the variation of the void fraction along the tube is small compared to the variation in x, the expression for $(dP/dz)_{\text{dec}}$ can be further simplified to the following approximate form:

$$\left(\frac{dP}{dz}\right)_{\text{dec}} = -\frac{2xDG^2}{\rho_v(D-2\delta)}\frac{dx}{dz} \tag{11.7}$$

We will also adopt the usual idealization that the pressure gradient in the vapor is equal to that felt by the liquid film. Analysis of the momentum transport in the liquid film is facilitated by defining a fictitious vapor density ρ_v^* such that

$$\rho_v^* g = \rho_v g - \frac{4\tau_i}{D-2\delta} - \frac{2xDG^2}{\rho_v(D-2\delta)}\left(\frac{dx}{dz}\right) \tag{11.8}$$

The gradient in the quality can be evaluated from an energy balance as

$$\frac{dx}{dz} = \frac{4q''}{DGh_{lv}} = \frac{4h(T_{\text{sat}} - T_w)}{DGh_{lv}} \tag{11.9}$$

where q'' is the local heat flux. Combining Eqs. (11.3) through (11.5), (11.7), and (11.8) yields

$$\frac{du}{dy} = \frac{(\delta - y)(\rho_l - \rho_v^*)g}{\mu_l} + \frac{\tau_i}{\mu_l} \tag{11.10}$$

In this form, the momentum balance can be integrated using $u = 0$ at $y = 0$ to obtain

$$u = \frac{(\rho_l - \rho_v^*)g}{\mu_l}\left(y\delta - \frac{y^2}{2}\right) + \frac{\tau_i y}{\mu_l} \tag{11.11}$$

Integrating this velocity profile across the liquid film, the following relation for the total liquid mass flow rate $\dot{m}_l = G(1 - x)\pi D^2/4$ is obtained:

$$\dot{m}_l = \pi D\left[\frac{\rho_l(\rho_l - \rho_v^*)g\delta^3}{3\mu_l} + \frac{\tau_i\rho_l\delta^2}{2\mu_l}\right] \tag{11.12}$$

Differentiating with respect to δ yields

$$\frac{d\dot{m}_l}{d\delta} = \frac{\pi D}{\mu_l}\left[\frac{\rho_l(\rho_l - \rho_v^*)g\delta^2}{\mu_l} + \frac{\tau_i\rho_l\delta}{\mu_l}\right] \tag{11.13}$$

As in the case of falling-film condensation, if subcooling of the film is neglected, an overall mass and energy balance requires that

$$\frac{d\dot{m}_l}{dz} = \frac{k_l(T_{\text{sat}} - T_w)\pi D}{h_{lv}\delta} \tag{11.14}$$

Combining this with Eq. (11.13) yields

$$\frac{d\delta}{dz} = \frac{k_l\mu_l(T_{\text{sat}} - T_w)}{\rho_l(\rho_l - \rho_v^*)gh_{lv}\delta^3 + \tau_i\rho_l h_{lv}\delta^2} \tag{11.15}$$

This equation can be integrated using the condition that $\delta = 0$ at $x = 0$ to obtain

$$\frac{4zk_l\mu_l(T_{\text{sat}} - T_w)}{\rho_l(\rho_l - \rho_v^*)gh_{lv}} = \delta^4 + \frac{4\tau_i\delta^3}{3(\rho_l - \rho_v^*)g} \tag{11.16}$$

If convective effects in the subcooled film are included, the results of the integral analysis are the same except that h_{lv} is replaced by h'_{lv}:

$$\frac{4zk_l\mu_l(T_{\text{sat}} - T_w)}{\rho_l(\rho_l - \rho_v^*)gh'_{lv}} = \delta^4 + \frac{4\tau_i\delta^3}{3(\rho_l - \rho_v^*)g} \tag{11.17}$$

where

$$h'_{lv} = h_{lv}\left[1 + \left(\frac{3}{8}\right)\frac{c_{pl}(T_{\text{sat}} - T_w)}{h_{lv}}\right] \tag{11.18}$$

For laminar conduction-dominated transport, the heat transfer coefficient is then given by

$$h = \frac{k_l}{\delta} \tag{11.19}$$

To close the computational scheme, a means of evaluating τ_i is required. Because the viscosity of the liquid is much larger than that of the vapor and the film is thin, the mean vapor velocity in the core is generally much larger than the liquid velocity at the interface. As a first approximation, the vapor core could be treated as a single-phase flow in a round tube, with the velocity of the vapor taken to be zero at the interface. The interfacial shear τ_i can then be computed using a conventional single-phase correlation for the friction factor f_v:

$$\tau_i = f_v\left(\frac{\rho_v u_v^2}{2}\right) = f_v\left(\frac{G^2 x^2}{2\rho_v(1 - 4\delta/D)}\right) \tag{11.20}$$

For round tubes, f_v can be evaluated from the correlation

$$f_v = 0.079\left[\frac{Gx(D - \delta)}{\mu_v(1 - 4\delta/D)}\right]^{-0.25} \tag{11.21a}$$

Although prediction of the interfacial shear in this idealized manner is very approximate, it does provide a means of closing the problem mathematically. Alternatively, we could use the following simple relation for the interfacial friction factor discussed in Chapter Ten:

$$f_v = f_i = 0.005\left(1 + \frac{300\delta}{D}\right) \tag{11.21b}$$

Even with the use of a simplified method of computing the interfacial shear, determination of the heat transfer coefficient is possible only with the use of an iterative technique. An appropriate scheme of this type (for specified G, T_w, P, and thermophysical properties) is as follows:

1. Guess a value of δ.
2. Use Eqs. (11.9) and (11.19) to evaluate dx/dz.
3. Use Eqs. (11.20) and (11.21) to compute τ_i.
4. Determine ρ_v^* using Eq. (11.8).
5. Substitute the guessed δ value and the computed ρ_v^* and τ_i values into Eq. (11.17). If this equation is satisfied to an acceptable level of accuracy, the guessed value of δ and the computed h value are correct. If this relation is not satisfied, a new δ value is guessed and the process is repeated, beginning with step 2. This sequence of computations is repeated until convergence.

Because the thermophysical properties are assumed to be constant, this analysis applies to circumstances in which the pressure drop along the tube is small compared to the absolute pressure.

When the film and vapor core Reynolds numbers are high enough, part or all of the flow may be turbulent. For such circumstances, the analysis described above is not appropriate, and it must be modified to account for turbulent flow in the film. In addition, at higher liquid and vapor flow rates, entrainment of liquid droplets into the vapor core may also be significant. Prediction of the transport for such circumstances makes use of the analytical treatment of annular two-phase turbulent flow described in Section 10.4.

Such a prediction can be achieved in the following manner. At a given downstream location, Eqs. (10.98), (10.100), (10.101*a*), (10.102), (10.104), (10.105), (10.114), and (10.115) are first solved iteratively, using the correlation represented by Eqs. (10.116) through (10.121) to determine the entrainment fraction E. Solution of these equations yields the axial velocity profile $u(y)$ across the liquid film and the film thickness δ at a specified location along the length of the channel.

For steady turbulent flow, the transport of thermal energy in the liquid film is governed by the equation

$$u \frac{\partial T}{\partial x} + v \frac{\partial T}{\partial y} = \frac{\partial}{\partial y}\left[(\alpha_{T,l} + \epsilon_H) \frac{\partial T}{\partial y} \right] \tag{11.22}$$

where ϵ_H is the eddy diffusivity of heat. Because the film is thin compared to the downstream flow length, the usual boundary-layer approximations have been adopted. The boundary conditions on the temperature profile are

at $y = \delta$: $$T = T_{sat}(P_v) \tag{11.23}$$

at the interface and

at $y = 0$: $$T = T_w \tag{11.24}$$

The convection terms in the energy equation are often neglected because the transport rate across the film is much greater than downstream convection. If these terms are neglected, the energy equation can be simplified to

$$\frac{\partial}{\partial y}\left[(\alpha_{T,l} + \epsilon_H) \frac{\partial T}{\partial y} \right] = 0 \tag{11.25}$$

Integrating the energy equation (11.25) first from 0 to y and then from 0 to δ yields

$$\frac{T_{sat} - T_w}{q''_w/(\rho_l c_{pl} \alpha_l)} = \frac{k_l}{h} = \int_0^{\delta} \frac{dy}{1 + (\mathrm{Pr}_l/\mathrm{Pr}_t)(\epsilon_M/\nu_l)} \tag{11.26}$$

where h is the local heat transfer coefficient, q'' is the heat flux applied at the wall, and Pr_t is the turbulent Prandtl number defined as $\mathrm{Pr}_t = \epsilon_M/\epsilon_H$.

Before the above form of the energy transport equation can be analyzed further, a means of determining the turbulent Prandtl number and the eddy diffusivity

ϵ_M as functions of y must be known. Appropriate relations for ϵ_M/ν_l that can be used to compute the heat transfer coefficient using Eq. (11.26) are discussed in Section 10.4. The manner in which such relations can be integrated into this type of analysis is illustrated in Example 11.2.

Example 11.2 Develop an approximate heat transfer correlation from an integral analysis of the co-current downward annular film condensation in a tube (as shown in Fig. 11.5) at moderate vapor flow rates.

For the purposes of this approximate analysis, it will be assumed that entrainment of liquid in the vapor core is negligible and that the liquid flows in a thin film with a smooth interface along the tube wall. Effects of interfacial waves are neglected. The thickness of the film δ is assumed to be very small compared to the tube inside diameter D. Downstream convection in the film is neglected compared to cross-film diffusion in the momentum and energy balances. The downstream momentum equation is therefore given by

$$-\frac{1}{\rho_l}\frac{\partial P}{\partial z} + \frac{\partial}{\partial y}\left[(\nu_l + \epsilon_M)\frac{\partial u}{\partial y}\right] + g = 0$$

where ϵ_M is the turbulent eddy diffusivity in the liquid film. This equation can be rearranged to obtain

$$\frac{\partial}{\partial y}\left[\rho_l(\nu_l + \epsilon_M)\frac{\partial u}{\partial y}\right] = \left(\frac{\partial P}{\partial z}\right) - \rho_l g$$

In general, the two-phase pressure gradient and body force terms on the right side of the above equation will have some effect on the momentum balance. However, in the present analysis it will be assumed that these terms are small compared with viscous and turbulent shear stresses (note that the term in square brackets represents the combined effects of these stresses). Setting the right side of the equation to zero, it can be integrated once to obtain

$$\rho_l(\nu_l + \epsilon_M)\frac{du}{dy} = \tau_0$$

The constant obtained in the integration is denoted as τ_0, because the left side is clearly the local shear stress in the layer. Thus, with these idealizations, the film becomes a constant shear layer. Defining

$$u^+ = \frac{u}{\sqrt{\tau_0/\rho_l}} \qquad y^+ = \frac{y\sqrt{\tau_0/\rho_l}}{\nu_l}$$

the above equation can be cast in the form

$$\left(1 + \frac{\epsilon_M}{\nu_l}\right)\frac{du^+}{dy^+} = 1$$

It is further assumed here that the film is weakly turbulent so that ϵ_M/ν_l is small compared to 1, but not 0. Hence, to a first approximation,

$$\frac{du^+}{dy^+} = 1 \qquad \text{which implies that} \qquad u^+ = y^+$$

(since $u = 0$ at $y = 0$). This is consistent with the assumption that δ is small, resulting in δ^+ being so small that most of the film is in the viscous sublayer. Mass conservation of liquid requires that

$$\left(\frac{\pi D^2}{4}\right) G(1 - x) = \pi D \int_0^{\delta} \rho_l u \, dy$$

This relation can be written in terms of u^+ and y^+ as

$$\frac{GD(1 - x)}{4\mu_l} = \int_0^{\delta^+} u^+ \, dy^+$$

Substituting $u^+ = y^+$ and integrating yields

$$\delta^+ = \sqrt{\frac{1}{2}} \, \mathrm{Re}_l^{1/2} \qquad \mathrm{Re}_l = \frac{GD(1 - x)}{\mu_l}$$

For transport of heat in the film, neglecting downstream convection compared to cross-stream diffusion, the governing equation becomes

$$\frac{\partial}{\partial y}\left[\rho_l c_{pl}(\alpha_l + \epsilon_H)\frac{\partial T}{\partial y}\right] = 0$$

where ϵ_H is the eddy diffusivity for heat. Integrating this equation once, we obtain

$$\rho_l c_{pl}(\alpha_l + \epsilon_H)\frac{dT}{dy} = q''_0$$

The constant obtained in the integration is denoted as q''_0, because the left side represents the local heat flux in the film. The film is therefore a constant-heat-flux layer. Rearranging and integrating across the film yields

$$\frac{T_{\mathrm{sat}} - T_w}{q''_0/(\rho_l c_{pl}\alpha_l)} = \int_0^{\delta} \frac{dy}{1 + \mathrm{Pr}_l \, \epsilon_H/\nu_l}$$

where T_w is the wall temperature. Noting that $\rho_l c_{pl}\alpha_l = k_l$ and $\epsilon_H = \epsilon_M/\mathrm{Pr}_t$ (where Pr_t is the turbulent Prandtl number), the above equation can be written in terms of y^+ as

$$\frac{k_l\sqrt{\tau_0/\rho_l}}{h\nu_l} = \int_0^{\delta^+} \frac{dy^+}{1 + (\mathrm{Pr}_l/\mathrm{Pr}_t)(\epsilon_M/\nu_l)}$$

Because the film is a constant-heat-flux layer, q_0'' is equal to the heat flux removed at the wall of the tube and the local heat transfer coefficient h is taken to be

$$h = \frac{q_0''}{T_{sat} - T_w}$$

To use the relation derived above to predict the heat transfer coefficient, means of evaluating τ_0 and ϵ_M are needed. In the momentum balance relation, ϵ_M/ν_l was assumed to be small compared to 1. Pr_t is taken to be 0.9. However, it is also assumed here that the liquid Prandtl number Pr_l is large enough that $(\mathrm{Pr}_l/\mathrm{Pr}_t)(\epsilon_M/\nu_l)$ is not negligible compared to 1. If Von Karman's mixing-length model,

$$\epsilon_M = l^2\left|\frac{du}{dy}\right| \qquad l = \kappa y$$

is combined with the result $u^+ = y^+$, it can be easily shown that

$$\frac{\epsilon_M}{\nu_l} = \kappa^2(y^+)^2$$

In section 10.4 it was noted that recent investigations involving analysis of liquid film turbulent transport have used variations of the eddy diffusivity that go to zero at both the wall and the interface to account for the damping effect of the interface on turbulence. Consistent with this line of reasoning, the following continuous variation of $\epsilon_M/\nu_l\,\mathrm{Pr}_t$ is postulated here:

$$\frac{\epsilon_M}{\nu_l\,\mathrm{Pr}_t} = \begin{cases} \dfrac{\kappa^2(y^+)^2}{\mathrm{Pr}_t} & \text{for } 0 \leq y^+ \leq \dfrac{\delta^+}{2} \\ \dfrac{\kappa^2(\delta^+ - y^+)^2}{\mathrm{Pr}_t} & \text{for } \dfrac{\delta^+}{2} \leq y^+ \leq \delta^+ \end{cases}$$

Using this variation to evaluate the integral in the relation for h derived above, the following result is obtained:

$$h = \frac{\kappa k_l\sqrt{\tau_0/\rho_l}}{2\nu_l I}\sqrt{\frac{\mathrm{Pr}_l}{\mathrm{Pr}_t}}, \qquad I = \tan^{-1}\left(\sqrt{\frac{\mathrm{Pr}_l}{\mathrm{Pr}_t}}\frac{\kappa\delta^+}{2}\right)$$

For large values of the argument in parentheses, $\tan^{-1}$ approaches $\pi/2$. Consistent with the assumption of large Pr_l, I will be replaced with $\pi/2$ here. Also substituting $\kappa = 0.4$ (the usual recommended value of the Von Karman constant) and $\mathrm{Pr}_t = 0.9$ yields

$$h = 0.134k_l \sqrt{\frac{\tau_0}{\rho_l}} \frac{\sqrt{\mathrm{Pr}_l}}{\nu_l}$$

The frictional component of the two-phase flow pressure gradient $(dP/dz)_F$ must be related to the wall shear stress τ_0 as

$$\pi D \tau_0 = -\left(\frac{dP}{dz}\right)_F \frac{\pi D^2}{4}$$

Using the Martinelli correlation to evaluate $(dP/dz)_F$ yields

$$-\left(\frac{dP}{dz}\right)_F = \phi_l^2 \frac{2f_l G^2(1-x)^2}{D\rho_l} \qquad f_l = 0.046\left[\frac{G(1-x)D}{\mu_l}\right]^{-0.2}$$

Using these relations to evaluate τ_0, the equation for h can be written as

$$\frac{hD}{k_l} = 0.134\phi_l \sqrt{f_l}\left[\frac{G(1-x)D}{\mu_l}\right]\sqrt{\mathrm{Pr}_l}$$

Substituting the above expression for f_l (for turbulent flow) and the Martinelli turbulent-turbulent correlation for ϕ_l:

$$\phi_l = \left(1 + \frac{20}{X_{tt}} + \frac{1}{X_{tt}^2}\right)^{1/2}$$

The correlation for h can be cast in the form

$$\frac{hD}{k_l} = 0.028\,\mathrm{Re}_l^{0.9}\,\mathrm{Pr}_l^{1/2}\left(1 + \frac{20}{X_{tt}} + \frac{1}{X_{tt}^2}\right)^{1/2}$$

where

$$\mathrm{Re}_l = \frac{G(1-x)D}{\mu_l}$$

Perhaps the most interesting aspect of this result is that this relation is very similar to that proposed by Traviss et al. [11.10]. Their relation, which is described in detail in the next section, has the form

$$\frac{hD}{k_l} = f_1(\mathrm{Pr}_l, \mathrm{Re}_l) f_2(X_{tt})\,\mathrm{Re}_l^{0.9}$$

where f_1 is a relatively weak function of Re_l. This model analysis also demonstrates the direct link between the heat transfer performance and the shear stress in the liquid film during annular-flow convective condensation.

11.3 CORRELATION METHODS FOR CONVECTIVE CONDENSATION HEAT TRANSFER

It was demonstrated in Section 11.1 that for many applications involving internal flow condensation, the flow is expected to be in the annular regime over most of the passage length. While detailed analytical models of annular flow, like those discussed in Section 11.2, can be used as a means of predicting the heat transfer performance for annular flow condensation, such models generally require a high level of computational effort, and they can be used only if reliable closure relations for the interfacial shear and entrainment are available. As a result, somewhat simpler empirical relations that more directly correlate condensation heat transfer data have also been developed by a number of investigators.

In an early study, Ananiev et al. [11.3] proposed to correlate the local heat transfer coefficient for convective condensation with the relation

$$h = h_0 \sqrt{\frac{\rho_l}{\rho_m}} \tag{11.27}$$

where

$$\frac{1}{\rho_m} = \left(\frac{1}{\rho_l}\right)(1 - x) + \left(\frac{1}{\rho_v}\right)x \tag{11.28}$$

and h_0 is given by one of the available correlations for the single-phase heat transfer coefficient for the entire flow as liquid. (We could, for example, use the well-known Dittus-Boelter equation to evluate h_0 for round tubes.) Boyko and Kruzhilin [11.4] evaluated this type of approach for condensation of steam using the relation proposed by Miropolosky [11.5] for the single-phase cofficient h_0:

$$\frac{h_0 D}{k} = 0.021 \text{ Re}^{0.8} \text{ Pr}_b^{0.43} \left(\frac{\text{Pr}_b}{\text{Pr}_w}\right)^{0.25} \tag{11.29}$$

where Re is the Reynolds number and Pr_b and Pr_w are values of the Prandtl number evaluated at the bulk and wall temperatures, respectively. Boyko et al. [11.4] found that the local condensation heat transfer data obtained by Miropolsky [11.5] for condensation of steam could be correlated reasonably well with this scheme. However, a best fit to data for a steel tube was obtained by replacing the constant 0.021 in Eq. (11.29) by 0.024. The data for condensation of steam in a copper tube differed from that for the steel tube, even though both had a diameter of 2 cm. A best fit to the copper tube data required that the constant in Eq. (11.29) be changed to 0.032. In both cases the data matched the correlation prediction within about ±20%.

While this approach is simple to use, the apparent need for a different constant value in Eq. (11.29) for each fluid–solid combination makes its general applicability somewhat questionable. In addition, this correlation method ignores the effect of flow regime changes that are expected to occur along the channel as the flow proceeds downstream. In more recent studies, investigators have proposed correlation techniques for the local condensing heat transfer coefficient that allow for variation of the two-phase flow regime along the passage.

In horizontal tubes at low vapor velocities, liquid that condenses on the upper portion of the inside tube wall tends to run down the wall toward the bottom, as indicated schematically in Fig. 11.6. This stratified annular flow condition is observed most commonly at low condensation rates and/or short tube lengths. In an early study, Chato [11.6] developed a detailed analytical model of the heat transfer for these circumstances. His results implied that this type of flow configuration exists for inlet vapor Reynolds numbers (Re_v) less than 35,000. Predictions of the analytical model and experimental data were found to agree well with an equation of the form

$$h = 0.728 K_C \left[\frac{g \rho_l (\rho_l - \rho_v) k_l^3 h'_{lv}}{\mu_l (T_{sat} - T_w) D} \right]^{1/4} \tag{11.30}$$

where h'_{lv} is given by

$$h'_{lv} = h_{lv} \left[1 + 0.68 \frac{c_{pl}(T_{sat} - T_w)}{h_{lv}} \right] \tag{11.31}$$

This relation for h is identical to the relation obtained from the classic Nusselt analysis for a vertical flat plate of height D, except that the multiplying prefactor has been changed. This relation is consistent with the interpretation that the condensation process over the top portion of the inside tube wall is very similar to falling-film condensation over a flat plate.

Chato [11.6] found that test data for R-113 agreed well with this relation for $K_C = 0.76$. He further argued that heat transfer through the liquid pool at the bottom of the tube (see Fig. 11.6) is negligible compared to transport across the thin film on the upper portion of the tube wall. It follows, therefore, that the coefficient K_C in Eq. (11.30) must vary with the void fraction. In a subsequent

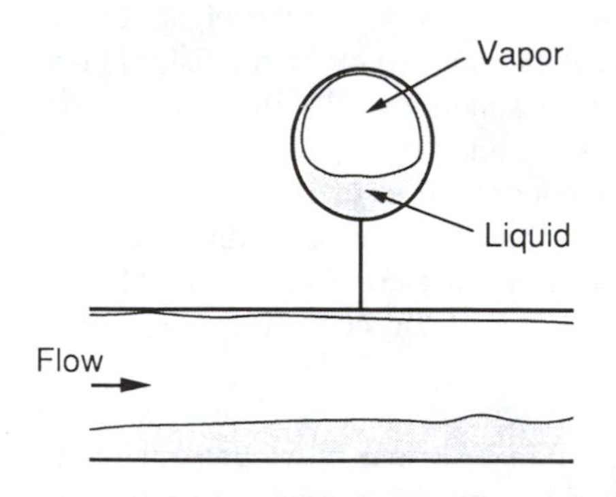

Figure 11.6 Horizontal co-current annular flow with condensation.

study, Jaster and Kosky [11.7] suggested the following simple relations as a means of predicting the void fraction and the factor K_C in Eq. (11.30):

$$K_C = \alpha^{3/4} \tag{11.32}$$

$$\alpha = \left[1 + \frac{1-x}{x}\left(\frac{\rho_v}{\rho_l}\right)\right]^{-1} \tag{11.33}$$

This correlation technique applies to the stratified annular flow conditions that may exist over much of the tube length if the inlet vapor velocity is low. This circumstance has also been examined in detail by Rufer and Kezios [11.8]. Flow circumstances of this type may be encountered near the end of the condensation process when the void fraction is still large, but deceleration of the flow has dropped the vapor velocity enough to produce stratified flow with a thin film of condensate on the upper portion of the tube wall.

At moderate to high inlet vapor velocities, annular flow is established almost immediately at the inlet and persists over most of the condensation process. A number of investigators have proposed ways of predicting the condensation heat transfer coefficient for annular flow conditions. Three of the more useful correlations are those developed by Soliman et al. [11.9], Traviss et al. [11.10], and Shah [11.11]. In the annular flow analysis described in the previous section, it became clear that the shear at the interface and at the tube wall were linked directly to the transport of heat across the liquid film. The correlation proposed by Soliman et al. [11.9] is cast in a form that explicitly acknowledges the importance of these shear parameters in determining the transport. Their correlation is given by

$$\frac{h\mu_l}{k_l\rho_l^{1/2}} = 0.036\,\mathrm{Pr}_l^{0.65}\,\tau_w^{1/2} \tag{11.34}$$

where

$$\tau_w = \tau_i + \tau_z + \tau_a \tag{11.35}$$

$$\tau_i = \frac{D}{4}\left(-\frac{dP}{dz}\right)_F, \qquad \left(-\frac{dP}{dz}\right)_F = \phi_g^2\left(-\frac{dP}{dz}\right)_v, \qquad \phi_g = 1 + 2.85X_{tt}^{0.523} \tag{11.36}$$

$$\tau_z = \frac{D}{4}(1-\alpha)(\rho_l - \rho_v)g\sin\theta, \qquad \alpha = \left[1 + \left(\frac{1-x}{x}\right)\left(\frac{\rho_v}{\rho_l}\right)^{2/3}\right]^{-1} \tag{11.37}$$

$$\tau_a = \frac{D}{4}\left(\frac{G^2}{\rho_v}\right)\left(\frac{dx}{dz}\right)\sum_{n=1}^{5} a_n\left(\frac{\rho_v}{\rho_l}\right)^{n/3} \tag{11.38}$$

$$a_1 = 2x - 1 - \beta x \tag{11.39a}$$

$$a_2 = 2(1-x) \tag{11.39b}$$

$$a_3 = 2(1 - x - \beta + \beta x) \tag{11.39c}$$

$$a_4 = \left(\frac{1}{x}\right) - 3 + 2x \tag{11.39d}$$

$$a_5 = \beta\left[2 - \left(\frac{1}{x}\right) - x\right] \tag{11.39e}$$

$$\beta = \frac{\text{interface velocity}}{\text{mean film velocity}}$$

$$= 1.25 \text{ for turbulent film flow}$$

$$= 2.0 \text{ for laminar film flow}$$

Note that in these correlation equations, $(dP/dz)_v$ is the single-phase pressure gradient for the vapor phase flowing alone, and X_{tt} is the turbulent-turbulent Martinelli parameter defined in Chapter Ten. For flow in a round tube these can be evaluated as

$$-\left(\frac{dP}{dz}\right)_v = \frac{2f_v G^2 x^2}{D\rho_v}, \qquad f_v = 0.046\left(\frac{GxD}{\mu_v}\right)^{-0.2} \tag{11.40a}$$

$$X_{tt} = \left(\frac{1-x}{x}\right)^{0.9}\left(\frac{\rho_v}{\rho_l}\right)^{0.5}\left(\frac{\mu_l}{\mu_v}\right)^{0.1} \tag{11.40b}$$

The value of β is determined based on the Reynolds number for the liquid phase flowing alone in the passage. If it is greater than 2000, the turbulent value is used. Otherwise the laminar value is used. Although the scheme proposed by Soliman et al. [11.9] is complex, it can easily be implemented as an explicit calculation in a simple computer program.

Traviss et al. [11.10] proposed the following relation for the local heat transfer coefficient for annular-flow convective condensation:

$$\frac{hD}{k_l} = \frac{0.15\,\mathrm{Pr}_l\,\mathrm{Re}_l^{0.9}}{F_T}\left[\frac{1}{X_{tt}} + \frac{2.85}{X_{tt}^{0.476}}\right], \qquad \mathrm{Re}_l = \frac{G(1-x)D}{\mu_l} \tag{11.41}$$

where X_{tt} is given by equation (11.40*b*) and F_T is given by

$$F_T = 5\,\mathrm{Pr}_l + 5\ln\{1 + 5\,\mathrm{Pr}_l\} + 2.5\ln(0.0031\,\mathrm{Re}_l^{0.812}) \quad \text{for } \mathrm{Re}_l > 1125 \tag{11.42a}$$

$$= 5\,\mathrm{Pr}_l + 5\ln[1 + \mathrm{Pr}_l(0.0964\,\mathrm{Re}_l^{0.585} - 1)] \quad \text{for } 50 < \mathrm{Re}_l < 1125 \tag{11.42b}$$

$$= 0.707\,\mathrm{Pr}_l\,\mathrm{Re}_l^{0.5} \quad \text{for } \mathrm{Re}_l < 50 \tag{11.42c}$$

Based on a purely empirical approach, Shah [11.11] has proposed the following correlation as a best fit to available convective condensation heat transfer data for round tubes:

$$\frac{h}{h_{lo}} = (1 - x)^{0.8} + \frac{3.8x^{0.76}(1 - x)^{0.04}}{(P/P_{cr})^{0.38}} \tag{11.43}$$

where

$$h_{lo} = 0.023\left(\frac{k_l}{D}\right)\left(\frac{GD}{\mu_l}\right)^{0.8} \mathrm{Pr}_l^{0.4} \tag{11.44}$$

and P and P_{cr} are the absolute local and critical pressures, respectively. Shah [11.11] found good agreement between this correlation and film condensation data for a wide variety of fluids and flow conditions. This correlation was recommended for $11 \leqslant G \leqslant 211$ kg/m² s, $0 \leqslant x \leqslant 1.0$, $1 \leqslant \mathrm{Pr}_l \leqslant 13$.

More recently, Chen et al. [11.12] have proposed a comprehensive film-condensation heat transfer correlation based on analytical and theoretical results from the literature. For co-current annular film flow they argued that for falling-film condensation in the absence of interfacial shear, the method of Churchill and Usagi [11.13] can be used to determine the local Nusselt number for an arbitrary condition from separate correlations for the laminar-wavy and turbulent film flow regimes. Based on such arguments, they postulated the following relation for the zero-shear Nusselt number Nu_0 at an arbitrary location:

$$\mathrm{Nu}_0 = [(\mathrm{Nu}_{lw})^{n_1} + (\mathrm{Nu}_t)^{n_1}]^{1/n_1} \tag{11.45}$$

where Nu_{lw} and Nu_t are the Nusselt numbers for wavy laminar and turbulent film flow, respectively. The Nusselt number in the correlations developed by these investigators is defined as $\mathrm{Nu} = h\nu_l^{2/3}/(k_l g^{1/3})$.

In a similar fashion, Chen et al. [11.12] proposed the following relation for the local Nusselt number Nu_x in the presence of vapor shear:

$$\mathrm{Nu}_x = [(\mathrm{Nu}_0)^{n_2} + (\mathrm{Nu}_{sd})^{n_2}]^{1/n_2} \tag{11.46}$$

where Nu_0 is the zero shear value as determined using Eq. (11.45), and Nu_{sd} is the value of the Nusselt number obtained from a prediction of shear-dominated film flow. For laminar-wavy flow with zero shear, Chen et al. [11.12] used the correlation recommended by Chun and Seban [11.14]:

$$\mathrm{Nu}_{lw} = 0.823\,\mathrm{Re}_x^{-0.22} \tag{11.47}$$

where

$$\mathrm{Re}_x = \frac{G(1 - x)D}{\mu_l} \tag{11.48}$$

Example 11.3 Steam condenses at 247 kPa as it flows inside a horizontal tube with an inside diameter of 2.0 cm. For $G = 200$ kg/m² s, $x = 0.7$, and a wall temperature of 100°C, compute the local heat transfer coefficient using the correlations of (*a*) Traviss et al. [11.10], (*b*) Shah [11.11], and (*c*) Boyko and Kruzhilin [11.4].

For saturated steam at 247 kPa, $T_{sat} = 400$ K = 127°C, $\rho_l = 938$ kg/m^3, $\rho_v = 1.37$ kg/m^3, $h_{lv} = 2183$ kJ/kg, $c_{pl} = 4.24$ kJ/kg K, $\mu_l = 219 \times 10^{-6}$ N s/m^2, $\mu_v = 13.6 \times 10^{-6}$ N s/m^2, $k_l = 0.686$ W/m K, and $\Pr_l = 1.35$. At 100°C, $\Pr_l = 1.72$.

(*a*) For the specified conditions,

$$\mathrm{Re}_v = \frac{GxD}{\mu_v} = \frac{200(0.7)(0.02)}{13.6 \times 10^{-6}} = 2.06 \times 10^5$$

$$\mathrm{Re}_l = \frac{G(1-x)D}{\mu_l} = \frac{200(0.3)(0.02)}{219 \times 10^{-6}} = 5480$$

and for turbulent-turbulent flow,

$$X_{tt} = \left(\frac{\rho_v}{\rho_l}\right)^{0.5}\left(\frac{\mu_l}{\mu_v}\right)^{0.1}\left(\frac{1-x}{x}\right)^{0.9}$$

$$= \left(\frac{1.37}{938}\right)^{0.5}\left(\frac{219}{13.6}\right)^{0.1}\left(\frac{0.3}{0.7}\right)^{0.9} = 0.0235$$

Substituting into Eq. (11.42*a*),

$$F_T = 5\,\Pr_l + 5\ln(1 + 5\,\Pr_l) + 2.5\ln(0.0031\,\mathrm{Re}_l^{0.812})$$

$$= 5(1.35) + 5\ln[1 + 5(1.35)] + 2.5\ln[0.0031(5480)^{0.812}] = 20.0$$

Rearranging equation (11.41) and substituting,

$$h_T = \left(\frac{k_l}{D}\right)\frac{0.15\,\Pr_l\,\mathrm{Re}_l^{0.8}}{F_T}\left(\frac{1}{X_{tt}} + \frac{2.85}{X_{tt}^{0.476}}\right)$$

$$= \left(\frac{0.686}{0.02}\right)\frac{0.15(1.35)(5480)^{0.9}}{20.0}\left[\frac{1}{0.0235} + \frac{2.85}{(0.0235)^{0.476}}\right]$$

$$= 47{,}900 \text{ W/m}^2\text{ K}$$

Note also that

$$F_{TD} = \left[\frac{G^2x^2}{\rho_v(\rho_l - \rho_v)Dg}\right]^{0.5} = \left[\frac{(200)^2(0.7)^2}{1.37(938 - 1)(0.02)9.8}\right]^{0.5} = 8.83$$

For the values of $X = X_{tt} = 0.0235$ and $F_{TD} = 8.83$, Fig. 10.5 indicates that the flow is in the annular flow regime.

(*b*) Substituting into Eq. (11.44),

$$h_{lo} = 0.023\left(\frac{k_l}{D}\right)\left(\frac{GD}{\mu_l}\right)^{0.8} \mathrm{Pr}_l^{0.4}$$

$$= 0.023\left(\frac{0.686}{0.02}\right)\left(\frac{200(0.02)}{219 \times 10^{-6}}\right)^{0.8} (1.35)^{0.4} = 2280 \ \mathrm{W/m^2\,K}$$

The Shah [11.11] prediction of the condensation heat transfer coefficient is given by Eq. (11.43) as

$$h_S = h_{lo}\left[(1-x)^{0.8} + \frac{3.8x^{0.76}(1-x)^{0.04}}{(P/P_{\mathrm{cr}})^{0.38}}\right]$$

Noting that P_{cr} for water is 22,124 kPa and substituting, we find that

$$h_S = (2280)\left[(0.3)^{0.8} + \frac{3.8(0.7)^{0.76}(0.3)^{0.04}}{(247/22{,}124)^{0.38}}\right] = 35{,}600 \ \mathrm{W/m^2\,K}$$

(*c*) The heat transfer coefficient is determined using the Boyko and Kruzhilin [11.4] correlation as follows:

$$h_0 = 0.021\left(\frac{k_l}{D}\right)\left(\frac{GD}{\mu_l}\right)^{0.8} \mathrm{Pr}_b^{0.43}\left(\frac{\mathrm{Pr}_b}{\mathrm{Pr}_w}\right)^{0.25}$$

Noting that at the wall and bulk flow temperatures $\mathrm{Pr} = \mathrm{Pr}_w = 1.72$ and $\mathrm{Pr} = \mathrm{Pr}_b = 1.35$, respectively, substitution yields

$$h_0 = 0.021\left(\frac{0.686}{0.02}\right)\left(\frac{200(0.02)}{219 \times 10^{-6}}\right)^{0.8} (1.35)^{0.43}\left(\frac{1.35}{1.75}\right)^{0.25} = 1970 \ \mathrm{W/m^2\,K}$$

Equations (11.27) and (11.28) are combined to obtain

$$h_{BK} = h_0\left[(1-x) + x\left(\frac{\rho_l}{\rho_v}\right)\right]^{1/2}$$

$$= 1970\left[(0.3) + 0.7\left(\frac{938}{1.37}\right)\right]^{1/2} = 43{,}100 \ \mathrm{W/m^2\,K}$$

Thus the predicted h values for these three correlations vary from 35.6 kW/m^2 K to 47.9 kW/m^2 K.

For the turbulent regime, Chen et al. [11.12] used the following relation to predict Nu_t, which is a curve-fit to the theoretical results of Blangetti and Schlunder [11.15]:

$$\mathrm{Nu}_t = 0.0040\ \mathrm{Re}_x^{0.4}\mathrm{Pr}_l^{0.65} \tag{11.49}$$

For the high interfacial shear stress regime, a modified form of the relation proposed by Soliman et al. [11.9] was used to predict Nu_{sd}:

$$\mathrm{Nu}_{sd} = 0.036\,\mathrm{Pr}_l^{0.65}(\tau_i^*)^{1/2} \tag{11.50}$$

where τ_i^* is a dimensionless interfacial shear stress defined as

$$\tau_i^* = \frac{\tau_i}{\rho_l(g\nu_l)^{2/3}} \tag{11.51}$$

In obtaining Eq. (11.50) from the original relation of Soliman et al. [11.9], the gravitational and acceleration effects have been neglected, as would be appropriate for the interfacial shear-dominated case.

For co-current flow, Chen et al. [11.12] used the following expression to determine τ_i^*, which was derived from the empirical two-phase pressure drop correlation developed by Dukler [11.16]:

$$\tau_i^* = A_D(\mathrm{Re}_{\mathrm{ter}} - \mathrm{Re}_x)^{1.4}\,\mathrm{Re}_x^{0.4} \tag{11.52}$$

where

$$A_D = \frac{0.252\mu_l^{1.177}\mu_v^{0.156}}{D^2 g^{2/3}\rho_l^{0.553}\rho_v^{0.78}} \tag{11.53}$$

$$\mathrm{Re}_{\mathrm{ter}} = \frac{GD}{\mu_l} \tag{11.54}$$

Using Eqs. (11.47) through (11.54) to evaluate the terms in Eqs. (11.45) and (11.46), Chen et al. [11.12] determined the values of n_1 and n_2 that provided a best fit to the measured local heat transfer coefficients obtained by Blangetti and Schlunder [11.15]. They found that $n_1 = 6$ and $n_2 = 2$ provided a very good fit to these data. Combining the above equations for these values of n_1 and n_2, the following correlation equation is obtained:

$$\mathrm{Nu}_x = \frac{h\nu_l^{2/3}}{k_l g^{1/3}} = \left[\left(0.31\,\mathrm{Re}_x^{-1.32} + \frac{\mathrm{Re}_x^{2.4}\,\mathrm{Pr}_l^{3.9}}{2.37\times 10^{14}}\right)^{1/3} + \frac{A_D\,\mathrm{Pr}_l^{1.3}}{771.6}(\mathrm{Re}_{\mathrm{ter}} - \mathrm{Re}_x)^{1.4}\,\mathrm{Re}_x^{0.4}\right]^{1/2} \tag{11.55}$$

This equation applies to annular flow condensation in vertical tubes, and includes the effects of gravity, interfacial waves, and interfacial shear. As illustrated in Fig. 11.7, Eq. (11.55) agrees fairly well with the local condensation heat transfer data obtained by Ueda et al. [11.17].

If the gravity force terms are neglected, the above equation reduces to

$$\mathrm{Nu}_x = \frac{h\nu_l^{2/3}}{k_l g^{1/3}} = 0.036A_D^{0.5}\,\mathrm{Pr}_l^{0.65}(\mathrm{Re}_{\mathrm{ter}} - \mathrm{Re}_x)^{0.7}\,\mathrm{Re}_x^{0.2} \tag{11.56}$$

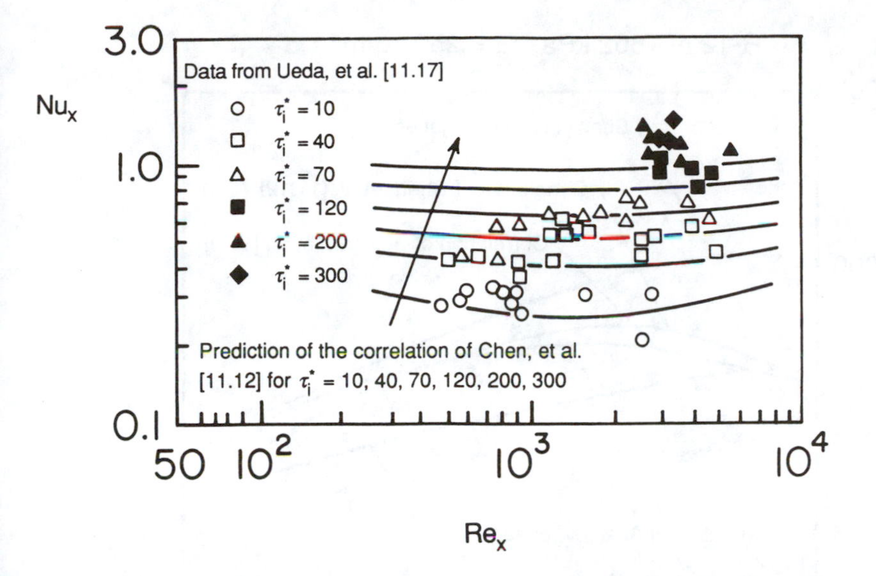

Figure 11.7 Comparison of the data of Ueda et al. [11.17] with the Nusselt number-versus-Reynolds number variation predicted by the correlation of Chen et al. [11.12] for internal flow condensation in a tube.

When shear forces are large compared to body forces, the tube orientation is unimportant. Equation (11.56) is therefore expected to apply to shear-dominated annular film condensation in either horizontal or vertical tubes. Although the gravitational acceleration g appears in the definitions of Nu_x and A_D, it cancels out of Eq. (11.56), leaving the heat transfer coefficient independent of g. Chen et al. [11.12] also used Eq. (11.55) to derive a relation for the average heat transfer coefficient for annular film condensation in vertical tubes, and they extended their analysis and developed a correlation for counterflow condensation.

For co-current flow, Chen et al. [11.12] noted that their correlation equation (11.55) may be inaccurate near the inlet of the tube and/or at low quality near the end of the condensation process. Near the inlet, high vapor shear may result in high liquid entrainment, resulting in a mist annular flow or breakdown of a continuous liquid film. Condensation heat transfer for such conditions would be significantly different from the annular flow postulated in their analysis. At the end of the condensation process, the flow pattern will change to slug flow for vertical tubes, or to stratified or intermittent (slug/plug) flow for horizontal tubes. For these nonannular regimes, the heat transfer mechanisms are expected to differ and the correlation for annular flow is not expected to apply.

A comparison of the heat transfer coefficient variations along the passage predicted by the above correlations for co-current annular flow is shown in Fig. 11.8. The variations have been computed for R-12 condensing at conditions typical of a vapor-compression refrigeration system.

For the slug flow often encountered at the end of the condensation process,

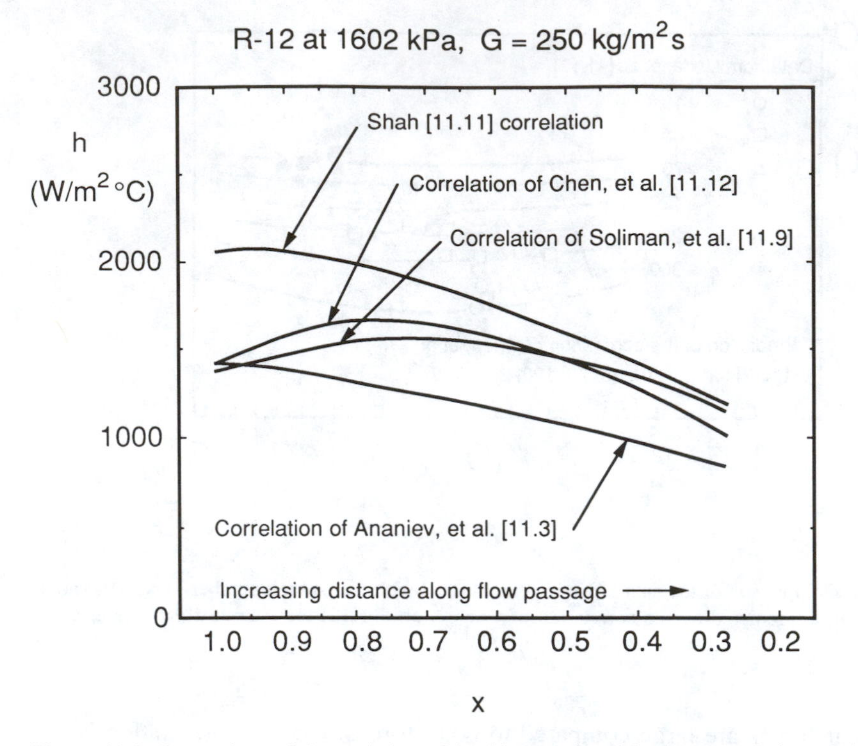

Figure 11.8 Comparison of the variation of h with x predicted by four correlation methods.

the intermittent nature of the flow makes analysis of the condensation heat transfer difficult at best. Based on point measurements of the heat transfer coefficient around the perimeter during slug, plug, and wavy (stratified) flow condensation in a horizontal tube, Rosson and Meyers [11.18] recommended that the upper and lower portions of the tube be treated separately. Over the upper portion of the tube, the condensation process is basically falling-film condensation with superimposed effects of vapor shear. They proposed to correlate the heat transfer coefficient over the upper portion of the tube with the relation

$$h_{\text{top}} = 0.31\left(\frac{GxD}{\mu_v}\right)^{0.12}\left[\frac{g\rho_l(\rho_l - \rho_v)k_l^3 h'_{lv}}{\mu_l(T_{\text{sat}} - T_w)D}\right]^{1/4} \tag{11.57}$$

where h'_{lv} is given by Eq. (11.31).

Based on arguments regarding the analogy between heat and mass transfer, Rosson and Meyers [11.18] suggested the following relation for the heat transfer coefficient over the bottom portion of the tube:

$$\frac{h_{\text{bot}}D}{k_l} = \frac{\phi_{l,vt}[8G(1 - x)D/\mu_l]^{1/2}}{5[1 + \text{Pr}_l^{-1}\ln(1 + 5\text{Pr}_l)]} \tag{11.58}$$

where $\phi_{l,vt}$ is the two-phase multiplier for viscous (laminar) liquid flow and turbulent vapor flow, as predicted by the Martinelli correlation (Eq. [10.58*a*]) with $C = 12$. The angular position θ_m at which the h value undergoes a transition from the higher value on the top to the lower value on the bottom must be determined to obtain an overall mean value for the condensing heat transfer coefficient at a given downstream location. Based on their data, they concluded that the angular position at which this transition occurs is dependent on the vapor and liquid Reynolds numbers Re_v and Re_l, respectively, and the Galileo number Ga, defined as

$$Re_v = \frac{GxD}{\mu_v} \tag{11.59}$$

$$Re_l = \frac{G(1-x)D}{\mu_l} \tag{11.60}$$

$$Ga = \frac{D^3 \rho_l(\rho_l - \rho_v)g}{\mu_l^2} \tag{11.61}$$

From their data, Rosson and Meyers [11.18] developed a correlation for θ_m as a function of the above parameters. This correlation was presented graphically as shown in Fig. 11.9. With the value of θ_m determined from this plot, the streamwise local condensation heat transfer coefficient, averaged over the tube perimeter, can be approximately computed as

$$h = \left(\frac{\theta_m}{\pi}\right) h_{top} + \left(\frac{\pi - \theta_m}{\pi}\right) h_{bot} \tag{11.62}$$

A similar but more detailed model of slug flow condensation has also been developed by Tien et al. [11.19].

Example 11.4 Use the correlation of Chen et al. [11.12] to predict the local heat transfer coefficient for condensation of R-12 at 1602 kPa in a horizontal tube with an inside diameter of 1.0 cm. The local flow conditions are such that $G = 500$ kg/m^2 s and $x = 0.6$. Compare the prediction with that obtained using the Shah [11.11] correlation.

For saturated R-12 at 1602 kPa, $T_{sat} = 335$ K $= 62$°C, $\rho_l = 1157$ kg/m^2 s, $\rho_v = 94.6$ kg/m^2 s, $\mu_l = 167 \times 10^{-6}$ N s/m^2, $\mu_v = 15.2 \times 10^{-6}$ N s/m^2, $k_l = 0.0530$ W/m K, $Pr_l = 3.40$, and $\sigma = 0.0042$ N/m. To determine the flow regime, we first compute Re_l and Re_v:

$$Re_l = \frac{G(1-x)D}{\mu_l} = \frac{500(0.4)(0.01)}{167 \times 10^{-6}} = 12{,}000$$

$$Re_v = \frac{GxD}{\mu_v} = \frac{500(0.6)(0.01)}{15.2 \times 10^{-6}} = 1.97 \times 10^5$$

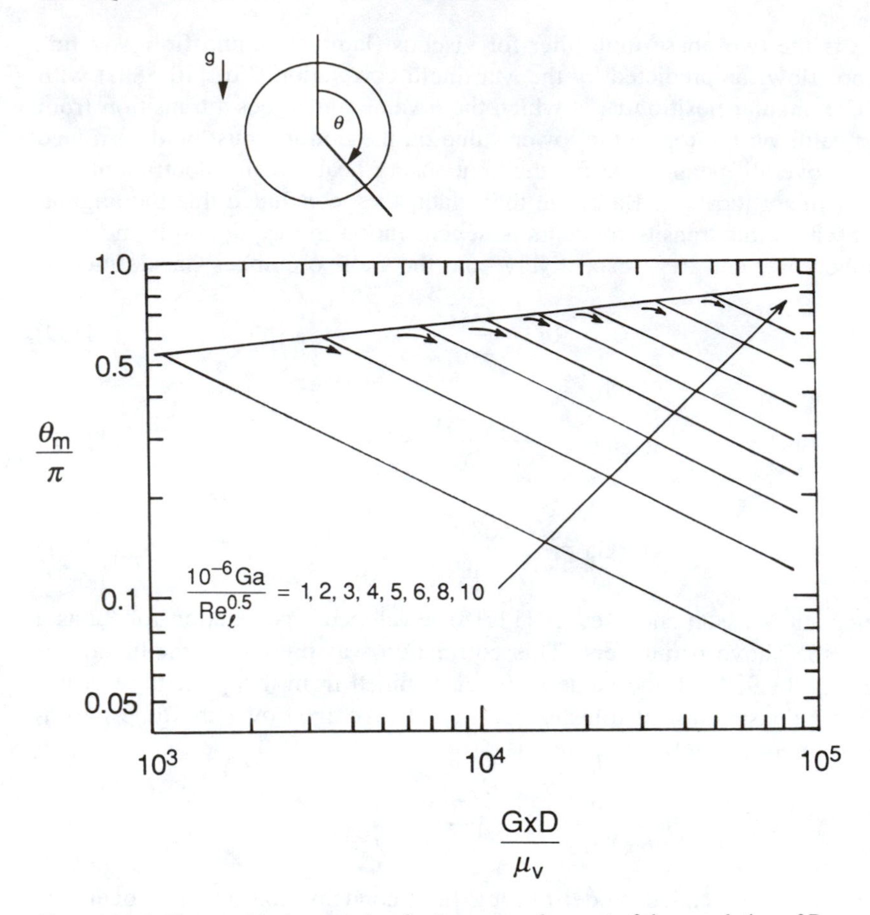

Figure 11.9 The graphical correlation for θ_m proposed as part of the correlation of Rosson and Meyers [11.18] for plug, slug, or wavy flow condensation heat transfer in a horizontal tube. (*Adapted from reference [11.18] with permission, copyright © 1965, American Institute of Chemical Engineers.*)

Thus the flow is turbulent-turbulent and

$$X_{tt} = \left(\frac{\rho_v}{\rho_l}\right)^{0.5}\left(\frac{\mu_l}{\mu_v}\right)^{0.125}\left(\frac{1-x}{x}\right)^{0.875}$$

$$= \left(\frac{94.6}{1157}\right)^{0.5}\left(\frac{167}{15.2}\right)^{0.125}\left(\frac{0.4}{0.6}\right)^{0.875} = 0.271$$

Evaluating F_{TD} using Eqs. (10.7*a*) and (10.16), we find

$$F_{TD} = \left[\frac{G^2x^2}{\rho_v(\rho_l - \rho_v)Dg}\right]^{0.5} = \left[\frac{(500)^2(0.6)^2}{94.6(1157 - 95)(0.1)9.8}\right]^{0.5} = 3.02$$

For these values of $X = X_{tt}$ and F_{TD}, Fig. 10.5 indicates that the flow is well into the annular flow regime. This implies that gravity effects on the annular film are small and may be neglected.

Evaluating A_D and $\mathrm{Re}_{\mathrm{ter}}$ using Eqs. (11.53) and (11.54), we obtain

$$A_D = \frac{0.252\mu_l^{1.177}\mu_v^{0.156}}{D^2 g^{2/3}\rho_l^{0.553}\rho_v^{0.78}}$$

$$= \frac{0.252(167 \times 10^{-6})^{1.177}(15.2 \times 10^{-6})^{0.156}}{(0.1)^2(9.8)^{2/3}(1157)^{0.553}(94.6)^{0.78}} = 2.03 \times 10^{-6}$$

$$\mathrm{Re}_{\mathrm{ter}} = \frac{GD}{\mu_l} = \frac{500(0.01)}{(167 \times 10^{-6})} = 30{,}000$$

Neglecting the effects of gravity, the heat transfer coefficient is given by Eq. (11.56) as

$$h = 0.036\left[\frac{k_l g^{1/3}}{(\mu_l/\rho_l)^{2/3}}\right] A_D^{0.5}\,\mathrm{Pr}_l^{0.65}(\mathrm{Re}_{\mathrm{ter}} - \mathrm{Re}_x)^{0.7}\,\mathrm{Re}_x^{0.2}$$

Using the fact that $\mathrm{Re}_x = \mathrm{Re}_l$ and substituting yields

$$h = 0.036\left[\frac{(0.0530)(9.8)^{1/3}}{(167 \times 10^{-6}/1157)^{2/3}}\right]$$

$$\cdot (2.03 \times 10^{-6})^{0.5}(3.40)^{0.65}(30{,}000 - 12{,}000)^{0.7}(12{,}000)^{0.2}$$

$$= 2920 \mathrm{\ W/m^2\ K}$$

Using the Shah correlation (noting that $P_{\mathrm{cr}} = 4132$ kPa for R-12), it follows from Eqs. (11.43) and (11.44) that

$$h_{lo} = 0.023\left(\frac{k_l}{D}\right)\left(\frac{GD}{\mu_l}\right)^{0.8} \mathrm{Pr}_l^{0.4}$$

$$= 0.023\left(\frac{0.0530}{0.01}\right)\left(\frac{500(0.01)}{167 \times 10^{-6}}\right)^{0.8}(3.40)^{0.4} = 758 \mathrm{\ W/m^2\ K}$$

$$h = h_{lo}\left[(1 - x)^{0.8} + \frac{3.8x^{0.76}(1 - x)^{0.04}}{(P/P_{\mathrm{cr}})^{0.38}}\right]$$

$$= 758\left[(0.4)^{0.8} + \frac{3.8(0.6)^{0.76}(0.4)^{0.04}}{(1602/4132)^{0.38}}\right] = 3060 \mathrm{\ W/m^2\ K}$$

This result is within 5% of the value obtained with the correlation of Chen et al. [11.12].

REFERENCES

11.1 Baker, O., Simultaneous flow of oil and gas, *Oil and Gas J.*, vol. 53, pp. 185–195, 1954.

11.2 Taitel, Y., and Dukler, A. E., A model for predicting flow regime transitions in horizontal and near horizontal gas–liquid flow, *AIChE J.*, vol. 22, pp. 47–55, 1976.

11.3 Annaiev, E. P., Boyko, L. D., and Kruzhilin, G. N., Heat transfer in the presence of steam condensation in a horizontal tube, *Proc. 1st Int. Heat Transfer Conf.*, part II, p. 290, 1961.

11.4 Boyko, L. D., and Kruzhilin, G. N., Heat transfer and hydraulic resistance during condensation of steam in a horizontal tube and in a bundle of tubes, *Int. J. Heat Mass Transfer*, vol. 10, pp. 361–373, 1967.

11.5 Miropolsky, Z. L., Heat transfer during condensation of high pressure steam inside a tube, *Teploenergetika*, vol. 3, pp. 79–83, 1962.

11.6 Chato, J., *ASHRAE J.*, p. 52, 1962.

11.7 Jaster, H., and Kosky, P. G., Condensation in a mixed flow regime, *Int. J. Heat Mass Transfer*, vol. 19, pp. 95–99, 1976.

11.8 Rufer, C. E., and Kezios, S. P., Analysis of stratified flow with condensation, *J. Heat Transfer*, vol. 88, pp. 265–275, 1966.

11.9 Soliman, M., Schuster, J. R., and Berenson, P. J., A general heat transfer correlation for annular flow condensation, *J. Heat Transfer*, vol. 90, pp. 267–276, 1968.

11.10 Traviss, D. P., Rohsenow, W. M., and Baron, A. B., Forced convection condensation in tubes: A heat transfer correlation for condenser design, *ASHRAE Trans.*, vol. 79, part I, pp. 157–165, 1973.

11.11 Shah, M. M., A general correlation for heat transfer during film condensation inside pipes, *Int. J. Heat Mass Transfer*, vol. 22, pp. 547–556, 1989.

11.12 Chen, S. L., Gerner, F. M., and Tien, C. L., General film condensation correlations, *Exp. Heat Transfer*, vol. 1, pp. 93–107, 1987.

11.13 Churchill, S. W., and Usagi, R., A general expression for the correlation of rates of transfer and other phenomena, *AIChE Journal*, Vol. 18, pp. 1121–1128, 1972.

11.14 Chun, K. R., and Seban, R. A., Heat transfer to evaporating liquid films, *J. Heat Transfer*, vol. 93, pp. 391–396, 1971.

11.15 Blangetti, F., and Schlunder, E. O., Local heat transfer coefficients of condensation in a vertical tube, *Proc. 6th Int. Heat Transfer Conf.*, vol. 2, pp. 437–442, 1978.

11.16 Dukler, A. E., Fluid mechanics and heat transfer in vertical falling-film systems, *Chem. Eng. Prog. Symp. Ser.*, vol. 56, no. 30, pp. 1–10, 1960.

11.17 Ueda, T., Kubo, T., and Inoue, M., Heat transfer for steam condensing inside a vertical tube, *Proc. 5th Int. Heat Transfer Conf.*, vol. 3, pp. 304–308, 1976.

11.18 Rosson, H. F., and Meyers, J. A. Point values of condensing film coefficients inside a horizontal tube, *Chem. Eng. Prog. Symp. Ser.*, vol. 61, no. 59, pp. 190–199, 1965.

11.19 Tien, C. L., Chen, S. L., and Peterson, P. F., Condensation inside tubes, Electric Power Research Institute Report no. EPRI NP-5700, January 1988.

PROBLEMS

11.1 Oxygen condenses at a nominal pressure of 196 kPa inside a horizontal tube. The tube inside diameter is 8 mm and the mass flux is 250 kg/m^2 s. Determine the condensation regime at qualities of 0.9, 0.5, and 0.05.

11.2 Mercury vapor condenses inside a horizontal tube with an inside diameter of 1.7 cm. The pressure is essentially constant along the tube at 145 kPa. For a mass flux of 2000 kg/m^2 s, determine the condensation regime at locations where the quality is 0.95, 0.4, and 0.01.

11.3 R-12 condenses in the horizontal tubes of an air-conditioning condenser at a pressure of 793 kPa. The tubes of the condenser are 0.9 cm in inside diameter and the mass flux through each tube is 500 kg/m^2 s. Determine the condensation regime at locations in the condenser where the quality is equal to 0.95, 0.4, and 0.02.

11.4 R-22 condenses at 2020 kPa in a vertical tube having an inside diameter of 5 mm. At a quality of 0.8, use the correlations of Traviss et al. [11.10] and Chen et al. [11.12] to predict the heat transfer coefficient for downflow mass flux values of 100, 200, 500, and 1000 kg/m^2 s. Plot both sets of results on log–log paper. How does the power-law dependence of h on G for these correlations compare?

11.5 R-12 condenses at 1602 kPa in a vertical tube having an inside diameter of 8 mm. For downward flow at a quality of 0.95 and a mass flux of 400 kg/m^2 s, use the correlations of Traviss et al. [11.10] and Chen et al. [11.12] to predict the heat transfer coefficient. Also use one of the correlations from Table 10.2 to predict the void fraction. From the heat transfer results, estimate the liquid film thickness as $\delta = k_l/h$. Compute the film thickness from the void fraction as $\delta = D(1 - \alpha)/4$ and compare the result to the values of δ computed from the heat transfer coefficients. How would you expect entrainment to affect the validity of these estimates for δ?

11.6 Ammonia condenses in the horizontal tubes of a refrigeration system condenser. The tubes have an inside diameter of 1.3 cm, and the mass flux through each tube is 600 kg/m^2 s. At a particular location where the pressure is 1425 kPa and the quality is 0.8, determine the void fraction and condensation regime. Also determine the condensation heat transfer coefficient using (*a*) the heat transfer relation obtained from the analysis in Example 11.2, (*b*) the correlation of Boyko et al. [11.4], and (*c*) the correlation of Traviss et al. [11.10].

11.7 Nitrogen condenses in downward co-current flow in a vertical tube. The tube has an inside diameter of 8 mm and the mass flux is 200 kg/m^2 s. At a location where the heat flux is 500 kW/m^2, the quality is 0.6, and the pressure is 540 kPa, determine the condensing heat transfer coefficient using the correlation of Soliman et al. [11.9]. Compare your result with the prediction of the correlation of Boyko et al. [11.4].

11.8 In a miniature condenser unit, it is proposed to condense ammonia at a mass flux of 100 kg/m^2 s in a round tube with an inside diameter of 0.2 mm. The tube wall temperature is 50°C. For a pressure of 2422 kPa and a quality of 0.8, estimate the heat transfer coefficient using Shah's [11.11] correlation. Estimate how much the saturation temperature shifts because of the curvature of the interface of the liquid film on the wall of the tube (Use results given in Chapter 5). If the h value is based on the normal (flat interface) saturation temperature $h = q''/(T_{sat} - T_w)$, how much (by how many percent) does this effectively alter the h value?

11.9 R-12 condenses inside a horizontal tube with an inside diameter of 1.2 cm. The mass flux is 280 kg/m^2 s, and the pressure is virtually constant at 1602 kPa along the tube. Determine the approximate value of quality at which the flow is expected to be in the slug, plug, or bubbly regime. For a quality of 0.02 and a wall subcooling $T_{sat} - T_w$ of 10°C, use the correlation of Rosson and Meyers [11.18] to estimate the heat transfer coefficient. How does this value compare with that predicted by the Dittus-Boelter equation for the entire flow as liquid?

11.10 Steam condenses at 247 kPa as it flows inside a horizontal tube with an inside diameter of 2.0 cm. For $G = 200$ kg/m^2 s and $x = 0.9$, 0.7, and 0.5, compute the local heat transfer coefficient using the correlation of Chen et al. [11.12]. Compare your results for $x = 0.7$ to the values obtained with other correlations in Example 11.3.

11.11 Ammonia condenses inside a horizontal tube with an inside diameter of 9 mm. The mass flux is 400 kg/m^2 s. For a location near the end of the condensation process, the quality is 0.01, the wall temperature is 65°C, and the pressure is 3870 kPa. (*a*) what are the void fraction and flow regime for these circumstances? (*b*) Use the correlation of Rosson and Meyers [11.18] to estimate the heat transfer coefficient at this location.

11.12 Nitrogen condenses during downflow in a round tube with an inside diameter of 1.5 cm. The pressure is nominally constant along the tube at 229 kPa, and the mass flux is 100 kg/m^2 s.

Use the correlation of Chen et al. [11.12] to determine the heat transfer coefficient at $x = 0.9$, 0.65, and 0.4. Are gravity effects important for these conditions? Explain briefly.

11.13 Steam condenses in a horizontal tube with an inside diameter of 25 mm. The mass flow rate through the tube is 5.9 g/s, the wall temperature of the tube is held constant at 112°C, and the pressure is essentially constant along the tube and equal to 247 kPa. Consider a location near the inlet where the quality is 0.9 and one near the outlet where the quality is 0.1. Determine (*a*) the void fraction for each location, (*b*) the two-phase flow regime at each location, and (*c*) the local heat transfer coefficient for each location. Repeat parts (*a*) through (*c*) for a flow rate of 25 g/s.

11.14 A downflow heat exchanger for condensing R-22 has a tube configuration like that shown in Fig. 11.2. Each tube in the condenser is 1.2 m long and has an inside diameter of 9 mm. The flow rate through the unit is such that the mass flux through each tube is 250 kg/m^2 s. The R-22 enters as a saturated vapor at 2800 kPa. If the walls of the tube are essentially at a constant and uniform temperature of 40°C, estimate the exit quality of the flow by dividing the tube into four segments and computing the heat transfer for each section assuming that the heat transfer coefficient is constant over the segment. Begin at the inlet segment and work progressively downstream, using the exit condition from the previous segment as the inlet condition to the next. (You may compute the heat transfer coefficient for each segment based on its inlet conditions as a first approximation.)

11.15 A cross-flow condenser like that shown in Fig. 11.1 is to be used for condensing ammonia at 1425 kPa. The tubes in the exchanger have an inside diameter of 8 mm, and the temperature of the tube wall is held essentially constant throughout the unit at 20°C. The ammonia enters as saturated vapor, and the mass flow rate is such that the mass flux through each tube is 300 kg/m^2 s. Estimate the tube length required in the exchanger to completely condense the vapor for these conditions. Do so by considering segments of the tube 20 cm long and assuming that the heat transfer coefficient is constant over each segment. Begin at the inlet and work progressively downstream until the exit quality from a segment equals zero or would be negative based on an energy balance.

CHAPTER
TWELVE

CONVECTIVE BOILING IN TUBES AND CHANNELS

12.1 REGIMES OF CONVECTIVE BOILING

Flow boiling in tubes and channels is perhaps the most complex convective phase-change process encountered in applications. In most evaporator and boiler applications, the flow is either vertically upward or horizontal. To make this discussion more concrete, we will therefore focus on these two specific flow circumstances.

Figure 12.1 depicts schematically a typical low-heat-flux vaporization process in a horizontal round tube. The widest possible range of flow conditions is encountered if the liquid enters as subcooled liquid and leaves as superheated vapor. As the vaporization process proceeds, the vapor content of the flow increases with distance along the tube. To maintain the specified mass flow rate as the mean density of the flow decreases, the mean flow velocity must increase substantially. As noted in Chapter Ten, the two-phase flow regime generally is strongly dependent on the relative velocities of the two phases. The acceleration of the flow results in an increasing difference between the mean liquid and vapor velocities, which produces a sequence of changes in the flow regime.

When boiling is first initiated, bubbly flow exists in the tube. Increasing quality typically produces transitions from bubbly to plug flow, plug to annular flow, and annular to mist flow, as indicated in Fig. 12.1. Regimes of slug flow, stratified flow, or wavy flow may also be observed at intermediate qualities, depending on the flow conditions.

Flow boiling generally is further complicated by the fact that, in addition to flow-regime changes, different vaporization mechanisms may be encountered at different locations along the tube. As indicated in Fig. 12.1, nucleate boiling is

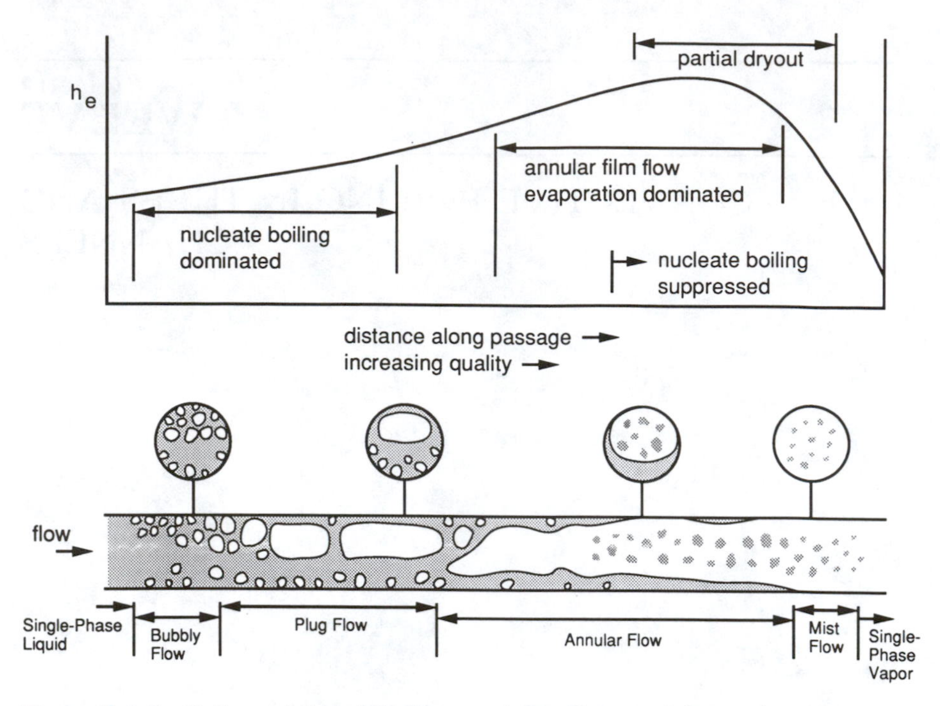

Figure 12.1 Qualitative variation of the heat transfer coefficient and flow regime with quality for internal convective boiling in a horizontal tube.

usually the dominant vaporization mechanism near the onset of boiling. As more vapor is generated in the flow and the void fraction increases, the flow undergoes a transition to an annular or nearly annular configuration, whereupon evaporation from the liquid–vapor interface becomes increasingly important. At low to moderate qualities, both mechanisms may be important. As the liquid film on the wall thins, however, film evaporation may become so effective that it becomes the dominant mechanism. In some cases nucleate boiling may be completely suppressed, leaving film evaporation as the only active vaporization mechanism.

Acceleration of the vapor core during the latter annular flow stages of the vaporization process very often produces entrainment of liquid droplets. This effect, together with direct vaporization of the film, tends to reduce the film thickness as it flows downstream. Eventually, the film may disappear completely from portions of the tube wall. This is usually referred to as *dryout*, or in this case, *partial dryout* of the tube wall. For horizontal tubes, because of the tendency of gravity to thin the liquid film on the top of the tube, partial dryout of the tube usually is first observed along the top portion of the tube wall, as indicated in Fig. 12.1. The lower portion of the tube wall may remain wetted with the liquid film for a substantial distance beyond the first partial dryout of the top portion. The portion of the tube wall covered by the liquid film generally decreases with downstream distance until the wall is completely dry around the entire perimeter of the tube.

Prior to the onset of partial dryout, transfer of heat across the liquid film becomes more efficient as the film becomes progressively thinner. As a result, the heat transfer coefficient associated with the combined nucleate boiling and film evaporation mechanisms often increases with downstream distance prior to the onset of dryout, as indicated in Fig. 12.1.

When the tube wall is partially dry, heat transfer from dry portions of the wall surface is negligible compared to that at locations wetted by liquid where film evaporation is occurring. Because these dry locations are relatively inactive, the local heat transfer coefficient, averaged around the perimeter of the tube, is lower than if the entire perimeter were covered by the liquid film. Furthermore, as the wetted fraction of the wall decreases with downstream distance, more of the wall becomes inactive and the local heat transfer coefficient, averaged over the perimeter, also progressively decreases, as indicated in Fig. 12.1.

At some point along the tube, the film may completely disappear from the wall of the tube. Liquid may still be present in the flow, however, as entrained droplets. Transport of heat from the tube wall to the droplets is necessary if the vaporization process is to continue. In general, the continued vaporization of the droplets may be accomplished by a combination of mechanisms, including convection through the gas, radiation, and collisions or near collisions of droplets with the wall. These mechanisms are not very effective, and the associated heat transfer coefficient is usually significantly lower than the values associated with nucleate boiling and/or liquid film evaporation. In the mist evaporation process, the heat transfer coefficient usually continues to decrease as the quality increases, until ultimately the single-phase vapor value is attained.

For convective boiling in vertical tubes at moderate heat flux levels, the typical sequence of flow regimes observed is shown in Fig. 12.2. Near the onset of boiling, bubbly flow exists. As the quality increases, progressive transitions from bubbly to slug, slug to churn, churn to annular, and annular to mist flow are generally observed. The progression from one regime to the next can be traced on the flow regime map like that of Hewitt and Roberts [12.1] for vertical upward co-current flow. As liquid is converted to vapor, $j_v = Gx/\rho_v$ increases and $j_l = G(1 - x)/\rho_l$ decreases. On the Hewitt and Roberts map, the sequence of flow regimes would trace a curve similar to that shown in Fig. 12.3. This specific curve represents the locus of state points for vaporization of R-12 at a pressure of 339 kPa and a mass flux of 300 kg/m^2 s.

Since the total mass flux is the same everywhere along the tube, the path of the system on this map is defined by the relation

$$\rho_v j_v + \rho_l j_l = G \tag{12.1}$$

which can be rearranged to the form

$$(\rho_v)^{1/2}(\rho_v j_v^2)^{1/2} + (\rho_l)^{1/2}(\rho_l j_l^2)^{1/2} = G \tag{12.2}$$

From the latter form of this relationship, it can be easily shown that for a specified value of G, the curve representing the system state points on the flow regime map will have a shape like that of the curve shown on the log–log plot in Fig. 12.3.

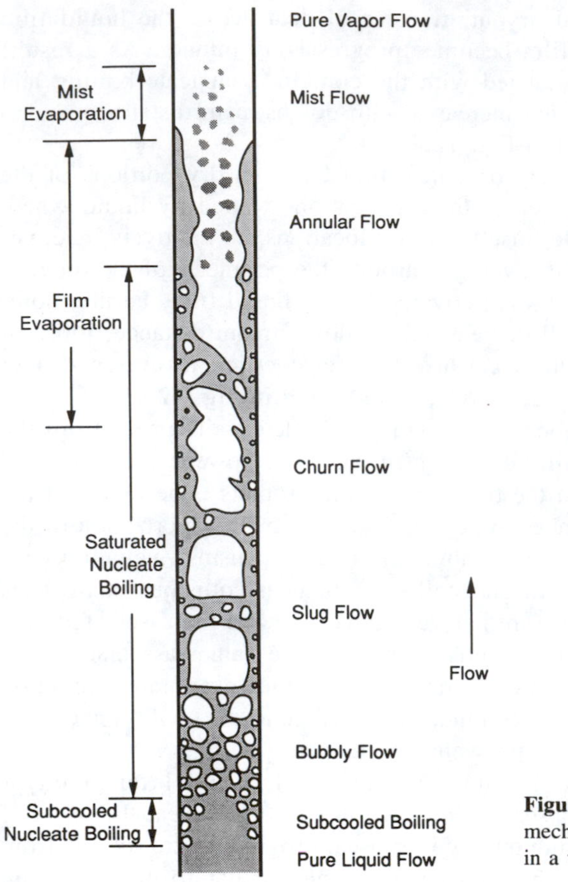

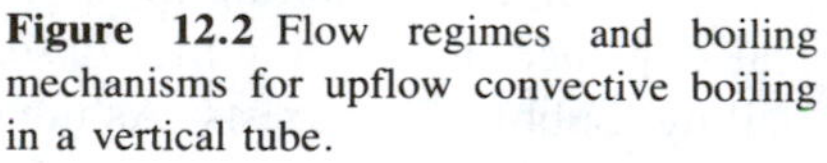
Figure 12.2 Flow regimes and boiling mechanisms for upflow convective boiling in a vertical tube.

While different values of G or fluid densities shift the curve somewhat, the overall shape will remain similar to the curve in Fig. 12.3.

Upward flow boiling in a vertical tube generally exhibits a sequence of vaporization mechanisms similar to those in the horizontal flow case described above. As indicated schematically in Fig. 12.2, nucleate boiling is a major factor in the early stages of the vaporization process, where bubbly and slug flow occurs. Once churn or annular flow is attained, film evaporation usually becomes important. If the vaporization process continues, dryout of the liquid film will eventually occur, leaving vaporization of entrained droplets of liquid as the final stage of the boiling process. As in the horizontal case, convection, radiation, and droplet collisions with the wall may all play a role in the mist evaporation that occurs during the final stage of vaporization in a vertical tube.

The sequences of flow regimes and boiling mechanisms described above correspond to low wall heat flux conditions and/or low wall superheat levels. The effect of varying heat flux or wall superheat on the flow and boiling mechanisms can be better understood by considering the boiling regime maps shown in Fig. 12.4 and Fig. 12.5. Figure 12.4 is a modified version of a boiling regime map presented by Collier [12.2]. This type of map applies specifically to flow boiling with a constant applied heat flux condition at the tube wall. The horizontal coordinate is the local bulk enthalpy of the boiling fluid. The vertical coordinate is the applied heat flux. For a specified constant heat flux, the system state traverses a horizontal line on this map.

The line labeled q''_1 in Fig. 12.4 is typical of the path followed by a system at a low applied heat flux. The sequence of boiling regimes is exactly that described above for the horizontal and vertical flows shown in Figs. 12.1 and 12.2. If, however, a higher heat flux is applied, the sequence of boiling regimes may be quite different. At the higher heat flux of q''_2 indicated in Fig. 12.4, nucleate boiling is initiated while the bulk fluid is still subcooled. Subcooled nucleate boiling gives way to saturated nucleate boiling when the bulk enthalpy reaches the value for saturated liquid. As indicated in this diagram, at this higher value of heat flux, as the quality of the flow increases, a departure from nucleate boiling (DNB) eventually occurs, resulting in a transition to saturated film boiling. For a constant-heat-flux wall condition, this transition is usually accompanied by a substantial rise in wall temperature.

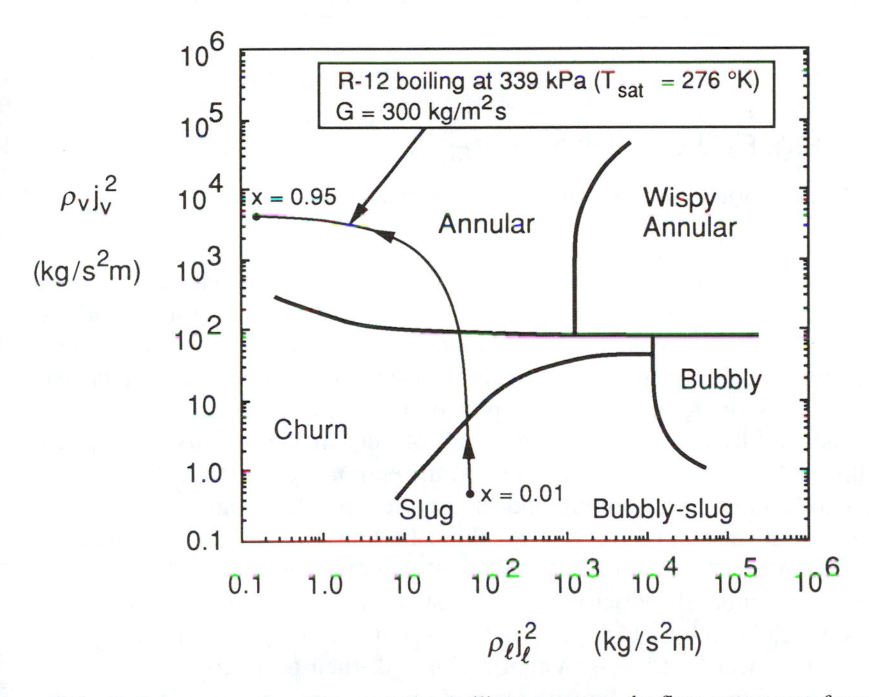

Figure 12.3 Representation of a convective boiling process on the flow pattern map for vertical cocurrent flow.

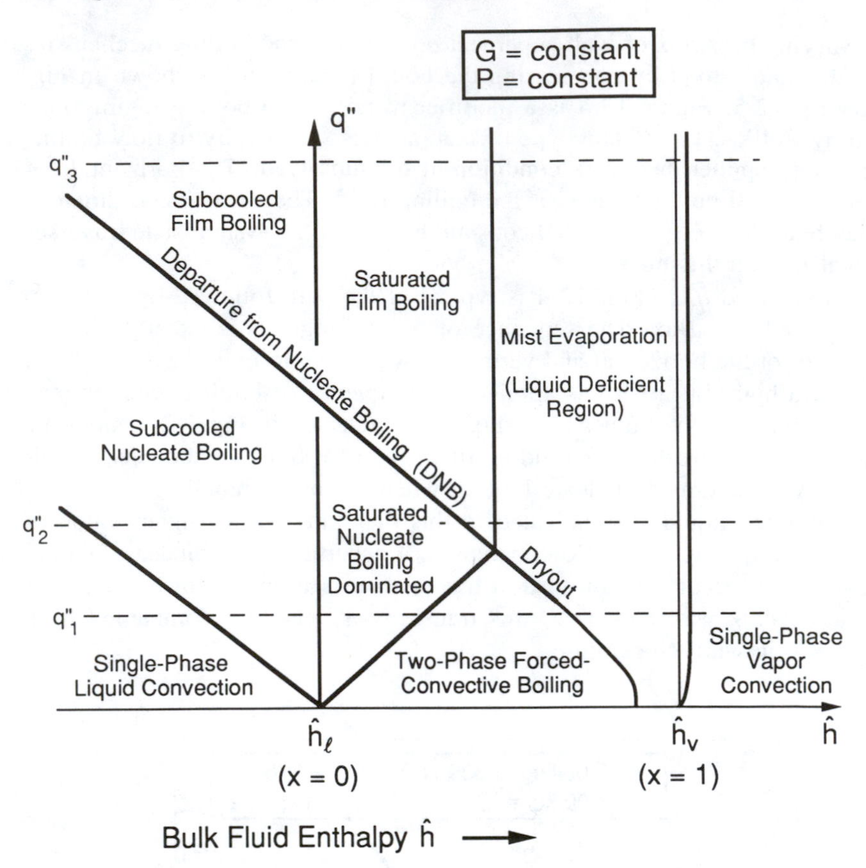

Figure 12.4 Boiling regimes for a constant wall heat flux condition.

If the vaporization process continues in the film boiling mode, the quantity of liquid in the core eventually becomes so small that it breaks up into droplets, thus resulting in a transition to the mist evaporation mode (liquid deficient region). As in the low-heat-flux case, when droplet vaporization is complete, further heat transfer occurs as single-phase vapor convection.

As illustrated by the q''_3 line in Fig. 12.4, at very high heat flux levels, subcooled film boiling may exist immediately at the entrance of the tube. Subsequent transitions to saturated film boiling and the mist evaporation region occur as the flow proceeds downstream. Because all three boiling regimes are characterized by inefficient heat transfer mechanisms, wall superheat is expected to be very high all along the tube. Becauses the heat transfer coefficient associated with these regimes is typically so low, heat transfer equipment is usually designed to operate at relatively low heat flux levels. Vaporization will then occur mostly in the nucleate boiling and two-phase, forced-convection regimes, in which the heat transfer coefficient is usually high.

An alternative view of boiling regime transitions can be seen in Fig. 12.5.

In this figure, the horizontal coordinate again is the bulk enthalpy of the fluid and the vertical coordinate is the imposed wall superheat. For a fixed wall superheat along the tube, the flow will exhibit the sequence of flow regimes that lie along the horizontal constant superheat line.

The bottom of this diagram reflects the fact that if the wall superheat is less than that required for the onset of boiling, no vaporization will take place, and only superheated liquid will leave the tube. At low wall superheats above the onset threshold, regimes of nucleate boiling, two-phase forced-convective boiling, and mist evaporation will be encountered, in that order, as the flow proceeds downstream.

Because the wall superheat is controlled, at high wall superheats, transition

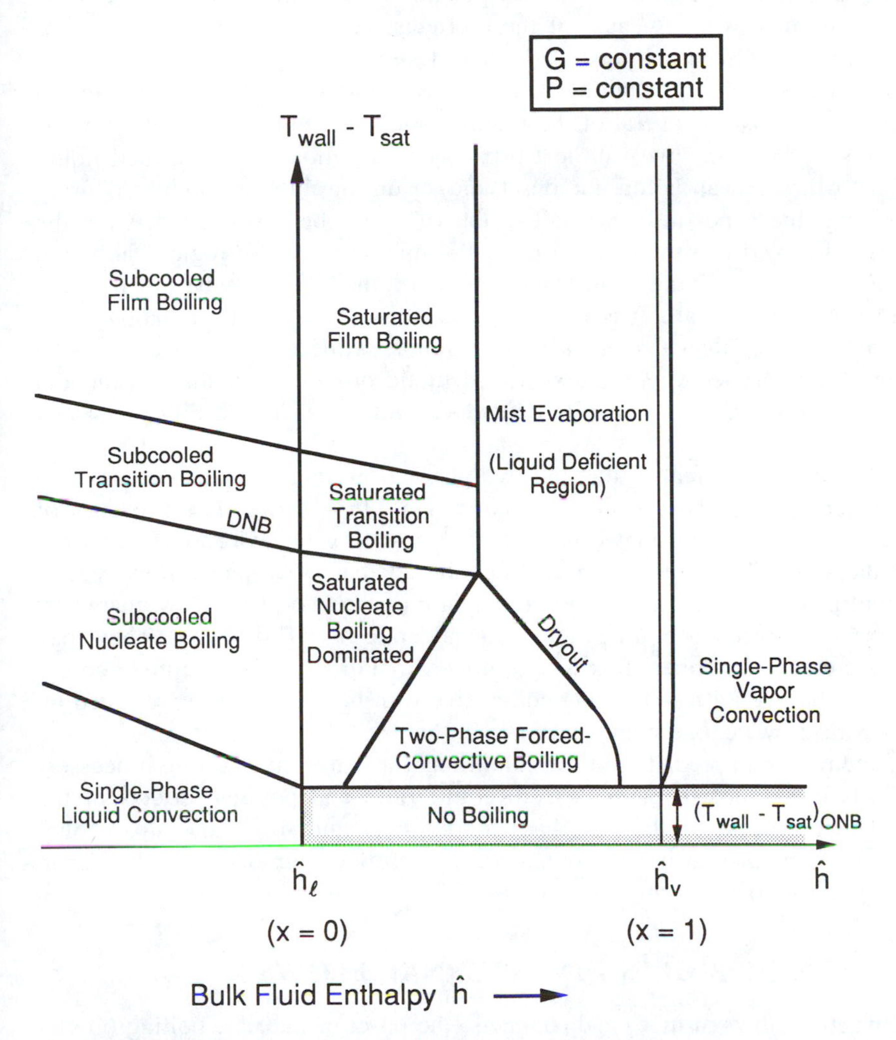

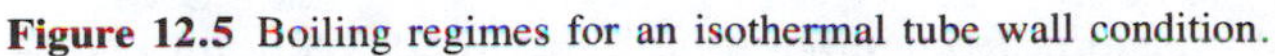

Figure 12.5 Boiling regimes for an isothermal tube wall condition.

boiling may occur beyond the departure from nucleate boiling. At even higher wall superheat levels, subcooled and saturated film boiling will occur in the early stages of the vaporization process, followed again by mist evaporation.

Heat transfer equipment is generally designed to operate at low wall superheat levels so as to avoid transition and film boiling. The high wall temperatures required to sustain film boiling can result in thermal stresses that may lead to failure of the equipment. In most real evaporators, an isothermal wall condition is not exactly maintained. Because the wall has finite conductivity and thermal capacity, its temperature can be perturbed by changes in the boiling-side heat transfer coefficient. Intermittent contact of the liquid with the tube wall during transition boiling may produce large, rapid temperature fluctuations in such cases. These fluctuations may, in turn, produce severe thermal stress fluctuations, which may cause failure of the tube wall. Because of these effects, conditions that result in transition boiling are also usually avoided when designing evaporators or boilers.

One final aspect of flow boiling processes worth noting is that nonequilibrium conditions can exist in several of the regimes shown in Figs. 12.4 and 12.5. At low wall superheats or low wall heat-flux levels, a region of superheated liquid may exist near the wall before the onset and/or during nucleate boiling. During film boiling, the vapor near the wall in the film may be superheated while the liquid is subcooled or saturated. Also, in the mist evaporation region, although the liquid droplets are at the saturation temperature, the vapor may be superheated, particularly near the wall. It is possible, therefore, for the bulk enthalpy to be equal to or greater than that for saturated vapor, while the system exists as a mixture of superheated vapor and saturated liquid droplets. For this reason, the mist evaporation region is shown to extend beyond the $\hat{h}_v$ line in Figs. 12.4 and 12.5.

It is important to realize that the two-phase flow regime maps described in Chapter Ten apply to equilibrium, adiabatic, two-phase flows. The existence of nonequilibrium conditions may cause the two-phase flow behavior to deviate from the predictions of these maps. In general, the adiabatic two-phase flow regime maps will provide a reasonably good prediction of the two-phase flow characteristics during convective boiling at low to moderate heat flux levels. They may not be accurate at high heat flux levels, however. For example, the inverted annular flow that would result during convective film boiling is not represented on the adiabatic flow regime map (see Fig. 12.3).

In attempting to predict the flow conditions for convective boiling processes, the effects of possible departures from equilibrium and possible effects of the phase-change process on the two-phase morphology must be taken into account if an accurate prediction is to be obtained. These effects are discussed further in the following sections of this chapter.

12.2 ONSET OF BOILING IN INTERNAL FLOWS

For convective flows in tubes and channels, the onset of nucleate boiling (ONB) marks the beginning of the transition from single-phase liquid convection to combined convection and nucleate boiling. In a convective flow of this type, the onset is usually defined as occurring at the location where active nucleation sites are first observed. The conditions at which potential nucleation sites may become active, and methods for predicting such conditions, are discussed at length in Chapter Six. The reader who is unfamiliar with these concepts is urged to review the relevant material in that chapter before reading further in this section. This section will focus on extension of those fundamental concepts to internal convective flows.

As discussed in Section 6.3, Hsu's [12.3] model analysis predicts that the range of active nucleation sites on a heated surface depends directly on the thermal boundary-layer thickness, subcooling, wall superheat, and the thermophysical properties of the liquid. It further predicts that a threshold value of wall superheat must be attained before any sites will become active. While this model analysis is approximate in many respects, qualitatively it agrees with trends observed in many boiling systems. The trends predicted by this analysis are particularly relevant to the internal flows considered here.

Hsu's analysis implies that for a given wall temperature, boundary-layer thickness, subcooling, and fluid properties, the range of active cavity sizes is determined. Because the temperature profile is assumed to be linear across the thermal boundary layer in that model, for a given wall temperature and set of fluid properties, specifying the bulk temperature and the boundary-layer thickness is equivalent to specifying the heat flux. This suggests that the onset conditions for a coolant with given properties corresponds to a specific combination of wall temperature (or, equivalently, wall superheat) and heat flux. This is, in fact, the way in which several commonly used onset correlations are cast.

For a constant-heat-flux boundary condition, if the single-phase heat transfer coefficient h_{le} associated with the pure liquid flow is constant along the tube, then it follows from the definition of h_{le} and conservation of energy that

$$q'' = h_{le}[T_w(z) - T_l(z)] \tag{12.3}$$

$$q'' = \left(\frac{d_h}{4z}\right) G c_{pl} [T_l(z) - T_{l,\text{in}}] \tag{12.4}$$

where d_h is the hydraulic diameter, $T_l(z)$ is the local bulk liquid temperature, $T_w(z)$ is the local wall temperature, and $T_{l,\text{in}}$ is the bulk temperature of the liquid at the inlet of the channel. Combining these relations to eliminate $T_l(z)$, it can be shown that

$$T_w(z) - T_{\text{sat}} = \frac{q''}{h_{le}} \left[1 + 4\left(\frac{h_{le}}{G c_{pl}}\right)\left(\frac{z}{d_h}\right) \right] - (T_{\text{sat}} - T_{l,\text{in}}) \tag{12.5}$$

Thus for a constant-heat-flux wall boundary condition, the wall superheat increases linearly with distance along the tube z, and at each location is proportional to the heat flux q''.

Equation (12.3) applies also to the case of an isothermal wall boundary condition, except that T_w is constant. For this circumstance, the heat flux will now vary along the tube because as the bulk temperature rises, the driving temperature difference will vary. The energy balance for this case is given by

$$q'' = \left(\frac{Gc_{pl}d_h}{4}\right)\frac{dT_l}{dz} \tag{12.6}$$

Combining Eqs. (12.3) and (12.6) yields the differential equation

$$\frac{dT_l}{dz} = \left(\frac{4h_{le}}{Gc_{pl}d_h}\right)(T_w - T_l) \tag{12.7}$$

which can be solved using the boundary condition $T_l = T_{l,\mathrm{in}}$ at $z = 0$ to obtain

$$T_w - T_l(z) = (T_w - T_{l,\mathrm{in}}) \exp\left(-\frac{4h_{le}z}{Gc_{pl}d_h}\right) \tag{12.8}$$

It follows directly from Eqs. (12.3) and (12.8) that, for the isothermal wall condition,

$$q''(z) = h_{le}[(T_w - T_{\mathrm{sat}}) + (T_{\mathrm{sat}} - T_{l,\mathrm{in}})] \exp\left(-\frac{4h_{le}z}{Gc_{pl}d_h}\right) \tag{12.9}$$

To further explore the necessary conditions for the onset of boiling, we now will consider a fixed z location along the tube. It can be seen from Eq. (12.5) that for a uniform applied heat flux, at a fixed value of z the wall superheat increases linearly with applied heat flux. Similarly, Eq. (12.9) indicates that for a constant wall temperature condition, at a given location, the resulting heat flux increases linearly with the specified wall superheat.

For either the isothermal or uniform flux wall condition, as the wall superheat or heat flux is increased, the onset of boiling will eventually occur, initiating a transition from single-phase liquid convection to fully developed nucleate boiling. The nature of this transition can be better understood by considering Fig. 12.6, which shows possible variations of the heat flux with wall superheat near the onset condition. For the isothermal wall condition, increasing the wall temperature causes the operating point at the specified z location to move up the single-phase operating curve shown in this diagram. If this process is continued, eventually the point representing the system operating condition approaches the intersection of the single-phase curve and the fully developed nucleate boiling curve.

As noted in Chapter Eight, the nucleate boiling curve is usually only weakly dependent on flow velocity and subcooling. Consequently, the fully developed nuleate boiling curve in the presence of forced convection and subcooling is generally taken to be identical to the ordinary saturated nucleate pool boiling curve at the same pressure. Correlating equations that relate the heat flux to the wall superheat for saturated nucleate boiling are generally of the form

$$q'' = \gamma[T_w - T_{\mathrm{sat}}(P)]^m \tag{12.10}$$

where γ is a factor that depends on surface and fluid properties, and the exponent m is typically between 2 and 4.

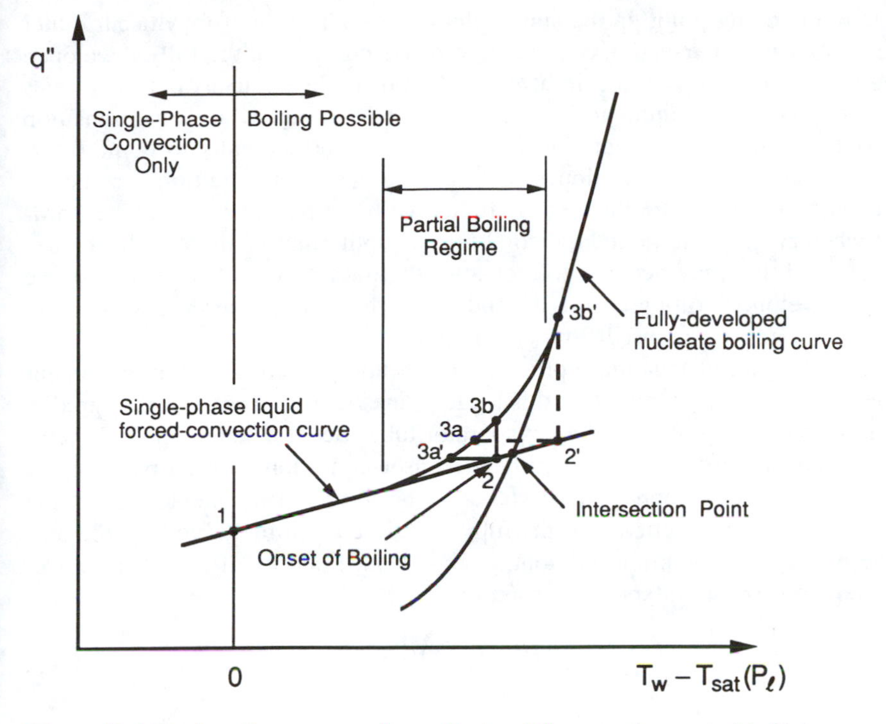

Figure 12.6 The heat flux-versus-wall superheat variation near the onset of boiling.

A necessary (but not sufficient) condition for the onset of boiling is that the wall temperature must be above T_{sat} at the local system pressure (i.e., the wall superheat must be greater than zero). If the wall superheat is gradually increased at a given downstream location, the onset of boiling generally occurs in the vicinity of the intersection of the single-phase convection and nucleate boiling curves. The onset may occur before the intersection is reached at, for example, point 2 in Fig. 12.6, or it may occur at a point 2′ beyond the intersection. If the former is true, and the wall is held isothermal, the system operating point jumps vertically to a curve that marks a smooth transition between the single-phase curve and the fully developed boiling curve. If the latter is true, the system operating point will similarly jump vertically from point 2′ to point $3b'$ on the fully developed boiling curve.

The smooth transition sometimes observed between the single-phase and fully developed boiling curves is usually referred to as the *partial boiling regime*. As the superheat is increased in this regime, the active nucleation site density increases and the nucleate boiling contribution to the total heat transfer increases until it essentially equals that for nucleate pool boiling at the same wall superheat and pressure conditions.

Similar arguments apply for a uniform heat flux wall condition. At a given location, increasing the heat flux would cause the wall superheat to rise, moving the system operating point up the single-phase curve. The onset may occur either before or after the intersection of the single-phase curve with the fully developed boiling curve. If the onset occurs at point 2, and the heat flux is held constant, the operating point may jump horizontally to point $3a$ in Fig. 12.6. If the transition is delayed to point $2'$, the operating point may jump horizontally to point $3a'$.

The transition to fully developed boiling once the onset condition is achieved will be discussed further in the next section. The main point here is that, regardless of whether the wall boundary condition is isothermal or uniform heat flux, the onset condition may occur before or after the intersection of the single-phase and fully developed boiling curves depending on the fluid properties, wall cavity size distribution, and imposed flow conditions.

Because the onset location represents the location where a new heat transfer mechanisms "turns on," predicting this location is essential to any effort to predict the heat transfer performance of an evaporator tube. Several semiempirical methods have been proposed for predicting the onset condition. As a first approximation, the onset condition could be taken to be the point of intersection of the single-phase convection curve and the fully developed boiling curve. For the uniform heat flux wall condition, combining Eqs. (12.5) and (12.10) yields the following relation for the intersection condition:

$$\left(\frac{q''}{\gamma}\right)^{1/m} - \frac{q''}{h_{le}}\left[1 + 4\left(\frac{h_{le}}{Gc_{pl}}\right)\left(\frac{z}{d_h}\right)\right] + [T_{\text{sat}} - T_{l,\text{in}}] = 0 \qquad (12.11)$$

Alternatively, for the isothermal wall condition, combining Eqs. (12.8) and (12.10), the following relation is obtained:

$$\gamma(T_w - T_{\text{sat}})^m - h_{le}[(T_w - T_{\text{sat}}) + (T_{\text{sat}} - T_{l,\text{in}})] \exp\left(-\frac{4h_{le}z}{Gc_{pl}d_h}\right) = 0 \qquad (12.12)$$

Solving Eq. (12.11) for q'' or Eq. (12.12) for $T_w - T_{\text{sat}}$ yields the values of these parameters for which the intercept condition is just met at the specified z location. Alternatively, if q'' or $T_w - T_{\text{sat}}$ is specified, the appropriate equation can be solved for the value of z at which the intercept condition is met. If the intercept is interpreted as the onset condition, these equations thus predict the onset conditions for these boundary conditions.

Collier [12.2] reports that this approach was, in fact, proposed in an early study by Bowring [12.4]. As noted in the discussion above, the actual onset condition is a function of a number of system parameters, and this idealization is, at best, a very crude approximation to the actual system behavior.

Several investigators have developed methods for predicting the onset conditions based on semitheoretical arguments similar to those used in the model analysis of Hsu [12.3] (see Chapter Six). Bergles and Rohsenow [12.5] developed a model analysis similar to Hsu's and used a graphical technique to solve the governing equations. Based on their computed results, they recommended the following relation as a means of predicting the onset of boiling for flow of water in a heated tube:

$$q''_{\text{ONB}} = 5.30P^{1.156}[1.80(T_w - T_{\text{sat}})_{\text{ONB}}]^{2.41/P^{0.0234}} \tag{12.13}$$

This is a dimensional relation in which the pressure, temperature, and heat flux are in units of kPa, °C, and W/m^2, respectively. This relation was reported to match the values calculated using the model analysis for $103 \leq P \leq 13{,}700$ kPa. As noted by Bergles and Rohsenow [12.5], their model analysis, like Hsu's [12.3], is plausible only for surfaces that have cavity sizes distributed over a wide range. They argue, however, that commercially produced surfaces usually do have cavities over a wide range of sizes, and therefore the results of this model should be applicable to many real systems.

Sato and Matsumura [12.6] developed an analytical treatment of the onset problem similar to that proposed by Bergles and Rohsenow [12.5]. They proposed the following relation as a means of predicting the onset condition:

$$q''_{\text{ONB}} = \frac{k_l h_{lv} \rho_v}{8\sigma T_{\text{sat}}} [(T_w - T_{\text{sat}})_{\text{ONB}}]^2 \tag{12.14}$$

Davis and Anderson [12.7] modified and extended the analytical treatment of Bergles and Rohsenow [12.5]. Their detailed analysis yielded the following expression for the onset condition:

$$(T_w - T_{\text{sat}})_{\text{ONB}} = \frac{(RT_{\text{sat}}^2/h_{lv}) \ln(1 + \xi')}{1 - (RT_{\text{sat}}/h_{lv}) \ln(1 + \xi')} + \frac{q''_{\text{ONB}} y'}{k_l} \tag{12.15}$$

where

$$y' = \frac{C_\theta \sigma}{P} + \sqrt{\left(\frac{C_\theta \sigma}{P}\right) + \frac{2C_\theta k_l \sigma T_{\text{sat}}}{q''_{\text{ONB}} h_{lv} \rho_v}} \tag{12.16}$$

$$\xi' = \frac{2C_\theta \sigma}{Py'} \tag{12.17}$$

$$C_\theta = 1 + \cos\theta \tag{12.18}$$

In these expressions θ is the contact angle of the liquid–vapor interface on the solid surface for the model bubble considered in the analysis.

Davis and Anderson [12.7] also argued that, for systems at higher pressures or for low surface tension, the relation for the onset condition could be simplified to

$$q''_{\text{ONB}} = \frac{k_l h_{lv} \rho_v}{8C_\theta \sigma T_{\text{sat}}} [(T_w - T_{\text{sat}})_{\text{ONB}}]^2 \tag{12.19}$$

It can readily be seen that for a hemispherical model bubble $\theta = 90°$, $C_\theta = 1$, and Eq. (12.19) becomes identical to the correlation of Sato and Matsumura [12.6]. Davis and Anderson [12.7] found that Eq. (12.19) with $C_\theta = 1$ (the Sato and Matsumura correlation) agreed well with experimentally reported onset conditions for water reported by Sato and Matsumura [12.6] and Rohsenow [12.8]. This equation was also shown to agree well with the correlation of Bergles and Rohsenow [12.5] over wide ranges of pressure and wall heat flux.

Somewhat later, Frost and Dzakowic [12.9] explored the applicability of the analytical treatment used by Davis and Anderson [12.7] to other liquids. Based on additional arguments regarding the effect of liquid Prandtl number on the onset condition, they recommended the following relation for the onset condition:

$$q''_{ONB} = \frac{k_l h_{lv} \rho_v}{8\sigma T_{sat}} [(T_w - T_{sat})_{ONB}]^2 \, Pr_l^2 \qquad (12.20)$$

This correlation, which is identical to the Sato and Matsumura [12.6] correlation except for the inclusion of the Pr_l^2 multiplier, was found to agree well with data for a wide variety of fluids, including water, various hydrocarbons, mercury, and cryogenic liquids.

It cannot be overemphasized that the onset correlations described above are expected to predict the onset conditions accurately only if there exists a sufficiently wide range of potential nucleation site sizes on the wall surface. In some instances, larger active sites may not be present in large numbers, due to the surface being polished or due to the highly wetting nature of the liquid, which may cause the liquid to displace vapor or air in all but the smallest cavities. If such conditions exist in the system of interest, the above relations should be used with caution.

The above predictive relations for the onset condition indicate the threshold condition at which nucleation is initiated or turned off. When analyzing internal convective boiling in a tube, the first step is usually to determine the portion of the tube over which boiling occurs. While the correlations described above indicate the boundary between the boiling and nonboiling regions, determining the region in which boiling occurs can be a little tricky.

The variation of the wall superheat with heat flux at the onset condition, as predicted by the Sato and Matsumura [12.6] correlation, is plotted in Fig. 12.7 for water at atmospheric pressure. The curve in this plot represents combinations of wall superheat and heat flux at which nucleation will just begin. The question to be answered, then, is: "Where along the tube will the actual operating conditions first cross this curve?" For a tube with a constant heat flux applied along the wall, the wall superheat increases linearly with downstream distance z, as indicated by Eq. (12.5). As the flow proceeds downstream, the system state point thus moves upward along a vertical (constant q'') line on Fig. 12.7. This has the effect of raising T_w while keeping the slope of the temperature profile near the wall constant. At a given distance y away from the tube wall, the temperature will be higher, and nucleation conditions will be more favorable at locations that are farther downstream. Moving upward across the onset curve thus carries the system from a region with no boiling present to one where nucleate boiling is more favored. State points above the onset curve in Fig. 12.7 are therefore in the nucleate boiling region, and points below the curve are in the single-phase region. Equation (12.5) can be combined with one of the onset correlations described above to eliminate the wall superheat and solve for the z location of the onset for the specified q'' value.

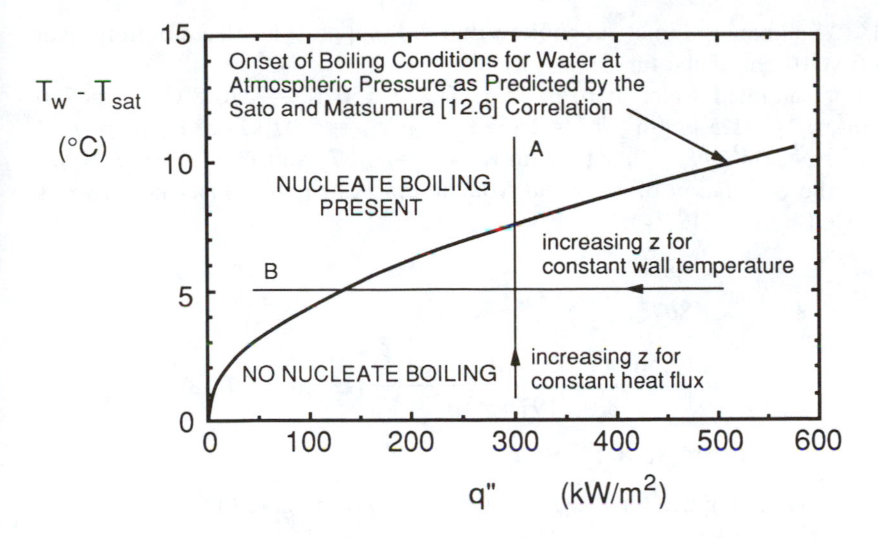

Figure 12.7 Onset of boiling conditions predicted for water at atmospheric pressure.

For a constant and uniform wall temperature condition, the wall heat flux decreases exponentially with downstream distance z, as indicated by Eq. (12.9). The system point thus moves horizontally to the left in Fig. 12.7 along a constant superheat line as the fluid proceeds downstream. Decreasing q'' has the effect of decreasing the temperature gradient at the wall with downstream distance. Hence, at a given y location (away from the wall), the fluid will be hotter, and nucleation will therefore be favored, at a z location that is farther downstream. Thus the decreasing heat flux with downstream distance will carry the system from a region of no boiling into the boiling region as its operating point crosses the onset curve. Combining Eq. (12.9) with one of the onset relations to eliminate q'', the z location at which the onset occurs can be determined for the specified wall superheat.

In applications involving convective boiling, the combined variations of heat flux and wall superheat with downstream distance may be more complicated than the variations that result for the constant q'' and T_w circumstances described above. However, once the variation of q'' with superheat along the tube has been determined (using energy balance requirements and the imposed boundary condition), the intersection of the resulting q''-versus-superheat relation with the onset curve will define onset and/or suppression locations. The most noteworthy trends are that increasing superheat and decreasing heat flux favor nucleation. The direction in which the onset curve is crossed, and the associated changes in heat flux and superheat, determine whether nucleation is initiated or suppressed.

Example 12.1 Water flows upward in a vertical round tube with an inside diameter of 1.0 cm. The pressure along the tube is virtually constant at 6124 kPa. The water enters as subcooled liquid at a flow rate that corresponds to a mass flux of 9000 kg/m^2 s. The wall is held at a uniform temperature of

281°C. Estimate the inlet subcooling that will suppress nucleate boiling over the first 10 cm of the tube.

For saturated water at 6124 kPa, $T_{sat} = 550$ K $= 276.8$°C, $\rho_l = 756$ kg/m^3, $\rho_v = 31.5$ kg/m^3, $h_{lv} = 1563$ kJ/kg, $c_{pl} = 5.07$ kJ/kg K, $\mu_l = 99.2 \times 10^{-6}$ N s/m^2, $k_l = 0.581$ W/m K, $\mathrm{Pr}_l = 0.87$, and $\sigma = 0.0197$ N/m. Using the correlation of Sato and Matsumura [12.6], the onset heat flux is predicted by Eq. (12.14):

$$q''_{\mathrm{ONB}} = \frac{k_l h_{lv} \rho_v}{8\sigma T_{sat}} (T_w - T_{sat})^2$$

$$= \frac{(0.581)(1563 \times 1000)(31.5)}{8(0.0197)(550)} (281.0 - 276.8)^2$$

$$= 5.82 \times 10^6 \text{ W/m}^2$$

The Reynolds number for the entire flow as liquid is given by

$$\mathrm{Re}_{le} = \frac{GD}{\mu_l} = \frac{900(0.01)}{99.2 \times 10^{-6}} = 9.07 \times 10^5$$

Assuming that fully developed turbulent flow is established just downstream of the tube inlet, the heat transfer coefficient for pure liquid flow is computed using the Dittus-Boelter equation:

$$h_{le} = 0.023 \left(\frac{k_l}{D} \right) \mathrm{Re}_{le}^{0.8} \, \mathrm{Pr}_l^{0.4}$$

$$= 0.023 \left(\frac{0.581}{0.01} \right) (9.07 \times 10^5)^{0.8} (0.87)^{0.4}$$

$$= 7.38 \times 10^4 \text{ W/m}^2 \text{ K}$$

Up to the onset of boiling, the heat flux variation is given by Eq. (12.9):

$$q'' = h_{le}[(T_w - T_{sat}) + (T_{sat} - T_{l,\mathrm{in}})] \exp\left(- \frac{4 h_{le} z}{G c_{pl} D} \right)$$

We want to find the value of $T_{sat} - T_{l,\mathrm{in}}$ that just makes $q'' = q''_{\mathrm{ONB}}$ at $z = 0.10$ m. Rearranging this equation to solve for $T_{sat} - T_{l,\mathrm{in}}$ yields

$$T_{sat} - T_{l,\mathrm{in}} = \frac{q''}{h_{le}} \exp\left(\frac{4 h_{le} z}{G c_{pl} D} \right) - (T_w - T_{sat})$$

Substituting $q'' = q''_{\mathrm{ONB}}$, $z = 0.10$ m, and the respective values of the other parameters, we obtain

$$T_{sat} - T_{l,\mathrm{in}} = \frac{5.82 \times 10^6}{7.38 \times 10^4} \exp\left[\frac{4(7.38 \times 10^4)(0.1)}{9000(5.07 \times 1000)(0.01)} \right] - (281.0 - 276.8)$$

$$= 79.9\text{°C}$$

Note that for $z < 10$ cm, Eq. (12.9) predicts that q'' will be greater than q''_{ONB} for this subcooling and from Fig. 12.7 it can be seen that no boiling will occur. Thus if the inlet subcooling is 80°C, the onset is predicted not to occur over the first 10 cm downstream of the inlet.

12.3 SUBCOOLED FLOW BOILING

Regimes of Subcooled Flow Boiling

Subcooled internal flow boiling can arise in a number of applications. Boilers and vapor generators are occasionally fed liquid that is somewhat subcooled. Subcooled flow boiling has also been of particular interest as a means of providing high-heat-flux cooling in some specialized thermal control applications. In electronic systems, for example, cooling of microelectronic chips may require removal of heat at flux levels exceeding 100 W/cm^2. At the present time, perhaps the easiest way to achieve flux levels of this magnitude is with a subcooled flow boiling process.

As noted in Chapter Eight, for external flow over a surface, the critical heat flux increases with subcooling and flow velocity over the surface. Similar trends are observed for internal flow in tubes, which makes subcooled flow boiling an attractive prospect for removing heat at high flux levels from coolant passage walls in special applications that require it.

Figure 12.8 indicates schematically the sequence of regimes associated with subcooled boiling. Immediately downstream of the onset of boiling, a region of *partial subcooled boiling* exists. This is a transition region in which both forced-convective effects and nucleate boiling effects are important, with nucleate boiling effects increasing in strength as the flow proceeds downstream. Near the onset, active nucleation sites are few and widely spaced. The effect of bubble growth and release on the overall heat transfer from the wall is small in this region.

For either the uniform wall temperature or uniform wall heat flux condition, the conditions become progressively more favorable to nucleation as the flow proceeds downstream (see the discussion in Section 12.2). This generally results in an increase in the nucleation site density with increasing downstream distance in the partial boiling region. Eventually the nucleation site density becomes so high that the nucleate boiling contribution to the heat transfer is essentially equivalent to that for saturated nucleate pool boiling. The flow is then said to have entered the *fully developed nucleate boiling regime*.

In the fully developed nucleate boiling regime, the nucleate boiling mechanism generally is so strong that it completely dominates the heat transfer process. At extremely high flow velocities and subcooling levels, convective effects may be important well beyond the point where the nucleation site density has attained a level equal to that for pool boiling at the same superheat. However, for most cases of practical interest, nucleate boiling completely dominates in the fully developed nucleate boiling regime.

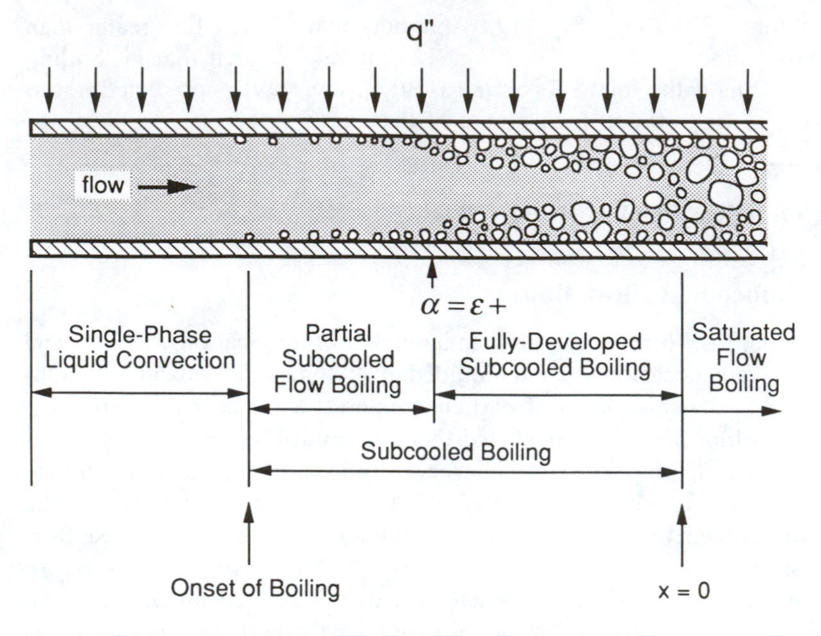

Figure 12.8 Transition of flow through regimes of subcooled boiling.

Once fully developed nucleate boiling dominates the transport, the heat transfer rate becomes virtually independent of the flow rate and subcooling. This is a direct consequence of the observation (discussed in Chapter Eight) that nucleate pool boiling at moderate to high wall superheats is affected only slightly by ambient motion and subcooling of the surrounding liquid. The heat flux-versus-wall superheat curve will vary with flow velocity or subcooling in the single-phase liquid convection regime and in the partial subcooled boiling regime. However, the above observations imply that in the different curves will all eventually merge into the fully developed boiling curve, as indicated in Fig. 12.9.

Methods of Predicting Partial Subcooled Boiling Heat Transfer

As noted in Chapter Eight, the mechanisms of subcooled nucleate boiling have been the subject of much debate over the past 40 years. For subcooled flow boiling in particular, a number of investigators have examined the nucleate boiling process experimentally and/or analytically (see, e.g., [12.10–12.14]). In the partial subcooled boiling regime, methods used to predict the heat transfer rate have typically been based on the premise that the forced-convection and nucleate boiling mechanisms act in parallel and independently. To account for the contributions of these two mechanisms, Kutateladze [12.15] proposed the following interpolation formula for the heat transfer coefficient associated with partial subcooled flow boiling:

$$h = (h_{spl}^2 + h_{snb}^2)^{1/2} \tag{12.21}$$

As the wall temperature increases, the subcooled nucleate boiling contribution h_{snb} generally becomes large compared to the single-phase liquid convection contribution h_{spl}, and this relation predicts that h approaches h_{snb}. Similarly, as the wall temperature decreases toward T_{sat}, h_{snb} goes to zero, and consequently, h approaches h_{spl}. This type of interpolation relation thus correctly predicts the limiting behavior of h as one mechanism turns on and the other turns off.

In more recent investigations, the total heat flux has more often been postulated to be the sum of contributions due to single-phase liquid q''_{spl} convection and nucleate boiling q''_{snb}:

$$q''_{total} = q''_{spl} + q''_{snb} \tag{12.22}$$

One possible superposition scheme of this type for predicting partial subcooled boiling heat transfer can be understood by considering Fig. 12.10. This figure is a plot of surface heat flux as a function of wall temperature T_w. In this plot, the region of partial subcooled boiling is idealized as beginning when the wall temperature just exceeds the saturation temperature. The objective is to predict the portion of the curve *BCD* in the partial boiling regime. The broken curve

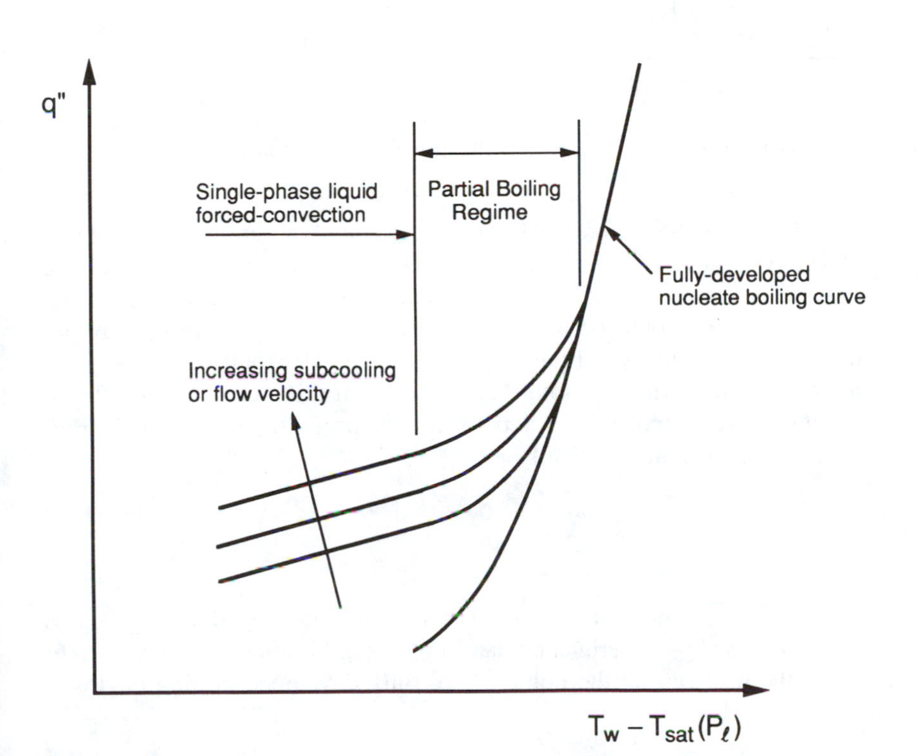

Figure 12.9 The partial boiling transition from single-phase, forced-convection to fully developed nucleate boiling.

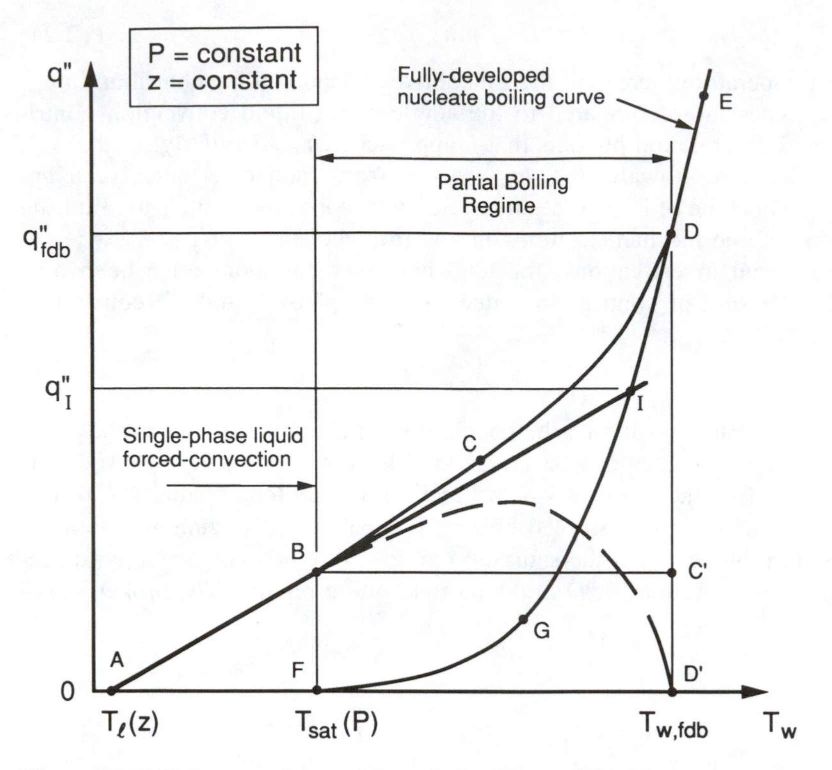

Figure 12.10 Reference points used in analysis of the partial boiling regime.

in this figure represents the difference between the actual system curve *BCD* and the fully developed nucleate boiling curve *FGD*. This difference curve can be interpreted as the covective contribution to the total heat flux in the partial boiling regime q''_{spl}. If this curve could be predicted somehow, adding it to the fully developed boiling curve would yield the desired curve for the total heat flux.

One method of predicting the total heat transfer in this manner would be to approximate the broken curve as the step function represented by $BC'D'$ in Fig. 12.10. This is specified mathematically as

$$T_{\text{sat}} \leq T_w \leq T_{w,fdb}: \qquad q''_{spl} = h_{le}[T_{\text{sat}} - T_l(z)] \tag{12.23a}$$

$$T_{w,fdb} < T_w: \qquad q''_{spl} = 0 \tag{12.23b}$$

To use this method, the condition at point *D* where fully developed boiling begins must be specified. Based on experimental data, Engelberg-Forester and Grief [12.16] concluded that the heat flux at the point where fully developed boiling begins is given approximately by

$$q''_{fdb} = 1.4q''_I \tag{12.24}$$

where q''_I is the heat flux at the intersection of the single-phase liquid convection curve and the fully developed nucleate boiling curve. If the nucleate boiling curve is given by Eq. (12.10), then the heat flux at the intercept point q_I'' for the uniform heat flux wall condition can be determined from Eq. (12.11), as described in Section 12.2. If, on the other hand, the wall is held at a uniform temperature, then q_I'' can be determined from Eqs. (12.10) and (12.12). Then, using Eq. (12.24), the wall temperature corresponding to the beginning of fully developed boiling can be determined as

$$T_{w,fdb} = \left(\frac{1.4q_I''}{\gamma}\right)^m + T_{\text{sat}} \tag{12.25}$$

With $T_{w,fdb}$ specified, the relations for q_{spl}'' can be evaluated and added to the fully developed nucleate boiling curve, as specified by Eq. (12.22), to determine the total heat flux in the partial subcooled regime. This scheme for predicting the heat transfer during partial subcooled boiling was proposed by Bowring [12.4].

Rohsenow [12.17] developed a slightly different method for predicting partial subcooled boiling heat transfer. He proposed to use a conventional single-phase correlation to predict the single-phase contribution to the total heat flux,

$$q_{spl}'' = h_{le}[T_w - T_l(z)] \tag{12.26}$$

and he further postulated that the remaining additional nucleate boiling contribution to the total heat flux could be computed using the Rohsenow correlation for nucleate pool boiling,

$$q_{snb}'' = \mu_l h_{lv}\left[\frac{g(\rho_l - \rho_v)}{\sigma}\right]^{1/2} \text{Pr}_l^{-s/r}\left[\frac{c_{pl}[T_w - T_{\text{sat}}(P_l)]}{C_{sf}h_{lv}}\right]^{1/r} \tag{12.27}$$

As described in Chapter Seven, for this correlation, $r = 0.33$, $s = 1.0$ for water and 1.7 for other fluids, and the constant C_{sf} may vary from one fluid–surface combination to another. To correlate partial subcooled boiling data with Rohsenow's method, the convective contribution to the total heat flux is first computed using Eq. (12.26) and subtracted from the total measured heat flux at each data point. The value of C_{sf} is then determined that best fits the data representing the implied nucleate boiling contributions. This method has been used to correlate partial subcooled boiling data obtained by Rohsenow and Clarke [12.18], Kreith and Sommerfield [12.19], Piret and Isbin [12.20], and Bergles and Rohsenow [12.5]. Values of C_{sf} obtained in this manner are listed in Table 7.2.

Yet another method of predicting the heat transfer during partial subcooled boiling was proposed by Bergles and Rohsenow [12.5]. This method can be best understood by considering the heat flux-versus-wall temperature plot shown in Fig. 12.11. For the partial subcooled boiling regime, the following relation was proposed to predict the heat transfer:

For $q'' > q_{\text{ONB}}''$:

$$q'' = q_{spl}''\left\{1 + \left[\frac{q_{snb}''}{q_{spl}''}\left(1 - \frac{q_D''}{q_{snb}''}\right)\right]^2\right\}^{1/2} \tag{12.28}$$

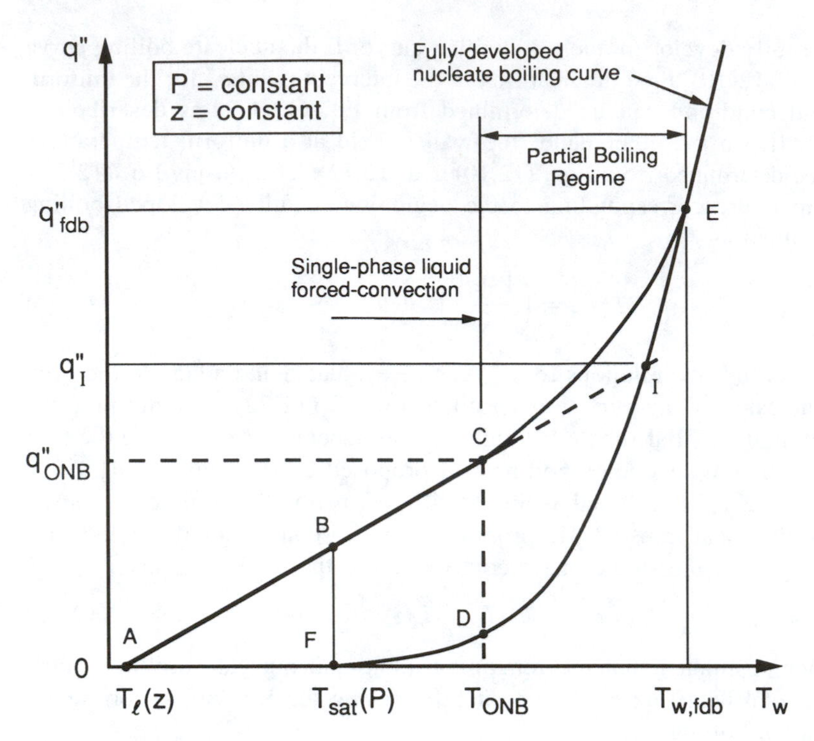

Figure 12.11 Reference points used in the methods of Bergles and Rohsenow [12.5] for prediction of partial subcooled boiling heat transfer.

In this relation the value of q''_{spl} is determined using an appropriate single-phase liquid convection correlation and q''_{snb} is determined using the fully developed subcooled boiling correlation at the local conditions. The value of q''_D is determined using the fully developed boiling correlation at the wall temperature corresponding to the onset of boiling as determined using the Bergles and Rohsenow onset correlation (12.13) described in the previous section. As the wall temperature increases, q''_{snb} becomes large compared to q''_{spl} and q''_D, and q'' approaches q''_{snb}. Similarly, as the wall temperature decreases toward T_{sat}, q''_{snb} goes to zero, and q'' approaches q''_{spl}.

In evaluating the fully developed nucleate boiling contribution in the above schemes, it has often been assumed that the heat flux contribution due to subcooled nucleate boiling in the presence of liquid convection is identical to that for saturated pool boiling. Bergles and Rohsenow [12.4] specifically explored this issue by testing a heated cylinder in a forced-convective boiling (as the inside wall of an annulus) and in pool boiling in a nearly saturated liquid pool. The forced-convection q''-versus-superheat curves for different velocities did approach a common limiting curve at high superheat levels. However, this limiting curve was not exactly an extension of the curve indicated by the pool boiling data.

Their results indicate that using a correlation based on pool boiling data to predict the fully developed nucleate boiling curve during subcooled convective boiling can result in significant errors in some cases. Better results can be obtained if the fully developed nucleate boiling curve is determined by correlation of the implied nucleate boiling effect back-calculated from measured data using one of the correlation schemes described above. If no better information is available, it may be necessary to use an extension of the saturated pool boiling curve to predict q''_{snb}. The predictions may be quite good in some cases. In general, however, the results of such a calculation should be used with caution.

Example 12.2 Subcooled liquid ethanol at 226 kPa flows through a vertical round tube having walls held at 140°C. The liquid enters the tube at 50°C and the onset of boiling occurs immediately at the entrance of the tube. The tube diameter is 1.2 cm and the mass flux is 600 kg/m^2 s. Determine the partial boiling heat transfer coefficient predicted by Rohsenow's method at a location downstream of the inlet where the fluid bulk temperature is 90°C.

For saturated ethanol at 226 kPa, T_{sat} = 373 K = 99.8°C, ρ_l = 734 kg/m^3, ρ_v = 3.18 kg/m^3, h_{lv} = 927 kJ/kg, c_{pl} = 3.30 kJ/kg K, μ_l = 314 × 10^{-6} N s/m^2, k_l = 0.151 W/m K, Pr_l = 6.88, and σ = 0.0157 N/m. Rohsenow's [12.17] method postulates that the total partial subcooled boiling heat flux is given by

$$q'' = q''_{spl} + q''_{snb}$$

where, for ethanol,

$$q''_{spl} = h_{le}[T_w - T_l(z)]$$

$$q''_{snb} = \mu_l h_{lv}\left[\frac{g(\rho_l - \rho_v)}{\sigma}\right]^{1/2} \mathrm{Pr}_l^{-5.15}\left[\frac{c_{pl}(T_w - T_{sat})}{C_{sf}h_{lv}}\right]^{3.0}$$

For the entire flow as liquid,

$$\mathrm{Re}_{le} = \frac{GD}{\mu_l} = \frac{600(0.012)}{314 \times 10^{-6}} = 2.24 \times 10^4$$

Assuming fully developed turbulent flow,

$$h_{le} = 0.023\left(\frac{k_l}{D}\right)\mathrm{Re}_{le}^{0.8}\,\mathrm{Pr}_l^{0.4}$$

$$= 0.023\left(\frac{0.151}{0.012}\right)(2.24 \times 10^4)^{0.8}(6.88)^{0.4}$$

$$= 1930 \text{ W/m}^2 \text{ K}$$

It follows that

$$q''_{spl} = 1930(140 - 90) = 9.65 \times 10^4 \text{ W/m}^2$$

Using $C_{sf} = 0.013$ and substituting the respective values of the fluid properties in the above equation for q_{snb} yields

$$q''_{snb} = (314 \times 10^{-6})(927 \times 1000)\left[\frac{9.8(734 - 3)}{0.0157}\right]^{1/2}(6.88)^{-5.15}\left[\frac{3.33(140 - 99.8)}{0.013(927)}\right]^{3}$$

$$= 1.44 \times 10^{4}\ \mathrm{W/m^2}$$

The total heat flux and overall heat transfer coefficient are then given by

$$q'' = q''_{spl} + q''_{snb} = 9.65 \times 10^4 + 1.44 \times 10^4 = 1.11 \times 10^5\ \mathrm{W/m^2}$$

$$h = \frac{q''}{T_w - T_l(z)} = \frac{1.11 \times 10^5}{140 - 90} = 2220\ \mathrm{W/m^2\,K}$$

Pressure Drop and Void Fraction

Determination of the two-phase pressure drop during subcooled flow boiling is an important aspect of the design of some power systems and heat exchangers. To determine the two-phase pressure drop during the subcooled boiling process, a means of predicting the void fraction is also needed. Predicting the void fraction and two-phase pressure drop for these conditions is a challenging problem, requiring methods that account for the fact that the two-phase flow is not one-dimensional. Fortunately, the simple methods described above can be used to predict the heat transfer for subcooled flow boiling in spite of the complexity of the two-phase flow. A full discussion of the two-phase flow characteristics for these circumstances is beyond the scope of this book. However, because two-phase pressure drop is important to heat exchanger design, a brief overview of the void fraction and pressure drop characteristics of subcooled boiling will be presented in the remaining portion of this section.

Prediction of the void fraction and two-phase pressure drop is generally even more complicated for convective subcooled boiling than for saturated convective boiling. The added complexity arises from the nonuniformity of the flow across the tube. As indicated schematically in Fig. 12.8, during partial subcooled boiling in a highly subcooled flow, vapor bubbles generated at the wall of the tube generally condense rapidly when they depart from the wall region and enter the subcooled bulk liquid. As a result, in a highly subcooled turbulent flow, vapor bubbles are confined to the thin thermal boundary layer (viscous sublayer) near the wall. The average void fraction in this regime is extremely small, but the impact of the bubbles on the wall region transport can be quite significant. An approximate model of the flow in this regime near the wall for these conditions was developed by Griffith, Clark, and Rohsenow [12.21]. From an analysis based on this model, the following relation for the void fraction was obtained:

$$\alpha = 3.73 \frac{q''_{snb}}{h_{le}[T_{sat} - T_l(z)]}\left(\frac{k_l}{h_{le} d_h}\right) \mathrm{Pr}_l \tag{12.29}$$

This relation is limited to highly subcooled flow boiling at moderate pressures.

In the partially subcooled boiling regime, as heat is added and the flow proceeds downstream, the bulk subcooling decreases. At lower subcooling levels, bubbles that leave the wall region and enter the bulk flow condense more slowly. As a result, vapor bubbles are present in a progressively larger portion of the bulk flow as the subcooling continues to decrease with downstream distance.

The point in the flow at which the subcooling is low enough that significant amounts of vapor begin to enter the bulk flow is of particular importance, since the pressure drop characteristics of the two-phase flow upstream and downstream of this point are quite different. This location is designated as the point where $\alpha = \epsilon+$ in Fig. 12.8. Upstream of this point (or at high subcooling), vapor is present only very near the wall. Downstream of this location, vapor exists near the wall and in a significant portion of the bulk flow near the wall. From a simple model, Levy [12.22] has developed a relation that can be used to predict the void fraction at the $\alpha = \epsilon+$ point between these two regimes.

It has also been argued by Griffith et al. [12.21] that the $\alpha = \epsilon+$ point corresponds to the onset of fully developed boiling (see Fig. 12.8). Based on this premise, it is possible to directly estimate the $\alpha = \epsilon+$ location from the boiling heat transfer relations described previously in this section. Collier [12.2] reports that an alternative methodology for predicting the $\alpha = \epsilon+$ location has also been developed by Bowring [12.4].

Levy [12.22] interpreted the $\alpha = \epsilon+$ point as being the location where bubbles are first able to detach from the surface. By considering the conditions necessary for growth and the force balance on an attached bubble, Levy [12.22] developed an analytical method for predicting the conditions at the $\alpha = \epsilon+$ point. The predictions of the Levy [12.22] model agree fairly well with the data of Egen et al. [12.23] and Mauer [12.24]. An analysis similar to Levy's has also be presented by Staub [12.25]. An alternative approach to predicting the $\alpha = \epsilon+$ point, interpreted as the point of net vapor generation, has also been developed by Saha and Zuber [12.26].

As noted above, in the low subcooling regime (downstream of the $\alpha = \epsilon+$ point), the vapor void fraction increases progressively with downstream distance (and decreasing subcooling). Methods that can be used to predict the void fraction in the flow for conditions in this regime have been developed by Griffith et al. [12.21], Bowring [12.4], Kroeger and Zuber [12.27], and Levy [12.22].

Understanding of the pressure drop characteristics for subcooled flow boiling is relatively limited at this time. In the low subcooling region, vapor bubbles are present over much of the passage cross section. Prediction of the two-phase pressure gradient using the separated flow model would therefore seem appropriate, since the flow deviates only slightly from saturated flow boiling. The main difficulty in applying this approach is determination of the vapor void fraction and the local mass fraction of vapor in the flow. Note that because the flow is far from equilibrium, the local vapor mass fraction is not equal to the quality, and thus cannot be calculated from simple thermodynamic considerations. Collier [12.2] describes a methodology proposed by Sher [12.28] that handles this difficulty.

For the highly subcooled region, the presence of vapor in the near wall region may have two possible effects: (1) The bubbles may act like surface roughness elements acting to increase the pressure gradient; or (2) the vapor in the layer near the wall may act to reduce the effective viscosity of the layer, thereby resulting in a lower wall shear stress and lower pressure gradient for a given bulk velocity. Because these are opposite effects, the magnitude of the pressure gradient relative to that for single-phase liquid at the same flow rate is not clear.

In an attempt to examine some of the mechanisms associated with momentum transfer during highly subcooled flow boiling, Hirata and Nishikawa [12.29] developed an analysis of the analogous circumstance of liquid flow over a porous plate through which gas was injected into the boundary layer. Their results suggest that generation of the gas at the wall can produce an increase in the wall shear stress. The increase in the shear stress was found to be strongly affected by the bubble (nucleation) site density and the size of the bubbles at departure.

Results of experimental investigations indicate that the pressure gradient associated with highly subcooled flow boiling may be either larger or smaller than that for pure liquid at the same flow rate. The lack of a clear trend in the data is not surprising given the complexity of the mechanisms described above. Measured values of the total pressure gradient for subcooled flow boiling obtained by Reynolds [12.30] imply that the pressure gradient just downstream of the onset of subcooled boiling may be lower than that for single-phase liquid flow. Farther downstream, however, the pressure gradient was observed to increase rapidly, becoming much larger than that for a single-phase liquid at the same flow rate. Data obtained by Dormer and Bergles [12.31] exhibit a similar trend. On the other hand, the pressure gradients for subcooled flow boiling measured by Buchberg et al. [12.32] were virtually always higher than that for the entire flow as liquid.

Additional experimental studies of the pressure gradient associated with subcooled flow boiling have been conducted by Sher [12.28], Owens and Schrock [12.33], Jicha and Frank [12.34] and Jordan and Leppert [12.35]. Based on the data obtained in his experiments, Reynolds [12.30] developed an empirical correlation for the total two-phase pressure gradient. Owens and Schrock [12.32] also developed a pressure gradient correlation based on a fit to their data. Additional information regarding empirical correlations for the two-phase pressure gradient for subcooled flow boiling can be found in Collier [12.2] and Tong [12.36].

12.4 SATURATED FLOW BOILING

Saturated internal flow boiling is most often encountered in applications where complete or nearly complete vaporization of the coolant is desired. Perhaps the most frequently encountered example is the evaporator in a refrigeration or air-conditioning system. Other examples include cryogenic processing applications, boilers in nuclear and conventional power plant systems, and chemical processing involving pure hydrocarbons. As seen in Figs. 12.4 and 12.5, to avoid the high wall temperatures and/or the poor heat transfer associated with the saturated film boiling regime, the vaporization must be accomplished at low superheat or low heat flux levels. For this reason, evaporators and boilers are usually designed to avoid the high heat flux and high wall superheat levels that may produce film boiling at some point during the process. Because most equipment operates in this range, this section will focus on flow boiling processes at low to moderate superheat and heat flux conditions.

Before discussing the boiling process itself, it is worth noting that at low wall superheat conditions, it is possible for the onset of nucleate boiling to be delayed until the mean coolant enthalpy is higher than that for saturated liquid $\hat{h}_l$. The required wall superheat for the onset of boiling for saturated or superheated bulk liquid can be predicted by the correlations of Bergles and Rohsenow [12.5], Sato and Matsumura [12.6], Davis and Anderson [12.7], or Frost and Dzakowic [12.9], as described in Section 12.2. In most (but not all) systems of practical interest, the onset of nucleate boiling is achieved at or just beyond the point where the bulk flow reaches the saturated liquid condition.

When boiling is initiated, both nucleate boiling and liquid convection may be active heat transfer mechanisms. Usually the walls of the passage have an abundance of active nucleation sites, and at low quality, the vapor void fraction is relatively low and the nucleate boiling mechanism is much stronger than the forced-convective effect. In general, however, the relative importance of these two mechanisms varies over the length of the passage. As the flow proceeds downstream and vaporization occurs, the void fraction rapidly increases at low to moderate pressures. As a result, the flow must accelerate, which tends to enhance the convective transport from the heated wall of the tube.

As described in Section 12.1, the increasing void fraction and acceleration of the flow also produce changes in the flow regime with downstream location. For vertical upward flow, bubbly flow at the onset location subsequently changes to slug, churn, and then annular flow. When there is a large difference in the liquid and vapor densities, the transition from bubbly to the annular configuration associated with churn or annular flow can occur over a very short portion of the tube length. Once such an annular configuration is achieved, convective transport of heat across the liquid film on the wall can directly vaporize liquid at the liquid–vapor interface of the film. Further, it is clear that, as vaporization continues, the thickness of the liquid film on the tube wall will decrease, reducing its thermal resistance and thereby enhancing the effectiveness of this mechanism.

In the case of a uniform applied heat flux, it can be seen that as annular film evaporation increases in effectiveness, if the nucleate boiling contribution is unchanged, the wall-to-interface temperature difference needed to drive the heat flux is reduced. However, the decrease in wall superheat resulting from this effect tends to extinguish the smaller active nucleation sites, reducing the effectiveness of the nucleate boiling mechanism. As the liquid film becomes very thin near the latter stages of the vaporization process, the required superheat to transport all the surface heat flux across the liquid film may become so low that nucleation is completely suppressed. This is indicated schematically in Fig. 12.12.

For a tube with a constant wall superheat condition, the variation of the transport mechanisms is similar to that for the uniform flux wall condition. As in the

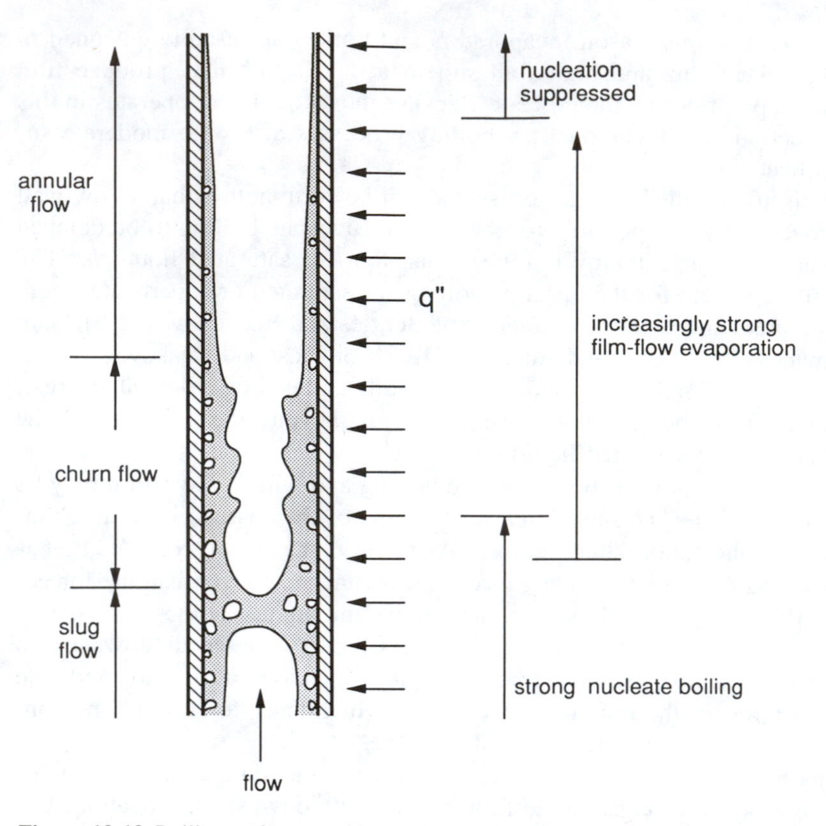

Figure 12.12 Boiling regime transitions at moderate qualities.

uniform heat flux case described above, the transition from bubbly to annular flow is generally expected to strengthen the convective transport mechanism. Although the wall superheat is fixed at a level adequate to result in the onset of nucleation initially, the enhancement of the convective effect increases the heat flux from the wall as the flow proceeds downstream. As indicated in Fig. 12.7, increasing the heat flux from the surface at a fixed wall superheat level may ultimately cause nucleate boiling to be suppressed. Thus, for flow in a round tube with an isothermal wall condition or a uniform applied heat flux, the forced-convective effect generally becomes stronger and the nucleate boiling effect tends to become weaker as the flow proceeds downstream.

Because of the trends described above, prediction of the convective boiling heat transfer coefficient requires an approach that accommodates a transition from a nucleate-pool-boiling-like condition at low qualities to a nearly pure film evaporation condition at higher qualities. Heat transfer in the latter case can be modeled in the manner described in Section 11.2 for annular film flow condensation. The only differences here are that the direction of the heat flux is reversed and the driving temperature difference $T_{sat} - T_{wall}$ used in the Section 11.2 analysis is replaced with $T_{wall} - T_{sat}$. In fact, with these changes, the integral analysis and resulting heat transfer relation presented in Example 11.2 apply equally well to annular film flow evaporation in a vertical round tube.

The approximate analysis presented in Example 11.2 indicates that, for convective transport across the liquid film, the resulting heat transfer coefficient can be correlated in terms of the turbulent-turbulent Martinelli parameter X_{tt}, the Reynolds number for the liquid flowing alone Re_l, and the liquid Prandtl number Pr_l:

$$\frac{hD}{k_l} = f(X_{tt}, Re_l, Pr_l) \tag{12.30}$$

Variations of the above form that have been used to correlate boiling heat transfer data have included relations of the form

$$\frac{h}{h_{le}} = f\left(\frac{1}{X_{tt}}\right) \quad \text{and} \quad \frac{h}{h_l} = f\left(\frac{1}{X_{tt}}\right) \tag{12.31}$$

In these relations h_{le} is the single-phase convection coefficient for the entire flow as liquid and h_l is the single-phase coefficient for the liquid phase flowing alone. X_{tt} is the turbulent-turbulent Martinelli parameter. If the single-phase turbulent-flow correlation

$$f = 0.046\ Re^{-0.2}$$

is used to evaluate the friction factors in the definition of the Martinelli parameter (Eqs. [10.11*b*] and [10.12]), it can be shown that X_{tt} is given by

$$X_{tt} = \left(\frac{1-x}{x}\right)^{0.9} \left(\frac{\rho_v}{\rho_l}\right)^{0.5} \left(\frac{\mu_l}{\mu_v}\right)^{0.1} \tag{12.32}$$

If the Dittus-Boelter correlation is used to evaulate h_{le} and h_l, relations of the type indicated in Eqs. (12.31) can be rearranged to show that h is a function of Re_l, Pr_l, and X_{tt}, making the form of these relations similar to that indicated by Eq. (12.30).

Several investigators have, in fact, correlated convective boiling data in the absence of strong nucleate boiling effects, using relations similar to the forms indicated above. Dengler and Addoms [12.37] proposed the following relation as a fit to heat transfer data for convective vaporization of water in a vertical tube:

$$\frac{h}{h_{le}} = 3.5\left(\frac{1}{X_{tt}}\right)^{0.5} \tag{12.33}$$

Based on a fit to data for convective boiling of organic liquids, Guerrieri and Talty [12.38] similarly proposed the correlation

$$\frac{h}{h_{le}} = 3.4\left(\frac{1}{X_{tt}}\right)^{0.45} \tag{12.34}$$

Equations (12.33) and (12.34) strictly apply only to conditions where nucleate boiling has been completely suppressed. Hence, their usefulness is very limited. However, if one of these relations is assumed to hold at the location where complete suppression is just achieved, it can be combined with one of the onset correlations described in Section 12.2 to obtain a relation for the conditions at which nucleation will be suppressed. Combining Eq. (12.33) with the Sato and Matsumura [12.6] correlation (12.14), for example, it can easily be shown that the quality at which nucleation is suppressed is given by the relation

$$x_{\text{SUP}} = \frac{\gamma}{1 + \gamma} \tag{12.35}$$

where for an isothermal wall boundary condition,

$$\gamma = \gamma_T = \left(\frac{\rho_v}{\rho_l}\right)^{0.56} \left(\frac{\mu_l}{\mu_v}\right)^{0.11} \left[\frac{k_l h_{lv} \rho_v (T_w - T_{\text{sat}})}{28 h_{le} \sigma T_{\text{sat}}}\right]^{2.22} \tag{12.36a}$$

and for a uniform heat flux applied to the wall,

$$\gamma = \gamma_H = \left(\frac{\rho_v}{\rho_l}\right)^{0.56} \left(\frac{\mu_l}{\mu_v}\right)^{0.11} \left(\frac{q'' k_l h_{lv} \rho_v}{98 \sigma T_{\text{sat}} h_{le}^2}\right)^{1.11} \tag{12.36b}$$

The variation of x_{SUP} with γ is shown in Fig. 12.13. Thus, if the coolant properties and the wall boundary condition are specified, the quality at which nucleate boiling is virtually completely suppressed can be estimated using these relations.

The above discussion indicates that nucleate boiling may dominate at low qualities, while at moderate to high qualities nucleate boiling may be completely suppressed and film evaporation may dominate. At intermediate qualities, both

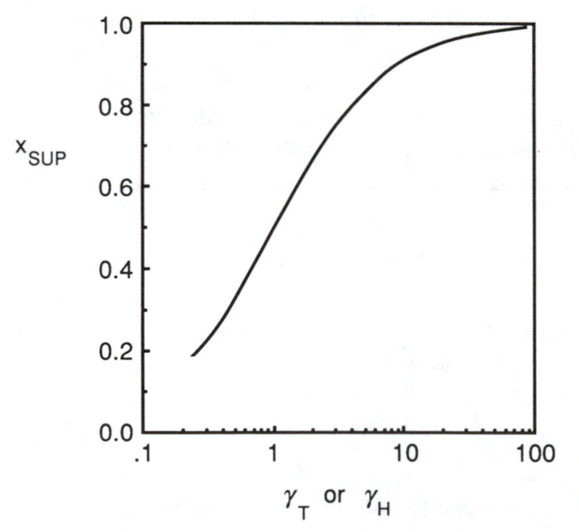

Figure 12.13 Variation of the quality at which nucleation is suppressed with the parameters defined by Eqs. (12.36).

nucleate boiling and film evaporation effects may be important. Several correlations have been proposed that attempt to use a superposition technique to account for the gradual transition between these limiting circumstances. Six correlations of this type are described and briefly discussed below.

The Chen Correlation

Chen [12.39] argued that the heat transfer coefficient h for saturated convective boiling is equal to the sum of a microscopic (nucleate boiling) contribution h_{mic} and a macroscopic (bulk convective) contribution h_{mac}:

$$h = h_{\text{mic}} + h_{\text{mac}} \tag{12.37}$$

Chen [12.39] proposed to evaluate the macroscopic contribution using a correlation similar in form to the Dengler and Addoms [12.37] correlation:

$$h_{\text{mac}} = h_l F(X_{tt}) \tag{12.38}$$

where the single-phase coefficient for the liquid alone, h_l, is evaluated using the Dittus-Boelter equation,

$$h_l = 0.023\left(\frac{k_l}{D}\right) \text{Re}_l^{0.8}\,\text{Pr}_l^{0.4} \tag{12.39a}$$

$$\text{Re}_l = \frac{G(1-x)D}{\mu_l} \tag{12.39b}$$

Example 12.3 Saturated flow boiling of R-22 at 218 kPa occurs in a small vertical round tube with a diameter of 0.7 cm. The mass flux is 200 $\text{kg/m}^2\,\text{s}$, and a uniform heat flux of 6000 W/m^2 is applied to the tube wall. Estimate the quality at which nucleation is expected to be completely suppressed.

For saturated R-22 at 218 kPa, $T_{\text{sat}} = 250\text{ K} = -23.2\text{ K}$, $\rho_l = 1360\text{ kg/m}^3$, $\rho_v = 9.59\text{ kg/m}^3$, $h_{lv} = 226\text{ kJ/kg}$, $c_{pl} = 1.13\text{ kJ/kg K}$, $\mu_l = 282 \times 10^{-6}\text{ N s/m}^2$, $\mu_v = 10.9 \times 10^{-6}\text{ N s/m}^2$, $k_l = 0.109\text{ W/m K}$, $\text{Pr}_l = 2.92$, and $\sigma = 0.0155\text{ N/m}$. For the entire flow as liquid,

$$\text{Re}_{le} = \frac{GD}{\mu_l} = \frac{2000(0.007)}{282 \times 10^{-6}} = 4960$$

So, for fully developed turbulent flow,

$$\begin{aligned} h_{le} &= 0.023\left(\frac{k_l}{D}\right) \text{Re}_{le}^{0.8}\,\text{Pr}_l^{0.4} \\ &= 0.023(0.109)/(0.007)4960^{0.8}(2.92)^{0.4} \\ &= 497\text{ W/m}^2\text{ K} \end{aligned}$$

Using Eq. (12.36*b*), γ is determined as

$$\gamma = \left(\frac{\rho_v}{\rho_l}\right)^{0.56} \left(\frac{\mu_l}{\mu_v}\right)^{0.11} \left(\frac{q'' k_l h_{lv} \rho_v}{98\sigma T_{\text{sat}} h_{le}^2}\right)^{1.11}$$

$$= \left(\frac{9.59}{1360}\right)^{0.56} \left(\frac{282}{10.9}\right)^{0.11} \left[\frac{6000(0.109)(226 \times 1000)(9.59)}{98(0.0155)(250)(497)^2}\right]^{1.11}$$

$$= 1.82$$

Substituting this value of γ into Eq. (12.35) yields

$$x_{\text{SUP}} = \frac{\gamma}{1 + \gamma} = \frac{1.82}{1 + 1.82} = 0.645$$

Thus, in this system, nucleation is predicted to exist almost to 65% quality.

The microscopic contribution to the overall heat transfer coefficient was determined by applying a correction to the Forster and Zuber [12.40] relation for the heat transfer coefficient for nucleate pool boiling:

$$h_{\text{mic}} = 0.00122\left[\frac{k_l^{0.79} c_{pl}^{0.45} \rho_l^{0.49}}{\sigma^{0.5} \mu_l^{0.29} h_{lv}^{0.24} \rho_v^{0.24}}\right][T_w - T_{\text{sat}}(P_l)]^{0.24}[P_{\text{sat}}(T_w) - P_l]^{0.75} S \quad (12.40)$$

The so-called suppression factor S corrects the fully developed nucleate boiling prediction of h_{mic} to account for the fact that as the macroscopic convective effect increases in strength, nucleation is more strongly suppressed. Chen argued that this suppression factor ought to be a function of an appropriately defined two-phase Reynolds number, Re_{tp}. He further conjectured that the macroscopic contribution h_{mac} should also be related to this two-phase Reynolds number via an extension of the Dittus-Boelter equation,

$$h_{\text{mac}} = 0.023\left(\frac{k_{tp}}{D}\right) \text{Re}_{tp}^{0.8} \,\text{Pr}_{tp}^{0.4} \quad (12.41)$$

Taking $k_{tp} = k_l$ and $\text{Pr}_{tp} = \text{Pr}_l$ and combining Eq. (12.41) with Eq. (12.39*a*), he concluded that

$$\text{Re}_{tp} = \text{Re}_l[F(X_{tt})]^{1.25} \quad (12.42)$$

From a regression analysis of available data, Chen obtained $F(X_{tt})$ and $S(\text{Re}_{tp})$ curves that provided a best fit for this correlation technique. The correlation curves for $F(X_{tt})$ and $S(\text{Re}_{tp})$ were originally presented only in graphical form. Somewhat later, Collier [12.41] proposed the following empirical relations as fits to Chen's original $F(X_{tt})$ and $S(\text{Re}_{tp})$ curves:

$$F(X_{tt}) = 1 \qquad \text{for } X_{tt}^{-1} \leq 0.1 \quad (12.43a)$$

$$F(X_{tt}) = 2.35\left(0.213 + \frac{1}{X_{tt}}\right)^{0.736} \qquad \text{for } X_{tt}^{-1} > 0.1 \quad (12.43b)$$

$$S(\text{Re}_{tp}) = (1 + 2.56 \times 10^{-6}\,\text{Re}_{tp}^{1.17})^{-1} \tag{12.44}$$

If the wall superheat, mass flux, fluid properties, and quality are specified, Chen's [12.39] correlation can then be used to calculate the heat transfer coefficient as follows:

1. For the specified G, x, and fluid properties, Eqs. (12.32), (12.39*a*), and (12.39*b*) can be used to calculate X_{tt}, Re_l, and h_l.
2. Equation (12.43) can be used to compute $F(X_{tt})$.
3. Equation (12.42) can be used to calculate Re_{tp}, which can then be inserted into Eq. (12.44) to determine S.
4. With the results of steps 1 through 3, Eqs. (12.38) and (12.40) can be used to determine h_{mac} and h_{mic}. The overall heat transfer coefficient is then computed using Eq. (12.37): $h = h_{\text{mic}} = h_{\text{mac}}$.

If a uniform heat flux is applied to the tube walls, the above calculation process must be iterated to determine the wall superheat that when multiplied by the resulting h value yields the specified heat flux.

Modifications to the original Chen correlation have also been proposed in more recent publications. Based on a model analysis of the thermal region near the wall and its effect on nucleation, Bennett et al. [12.42] proposed that the suppression factor S in the Chen correlation be computed using the relation

$$S = \frac{[1 - \exp\{-F(X_{tt})h_l X_0/k_l\}]}{F(X_{tt})h_l X_0/k_l} \tag{12.45}$$

where

$$X_0 = 0.041\left[\frac{\sigma}{g(\rho_l - \rho_v)}\right]^{0.5} \tag{12.46}$$

In most cases, use of this correlation for S yields values of h that are comparable to those obtained using the empirical relation (12.44).

Noting that the original Chen correlation was developed mainly to fit flow boiling data for water, Bennett and Chen [12.43] modified the correlation to account for the effect of the liquid Prandtl number being significantly different from 1. To generalize the Chen correlation for all nonmetallic liquids, they recommended replacing Eq. (12.38) with the relation

$$h_{\text{mac}} = h_l F(X_{tt})\,\text{Pr}_l^{0.296} \tag{12.47}$$

The variation of h with x predicted using the Bennett and Chen [12.43] correlation for upward vertical flow boiling of R-12 at a saturation pressure of 384.5 kPa is shown in Fig. 12.14. Computed results are shown in this figure for G = 300 kg/m^2 s and a wall superheat of 10°C. In addition to the overall h variation, the variations of h_{mic} and h_{mac} are also plotted in this figure. As the quality increases, trends of increasing h_{mac} and decreasing h_{mic} are clearly discernible in Fig. 12.14. Combining the two values, the net trend is a slight but steady increase in h as x increases.

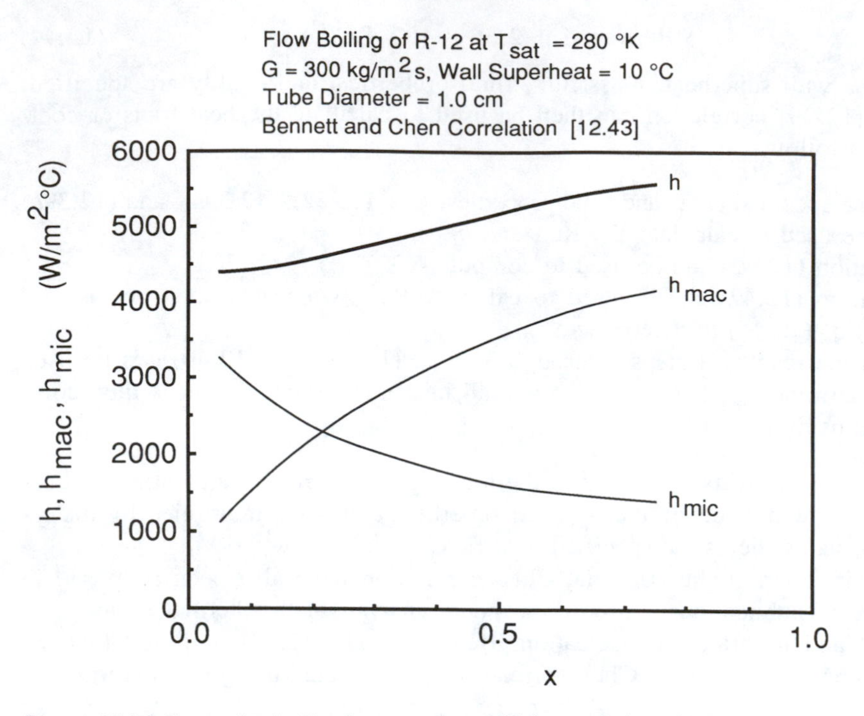

Figure 12.14 Convective boiling heat transfer coefficient variation with quality as predicted for R-12 using the Bennett and Chen correlation [12.43].

The Shah Correlation

For saturated flow boiling in vertical and horizontal tubes, Shah [12.44] has proposed a correlation for the heat transfer coefficient in the form

$$\psi_S = \frac{h}{h_l} = f(\text{Co}, \text{Bo}, \text{Fr}_{le}) \tag{12.48}$$

where

$$\text{Co} = \left(\frac{1-x}{x}\right)^{0.8} \left(\frac{\rho_v}{\rho_l}\right)^{0.5} \tag{12.49}$$

$$\text{Bo} = \frac{q''}{G h_{lv}} \tag{12.50}$$

$$\text{Fr}_{le} = \frac{G^2}{\rho_l^2 g D} \tag{12.51}$$

and h_l is the single-phase coefficient for the liquid phase flowing alone in the tube, as predicted by equations (12.39). The correlation in terms of these variables was specified in graphical form. The graphical representation of Shah's correlation is shown in Fig. 12.15. For specified quality, mass flux, and heat flux values, the dimensionless parameters Co, Bo, and Fr_{le} can be computed and the ratio h/h_l can be determined from the plot in Fig. 12.15.

For vertical tubes, the value of the Froude number, Fr_{le}, is ignored. The user reads up along the vertical line corresponding to the computed Co value to its intersection with the appropriate Bo line and then horizontally across to the vertical axis to determine the h/h_l value. As Bo decreases, the constant Bo lines merge into line AB.

For horizontal tubes, the use of the plot is somewhat different. In this case, the user reads vertically up along the appropriate constant Co line to its intersection with the Fr_{le} line corresponding to the computed value. One then proceeds horizontally to line AB, vertically to the appropriate Bo line, and then horizontally to the left to read the h/h_l value from the vertical axis.

More recently, Shah [12.45] has recommended the following computational representation of his correlation:

$$N_S = \mathrm{Co} \qquad \text{for } \mathrm{Fr}_{le} \geqslant 0.04 \tag{12.52a}$$

$$N_S = 0.38\,\mathrm{Fr}_{le}^{-0.3}\mathrm{Co} \qquad \text{for } \mathrm{Fr}_{le} < 0.04 \tag{12.52b}$$

$$F_S = 14.7 \qquad \text{for } \mathrm{Bo} \geqslant 11 \times 10^{-4} \tag{12.53a}$$

$$F_S = 15.4 \qquad \text{for } \mathrm{Bo} < 11 \times 10^{-4} \tag{12.53b}$$

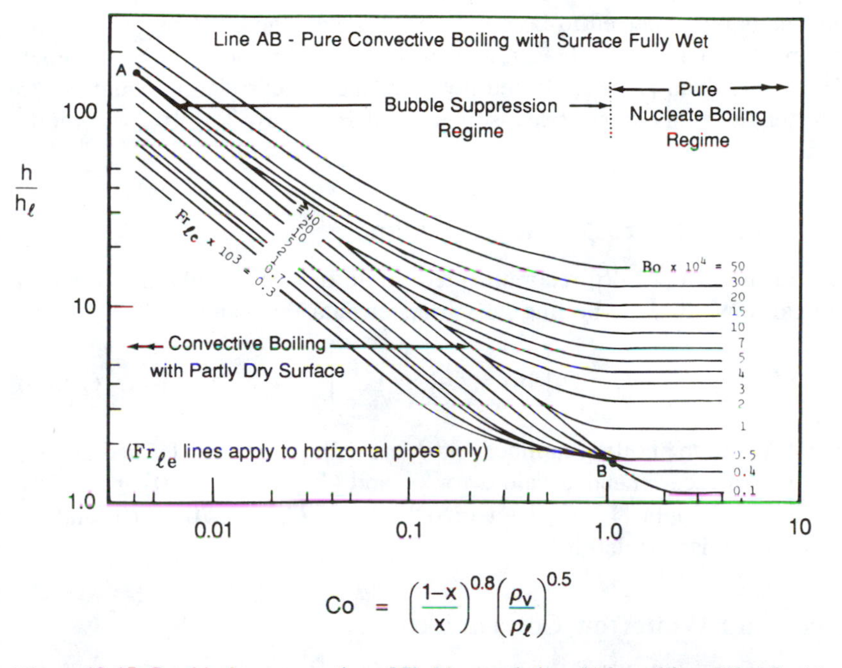

Figure 12.15 Graphical representation of Shah's correlation. (*Adapted from* [12.44] *with permission, copyright © 1976, American Society of Heating Refrigerating and Air-Conditioning Engineers.*)

$$\psi_{cb} = 1.8 N_S^{-0.8} \tag{12.54}$$

For $N_S > 1.0$:

$$\psi_{nb} = 230 \text{ Bo}^{0.5} \qquad \text{for Bo} > 0.3 \times 10^{-4} \tag{12.55a}$$

$$\psi_{nb} = 1 + 46 \text{ Bo}^{0.5} \qquad \text{for Bo} \leqslant 0.3 \times 10^{-4} \tag{12.55b}$$

$$\psi_S = \text{the larger of } \psi_{nb} \text{ and } \psi_{cb} \tag{12.56}$$

For $N_S \leqslant 1.0$:

$$\psi_{bs} = F_S \text{ Bo}^{0.5} \exp(2.74 N_S^{-0.1}) \qquad \text{for } 0.1 < N_S \leqslant 1.0 \tag{12.57a}$$

$$\psi_{bs} = F_S \text{ Bo}^{0.5} \exp(2.47 N_S^{-0.15}) \qquad \text{for } N_S \leqslant 0.1 \tag{12.57b}$$

$$\psi_S = \text{the larger of } \psi_{bs} \text{ and } \psi_{cb} \tag{12.58}$$

Shah [12.45] indicates that the equations given above agree with the curves in the graphical representation within ±6% over most of the chart except for two regions:

1. Near Co = 0.004 and Bo = 50×10^{-4}
2. For horizontal tubes at $\text{Fr}_{le} < 0.04$ and Bo $< 1 \times 10^{-4}$

Shah [12.45] notes that these equations overpredict h by about 11% on the first region, but this is not expected to be a problem because these conditions usually fall into the postdryout region. Inaccuracy in the second region can be as much as 20%. However, Shah [12.45] elected not to refine the correlating equations to achieve a better fit in this range, because values of Bo below 1×10^{-4} are rarely encountered.

The Correlation of Schrock and Grossman

Schrock and Grossman [12.46] recommended the following correlation based on a fit to vertical upward, flow-boiling heat transfer data for water:

$$\frac{h}{h_l} = C_1\left[\text{Bo} + C_2\left(\frac{1}{X_{tt}}\right)^{0.66}\right] \tag{12.59}$$

where Bo and X_{tt} are the boiling number and Martinelli parameter defined above. These investigators recommended values for C_1 and C_2 of 7.39×10^3 and 1.5×10^{-4}, respectively. Wright [12.47] later recommended $C_1 = 6.70 \times 10^3$ and $C_2 = 3.5 \times 10^{-4}$ for this correlation.

The Gungor and Winterton Correlation

Gungor and Winterton [12.48] proposed the following correlation for heat transfer during convective flow boiling in vertical tubes:

$$h = h_l \left[1 + 3000 \text{ Bo}^{0.86} + \left(\frac{x}{1-x} \right)^{0.75} \left(\frac{\rho_l}{\rho_v} \right)^{0.41} \right] \tag{12.60}$$

Here again, h_l is the heat transfer coefficient for the liquid phase flowing alone, and Bo is the boiling number defined by Eq. (12.50).

The Correlation of Bjorge, Hall, and Rohsenow

For saturated upward flow boiling in vertical tubes at qualities above 0.05, Bjorge, Hall, and Rohsenow [12.49] recommended the following correlation technique for predicting h:

$$h = \frac{q''_{\text{tot}}}{T_w - T_{\text{sat}}} \tag{12.61}$$

where

$$q''_{\text{tot}} = q''_{fc} + q''_{fdb} \left[1 - \left(\frac{(T_w - T_{\text{sat}})_i}{T_w - T_{\text{sat}}} \right)^3 \right] \tag{12.62}$$

$$q''_{fc} = F_B \text{ Pr}_l (k_l/D)(T_w - T_{\text{sat}}) \frac{\text{Re}_l^{0.9}}{C_2} \tag{12.63}$$

$$F_B = 0.15 \left(\frac{1}{X_{tt}} + \frac{2}{X_{tt}^{0.32}} \right) \tag{12.64}$$

$$C_2 = 5 \text{ Pr}_l + 5 \text{ Pr}_l \ln(1 + 5 \text{ Pr}_l) + 2.5 \ln(0.0031 \text{ Re}_l^{0.812}) \qquad \text{for Re}_l > 1125 \tag{12.65a}$$

$$C_2 = 5 \text{ Pr}_l + 5 \ln[1 + \text{Pr}_l(0.0964 \text{ Re}_l^{0.585} - 1)] \qquad \text{for } 50 < \text{Re}_l \leq 1125 \tag{12.65b}$$

$$C_2 = 0.0707 \text{ Pr}_l \text{ Re}_l^{0.5} \qquad \text{for Re}_l \leq 50 \tag{12.65c}$$

$$q''_{fdb} = B_M \mu_l h_{lv} \left[\frac{g(\rho_l - \rho_v)}{\sigma} \right]^{1/2} \left[\frac{k_l^{1/2} \rho_l^{17/8} c_{pl}^{19/8} \rho_v^{1/8} (T_w - T_{\text{sat}})^3}{\mu_l h_{lv}^{7/8} (\rho_l - \rho_v)^{9/8} \sigma^{5/8} T_{\text{sat}}^{1/8}} \right] \tag{12.66}$$

$$(T_w - T_{\text{sat}})_i = \frac{8 \sigma T_{\text{sat}} h_{fc}}{k_l h_{lv}} \left(\frac{1}{\rho_v} - \frac{1}{\rho_l} \right) \tag{12.67}$$

This correlation technique postulates a superposition of heat fluxes rather than a superposition of heat transfer coefficients. Correlation of the forced-convective (macroscopic) contribution to the heat flux again presumes that the corresponding heat transfer coefficient is primarily a function of Re_l, X_{tt}, and Pr_l. The nucleate boiling (microscopic) contribution is predicted using the Mikic and Rohsenow [12.50] correlation (12.66) for fully developed nucleate boiling. B_M in this relation is a dimensional constant that depends on the solid-surface cavity size distribution and the fluid properties. For forced convection of water, they recommend $B_M =$

Table 12.1

	Co < 0.65 (convective region)	Co ⩾ 0.65 (nucleate boiling region)
C_1	1.1360	0.6683
C_2	−0.9	−0.2
C_3	667.2	1058.0
C_4	0.7	0.7
C_5[a]	0.3	0.3

[a]$C_5 = 0$ for vertical tubes and horizontal tubes with $\mathrm{Fr}_{le} > 0.04$.

1.89×10^{-14} for properties in SI units. $(T_w - T_{sat})_i$ in Eq. (12.62) is the wall superheat at the incipient boiling (onset) condition.

Kandlikar's Correlation

Very recently, Kandlikar [12.51] proposed the following correlation as a fit to a very broad spectrum of data for flow boiling heat transfer in vertical and horizontal tubes:

$$h = h_l[C_1\,\mathrm{Co}^{C_2}(25\,\mathrm{Fr}_{le})^{C_5} + C_3\,\mathrm{Bo}^{C_4}\,F_K] \tag{12.68}$$

The constants C_1 through C_5 are given in Table 12.1. The factor F_K is a fluid-dependent parameter, values of which are listed for various fluids in Table 12.2. For fluids other than those listed in Table 12.2, Kandlikar [12.51] recommends that F_K be estimated as the multiplier that must be applied to the Forster and Zuber [12.40] correlation to correlate pool boiling data for the fluid of interest.

Comparison of Correlations

Like most of the other correlations described above, Kandlikar's [12.51] correlation sums the contributions of terms representing nucleate boiling and forced-convective effects. The variation of these contributions and that of the overall

Table 12.2

Fluid	F_K
Water	1.00
R-11	1.30
R-12	1.50
R-13B1	1.31
R-22	2.20
R-113	1.30
R-114	1.24
R-152a	1.10
Nitrogen	4.70
Neon	3.50

heat transfer coefficient are shown in Fig. 12.16 for flow boiling of R-12 at 384.5 kPa in a vertical tube. The tube inside diameter is 1.0 cm, $G = 300\ \text{kg/m}^2\ \text{s}$, and the wall superheat is 10°C. The conditions are identical to those for Fig. 12.14, which shows similar variations as predicted by the Bennett and Chen [12.43] correlation.

Comparing Figs. 12.14 and 12.16, it can be seen that both correlations predict that, as the quality increases, the nucleate boiling contribution diminishes while the forced-convective effect increases. However, the net effect is different in the two cases. As quality increases, the Kandlikar correlation predicts a steadily decreasing heat transfer coefficient, whereas the Bennett and Chen correlation predicts a gradual increase in overall h. Kandlikar [12.51] points out that data for refrigerants often exhibit a decrease in h with quality at a specified mass flux.

Figure 12.17 shows the variation of overall h predicted by all of the correlations described above for the same conditions as Figs. 12.14 and 12.16. While the Bennett and Chen [12.43] correlation and the correlation of Bjorge et al. [12.49] show a slight monotonic increase with quality, the other correlations exhibit more complicated trends. The Shah [12.44] and Schrock-Grossman [12.46] correlations generally indicate that h first increases and then decreases slightly with quality. The Gungor and Winterton [12.48] and Kandlikar [12.51] correlations indicate a steady drop in heat transfer coefficient with quality.

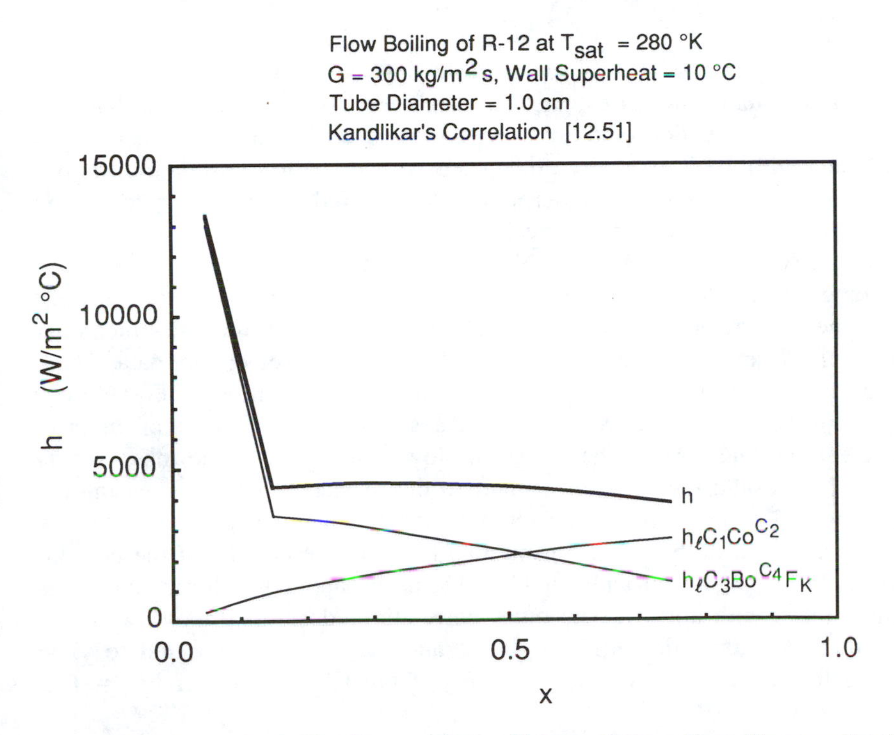

Figure 12.16 Convective boiling heat transfer coefficient variation with quality as predicted for R-12 using Kandlikar's correlation.

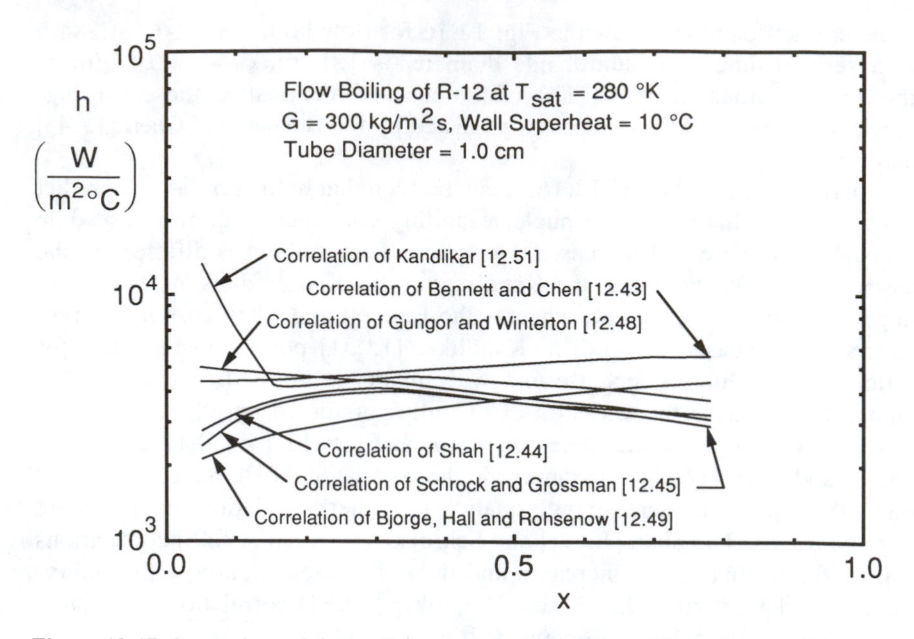

Figure 12.17 Comparison of the predictions of six different correlations for convective boiling heat transfer.

While there seems to be a consensus regarding the behavior of the individual mechanisms as quality increases, the net combined effect of these mechanisms varies depending on which of these correlations is used. The agreement of refrigerant data with the low-quality predictions of Kandlikar's correlation implies that the other correlations may underpredict the strength of the nucleate boiling effect in hydrocarbon fluids at low quality.

The relatively good agreement with data for a broad spectrum of fluids over a wide range of conditions suggests that Kandlikar's [12.51] may be the most reliable general correlation. It was found to agree with water data to a mean deviation of 15.9% and to a mean deviation of 18.8% for all refrigerant data. There is, however, one curious aspect of this correlation. Examination of Eq. (12.68) together with the relation (12.39) for h_l indicates that the contribution of the (second) nucleate boiling term to the overall h slowly increases as the mass flux G increases. This would seem to run counter to the expectation that increasing the flow rate would increase bulk convection, suppressing nucleation and thereby weakening the nucleate boiling contribution. It should be noted that the correlations of Shah [12.44] and Kandlikar [12.51] can be applied to horizontal tubes or vertical tubes with upflow. The other correlations described above were developed to match data for vertical upflow, and they can be expected to yield reasonable results for other orientations only if the Froude number $\mathrm{Fr}_{le} = G^2/(\rho_l^2 g D)$ is large (>.05)

All the correlations described above have been used with some success to correlate flow-boiling heat transfer data over a finite range of flow conditions. In selecting one of the above correlations to predict the flow-boiling heat transfer coefficient for a specific set of circumstances, perhaps the best advice is to select one that has been verified against data for fluid and flow conditions that are as similar as possible to those under consideration. It should be noted that the correlations described above generally will not be accurate at qualities beyond that at which dryout of the liquid film on the wall occurs. For round tubes, this is usually estimated to occur at a quality somewhere above $x = 0.7$. Determination of the exact condition at which dryout occurs will be discussed further in the next section.

Example 12.4 Use Kandlikar's [12.51] correlation to predict the heat transfer coefficient for flow boiling of nitrogen in a vertical tube at a pressure of 778 kPa and qualities of 0.20 and 0.60. The tube diameter is 0.9 cm, the flow rate is such that the mass flux is 200 kg/m^2 s, and a uniform heat flux of 20 kW/m^2 is applied to the tube wall. Compare the results with the predictions of the Gungor and Winterton [12.48] correlation for the same conditions.

For saturated nitrogen at 778 kPa, $T_{sat} = 100$ K $= -173.2$°C, $\rho_l = 691$ kg/m^3, $\rho_v = 32.0$ kg/m^3, $h_{lv} = 162.2$ kJ/kg, $c_{pl} = 2.31$ kJ/kg K, $\mu_l = 86.9 \times 10^{-6}$ N s/m^2, $\mu_v = 7.28 \times 10^{-6}$ N s/m^2, $k_l = 0.0955$ W/m K, $\mathrm{Pr}_l = 2.10$, and $\sigma = 0.00367$ N/m. At $x = 0.20$,

$$\mathrm{Re}_l = \frac{G(1-x)D}{\mu_l} = \frac{200(0.8)(0.009)}{86.9 \times 10^{-6}} = 16{,}600$$

Because the flow is turbulent, the Dittus-Boelter equation is used to determine h_l:

$$\begin{aligned} h_l &= 0.023\left(\frac{k_l}{D}\right) \mathrm{Re}_l^{0.8}\, \mathrm{Pr}_l^{0.4} \\ &= 0.023\left(\frac{0.0955}{0.009}\right)(16{,}600)^{0.8}(2.10)^{0.4} = 780 \text{ W/m}^2\text{ K} \end{aligned}$$

A similar set of calculations indicates that, for $x = 0.60$, $\mathrm{Re}_l = 8400$ and $h_l = 448$ W/m^2 K. The convection and boiling numbers are given by Eqs. (12.49) and (12.50):

$$\mathrm{Co} = \left(\frac{1-x}{x}\right)^{0.8}\left(\frac{\rho_v}{\rho_l}\right)^{0.5}$$

$$\mathrm{Bo} = \frac{q''}{Gh_{lv}}$$

Substituting into the equation for Bo yields

$$\mathrm{Bo} = \frac{20}{(200 \times 162.2)} = 6.17 \times 10^{-4}$$

For $x = 0.2$,

$$\mathrm{Co} = \left(\frac{0.8}{0.2}\right)^{0.8}\left(\frac{32.0}{691}\right)^{0.5} = 0.652$$

and for $x = 0.6$ one similarly finds that $\mathrm{Co} = 0.156$. From Tables 12.1 and 12.2, it follows that for $x = 0.2$, the constants for $\mathrm{Co} > 0.65$ apply and $F_K = 4.70$ for nitrogen. We further note that $C_5 = 0$ since the tube is vertical. Kandlikar's relation (12.68) for the heat transfer coefficient then reduces to

$$h = h_l(C_1\,\mathrm{Co}^{C_2} + C_3\,\mathrm{Bo}^{C_4}\,F_K)$$

Substituting the appropriate constant values,

$$h = 780[0.6683(0.652)^{-0.2} + 1058(6.17 \times 10^{-4})^{0.7}(4.70)]$$

$$= 2.25 \times 10^4\ \mathrm{W/m^2\,K} \qquad (\text{for } x = 0.2)$$

Using the constants from Table 12.1 for $\mathrm{Co} < 0.65$, substituting the appropriate parameter values for $x = 0.6$ yields

$$h = 448[1.1360(0.156)^{-0.9} + 667.2(6.17 \times 10^{-4})^{0.7}(4.70)]$$

$$= 1.07 \times 10^4\ \mathrm{W/m^2\,K} \qquad (\text{for } x = 0.6)$$

The correlation of Gungor and Winterton [12.48] is given by equation (12.60):

$$h = h_l\left[1 + 3000\,\mathrm{Bo}^{0.86} + \left(\frac{x}{1-x}\right)^{0.75}\left(\frac{\rho_l}{\rho_v}\right)^{0.41}\right]$$

Substituting the appropriate h_l and x values for $x = 0.2$ and $x = 0.6$ yields

$$h = 780\left[1 + 3000(6.17 \times 10^{-4})^{0.86} + \left(\frac{0.2}{0.8}\right)^{0.75}\left(\frac{691}{32.0}\right)^{0.41}\right]$$

$$= 5.82 \times 10^3\ \mathrm{W/m^2\,K} \qquad \text{for } x = 0.2$$

$$h = 448\left[1 + 3000(6.17 \times 10^{-4})^{0.86} + \left(\frac{0.6}{0.4}\right)^{0.75}\left(\frac{691}{32.0}\right)^{0.41}\right]$$

$$= 4.92 \times 10^3\ \mathrm{W/m^2\,K} \qquad \text{for } x = 0.6$$

Thus, it can be seen that for these circumstances, the values of h predicted by these schemes differ by a factor of 3.9 at $x = 0.2$ and a factor of 2.2 at $x = 0.6$.

12.5 CRITICAL HEAT FLUX CONDITIONS FOR INTERNAL FLOW BOILING

The terms *critical heat flux condition* (CHF) and boiling *burnout* are used to describe the conditions at which the wall temperature rises and/or the heat transfer coefficient decreases sharply due to a change in the heat transfer mechanism. The term "burnout" is used even when failure of the passage wall does not occur due to overheating and melting, although clearly its origins can be traced back to circumstances when the tube wall does fail in this manner.

The critical heat flux condition is indicated schematically in Figs. 12.4 and 12.5 as diagonal (or, in the limit, vertical) lines. The nature of the transition indicated by these lines varies with the enthalpy of the flow. At subcooled (bulk flow) conditions and low qualities, this transition corresponds to a change in the boiling mechanism from nucleate to film boiling. For this reason, the critical heat flux condition for these circumstances is often referred to as the *departure from nucleate boiling* (DNB). In the Soviet literature, this transition is referred to as *burnout of the first kind.*

At moderate to high qualities, the flow is almost invariably in an annular configuration, and the transition corresponds to dryout of the liquid film on the tube wall (see Fig. 12.18). For this range of conditions, this transition is usually referred to simply as *dryout*. Most Soviet investigators have used the terminology *burnout of the second kind* for this transition. As indicated in Figs. 12.4 and 12.5, once dryout occurs, the flow enters the so-called liquid-deficient region, in which the remaining liquid exists as entrained droplets. The heat transfer associated with this process will be discussed in more detail in Section 12.6. However, it is worth noting at this stage that because of the high vapor velocity typical of these conditions, the convection of heat from the tube wall to the vapor is generally strong. As a result, the drop in the heat transfer coefficient for dryout is usually not as severe as that which accompanies departure from nucleate boiling at the same total mass flux.

Before discussing measurements of the critical heat flux conditions and methods for predicting them, it is useful first to consider the mechanisms responsible for part or all of the tube wall becoming dry. Although numerous investigations of critical heat flux conditions have been conducted, the mechanisms responsible for the critical heat flux transitions are not well understood at the present time. However, some light has been shed on the nature of these mechanisms. At subcooled or very low quality conditions, bubbly or slug flow is typically encountered. For such conditions, three potential mechanisms have been discussed by Tong and Hewitt [12.52]. These mechanisms are illustrated in Fig. 12.19.

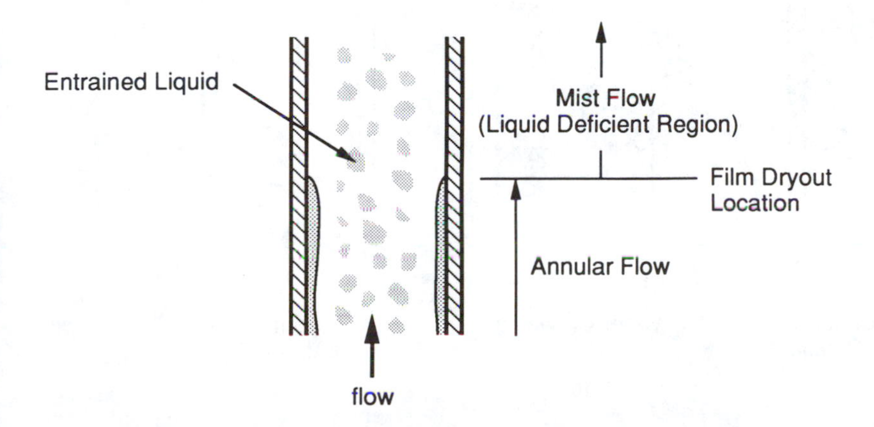

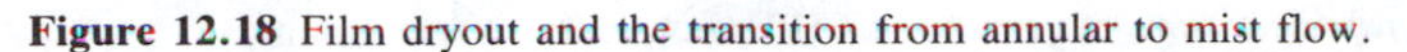

Figure 12.18 Film dryout and the transition from annular to mist flow.

The first of these mechanisms is associated with the evaporation of the liquid microlayer under a growing vapor bubble on the heated tube wall (Fig. 12.19*a*). Just prior to release of the bubble, evaporation of the microlayer may leave a portion of the wall under the bubble completely dry. If a constant-heat-flux condition is applied to the wall, the temperature of the dry patch may rise above the Leidenfrost temperature, preventing the patch, or a portion of it, from rewetting. Continued evaporation of the microfilm at the perimeter of the dry patch may cause it to grow. This may eventually lead to the entire wall becoming dry, whereupon a transition to convective film boiling occurs.

The second mechanism identified by Tong and Hewitt [12.52] may be encountered at moderate bulk subcooling levels. For such conditions, bubbles may be concentrated in a boundary layer near the wall (see Fig. 12.19*b*). If the number density of the bubbles and size of the bubble boundary layer become large enough, liquid flow to the surface may be impeded. Liquid under and between bubbles at the surface may then be evaporated away, producing dry patches at the wall. Dry patches produced in this fashion may continue to grow as liquid near the surface is evaporated away, ultimately producing a transition to convective film boiling.

The third mechanism described by Tong and Hewitt [12.52] is associated with slug flow encountered for low-quality saturated flows. As a slug flow bubble moves downstream, if the heat flux from the wall is high enough, the liquid film between the bubble interface and the wall may completely evaporate at a particular location, forming a dry patch, as indicated in Fig. 12.19*c*. For a sufficiently high applied heat flux, the surface temperature may exceed the Leidenfrost temperature and prevent the patch from rewetting. The dry patch may subsequently grow in size, leading to a transition to convective film boiling.

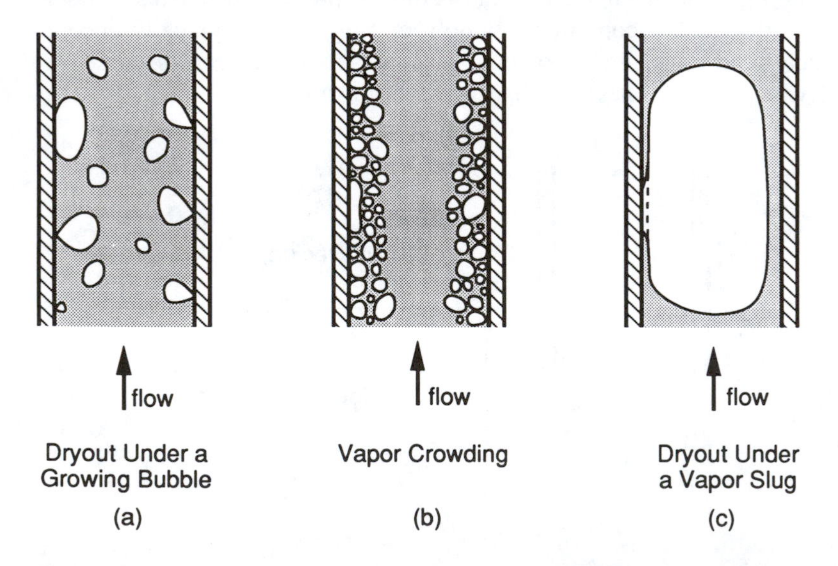

Figure 12.19 Schematic representations of postulated CHF mechanisms at low quality.

Despite numerous studies of critical heat flux (CHF) conditions, the mechanisms described above are not well understood. The complexity of these mechanisms has also made analytical modeling generally difficult. As a result, most proposed methods for predicting CHF conditions have been correlations based on experimental data.

The body of CHF data generated in prior experimental studies is extremely large. Bergles [12.53] has estimated that several hundred thousand CHF data points have been obtained in such studies, and that over 200 correlations have been developed in attempts to correlate the data. Most of the available data have been obtained for upward flow boiling of water in a uniformly heated tube. A review of all the available CHF data is clearly beyond the scope of this text. The interested reader may wish to consult the summary of CHF data for water in vertical tubes compiled by Thompson and Macbeth [12.54] for further information.

With the large available data base, evaluation of the data to eliminate questionable points is a formidable task. In the late 1960s, a group in the Soviet Union at the All-Union Heat Engineering Institute (VTI) addressed this issue by reviewing available data. They subsequently tabulated recommended values of CHF conditions for upflow of water in a vertical tube with an inside diameter of 8 mm. These tabulated values were listed in a paper by Doroschuk and Lantsman [12.55]. This paper also described methods for extrapolating the CHF conditions to round tubes with different diameters. The CHF values summarized in this paper are for the most part departures from nucleate boiling (DNB) transitions at low quality or subcooled conditions.

A more extensive tabulation, including DNB as well as dryout transitions, was published somewhat later by the Heat and Mass Transfer Section of the Scientific Council, USSR Academy of Sciences [12.56]. The tabulated CHF conditions were for water in a vertical round tube with an inside diameter of 8 mm, and the following method was recommended for predicting the critical heat flux for other tube diameters:

$$q''_{\text{crit}} = (q''_{\text{crit}})_{8\ \text{mm}} \left(\frac{8}{D}\right)^{1/2} \tag{12.69}$$

where D is the tube inside diameter in millimeters.

Values of the CHF presented in tabulated form in the USSR Academy of Sciences report [12.56] are plotted for two pressure levels in Figs. 12.20 and 12.21. These figures clearly indicate the effects of varying system conditions on the critical heat flux.

In general, for upward flow of water in a round tube, the parameters that may affect the critical heat flux include the mass flux G, the pressure P, the tube diameter D, the location z downstream of the tube inlet, and either the quality x or subcooling $T_{\text{sat}} - T_{\text{bulk}}$. The bulk of the available data indicate that for $z/D > 20$, the downstream location z has no significant effect on the CHF condition, implying that the critical heat flux is dictated by local conditions alone. For sufficiently long tubes, it follows that the critical heat flux depends mainly on the

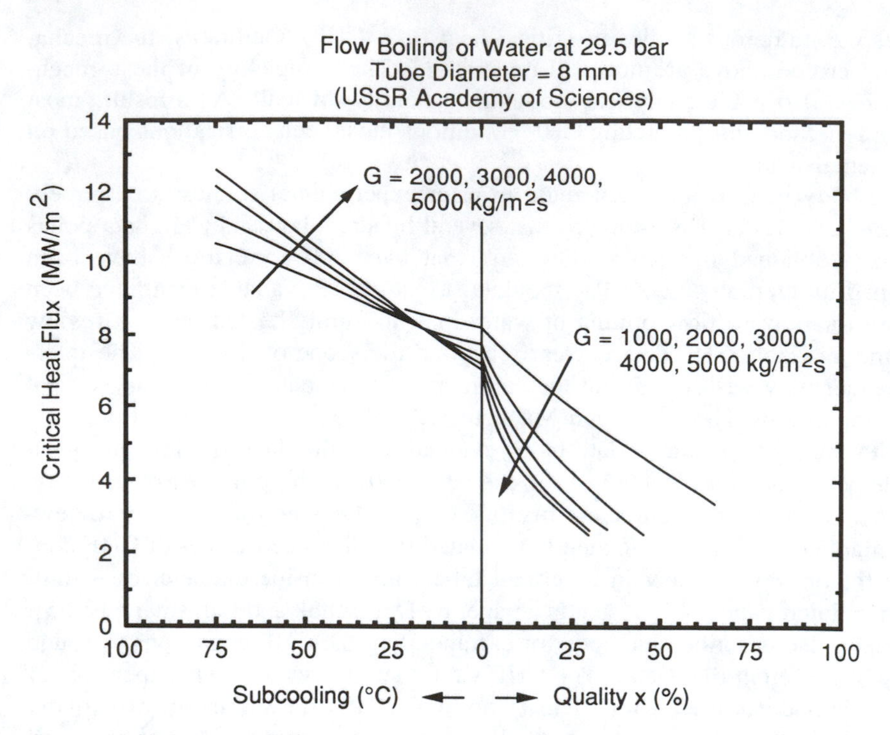

Figure 12.20 Predicted CHF conditions for water at 29.5 bar based on data collected by the USSR Academy of Sciences.

four remaining variables:

$$q''_{\text{crit}} = f(D, P, G, x_{\text{crit}}) \qquad \text{for saturated flow} \tag{12.70a}$$

or

$$q''_{\text{crit}} = f(D, P, G, T_{\text{sat}} - T_{\text{bulk}}) \qquad \text{for subcooled flow} \tag{12.70b}$$

The effects of each of these parameters on the critical heat flux can be observed by examining Figs. 12.20 and 12.21. Figure 12.20 shows the variation of the critical heat flux q''_{crit} with subcooling and quality for various mass flux levels at a pressure of 29.5 bar. Figure 12.21 show similar variations of the critical heat flux at a pressure of 69 bar. Comparison of these two figures indicates that, other parameters being the same, q''_{crit} generally increases as the pressure decreases over this pressure range. For subcooled conditions, this trend persists throughout the tabulated conditions. For saturated conditions, the trend weakly reverses at high pressures and moderate to high qualities.

It is clear from the plots that the critical heat flux decreases as the bulk enthalpy of the coolant increases: q''_{crit} decreases as subcooling decreases or quality increases. It can be seen that if the subcooled and saturated conditions are considered separately, the tabulated values are often well fit by a linear variation of subcooling or quality. As will be seen below, this provides some guidance in the development of correlations for the CHF condition.

The effects of increasing mass flux are clearly different for subcooled and saturated flows. For strongly subcooled conditions, increasing the flow rate increases the critical heat flux, apparently because doing so strengthens convective transport. For saturated flow, increasing the mass flux decreases the critical heat flux at a given quality. When the flow is in the churn or annular regime, this trend is apparently a consequence of the increase in entrainment that generally accompanies increasing mass flux. Since a greater portion of the liquid inventory flows as entrained droplets on the core, less flows in the liquid film, making it more difficult to keep the walls fully wetted.

The CHF conditions indicated in Figs. 12.20 and 12.21 are all for relatively high heat flux levels. Figure 12.22 shows a sample of CHF data reported by Levitan and Lantsman [12.57] for lower heat flux levels. At these lower heat flux levels, their data indicate that below a certain heat flux, the critical quality becomes essentially independent of heat flux. In considering this type of behavior in low-heat-flux CHF data, Doroshchuk et al. [12.58] interpreted this range as corresponding to dryout of the liquid film on the wall of the tube (burnout of the second kind). Prior to this transition, the CHF transition was interpreted to be a departure from nucleate boiling.

Based on an assessment of the available data base, values of the limiting

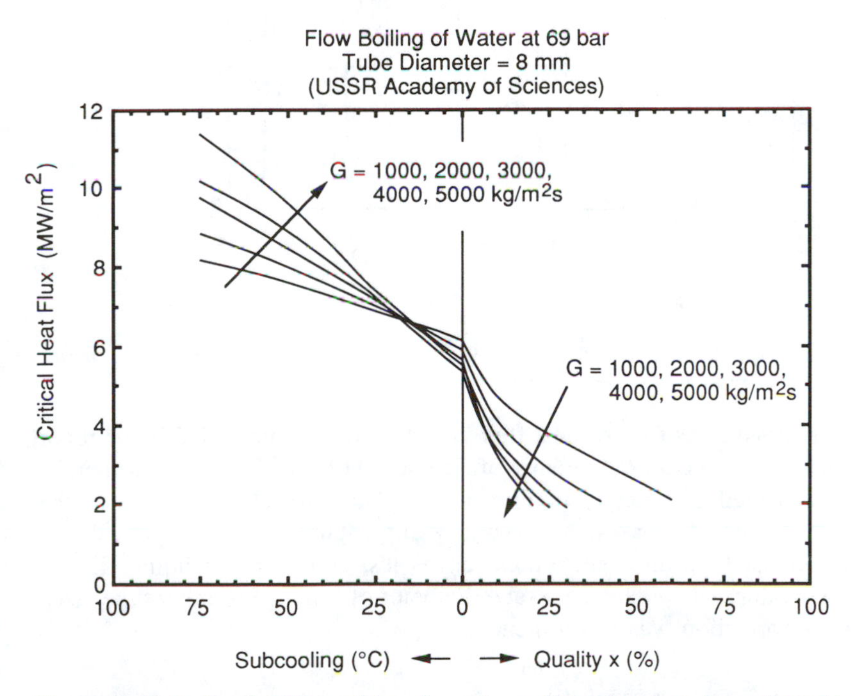

Figure 12.21 Predicted CHF conditions for water at 69 bar based on data collected by the USSR Academy of Sciences.

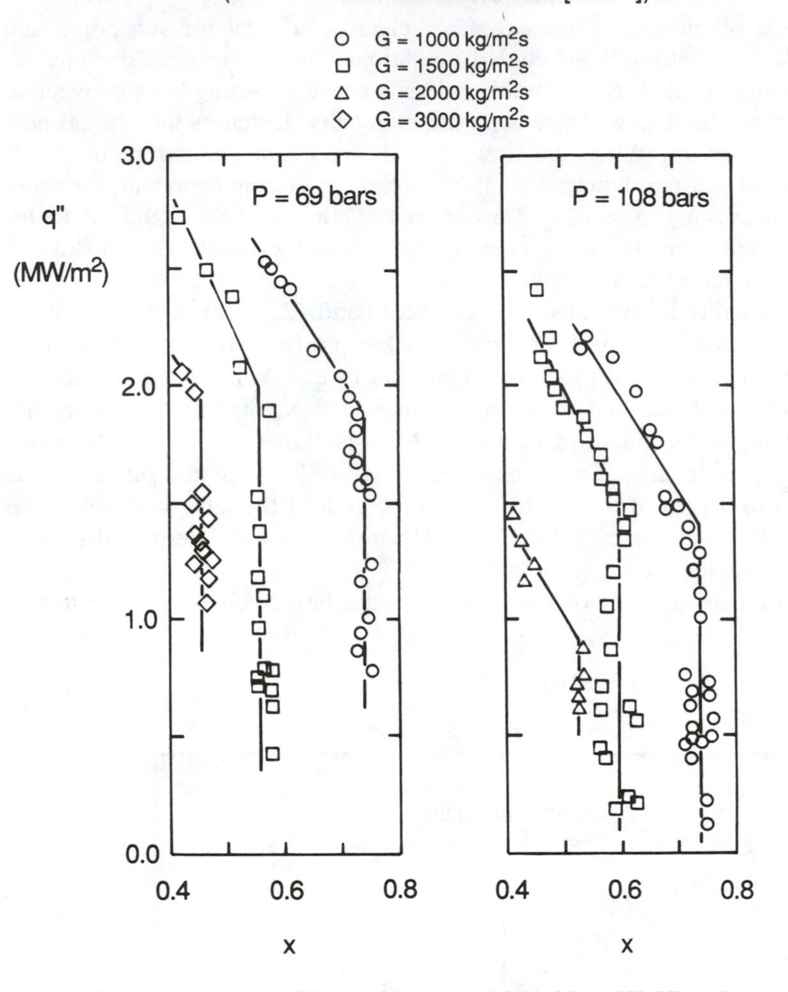

Figure 12.22 Low-heat-flux CHF data for water. (*Adapted from* [*12.57*] *with permission, copyright © 1975, British Library Document Supply Centre.*)

dryout quality conditions for low heat flux values were recommended in the report by the members of the USSR Academy of Sciences [12.56]. These recommended values are presented graphically in Fig. 12.23. This plot clearly indicates that increasing mass flux decreases the dryout quality. This apparently is a consequence of increased entrainment, which leaves less liquid in the liquid film on the walls. All values shown are for a tube diameter of 8 mm. For other tube sizes, the following correction was recommended:

$$x_{\text{crit}} = (x_{\text{crit}})_{8\text{mm}} \left(\frac{8}{D} \right)^{0.15} \qquad \text{(where } D \text{ is in mm)} \tag{12.71}$$

The lack of heat flux dependence on the measured dryout quality values for these conditions was attributed by Doroshchuk et al. [12.58] to the nonprecipitation of entrained droplets on the wall at low mass flux levels and moderate heat flux levels. Doroshchuk et al. [12.58] also presented CHF data at high mass flux levels that suggest that the dryout quality may exhibit some renewed heat flux dependence at very low heat flux levels and high mass flux levels. These investigators attributed this dependence on heat flux to the increased precipitation (deposition) of droplets on the tube wall at higher mass flux levels. The model of the entrainment and deposition mechanisms discussed in Chapter Ten does imply that the deposition rate increases as the mass flux increases, which tends to support this hypothesis. This matter apparently was explored in only a limited fashion, because the heat flux levels at which it occurs are low compared to those of interest in power system boilers. These low heat flux trends in the dryout quality may be of importance, however, to refrigeration-system evaporators and other applications where low-heat-flux evaporation processes are encountered.

The effect of tube diameter on critical heat flux may be surmised from in-

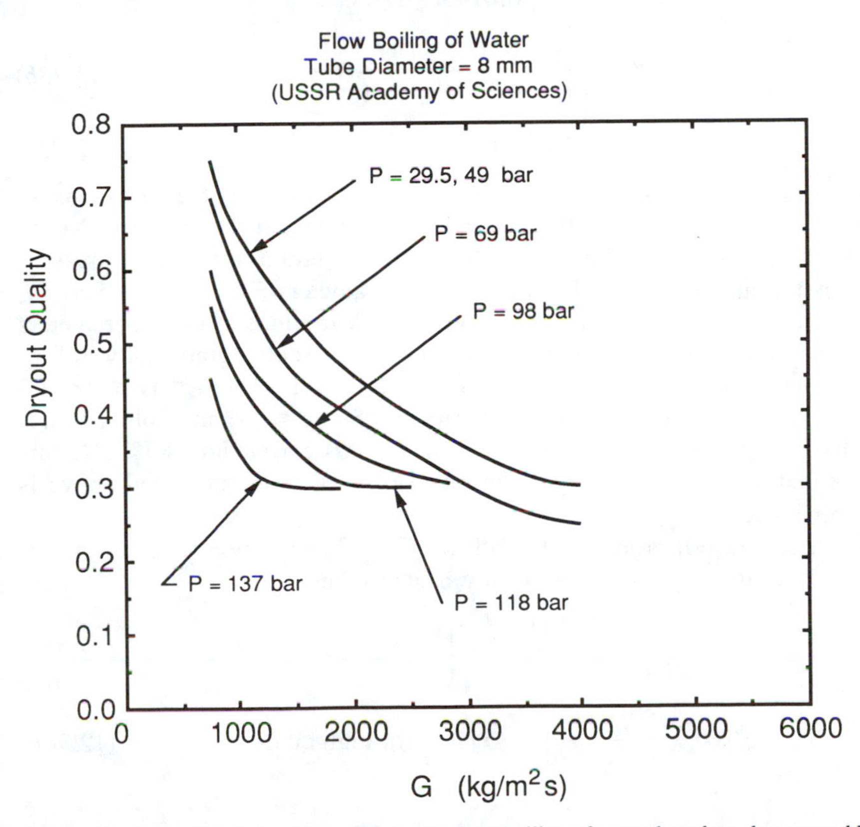

Figure 12.23 Predicted dryout conditions for flow boiling of water based on data assembled by the USSR Academy of Sciences.

spection of the recommended correction relation (12.69). It implies that tubes of larger diameter will have lower critical heat flux values.

Correlations for Predicting the Critical Heat Flux

As noted above, a very large number of correlation methods for predicting the CHF condition have been described in the literature. Discussion of all such correlations here is clearly impossible. However, a representative sample of these correlations will be described in this subsection.

Building on earlier investigations by Thompson and Macbeth [12.54], Bowring [12.59] proposed the following empirical correlation for the CHF condition:

$$q''_{\text{crit}} = \frac{A' + DG(\hat{h}_{l,\text{sat}} - \hat{h}_{\text{in}})/4}{C' + z} \tag{12.72}$$

where

$$A' = \frac{0.579F_{B1}DGh_{lv}}{1.0 + 0.0143F_{B2}D^{1/2}G} \tag{12.73a}$$

$$C' = \frac{0.077F_{B3}DG}{1.0 + 0.347F_{B4}(G/1356)^n} \tag{12.73b}$$

$$n = 2.0 - 0.00725P \tag{12.73c}$$

In these relations, D is the internal tube diameter in m, G is the mass flux in kg/m^2 s, h_{lv} is the latent heat of vaporization in J/kg, and the pressure P is in bar. In Eq. (12.72), $\hat{h}_{l,\text{sat}}$ is the enthalpy of saturated liquid at the local pressure P and $\hat{h}_{\text{in}}$ is the inlet enthalpy. The correlation parameters F_{B1}, F_{B2}, F_{B3}, and F_{B4} (for water) were taken to be functions of pressure. Variations of these parameters recommended for use with this correlation scheme are shown graphically in Fig. 12.24. This correlation is based on a fit to data in the ranges $136 \leqslant G \leqslant 18{,}600$ kg/m^2 s, $2 \leqslant P \leqslant 190$ bar, $2 \leqslant D \leqslant 45$ mm, $0.15 \leqslant z \leqslant 3.7$ m. Collier [12.4] reports that the root-mean-square (RMS) error for this correlation is 7%. He further notes that extrapolation outside the range of parameters indicated above is not recommended.

Biasi et al. [12.60] proposed the following CHF correlation scheme for vertical upflow boiling of water in uniformly heated tubes:

$$q''_{\text{crit}} = \frac{1883}{D^m G^{1/6}}\left[\frac{F_{Bi}(P)}{G^{1/6}} - x\right] \quad \text{for low qualities} \tag{12.74a}$$

$$q''_{\text{crit}} = \frac{3780H_{Bi}(P)}{D^m G^{0.6}}[1 - x] \quad \text{for high qualities} \tag{12.74b}$$

where

$$m = 0.4 \qquad \text{for } D \geqslant 1 \text{ cm} \tag{12.75a}$$

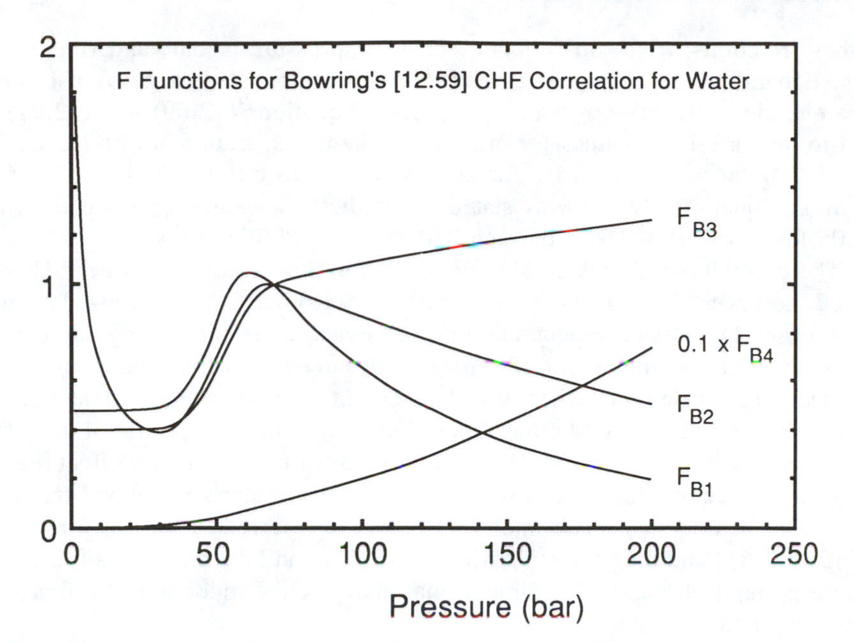

Figure 12.24 Pressure dependence of constants in Bowring's [12.59] CHF correlation for water.

$$m = 0.6 \qquad \text{for } D < 1 \text{ cm} \tag{12.75b}$$

$$F_{Bi}(P) = 0.7249 + 0.099P \exp(-0.032P) \tag{12.76a}$$

$$H_{Bi}(P) = -1.159 + 0.149P \exp(-0.019P) + \frac{8.99P}{10 + P^2} \tag{12.76b}$$

In these dimensional relations, D is in cm, P is in bar, and G is in g/cm^2 s. The correlation is reported to be valid over the ranges $10 < G < 600$ g/cm^2 s, $2.7 < P < 140$ bar, $0.3 < D < 3.75$, $20 < z < 600$ cm, $\rho_v/(\rho_l - \rho_v) < x < 1$. Operationally, one calculates the critical heat flux using both Eqs. (12.74*a*) and (12.74*b*), and the larger of the two is taken as the correct value. The RMS error of this correlation against a data base of over 4500 points was reported to be 7.3%.

For upflow boiling of water in a vertical tube with constant applied heat flux, Levitan and Lantsman [12.57] recommended the following correlation for burnout of the first kind (DNB) in an 8-mm-diameter tube:

$$q''_{\text{crit}} = \left[10.3 - 7.8\left(\frac{P}{98}\right) + 1.6\left(\frac{P}{98}\right)^2\right]\left(\frac{G}{1000}\right)^{1.2\{[0.25(P-98)/98]-x\}} e^{-1.5x} \tag{12.77}$$

For burnout of the second kind (dryout), they recommend the following relation for predicting the dryout quality for an 8-mm tube:

$$x_{\text{crit}} = \left[0.39 + 1.57\left(\frac{P}{98}\right) - 2.04\left(\frac{P}{98}\right)^2 + 0.68\left(\frac{P}{98}\right)^3\right]\left(\frac{G}{1000}\right)^{-0.5} \tag{12.78}$$

In these relations, q''_{crit} is in MW/m^2, P is the pressure in bar, and G is in kg/m^2 s. Equation (12.77) may be extended into the subcooled range by using negative qualities defined as $x = (\hat{h} - \hat{h}_{l,\text{sat}})/h_{lv}$. Equations (12.69) and (12.71) are used to obtain critical values for other tube diameters. Equation (12.77) was reported to be accurate to $\pm 15\%$ for $29.4 \leqslant P \leqslant 196$ bar and $750 \leqslant G \leqslant 5000$ kg/m^2 s. Equation (12.78) was stated to predict x_{crit} values accurate to within ± 0.05 for $9.8 \leqslant P \leqslant 166.6$ bar and $750 \leqslant G \leqslant 3000$ kg/m^2 s.

The correlations discussed above are all for flow boiling of water. There is a need, however, for CHF prediction methods for other fluids as well. This need stems from (1) a desire to accurately design evaporators for boiling applications involving other coolants such as fluorocarbons and hydrocarbons, and (2) a desire to model large-scale boiling heat transfer equipment using water with smaller experimental simulations using other fluids. Because of the complexity of most flow boiling systems, a number of dimensionless groups may potentially affect the critical heat flux conditions. A full discussion of these matters is beyond the scope of the current treatment. The interested reader is referred to the early report by Barnett [12.61] and the papers by Ahmad [12.62] and Ishii and Jones [12.63] for a more in-depth discussion of dimensional analysis and modeling of critical heat flux phenomena.

Perhaps the best generalized correlation of CHF conditions in vertical, uniformly heated tubes is that recently proposed by Katto and Ohno [12.65]. This correlation is an updated version of a correlation proposed earlier by Katto [12.66, 12.67]. The updated version attempts to correct some inconsistencies and improve the accuracy of the earlier version. This correlation takes the form of the following algorithm:

$$\gamma = \frac{\rho_v}{\rho_l} \tag{12.79}$$

$$\text{We}_k = \frac{G^2 L}{\rho_l \sigma} \tag{12.80}$$

$$C_K = 0.25 \qquad \text{for } \frac{L}{D} < 50 \tag{12.81a}$$

$$C_K = 0.25 + 0.0009\left[\left(\frac{L}{D}\right) - 50\right] \qquad \text{for } 50 \leqslant \frac{L}{D} \leqslant 150 \tag{12.81b}$$

$$C_K = 0.34 \qquad \text{for } \frac{L}{D} > 150 \tag{12.81c}$$

$$q''_{\text{co1}} = C_K G h_{lv}\, \text{We}_K^{-0.043} \left(\frac{L}{D}\right)^{-1} \tag{12.82a}$$

$$q''_{\text{co2}} = 0.10 G h_{lv} \gamma^{0.133}\, \text{We}_K^{-1/3} \left[\frac{1}{1 + 0.0031(L/D)}\right] \tag{12.82b}$$

$$q''_{co3} = 0.098Gh_{lv}\gamma^{0.133} \text{ We}_K^{-0.433}\left[\frac{(L/D)^{0.27}}{1 + 0.0031(L/D)}\right] \qquad (12.82c)$$

$$q''_{co4} = 0.0384Gh_{lv}\gamma^{0.60} \text{ We}_K^{-0.173}\left[\frac{1}{1 + 0.280 \text{ We}_K^{-0.233}(L/D)}\right] \qquad (12.82d)$$

$$q''_{co5} = 0.234Gh_{lv}\gamma^{0.513} \text{ We}_K^{-0.433}\left[\frac{(L/D)^{0.27}}{1 + 0.0031(L/D)}\right] \qquad (12.82e)$$

$$K_{K1} = \frac{1.043}{4C_K \text{ We}_K^{-0.043}} \qquad (12.83a)$$

$$K_{K2} = \left(\frac{5}{6}\right)\frac{0.0124 + D/L}{\gamma^{0.133} \text{ We}_K^{-1/3}} \qquad (12.83b)$$

$$K_{K3} = (1.12)\frac{1.52 \text{ We}_K^{-0.233} + D/L}{\gamma^{0.6} \text{ We}_K^{-0.173}} \qquad (12.83c)$$

For $\gamma < 0.15$:

$$\text{for } q''_{co1} < q''_{co2}: \quad q''_{co} = q''_{co1} \qquad (12.84a)$$

$$\text{for } q''_{co1} > q''_{co2}: \quad q''_{co} = q''_{co2} \quad \text{for } q''_{co2} < q''_{co3} \qquad (12.84b)$$

$$q''_{co} = q''_{co3} \quad \text{for } q''_{co2} \geqslant q''_{co3} \qquad (12.84c)$$

$$\text{for } K_{K1} > K_{K2}: \quad K_K = K_{K1} \qquad (12.85a)$$

$$\text{for } K_{K1} \leqslant K_{K2}: \quad K_K = K_{K2} \qquad (12.85b)$$

For $\gamma \geqslant 0.15$:

$$\text{for } q''_{co1} < q''_{co5}: \quad q''_{co} = q''_{co1} \qquad (12.86a)$$

$$\text{for } q''_{co1} > q''_{co5}: \quad q''_{co} = q''_{co5} \quad \text{for } q''_{co5} > q''_{co4} \qquad (12.86b)$$

$$q''_{co} = q''_{co4} \quad \text{for } q''_{co5} \leqslant q''_{co4} \qquad (12.86c)$$

$$\text{for } K_{K1} > K_{K2}: \quad K_K = K_{K1} \qquad (12.87a)$$

$$\text{for } K_{K1} \leqslant K_{K2}: \quad K_K = K_{K2} \quad \text{for } K_{K2} < K_{K3} \qquad (12.87b)$$

$$K_K = K_{K3} \quad \text{for } K_{K2} \geqslant K_{K3} \qquad (12.87c)$$

After executing the above calculations, the critical heat flux q''_{crit} is computed as

$$q''_{crit} = q''_{co}\left(1 + K_K \frac{\hat{h}_{l,sat} - \hat{h}_{in}}{h_{lv}}\right) \qquad (12.88)$$

This correlation method has been found to agree reasonably well with available data for flow boiling of water, R-12, R-22, and liquid helium for tube diameters near 1 cm, heated lengths near 1 m, and mass flux levels from 120 to 2100 kg/m^2 s.

Additional CHF prediction methods have also been proposed recently by Shah [12.67], Levy et al. [12.68], and Groeneveld et al. [12.69]. The correlation of Groeneveld et al. [12.69] is a simple tabulated correlation cast in terms of the following dimensionless variables:

$$\mathrm{Bo} = \frac{q''_{\mathrm{crit}}}{Gh_{lv}} \qquad \frac{\rho_l}{\rho_v} \qquad \psi_G = \left(\frac{G^2 D}{\sigma \rho_l}\right)^{1/2} \tag{12.89}$$

This relatively simple correlation provided a reasonably good fit to CHF data for water, R-11, R-12, R-21, R-113, R-114, CO_2, and N_2.

Dryout

As mentioned above, for saturated conditions at moderate to high qualities, the critical heat flux condition most often corresponds to dryout of the liquid film on the tube wall. As suggested by Figs. 12.4 and 12.5, this variation of the critical heat flux condition is encountered at low-heat-flux and/or low-wall-superheat conditions. Perhaps the most common application where such conditions are encountered is in the tubes of evaporators used in air-conditioning and refrigeration systems. In such systems, the two-phase flow leaving the expansion valve generally enters the evaporator at about 20–30% quality, and the flow leaving the unit is usually near 100% quality or slightly superheated. Dryout is generally encountered somewhere along the refrigerant flow path in the evaporator.

To predict the conditions at which dryout occurs, one approach would be to model the annular film flow evaporation process in the manner described in Sections 10.4 and 12.4 and computationally predict the conditions at which the film thickness or the film flow rate goes to zero. As indicated in these previous sections, this would require analysis of the mass and momentum transport in the core flow and the liquid film on the tube wall and transport of heat across the liquid film. As noted in Section 10.4, this usually requires including the effect of entrainment of liquid droplets in the vapor core.

The condition at which the film dries out is largely dictated by the transport in the core and the film regions and the interaction of the mechanisms of deposition, entrainment, and vaporization. This approach to predicting the dryout condition is plausible because, at low to moderate heat flux levels, the entrainment and deposition mechanisms apparently are largely unaffected by the presence of the applied heat flux. This makes it possible to apply the entrainment/deposition models developed for adiabatic equilibrium annular flow to annular flow with evaporation of the liquid film.

An analysis of this type proceeds by integrating the transport equations in the core flow and in the film along the channel from known or assumed boundary conditions at the starting point to the point where the film thickness goes to zero.

Dryout is thus determined as the point where the processes of evaporation, deposition, and entrainment lead to a condition in which the film flow rate becomes zero. Analytical treatments of this type have been explored in depth by Whalley, Hewitt, and co-workers [12.70–12.73]. The results of these investigations indicate that this type of method can be used successfully to predict dryout conditions for forced-convective boiling in tubes, annuli, and rod bundles. For a further discussion of this method of predicting dryout conditions, the interested reader may wish to consult references [12.74–12.76].

Other Factors

The discussion of CHF conditions in this section has focused on the relatively idealized circumstances of vertical upward flow boiling in a round tube with a uniform applied heat flux. Obviously, the conditions of interest in applications involving flow boiling may deviate in a number of ways from these conditions. Differences in the flow conditions will usually alter the CHF conditions somewhat from those for vertical upflow boiling in a round tube. Previous investigations have examined the effects of a nonuniform heat flux profile (axially and cirumferentially) [12.77–12.83], step changes in heat flux [12.84], noncircular geometries [12.85–12.92], pulsating or transient flow [12.93–12.96], and channel orientation [12.97]. Further information on these specialized aspects of CHF phenomena may be obtained in the indicated references or in the summary articles by Bergles [12.53] and Hewitt [12.75].

While the studies cited above provide some insight into the nature of CHF transitions, many aspects of CHF phenomena are still not well understood. Continued research in this area is clearly needed to fill the gaps in current knowledge of this aspect of flow boiling processes.

Example 12.5 Saturated flow boiling of water at low applied heat flux occurs in a vertical round tube with an inside diameter of 1.2 cm. The flow rate is such that the mass flux is 2000 kg/m^2 s, and the pressure along the tube is virtually uniform at 3773 kPa. Use the correlation of Levitan and Lantsman [12.57] to determine the dryout quality for these conditions.

For an 8-mm tube, the correlation of Levitan and Lantsman [12.57] is given by Eq. (12.78):

$$(x_{crit})_{8\ \mathrm{mm}} = \left[0.39 - 1.57\left(\frac{P}{98}\right) + 2.04\left(\frac{P}{98}\right)^2 + 0.68\left(\frac{P}{98}\right)^3\right]\left(\frac{G}{1000}\right)^{-0.5}$$

where P and G must be in bar and kg/m^2 s, respectively. Here

$$P = 3773\left(\frac{1.0}{101}\right) = 37.4 \text{ bar}$$

Substituting into the above equation yields

$$(x_{\text{crit}})_{8\text{ mm}} = \left[0.39 + 1.57\left(\frac{37.4}{98}\right) - 2.04\left(\frac{37.4}{98}\right)^2 + 0.68\left(\frac{37.4}{98}\right)^3\right]\left(\frac{2000}{1000}\right)^{-0.5}$$

$$= 0.516$$

Using Eq. (12.71) to correct for the diameter effect yields

$$x_{\text{crit}} = (x_{\text{crit}})_{8\text{ mm}}\left(\frac{8}{D}\right)^{0.15} \qquad (D \text{ in mm})$$

$$= 0.516\left(\frac{8}{12}\right)^{0.15} = 0.486$$

Thus dryout is predicted to occur at a quality of about 48%.

Example 12.6 Flow boiling of R-12 occurs in a vertical tube with a diameter and length of 1.1 cm and 30 cm, respectively. The pressure is virtually uniform along the tube at 333 kPa. The mass flow rate is such that the mass flux is 500 kg/m^2 s. Use the correlation of Katto and Ohno [12.65] to predict the critical heat flux for these circumstances if (*a*) the inlet condition is saturated liquid, and (*b*) the inlet flow is subcooled by 20°C.

For saturated R-12 at 333 kPa, $T_{\text{sat}} = 275$ K $= 1.8$°C, $\rho_l = 1388$ kg/m^3, $\rho_v = 19.2$ kg/m^3, $h_{lv} = 154.7$ kJ/kg, $c_{pl} = 0.932$ kJ/kg K, and $\sigma = 0.0114$ N/m.

(*a*) Using the embodiment of the Katto and Ohno correlation in Eqs. (12.79) through (12.88), we proceed as follows:

$$\gamma = \frac{\rho_v}{\rho_l} = \frac{19.2}{1388} = 0.0138$$

$$\text{We}_K = \frac{G^2L}{\rho_l\sigma} = \frac{500(0.30)}{1388(0.0114)} = 9.48$$

$L/D = 30/1.1 = 27.3$, which implies that $C_K = 0.25$.

$$q''_{\text{co1}} = C_K G h_{lv}\,\text{We}_K^{-0.043}\left(\frac{L}{D}\right)^{-1}$$

$$= (0.25)(500)(154.7)(9.48)^{-0.043}(27.3)^{-1} = 644 \text{ kW/m}^2$$

$$q''_{\text{co2}} = 0.10 G h_{lv}\gamma^{0.133}\,\text{We}_K^{-1/3}\left[1 + 0.0031\left(\frac{L}{D}\right)\right]^{-1}$$

$$= 0.10(500)(154.7)(0.0138)^{0.133}(9.48)^{-1/3}[1 + 0.0031(27.3)]^{-1}$$

$$= 1906 \text{ kW/m}^2$$

Because $q''_{co1} < q''_{co2}$, $q''_{co} = q''_{co1} = 644\ \text{kW/m}^2$.

$$K_{K1} = \frac{1.043}{4C_K\,\text{We}_K^{-0.043}} = \frac{1.043}{4(0.25)(9.48)^{-0.043}} = 1.149$$

$$K_{K2} = \left(\frac{5}{6}\right)\frac{0.0124 + D/L}{\gamma^{0.133}\,\text{We}_K^{-1/3}} = \left(\frac{5}{6}\right)\frac{0.0124 + 1/27.3}{(0.0138)^{0.133}(9.48)^{-1/3}} = 0.153$$

Because $K_{K1} > K_{K2}$, $K_K = K_{K1} = 1.149$. For saturated liquid at the inlet, Eq. (12.88) reduces to

$$q''_{\text{crit}} = q''_{co}$$

And it follows from the above results that for these conditions,

$$q''_{\text{crit}} = 644\ \text{kW/m}^2$$

(*b*) Evaluation of q''_{co} for the subcooled inlet condition is identical to the saturated case. However, for the subcooled case,

$$q''_{\text{crit}} = q''_{co}\left(1 + K_K\frac{\hat{h}_{l,\text{sat}} - \hat{h}_{\text{in}}}{h_{lv}}\right)$$

Computing $\hat{h}_{l,\text{sat}} - \hat{h}_{\text{in}}$ as $c_{pl}\,(T_{\text{sat}} - T_{\text{in}})$, this becomes

$$q''_{\text{crit}} = q''_{co}\left[1 + K_K\frac{c_{pl}(T_{\text{sat}} - T_{\text{in}})}{h_{lv}}\right]$$

Substituting yields

$$q''_{\text{crit}} = 644\left[1 + (1.149)\frac{(0.932)(20)}{154.7}\right]$$

$$= 733\ \text{kW/m}^2$$

Thus the 20°C subcooling is estimated to increase the critical heat flux by about 14%.

12.6 POST-CHF INTERNAL FLOW BOILING

As indicated in Figs. 12.4 and 12.5, if the flow boiling process exceeds the burnout condition, the boiling process may enter the transition boiling, film boiling, or mist evaporation regime. Usually, heat transfer equipment is designed so that transition boiling and film boiling are avoided to prevent the associated high wall temperatures from damaging the walls of the flow passages in the unit. Mist flow evaporation is often encountered in the latter stages of vaporization processes at low to moderate heat flux or wall superheat levels. It may be particularly important in refrigeration and air-conditioning evaporators, where complete vaporization of the working fluid must be achieved. Each of the three possible regimes of vaporization is discussed in this section.

Transition Flow Boiling

Of the three regimes considered in this section, transition boiling is perhaps the least understood. In order to encounter this regime, the wall temperature of the passage must be controlled in the physical system so that it remains in the transition boiling range. While this regime is rarely encountered under normal operating conditions, it may potentially arise during loss-of-coolant accident scenarios for nuclear power plants. Reflooding the core of the reactor with coolant may produce the forced-flow version of the quenching process described in Chapter Seven (for pool boiling). Because of the need to understand how transition flow boiling may affect emergency core cooling, many of the studies of flow transition boiling have been conducted in connection with nuclear power applications.

Experimental investigations of forced-convection transition boiling have been conducted by McDonough et al. [12.98], Iloeje et al. [12.99] and Cheng and Ng [12.100]. Work on forced-flow transition boiling up to 1976 has been summarized by Groeneveld and Fung [12.101]. Attempts to develop heat transfer correlations for forced-flow transition boiling were made by McDonough et al. [12.98], Tong [12.102], and Ramu and Weisman [12.103]. Tong and Young [12.104] proposed the following correlation for transition boiling of water:

$$q''_{tb} = q''_{fb} + q''_{nb} \exp\left[-0.0394 \frac{x_e^{2/3}}{dx_e/dz}\left(\frac{T_w - T_{\text{sat}}}{55.6}\right)^{1+0.00288(T_w - T_{\text{sat}})}\right] \tag{12.90}$$

where q''_{tb} is the total transition boiling heat flux, q''_{nb} is the nucleate boiling heat flux based on the instantaneous local conditions, q''_{fb} is the film boiling heat flux based on the instantaneous local conditions, x_e is the equilibrium quality, $T_w - T_{\text{sat}}$ is the wall superheat in °C, and dx_e/dz is the quality gradient in m^{-1}. This is a dimensional correlation, and terms must be evaluated in the proper units. Collier [12.105] questioned the manner in which this correlation was derived, but nevertheless acknowledged its usefulness as a predictive tool.

Convective Film Boiling

As seen in Figs. 12.4 and 12.5, at low qualities and subcooled conditions, the CHF transition may lead to film boiling at the walls of the tube. Knowledge of the transport for this type of flow condition is relatively limited. For flow boiling of water, which is of central interest to the power industry, to achieve sustained film boiling at high mass flux levels requires wall temperatures so high that most conventional materials would melt. Thus, conducting controlled experiments is extremely difficult for such circumstances. It is not surprising, therefore, that few extensive experimental investigations of convective film boiling have been conducted.

Experimental investigations by Dougall and Rohsenow [12.106] and others indicate that for convective film boiling at low to moderate flow rates, the flow takes on the so-called *inverted annular* flow configuration shown in Fig. 12.25

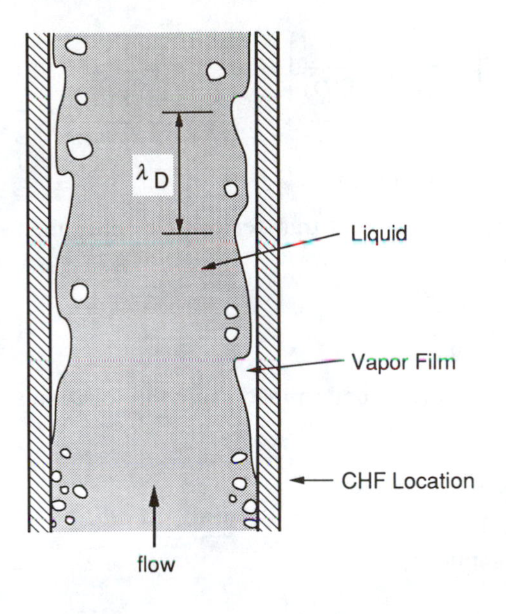

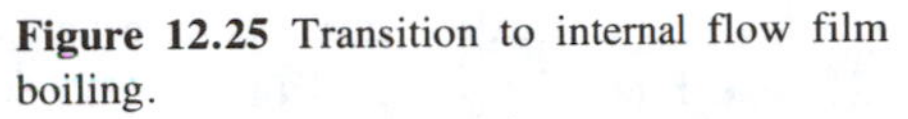
Figure 12.25 Transition to internal flow film boiling.

(regular annular flow corresponds to a liquid film on the wall with vapor flowing in the core). The vapor film along the tube wall is generally not smooth but exhibits irregularities at random locations. This behavior is similar to the irregular film behavior for buoyancy-driven film boiling on a vertical surface described in Chapter Seven. These irregular bulges or bubbles in the film tend to retain their identity, moving downstream at a velocity slightly slower than the peak velocity in the vapor film.

It is tempting to develop an annular flow model of the inverted annular flow similar to those described in Section 11.2 for annular flow condensation. This can, in fact, be easily done by following the same steps indicated in that section, but inverting the role of the liquid and vapor and the direction of heat flow. Doing so leads to the following set of relations:

$$\rho_v^* g = \rho_v g - \frac{4\tau_i}{D - 2\delta} - \frac{2xDG^2}{\rho_v(D - 2\delta)}\left(\frac{dx}{dz}\right) \tag{12.91}$$

$$\frac{dx}{dz} = \frac{4q''}{DGh_{lv}} = \frac{4h(T_w - T_{sat})}{DGh_{lv}} \tag{12.92}$$

$$\frac{4zk_v\mu_v(T_w - T_{sat})}{\rho_v(\rho_l - \rho_v^*)gh'_{lv}} = \delta^4 + \frac{4\tau_i\delta^3}{3(\rho_l - \rho_v^*)g} \tag{12.93}$$

where

$$h'_{lv} = h_{lv}\left[1 + \left(\frac{3}{8}\right)\frac{c_{pv}(T_w - T_{sat})}{h_{lv}}\right] \tag{12.94}$$

$$\tau_i = f_l\left(\frac{\rho_l u_l^2}{2}\right) = f_l\left[\frac{G^2(1-x)^2}{2\rho_l(1-4\delta/D)}\right] \tag{12.95}$$

$$f_l = 0.079\left[\frac{G(1-x)(D-\delta)}{\mu_l(1-4\delta/D)}\right]^{-0.25} \tag{12.96}$$

For laminar conduction-dominated transport, the heat transfer coefficient is then given by

$$h = \frac{k_v}{\delta} \tag{12.97}$$

The heat transfer coefficient can then be iteratively determined using the following algorithm:

1. Guess a value of δ.
2. Use Eqs. (12.92) and (12.97) to evaluate dx/dz.
3. Use Eqs. (12.95) and (12.96) to compute τ_i.
4. Determine ρ_v^* using Eq. (12.91).
5. Substitute the guessed δ value and the computed ρ_v^* and τ_i values into Eq. (12.93). If this equation is satisfied to an acceptable level of accuracy, the guessed value of δ and the computed h value are correct. If this relation is not satisfied, a new δ value is guessed and the process is repeated, beginning with step 2. This sequence of computations is repeated until convergence.

If the interfacial shear and acceleration pressure gradient are small, this model yields the following Nusselt-type relation for h:

$$h = \left[\frac{\rho_v(\rho_l - \rho_v)gh'_{lv}k_v^3}{4(z - z_{\mathrm{CHF}})\mu_v(T_w - T_{\mathrm{sat}})}\right]^{1/4} \tag{12.98}$$

The problem with this model is, of course, that it does not account for the bubbles of vapor that rise with the film. Observations suggest that vapor accumulates in these bubbles, which tends to thin the vapor film between them. This, in turn, results in a generally higher time-averaged heat transfer coefficient than is predicted by the smooth-film steady-flow analysis described above.

There apparently have been very few efforts to examine the detailed aspects of convective film boiling inside tubes. However, several investigations have explored the effects of a nonsmooth interface in the broader context of film boiling on a surface in a motionless pool or a convective liquid flow. The model developed by Suryanarayana and Merte [12.107] for turbulent film boiling on vertical surfaces immersed in a motionless liquid pool also provides insight into the mechanisms of convective film boiling inside tubes (see Section 7.6).

Several investigators have linked the appearance of regular distortions of the interface to the growth of waves due to interface instability. Collier [12.2] reported the suggestion of Bailey [12.108] that these bulging portions of the film are due to a varicose instability of the vapor film during flow boiling over a vertical cylinder. This argument is supported by the analogous instability of a hollow gas cylinder within a denser liquid, as described by Chandrasekhar [12.109]. Based on the analysis presented by Chandrasekhar [12.109], the most rapidly growing wavelength associated with this type of instability is given by the simple relation

$$\lambda_D = 2\pi\left(\frac{D}{2}\right) \tag{12.99}$$

Collier [12.2] further suggests that this wavelength, which is expected to approximate the mean spacing between the bulging sections of the film, would be a more appropriate length scale than $z - z_{\text{CHF}}$ in Eq. (12.98).

Experimental evidence does indicate that classical laminar film boiling exists for only a short distance downstream of the CHF location on the channel wall. Beyond this initial region the interface exhibits the bulging irregular structure discussed above, which results in a time-varying thickness of the film that, in the mean, is thinner than the prediction of the steady laminar flow model. In addition, heat transfer data do indicate that for these conditions there is no detectable z dependence of the measured heat transfer coefficient, implying that $z - z_{\text{CHF}}$ is not the appropriate length scale for the heat transfer correlation. Use of λ_D as a length scale in the correlation obtained from the laminar flow model would appear to be an improvement. However, this correlation methodology has not been tested against a wide data base.

For film boiling over a vertical surface under emergency core reflooding conditions in nuclear reactors, Leonard et al. [12.110] suggested that the spacing of the wave crests or bulges corresponded to the Helmholtz-unstable wavelength λ_H, defined as

$$\lambda_H = 16.24\left[\frac{\sigma^4 h_{lv}^3 \mu_v^5}{\rho_v(\rho_l - \rho_v)^5 g^5 k_v^3 (T_w - T_{\text{sat}})^2}\right]^{1/2} \tag{12.100}$$

To predict the heat transfer coefficient, they proposed substituting the Helmholtz length scale defined above into Bromley's equation (see Chapter 7) for film boiling from a horizontal cylinder:

$$h = 0.62\left[\frac{k_v^3 \rho_v(\rho_l - \rho_v) g h_{lv}}{\mu_v (T_w - T_{\text{sat}})\lambda_H}\right]^{1/4} \tag{12.101}$$

Additional information on film boiling heat transfer associated with reflooding of a hot tube is provided by the study of Groeneveld and Young [12.111].

Mist Flow Evaporation

At low to moderate heat flux and wall superheat levels, the CHF transition corresponds to dryout of the liquid film and a transition to mist flow evaporation. This regime is also sometimes referred to as the *liquid-deficient region* and the regime of *dispersed-flow heat transfer*. The mechanisms that facilitate the continued vaporization are unique to this regime.

Six different heat transfer mechanisms may play a role in the mist-flow evaporation process. These mechanisms are itemized below and indicated schematically in Fig. 12.26.

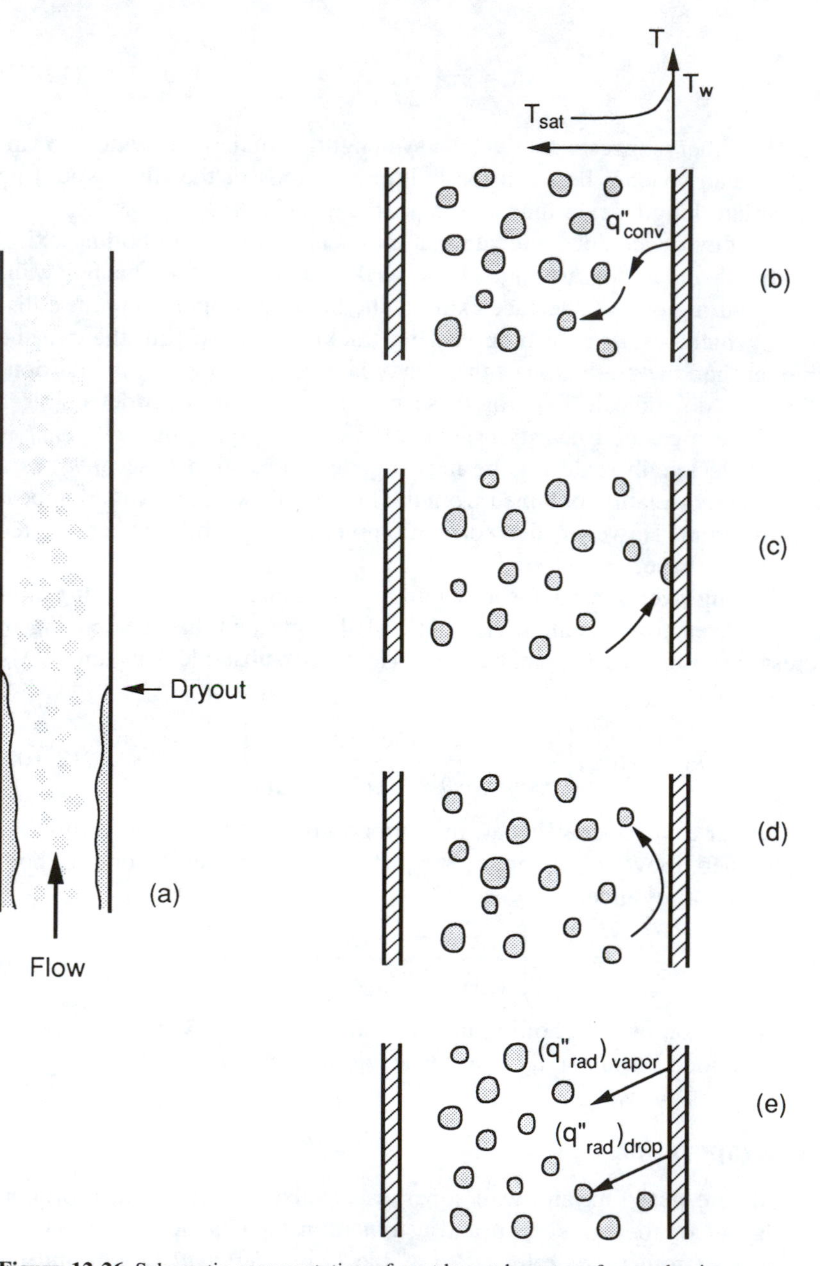

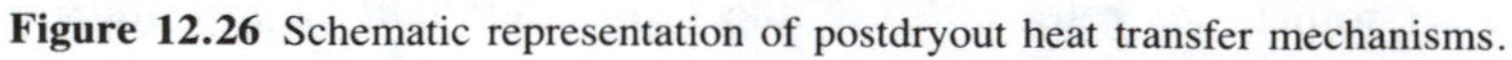
Figure 12.26 Schematic representation of postdryout heat transfer mechanisms.

1. Convective heat transfer from the wall to the vapor (Fig. 12.26*b*)
2. Convective heat transfer from the vapor to the entrained droplets (Fig. 12.26*b*)
3. Evaporation of droplets that collide with the wall and wet its surface (Fig. 12.26*c*)
4. Evaporation of droplets that come into close proximity to the wall but do not wet the surface (Fig. 12.26*d*)
5. Radiation heat transfer from the wall to the droplets (Fig. 12.26*e*)
6. Radiation heat transfer from the wall to the vapor (Fig. 12.26*e*)

The first two mechanisms function in tandem to transport heat from the wall to the liquid droplets. Convective transport from the vapor to the droplets requires that the vapor in the core must be at least slightly superheated. The nature of droplet interactions with the wall generally depends on the wall temperature and the wetting characteristics of the liquid. If the wall temperature is below the Leidenfrost temperature (see Chapter Eight), droplets that approach the wall may collide with it, wetting the surface. Some portion of the liquid may rebound as smaller droplets back into the vapor, with the remainder staying on the surface. The liquid left on the surface may vaporize by nucleate boiling or by conduction of heat across the liquid and vaporization at the liquid–vapor interface.

If the wall temperature is above the Leidenfrost temperature, the wall side of the droplet vaporizes rapidly as the droplet nears the wall, producing a significant repelling force on the droplet. At high wall temperatures, only droplets having a sufficiently high surface-normal component of momentum will be able to overcome the repelling effect and collide with the wall. Some may then only closely approach the wall and be repelled. However, the vaporization of these droplets also serves to remove heat from the wall and continue the vaporization process.

The radiation heat transfer mechanisms listed above are generally significant contributors to the overall transport only for high wall temperatures. Radiant transfer from the wall to the vapor is important only if the vapor has absorption bands in the wavelength range emitted from the surface. Most hydrocarbons and vapors encountered in cryogenic applications usually do not absorb radiation in the range of wavelengths emitted from the tube wall at temperatures encountered in typical applications. A notable exception to this generalization is steam, which does absorb and emit in wavelength bands in the infrared range.

Because the vapor must be superheated slightly to convectively transport heat from the vapor to the droplets, the flow will generally be in a nonequilibrium state. In some cases the departure from equilibrium may be so small as to be negligible. In other cases the flow may be far from equilibrium. The flow may be thought of as existing somewhere between two limiting cases. If the heat transfer between the droplets and the vapor is infinitely fast, the liquid and vapor in the flow will be at thermodynamic equilibrium. At the other extreme, if the heat transfer rate between the droplets and the vapor is virtually zero, all of the heat transferred to the flow will go into superheating the gas, and the flow will be far from thermodynamic equilibrium.

For a real dispersed flow, the fluid is generally somewhere between the two limiting cases indicated above. The assumption that full thermodynamic equilibrium exists in the flow is usually a valid idealization only in flows where there is expected to be a significant relative motion between the droplets and the vapor and/or turbulence that will enhance convective transport between the phases. This is generally expected to be true at high mass flux levels or in flows with high turbulence or secondary flows generated by recirculations in complex flow passages.

Model analyses of dispersed flow heat transfer have generally fallen into two categories: (1) those that assume complete thermodynamic equilibrium in the flow, and (2) those that account for some departure from equilibrium. We shall begin here by first considering a very idealized model of the first type.

A dispersed flow of droplets and vapor under saturated conditions in a round tube, like that shown in Fig. 12.26*a*, is assumed to exist in a heated tube. The flow is postulated to be turbulent and heat transfer between the vapor and the droplets is assumed to be sufficiently fast that equilibrium is maintained. We will treat the two-phase flow using an extension of the homogeneous flow model (see Chapter Ten). It is therefore postulated that the two-phase flow behaves like single-phase flow of a pseudo-fluid having properties that are some appropriately defined average of the vapor and liquid properties. The heat transfer coefficient could then be computed using the Dittus-Boelter equation with appropriately defined mean properties:

$$\mathrm{Nu} = \frac{hD}{k_{tp}} = 0.023\left(\frac{GD}{\mu_{tp}}\right)^{0.8} \mathrm{Pr}_{tp}^{0.4} \tag{12.102}$$

Rearranging, this relation can be cast in the form

$$\mathrm{Nu}_v = \frac{hD}{k_v} = 0.023\left(\frac{GD}{\mu_v}\right)^{0.8} \mathrm{Pr}_v^{0.4}\left(\frac{k_{tp}}{k_v}\right)\left(\frac{\mu_v}{\mu_{tp}}\right)^{0.8}\left(\frac{\mathrm{Pr}_v}{\mathrm{Pr}_{tp}}\right)^{0.4} \tag{12.103}$$

As described in Section 10.2, for the homogeneous flow model, one possible relation for the effective two-phase viscosity is

$$\mu_{tp} = \frac{xv_v\mu_v + (1-x)v_l\mu_l}{xv_v + (1-x)v_l} \tag{12.104a}$$

If this relation is similarly applied to the other properties,

$$k_{tp} = \frac{xv_v k_v + (1-x)v_l k_l}{xv_v + (1-x)v_l} \tag{12.104b}$$

$$\mathrm{Pr}_{tp} = \frac{xv_v \,\mathrm{Pr}_v + (1-x)v_l \,\mathrm{Pr}_l}{xv_v + (1-x)v_l} \tag{12.104c}$$

then Eq. (12.103) can be written as

$$\mathrm{Nu}_v = \frac{hD}{k_v} = 0.023\left(\frac{GD}{\mu_v}\right)^{0.8} \mathrm{Pr}_v^{0.4}\Gamma\left(x, \frac{\rho_l}{\rho_v}, \frac{k_l}{k_v}, \frac{\mu_v}{\mu_l}, \frac{\mathrm{Pr}_v}{\mathrm{Pr}_l}\right) \tag{12.105}$$

where

$$\Gamma = \frac{x + (1 - x)(\rho_v/\rho_l)(k_l/k_v)}{x + (1 - x)(\rho_v/\rho_l)(\mu_l/\mu_v)} \left[\frac{x + (1 - x)(\rho_v/\rho_l)}{x + (1 - x)(\rho_v/\rho_l)(\mathrm{Pr}_l/\mathrm{Pr}_v)} \right]^{0.4} \tag{12.106}$$

While this model is overly simplistic, it does suggest that one possible approach to correlating mist-flow evaporation heat transfer data is to modify a single-phase correlation for the entire flow as vapor with a correction factor Γ that may, in general, be a function of x and the property ratios ρ_v/ρ_l, k_l/k_v, μ_l/μ_v, and $\mathrm{Pr}_v/\mathrm{Pr}_l$. A number of correlations of this general type have been proposed. Collier [12.105] and Mayinger and Langer [12.112] describe 15 correlations having this general form. An extensive summary of such correlation methods has also been presented by Groeneveld [12.113]. Three representative correlations of this type will be briefly discussed here.

Groeneveld's Correlation

Based on a best fit to data from a number of previous investigations, Groeneveld [12.113] proposed the following correlation for heat transfer in the dispersed flow regime:

$$\mathrm{Nu}_v = \frac{hD}{k_v} = a\left[\left(\frac{GD}{\mu_v}\right)\left(x + \frac{\rho_v}{\rho_l}(1 - x)\right)\right]^b \mathrm{Pr}_{v,w}^c Y^d \tag{12.107}$$

where

$$Y = 1 - 0.1\left(\frac{\rho_l}{\rho_v} - 1\right)^{0.4} (1 - x)^{0.4} \tag{12.108}$$

Note that $\mathrm{Pr}_{v,w}$ is the vapor Prandtl number evaluated at the wall temperature. All other properties are evaluated at the saturation condition. This correlation is very similar to the correlation proposed earlier by Miropolsky [12.114]. Values of the constants a, b, c, and d were provided for round tubes and annuli. Recommended values are listed in Table 12.3.

For tubes, this correlation was based on data for water in horizontal and vertical tubes in the following ranges: $0.25 < D < 2.5$ cm, $68 < P < 215$ bar, $700 < G < 5300$ kg/m^2 s, $0.1 < x < 0.9$, $120 < q'' < 2100$ kW/m^2, $0.88 < \mathrm{Pr}_{v,w} < 2.21$, $6.6 \times 10^4 < GD[x + (1 - x)\rho_v/\rho_l]/\mu_v < 1.3 \times 10^6$, $0.706 < Y < 0.976$. This correlation reportedly matches the data to a RMS error of 11.5%. A modified

Table 12.3

Geometry	a	b	c	d
Tubes	0.00109	0.989	1.41	−1.15
Annuli	0.0520	0.688	1.26	−1.06
Tubes and annuli	0.00327	0.901	1.32	−1.50

version of Groeneveld's correlation has also been developed by Slaughterback et al. [12.115] in an effort to better match heat transfer data at low pressures.

The Correlation of Dougall and Rohsenow

Along similar lines, Dougall and Rohsenow [12.106] proposed the correlation

$$\mathrm{Nu}_v = \frac{hD}{k_v} = 0.023\left[\left(\frac{GD}{\mu_v}\right)\left(x + \frac{\rho_v}{\rho_l}(1 - x)\right)\right]^{0.8} \mathrm{Pr}_{v,\mathrm{sat}}^{0.4} \tag{12.109}$$

It can be seen that the above correlations are similar in form to the relation that emerged from the homogeneous flow model described above. As noted above, a major shortcoming of these simple correlations is that they do not account for nonequilibrium conditions in the flow. More intricate prediction techniques that account for the nonequilibrium condition as well as mechanisms other than vapor convection have been proposed by a number of investigators [12.99, 12.116–12.125]. Generally these models fall into one of two categories: (1) models from which correlations are developed that account for nonequilibrium effects; and (2) more complex models that attempt to treat the transport mechanisms and the nonequilibrium effects in a more fundamental theoretical manner. Two examples of models of these types are described in detail below.

The Correlation of Groeneveld and Delorme

Groeneveld and Delorme [12.119] proposed a modified version of the equilibrium correlation model described above that accounts for nonequilibrium effects. The methodology proposed by these investigators makes use of a single-phase heat transfer correlation, modified for two-phase homogeneous flow, having the following general form:

$$\frac{hD}{k_{v,f}} = \frac{q''D}{(T_w - T_{v,a})k_{v,f}} = a\left[\left(\frac{GD}{\mu_{v,f}}\right)\left(x_a + \frac{\rho_v}{\rho_l}(1 - x_a)\right)\right]^b \mathrm{Pr}_{v,f}^c\left(e + f\frac{D}{L}\right)^g \tag{12.110}$$

Although a general relation of this form may be used, these authors obtained a particularly good fit to available data by using the following relation, which is a modified form of a single-phase correlation proposed by Hadaller and Banerjee [12.126]:

$$\frac{hD}{k_{v,f}} = \frac{q''D}{(T_w - T_{v,a})k_{v,f}}$$
$$= 0.008348\left[\left(\frac{GD}{\mu_{v,f}}\right)\left(x_a + \frac{\rho_v}{\rho_l}(1 - x_a)\right)\right]^{0.8774} \mathrm{Pr}_{v,f}^{0.6112} \tag{12.111}$$

(This corresponds to $e = f = g = 0$ in Eq. [12.110].) The v, f subscripts in the above relations indicate vapor properties to be evaluated at the film temperature

$T_{v,f} = (T_w - T_{v,a})/2$. To facilitate determination of the actual vapor quality x_a and temperature $T_{v,a}$, the following relations were used:

$$x_a = \frac{h_{lv} x_e}{\hat{h}_{v,a} - \hat{h}_{l,\text{sat}}} \tag{12.112}$$

$$\hat{h}_{v,a} - \hat{h}_{l,\text{sat}} = h_{lv} + \int_{T_{\text{sat}}}^{T_{v,a}} c_{pv}\, dT_v \tag{12.113}$$

$$\text{Re}_{\text{hom}} = \left(\frac{GD}{\mu_v}\right)\left(x_e^* + \frac{\rho_v}{\rho_l}(1 - x_e^*)\right) \tag{12.114}$$

$$x_e^* = x_e \text{ for } 0 \leqslant x_e \leqslant 1 \qquad x_e^* = 1 \text{ for } x_e > 1 \tag{12.115}$$

$$\psi = a_1 \, \text{Pr}_v^{a_2} \, \text{Re}_{\text{hom}}^{a_3} \left(\frac{q''_{D} c_{pv,e}}{k_v h_{lv}}\right)^{a_4} \sum_{i=0}^{2} b_i x_e^i \qquad \text{for } 0 \leqslant \psi \leqslant \frac{\pi}{2} \tag{12.116}$$

$$\left(\text{if computed } \psi < 0, \text{ set } \psi = 0, \text{ if computed } \psi > \frac{\pi}{2}, \text{ set } \psi = \frac{\pi}{2}\right)$$

$$\frac{\hat{h}_{v,a} - \hat{h}_{v,e}}{h_{lv}} = \exp(-\tan\psi) \tag{12.117}$$

$$\hat{h}_{v,e} = \hat{h}_{v,\text{sat}} \text{ for } 0 \leqslant x_e \leqslant 1 \qquad \hat{h}_{v,e} = \hat{h}_{v,sat} + (x_e - 1)h_{lv} \text{ for } x_e > 1 \tag{12.118}$$

Equations (12.112) and (12.113) are a consequence of conservation of energy and the definitions of the actual and equilibrium qualities. Equations (12.114)–(12.117) embody an empirical correlation for the degree of nonequilibrium in the flow. In Eq. (12.117), $\hat{h}_{v,a}$ is the actual vapor enthalpy and $\hat{h}_{v,e}$ is the equilibrium vapor enthalpy. Values of the constants a_i and b_i were determined to provide a best fit to a data base of over 1400 data points for water. The recommended values of these constants are

$$\begin{aligned} a_1 &= 0.13864 & b_0 &= 1.3072 \\ a_2 &= 0.2031 & b_1 &= -1.0833 \\ a_3 &= 0.20006 & b_2 &= 0.8455 \\ a_4 &= -0.09232 & & \end{aligned}$$

Example 12.7 Example 12.5 considered saturated flow boiling of water in a vertical tube for $D = 1.2$ cm, $G = 2000$ kg/m^2 s, and P = 3773 kPa. In that example it was estimated that dryout occurred at $x_{\text{crit}} = 0.49$ for these circumstances. For these same flow conditions and an applied heat flux of 600 kW/m^2, use the correlation of Groeneveld [12.113] to estimate the heat transfer coefficient for $x = 0.6$ and $x = 0.9$. Compare the results with those predicted by the correlation of Dougall and Rohsenow [12.106].

For saturated liquid water at 3773 kPa, $T_{sat} = 520$ K $= 246.8$°C, $\rho_l = 804$ kg/m^3, $\rho_v = 18.9$ kg/m^3, $\mu_v = 18.1 \times 10^{-9}$ N s/m^2, $k_v = 0.0501$ W/m K, and $Pr_v = 1.39$. In addition, at 3773 kPa, $Pr_v = 1.05$ at 623 K and $Pr_v = 0.93$ at 773 K. Inserting the recommended round-tube constants into the correlation (12.107) proposed by Groeneveld [12.113], the following relation is obtained for the heat transfer coefficient:

$$h = 0.00109\left(\frac{k_v}{D}\right)\left[\frac{GD}{\mu_v}\left(x + \frac{\rho_v}{\rho_l}(1 - x)\right)\right]^{0.989} Pr_{v,w}^{1.41} Y^{-1.15}$$

where

$$Y = 1 - 0.1\left(\frac{\rho_l}{\rho_v} - 1\right)^{0.4} (1 - x)^{0.4}$$

Taking $Pr_{v,w} = 1.05$ as a first approximation, substituting for $x = 0.6$ yields

$$Y = 1 - 0.1\left(\frac{804}{18.9} - 1\right)^{0.4} (1 - 0.6)^{0.4} = 0.692$$

$$\begin{aligned} h &= 0.00109\left(\frac{0.0501}{0.012}\right)\left[\frac{2000(0.012)}{18.1 \times 10^{-6}}\left(0.6 + \frac{18.9}{804}(1 - 0.6)\right)\right]^{0.989} \\ &\quad (1.05)^{1.41}(0.692)^{-1.15} \\ &= 5180 \text{ W/m}^2\text{ K} \end{aligned}$$

This implies that the wall temperature T_w is

$$T_w = T_{sat} + \frac{q''}{h} = 520 + 600{,}000/5{,}180 = 636 \text{ K}$$

Interpolating to estimate the vapor Prandtl number at $T_w = 636$ K, we obtain

$$\begin{aligned} Pr_{v,w} &= Pr_{v,623} + \frac{(T_w - 623)(Pr_{v,773} - Pr_{v,623})}{773 - 623} \\ &= 1.05 + \frac{(636 - 623)(0.93 - 1.05)}{773 - 623} \\ &= 1.04 \end{aligned}$$

Correcting the computed h value for the different $Pr_{v,w}$ yields

$$h = 5180\left(\frac{1.04}{1.05}\right)^{1.41} = 5110 \text{ W/m}^2\text{ K}$$

This new value of h is then used to determine T_w:

$$T_w = 520 + \frac{600{,}000}{5{,}110} = 637 \text{ K}$$

After one more iteration, this process converges to

$$T_w = 637 \text{ K}, \qquad \text{Pr}_{v,w} = 1.04, \qquad h = 5110 \text{ W/m}^2 \text{ K}$$

A similar iterative calculation for $x = 0.9$ yields

$$T_w = 584 \text{ K}, \qquad \text{Pr}_{v,w} = 1.08, \qquad h = 9450 \text{ W/m}^2 \text{ K} \qquad (\text{for } x = 0.9),$$

Using the correlation of Dougall and Rohsenow [12.106], the heat transfer coefficient is given by Eq. (12.109) as

$$h = 0.023\left(\frac{k_v}{D}\right)\left[\left(\frac{GD}{\mu_v}\right)\left(x + \frac{\rho_v}{\rho_l}(1-x)\right)\right]^{0.8} \text{Pr}_{v,\text{sat}}^{0.4}$$

Substituting for $x = 0.6$,

$$h = 0.023\left(\frac{0.0501}{0.012}\right)\left[\frac{2000(0.012)}{18.1 \times 10^{-6}}\left(0.6 + \frac{18.9}{804}(1-0.6)\right)\right]^{0.8} (1.39)^{0.4}$$

$$= 5830 \text{ W/m}^2 \text{ K}$$

Similarly, for $x = 0.9$,

$$h = 0.023\left(\frac{0.0501}{0.012}\right)\left[\frac{2000(0.012)}{18.1 \times 10^{-6}}\left(0.9 + \frac{18.9}{804}(1-0.9)\right)\right]^{0.8} (1.39)^{0.4}$$

$$= 7980 \text{ W/m}^2 \text{ K}$$

Thus the h values predicted by these two correlations agree within about 18% for these conditions.

To compute the local wall temperature for a specified wall heat flux, the Groeneveld and Delorme [12.119] correlation is used in the following manner:

1. Assuming that the local equilibrium quality x_e and mass flux are given, a value of the film temperature is guessed so that properties may be evaluated.
2. Equations (12.114) through (12.117) are used to determine Re_{hom} and ψ, which are then substituted into Eqs. (12.117) and (12.118) to determine $\hat{h}_{v,a}$.
3. Using the value of $\hat{h}_{v,a}$ determined in step 2, Eq. (12.112) is used to determine x_a, and Eq. (12.113) is solved iteratively to determine $T_{v,a}$.
4. Using the values of $\hat{h}_{v,a}$ and x_a determined in step 3, Eq. (12.111) can be solved for the wall temperature T_w for the specified heat flux.
5. For the computed values of $T_{v,a}$ and T_w, a new value of the film temperature is computed as the arithmetic average of the two. If this value agrees with the previous guess of $T_{v,f}$, the computation is complete. If not, a new value of $T_{v,f}$ is guessed and the process is repeated beginning with step 2.

Although iteration is required, this correlation procedure is straightforward and can be easily programmed on a computer. While this correlation accounts for nonequilibrium conditions, it does not account for direct transfer of heat from the wall to liquid droplets or for radiation heat transfer. These investigators argue, however, that these effects are small for many circumstances of practical interest. They specifically cite the visual observations of the postdryout region made by Cumo et al. [12.127], which suggest that droplet collisions with the wall were infrequent and contributed little to the overall heat transfer. They also note that the droplet velocity-to-vapor velocity ratios determined experimentally by Cumo et al. [12.127] were typically close to 1, supporting the homogeneous flow idealization implicit in this model.

Cumo et al. [12.127] also argued that calculations indicate that the radiative heat flux component is very small compared to the convective component except for low flows and very high surface temperatures. The generally good agreement of the predictions of this correlation with available data supports arguments regarding the validity of the idealization in this model. In using it, however, one must be sure that the conditions of interest conform to the assumptions in this model.

Figure 12.27 shows a comparison of the predicted wall temperature variation with equilibrium quality and measured data obtained by Era et al. [12.128]. Also shown is the variation predicted by Groeneveld's [12.113] empirical correlation described by Eqs. (12.107) and (12.108). It can be seen that the nonequilibrium correlation exhibits trends similar to the data, whereas the empirical correlation generally does not, particularly at equilibrium qualities greater than 1.

The Model of Ganic and Rohsenow

In contrast to the model of Groeneveld and Delorme [12.119], the model proposed by Ganic and Rohsenow [12.121] attempts to account for more of the mechanisms. The total heat flux was postulated to consist of contributions due to vapor convection q''_v, droplet heat transfer q''_d, and radiation q''_r:

$$q'' = q''_v + q''_d + q''_r \tag{12.119}$$

To predict the convective contribution, the MacAdams correlation was specified,

$$q''_v = 0.0023\left(\frac{k_v}{D}\right)\left(\frac{GxD}{\alpha\mu_v}\right)^{0.8} \mathrm{Pr}_v^{0.4}(T_w - T_{\mathrm{sat}}) \tag{12.120}$$

in which properties were evaluated at saturation conditions. Note that in the above definition of the vapor Reynolds number, α is the void fraction.

The radiation contribution was postulated to be given by the sum of radiation from the surface to the liquid drops and the vapor:

$$q''_r = F_{wl}\sigma(T_w^4 - T_{\mathrm{sat}}^4) + F_{wv}\sigma(T_w^4 - T_{\mathrm{sat}}^4) \tag{12.121}$$

Implicit in the above relation is the assumption that the vapor and the liquid are at the saturation temperature. To evaluate the view factors F_{wl} and F_{wv}, the method

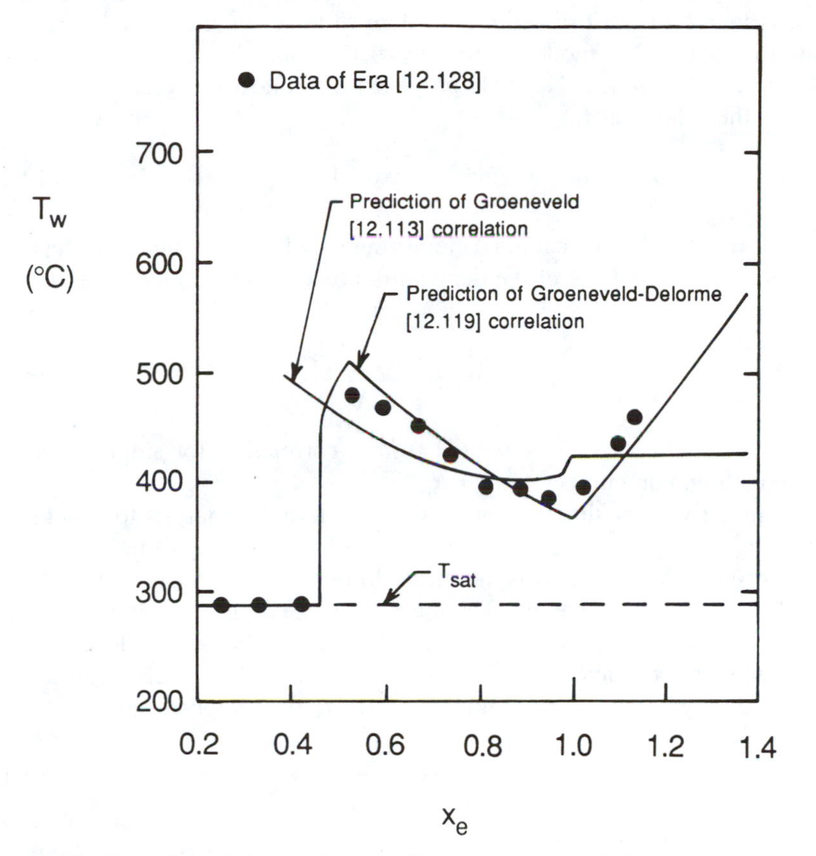

Figure 12.27 Comparison of the postdryout wall temperature data of Era et al. [12.128] with wall temperature variations predicted by two correlation techniques. (*Adapted from [12.119] with permission, copyright © 1976, Elsevier Science Publishers.*)

described by Sun et al. [12.129] for a dispersed system was recommended. For the specific circumstances of convective boiling of nitrogen, Ganic and Rohsenow [12.121] concluded that the radiation flux contribution was negligible. When comparing the predictions of their model with experimental data for nitrogen, q_r'' was therefore taken to be zero.

A key element of this model is its very detailed analytical treatment of the wall-to-droplet heat transfer process. The Groeneveld and Delorme [12.119] model neglects this mechanism, which is appropriate for high-quality conditions where the liquid inventory is low and droplet collisions with the wall are infrequent. At lower qualities, where more liquid is present in the flow, liquid is likely to contact the wall more frequently, and the importance of this mechanism to the overall heat transfer is expected to be greater. Also, because more liquid and less vapor is present at lower qualities, there is a reduced tendency for the vapor to superheat, because it contacts liquid more frequently. Ganic and Rohsenow's model, which neglects nonequilibrium effects but includes the effects of heat transfer to the droplets, is therefore most directly applicable to lower-quality flows.

Based on a detailed model of the interaction between liquid drops and the wall, Ganic and Rohsenow [12.121] proposed the following relation for its contribution to the total heat flux:

$$q''_d = v_0(1 - \alpha)\rho_l h_{lv} f_{cd} \exp\left[1 - \left(\frac{T_w}{T_{sat}}\right)^2\right] \tag{12.122}$$

where v_0 is the droplet deposition velocity and f_{cd} is the cumulative deposition factor. Based on an analysis of the deposition process, the following relation was developed to predict v_0:

$$v_0 = 0.15 \frac{Gx}{\rho_l \alpha} \sqrt{\frac{f_v}{2}} \tag{12.123}$$

where f_v is the friction factor computed from a correlation for single-phase flow at the vapor Reynolds number $GxD/\alpha\mu_v$.

The cumulative deposition factor f_{cd} was shown in the analysis to be a function only of the ratios $a_c/\bar{a}$ and $a_m/\bar{a}$, where a_c is the critical droplet radius, a_m is the maximum droplet radius, and $\bar{a}$ is the mean droplet radius. In any dispersed flow, a range of droplet sizes may exist. The maximum size a_m is dictated by the critical Weber number, at which dynamic pressure forces in the vapor flow overcome surface tension forces, leading to the breakup of liquid droplets. Based on arguments of this type, it was recommended that a_m be computed as

$$a_m = \frac{7.5\sigma}{\rho_v(u_v - u_l)^2} \tag{12.124}$$

where

$$u_v = \frac{Gx}{\rho_v \alpha} \qquad u_l = \frac{G(1 - x)}{\rho_l(1 - \alpha)} \tag{12.125}$$

$$\alpha = \frac{x}{x + (\rho_v/\rho_l)S(1 - x)} \tag{12.126}$$

In Eq. (12.126), S is the slip ratio u_v/u_l. Although Ganic and Rohsenow developed a specific correlation for S, they suggested that other correlations, such as those described by Tong and Young [12.104], may also be used. To predict the mean droplet size $\bar{a}$, the following empirical correlation was proposed:

$$\bar{a} = \frac{0.732}{u_v - u_l} \sqrt{\frac{\sigma}{\rho_l}} \left[\frac{\mu_v}{(u_v - u_l)D\rho_v}\right]^{1/2} \tag{12.127}$$

This a dimensional relation in which $\bar{a}$ and D are in m, u_v and u_l are in m/s, σ is in N/s, ρ_l and ρ_v are in kg/m^3, and μ_v is in N s/m^2.

The critical droplet radius is a parameter that emerges from analysis of the motion of droplets near the wall. Ganic and Rohsenow [12.121] solved the differential equations governing the motion of droplets near the wall, including the effects of lift and drag forces, buoyancy, and the repelling effect of rapid evaporation on the wall side of the drop. They found that for a given set of flow and wall temperature conditions, droplets below a critical size were returned to the main flow without striking the wall. Above this critical size, droplets did strike the wall. Their computed results indicated that, as expected, for a given drop size, the tendency to hit the wall was increased as the wall temperature was decreased and as the deposition velocity v_0 increased.

The computational scheme used to determine a_c is rather lengthy and will not be described in detail here. A detailed description of this scheme can be found in reference [12.121]. Once a_c is obtained for the circumstances of interest, and a_m and $\bar{a}$ are computed using the equations given above, the ratios $a_c/\bar{a}$ and $a_m/\bar{a}$ can be computed, and f_{cd} can be computed from an explicit relation given by Ganic and Rohsenow [12.121]. Equation (12.122) can then be evaluated, using Eq. (12.123) and (12.126), to determined q''_d.

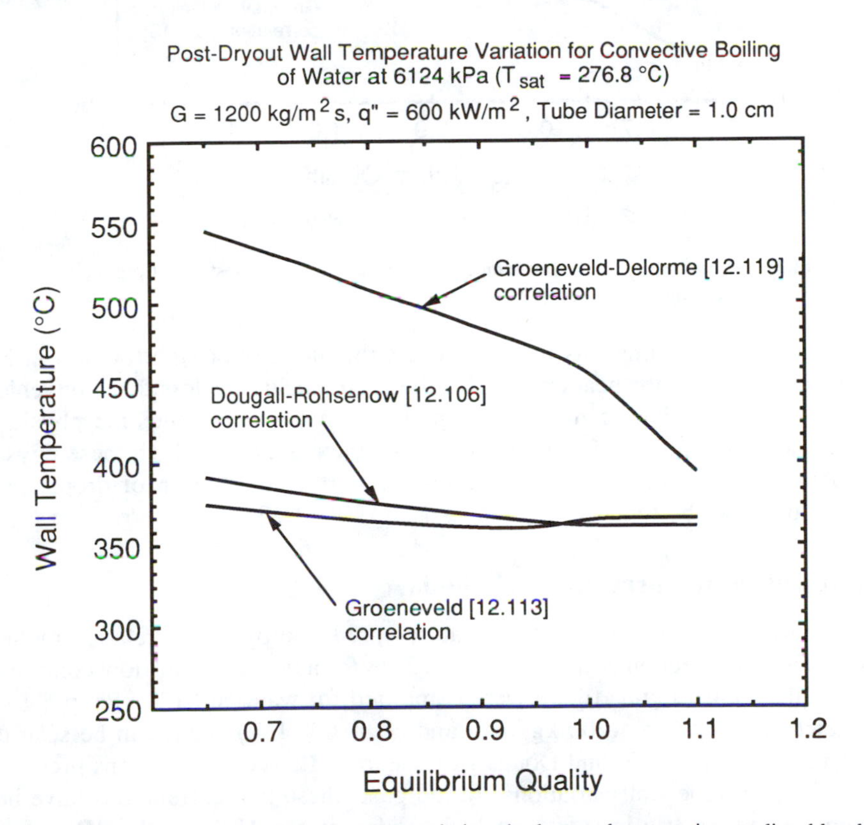

Figure 12.28 Comparison of wall temperature variations in the postdryout region predicted by three correlation methods.

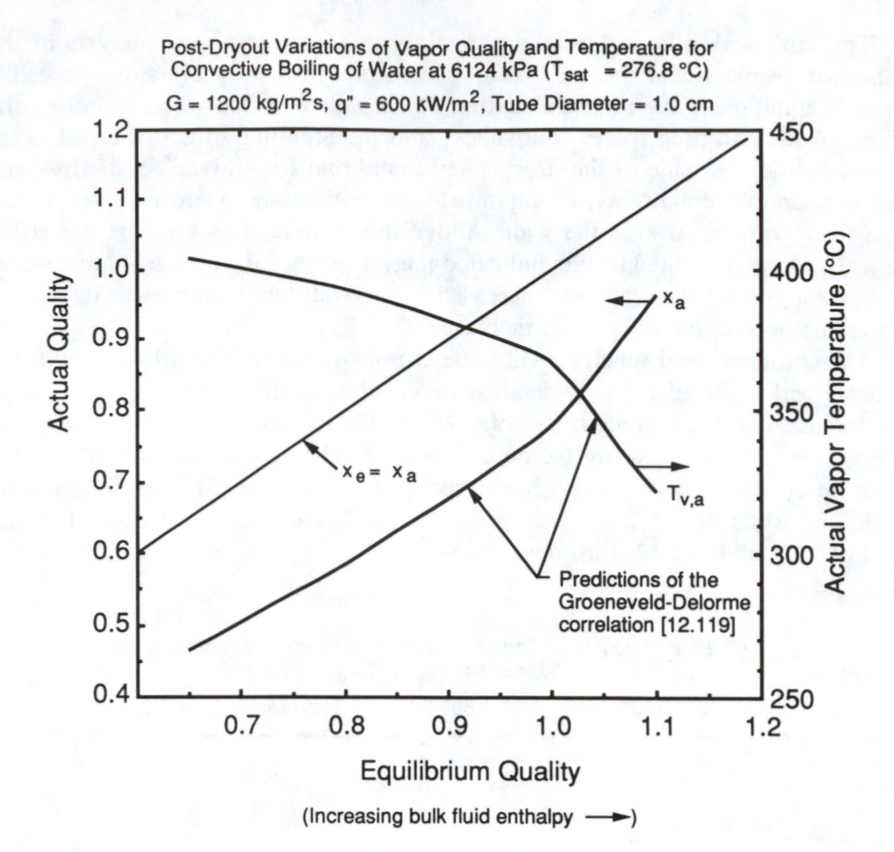

Figure 12.29 Actual quality and vapor temperature variations in the postdryout regime predicted by the Groeneveld-Delorme correlation [12.119].

While programming this model into a computer is a straightforward task, it makes prediction of the heat transfer for these circumstances less than convenient. Perhaps the main significance of this model is that it illuminates the physics of the droplet–wall interaction and demonstrates the manner in which these physical mechanisms affect the overall heat transfer. Further discussion of droplet–wall interactions may be found in references [12.130–12.133].

Comparison of Correlation Methods

A comparison of the tube wall temperatures predicted by the correlation methods described in this section is shown in Fig. 12.28 for a typical set of flow conditions. The results shown in this figure were computed for water at 6124 kPa in a 1-cm-diameter tube for $G = 1200$ kg/m^2 s and $q'' = 600$ kW/m^2. It can be seen that the Groeneveld [12.113] and Dougall-Rohsenow [12.106] correlations predict almost identical tube wall variations. In the past, these two correlations have been recommended for use in safety analysis codes by the U.S. Nuclear Regulatory Commission.

The temperature variation predicted by the nonequilibrium correlation of Groeneveld and Delorme [12.119] is substantially higher than those for the other equilibrium correlations. Plots of the actual vapor temperature and quality as predicted by the nonequilibrium correlation are shown in Fig. 12.29. It is clear from these plots that this correlation predicts a higher wall temperature because it predicts a significant departure from equilibrium, with the vapor being superheated by as much as 100°C and the actual quality being lower than the equilibrium value. This model further predicts that all droplets will not be evaporated until the (effective) equilibrium quality has reached 1.11.

As noted earlier in this section, additional semitheoretical models of postdryout heat transfer have been proposed by other investigators. While the models discussed in this section are useful, further experimental investigation of the mechanisms and verification of the predictions of these models is needed before any of them can be used with confidence over a wide range of conditions. In particular, it is worth noting that most of the experimental work in this area has been done with water. Much less information on postdryout heat transfer exists for hydrocarbons, refrigerants, and cryogenic fluids. Further general information on postdryout heat transfer can be obtained from the review articles by Collier [12.105] and Mayinger and Langner [12.112].

REFERENCES

12.1 Hewitt, G. F., and Roberts, D. N., Studies of two-phase flow patterns by simultaneous X-ray and flash photography, AERE-M 2159, Her Majesty's Stationery Office, London, 1969.

12.2 Collier, J. G., *Convective Boiling and Condensation,* 2nd ed., McGraw-Hill, New York, 1981.

12.3 Hsu, Y. Y., On the size range of active nucleation cavities on a heating surface, *J. Heat Transfer,* vol. 84, p. 207, 1962.

12.4 Bowring, R. W., Physical model based on bubble detachment and calculation of steam voidage in the subcooled region of a heated channel, OECD Halden Reactor Project Report HPR-10, 1962.

12.5 Bergles, A. E., and Rohsenow, W. M. The determination of forced-convection surface-boiling heat transfer, *J. Heat Transfer,* vol. 86, pp. 365–372, 1964.

12.6 Sato, T., and Matsumura, H, *Bull. Japan. Soc. Mech. Eng.*, vol. 7, p. 392, 1964.

12.7 Davis, E. J., and Anderson, G. H., The incipience of nucleate boiling in forced convection flow, *AIChE J.*, vol. 12, pp. 774–780, 1966.

12.8 Rohsenow, W. M., Department of Mechanical Engineering Technical Report No. 8767-21, MIT, Cambridge, MA, 1962.

12.9 Frost, W., and Dzakowic, G. S., An extension of the methods of predicting incipient boiling on commercially finished surfaces, ASME paper 67-HT-61, presented at the 1967 National Heat Transfer Conf., Seattle, Washington, 1967.

12.10 Clark, J. A., and Rohsenow, W. M., A study of the mechanism of boiling heat transfer, *Trans. ASME,* vol. 23, p. 603, 1951.

12.11 Bankoff, S. G., On the mechanism of subcooled nucleate boiling, Parts I and II, *Chem. Eng. Prog. Symp. Ser.*, vol. 57, no. 32, pp. 156–172, 1960.

12.12 Jiji, L. M., and Clark, J. A., Bubble boundary layer and temperature profiles for forced convective boiling in channel flow, *J. Heat Transfer,* vol. 76, pp. 50–58, 1964.

12.13 Hsu, S. T., and Ing, P. W., Experiments in forced convection subcooled nucleate boiling heat transfer, Paper 62-HT-38, presented at the 1962 Nat. Heat Transfer Conf., Houston, Texas, 1962.

12.14 Bernath, L., and Begell, W., Forced convection local boiling heat transfer in narrow annuli, *Chem. Eng. Prog. Symp. Ser.*, vol. 55, no. 29, pp. 59–65, 1959.

12.15 Kutateladze, S. S., Boiling heat transfer, *Int. J. Heat Mass Transfer,* vol. 4, pp. 31–45, 1961.

12.16 Engelberg-Forester, K., and Greif, R., Heat transfer to a boiling liquid—Mechanism and correlations, *J. Heat Transfer,* vol. 81, pp. 43–53, 1959.

12.17 Rohsenow, W. M., Heat transfer with evaporation, *Heat Transfer—A Symposium Held at the University of Michigan During the Summer of 1952,* Univ. of Michigan Press, Ann Arbor, pp. 101–150, 1953.

12.18 Rohsenow, W. M., and Clarke, J. A., Heat transfer and pressure drop data for high heat flux densities to water at high subcritical pressure, 1951 Heat Transfer and Fluid Mechanics Inst., Stanford University Press, Stanford, CA, 1951.

12.19 Kreith, F., and Sommerfield, M., Heat transfer to water at high flux densities with and without surface boiling, *Trans. ASME,* vol. 71, pp. 805–815, 1949.

12.20 Piret, E. L., and Isbin, H. S., Two-phase heat transfer in natural circulation evaporators, *Chem. Eng. Prog. Symp. Ser.*, vol. 50, no. 6, p. 305, 1953.

12.21 Griffith, P., Clark, J. A., and Rohsenow, W. M. Void volumes in subcooled boiling systems, Paper 58-HT-19, presented at the 1958 Nat. Heat Transfer Conf., Chicago, 1958.

12.22 Levy, S., Forced convection subcooled boiling prediction of vapor volumetric fraction, *Int. J. Heat Mass Transfer,* vol. 10, pp. 951–965, 1967.

12.23 Egen, R. A., Dingee, D. A., and Chastain, J. W., Vapor formation and behavior in boiling heat transfer, ASME Paper 57-A-74, 1957.

12.24 Maurer, G. W., A method of predicting steady state boiling vapor fraction in reactor coolant channels, Bettis Tech. Rev., WAPD-BT-19, 1960.

12.25 Staub, F. W., The void fraction in subcooled boiling—Prediction of the point of net vapor generation, ASME Paper 67-HT-36, presented at the 1967 Nat. Heat Transfer Conf., Seattle, Washington, 1967.

12.26 Saha, P., and Zuber, N., Point of net vapor generation and vapor void fraction in subcooled boiling, Proc. of the 5th Int. Heat Transfer Conf., Tokyo, Paper B4.7, 1974.

12.27 Kroeger, P. G., and Zuber, N., An analysis of the effects of various parameters on the average void fractions in subcooled boiling, *Int. J. Heat Mass Transfer,* vol. 11, pp. 211–233, 1968.

12.28 Sher, N. C., Estimation of boiling and non-boiling pressure drop in rectangular channels at 2000 psia, USAEC Report WAPD-TH-300, 1957.

12.29 Hirata, M., and Nishiwaka, N., Skin friction and heat transfer for liquid flow over a porous wall with gas injection, *Int. J. Heat Mass Transfer,* vol. 6, pp. 941–949, 1963.

12.30 Reynolds, J. B., Local boiling pressure drop, USAEC Report ANL-5178, 1954.

12.31 Dormer, J., Jr., and Bergles, A. E., Pressure drop with surface boiling in small diameter tubes, Report no. 8767-31, Department of Mechanical Engineering, MIT, Cambridge, MA, 1964.

12.32 Buchberg, H., Romie, F., Lipkis, R., and Greenfield, M., Heat transfer, pressure drop and burnout studies with and without surface boiling for de-aerated and gassed water at elevated pressures in a forced flow system, *Proc. 1951 Heat Transfer and Fluid Mechanics Inst.*, Stanford University, pp. 171–191, 1951.

12.33 Owens, W. L., and Schrock, V. E., Local pressure gradients for subcooled boiling of water in vertical tubes, ASME Paper no. 60-WA-249, 1960.

12.34 Jicha, J. J., and Frank, S., An experimental local boiling heat transfer and pressure drop study of a round tube, Paper 62-HT-48, presented at the 1962 Nat. Heat Transfer Conf., Houston, Texas, 1962.

12.35 Jordan, D. P., and Leppert, G., Pressure drop and vapor volume with subcooled nucleate boiling, *Int. J. Heat Mass Transfer,* vol. 5, pp. 751–761, 1962.

12.36 Tong, L. S., *Boiling Heat Transfer and Two-Phase Flow,* Robert Krieger, 1975.

12.37 Dengler, C. E., and Addoms, J. N., Heat transfer mechanism for vaporization of water in a vertical tube, *Chem. Eng. Prog. Symp. Ser.*, vol. 52, no. 18, pp. 95–103, 1956.

12.38 Guerrieri, S. A., and Talty, R. D., A study of heat transfer to organic liquids in single-tube natural circulation vertical tube boilers, *Chem. Eng. Prog. Symp. Ser.*, vol. 52, no. 18, pp. 69–77, 1956.

12.39 Chen, J. C., Correlation for boiling heat transfer to saturated fluids in convective flow, *Ind. Eng. Chem. Proc. Design and Dev.*, vol. 5, no. 3, pp. 322–339, 1966.

12.40 Forster, H. K., and Zuber, N., Dynamics of vapor bubbles and boiling heat transfer, *AIChE J.*, vol. 1, pp. 531–535, 1955.

12.41 Collier, J. G., Forced convective boiling, in *Two-Phase Flow and Heat Transfer in the Power and Process Industries,* A. E. Bergles, J. G. Collier, J. M. Delhaye, G. F. Hewitt, and F. Mayinger, Eds., Hemisphere, New York, 1981.

12.42 Bennett, D. L., Davis, M. W., and Hertzler, B. L., The suppression of saturated nucleate boiling by forced convective flow, *AIChE Symp. Ser.*, vol. 76, no. 199, pp. 91–103, 1980.

12.43 Bennett, D. L., and Chen, J. C., Forced convective boiling in vertical tubes for saturated pure components and binary mixtures, *AIChE J.*, vol. 26, pp. 454–461, 1980.

12.44 Shah, M. M., A new correlation for heat transfer during boiling flow through pipes, *ASHRAE Trans.*, vol. 82, part 2, pp. 66–86, 1976.

12.45 Shah, M. M., Chart correlation for saturated boiling heat transfer: Equations and further study, *ASHRAE Trans.*, vol. 88, part 1, pp. 185–196, 1982.

12.46 Schrock, V. E., and Grossman, L. M., Forced convection boiling studies, Univ. of California, Inst. of Engineering Research, Report no. 73308-UCX-2182, Berkeley, California, 1959.

12.47 Wright, R. M., Downflow forced convection boiling of water in uniformly heated tubes, Univ. of California, Report no. UCRL-9744, Berkeley, California, 1961.

12.48 Gungor, K. E., and Winterton, R. H. S., A general correlation for flow boiling in tubes and annuli, *Int. J. Heat Mass Transfer,* vol. 29, pp. 351–358, 1986.

12.49 Bjorge, R. W., Hall, G. R., and Rohsenow, W. M., Correlation of forced convection boiling heat transfer data, *Int. J. Heat Mass Transfer,* vol. 25, pp. 753–757, 1982.

12.50 Mikic, B. B., and Rohsenow, W. M., A new correlation of pool boiling data including the effect of heating surface characteristics, *J. Heat Transfer,* vol. 91, pp. 245–251, 1969.

12.51 Kandlikar, S. G., A general correlation for saturated two-phase flow boiling heat transfer inside horizontal and vertical tubes, *J. Heat Transfer,* vol. 112, pp. 219–228, 1989.

12.52 Tong, L. S., and Hewitt, G. F., Overall viewpoint of flow boiling CHF mechanisms, ASME Paper 72-HT-54, presented at the Nat. Heat Transfer Conf., 1972.

12.53 Bergles, A. E., Burnout in boiling heat transfer: High-quality forced-convection systems, in *Two-Phase Flow and Heat Transfer, China–U.S. Progress,* X.-J. Chen and T. N. Veziroglu, Eds., Hemisphere, New York, pp. 177–206, 1985.

12.54 Thompson, B., and Macbeth, R. V., Boiling water heat transfer—Burnout in uniformly heated round tubes: A compilation of world data with accurate correlations, Br. Report AEEW-R356, Winfrith, U.K., 1964.

12.55 Doroshchuk, V. E., and Lantsman, F. P., Selecting magnitudes of critical heat fluxes with water boiling in vertical uniformly heated tubes, *Thermal Eng.* (USSR), English transl., vol. 17, no. 12, pp. 18–21, 1970.

12.56 Heat and Mass Transfer Section, Scientific Council, USSR Academy of Sciences, Tabular data for calculating burnout when boiling water in uniformly heated round tubes, *Thermal Eng.* (USSR), English transl., vol. 23, no. 9, pp. 90–92, 1976.

12.57 Levitan, L. L., and Lantsman, F. P., Investigating burnout with flow of a steam–water mixture in a round tube, *Thermal Eng.* (USSR), English trans., vol. 22, no. 1, pp. 102–105, 1975.

12.58 Doroshchuk, V. E., Kon'kov, A. S., Lantsman, F. P., Levitan, L. L., and Sinitsyn, I. T., Burnout of the second kind with high mass velocities, *Thermal Eng.* (USSR), English transl., vol. 19, no. 3, pp. 105–107, 1972.

12.59 Bowring, R. W., A simple but accurate round tube uniform heat flux dryout correlation over the pressure range 0.7–17 MN/m^2 (100–2500 psia), Br. Report AEEW-R789, Winfrith, U.K., 1972.

12.60 Biasi, L., Clerici, G.C., Gariloben, S., Sala, R., and Tozzi, A., Studies on burnout, Part 3, A new correlation for round ducts and uniform heating and its comparison with world data, *Energia Nucleare,* vol. 14, no. 9, pp. 530–536, 1967.

12.61 Barnett, P. G., The scaling of forced convection boiling heat transfer, Br. Report no. AEEW-R134, Winfrith, U.K., 1963.

12.62 Ahmad, S. Y., Fluid to fluid modeling of critical heat flux: A compensated distortion model, *Int. J. Heat Mass Transfer,* vol. 16, pp. 641–662, 1973.

12.63 Ishii, M., and Jones, O. C., Jr., Derivation and application of scaling criteria for two-phase flows, in *Two-Phase Flow and Heat Transfer, Proc. of NATO Advanced Study Institute,* Istanbul, S. Kakac and F. Mayinger, Eds., vol. 1, Hemisphere, New York, pp. 163–186, 1977.

12.64 Katto, Y., and Ohno, H., An improved version of the generalized correlation of critical heat flux for the forced convective boiling in uniformly heat vertical tubes, *Int. J. Heat Mass Transfer,* vol. 27, pp. 1641–1648, 1984.

12.65 Katto, Y., A generalized correlation of critical heat flux for the forced convective boiling in vertical uniformly heated round tubes, *Int. J. Heat Mass Transfer,* vol. 21, pp. 1527–1542, 1978.

12.66 Katto, Y., A generalized correlation of critical heat flux for the forced convective boiling in vertical uniformly heated round tubes—A supplementary report, *Int. J. Heat Mass Transfer,* vol. 22, pp. 783–794, 1979.

12.67 Shah, M. M., A generalized graphical method for predicting CHF in uniformly heated vertical tubes, *Int. J. Heat Mass Transfer,* vol. 22, pp. 557–568, 1979.

12.68 Levy, S., Healzer, J. M., and Abdollahian, D., Prediction of critical heat flux for annular flow in vertical pipes, EPRI Report no. NP-1619, 1980.

12.69 Groeneveld, D. C., Kiameh, B. P., and Chang, S. C., Prediction of critical heat flux (CHF) for non-aqueous fluids in forced convective boiling, *Proc. 8th Int. Heat Transfer Conf.,* Hemisphere, New York, vol. 5, pp. 2209–2214, 1986.

12.70 Whalley, P. B., Hutchinson, P., and Hewitt, G. F., The calculation of critical heat flux in forced convective boiling, *Proc. 5th Int. Heat Transfer Conf.,* Tokyo, Paper B6.11, 1974.

12.71 Whalley, P. B., The calculation of dryout in a rod bundle, Br. Report AERE-R8319, 1976.

12.72 Whalley, P. B., Hutchinson, P., and Hewitt, G. F., Prediction of annular flow parameters for transient flows and for complex geometries, Br. Report AERE-M2661, 1974.

12.73 Govan, A. H., Hewitt, G. F., Owen, D. G., and Bott, T. R., An improved CHF modeling code, *Proc. 2nd U.K. Conf. Heat Transfer,* Institute of Mechanical Engineers, London, vol. 1, pp. 33–48, 1988.

12.74 Hewitt, G. F., and Hall-Taylor, N. S., *Annular Two-Phase Flow,* Pergamon Press, Oxford, chap. 11, 1970.

12.75 Hewitt, G. F., Critical heat flux in flow boiling, Keynote lecture, *Proc. 6th Int. Heat Transfer Conf.,* Toronto, vol. 6, pp. 143–172, 1978.

12.76 Hewitt, G. F., Burnout, in *Handbook of Multiphase Systems,* G. Hetsroni, Ed., Hemisphere, New York, sec. 6.4, 1982.

12.77 Swenson, H. S., Carver, J. R., and Karkarla, C. R., The influence of axial heat flux distribution on the departure from nucleate boiling in a water-cooled tube, ASME Paper 62-WA-297, presented at the ASME Winter Annual Meeting, New York, 1962.

12.78 Lee, D. H., and Obertelli, J. D., An experimental investigation of forced convection burnout in high pressure water—Part II, Br. Report AEEW-R 309, Winfrith, U.K., 1963.

12.79 Barnett, P. G., The prediction of burnout in non-uniformity heated rod clusters from burnout data for uniformly heated round tubes, Br. Report AEEW-R 362, Winfrith, U.K., 1964.

12.80 Tong, L. S., Currin, H. B., Larsen, P. S., and Smith, O. G., Influence of axially non-uniform heat flux on DNB, *Chem. Eng. Prog. Symp. Ser,* vol. 62, no. 64, pp. 35–40, 1966.

12.81 Kirby, G. J., A new correlation of non-uniformly heated round tube burnout data, Br. Report AEEW-R 500, Winfrith, U.K., 1966.

12.82 Tong, L. S., Prediction of departure from nucleate boiling for an axially non-uniform heat flux distribution, *J. Nuclear Energy,* vol. 21, no. 3, pp. 241–248, 1967.

12.83 Lee, D. H., Burnout in a channel with non-uniform circumferential heat flux, Br. Report AEEW-R 477, Winfrith, U.K., 1966.

12.84 Bertoletti, S., Gaspar, G. P., Lombardi, C., Peterlongo, G., Silvestri, M., and Tacconi, E. A., Heat transfer crisis with steam–water mixtures, *Energia Nucleare*, vol. 12, no. 3, pp. 121–172, 1965.

12.85 Tippets, F. E., Critical heat fluxes and flow patterns in high pressure boiling water flows, ASME Paper 62-WA-162, presented at the ASME Winter Annual Meeting, New York, 1962.

12.86 Levy, S., Fuller, R. A., and Niemi, R. O., Heat transfer to water in thin rectangular channels, *J. Heat Transfer*, vol. 81, p. 129, 1959.

12.87 Gambill, W. R., and Bundy, R. D., HFIR heat transfer studies of turbulent water flow in thin regular channels, Report ORNL 3079, Oak Ridge, Tenn., 1961.

12.88 Miropolsky, Z. L., and Pikus, V. Y., Critical boiling heat fluxes in curved channels, *Heat Transfer—Sov. Res.*, vol. 1, no. 1, pp. 74–79, 1969.

12.89 Janssen, E., Levy, S., and Kervinene, J. A., Investigation of burnout in an internally heated annulus cooled by water at 600 to 1400 psia, ASME Paper 63-WA-149, presented at the ASME Winter Annular Meeting, Philadelphia, 1963.

12.90 Becker K. M., and Hernborg, G., Measurements of burnout conditions for flow boiling water in an annulus, ASME Paper 63-HT-25, presented at the 6th Nat. Heat Transfer Conf., Boston, 1963.

12.91 Bennett, A. W., Collier, J. G., and Kersey, H. A., Heat transfer to mixtures of high pressure steam and water in an annulus. Part IV, Br. Report AERE-R 3961, 1964.

12.92 Macbeth, R. V., Burnout analysis—Part 5. Examination of published world data for rod bundles, Br. Report AEEW-R 358, Winfrith, U.K., 1964.

12.93 Stryrikovich, M. A., Miropolsky, Z. L., Shitsman, M. Y., Mostinski, I. L., Stavrovski, A. A., and Factorovich, L. E., The effect of prefixed units on the occurrence of critical boiling in steam generating tubes, *Teploenergetika*, vol. 6, p. 81, 1960.

12.94 Moxon, D., and Edwards, P. A., Dryout during flow and power transients, Br. Report AEEW-R 553, Winfrith, U.K., 1967.

12.95 Leung, J. C. M., Critical heat flux under transient conditions: A literature survey, Argonne National Lab., Report ANL-78-39, 1978.

12.96 Ishigai, S., Nakanishi, S., Yamanchi, S., and Masuda, T., Effect of transient flow on premature dryout in tube, *Proc. 5th Int. Heat Transfer Conf.*, Tokyo, vol. 4, pp. 300–304, 1974.

12.97 Kefer, V., Kohler, W., and Kastner, W., Critical heat flux (CHF) and post-CHF heat transfer in horizontal and inclined evaporator tubes, *Int. J. Multiphase Flow*, vol. 15, pp. 385–392, 1989.

12.98 McDonough, J. B., Milich, W., and King, E. C., An experimental study of partial film boiling region with water at elevated pressures in a round vertical tube, *Chem. Eng. Prog. Symp. Ser.*, vol. 57, no. 32, pp. 197–208, 1961.

12.99 Iloeje, O. C., Plummer, D. N., Rohsenow, W. M., and Griffith, P., An investigation of the collapse and surface rewet in film boiling in forced vertical flow, *J. Heat Transfer*, vol. 97, pp. 166–172, 1975.

12.100 Cheng, S. C., and Ng, W., Transition boiling heat transfer in forced vertical flow via a high thermal capacity heating process, *Lett. Heat Transfer*, vol. 3, pp. 333–342, 1976.

12.101 Groeneveld, D. C., and Fung, K. K. Forced convective transition boiling—A review of literature and comparison of prediction methods, Report AECL-5543, 1976.

12.102 Tong, L. S., Heat transfer in water-cooled nuclear reactors, *Nuclear Eng. and Design*, vol. 6, p. 301, 1967.

12.103 Ramu, K., and Weisman, J., A method for the correlation of transition boiling heat transfer data, *Proc. 5th Int. Heat Transfer Conf.*, Tokyo, Paper B.4.4, vol. IV, pp. 160–164, 1974.

12.104 Tong, L. S., and Young, J. D., A phenomenological transition and film boiling heat transfer correlation, *Proc. 5th Int. Heat Transfer Conf.*, Tokyo, Paper B.3.9, vol. IV, pp. 120–124, 1974.

12.105 Collier, J. G., Heat transfer in the postburnout region and during quenching and reflooding, in *Handbook of Multiphase Systems*, G. Hetsroni, Ed., Hemisphere, New York, Sect. 6.5, pp. 6-142 to 6-188, 1982.

12.106 Dougall, R. S., and Rohsenow, W. M., Film boiling on the inside of vertical tubes with upward flow of the fluid at low qualities, MIT report no. 9079-26, MIT, Cambridge, MA, 1963.

12.107 Suryanarayana, N. V., and Merte, H., Film boiling on vertical surfaces, *J. Heat Transfer*, vol. 94, pp. 377–384, 1972.

12.108 Bailey, N. A., Film boiling on submerged vertical cylinders, Report AEEW-M1051, Winfrith, U.K., 1971.

12.109 Chandrasekhar, S., *Hydrodynamic and Hydromagnetic Stability*, Dover, 1981.

12.110 Leonard, J. E., Sun, K. H., and Dix, G. E., Solar and nuclear heat transfer, *AIChE Symp. Ser.*, vol. 73, no. 164, p. 7, 1977.

12.111 Groeneveld, D. C., and Young, J. M., Film boiling and rewetting heat transfer during bottom flooding of a hot tube, *Proc. 6th Int. Heat Transfer Conf.*, Toronto, vol. 5, pp. 89–94, 1978.

12.112 Mayinger, F., and Langer, H., Post-dryout heat transfer, *Proc. 6th Int. Heat Transfer Conf.*, Toronto, vol. 6, pp. 181–198, 1978.

12.113 Groeneveld, D. C., Post-dryout heat transfer at reactor operating conditions, Report AECL-4513, 1973.

12.114 Miropolsky, Z. L., Heat transfer in film boiling of a steam water mixture in steam generating tube, *Teploenergetika*, vol. 10, pp. 49–52, 1963 (translated as report AEC-tr-6252, 1964).

12.115 Slaughterback, D. C., Veseley, W. E., Ybarrando, L. J., Condie, K. G., and Mattson, R. J., Statistical regression analysis of experimental data for flow film boiling heat transfer, ASME Paper 73-HT-20 presented at the National Heat Transfer Conference, Atlanta, 1973.

12.116 Laverty, W. F., and Rohsenow, W. M., Film boiling of saturated nitrogen flowing in a vertical tube, *J. Heat Transfer*, vol. 89, pp. 90–98, 1967.

12.117 Brevi, R., and Cumo, M., Quality influence in post burnout heat transfer, *Int. J. Heat Mass Transfer*, vol. 14, pp. 483–489, 1971.

12.118 Plummer, D. N., *Post critical heat transfer to flowing liquid in a vertical tube*, Ph.D. dissertation, MIT, Cambridge, MA, 1974.

12.119 Groeneveld, D. C., and Delorme, G. G. J., *Nuc. Eng. and Design*, vol. 36, pp. 17–26, 1976.

12.120 Chen, J. C., Ozkaynak, F. T., and Sundaram, R. K., Vapor heat transfer in the post-CHF region including the effect of thermodynamic non-equilibrium, *Nucl. Eng. and Design*, vol. 51, pp. 143–155, 1979.

12.121 Ganic, E. N., and Rohsenow, W. M., Dispersed flow heat transfer, *Int. J. Heat Mass Transfer*, vol. 20, pp. 855–866, 1977.

12.122 Hein, D., and Kohler, W., The role of thermal non-equilibrium in post-dryout heat transfer, Paper presented at the European Two-Phase Flow Group Meeting, Grenoble, France, 1976.

12.123 Saha, P., Shiralkar, B. S., and Dix, G. E., A post-dryout heat transfer model based on actual vapour generation rate in the dispersed droplet regime, ASME Paper 77-HT-80 presented at the 17th National Heat Transfer Conference, Salt Lake City, 1980.

12.124 Marinelli, V., and Sabatto, A., A correlation for convective heat transfer coefficients in post dryout regime, Paper presented at the European Two-Phase Flow Group Meeting, Grenoble, France, 1976.

12.125 Varma, H. K., A model for heat transfer coefficients in dryout region of forced convective evaporation, *Proc. 6th Int. Heat Transfer Conf.*, Toronto, vol. 1, pp. 417–422, 1978.

12.126 Hadaller, G., and Banerjee, S., Heat transfer to superheated steam in round tubes, AECL unpublished report, 1969.

12.127 Cumo, M., Ferrari, G., and Farello, G. E., A photographic study of two-phase, highly dispersed flows, CNEN-RT/ING (72), vol. 19, pp. 241–268, 1972.

12.128 Era, A., Gaspari, G. P., Hassid, A., Milani, A., and Zavattarelli, R., Heat transfer data in the liquid deficient region for steam-water mixtures at 70 bar flowing in tubular and annular conduits, CISE Report CISE-R-184, 1967.

12.129 Sun, K. H., Gonzalez, J. M., and Tien, C. L., Calculations of combined radiation and convection heat transfer in rod bundles under emergency cooling conditions, ASME Paper 75-HT-64, 1975.
12.130 Cumo, M., and Farello, G. E., Heated wall-droplet interaction for two-phase flow heat transfer in liquid deficient region, CNEN-RT/ING (72), vol. 19, pp. 146–178, 1972.
12.131 Cumo, M., Farello, G. E. and Ferrari, G., Notes on droplet heat transfer, CNEN-RT/ING (72), vol. 19, pp. 180–202, 1972.
12.132 McGinnis, F. K., and Holman, J. P., Individual droplet heat transfer rates for splattering on hot surfaces, *Int. J. Heat Mass Transfer,* vol. 12, pp. 95–108, 1969.
12.133 Toda, S., A study of mist cooling (2nd report: Theory of mist cooling and its fundamental experiments), *Heat Transfer—Jap. Res.,* vol. 1, no. 3, pp. 39–307, 1972.

PROBLEMS

12.1 Liquid nitrogen flows upward in a vertical round tube with an inside diameter of 0.7 cm. The pressure along the tube is virtually constant at 360 kPa. The nitrogen enters subcooled at 80 K and a flow rate corresponding to a mass flux of 800 kg/m^2 s. Assuming that the flow of liquid becomes fully developed immediately at the entrance, estimate the location (distance downstream of the inlet) at which the onset of boiling occurs.

12.2 Subcooled liquid ammonia at a temperature of 250 K and a pressure of 382 kPa enters a horizontal tube with an inside diameter of 1.2 cm. The mass flux is 1000 kg/m^2 s. Estimate the wall superheat necessary to initiate nucleate boiling at a location 0.5 m downstream of the inlet.

12.3 Subcooled water at 9460 kPa and 550 K enters a vertical evaporator tube with an inside diameter of 1.3 cm. The tube wall temperature is constant along the length of the tube at 600 K. Estimate the mass flux at which the onset of boiling just occurs at a distance 0.3 m downstream of the inlet. If the mass flux is decreased below the onset level, does nucleate boiling persist, or is it suppressed?

12.4 Subcooled liquid water at 571 kPa flows through a vertical round tube having walls held at 460 K. The liquid enters the tube at 400 K, and the onset of boiling occurs immediately at the entrance of the tube. The tube diameter is 0.9 cm and the mass flux is 1200 kg/m^2 s. Determine the partial boiling heat transfer coefficient predicted by Rohsenow's method [12.17] at a location downstream of the inlet where the bulk fluid temperature is 420 K.

12.5 Subcooled liquid nitrogen at 229 kPa and 77.4 K enters a round tube with an inside diameter of 1.0 cm. The wall of the tube is held at a constant temperature of 90 K. The mass flux through the tube is 300 kg/m^2 s. At a location along the tube where the bulk fluid temperature is 81 K, estimate the partial boiling heat transfer coefficient using (*a*) Rohsenow's method [12.17] and (*b*) the method of Bergles and Rohsenow [12.5].

12.6 Saturated flow boiling of R-12 at 528 kPa occurs in a vertical round tube with a diameter of 0.9 cm. The mass flux is 300 kg/m^2 s, and a uniform heat flux of 7000 W/m^2 is applied to the wall. Estimate the quality at which nucleation is expected to be completely suppressed.

12.7 Saturated flow boiling of ammonia at 775 kPa occurs in a vertical round tube with a diameter of 1.3 cm. The mass flux is 200 kg/m^2 s, and a uniform heat flux of 50,000 W/m^2 is applied to the wall. Estimate the quality at which nucleation is expected to be completely suppressed.

12.8 Use Kandlikar's [12.51] correlation to predict the heat transfer coefficient for flow boiling of R-12 in a vertical tube at a pressure of 333 kPa and qualities of 0.3 and 0.7. The tube diameter is 1.2 cm, the mass flux is 250 kg/m^2 s, and a uniform heat flux is 15 kW/m^2 is applied to the tube wall. Compare the results to the predictions of the Shah [12.44] correlation for the same conditions.

12.9 Use the correlation of Bjorge et al. [12.49] to predict the heat transfer coefficient for flow boiling of water in a vertical tube at a pressure of 6124 kPa and qualities of 0.3 and 0.7. The tube diameter is 1.4 cm, the mass flux is 900 kg/m^2 s, and the wall superheat is 25°C. Compare the results to the predictions of the Bennett and Chen [12.43] correlation for the same conditions.

12.10 Saturated flow boiling of water at low wall superheat occurs in a vertical round tube with an inside diameter of 1.8 cm. The mass flux through the tube is 3000 kg/m^2 s, and the pressure along the tube is essentially constant at 6124 kPa. Estimate the dryout quality for these conditions.

12.11 Saturated flow boiling of water occurs in a vertical round tube with an inside diameter of 0.9 cm. The mass flux through the tube is 4000 kg/m^2 s, and the pressure along the tube is essentially constant at 9460 kPa. Estimate the critical heat flux at qualities of 0.1, 0.3, and 0.5 using the Levitan and Lantsman [12.57] correlation. Compare the results with the predictions of the correlation of Biasi et al. [12.60] for the same conditions.

12.12 Flow boiling of nitrogen occurs in a vertical tube with a diameter of 0.8 cm and a length of 50 cm. The flow enters the tube as saturated liquid. The pressure is virtually uniform along the tube at 540 kPa. The mass flux through the tube is 300 kg/m^2 s. Use the correlation of Katto and Ohno [12.64] to predict the critical heat flux for these conditions.

12.13 Convective vaporization of R-12 occurs at low wall superheat in a vertical tube with a diameter of 0.6 cm. The mass flux is 300 kg/m^2 s, and the pressure is essentially constant at 333 kPa. For these conditions and an applied heat flux of 100 kW/m^2, use the correlation of Groeneveld [12.113] to estimate the heat transfer coefficient for $x = 0.7$ and $x = 0.9$. Compare the results with those predicted by the correlation of Dougall and Rohsenow [12.106].

12.14 Write a computer program to compute the heat transfer coefficient for mist flow evaporation using the correlation of Groeneveld and Delorme [12.119]. Use the program to compute the heat transfer coefficient value for the conditions specified in Problem 12.13, and compare the computed values of the heat transfer coefficient and wall temperature with those predicted by the Groeneveld [12.113] correlation.

12.15 Saturated liquid water at 9460 kPa enters a vertical tube 2.4 m long and having an inside diameter of 1.27 cm. The mass flow rate through the tube is 0.127 kg/s. A constant heat flux of 1.27 MW/m^2 is applied all along the tube. The onset of boiling occurs immediately at the entrance of the tube ($z = 0$). The pressure is essentially constant along the tube. (*a*) Determine the regimes of convective boiling that exist along the tube and the portion of the tube (range of z) that corresponds to each regime. (*b*) Estimate the wall temperature at $z = 1.2$ m.

12.16 Water at 1172 kPa and 100°C enters (at $z = 0$) a horizontal copper tube that is 25 mm in inside diameter and 5.0 m long (the exit corresponds to $z = 5.0$ m). A uniform heat flux of 200 kW/m^2 is applied to the tube. The water mass flow rate is 0.25 kg/s. The last 2 m of the tube contain a twisted-tape insert that for single-phase flow produces a heat transfer coefficient of 6000 W/m^2 K. The first 2 m of the tube contain no insert and have a single-phase h value of 3000 W/m^2 K. Does nucleate boiling occur on the wall of the tube? If so, determine the portion, or portions of the tube (range of z) over which nucleate boiling is expected to occur.

12.17 Saturated liquid water at 6124 kPa enters a vertical tube 1.5 m long and having an inside diameter of 1.27 cm. The mass flow rate through the tube is 0.191 kg/s. The applied heat flux to the tube is 3.65 MW/m^2 over the first 0.5 m of the tube ($0 < z \leq 0.5$ m) and 2.40 MW/m^2 over the last 1.0 m of the tube ($0.5 < z \leq 1.5$ m). The onset of boiling occurs immediately at the entrance of the tube ($z = 0$). The pressure is essentially constant along the tube. (*a*) Determine the regimes of convective boiling that exist along the tube, and the portion of the tube (range of z) that corresponds to each regime. (*b*) Estimate the wall temperature at $z = 0.8$ m. (*c*) At $z = 1.45$ m, will the Hewitt and Roberts flow regime map shown in Fig. 12.3 predict the correct flow regime? Briefly explain your answer.

PART FOUR

SPECIAL TOPICS

CHAPTER
THIRTEEN

SPECIAL TOPICS AND APPLICATIONS

13.1 TWO-PHASE FLOW INSTABILITIES

Many two-phase flow circumstances that arise in technological applications can exhibit modes of instability. The physical mechanisms that bring about such instabilities are often quite complex. A full treatment of the various instability mechanisms that can arise and efforts to model them is beyond the scope of this text. However, instabilities in the two-phase flow can alter vaporization or condensation processes to the point that the equipment fails to function as designed or is mechanically damaged. Some knowledge of two-phase instability mechanisms is therefore essential to avoid such problems in the design of equipment for phase-change applications.

Phase-change systems are almost always subjected to small-scale fluctuations. Examples of such fluctuations include the velocity fluctuations associated with turbulent flow, waves on the liquid–vapor interface, and the ebullition process associated with nucleate boiling. Thus, even when a flow is nominally steady, small-scale fluctuations in the system can amplify and trigger modes of instability. Despite these fluctuations, a flow is considered *stable* if, when perturbed, the system tends—at least asymptotically—to return to the original operating conditions.

Two-phase flow instabilities generally fall into one of two categories. A flow is said to be subject to a *static instability* if the source of the instability is intrinsic to the steady-state operating characteristics of the system. Since the instability is a consequence of the steady-state characteristics of the system, it is expected that the onset of instability can be predicted from knowledge of its steady-state behavior alone. A static instability most often leads to a different steady-state operating point for the system, or a periodic oscillating behavior.

When thermal or hydrodynamic inertia effects play a major role in causing the flow to become unstable, the flow is said to be subject to a *dynamic instability*. Since such inertia effects do not usually affect the steady-state flow, knowledge of the steady-state behavior alone is generally not sufficient to predict the onset of instability. Methods of predicting the onset condition and the subsequent behavior of the system must therefore account for additional mechanisms of this type.

Static Instabilities

Perhaps the two most commonly encountered static instabilities are the critical heat flux transition and the so-called Ledinegg instability. As in the case of pool boiling, the critical heat flux transition during convective vaporization processes may be associated with a transition from a nucleate-boiling vaporization process to a convective film boiling process. This transition was discussed extensively in Chapter Twelve and therefore will not be considered further here.

The *Ledinegg instability* [13.1] results from the pressure drop-versus-flow response associated with some two-phase flows. The circumstances of interest are indicated in Fig. 13.1. The Ledinegg instability comes about because there are some physical situations where the sum of the friction, acceleration, and gravity contributions to the total pressure drop decreases with increasing flow. Figure

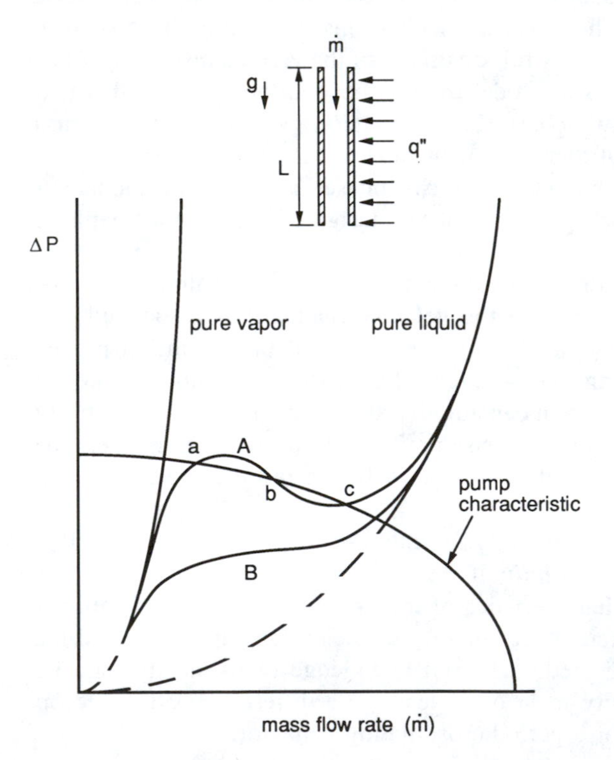

Figure 13.1 Interaction of a typical pump characteristic with the pressure drop-versus-flow behavior for downward vaporizing two-phase flow in a vertical tube.

13.1 indicates the ΔP versus mass flow rate ($\dot{m}$) for a downward boiling flow in a tube of finite length L, with the flow entering as subcooled liquid. At very high flow rates the flow remains all liquid, and the ΔP value is expected to lie on the pure liquid curve in Fig. 13.1. As the flow decreases, some vaporization will occur, whereupon the operating point ($\dot{m}$, ΔP) must deviate from the pure liquid curve. At very low flows, the operating point merges with the curve for pure vapor flow, since complete vaporization will occur almost immediately at the inlet, and flow along most of the tube will be pure vapor.

Depending on the imposed conditions, the ΔP-versus-m curve may have a form either like curve A or curve B in Fig. 13.1. If the curve has a region of negative slope, like curve A, it is possible that the characteristic curve for the pump may intersect the curve at more than one point. The pump curve shown in Fig. 13.1, for example, intersects curve A at three points.

The slope of the flow characteristic curve for the tube is positive at points a and c and negative at point b. At locations where the slope of the flow curve is more negative than the pump curve, the operating point is unstable. A perturation that increases the flow at point b decreases the ΔP required to drive the flow while decreasing the ΔP produced by the pump by a smaller amount. This disparity between the applied and required ΔP tends to increase the flow further. The flow will therefore continue to increase until the slope of the system curve becomes positive, with the ultimate effect of driving the system to point c. A similar set of arguments can be made that indicate that a perturbation that reduces the mass flow rate at point b will cause the operating point to progressively move to point a.

Points a and c, on the other hand, are stable operating conditions, because changes in the supplied and required pressure drop associated with small perturbations in the mass flow rate act to restore the system to the original operating point. The point at which the system will operate depends on the startup conditions and/or operating transients experienced by the system. Thus, the system may exist stably at a or c. Large enough perturbations can cause the system to shift from one point to the other. The capability of the flow to jump from one operating point to another, when subjected to a large enough perturbation, is the characteristic feature of the Ledinegg instability. Addition of a control valve upstream of the channel is a simple means of altering the pressure drop characteristics of the channel so as to avoid Ledinegg instability.

Other static instabilities that result in periodic variations in the flow can also arise. These are sometimes referred to as *relaxation instabilities*. This type of instability is postulated to occur in evaporating flows when flow conditions are close to the transition between bubbly and annular flow. A temporary reduction in flow rate may result in a temporary increase in the rate of vapor generation as a percentage of the total flow. This, in turn, may cause the flow to undergo a transition from bubbly to annular flow.

Because the pressure drop for annular flow is generally lower than that for a bubbly flow under the same quality conditions, the transition from bubbly to annular flow caused by the perturbation may result in a reduction in the pressure drop. The disparity between the required and supplied pressure drop will then cause the flow to accelerate. The relatively lower vaporization rate then causes the flow to revert to the bubbly regime. Usually the delay associated with acceleration and deceleration of the flow is enough to allow this cycle to repeat with a regular frequency. Oscillatory behavior of this type that is due to a flow-regime transition only is sometimes referred to as a *fundamental relaxation instability*.

For boiling systems, relaxation instabilities can also be directly linked to variations in the vaporization process. The phenomena of bumping, geysering, and chugging are directly linked to characteristics of the boiling process. *Bumping* is associated with the switching between natural convection and violent nucleate boiling. This phenomenon is typically observed in systems containing a highly wetting liquid, which tends to eliminate entrapment of vapor in larger crevices on the solid heated surface. At low applied heat flux levels, a substantial superheat must be achieved to activate very small sites that do contain entrapped vapor. Once these sites become active, the high superheat level produces very violent boiling, which dissipates the superheat very rapidly.

As the superheat drops, the small sites on the surface are deactivated and the nucleate boiling mechanism turns off, leaving only natural convection as a means of dissipating heat from the surface. Because natural convection alone cannot remove heat from the surface fast enough, the superheat builds up again and the cycle repeats. The cycle associated with this phenomenon is generally quite irregular, and it usually disappears at higher heat fluxes where the high wall superheat necessary to activate small cavities is sustained during steady boiling.

Geysering is typically observed in a vertical column of liquid with a closed bottom end that is heated. For sufficiently high heat flux levels, boiling occurs at the heated base of the column. Production of vapor in the column reduces the hydrostatic head at the base where boiling is occurring. For low-pressure systems, this reduction in pressure produces a very rapid increase in the vaporization rate, producing an explosion of vapor from the channel that often carries some liquid with it. Escape of the vapor from the channel allows subcooled liquid to flow back downward into the base region, thereby establishing a subcooled, nonboiling condition. In time, sufficient superheat is developed to initiate boiling again, and the cycle then repeats.

Cyclic expulsion of coolant from a flow channel is often referred to as *chugging*. This type of phenomenon may vary from moderate variations in the channel inlet and exit flow rates to violent ejection of fluid from the channel. For boiling systems, the nature of the process typically is similar to that for geysering described above. The cycle generally includes a waiting period, nucleation, ejection, and then reentry of the coolant. Thermosiphon reboilers frequently exhibit instabilities similar to the geysering and chugging types described above.

Chugging can also be observed in circumstances where condensation occurs. In nuclear power plants, for example, provisions are sometimes made to vent air and steam into a water pool to limit the rise in pressure during a malfunction. If the gas flow rate is low and/or the water temperature is low, a bubble may grow at the submerged exit of the vent pipe and suddenly collapse as direct-contact condensation of the steam occurs. When the sudden condensation takes place, water from the pool may surge up the vent pipe until it is stopped by increasing pressure in the line. The repeated occurrence of this sequence of events is referred to as *condensation chugging*.

Dynamic Instabilities

Disturbances in gas–liquid two-phase flows may be transported by either of two mechanisms: (1) acoustic (pressure) waves, or (2) density waves (resulting from fluctuations in void fraction). While these are both wave-type phenomena, the wave propagation velocities are very different. Acoustic waves are characterized by high frequencies, while density wave oscillations are typically much lower in frequency.

Acoustic instability is simply the growth of acoustic waves in the two-phase flow. Acoustic oscillations can occur in subcooled flow boiling, in saturated flow boiling below the CHF condition, and in convective film boiling. While the acoustic oscillations often have little effect on the flow, an early study by Bergles et al. [13.2] indicated that the amplitude of acoustic pressure fluctuations can be large compared to the mean value in some cases. Frequencies in the range of 10–10,000 Hz have been observed experimentally for oscillations of this type.

The time scale associated with density waves is about the time required for a fluid particle to travel through the channel. Waves of this type are most often observed in flow boiling processes in which the flow enters as subcooled liquid. The oscillations are a direct consequence of the link between the vaporization process and the two-phase flow behavior. A momentary drop in the inlet flow rate results in an increase in the specific enthalpy of the fluid in the inlet region. In the subcooled portion of the flow, this higher enthalpy is manifested as a higher local bulk temperature. As the higher-enthalpy fluid is convected downstream, the reduced subcooling causes the onset of boiling to occur earlier. As a result, the location of the onset of boiling shifts in the upstream direction as a result of the perturbation. Beyond the onset location, the increased enthalpy will be manifested as increased quality and increased void fraction.

The locally increased quality and void fraction resulting from the perturbation may thin the liquid film where annular flow occurs, increasing the vaporization rate momentarily, which in turn leads to further acceleration of the flow. The resulting increased acceleration of the flow will increase the local pressure gradient, resulting in a higher overall pressure drop for the two-phase portion of the flow in the channel.

If the flow is driven by a fixed head, the sum of the pressure drop in the subcooled and two-phsae regions must be a constant. This implies that an increase in the pressure drop for the two-phase portion of the flow can be sustained only if the pressure drop in the single-phase portion decreases. A decrease in the single-phase pressure drop can be achieved only if the flow rate decreases, which thus reinforces the original perturbation at the inlet.

Because pressure (acoustic) waves travel very rapidly in the flow, the effects of pressure drop changes are felt throughout the channel immediately. However, due to the finite speed with which the density perturbation travels downstream, there is a time lag between the original perturbation at the inlet and the time at which the reinforcing effect described above will be felt at the inlet. For just the right set of imposed conditions, sinusoidal fluctuations of the inlet flow rate will be in phase with the subsequent reinforcing feedback effects due to the mechanisms described above. (Random disturbances can be thought of as being composed of a linear combination of their sinusoidal Fourier series components.) Small fluctuations in the inlet flow may then amplify into finite-amplitude density wave disturbances.

Experimental studies, such as those by Saha et al. [13.3], have documented the characteristic features of density wave oscillations. Studies of this type have indicated that density wave oscillations are strongly dependent on the heat flux variation, inlet and exit flow restrictions, single- and two-phase frictional pressure drop characteristics of the channel, subcooling, inlet flow rate, and system pressure.

A number of studies [13.4–13.10] have analytically explored the stability of vaporizing flows with respect to density wave oscillations. A detailed discussion of this type of analysis is beyond the scope of this book. It is worth noting, however, that these treatments indicate that the criteria for the onset of instability can be cast in terms of specific dimensionless groups. A stability plane of the type presented by Ishii and Zuber [13.8] is shown in Fig. 13.2. The predicted stability conditions are plotted in terms of the subcooling number Su and the dimensionless heat flux Q^*, defined as

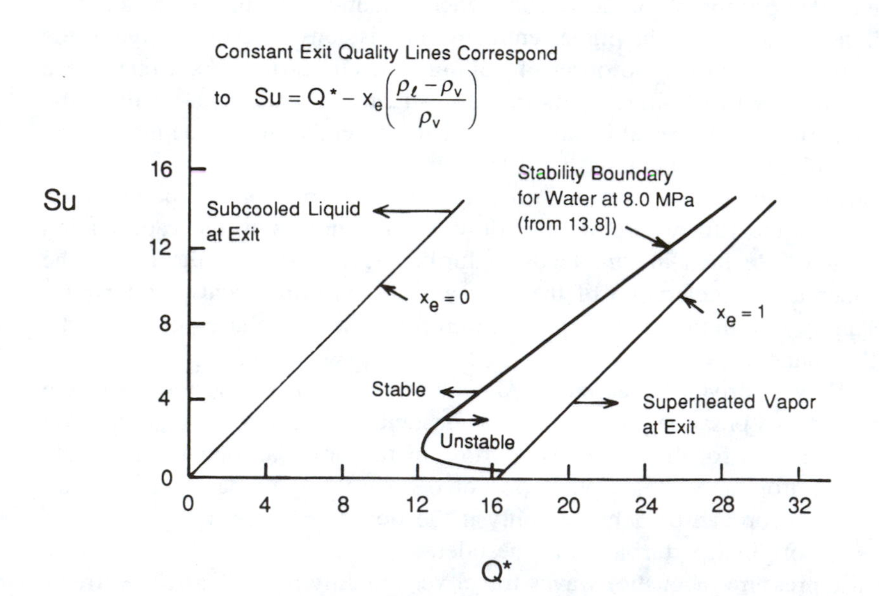

Figure 13.2 Stability plane obtained from the analysis of Ishii and Zuber [13.8].

$$\mathrm{Su} = \frac{c_{pl}(T_{\mathrm{sat}} - T_{\mathrm{in}})}{h_{lv}} \left(\frac{\rho_l - \rho_v}{\rho_v} \right) \tag{13.1}$$

$$Q^* = \frac{4q''L}{d_h(G/\rho_{l,\mathrm{in}})h_{lv}} \left(\frac{\rho_l - \rho_v}{\rho_v \rho_l} \right) \tag{13.2}$$

This stability map was developed from a linear stability analysis in which the disturbances are assumed to be small perturbations about steady state. The predictions of the analysis agree reasonably well with experimental determinations of the onset of density wave instability. Perhaps more important, the results of such analyses give insight into the physical mechanisms and parametric effects of system operating conditions on the onset of instability.

It can be seen from Fig. 13.2 that, at a given inlet subcooling and flow rate, increasing the heat flux will tend to make a stable flow become unstable. Likewise, for fixed heat flux and inlet subcooling, decreasing the inlet velocity can cause a flow to become unstable. It is also known that the frictional pressure drop in the liquid region has a stabilizing effect, since this pressure drop is in phase with the inlet flow and has a damping effect on oscillations. Adding a flow restriction at the inlet can therefore serve as a means of suppressing density wave instabilities.

Further discussion of static and dynamic two-phase flow instabilities can be found in the review by Bergles [13.11] and in reference [13.12].

Example 13.1 Water flows upward in a vertical round heated tube at a mass flux of 1000 kg/m^2 s. The tube length and diameter are 20 cm and 0.8 cm, respectively. The pressure is virtually constant along the tube at 7800 kPa. If the liquid entering the bottom of the tube is at 95°C, estimate the range of heat flux for which the flow is expected to be stable.

For saturated water at 7800 kPa, $T_{\mathrm{sat}} = 292$°C, $\rho_l = 727$ kg/m^3, $h_{lv} = 1456$ kJ/kg, and $c_{pl} = 5.39$ kJ/kg K. For the specified conditions,

$$\mathrm{Su} = \frac{c_{pl}(T_{\mathrm{sat}} - T_{\mathrm{in}})}{h_{lv}} \left(\frac{\rho_l - \rho_v}{\rho_v} \right)$$

$$= \frac{5.39(292 - 95)}{1456} \left[\frac{727 - 41.7}{41.7} \right] = 12.0$$

For this value of the subcooling number, assuming Fig. 13.2 is applicable at a pressure of 7800 kPa, the flow is seen to be stable for $Q^* < 24$, or equivalently, for

$$\frac{4q''L}{d_h(G/\rho_{\mathrm{in}})h_{lv}} \left(\frac{\rho_l - \rho_v}{\rho_v} \right) < 24$$

This can be rearranged to obtain

$$q'' < \frac{6.0 d_h(G/\rho_{\mathrm{in}})h_{lv}}{L} \left(\frac{\rho_v}{\rho_l - \rho_v} \right)$$

Approximating $\rho_{in} = \rho_l$ and substituting for the other variables yields

$$q'' < \frac{6.0(0.008)(1000/727)(1456)}{0.20}\left(\frac{41.7}{727 - 41.7}\right)$$

$$q < 29.2 \text{ kW/m}^2$$

Thus, the flow for these conditions is expected to become unstable for heat flux levels above 30 kW/m^2. It can be seen, however, that the flux can be made stable at higher heat fluxes by increasing the subcooling or the mass flux.

13.2 POOL BOILING OF BINARY MIXTURES

Despite the common occurrence and importance of boiling of multicomponent liquid mixtures in petrochemical and cryogenic processing systems, its mechanisms are not well understood at the present time. The simplest and most widely studied example of such a process is the boiling of a binary mixture. Even for this simplest type of mixture, the mechanisms may be quite complex, and accurate prediction of the pool boiling heat transfer coefficient is difficult. As a prelude to considering the overall boiling process, we will therefore first discuss some of the fundamental aspects of the thermodynamics of binary systems.

Thermodynamics of Binary Mixtures

For a binary mixture, the equilibrium conditions at which a phase change is initiated are represented in an equilibrium phase diagram. For a simple binary mixture with no azeotropic points, the phase diagram looks qualitatively like that shown in Fig. 13.3. For a mixture that forms an azeotrope at one concentration, the phase diagram looks like that shown in Fig. 13.4. The phase diagrams in these figures represent the mixture behavior at a fixed pressure, designated as P_l. In these diagrams, X_l and X_v are the concentrations of the more volatile component (lower pure component boiling temperature) in the liquid and vapor, respectively.

The *dew-point curve* shown in Fig. 13.3 represents the locus of points at which condensation is first observed as the binary mixture, with a specified concentration, is cooled at constant pressure. Similarly, the *bubble-point curve* is the locus of point at which vaporization begins as the binary mixture, with a given concentration, is heated at constant pressure. These curves represent the observed transitions for a system cooled or heated (quasi-statically) so that it passes through a sequence of equilibrium states.

The phase diagram shown in Fig. 13.3 also reflects the fact that for a system with fixed temperature and pressure, at saturation, the concentration of the more volatile component is higher in the vapor than in the liquid. If the vapor is idealized as a mixture of independent ideal gases, then *Dalton's law* dictates that the partial pressure of each component P_{vi} is equal to the total pressure multiplied by the mole fraction X_{vi} of component i:

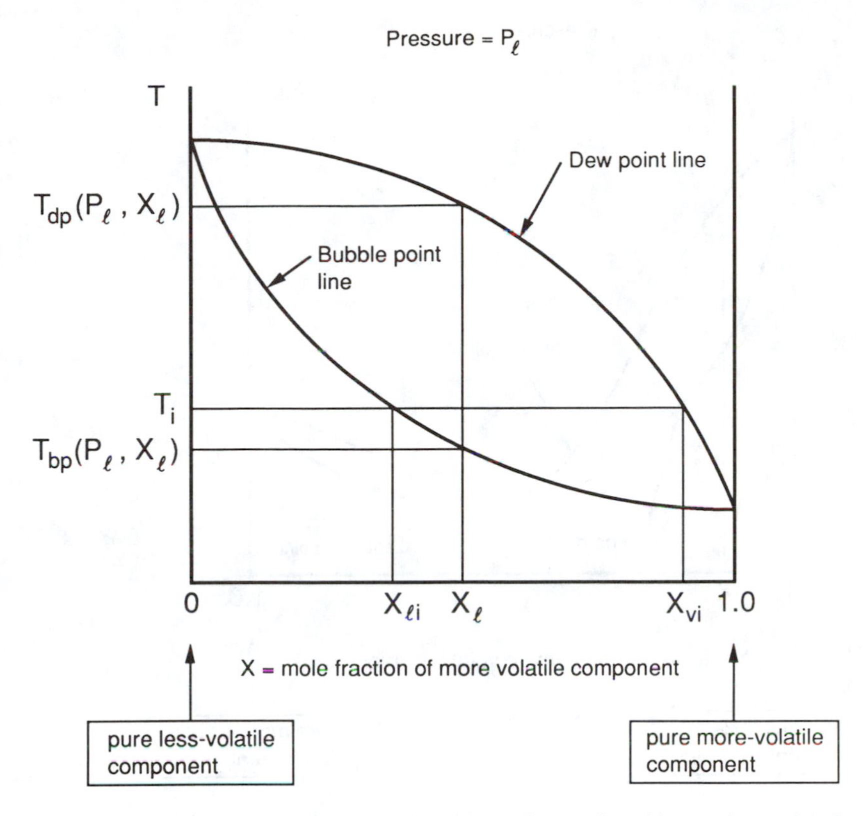

Figure 13.3 Typical form of the equilibrium phase diagram for a binary mixture that does not form an azeotrope.

$$P_{vi} = PX_{vi} \tag{13.3}$$

Two idealized models that relate the vapor partial pressure of a component to its concentration in the liquid phase are *Henry's law* and *Raoult's law*. Henry's law states that the partial pressure P_{vi} is proportional to the mole fraction of the component in the liquid X_{li}:

$$P_{vi} = C_H X_{li} \tag{13.4}$$

In this relation C_H designates the Henry's law proportionality constant. Henry's law is usually a good approximation for components having low concentrations (i.e., small X_{li}).

Raoult's law states that the partial pressure for the *i*th component is given by

$$P_{vi} = P_{pi} X_{li} \tag{13.5}$$

where P_{pi} is the saturation pressure for pure component i at the specified system temperature T, and X_{li} is the mole fraction of component i in the liquid. Raoult's law is generally a good approximation for values of X_{li} near 1.

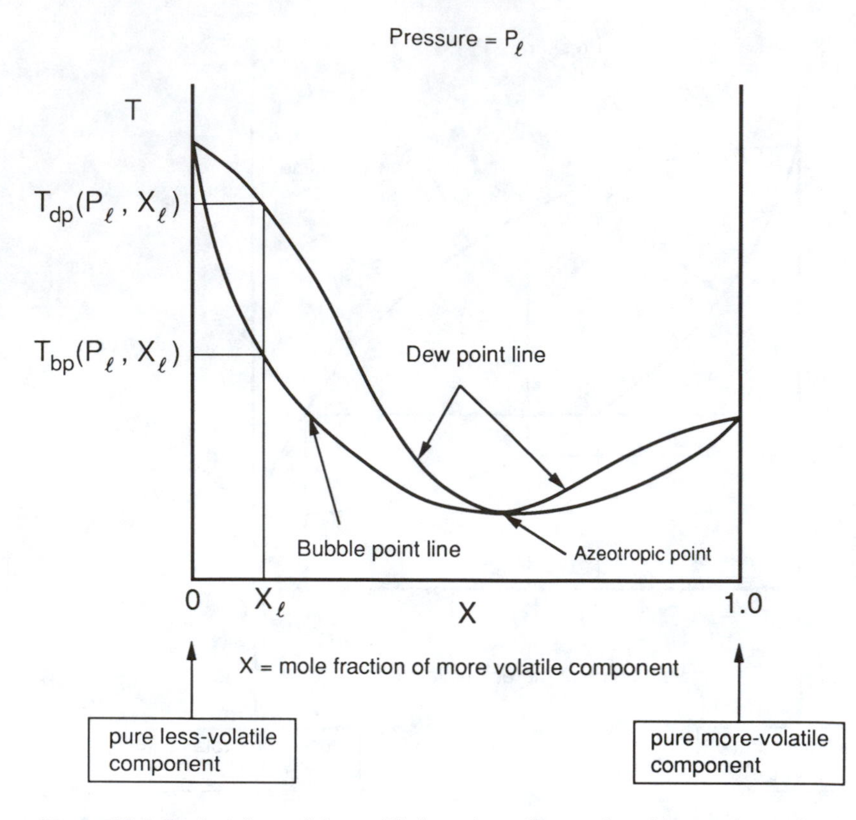

Figure 13.4 Typical form of the equilibrium phase diagram for a binary mixture that forms an azeotrope at one concentration.

If Raoult's law is assumed to apply to a binary system, it follows that

$$P_{v1} = P_{p1}X_{l1} \qquad P_{v2} = P_{p2}X_{l2} \tag{13.6}$$

where the 1 subscript designates the more volatile component. Since, by definition, the sum of the partial pressures must equal the total system pressure,

$$P = P_{v1} + P_{v2} = P_{p1}X_{l1} + P_{p2}X_{l2} \tag{13.7}$$

Rearranging and combining these relations and using the fact that $X_{l1} + X_{l2} = 1$, the following relation for $X_l = X_{l1}$ can be obtained:

$$X_l = X_{l1} = \frac{P - P_{p2}(T)}{P_{p1}(T) - P_{p2}(T)} \tag{13.8}$$

In this expression, the functional dependence of the pure component vapor pressures on temperature have been indicated. Thus, if the binary mixture conforms to Raoult's law, the variation of $X_l = X_{l1}$ with T on the phase diagram can be predicted from the pure-component vapor curves using Eq. (13.8). If, in addition, Dalton's law applies to the vapor mixture at equilibrium, then

$$P_{v1} = P_{p1}X_l = PX_v \tag{13.9}$$

Rearranging the above result yields

$$X_v = \frac{P_{p1}(T)X_l}{P} \tag{13.10}$$

Thus, having determined $X_l(T)$ as described above, Eq. (13.10) can be used to determine $X_v(T)$, completing the phase diagram shown in Fig. 13.3.

The phase diagram shown in Fig. 13.5 was constructed by applying this idealized method to a mixture of R-11 and R-113 at atmospheric pressure. The predictions of this model are actually fairly close to the actual observed mixture behavior for these fluids at atmospheric pressure. However, not all binary mixtures conform to the Raoult's law and Dalton's law idealizations described above. This idealized model does, however, provide some insight into the reasons for the qualitative trends observed in phase diagrams like that shown in Fig. 13.3.

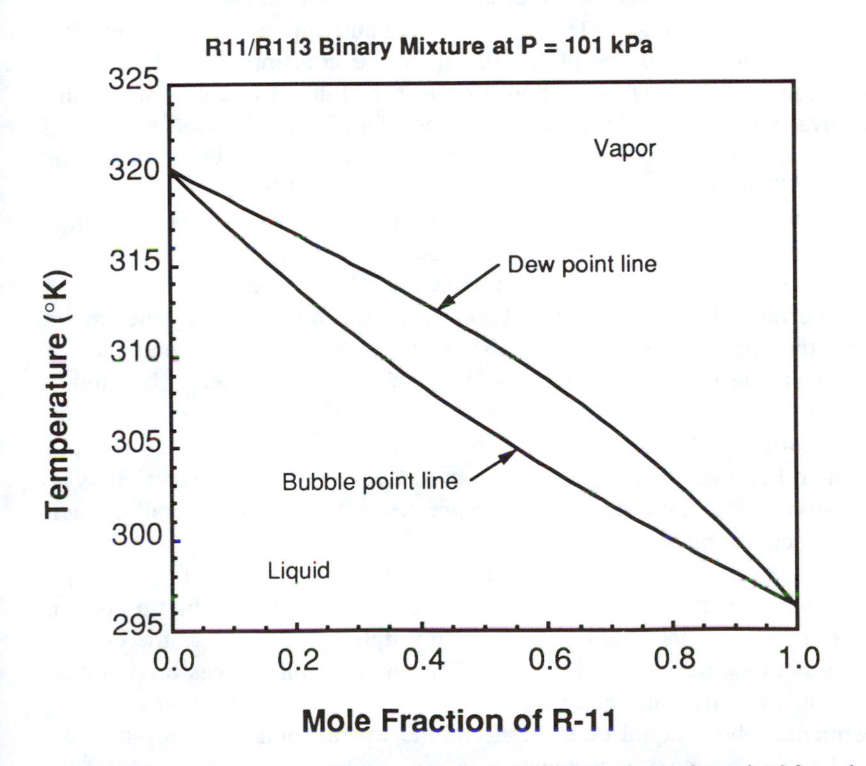

Figure 13.5 Equilibrium phase diagram for an R-11/R-113 mixture determined from the ideal mixture model.

Nucleate Boiling Heat Transfer

A number of investigators have experimentally determined the heat transfer coefficients associated with nucleate boiling of a binary mixture [13.13–13.20]. Such experiments usually consist of a series of steady-state boiling tests at liquid compositions ranging from all one pure fluid to all of the other while keeping other conditions constant. For these experiments, a heat transfer coefficient h_{bl} is often defined as

$$h_{bl} = \frac{q''}{[T_w - T_{bp}(P_l, X_l)]} \tag{13.11}$$

where T_w is the wall temperature and $T_{bp}(P_l, X_l)$ is the bubble-point temperature at the liquid pressure P_l and bulk concentration X_l as indicated in the equilibrium binary phase diagram (see Fig. 13.3).

When vaporization occurs in a nonazeotropic binary mixture, the vapor generated is richer in the more volatile component than the bulk liquid, and the remaining liquid in the vicinity of the interface has a correspondingly lower concentration of the more volatile component. Consequently, in the liquid phase, the more volatile component diffuses toward the interface, and the excess less volatile component diffuses away from the interface into the bulk liquid.

As a result of the behavior noted above, the actual temperature at the interface must be equal to the bubble point for the interface concentration of the more volatile component, which is lower than that in the ambient liquid. This implies that the interface temperature is somewhat higher than $T_{bp}(P_l, X_l)$ (see Fig. 13.3). It follows that the actual driving temperature difference for supplying heat to the interface $T_w - T_i$ is somewhat less than the temperature difference $T_w - T_{bp}(P_l, X_l)$ on which the heat transfer coefficient is often based. The amount by which $T_w - T_i$ is less than $T_w - T_{bp}(P_l, X_l)$ increases as the difference between the equilibrium vapor and liquid concentrations $|X_v - X_l|$ increases.

At the interface, the vapor and liquid concentrations must differ by the amount specified by the phase diagram (Fig. 13.3) at the interface temperature. The difference between the bulk concentrations $|X_v - X_l|$ may, in general, be slightly different than the concentration difference $|X_{vi} - X_{li}|$ at the interface, due to additional concentration differences established to facilitate transport of one component or the other away from the interface by diffusion and/or convection. Very often, diffusion in the liquid phase has the greatest effect on the overall concentration difference during vaporization processes.

It can be seen in Fig. 13.3 that the difference between the liquid and vapor concentrations at the interface $|X_{vi} - X_{li}|$ must go to zero at bulk liquid concentrations of 0 and 1, but increases and passes through a maximum as the concentration varies between the pure fluid limits. Concentration differences driving mass transfer to and from the interface will also go to 0 for these limiting cases. At some intermediate bulk liquid concentration, the overall bulk concentration difference will achieve a maximum value.

It follows directly from the above arguments that the amount by which $T_w -$

T_i is less than $T_w - T_{bp}(P_l, X_l)$ is 0 at bulk liquid concentrations of 0 and 1, achieving a maximum value at some intermediate concentration. Consequently, for nucleate boiling at a given pressure and wall temperature, the heat transfer coefficient h_{bl} defined by Eq. (13.11) generally varies in the manner indicated in Fig. 13.6 with bulk liquid concentration. The minimum of the h_{bl} variation is a direct consequence of the maximum in the difference between the liquid and vapor concentrations at some intermediate bulk liquid concentration, as described above.

If the diffusion of the more volatile component in the liquid were very rapid during the vaporization process, its concentration at the interface would be virtually identical to that of the bulk liquid, and the equilibrium temperature of the interface at the given system pressure would simply equal $T_{bp}(P_l, X_l)$. Thus the fact that $T_w - T_i$ is less than $T_w - T_{bp}(P_l, X_l)$ is a direct consequence of the resistance to mass diffusion in the liquid. For this reason, the depression of the heat transfer coefficient when $|X_v - X_l|$ is nonzero is sometimes referred to as resulting from a *diffusion resistance* to heat transfer.

One obvious alternative to the above approach would be to base the heat transfer coefficient on the actual temperature difference $T_w - T_i$ between the wall and the interface. However, the value of T_i established in a given circumstance depends on the mass transport from the interface, which is difficult to accurately predict even in the simplest systems (see, for example, the discussion by Cooper and Stone [13.21]).

If diffusion in the liquid were infinitely fast, the interface concentration would equal that in the bulk liquid. As far as heat transfer is concerned, the mixture would then behave essentially like an azeotropic mixture (i.e., like a pure fluid with properties that are averages of pure-fluid properties for components in the mixture). For such idealized circumstances, a plausible estimate of the heat trans-

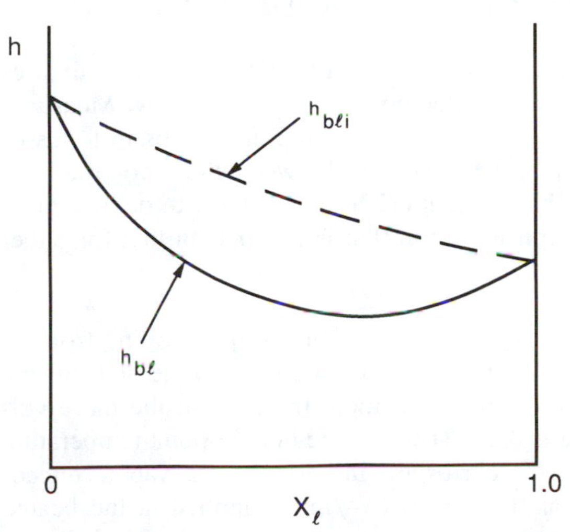

Figure 13.6 Qualitative variation of the nucleate boiling heat transfer coefficient with the bulk liquid concentration of a binary mixture at a fixed pressure.

fer coefficient h_{bli} would be a weighted average of the heat transfer coefficients for the pure components:

$$h_{bli} = h_{l0}X_l + h_{h0}(1 - X_l) \tag{13.12}$$

In this relation h_{l0} and h_{h0} are the pure-component heat transfer coefficients at the same pressure and heat flux for the low-boiling-point and high-boiling-point constituents, respectively. The arguments imply a linear variation of the heat transfer coefficient with concentration, as shown in Fig. 13.6, if mass diffusion is infinitely fast. As described above, the deviation from this idealized behavior due to finite mass transfer rates will be greatest where $|X_v - X_l|$ is largest. This suggests that the actual variation of h_{bl} with concentration can be represented by a relation of the form

$$\frac{h_{bl}}{h_{bli}} = 1 - C_C|X_v - X_l|^n \tag{13.13}$$

where C_c and n are constants that may vary depending on the substances and system pressure.

For nucleate boiling of a benzene-toluene mixture at atmospheric pressure and an applied heat flux of 100 kW/m^2, Happel [13.17] obtained good agreement between this relation and experimental data for $C_c = 1.5$ and $n = 1.4$. Although this appears to be a promising approach, appropriate values of C_c and n for a given set of conditions are generally unknown. They could, however, be determined from experiments for a given system. Trends in the data considered by Happel [13.17] also suggest that C_c and/or n will likely be larger at higher pressures. Happel [13.17] further observed that for transition boiling of binary mixtures, the heat transfer reduction was dependent on $|X_v - X_l|$, as in nucleate boiling, but its dependence on pressure was weaker. Alternative approaches to correlation of nucleate boiling data have also been proposed by Thome [13.18] and Stephan [13.19].

While precise prediction of heat transfer for pool boiling of binary mixtures is difficult, some general trends have been documented experimentally. Measured data reported by van Stralen [13.20] indicate that the boiling curves in terms of q'' and $T_w - T_{bp}(P_l, X_l)$ for mixtures generally lie between those for the pure components at the same pressure, but the critical heat flux goes through a maximum value at some intermediate concentration that is larger than that for either of the pure components.

Example 13.2 Estimate the heat transfer coefficient for pool boiling from an upward-facing large flat surface in a binary mixture of benzene and toluene at atmospheric pressure. The concentration (mole fraction) of the more volatile benzene in the bulk liquid is 0.25, at which the bubble-point temperature is 102°C. At this temperature, the corresponding equilibrium vapor concentration is 0.48. A uniform heat flux of 10 kW/m^2 is applied at the heated surface.

For saturated pure toluene at atmospheric pressure, $T_{sat} = 110.6$°C, $\rho_l =$

778 kg/m^3, ρ_v = 2.91 kg/m^3, h_{lv} = 359 kJ/kg, c_{pl} = 1.81 kJ/kg K, μ_l = 251 × 10^{-6} N s/m^2, $\Pr_l$ = 4.02, and σ = 0.0180 N/m. Rearranging the general form of the Rohsenow correlation (7.43) (with C_{sf} = 0.013, r = 0.33 and s = 1.7) yields

$$h = \frac{q''}{T_w - T_{sat}} = \frac{q'' c_{pl}}{C_{sf} \Pr_l^{1.7} h_{lv}} \left[\frac{g(\rho_l - \rho_v)}{\sigma}\right]^{1/6} \left(\frac{\mu_l h_{lv}}{q''}\right)^{1/3}$$

Substituting the properties for pure toluene yields

$$h = \frac{10.0(1.81)}{(0.013)(4.02)^{1.7}(359)} \left[\frac{9.8(778 - 2.9)}{0.0180}\right]^{1/6} \left(\frac{251 \times 10^{-6}(359)}{10.0}\right)^{1/3}$$
$$= 0.675 \text{ kW/m}^2 \text{ K} = 657 \text{ W/m}^2 \text{ K}$$

For saturated pure benzene at atmospheric pressure, T_{sat} = 80.1°C, ρ_l = 823 kg/m^3, ρ_v = 2.74 kg/m^3, h_{lv} = 398 kJ/kg, c_{pl} = 1.88 kJ/kg K, μ_l = 321 × 10^{-6} N s/m^2, $\Pr_l$ = 4.61, and σ = 0.0212 N/m. Substituting these properties into the relation above for h yields

$$h = \frac{(10.0)(1.88)}{(.013)(4.61)^{1.7}(398)} \left[\frac{9.8(823 - 2.74)}{0.0212}\right]^{1/6} \left(\frac{321 \times 10^{-6}(398)}{10.0}\right)^{1/3}$$
$$= 0.518 \text{ kW/m}^2 \text{ K} = 518 \text{ W/m}^2 \text{ K}$$

Using Eq. (13.12), h_{bli} is computed as

$$h_{bli} = h_{h0}(1 - X_l) + h_{l0} X_l$$

where h_{l0} and h_{h0} are the pure component heat transfer coefficients at the same pressure and wall heat flux for the low-boiling-point and high-boiling-point constituents, respectively. Substituting the h values computed above and X_l = 0.25 yields

$$h_{bli} = 657(1 - 0.25) + 518(0.25) = 622 \text{ W/m}^2 \text{ K}$$

Computing the heat transfer coefficient using Eq. (13.13) with C_c = 1.5 and n = 1.4, we find

$$h_{bl} = h_{bli}(1 - C_c|X_v - X_l|^n)$$
$$= 622(1 - 1.5|0.48 - 0.25|^{1.4})$$
$$= 503 \text{ W/m}^2 \text{ K}$$

Critical Heat Flux

A number of investigations have examined the critical heat flux conditions for pool boiling of binary mixtures. Early studies by van Stralen and co-workers [13.22–13.25] examined pool boiling in binary mixtures extensively. Among the more interesting findings of these studies was the observation that the critical heat fluxes for water–alcohol mixtures at low alcohol concentrations were significantly higher than the corresponding values for pure water under comparable conditions.

Subsequent investigations by a number of investigators [13.26–13.31] added to the body of CHF data for pool boiling of various binary liquid mixtures with a variety of heater geometries. Many of these investigators also found that, in some instances, variation of the critical heat flux with concentration exhibits a maximum at low concentrations of one component. Reddy and Lienhard [13.32] point out that overall trends in these CHF data were almost impossible to identify at the time these data were obtained, because effects of heater geometry variations were not well understood until more recently.

Correlations to predict the critical heat flux for heaters in binary liquid mixtures have been proposed by Kutateladze et al. [13.33], Matorin [13.34], Gaidarov [13.35], Stephen and Preusser [13.36], and Yang [13.37]. A particularly illuminating approach to correlation CHF data for horizontal wires has also been developed by Reddy and Lienhard [13.32]. They focused on two factors recognized by earlier investigators:

1. The observation by van Stralen that q''_{cr} is a maximum where the quantity $(X_l - X_v)(\Delta T_{sat}/\Delta X_l)$ achieves its maximum value (as it varies with concentration). In this expression, $\Delta T_{sat}/\Delta X_l$ is the ratio of the change in liquid saturation (bubble-point) temperature to the change in liquid concentration.
2. McEligot's [13.38] observation that evaporation from a liquid–vapor interface in a binary system leaves the liquid phase at a higher temperature than the dew-point temperature of the vapor generated. The surrounding liquid, at the bulk concentration, is colder than the temperature at which the liquid evaporates, effectively being subcooled relative to the vaporization condition. Reddy and Lienhard [13.32] refer to this as *induced subcooling*. Because of the shape of the equilibrium phase diagram, this induced subcooling is greatest where the quantity $X_l - X_v$ is highest.

Given that subcooling of the surrounding bulk liquid pool is known to increase the critical heat flux for pool boiling of pure fluids, the above comments suggest that the peak in q''_{cr} is at least partially a consequence of the high induced subcooling that occurs where $|X_l - X_v|$ is largest. Reddy and Lienhard [13.32] used pure fluid correlations for the critical heat flux during saturated and subcooled pool boiling to assess quantitatively the induced subcooling in their experiments. For the ethanol–water mixture considered in their study, they noted that $X_l - X_v$ varies about linearly with $T_{dp}(X_l, P) - T_{bp}(X_l, P)$. They were consequently able to correlate their data in terms of this temperature difference instead of the concentration difference $X_l - X_v$ used by Kutateladze et al. [13.33] and others.

Based on the above arguments and some dimensional analysis considerations, Reddy and Lienhard [13.32] recommended a correlation of the following form for predicting the CHF condition for horizontal cylinders:

$$\frac{q''_{cr}}{(q''_{cr})_{SL}} = (1 + 0.10X_l)^{-1}(1 - 0.170\,\mathrm{Ja}_e^{0.308})^{-1} \tag{13.14}$$

where

$$(q''_{\text{cr}})_{\text{SL}} = \frac{\pi}{24} \rho_v^{1/2} h_{lv} [\sigma g (\rho_l - \rho_v)]^{1/4} \cdot (0.89 + 2.27 \exp\{-3.44 R [g(\rho_l - \rho_v)/\sigma]^{1/2}\}) \tag{13.15a}$$

$$\text{Ja}_e = \frac{\rho_l c_{pl} [T_{dp}(X_l, P) - T_{bp}(X_l, P)]}{\rho_v h_{lv}} \tag{13.15b}$$

The numerical constants in Eq. (13.14) were chosen to obtain a best fit to their data for saturated ethanol–water mixtures. In the above relations, T_{dp} and T_{bp} are the dew-point and bubble-point temperatures (see Fig. 13.3), respectively, and h_{lv} is the equilibrium latent heat at constant temperature. Equation (13.15*a*) is the pure-fluid correlation of Sun and Lienhard [13.39] for the critical heat flux for a horizontal cylinder of radius R. This correlation matched their data for wires and tubes to a RMS deviation of 15%. Although the agreement with their data is quite good, Reddy and Lienhard [13.32] acknowledge that additional research is needed to examine the effects of heat diffusion, heater geometry, and heater boundary condition on the critical heat flux condition.

Other Features

Nucleation, bubble growth, and other detailed facets of nucleate boiling in binary liquids are, as might be expected, quite a bit more complex than the corresponding phenomenon in pure-component systems. The interested reader is referred to the studies of van Stralen [13.14, 13.15], van Ouwerkerk [13.40], Calus et al. [13.16], Cooper [13.41], or Cooper and Stone [13.21] for more information on these aspects of boiling of binary mixtures.

Film boiling of binary mixtures on a horizontal surface was considered in an early study by Kautzky and Westwater [13.42]. These investigators obtained film boiling curves for pure CCl_4, for pure R-113, and for mixtures of these two components. Their results clearly demonstrate that a simple interpolation between the boiling curves for the pure components does not adequately represent the results for binary mixtures of these components.

More recently, Yue and Weber [13.43, 13.44] and Marschall [13.45] have presented analyses of film boiling of binary mixtures over a vertical surface. It is noteworthy that these analyses incorporate the effects of induced (or effective) subcooling on heat transfer. Further work is clearly needed before generally applicable methods of predicting pool boiling heat transfer for mixtures are available.

13.3 CONVECTIVE BOILING OF BINARY MIXTURES

All of Chapter Twelve in this book is dedicated to flow boiling of pure fluids. There are, however, numerous applications in which flow boiling of multicomponent fluid mixtures is important. Vaporization of hydrocarbon mixtures in the petrochemical industry is perhaps the most conspicuous example. In addition to the complex transport mechanisms found in pure fluid vaporization processes, vaporization of multicomponent fluid mixtures is further complicated by more complex thermodynamics and the additional effects of species mass transfer. In general, the complexity of the transport rapidly increases as the number of mixture components increases.

To examine multicomponent convective boiling processes with minimal added complexity (beyond that for pure fluids), this section will focus on flow boiling of binary mixtures. The two-phase flow characteristics for convective boiling of binary mixtures are generally similar to those for flow boiling of pure fluids. In considering the flow, it is useful to extend the definition of the quality x to be the ratio of the vapor mass flow rate W_v to the total mass flow rate,

$$x = \frac{W_v}{W_v + W_l} \tag{13.16}$$

The boiling mechanisms that arise during flow boiling of binary mixtures are similar to those for flow boiling of pure fluids. At low-quality conditions, where the vapor void fraction is low, nucleate boiling at the wall of the tube is the dominant mechanism of vaporization. As the flow proceeds downstream and the quality and void fraction increase, a transition to annular flow is expected to occur, which facilitates the increasing importance of film flow evaporation. As in the case of pure fluid convective boiling, nucleate boiling may be largely suppressed at high qualities, and dryout of the liquid film on the tube wall may result in a transition into the mist flow evaporation regime.

Prediction of Convective Boiling Heat Transfer

At low qualities, where nucleate boiling is the dominant vaporization mechanism, methodologies for predicting heat transfer for nucleate pool boiling of binary mixtures can be adapted to flow boiling circumstances. As described in Section 13.2, one proposed method for predicting the heat transfer coefficient for nucleate boiling of binary mixtures is embodied in the following relation:

$$h_{bl} = [h_{h0}(1 - X_l) + h_{l0}X_l](1 - C_c|X_v - X_l|^n) \tag{13.17}$$

where h_{l0} and h_{h0} are the pure-component heat transfer coefficients at the same pressure and heat flux for the low-boiling-point and high-boiling-point constituents, respectively. X_v and X_l are the mole fractions of the low-boiling-point constituent in the vapor and liquid phases, respectively. The constants C_c and n in Eq. (13.17) may vary depending on the components and the system pressure.

As discussed in Section 12.4, the increasing void fraction in vaporizing flows results in acceleration of the core flow and a transition to an annular configuration, both of which tend to enhance the forced-convective effect and suppress nucleate boiling. To account for this transition a superposition correlation method, similar to the Chen [13.46] correlation, could also be used for convective vaporization of binary mixtures. Some modification of the nucleate boiling contribution predicted, for example, by Eq. (13.17) would be necessary to account for the suppression of nucleation with increasing forced-convection effect. One approach would be to modify the nucleate boiling heat transfer coefficient predicted by Eq. (13.17) with a suppression factor S. The resulting relation for the nucleate boiling contribution would be

$$h_{bl} = S[h_{h0}(1 - X_l) + h_{l0}X_l](1 - C_c|X_v - X_l|^n) \tag{13.18}$$

A similar type of correlation based on a slightly different nucleate boiling heat transfer relation for a binary mixture has been proposed by Stephan and Korner [13.47].

As noted above, at moderate to high qualities, for vertical upward flow (and horizontal flow at high flow rates) the flow is expected to take on an annular morphology. For heat pump systems using a nonazeotropic refrigerant blend, the throttling process in the expansion valve may result in 20–30% quality at the inlet of the evaporator. The vaporization process may be annular film flow evaporation at the inlet and over most of the refrigerant flow passage in such circumstances. Because this type of vaporization process is important in this and other applications, we will examine it in more detail.

To facilitate an analytical examination of annular flow evaporation of a binary mixture in a round tube, the following idealizations will be adopted:

1. The two components of the binary mixture are miscible at all concentrations.
2. The sensible heat gain (or loss) of the liquid film is negligible compared to the latent heat of vaporization.
3. Thermodynamic equilibrium exists at the liquid–vapor interface.
4. Steady-state conditions exist throughout the flow.
5. The interface is smooth and the flow field is axisymmetric.
6. All liquid flows in the annular film on the tube wall (no entrainment).
7. Thermo-diffusion and diffusion-thermo effects are negligible.
8. Rates of axial diffusion and conduction are negligible.

The idealized physical system to be modeled is shown schematically in Figs. 13.7 and 13.8. To gain an overall sense of the nature of the analysis, the governing conservation equations will be developed assuming that heat transfer coefficients and mass transfer coefficients associated with the process can be determined from available correlations.

Considering the transport at the interface (see Fig. 13.8), conservation of the more volatile species requires that

$$\rho_l h_{li}^*(C_{bl} - C_{il}) = \rho_v h_{iv}^*(C_{iv} - C_{bv}) + \frac{C_{iv}}{\pi D}\left(\frac{dW_v}{dz}\right) \tag{13.19}$$

In this relation C_{bl} and C_{il} are the mass concentrations of the more volatile species in the bulk liquid and in the liquid at the interface. Similarly, C_{bv} and C_{iv} are the

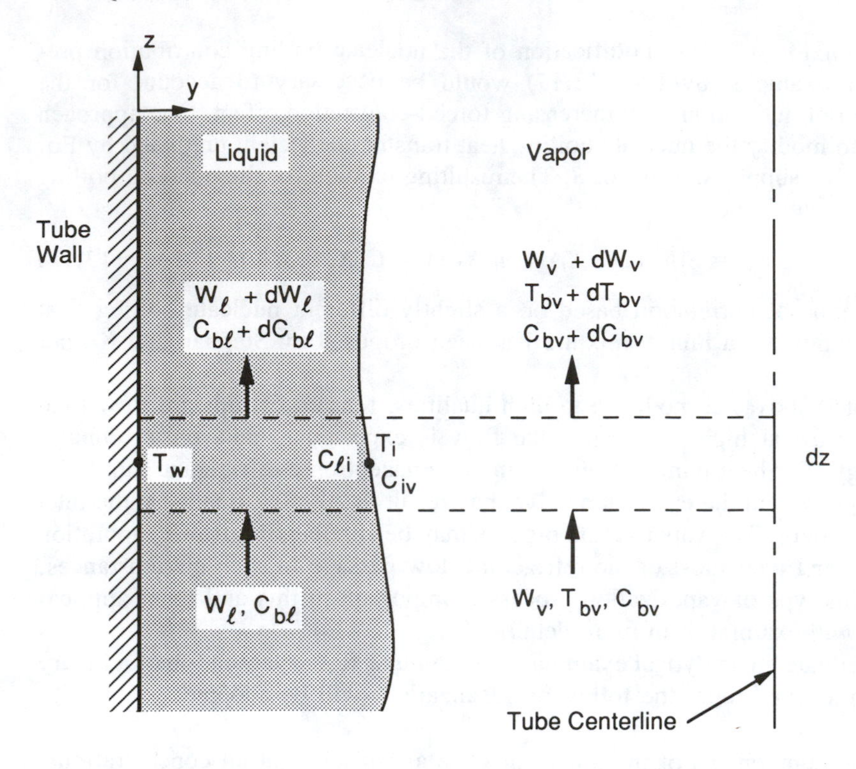

Figure 13.7 System model for analysis of convective vaporization of a binary mixture inside a tube.

mass concentrations in the bulk vapor and in the vapor at the interface. As indicated in Fig. 13.3, the binary phase diagram (for a mixture that does not form an azeotrope) dictates that the vapor concentration at a specified interface temperature T_i will be higher than the liquid concentration by a finite amount. The mass transfer coefficients associated with transport from the bulk liquid to the interface, and from the interface to the bulk vapor, are designated as h_{li}^* and h_{iv}^*, respectively. (Note that the units on these coefficients are m/s.)

A thermal energy balance at the interface similarly requires

$$h_{wi}(T_w - T_i) = h_{iv}(T_i - T_{bv}) + \frac{h_{lv}}{\pi D}\left(\frac{dW_v}{dz}\right) \tag{13.20}$$

This relation quantifies the transfer of heat from the wall to the interface across the liquid film in terms of the heat transfer coefficient h_{wi}. On the right side of Eq. (13.20), the first term represents convective transport from the interface to the bulk vapor. The second term represents the enthalpy contribution of vapor generated at the interface by the vaporization process.

Combining Eqs. (13.19) and (13.20) to eliminate dW_v/dz yields the relation

$$\frac{1}{C_{iv}}[\rho_l h_{li}^*(C_{bl} - C_{il}) - \rho_v h_{iv}^*(C_{iv} - C_{bv})] = \frac{1}{h_{lv}}[h_{wi}(T_w - T_i) - h_{iv}(T_i - T_{bv})] \tag{13.21}$$

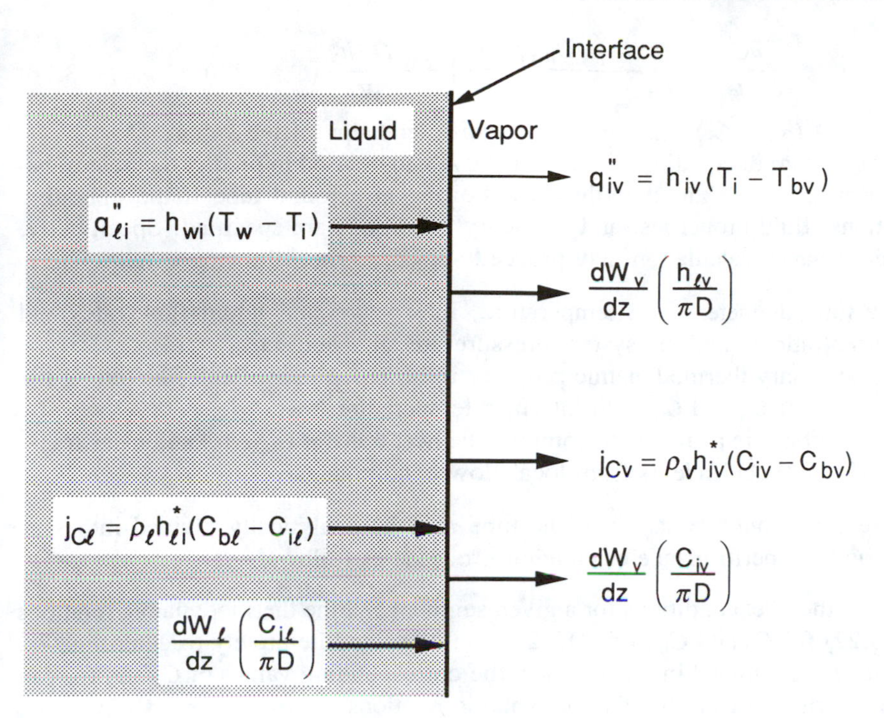

Figure 13.8 Mass, species, and energy fluxes at the liquid–vapor interface.

As indicated in the binary phase diagram (Fig. 13.3), at a given pressure, the thermodynamics provides relations that can be used to determine C_{il} and C_{iv} as a function of the interface temperature T_i:

$$C_{il} = C_{il}(T_i, P) \qquad C_{iv} = C_{iv}(T_i, P) \qquad \text{(from thermodynamics)} \tag{13.22}$$

Overall mass conservation at the interface further requires that

$$\frac{dW_l}{dz} = -\frac{dW_v}{dz} \tag{13.23}$$

Considering the differential elements of the flow indicated in Fig. 13.7, thermal energy and species conservation in the vapor core and species conservation in the liquid can be expressed as the following equations:

$$\frac{dT_{bv}}{dz} = \frac{\pi D h_{iv}}{W_v c_{pv}}(T_i - T_{bv}) + \frac{dW_v}{dz}\left[\frac{T_i - T_{bv}}{W_v} + \frac{\gamma_v(C_{vi} - C_{bv})}{W_v c_{pv}}\right] - \frac{\gamma_v}{W_v c_{pv}}\left(\frac{dW_v}{dz}\right) \tag{13.24}$$

$$\frac{dC_{bv}}{dz} = \left(\frac{C_{iv} - C_{bv}}{W_v}\right)\left(\frac{dW_v}{dz}\right) + \frac{\pi D \rho_v h^*_{vi}}{W_v}(C_{vi} - C_{bv}) \tag{13.25}$$

$$\frac{dC_{bl}}{dz} = -\left(\frac{C_{bl} - C_{il}}{W_l}\right)\left(\frac{dW_l}{dz}\right) - \frac{\pi D \rho_l h_{li}^*}{W_l}(C_{bl} - C_{il}) \tag{13.26}$$

where $\gamma_v = (\partial \hat{h}_v / \partial C_v)_T$, $\hat{h}_v$ being the specific enthalpy of the vapor.

The system of equations described above can be used to predict the heat transfer performance for annular film flow boiling in a round tube if the operating conditions, fluid properties, and boundary conditions are specified. Specifically, the performance calculation may proceed subject to the following conditions:

1. The tube diameter, wall temperature, inlet vapor and liquid flow rates and concentrations, and the system pressure are all specified.
2. All necessary thermodynamic property information is available, including the variation of C_{il} and C_{iv} with interface temperature.
3. Correlations are available to compute the heat transfer (h_{wi}, h_{iv}) and mass transfer (h_{li}^*, h_{iv}^*) coefficients from local flow conditions.

If these conditions are satisfied, the tube may be analytically divided into segments and the performance calculation proceeds as follows:

1. Using the inlet conditions for a given segment and the thermodynamic relations (13.22) for C_{il} and C_{iv}, Eq. (13.21) can be solved iteratively to determine T_i. With T_i determined in this manner, the corresponding values of C_{il} and C_{iv} can be determined from the thermodynamic relations.
2. The heat transfer and mass transfer coefficients are determined from the local flow conditions (at the inlet of the segment) and, if necessary, the interface conditions obtained in step 1.
3. Using the results of steps 1 and 2, Eqs. (13.20) and (13.23) are used to determine dW_v/dz and dW_l/dz.
4. Equations (13.24), (13.25), and (13.26) are used to determine dT_{bv}/dz, dC_{bv}/dz, and dC_{bl}/dz.
5. Values of W_v, W_l, x, T_{bv}, C_{bv}, and C_{bl} at the inlet to the next $(n + 1)$ segment are computed as

$$(W_v)_{n+1} = (W_v)_n + \left(\frac{dW_v}{dz}\right)\Delta z$$

$$(W_l)_{n+1} = (W_l)_n + \left(\frac{dW_l}{dz}\right)\Delta z$$

$$x_{n+1} = \frac{(W_v)_{n+1}}{(W_v)_{n+1} + (W_l)_{n+1}}$$

$$(T_{bv})_{n+1} = (T_{bv})_n + \left(\frac{dT_{bv}}{dz}\right)\Delta z$$

$$(C_{bv})_{n+1} = (C_{bv})_n + \left(\frac{dC_{bv}}{dz}\right)\Delta z$$

$$(C_{bl})_{n+1} = (C_{bl})_n + \left(\frac{dC_{bl}}{dz}\right)\Delta z$$

where Δz is the length of the tube segment. In addition, the heat flux from the wall is computed as

$$(q'')_n = h_{wi}(T_w - T_i)$$

6. If this is the last segment of the tube, the exit condition is known, as is the heat flux at each segment, and the computation is complete. If this is not the last segment, the algorithm is repeated, beginning with step 1.

The computational scheme described above will thus predict the variation of the heat flux along the flow passage. As specified, the integration of the equations for the bulk temperature of the vapor and the concentrations is obviously very simplistic, and the algorithm could be improved by the use of a more sophisticated numerical scheme. However, the basic scheme described above does indicate the nature of the computation required for this type of system.

A major issue in doing this type of computation is how to predict the heat and mass transfer coefficients at each downstream location. If the liquid film is laminar, a relatively crude approximation for the heat transfer coefficient would be

$$h_{wi} = \frac{k_l}{\delta} \tag{13.27}$$

where δ is the local liquid film thickness. If the film thickness is small compared to the tube diameter, it could be approximately computed from the void fraction, as

$$\delta = 4(1 - \alpha)D \tag{13.28}$$

The void fraction α could, in turn, be computed from the local properties and quality using one of the correlations represented in Table 10.2.

If the mass transfer from the bulk liquid film to the interface is idealized as being uniform along the interface, a simple laminar flow analysis for a shear-driven thin liquid film leads to the result

$$h_{li}^* = \frac{5.0 D_C^*}{\delta} \tag{13.29}$$

where D_C^* is the binary diffusion coefficient. Thus, by determining the film thickness from the void fraction as described above, the heat and mass transfer coefficients for laminar film flow can be computed approximately using Eqs. (13.27) and (13.29). For turbulent flow, treatment of the heat and mass transfer would require use of a model analysis like that described in Section 11.2. While this would make the computation more complex, conceptually, the adaptation of the turbulent film flow model described in Section 11.2 to these circumstances is relatively straightforward.

In the vapor core, the flow is most often turbulent. A crude estimate of the heat and mass transfer coefficients associated with transport between the bulk flow and the interface can be obtained by ignoring the presence of the liquid film and idealizing the interface as a smooth solid wall. The Dittus-Boelter equation can then be used to predict the heat transfer coefficient,

$$\mathrm{Nu} = \frac{hD}{k_v} = 0.023\left(\frac{W_v D}{(\pi D^2/4)\mu_v}\right)^{0.8} \mathrm{Pr}_v^{0.4} \tag{13.30}$$

and, using the heat and mass transfer analogy, this relation can be converted to a correlation that may be used to compute the mass transfer coefficient h^*:

$$\mathrm{Sh} = \frac{h^* D}{D_C^*} = 0.023\left(\frac{W_v D}{(\pi D^2/4)\mu_v}\right)^{0.8} \mathrm{Sc}_v^{0.4} \tag{13.31}$$

In this relation, Sh is the Sherwood number, D_C^* is the binary diffusion coefficient, and Sc is the Schmidt number, defined as the ratio of the kinematic viscosity to binary diffusion coefficient ν/D_C^*. There is, however, one note of caution that should be observed. Equations (13.30) and (13.31) apply strictly to heat and mass transfer to an impermeable surface. Because of the vaporization at the interface, there is an effective "blowing" velocity component at the interface in the vaporization process considered here. This blowing velocity may significantly alter the transport from that which occurs at an impermeable surface under comparable conditions. Consequently, the above relations are likely to yield useful predictions only if this blowing velocity is very small compared to the bulk velocity of the flow. Further discussion of methods for predicting the heat and mass transfer coefficients can be found in the paper by Shock [13.48].

The above idealized relations for the heat and mass transfer coefficients clearly will provide only a crude theoretical estimate of the transport during the convective vaporization of a binary mixture. Correlations based on experimental data or models tested against data would clearly be preferable. However, at the present time little information of this type is available in the open literature.

The analysis of annular film flow evaporation of a binary mixture described above considers only the film evaporation process. In an actual flow boiling circumstance, it is possible for both forced-convective evaporation and nucleate boiling effects to be present. This suggests that the above analysis could be modified to include a superimposed nucleate boiling contribution to the total heat flux at each location along the tube.

Along similar lines, Bennett and Chen [13.49] have extended the original Chen correlation to convective evaporation of a binary mixture. The modified Chen correlation developed by Bennett and Chen [13.49] retains the postulated form

$$h_{\mathrm{tot}} = h_{\mathrm{mic}} + h_{\mathrm{mac}} \tag{13.32}$$

As in the original correlation, the microscopic contribution h_{mic} is given by

$$\begin{aligned} h_{\mathrm{mic}} = 0.00122&\left[\frac{k_l^{0.79} c_{pl}^{0.45} \rho_l^{0.49}}{\sigma^{0.5} \mu_l^{0.29} h_{lv}^{0.24} \rho_v^{0.24}}\right][T_w - T_{bp}(C_{bl}, P_l)]^{0.24} \\ &\times [P_{\mathrm{sat}}(C_{bl}, T_w) - P_l]^{0.75} S_{\mathrm{bin}} \end{aligned} \tag{13.33}$$

where

$$S_{\text{bin}} = [S_{\text{pure}}(\text{Re}_{tp})]\left[1 - \frac{c_{pl}(C_{bv} - C_{bl})}{h_{lv}}\left(\frac{\partial T_{bp}}{\partial C_l}\right)_P\left(\frac{\rho_l c_{pl}}{k_l D_l^*}\right)^{1/2}\right]^{-1} \tag{13.34}$$

In the above relations, S_{pure} is the suppression factor for convective boiling of a pure fluid at the same two-phase Reynolds number (see Section 12.4), and T_{bp} is the bubble point of the binary mixture, which is a function of pressure and liquid concentration. The correction factor S_{bin} accounts for suppression of nucleate boiling and corrects the driving potential for bubble growth in the manner suggested by Florshuetz and Kahn [13.50] based on their study of bubble growth in binary liquid mixtures.

Bennett and Chen [13.49] further postulated that mass transfer in the flow does not affect the heat transfer coefficient, but that it does affect the driving potential for heat transfer. Based on this hypothesis and a simple model analysis, they recommended the following relations for the macroscopic contribution to the heat transfer coefficient:

$$h_{\text{mac}} = 0.023\left(\frac{k_l}{D}\right)\left(\frac{W_l D}{(\pi D^2/4)\mu_l}\right)^{0.8} \text{Pr}_l^{0.4} F_{\text{BC}}\left(\frac{\text{Pr}_l + 1}{2}\right)^{0.444} \Gamma_{\Delta T} \tag{13.35}$$

where

$$F_{\text{BC}} = \left[\frac{(dP/dz)_{Ftp}}{(dP/dz)_{Fl}}\right]^{0.444} \tag{13.36}$$

$$\Gamma_{\Delta T} = 1 + \frac{C_{bv} q''(\partial T_{bp}/\partial C_l)_P}{\rho_l h_{lv} h^*[T_w - T_{bp}(C_{bl}, P)]} \tag{13.37}$$

$$h^* = 0.023\left(\frac{D_C^*}{D}\right) \text{Re}_{tp}^{0.8} \text{Sc}_l^{0.4} \tag{13.38}$$

$$\text{Re}_{tp} = \left[\frac{W_l D}{(\pi D^2/4)\mu_l}\right]\left[F_{\text{BC}}\left(\frac{\text{Pr}_l + 1}{2}\right)^{0.444}\right]^{1.25} \tag{13.39}$$

Note that F_{BC} is basically the Chen F parameter, except that rather than explicitly correlating it with the Martinelli parameter, they have related it directly to the friction pressure gradient ratio in Eq. (13.36). The subscripts Ftp and Fl indicate the frictional pressure gradients for the two-phase flow and the liquid flowing alone, respectively. The Martinelli correlation or any other suitable correlation method can be used to determine the two-phase pressure gradient in Eq. (13.36). The factor $\Gamma_{\Delta T}$ corrects for mass transfer effects on the interface temperature.

The heat transfer coefficient must be computed iteratively using this correlation because the relation for $\Gamma_{\Delta T}$ contains the total heat flux $q'' = h_{\text{tot}}[T_w - T_{bp}(C_{bl}, P)]$. This correlation method was found to agree well with data for flow boiling of water and ethylene glycol mixtures to a mean deviation of 14.9%. It has the further advantage of reducing to a form that is virtually identical to the original Chen correlation when the concentration of the more volatile component approaches zero. This correlation shows promise as a generally applicable correlation method for predicting heat transfer during convective boiling of binary mixtures. However, this approach needs to be tested against a data base that spans a broad range of conditions before it can be widely used with confidence. Unfortunately, data available in the open literature that can be used for such a comparison is currently very limited.

CHF Conditions

The open literature provides little guidance regarding methods for predicting conditions for dryout during vaporization of binary mixtures. Collier [13.51] recommends the method proposed by Hewitt and co-workers [13.52, 13.53] for pure-fluid convective vaporization processes. In binary systems, the interfacial tension may vary strongly with concentration. Because entrainment is directly dependent on interfacial tension (see Section 10.4), the entrainment characteristics in a binary system may differ from those in a pure fluid system under comparable conditions. Shock [13.48] presented evidence that suggests that the entrainment characteristics in convective evaporation of binary mixtures often are not strongly affected by mass transfer effects. Based on this evidence, Collier [13.51] concluded that the annular flow model of Hewitt and co-workers [13.52, 13.53] should yield reasonably good results. It is possible, however, that during convective evaporation of a binary mixture, Marangoni effects at the interface of the liquid film may cause the film to become unstable and break down into rivulets if the more volatile component has a higher surface tension (see Section 2.5). This may lead to the onset of dryout prior to the point where dryout is predicted by models that ignore such effects.

A number of experimental investigations of the departure from nucleate boiling (DNB) during flow boiling of a binary mixture have been conducted (see, e.g., references [13.26, 13.54–13.57]). In general, the DNB heat flux for flow boiling of a binary mixture increases with increasing flow velocity and subcooling (below the bubble-point temperature) in much the same way as for a pure fluid. As a simplistic first estimate, we might expect that the DNB heat flux would vary linearly with liquid concentration between the pure fluid values (at the same subcooling and flow velocity) for $X_l = 0$ and $X_l = 1$:

$$(q''_{\rm cr})_i = (q''_{\rm cr})_{X=0}(1 - X_l) + (q''_{\rm cr})_{X=1}X_l \tag{13.40}$$

This variation is indicated by the broken line in Fig. 13.9. The solid curve in this figure indicates the trend typically observed in experimental data for a binary mixture that forms an azeotrope at one concentration. As in the case of pool boiling, maxima in the critical heat flux values are observed at or near locations where $|X_v - X_l|$ attains a maximum. For a binary mixture that does not form an azeotrope, $|X_v - X_l|$ attains a maximum value at one concentration, and only one maximum in the critical heat flux is expected.

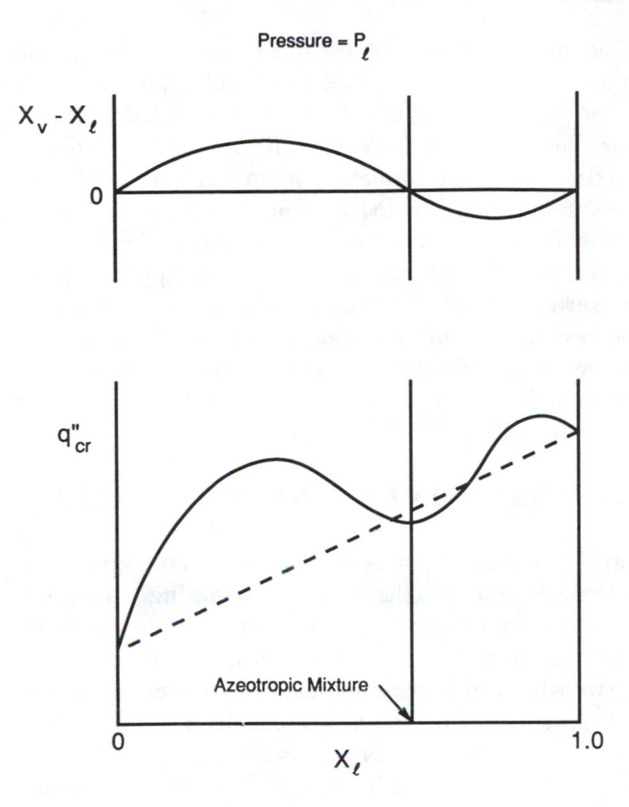

Figure 13.9 Qualitative variation of the flow boiling critical heat flux observed in a binary mixture that forms an azeotrope at one concentration.

To obtain a better fit to available data, a correlation having the following form has often been postulated:

$$q''_{cr} = (q''_{cr})_i(1 + \chi) \tag{13.41}$$

where $(q''_{cr})_i$ is given by Eq. (13.40). Note that χ is a correction term that accounts for the deviation from the idealized linear relation (13.40). Tolubinsky and Matorin [13.54] proposed the following relation for χ:

$$\chi = 1.5|X_v - X_l|^{1.8} + 6.8|X_v - X_l|\left[\frac{T_{sat}(X_l, P) - T_{sat}(1, P)}{T_{sat}(1, P)}\right] \tag{13.42}$$

where $T_{sat}(X_l, P)$ is the saturation (bubble point), temperature of the mixture and $T_{sat}(1, P)$ is the pure-component saturation temperature of the more volatile component. X_v and X_l are the bulk mole fractions in the vapor and liquid phases, respectively. This relation fit DNB data for flow boiling of ethanol–water, acetone–water, ethanol–benzene, and water–ethylene glycol mixtures to within ±20%. Note that this relation reflects the fact that the maxima in the critical heat flux variation with concentration roughly corresponds to the maxima in $|X_v - X_l|$.

The trends described above suggest that addition of small amounts of a more volatile liquid to another pure fluid may increase the critical heat flux above the pure fluid value for the same conditions. However, for subcooled flow boiling of mixtures of water with small amounts of 1-pentanol, Bergles and Scarola [13.57] observed a trend quite different from that suggested above. For added amounts of 1-pentanol (about 2% by weight), they found that the critical heat flux actually decreased. They attributed the decrease to the formation of smaller bubbles during the boiling process in the mixture, producing local liquid velocities that were smaller than those associated with the pure fluid boiling process, where larger bubbles are produced. The reduction in the local liquid velocities apparently allows vapor blanketing of the surface to occur more easily at a lower heat flux level. Taken collectively, the results of studies to date imply that the effect of adding small amounts of a more volatile fluid on the DNB condition may vary, depending on the component fluids involved.

13.4 CONVECTIVE CONDENSATION OF BINARY MIXTURES

While convective condensation of a pure vapor is perhaps more commonly encountered in refrigeration and power systems, there are an increasing number of important circumstances in which convective condensation of a multicomponent mixture occurs. In the petrochemical processing industries, processes of this types are, in fact, common. Binary working fluids are also being considered for use in power generation cycles, heat pump systems, and heat pipes, as a means of improving their thermal performance. In these types of systems, convective condensation of the mixture plays an important role in the operation of the system.

The general treatment of condensation of multicomponent mixtures is beyond the scope of this text. However, most of the important features of the general multicomponent case can be seen in the simpler circumstance of convective condensation of a binary mixture. With some extension, many of the concepts developed for convective condensation of pure vapors can be used for binary systems.

To gain some insight into the physics of multicomponent convective condensation, in this section we will therefore examine convective condensation in a binary system in some detail. To limit the complexity of the problem, we will confine our attention to simple binary mixtures without azeotropic points. As in the case of vaporization, the added complexity associated with condensation in a multicomponent system results primarily from (1) the more complicated thermodynamics associated with the binary system, and (2) the added effect of mass (species) transport during the condensation process.

The thermodynamic characteristics of convective condensation of a binary vapor mixture can best be understood by simultaneously considering the binary phase diagram in Fig. 13.10 and the flow configuration shown in Fig. 13.11. (See Section 13.2 for a discussion of the basic thermodynamic considerations.) Cocurrent downward flow condensation in a round tube is depicted in Fig. 13.12. If the vapor entering the tube has a concentration $X_{v,\text{in}}$ (of the more volatile component), the equilibrium saturation temperature at which condensation first occurs

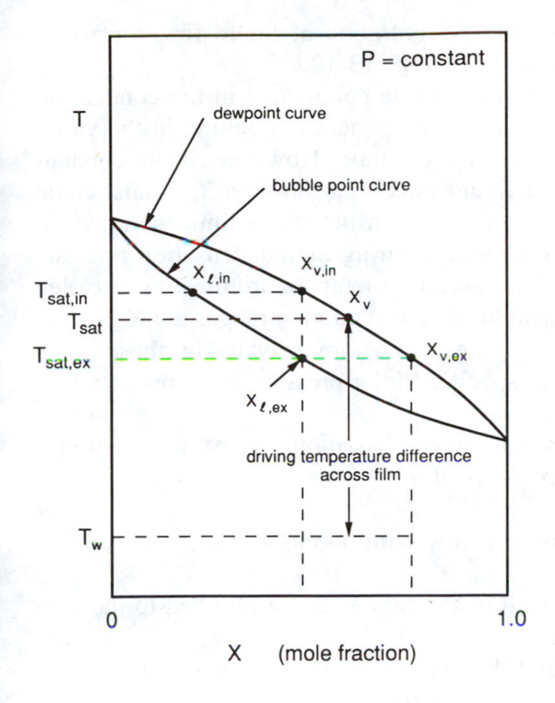

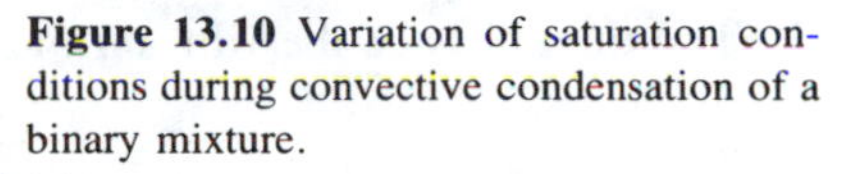
Figure 13.10 Variation of saturation conditions during convective condensation of a binary mixture.

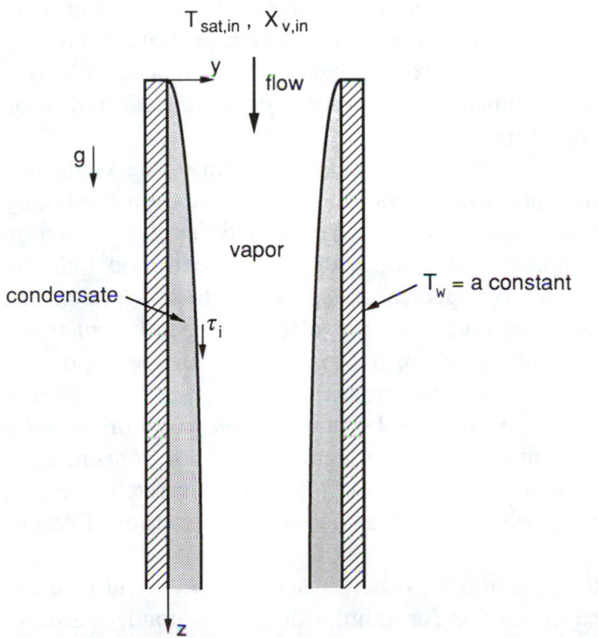

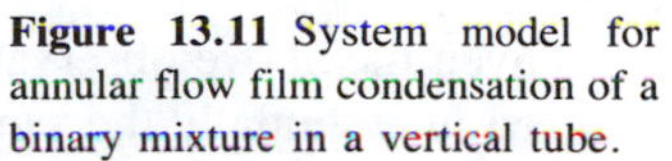
Figure 13.11 System model for annular flow film condensation of a binary mixture in a vertical tube.

is $T_{sat,in}$, as indicated in Fig. 13.10. The concentration of liquid first formed is $X_{l,in}$, as dictated by the bubble-point curve in Fig. 13.10.

Because the concentration of the more volatile component in the condensate is lower than in the vapor, the excess of this component left behind in the vapor serves to raise X_v as the flow proceeds down the tube. However, if the condensation process goes to completion, the liquid bulk concentration $X_{l,ex}$ must equal the inlet concentration in the vapor $X_{v,in}$. These considerations limit the range of saturation temperature exhibited by the system during the condensation process. From Fig. 13.10, it can be seen that the saturation temperature must be between $T_{sat,in}$ and $T_{sat,ex}$. Similarly, the vapor and liquid concentration ranges are $X_{v,in} < X_v \leq X_{v,ex}$ and $X_{l,in} < X_l \leq X_{l,ex}$. If the binary phase diagram is available, these limits can be determined explicitly from the specified inlet pressure and concentration in the manner described above.

For the purposes of this discussion of the condensation process shown in Fig. 13.11, the following idealizations are adopted:

1. There is annular co-current downward flow with no entrainment.
2. Saturated vapor enters the tube.
3. The thickness of the film on the wall of the tube is thin compared to the tube diameter.
4. The liquid film flow is laminar and the vapor core flow is turbulent.
5. Sensible cooling of the vapor and liquid is negligible compared to the latent heat energy associated with the condensation process.

These idealizations simplify analysis of the transport somewhat. They can be relaxed when necessary to provide a more accurate treatment. However, the main effects of the binary system thermodynamics and species transport are often only slightly different from those for these more idealized circumstances. We will therefore focus on these simpler circumstances in order to obtain a clear indication of the qualitative behavior of this type of system.

If the local concentration varies with cross-stream position in the vapor and liquid portions of the flow, the equilibrium conditions represented on the binary phase diagram are specified at the liquid–vapor interface. Satisfying this diagram at the interface becomes a boundary condition, with the concentration fields in the two phases then being dictated by species transport due to convection and diffusion. However, for the turbulent core flow postulated here, turbulent transport is expected to result in a virtually uniform concentration in the vapor and rapid transport to the interface. The temperature at the liquid–vapor interface at any downstream location is therefore taken to be equal to the equilibrium temperature specified by the dew-point curve at the local bulk vapor concentration X_v. With the interface temperature specified, the local heat transfer rate is dictated by transport of heat across the liquid film, driven by the temperature difference $T_{dp}(X_v, P) - T_w$.

Given the above observations, analysis of the transport in the liquid film can proceed in essentially the same manner as for annular downflow condensation of a pure vapor. Neglecting convective terms and downstream diffusion in the film, and invoking conservation of momentum, mass, and energy in the film, the same analytical arguments described in Section 11.2 lead to the following relation for the rate of change of the film thickness with downstream distance:

$$\frac{d\delta}{dz} = \frac{k_l \mu_l (T_{\text{sat}} - T_w)}{\rho_l(\rho_l - \rho_v^*) g h'_{lv} \delta^3 + \tau_i \rho_l h'_{lv} \delta^2} \tag{13.43}$$

where τ_i is the interfacial shear stress and ρ_v^* is defined by

$$\rho_v^* g = \rho_v g - \frac{4\tau_i}{D - 2\delta} - \frac{2xDG^2}{\rho_v(D - 2\delta)} \left(\frac{dx}{dz}\right) \tag{13.44}$$

In Eq. (13.43), h'_{lv} is the effective latent heat defined by Eq. (11.18), which accounts for latent heat and downstream convection effects. Because this is a binary system, x in Eq. (13.44) is not the quality in the usual thermodynamic sense. Consistent with its interpretation for two-phase flow, x is interpreted here as simply being the ratio of the vapor mass flow rate to the total mass flow rate.

In Section 11.2 it was noted that if τ_i and ρ_v^* are assumed to be constant, and 2δ is neglected compared to the tube diameter D, then Equation (13.43) can be analytically integrated to obtain an implicit relation for the variation of δ with z. However, for the binary system considered here, T_{sat} T_{dp} also varies with downstream position, even if these other variables are constant. Consequently, closed-form integration of Eq. (13.43) is not generally possible.

Perhaps the best approach is a numerical integration of Eq. (13.43). The following additional relations, developed in Section 11.2, are necessary for mathematical closure:

$$h = \frac{k_l}{\delta} \tag{13.45}$$

$$\frac{dx}{dz} = \frac{4q''}{DGh_{lv}} = \frac{4h(T_{dp} - T_w)}{DGh_{lv}} \tag{13.46}$$

$$\tau_i = f_v\left(\frac{\rho_v u_v^2}{2}\right) = f_v\left[\frac{G^2x^2}{2\rho_v(1 - 4\delta/D)}\right] \tag{13.47}$$

$$f_v = 0.005\left(1 - \frac{300\delta}{D}\right) \tag{13.48}$$

In addition to Eqs. (13.45) through (13.48), the following relations are also required to facilitate the numerical solution:

$$W_{vMV} = \frac{G(\pi D^2/4)x(M_{MV}/M_{LV})X_v}{1 - X_v + (X_v M_{MV}/M_{LV})} \tag{13.49}$$

$$W_{vLV} = G(\pi D^2/4)x - W_{vMV} \tag{13.50}$$

$$\frac{dW_{vMV}}{dz} = \frac{\pi D h (T_{dp} - T_w)}{(1 + \gamma) h'_{lv}} \tag{13.51}$$

$$X_v = \frac{W_{vMV}/M_{MV}}{W_{vMV}/M_{MV} + W_{vLV}/M_{LV}} \tag{13.52}$$

where

$$\gamma = \left(\frac{M_{LV}}{M_{MV}}\right)\left(\frac{1 - X_v}{X_v}\right) \tag{13.53}$$

In these relations, M_{MV} and M_{LV} are the molecular weights of the two species, and W_{vMV} and W_{vLV} are the mass flow rates of the two species in the vapor. The subscripts MV and LV designate the more volatile and less volatile components, respectively. These relations are easily obtained from conservation of energy, overall mass, and species. Equation (13.51) embodies the additional idealization that X_v is constant over the short segment of the tube ΔL considered in each numerical integration step.

The numerical integration is facilitated by first dividing the length of the tube into N segments. The segments are sequentially assigned a number n. With the above relations, the computational algorithm may then proceed as follows.

1. At the inlet to the first segment ($z = 0$), set $n = 1$, $\delta = 0$, $x = 1.0$, and $X_v = X_{v,\text{in}}$. Also determine W_{vMV} and W_{vLV} at the inlet to the first section using Eqs. (13.49) and (13.50).
2. If $n \geq N$, stop; the overall algorithm is complete.
3. Guess $\delta_{\text{avg},n}$ (the average film thickness) for the next segment to be analyzed.
4. For X_v at inlet of segment, use thermodynamic information to determine the equilibrium X_l corresponding to X_v, and $T_{dp}(X_v, P)$.
5. Use the guessed value of $\delta_{\text{avg},n}$ and the value of $T_{dp}(X_v, P)$ from step 4 to compute the heat transfer coefficient for the segment $h_{\text{avg},n}$.
6. Use Eqs. (13.47) and (13.48) to calculate $(\tau_i)_n$ using $\delta_{\text{avg},n}$ and the value of x at the inlet of the segment. Also compute dx/dz from Eq. (13.46) and the results of steps 4 and 5.
7. Use Eq. (13.44) with the results of previous steps to evaluate ρ_v^*.
8. Determine $(d\delta/dz)_n$ using Eq. (13.43) and the results of previous steps.
9. Compute:

$$\delta_{\text{out},n} = \delta_{\text{in},n} + \left(\frac{d\delta}{dz}\right)_n \Delta L$$

$$(\delta_{\text{avg},n}) = \frac{\delta_{\text{in},n} + \delta_{\text{out},n}}{2}$$

10. If $|\delta_{\text{avg}} - \delta_{\text{avg},n}|/D >$ error limit, iteration for this segment is not complete. Take $\delta_{\text{avg}} = (\delta_{\text{avg},n})$ (or otherwise generate a new guess for δ_{avg}), go to step

5, and repeat steps 5 through 10. If $|\delta_{avg} - \delta_{avg,n}|/D \leq$ error limit, continue to step 11.

11. Increment n and compute the inlet values of x, δ, and X_v for the next segment:

$$n = n + 1$$

$$x_{in} = x_{in} + \left(\frac{dx}{dz}\right) \Delta L$$

$$\delta_{in,n} = \delta_{out,n-1}$$

Using Eqs. (13.51) and (13.53) to evaluate dW_{vMV}/dz,

$$(W_{vMV})_{in,n} = (W_{vMV})_{in,n-1} + \left(\frac{dW_{vMV}}{dz}\right) \Delta L$$

Use Eq. (13.50) to determine $(W_{vLV})_{in,n}$.
Use Eq. (13.52) to determine $(X_{v,in})_n$.

12. Go to step 2.

Inspection of the above algorithm reveals that even for the simplified model considered here, the level of computational effort required to predict the heat transfer performance is high. Nevertheless, this type of algorithm can be easily programmed, even on a personal computer, if the thermophysical property data are available.

It cannot be overemphasized that the above analysis for downflow annular film condensation in a round tube is only approximate in many respects. It is presented here to illustrate the nature of this type of computation. Before attempting to use this scheme, the reader would be wise to carefully examine the idealizations in the analysis in order to determine the appropriateness of each to the application of interest. The determination of the interfacial shear is particularly crude in this model, and could be upgraded by using one of the more sophisticated methods available in the two-phase flow literature. Incorporating the effects of possible entrainment would also improve the model (see Chapter Ten).

It should also be noted that in the above analysis, the fully mixed turbulent core flow allows specification of the local interface temperature in a relatively simple manner. If this condition does not exist, the species transport in the core flow must also be computed to determine the interface condition at each downstream location. The species transport equation must then be numerically integrated downstream from the inlet condition; and for each finite tube segment, iterative determination of the local film thickness and interface condition is more complicated. Although this adds considerable complexity to the computational scheme, calculations of this type are certainly within the capabilities of currently available engineering workstation computers.

Relatively little information about internal convective condensation of binary mixtures is available in the open literature. In an early study, Van Es and Heertjes [13.58] theoretically and experimentally investigated the natural-convection condensation of nonazeotropic vapor mixtures of benzene and toluene in a vertical tube. In their experiments, the vapor was only partially condensed in the tube. The theory developed by these investigators was applied to flows driven by gravity and by the interfacial shear associated with the turbulent downward flow of the vapor mixture.

More recently, experimental and theoretical investigations of convective filmwise condensation of nonazeotropic mixtures of R-11 and R-114 inside a vertical tube were conducted by Mochizuki et al. [13.59]. Experiments were conducted in which local and overall heat transfer coefficients were determined for co-current downflow condensation of the mixture. A theoretical model of the condensation process, similar to that outlined above, was developed, and its predictions were compared to the experimentally determined variation of the heat transfer coefficient with downstream distance along the tube.

The predictions of the model for conditions resulting in laminar film flow agreed reasonably well with measured data. However, for turbulent flow, the model predictions were substantially different from the measured h profiles along the tube. For the overall mean heat transfer coefficient for the entire condensation process, these investigators developed the following correlation as a best fit to their data:

$$\frac{\overline{h}_L}{k_l} = 0.38\left[\mathrm{Ja}\left(\frac{L}{D}\right)\right]^{-0.3}\left(\frac{\mathrm{Re}^*\,\mathrm{Pr}_l}{R}\right)^{0.8} \tag{13.54}$$

where

$$\mathrm{Ja} = \frac{c_{pl}\,\overline{\Delta T}}{h_{lv}} \tag{13.55}$$

$$\mathrm{Re}^* = \frac{U_{\mathrm{in}}L}{\nu_l} \tag{13.56}$$

$$R = \left(\frac{\rho_l\mu_l}{\rho_v\mu_v}\right)^{1/2} \tag{13.57}$$

In these relations, U_{in} is the inlet vapor mean velocity and $\overline{\Delta T}$ is the mean difference between the interface and wall temperatures. This relation was reported to be accurate to $\pm 10\%$ for $5.8 \times 10^6 < \mathrm{Re}^* < 7.5 \times 10^7$ and $60 < L/D < 222$. While this correlation scheme seems to work well for this specific binary mixture, further testing of its predictions against data for other fluids is necessary before it can be widely used with confidence.

As noted by these investigators, further work is needed to develop better theoretical models of annular film condensation of binary mixtures. The last stages of convective condensation processes are generally in the slug and bubbly regimes (see Section 11.1). The transport for such circumstances is complicated by the intermittent nature of the two-phase flow and the coupled heat and mass transfer associated with the phase-change process. A survey of the literature at the present time reveals a virtual absence of information regarding bubbly or slug slow condensation of binary mixtures. Additional work is unquestionably needed to better define the convective condensation heat transfer characteristics for these complex circumstances.

13.5 ENHANCED FLOW PASSAGES FOR CONDENSERS AND EVAPORATORS

General Observations

While heat transfer coefficients associated with convective boiling and condensation in round tubes are generally quite high, there are some applications in which further enhancement of the heat transfer is desirable. Such applications may include high-performance and/or ultracompact evaporators and condensers for automotive air-conditioning systems, cryogenic processing systems for separation and liquefaction of oxygen and nitrogen, liquefaction of natural gas, electronics cooling, and thermal control systems for aircraft and spacecraft. The importance of these applications has stimulated a considerable amount of recent work in this area. A complete description of the recent work on enhanced surfaces for two-phase applications is beyond the scope of this text. However, in this section, a brief summary of recent developments in this area will be presented.

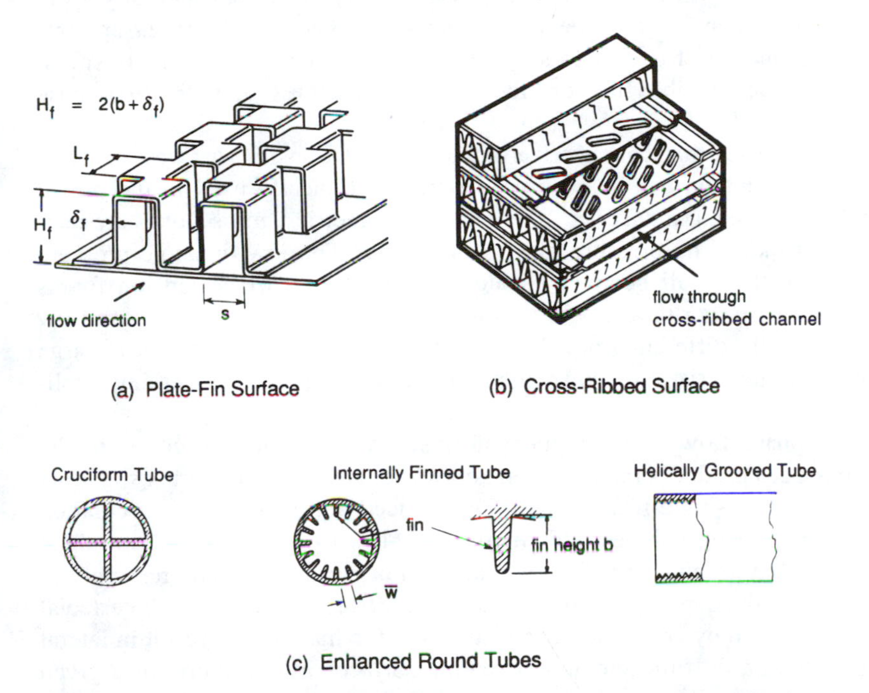

Figure 13.12 Enhanced passage geometries that are sometimes used for applications involving convective boiling or condensation.

In some compact evaporator and condenser designs, the phase change occurs in round tubes. For such circumstances, the methods described in Chapter Eleven or Twelve can be used to predict the local heat transfer coefficient, and the methods described in Chapter Ten can be applied to predict the two-phase pressure gradient. This section will focus primarily on more complicated matrix-type geometries, such as offset strip-fin and cross-ribbed channel configurations, and enhanced round tube configurations such as tubes with ribbed walls and tubes with twisted-tape inserts. Figure 13.12 shows two embodiments of the matrix-type geometries of interest here, and several common types of enhanced round tubes.

Effects of Geometry Variations on Two-Phase Flow

As noted in Chapter Ten, two-phase flow in round tubes has been studied extensively, and a considerable body of flow regime information exists in the literature. Two-phase flow in more complicated enhanced geometries is similar in many respects to two-phase flow in round tubes. The results of several previous studies [13.60–13.63] indicate that, for comparable conditions, the predictions of flow regime maps for round tube flows agree fairly well with the observed behavior of two-phase flows in simple noncircular geometries. This seems to suggest that the flow regime maps for round tubes may also provide a reasonably accurate estimate of the flow regime for two-phase flow in the more complex geometries of interest here. The flow in enhanced passages may differ substantially from round-tube flow behavior in some cases, however. These differences are frequently a consequence of two factors: (1) The flow passages are relatively small; and (2) the presence of ribs, fins, or other internal structures alters the two-phase flow behavior.

The length scales that characterize enhanced heat transfer surfaces are typically much smaller than the round tube geometries typically used in the power and chemical processing industries. In enhanced geometries, mass fluxes are usually kept low to keep the pressure drop penalty low. The low dynamic pressure of the flow and the small geometry scales can result in surface tension forces being comparable to inertia forces in the flow. In such cases, the two-phase flow can be significantly different from the high-mass-flux two-phase flows in large tubes that are characteristic of power generation and chemical processing applications.

In a two-phase flow, the presence of ribs, fins, or other structures in the enhanced passages produces recirculation zones and may promote vortex shedding and mixing just as in single-phase flows. The induced turbulence and recirculation can break up large slug bubbles into smaller bubbles. In annular flow, entrained liquid droplets may impinge on the upstream end of rib or fin structures, only to be shed back into the core flow at the trailing edge. Irregularities in the interfacial shear stress produced by core velocity variations in the matrix may result in lateral variations of the liquid film thickness over the surfaces in the matrix at a given downstream location. This, in turn, may affect the heat transfer performance and dryout of the liquid film during the later stages of vaporization.

The results of recent studies [13.64–13.67] of two-phase flow in more complex finned and ribbed channel geometries generally indicate that although the flow regimes encountered are basically the same as for round tubes under comparable conditions, the exact transition conditions between regimes can be significantly different. The results of these studies imply that use of round-tube flow regime maps to predict flow regimes in more complex passage geometries may result in significant errors. If no flow regime information for the geometry of interest is available, flow regime maps for round tubes may be the only recourse. In such circumstances, the round-tube predictions should be used with caution.

Enhanced Surfaces for Convective Condensation

The ability of different passage geometries to enhance convective condensation heat transfer has been the subject of several recent investigations. Geometries considered for this type of application include offset strip-fin and plate-fin surfaces [13.68–13.71], ribbed-plate surfaces [13.72–13.73], round tubes with twisted-tape inserts [13.74–13.76], and round tubes with internal fins [13.77–13.79].

As in the case of simpler passages, proposed techniques for predicting convective condensation in complex channel geometries typically involve either a numerical solution of basic governing equations, or use of dimensionless correlations developed to fit experimental data. For many condensing circumstances, the difference between the liquid and vapor densities is large and the void fraction is above 0.5 even at relatively low qualities. At moderate to high flow rates, an annular flow configuration may then exist over most of the channel length where condensation occurs. Analysis techniques and correlation methods that are applicable to annular flow are therefore of particular interest.

Correlations and analytical models developed in recent studies are, to a large degree, just extensions of similar methods developed for round tubes. A typical example is the correlation developed by Luu and Bergles [13.75] as a fit to their data for condensation of R-113 inside horizontal tubes with twisted tape inserts:

$$\bar{h} = 0.024\left(\frac{k_l}{d_h}\right)F_{tt}^*\left(\frac{F_t G d_h}{\mu_l}\right)^{0.8} \mathrm{Pr}_l^{0.43}\frac{[(\rho/\rho_m)_{\mathrm{in}}^{0.5} + (\rho/\rho_m)_{\mathrm{out}}^{0.5}]}{2} \tag{13.58}$$

where

$$\frac{\rho}{\rho_m} = \left[1 + \bar{x}\left(\frac{\rho_l}{\rho_v} - 1\right)\right] \tag{13.59}$$

$$F_t = \left(\frac{8y^2}{3\pi^2}\right)\left\{\left[\left(\frac{\pi}{2y}\right)^2 + 1\right]^{1.5} - 1\right\} \tag{13.60}$$

$$F_{tt}^* = 1 + \left(\frac{1}{\pi}\right)[\eta_f(4y + \pi^2)^{0.5}/2y - \delta_f/d_i] \tag{13.61}$$

In these equations, y is the ratio of twisted-tape pitch for 180° rotation to tube inside diameter, η_f is the fin efficiency of the tape (assuming perfect contact with the wall), δ_f is the tape thickness, $\bar{x}$ is the mean quality over the length of channel considered, d_i is the tube inside diameter, and d_h is the hydraulic diameter, given by

$$d_h = \frac{\pi d_i^2 - 4\delta_f d_i}{\pi d_i + 2(d_i - \delta_f)} \tag{13.62}$$

The above correlation scheme, which is basically an extension of the round-tube correlation proposed by Boyko and Kruzhilin [13.80], was found to agree well with both the localized h and overall mean h data obtained by Luu and Bergles [13.75].

For some geometries, correlations have been developed that predict a mean heat transfer coefficient for the entire condensation process [13.68, 13.72, 13.76, 13.78]. In some cases, correlations or analytical models have also been developed that make it possible to predict local values of the condensation heat transfer coefficient from the local quality and flow conditions [13.71, 13.73, 13.75, 13.77]. These different types of correlations can be used in different ways to predict the overall performance of a compact condenser unit. Correlations for the mean heat transfer coefficient can be used with a mean coefficient for the noncondensing side of the heat exchanger to predict the overall UA product for the unit (U being the overall heat transfer coefficient), and, subsequently, its overall heat transfer performance.

If methods of predicting local heat transfer coefficients on both sides of the condenser are available, the unit can be analytically broken down into an array of small elements in which the heat transfer coefficient on each side may be idealized as being constant. The performance of each element can be determined sequentially, and the overall heat transfer performance is the sum of the contributions of each of the elements. If the elements are chosen small enough, computerized calculations of this type can be more accurate and provide more detailed information about the performance of the unit than overall calculations based on mean heat transfer coefficients. Unfortunately, however, accurate, widely applicable methods of predicting local condensation heat transfer coefficients for most enhanced compact condenser surfaces are not yet available.

Example 13.3 Convective condensation of R-12 occurs in a horizontal tube having an aluminum twisted-tape insert. The tube inside diameter is 1.0 cm and the twist pitch-to-diameter ratio y for the twisted-tape insert is 2.0. The tape thickness is 0.5 mm. At a particular location, the pressure is 333 kPa and the quality is 0.8. The mass flux through the tube is 400 kg/m^2 s. Use the correlation of Luu and Bergles [13.75] to estimate the condensation heat transfer coefficient.

For saturated R-12 at 333 kPa, $T_{sat} = 275$ K $= 1.8°$C, $\rho_l = 1388$ kg/m^3, $\rho_v = 19.2$ kg/m^3, $h_{lv} = 154.7$ kJ/kg, $\mu_l = 262 \times 10^{-6}$ N s/m^2, $k_l = 0.0805$ W/m K, and $\mathrm{Pr}_l = 3.51$. Using Eqs. (13.58) through (13.62), we first compute $(\rho/\rho_m)_{in}$, $(\rho/\rho_m)_{out}$, F_t, F_{tt}^*, and d_h for the specified conditions. Since we are interested in a localized h value at the specified quality, we take

$$\left(\frac{\rho}{\rho_m}\right)_{in} = \left(\frac{\rho}{\rho_m}\right)_{out} = 1 + x\left(\frac{\rho_l}{\rho_v}\right)$$

$$= 1 + 0.8\left(\frac{1388}{19.2} - 1\right)$$

$$= 58.0$$

F_t is then computed as

$$F_t = \left(\frac{8y^2}{3\pi^2}\right)\left\{\left[\left(\frac{\pi}{2y}\right)^2 + 1\right]^{1.5} - 1\right\}$$

$$= \left[\frac{8(2)^2}{3\pi^2}\right]\left(\left\{\left[\frac{\pi}{2(2)}\right]^2 + 1\right\}^{1.5} - 1\right) = 1.14$$

Because the tape is made of aluminum, we will initially take the fin efficiency of the tape equal to 1 in computing F_{tt}^*:

$$F_{tt}^* = 1 + \left(\frac{1}{\pi}\right)\left[\frac{\eta_f(4y + \pi^2)^{0.5}}{2y} - \frac{\delta_f}{d_i}\right]$$

$$= 1 + \left(\frac{1}{\pi}\right)\left\{\frac{(1)[4(2) + \pi^2]^{0.5}}{2(2)} - \frac{0.0005}{0.01}\right\}$$

$$= 1.32$$

$$d_h = \frac{\pi d_i^2 - 4\delta_f d_i}{\pi d_i + 2(d_i - \delta_f)}$$

$$= \frac{\pi(0.01)^2 - 4(0.0005)(0.01)}{\pi(0.01) + 2(0.01 - 0.0005)}$$

$$= 0.00583 \text{ m} = 5.83 \text{ mm}$$

The localized $\bar{h}$ value is then given by Eq. (13.58) as

$$\bar{h} = 0.024\left(\frac{k_l}{d_h}\right)F_{tt}^*\left(\frac{F_t G d_h}{\mu_l}\right)^{0.8} \mathrm{Pr}_l^{0.43}\left[\frac{(\rho/\rho_m)_{in}^{0.5} + (\rho/\rho_m)_{out}^{0.5}}{2}\right]$$

Noting that $(\rho/\rho_m)_{in} = (\rho/\rho_m)_{out}$ and substituting yields

$$\bar{h} = 0.024\left(\frac{0.0805}{0.00583}\right)(1.32)\left[\frac{1.14(400)(0.00583)}{262 \times 10^{-6}}\right]^{0.8}(3.51)^{0.43}(58.0)^{0.5}$$

$$= 9140 \text{ W/m}^2\text{ K}$$

For this h value the fin efficiency for the twisted tape is computed as

$$\eta_f = \frac{\tanh \sqrt{2\bar{h}L^2/k_f\delta_f}}{\sqrt{2\bar{h}L^2/k_f\delta_f}}$$

Substituting with $k_f = 236$ W/m K for aluminum,

$$\eta_f = \frac{\tanh[2(9140)(0.01/2)^2/236(0.0005)]^{1/2}}{[2(9140)(0.01/2)^2/236(0.0005)]^{1/2}} = 0.489$$

Substituting $\eta_f = 0.489$ into the equation for F_{tt}^* yields $F_{tt}^* = 1.15$. An improved $\bar{h}$ is then computed as

$$\bar{h} = 9140\left(\frac{1.15}{1.32}\right) = 7963 \text{ W/m}^2 \text{ K}$$

Substituting this new $\bar{h}$ into the fin efficiency relation yields $\eta_f = 0.517$. After a few iterations this procedure converges to

$$\bar{h} = 8020 \text{ W/m}^2 \text{ K} \qquad \eta_f = 0.516$$

For a plain round tube with the same inside diameter, Shah's [11.11] correlation predicts $h = 5520$ W/m^2 K for these same conditions, about 30% less.

Enhanced Surfaces for Convective Boiling

The results of recent fundamental studies [13.64–13.66] indicate that the sequence of flow regimes encountered during flow boiling in most enhanced geometries is basically the same as that observed in round tubes under similar conditions. As for round tubes, the morphology of the two-phase flow in these geometries is expected to strongly influence the heat transfer performance. These observations imply that the boiling regime maps (Figs. 12.4 and 12.5) described in Chapter Twelve also apply in a qualititative sense to flow boiling in the enhanced geometries considered here. The interpretation of these diagrams may be somewhat different for the surface geometries considered here, however.

In the case of a vertical round tube, it is reasonable to expect that the applied heat flux or wall temperature at a given downstream location was uniform over the perimeter of the tube. For finned and ribbed surfaces, the wall boundary condition may, in fact, vary over the perimeter of the channel. In finned channels, for example, the inevitable variation of temperature from the tip to the root of the fin could, in extreme cases, cause film boiling to occur near the hotter root, whereas nucleate boiling or (liquid) film flow evaporation may exist near the cooler tip. At lower heat flux levels, the degree of nucleation suppression may vary over the surface because of the variation of the convective transport effect. The possibility of local variations in the transport conditions must be considered when interpreting this type of diagram for the more complicated channel geometries shown in Fig. 13.12.

The regime maps in Figs. 12.4 and 12.5 nevertheless suggest that, at low to moderate heat flux levels, a transition from nucleate boiling to forced-convective boiling and ultimately to the liquid-deficient regime is expected. Accompanying this sequence of boiling regimes is a sequence of flow regime transitions from slug to churn, from churn to annular, and from annular to mist flow. In Fig. 13.13, the data of Mandrusiak and Carey [13.81] for low-heat-flux flow boiling in a channel with offset strip fins clearly reflects these sequences of flow and vaporization regimes.

If the heat flux is high enough to produce film boiling, inverted annular flow or a transition to dispersed flow may result. Deviations from the sequences described above may also occur for downflow boiling. For high heat flux and downflow circumstances, flow regime information for round tube flows under similar conditions may provide some useful insight into the two-phase flow behavior for more complicated geometries. Unfortunately, relatively little is known at the present time about the behavior of these types of flow in more complex channel geometries, and the accuracy of using round-tube predictions for other, more complicated geometries is unclear.

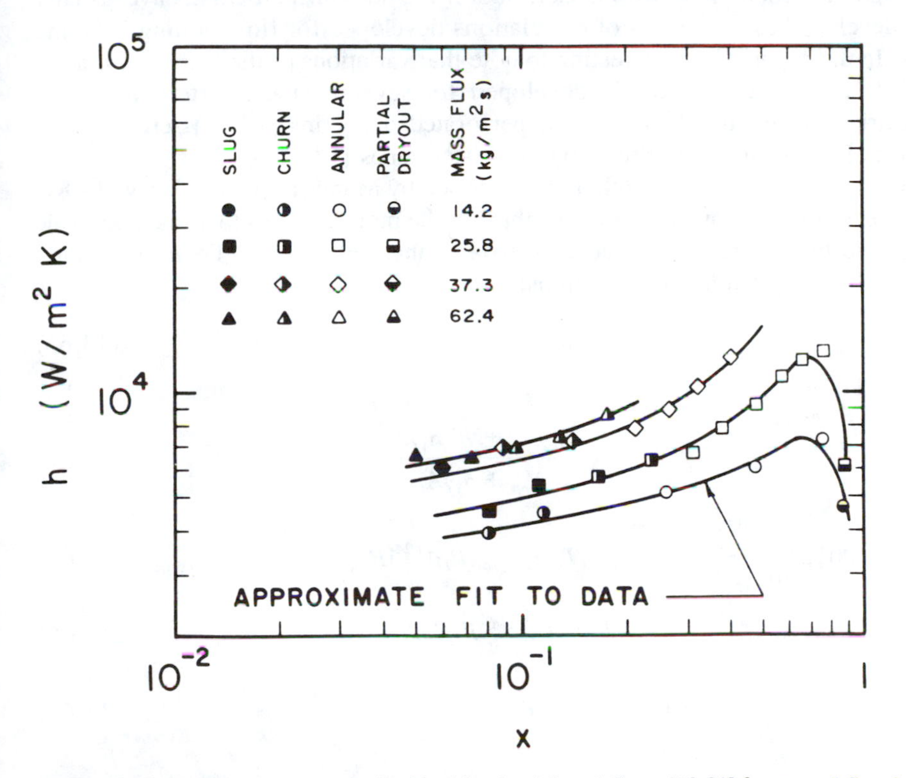

Figure 13.13 Heat transfer data obtained by Mandrusiak and Carey [13.81] for upward flow boiling of methanol in a partially heated vertical channel with offset strip fins (L_f = 12.7 mm, H_f = 9.52 mm, δ_f = 1.91 mm, fin pitch = 0.98 fins/cm).

Previous investigations of flow boiling in enhanced channel geometries have examined a number of different configurations, including channels with offset strip fins, [13.64, 13.65, 13.81–13.94], channels with perforated plate fins [13.95], cross-ribbed channels [13.66, 13.96–13.98], tubes with ribbed or grooved walls [13.99–13.104], and tubes with twisted-tape inserts [13.105–13.113]. The methods developed in these studies to predict the two-phase pressure drop and flow boiling heat transfer performance have varied somewhat with the channel geometry and imposed flow conditions. These techniques generally fall into one of two categories: (1) correlation techniques in terms of dimensionless groups, or (2) numerical solution of basic governing equations. Methods that include solution of basic governing equations may appear to be more fundamental. However, the difference between the sophistication level of this type of approach and correlation techniques may not be great, because correlation schemes often have a semitheoretical basis, and numerical solution of basic governing equations usually requires empirical relations to close the numerical scheme.

In developing methods to predict the heat transfer performance and pressure drop for convective boiling in enhanced geometries, researchers in the studies described above almost invariably have built upon concepts and methodologies developed for round-tube flows. Correlation methods in particular have usually been developed as extensions of correlations developed for flow boiling in round tubes. In this regard, it is interesting to note that variations of the Chen correlation [13.46] have been successfully developed for several widely different channel geometries: offset strip fins [13.81], perforated plate fins [13.95], cross-ribbed channels [13.98] and round tubes with twisted tapes [13.113].

The form of the Chen correlation developed by Mandrusiak and Carey [13.81] for offset strip fins provides some insight into the modifications that must be made to a round-tube correlation to adapt it to an enhanced surface. Their correlation is embodied in the following relations:

$$h = h_c + h_{ne} \tag{13.63}$$

where

$$h_{ne} = \frac{h_{FZ}A_p + h'_{FZ}\eta_f A_f}{A_p + \eta_f A_f} S \tag{13.64}$$

$$h_{FZ} = 0.00122 \frac{k_l^{0.79} c_{pl}^{0.45} \rho_l^{0.49}}{\sigma^{0.5} \mu_l^{0.29} h_{lv}^{0.45} \rho_v^{0.24}} \{[T_w - T_{sat}(P_l)]^{0.24} [P_{sat}(T_w) - P_l]^{0.75}\}_{prime} \tag{13.65}$$

$$h_c = \mathrm{Pr}_l^{0.296} F(X_{tt}) h_l \tag{13.66}$$

$$h_l = G(1-x)c_{pl} A \,\mathrm{Re}_l^{-n}\, \mathrm{Pr}_l^{-2/3}, \qquad \mathrm{Re}_l = \frac{G(1-x)d_{hp}}{\mu_l} \tag{13.67}$$

$$X_{tt} = \left(\frac{\rho_v}{\rho_l}\right)^{1/2} \left(\frac{\mu_l}{\mu_v}\right)^{n/2} \left(\frac{1-x}{x}\right)^{1-n/2} \tag{13.68}$$

$$S = \left(\frac{24.4}{N_b}\right)[1 - \exp(-0.041 N_b)], \qquad N_b = \left(\frac{h_c}{k_l}\right)\left[\frac{\sigma}{g(\rho_l - \rho_v)}\right]^{1/2} \tag{13.69}$$

$$F = \left[1 + \frac{28.0}{X_{tt}^2}\right]^{0.372} \tag{13.70}$$

In the above relations, A and n are constants in the turbulent-flow, single-phase heat transfer correlation for the geometry of interest:

$$\mathrm{St}_l \, \mathrm{Pr}_l^{2/3} = A \, \mathrm{Re}_l^{-n} \tag{13.71}$$

The subscript "prime" denotes that T_w is the temperature of the prime surface walls of the channel, and η_f denotes the fin efficiency for the offset strip fins. In Eq. (13.67), d_{hp} is the hydraulic diameter of the channel based on the heated perimeter.

Equation (13.63) reflects the usual assumption that the overall heat transfer coefficient is equal to the sum of contributions due to forced-convective (macroscopic) and nucleate boiling (microscopic) effects. Calculation of the nucleate boiling contribution is more complicated for this geometry than for a round tube, because of the temperature variation along the fins. Equation (13.64) was proposed as an approximate means of accounting for this variation at low to moderate superheat levels. Note that h_{FZ} is the nucleate boiling heat transfer coefficient predicted by the Forster and Zuber correlation [13.114]. In Eq. (13.64), h'_{FZ} is given by the right side of Eq. (13.65) after $(T_w)_{\mathrm{prime}}$ is replaced with $T_{\mathrm{sat}} + \eta_f[(T_w)_{\mathrm{prime}} - T_{\mathrm{sat}}]$.

In Eq. (13.64), S is the suppression factor. Even though the effects of convection on nucleation for the fins and the prime surface may be somewhat different, in this treatment, a single mean suppression factor has been used for the entire heated surface. To predict the values of S for their offset strip-fin geometries, Mandrusiak and Carey [13.81] used the general relation (13.69) for the suppression factor proposed by Bennett et al. [13.115] for round tubes.

Example 13.4 A vertical channel with aluminum offset strip fins is to be used to vaporize a flow of liquid nitrogen. The geometry parameters characterizing the surface are (see Fig. 13.12) $L_f = 3.18$ mm, $H_f = 6.35$ mm, $\delta_f = 0.20$ mm, $s = 1.49$ mm, and the hydraulic diameter d_h is 2.34 mm. The ratio of fin surface area to prime surface area A_f/A_p is 3.52 for this surface. The mass flux through the channel is 150 kg/m^2 s. At a particular location, the quality is 0.30, the pressure is 229 kPa, and the prime surface temperature is 87 K. Use the method of Mandrusiak and Carey [13.81] to estimate the local heat transfer coefficient. For this geometry, the single-phase heat transfer coefficient is given by Eq. (13.71) with $A = 0.245$ and $n = 0.38$, and F is given by $F = [1 + 3.53/X_{tt}^{0.5} + 1.05/X_{tt}^2]^{1/2}$.

For saturated nitrogen at 229 kPa, $T_{\mathrm{sat}} = 85$ K, $\rho_l = 771$ kg/m^3, $\rho_v = 9.83$ kg/m^3, $h_{lv} = 188$ kJ/kg, $c_{pl} = 2.096$ kJ/kg K, $\mu_l = 127 \times 10^{-6}$ N s/

m^2, μ_v = 5.60 × 10^{-6} N s/m^2, k_l = 0.1229 W/m K, Pr_l = 2.17, and σ = 0.00720 N/m. We begin by using Eqs. (13.66) through (13.68) to determine h_c:

$$Re_l = \frac{G(1-x)d_h}{\mu_l} = \frac{150(1-0.3)(0.00234)}{127 \times 10^{-6}} = 1935$$

$$\begin{aligned} h_l &= G(1-x)c_{pl}A\, Re_l^{-n}\, Pr_l^{-2/3} \\ &= 150(0.7)(2.096)(0.245)(1935)^{-0.38}(2.17)^{-2/3} \\ &= 1.81 \text{ kW/m}^2\text{ K} \end{aligned}$$

$$\begin{aligned} X_{tt} &= \left(\frac{\rho_v}{\rho_l}\right)^{1/2} \left(\frac{\mu_l}{\mu_v}\right)^{n/2} \left(\frac{1-x}{x}\right)^{1-n/2} \\ &= \left(\frac{9.83}{771}\right)^{1/2} \left(\frac{127}{5.60}\right)^{.38/2} \left(\frac{0.7}{0.3}\right)^{1-.38/2} = 0.406 \end{aligned}$$

$$\begin{aligned} F &= \left(1 + \frac{3.53}{X_{tt}^{0.5}} + \frac{1.05}{X_{tt}^2}\right)^{1/2} \\ &= \left[1 + \frac{3.53}{(0.406)^{0.5}} + \frac{1.05}{(0.406)^2}\right]^{1/2} = 3.59 \end{aligned}$$

$$\begin{aligned} h_c &= Pr_l^{0.296} F h_l = (2.17)^{0.296}(3.59)(1.81) \\ &= 8.17 \text{ kW/m}^2\text{ K} \end{aligned}$$

As a first estimate, we compute the fin efficiency taking $h = h_c$ and using k_f = 238 W/m K for aluminum:

$$\begin{aligned} \eta_f &= \frac{\tanh (2hb^2/k_f\delta_f)^{1/2}}{(2hb^2/k_f\delta_f)^{1/2}} \\ &= \frac{\tanh [2(8170)(0.00297)^2/(238)(0.0002)]^{1/2}}{[2(8170)(0.00297)^2/(238)(0.0002)]^{1/2}} = 0.526 \end{aligned}$$

For nitrogen near 229 kPa, $(\Delta P/\Delta T)_{sat}$ is approximately equal to 2.62 × 10^4 Pa/K. Thus, using the fact that

$$P_{sat}(T_w) - P_l = 2.62 \times 10^4[T_w - T_{sat}(P_l)]$$

Eq. (13.65) can be written as

$$h_{FZ} = 0.00122 \frac{k_l^{0.79} c_{pl}^{0.45} \rho_l^{0.49}}{\sigma^{0.5} \mu_l^{0.29} h_{lv}^{0.45} \rho_v^{0.24}} (2.62 \times 10^4)^{0.75} [T_w - T_{sat}(P_l)]_{prime}^{0.99}$$

Substituting,

$$h_{\mathrm{FZ}} = 0.00122 \frac{(0.1229 \times 10^{-3})^{0.79}(2.096)^{0.45}(771)^{0.49}}{(0.00720)^{0.5}(127 \times 10^{-6})^{0.29}(188)^{0.45}(9.83)^{0.24}}$$

$$\cdot (2.62 \times 10^4)^{0.75}(2)^{0.99}$$

$$= 1.28 \text{ kW/m}^2 \text{ K}$$

Replacing $(T_w - T_{\mathrm{sat}})_{\mathrm{prime}}$ with $\eta_f(T_w - T_{\mathrm{sat}})_{\mathrm{prime}}$, h'_{FZ} may be calculated as

$$h'_{\mathrm{FZ}} = (\eta_f)^{0.99} h_{\mathrm{FZ}} = (0.526)^{0.99}(1.28)$$

$$= 0.678 \text{ kW/m}^2 \text{ K}$$

The suppression factor is then calculated using Eqs. (13.69):

$$N_b = \left(\frac{h_c}{k_l}\right)\left[\frac{\sigma}{g(\rho_l - \rho_v)}\right]^{1/2} = \left(\frac{8170}{0.1229}\right)\left[\frac{0.00720}{9.8(771 - 9.83)}\right]^{1/2}$$

$$= 65.3$$

$$S = \left(\frac{24.4}{N_b}\right)[1 - \exp(-0.041 N_b)]$$

$$= \left(\frac{24.4}{65.3}\right)\{1 - \exp[-0.041(65.3)]\} = 0.348$$

With some rearrangement, Eq. (13.64) can be written in the form

$$h_{ne} = \frac{h_{\mathrm{FZ}} + h'_{\mathrm{FZ}}\eta_f(A_f/A_p)}{1 + \eta_f(A_f/A_p)} S$$

Substituting yields

$$h_{ne} = \frac{1.28 + 0.678(0.526)(3.52)}{1 + 0.526(3.52)}(0.348) = 0.309 \text{ kW/m}^2 \text{ K}$$

From Eq. (13.63) we then find that

$$h = h_c + h_{ne} = 8.17 + 0.309 = 8.48 \text{ kW/m}^2 \text{ K}$$

Substituting this new h value into the equation for η_f yields $\eta_f = 0.518$. With this η_f value, recomputing yields $h'_{\mathrm{FZ}} = 0.668$ kW/m^2 K, $h_{ne} = 0.308$ kW/m^2 K, and $h = 8.48$ kW/m^2 K. Thus h and η_f are estimated to be

$$h = 8.48 \text{ kW/m}^2 \text{ K} \qquad \eta_f = 0.52$$

After removing the contribution due to nucleate boiling, Mandrusiak and Carey [13.81] found that the remaining convective contribution to the overall heat transfer coefficient was well correlated in terms of the X_{tt} and F parameters defined by Eqs. (13.66) and (13.68). About 500 data points for annular film flow boiling in three different offset strip-fin geometries are well represented by Eq. (13.70). This correlation is slightly higher than the curve predicted by the classic Chen correlation (see Chapter Twelve).

For the overall correlation scheme described above, the predicted h values agreed with the measured values for the three geometries to a mean absolute deviation of 18.2%. Somewhat better agreement was obtained by fitting a separate $F(X_{tt})$ correlation to the data for each geometry. Mandrusiak and Carey [13.81] also found that the data of Robertson [13.86] and Robertson and Lovegrove [13.88] for a different offset fin geometry could also be correlated with good accuracy using this method.

Overall, the results of these investigations suggest that a superposition approach of this type can be used with reasonable success to predict the flow-boiling heat transfer performance of offset strip-fin geometries at low to moderate wall superheat levels. It may be possible to use the same F correlation for a family of geometrically similar surfaces. However, to get the greatest accuracy from this technique, it is necessary to determine an F correlation for the specific geometry of interest, in much the same way that a Colburn j factor correlation is determined for each compact heat exchanger geometry used for single-phase applications.

It is also noteworthy that the generalized suppression-factor relation proposed by Bennett et al. [13.115] seemed to work well with all the geometries tested in these studies. However, because the suppression factor is expected to depend on the nature of the convective mechanism and nucleation characteristics of the surface, it seems likely that the best results would be achieved by using a suppression-factor correlation determined experimentally for the specific geometry and materials of interest. In particular, this would help ensure that the suppression-factor correlation is consistent with the size distribution, liquid wetting angle, and other fluid properties that affect nucleation but do not appear explicitly in this correlation.

As noted above, variations of the Chen correlation have also been developed for several other flow-passage geometries. The broad success of this type of superposition technique suggests that, with appropriate modifications, it may be possible to apply this approach to other complex geometries as well.

For all geometries of interest, we desire the ability to predict the heat transfer coefficient and pressure gradient over the entire range of conditions encountered during the vaporization process: from low qualities, where nucleate boiling dominates, through annular film flow boiling at intermediate qualities, and for partial dryout and/or mist flow circumstances at high qualities. However, at the present time, accurate prediction of the heat transfer coefficient and pressure gradient for all flow regimes and geometries encountered in compact evaporators is not yet possible.

Even for offset strip fin configurations, which have been the subject of a number of studies, a widely applicable method for predicting the heat transfer coefficient in the mist flow (partial dryout) regime is not yet available. The flow boiling studies summarized above provide useful starting points for the development of better predictive methods. However, there clearly exists a need for further work if enhanced flow boiling surfaces are to be used to their fullest potential.

Further information on the performance characteristics of enhanced surfaces for flow boiling and condensation can be found in the reference [13.116].

REFERENCES

13.1 Ledinegg, M., Instabilitat der Stromung bei Naturlichen und Zwangumlauf, *Warme,* vol. 61, pp. 891–898, 1938.

13.2 Bergles, A. E., Goldberg, P., and Maulbetsch, J. S., Acoustic oscillations in a high pressure single channel boiling system, *EURATOM Report, Proc. Symp. on Two-Phase Flow Dynamics,* Eindhoven, EUR 4288e, pp. 535–550, 1967.

13.3 Saha, P., Ishii, M., and Zuber, N., An experimental investigation of the thermally induced flow oscillations in two-phase systems, *J. Heat Transfer,* vol. 98, pp. 616–622, 1976.

13.4 Terano, T., Kinetic behavior of monotube boiler, *Bull. JSME,* vol. 3, pp. 540–546, 1960.

13.5 Wallis, G. B., and Heasley, J. H., Oscillation in two-phase flow system, *J. Heat Transfer,* vol. 83, 363–369, 1961.

13.6 Serov, E. P., Analytical investigation of the boundary conditions for the formation of pulsation in steaming pipes during forced circulation, *High Temp.,* vol. 3, pp. 545–549, 1965.

13.7 Boure, J., and Mikaila, A., Oscillatory behavior of heated channels, in *Proc. Symp. on Two-Phase Flow Dynamics,* Eindhoven, vol. 1., pp. 695–720, 1967.

13.8 Ishii, M., and Zuber, N., Thermally induced flow instabilities in two-phase mixtures, *Proc. 4th Int. Heat Transfer Conf.,* Paris, vol. 5, paper B5.11, 1970.

13.9 Yadigaroglu, G., and Bergles, A. E., Fundamental and higher-mode density-wave oscillations in two-phase flow, *J. Heat Transfer,* vol. 94, pp. 189–195, 1972.

13.10 Saha, P., Thermally induced flow instabilities including the effect of thermal non-equilibrium between phases, Ph.D. thesis, Georgia Institute of Technology, 1974.

13.11 Bergles, A. E., Instabilities in two-phase systems, in *Two-Phase Flows and Heat Transfer in the Power and Process Industries,* Hemisphere, New York, Chap. 13, pp. 383–422, 1978.

13.12 Ishii, M., Wave phenomena and two-phase flow instabilities, in *Handbook of Multiphase Systems,* G. Hetsroni, Ed., Hemisphere, New York, sec. 2.4, 1982.

13.13 Bonnilla, C. F., and Perry, C. W., Heat transmission to boiling binary liquid mixtures, *Trans. AIChE,* vol. 37, pp. 269–290, 1941.

13.14 van Stralen, S., The mechanism of nucleate boiling in pure liquids and binary mixtures, Parts I & II, *Int. J. Heat Mass Transfer,* vol. 9, pp. 995–1046, 1966.

13.15 van Stralen, S., The mechanism of nucleate boiling in pure liquids and binary mixtures, Parts III & IV, *Int. J. Heat Mass Transfer,* vol. 10, pp. 1469–1498, 1967.

13.16 Calus, W. F., and Leonodopoulus, D. J., Pool boiling—Binary liquid mixtures, *Int. J. Heat Mass Transfer,* vol. 17, pp. 249–256, 1974.

13.17 Happel, O., Heat transfer during boiling of binary mixtures in the nucleate and film boiling ranges, in *Heat Transfer in Boiling,* E. Hane and U. Grigull, Eds., Academic Press/Hemisphere, New York, chap. 9, 1977.

13.18 Thome, J. R., Nucleate boiling of binary liquids, *AIChE Prog. Symp. Ser.,* vol. 77, pp. 238–250, 1981.

13.19 Stephan, K., Natural convection boiling in multicomponent mixtures, in *Heat Exchangers,* S. Kakac, A. E. Bergles, and F. Mayinger Eds., Hemisphere, New York, pp. 315–336, 1981.

13.20 van Stralen, S. J. D., Heat transfer to boiling binary and ternary systems, in *Boiling Phenomena,* S. van Stralen and R. Cole, Eds., Hemisphere/McGraw-Hill, New York, pp. 33–65, 1979.

13.21 Cooper, M. G., and Stone, C. R., Boiling of binary mixtures—Study of individual bubbles, *Int. J. Heat Mass Transfer,* vol. 24, pp. 1937–1950, 1981.

13.22 van Stralen, S. J. D., Heat transfer to boiling binary liquid mixtures at atmospheric and subatmospheric pressures, *Chem. Eng. Sci.,* vol. 5, pp. 290–296, 1956.

13.23 van Stralen, S. J. D., Heat transfer to boiling liquid mixtures, *Br. Chem. J.,* part I, vol. 4, pp. 8–17; part II, vol. 4, pp. 78–82; part III, vol. 6, pp. 834–840; part IV, vol. 7, pp. 90–97, 1959.

13.24 Van Wijk, W. R., Vos, A. S., and van Stralen, S. J. D., Heat transfer to boiling binary liquid mixtures, *Chem. Eng. Sci.,* vol. 5, pp. 68–80, 1956.

13.25 Vos, A. S., and van Stralen, S. J. D., Heat transfer to boiling water-methylethylketone mixtures, *Chem. Eng. Sci.,* vol. 5, pp. 50–56, 1956.

13.26 Carne, M., Studies of the critical heat flux for some binary mixtures and their components, *Can. J. Chem. Eng.,* vol. 41, pp. 235–241, 1963.

13.27 Dunskus, T., and Westwater, J. W., The effect of trace additives on the heat transfer to boiling isopropanol, *Chem. Eng. Symp. Ser.,* vol. 57, no. 32, pp. 173–181, 1961.

13.28 Grigoriev, L. N., Khairullin, I. Kh., and Usmanov, A. G., An experimental study of the critical heat flux in the boiling of binary mixtures. *Int. Chem. Eng.,* vol. 8, no. 1, pp. 39–42, 1968.

13.29 Jordan, D. P., and Leppert, G., Nucleate boiling characteristics of organic reactor coolants, *Nuc. Sci. and Eng.,* vol. 5, pp. 349–359, 1959.

13.30 Sterman, L. S., Vilemas, J. V., and Abramov, A. I., On heat transfer and critical heat flux in organic coolants and their mixtures, *Proc. 3rd Int. Heat Transfer Conf.,* Chicago, vol. 4, pp. 258–270, 1966.

13.31 van Stralen, S. J. D., and Sluyter, W. M., Investigations on the critical heat flux of pure liquids and mixtures under various conditions, *Int. J. Heat and Mass Transfer,* vol. 12, pp. 1353–1384, 1969.

13.32 Reddy, R. P., and Lienhard, J. H., The peak boiling heat flux in saturated ethanol–water mixtures, *J. Heat Transfer,* vol. 111, pp. 480–486, 1989.

13.33 Kutateladze, S. S., Brobrovich, G. I., Gogonin, I. I., Mamontova, N. N., and Moskvichova, V. N., The critical heat flux at the pool boiling of some binary liquid mixtures, *Proc. 3rd Int. Heat Transf. Conf.,* Chicago, vol. 3, pp. 149–159, 1966.

13.34 Matorin, A. S., Correlation of experimental data on heat transfer crisis in pool boiling of pure liquids and binary mixtures, *Heat Transfer—Soviet Research,* vol. 5, pp. 85–89, 1973.

13.35 Gaidorov, S. A., Evaluation of the critical heat flow in the case of a boiling mixture of large volume, *J. Appl. Mech. and Tech. Phys.,* vol. 16, pp. 601–603, 1975.

13.36 Stephen, K., and Preusser, P., Heat transfer and critical heat flux in pool boiling of binary and ternary mixtures, *Ger. Chem. Eng.,* vol. 2, pp. 161–169, 1979.

13.37 Yang, Y. M., An estimation of pool boiling critical heat flux for binary mixtures, *Proc. 2nd ASME/JSME Joint Thermal Eng. Conf.,* Honolulu, vol. 5, pp. 439–446, 1987.

13.38 McEligot, D. M., Generalized peak heat flux for dilute binary mixtures, *AIChE J.,* vol. 10, pp. 130–131, 1964.

13.39 Sun, K. H., and Lienhard, J. H., The peak pool boiling heat flux on horizontal cylinders, *Int. J. Heat Mass Transfer,* vol. 13, pp. 1425–1439, 1970.

13.40 van Ouwerkerk, H. J., Hemispherical bubble growth in a binary mixture, *Chem. Eng. Sci.,* vol. 27, pp. 1957–1960, 1972.

13.41 Cooper, M. G., The binary microlayer—A double diffusion problem, *Chem. Eng. Sci.,* vol. 37, pp. 27–35, 1982.

13.42 Kautzky, D. E., and Westwater, J. W., Film boiling of a mixture on a horizontal plate, *Int. J. Heat Mass Transfer,* vol. 10, pp. 253–256, 1967.

13.43 Yue, P.-L., and Weber, M. E., Film boiling of saturated binary mixtures, *Int. J. Heat Mass Transfer,* vol. 16, pp. 1877–1888, 1973.

13.44 Yue, P.-L., and Weber, M. E., Minimum film boiling flux of binary mixtures, *Trans. Inst. Chem. Engs.,* vol. 52, pp. 217–222, 1974.

13.45 Marschall, E., and Moresco, L. L., Analysis of binary film boiling, *Int. J. Heat Mass Transfer,* vol. 20, pp. 1013–1018, 1977.

13.46 Chen, J. C., Correlation for boiling heat transfer to saturated fluids in convective flow, *Ind. Eng. Chem. Proc. Design and Dev.*, vol. 5, no. 3, pp. 322–339, 1966.

13.47 Stephan, K., and Korner, M., Calculation of heat transfer in evaporating binary liquid mixtures, *Chemie-Ingenieur Technik,* vol. 41, pp. 409–417, 1969.

13.48 Shock, R. A. W., Evaporation of binary mixtures in upward annular flow, *Int. J. Multiphase Flow,* vol. 2, pp. 411–433, 1976.

13.49 Bennett, D. L., and Chen, J. C., Forced convective boiling in vertical tubes for saturated pure components and binary mixtures, *AIChE J.*, vol. 26, pp. 454–461, 1980.

13.50 Florshuetz, L. W., and Khan, A. R., Growth rates of free vapor bubbles in binary liquid mixtures at uniform superheats, *Proc. 4th Int. Heat Transfer Conf.*, paper B7.3, Paris, 1970.

13.51 Collier, J. G., *Convective Boiling and Condensation,* 2nd ed., McGraw-Hill, New York, 1981.

13.52 Govan, A. H., Hewitt, G. F., Owen, D. G., and Bott, T. R., An improved CHF modeling code, *Proc. 2nd U.K. Conf. on Heat Transfer,* Institute of Mechanical Engineers, London, vol. 1, pp. 33–48, 1988.

13.53 Hewitt, G. F., and Hall-Taylor, N. S., *Annular Two-Phase Flow,* Pergamon, Oxford, Chap. 11, 1970.

13.54 Tolubinsky, V. I., and Matorin, P. S., Forced convection boiling heat transfer crisis with binary mixtures, *Heat Transfer—Soviet Research,* vol. 5, pp. 98–101, 1973.

13.55 Andrews, D. G., Hooper, F. C., and Butt, P., Velocity, subcooling and surface effects in the departure from nucleate boiling of organic binaries, *Can. J. of Chem. Eng.*, vol. 46, pp. 194–199, 1968.

13.56 Naboichenko, K. V., Kiryutin, A. A., and Gribov, B. S., A study of critical heat flux with forced flow of monoisopropyldeiphenyl-benzene mixture, *Thermal Eng.*, vol. 12, pp. 107–114, 1965.

13.57 Bergles, A. E., and Scarola, L. S., Effect of a volative additive on the critical heat flux for surface boiling of water in tubes, *Chem. Eng. Sci.*, vol. 21, pp. 721–723, 1966.

13.58 Van Es, J. P., and Heertjes, P. M., On the condensation of a vapor of a binary mixture in a vertical tube, *Chem. Eng. Sci.*, vol. 5, pp. 217–225, 1956.

13.59 Mochizuki, S., Yagi, Y., Tadano, R., and Yang, W-J., Convective filmwise condensation of nonazeotropic binary mixtures in a vertical tube, *J. Heat Transfer,* vol. 106, pp. 531–538, 1984.

13.60 Quandt, E. R., Measurement of some basic parameters in two-phase annular flow, *AIChE J.*, vol. 11, pp. 31–38, 1965.

13.61 Hoopes, J. W., Jr., Flow of steam–water mixtures in a heated annulus and through orifices, *AIChE J.*, vol. 3, pp. 268–275, 1957.

13.62 Davis, E. J., and David, M. M., Heat transfer to high-quality steam–water mixtures flowing in a horizontal rectangular duct, *Can. J. Chem. Eng.*, vol. 39, pp. 99–105, 1961.

13.63 Holser, E. R., Flow patterns in high pressure two-phase (steam–water) flow with heat addition, *Chem. Eng. Prog. Symp. Ser.*, vol. 64, no. 82, pp. 54–66, 1968.

13.64 Carey, V. P., and Mandrusiak, G. D., Annular film-flow boiling of liquids in a partially heated vertical channel with offset strip fins, *Int. J. Heat Mass Transfer,* vol. 29, pp. 927–939, 1986.

13.65 Mandrusiak, G. D., Carey, V. P., and Xu, X., An experimental study of convective boiling in a partially-heated horizontal channel with offset strip fins, *J. Heat Transfer,* vol. 110, pp. 229–236, 1988.

13.66 Xu, X., and Carey, V. P., Heat transfer and two-phase flow during convective boiling in a partially-heated cross-ribbed channel, *Int. J. Heat Mass Transfer,* vol. 30, pp. 2385–2397, 1987.

13.67 Damianides, C. A., and Westwater, J. W., Two-phase flow patterns in a compact heat exchanger and in small tubes, *Proc. 2nd UK National Conf. on Heat Transfer,* vol. II, pp. 1257–1268, 1988.

13.68 Gopin, S. R., Usyukin, I. P., and Aver'yanov, I. G., Heat transfer in condensation of freons on finned surfaces, *Heat Transfer—Soviet Research,* vol. 8, no. 6, pp. 114–119, 1976.

13.69 Yung, D., Lorenz, J. J., and Panchal, C., Convective vaporization and condensation in serrated-fin channels, in *Heat Transfer in Ocean Thermal Energy Conversion Systems,* W. L. Owen, ed., ASME, New York, HTD-vol. 23, pp. 29–37, 1980.

13.70 Panchal, C. B., Heat transfer with phase change in plate-fin heat exchangers, *AIChE Symp. Ser.,* vol. 80, no. 236, pp. 90–97, 1984.

13.71 Robertson, J. M., Blundell, N., and Clarke, R. H., The condensing characteristics of nitrogen in plain brazed aluminum plate-fin heat-exchanger passages, *Heat Transfer 1986,* Hemisphere, New York, vol. 3, pp. 1719–1724, 1986.

13.72 Tovazhnyanskiy, L. L., and Kapustenko, P. A., Heat transfer from steam condensing in an extended-surface slot channel, *Heat Transfer—Sov. Res.,* vol. 12, no. 4, pp. 34–36, 1980.

13.73 Panchal, C. B., Condensation heat transfer in plate heat exchangers, in *Two-Phase Heat Exchanger Symposium,* J. T. Pearson and J. B. Kitto, Eds., ASME, New York, HTD-vol. 24, pp. 45–52, 1985.

13.74 Royal, J. H., and Bergles, A. E., Augmentation of horizontal in-tube condensation by means of twisted-tape inserts and internally finned tubes, *J. Heat Transfer,* vol. 100, pp. 17–24, 1978.

13.75 Luu, M., and Bergles, A. E., Enhancement of horizontal in-tube condensation of R-113, *ASHRAE Trans.,* vol. 86, part I, pp. 293–312, 1980.

13.76 Said, S. A., and Azer, N. Z., Heat transfer and pressure drop during condensation inside horizontal tubes with twisted tape inserts, *ASHRAE Trans.,* vol. 89, part I, pp. 96–113, 1983.

13.77 Vrable, D. L., Yang, W. J., and Clark, J. A., Condensation of refrigerant-12 inside horizontal tubes with internal axial fins, *Heat Transfer 1974,* vol. 3, pp. 250–254, 1974.

13.78 Said, S. A., and Azer, N. Z., Heat transfer and pressure drop during condensation inside horizontal internally finned tubes, *ASHRAE Trans.,* vol. 89, part I, pp. 114–134, 1983.

13.79 Venkatesh, S. K., and Azer, N. Z., Enhancement of condensation heat transfer of R-11 by internally finned tubes, *ASHRAE Trans.,* vol. 91, part II, pp. 128–144, 1985.

13.80 Boyko, L. D., and Kruzhilin, G. N., Heat transfer and hydraulic resistance during condensation of steam in a horizontal tube and in a bundle of tubes, *Int. J. Heat Mass Transfer,* vol. 10, pp. 361–373, 1967.

13.81 Mandrusiak, G. D., and Carey, V. P., Convective boiling in vertical channels with different offset strip fin geometries, *J. Heat Transfer,* vol. 111, pp. 156–165, 1989.

13.82 Panitsidis, H., Gresham, R. D., and Westwater, J. W., Boiling of liquids in a compact plate-fin heat exchanger, *Int. J. Heat Mass Transfer,* vol. 18, pp. 37–42, 1975.

13.83 Chen, C. C., Loh, J. V., and Westwater, J. W., Prediction of boiling heat transfer duty in a compact plate-fin heat exchanger using the improved local assumption, *Int. J. Heat Mass Transfer,* vol. 24, pp. 1907–1912, 1981.

13.84 Chen, C. C., and Westwater, J. W., Application of the local assumption for the design of compact heat exchangers for boiling heat transfer, *J. Heat Transfer,* vol. 106, pp. 204–209, 1984.

13.85 Galezha, V. B., Usyukin, I. P., and Kan, K. D., Boiling heat transfer with freons in finned-plate heat exchangers, *Heat Transfer—Sov. Res.,* vol. 8, no. 3, pp. 103–110, 1976.

13.86 Robertson, J. M., Boiling heat transfer with liquid nitrogen in brazed-aluminum plate-fin heat exchangers, *AIChE Symp. Ser.,* vol. 75, no. 189, pp. 151–154, 1979.

13.87 Robertson, J. M., and Clarke, R. H., The onset of boiling of liquid nitrogen in plate-fin heat exchangers, *AIChE Symp. Ser.,* vol. 77, no. 208, pp. 86–95, 1981.

13.88 Robertson, J. M., and Lovegrove, P. C., Boiling heat transfer with freon 11 (R11) in brazed aluminum plate-fin heat exchangers, *J. Heat Transfer,* vol. 105, pp. 605–610, 1983.

13.89 Robertson, J. M., The correlation of boiling coefficients in plate-fin heat exchanger passages with a film-flow model, *Heat Transfer 1982,* Munich, vol. 6, pp. 341–345, 1982.

13.90 Robertson, J. M., The prediction of convective boiling coefficients in serrated plate-fin passages using an interrupted liquid–film flow model, in *Basic Aspects of Two-Phase Flow and Heat Transfer,* ASME, New York, HTD-vol. 34, pp. 163–171, 1984.

13.91 Robertson, J. M., and Clarke, R. H., Investigations into the onset of convective boiling with liquid nitrogen in plate-fin heat exchanger passages under constant wall temperature boundary conditions, *AIChE Symp. Ser.,* vol. 80, no. 236, pp. 98–103, 1984.

13.92 Robertson, J. M., and Clarke, R. H., The general prediction of convective boiling coefficients in plate-fin heat exchanger passages. *AIChE Symp. Ser.* vol. 81, no. 245, pp. 129–134, 1985.

13.93 Carey, V. P., Criteria for evaluating the performance of compact heat exchanger surfaces in compact evaporators, in *Compact Heat Exchangers,* R. K. Shah, A. D. Kraus and D. Metzger, editors, pp. 287–310, Hemisphere, New York, 1990.

13.94 Mandrusiak, G. D., and Carey, V. P., A finite difference model of annular film-flow evaporation in a vertical channel with offset strip fins, *Int. J. Multiphase Flow,* vol. 16, pp. 1071–1098, 1990.

13.95 Robertson, J. M., The boiling characteristics of perforated plate-fin channels with liquid nitrogen in upflow, in *Heat Exchangers for Two-Phase Flow Applications,* ASME, New York, ASME HTD-vol. 27, pp. 35–40, 1983.

13.96 Panchal, C. B., Hillis, D. L., and Thomas, A., Convective boiling of ammonia and freon 22 in plate heat exchangers, *Proc. ASME/JSME Therm. Eng. Joint Conf.,* ASME, New York, vol. 2, pp. 261–268, 1983.

13.97 Marseille, T. J., Carey, V. P., and Estergreen, S. L., Full-core test method for experimental determination of convective boiling heat transfer coefficients in tubes of crossflow compact evaporators, *Exp. Thermal and Fluid Sci.,* vol. 1, pp. 395–404, 1988.

13.98 Cohen, M., and Carey, V. P., Convective boiling in a partially-heated channels with different cross-ribbed geometries, *Int. J. Heat Mass Transfer,* vol. 32, pp. 2459–2474, 1989.

13.99 Lavin, J. G., and Young, E. H., Heat transfer to evaporating refrigerants in two-phase flow, *AIChE J.,* vol. 11, pp. 1124–1132, 1965.

13.100 Withers, J. G., and Habdas, E. P., Heat transfer characteristics of helical-corrugated tubes for in tube boiling of refrigerant R-12, *AIChE Symp. Ser.,* vol. 70, no. 138, pp. 98–106, 1974.

13.101 D'yachkov, F. N., Investigation of heat transfer and hydraulics for boiling of freon-22 in internally-finned tubes, *Heat Transfer—Sov. Res.,* vol. 10, no. 2, pp. 10–19, 1978.

13.102 Ito, M., and Kimura, H., Boiling heat transfer and pressure drop in internal spiral-grooved tubes, *Bull. JSME,* vol. 221, pp. 1251–1257, 1979.

13.103 Kubanek, G. R., and Miletti, D. L., Evaporative heat transfer and pressure drop performance of internally-finned tubes with refrigerant 22, *J. Heat Transfer,* vol. 101, pp. 447–452, 1979.

13.104 Akhanda, M. A. R., and James, D. D., An experimental study of the relative effects of transverse and longitudinal ribbing of heat transfer surface in forced convective boiling, in *Two-Phase Heat Exchanger Symposium,* ASME, New York, HTD-vol. 44, pp. 83–90, 1985.

13.105 Gambill, W. R., Bundy, R. D., and Wansbrough, R. W., Heat transfer, burnout, and pressure drop for water in swirl flow through tubes with internal twisted tapes, *Chem. Eng. Prog. Symp. Ser.,* vol. 57, no. 32, pp. 127–137, 1961.

13.106 Feinstein, L., and Lundberg, R. E., Fluid friction and boiling heat transfer with water in vortex flow in tubes containing an internal twisted tape, Report RADC-TRR-63-451, AD430889, 1963.

13.107 Lopina, R. F., and Bergles, A. E., Subcooled boiling of water in tape-generated swirl flow, *J. Heat Transfer,* vol. 95, pp. 281–283, 1973.

13.108 Blatt, T. A., and Adt, R. R., Jr., The effects of twisted tape swirl generation on the heat transfer rate and pressure drop of boiling freon 11 and water, ASME Paper no. 63-WA-42, 1963.

13.109 Viskanta, R., Critical heat flux for water in swirling flow, *Nucl. Sci. and Eng.,* vol. 10, pp. 202–203, 1961.

13.110 Bergles, A. E., Fuller, W. D., and Hynek, S. J., Dispersed flow film boiling of nitrogen with swirl flow, *Int. J. Heat Mass Transfer,* vol. 14, pp. 1343–1354, 1971.

13.111 Cumo, M., Farello, G. E., Ferrari, G., and Palazzi, G., The influence of twisted tapes in subcritical, once-through vapor generators in counterflow, *J. Heat Transfer,* vol. 96, pp. 365–370, 1974.

13.112 Pai, R. H., and Pasint, D., Research at foster wheeler: Advanced once-through boiler design, *Electric Light and Power,* pp. 66–70, January 1965.

13.113 Jensen, M. K., and Beseler, H. P., Saturated forced-convective boiling heat transfer with twisted-tape inserts, *J. Heat Transfer*, vol. 108, pp. 93–99, 1986.

13.114 Forster, H. K., and Zuber, N., Dynamics of vapor bubbles and boiling heat transfer, *AIChE J.*, vol. 1, pp. 531–535, 1955.

13.115 Bennett, D. L., Davis, M. W., and Herteler, B. L., The suppression of saturated nucleate boiling by forced convective flow, *AIChE Symp. Ser.*, vol. 76, no. 199, pp. 91–103, 1980.

13.116 Carey, V. P., and Shah, R. K., Design of compact and enhanced heat exchangers for liquid–vapor phase-change applications, in *Two-Phase Heat Exchangers: Thermal Hydraulic Fundamentals and Design,* S. Kakac, A. E. Bergles, and E. O. Fernandes, Eds., Kluwer, Dordrecht, The Netherlands, pp. 909–968, 1988.

PROBLEMS

13.1 For the conditions in Example 13.1, determine and plot the variation of the heat flux at the onset of instability for mass flux values between 200 kg/m^2 s and 9000 kg/m^2 s. From an energy balance, also plot the exit quality from the tube when complete vaporization is not achieved. What ranges of heat flux and mass flux values are possible if complete vaporization (x = 1.0 at the end of the tube) and a stable flow are desired?

13.2 Convective condensation of nitrogen vapor occurs in a horizontal tube having an aluminum twisted-tape insert. The tube inside diameter is 8 mm and the twist pitch-to-diameter ratio y for the insert is 1.5. The tape thickness is 0.6 mm. At a particular location, the pressure is 229 kPa and the quality is 0.7. The total mass flux through the tube is 200 kg/m^2 s. Estimate the condensation heat transfer coefficient at this location. Using a suitable correlation from Chapter Eleven for a plain round tube, also compute the heat transfer coefficient for the round tube without the insert. How does the heat transfer coefficient change if the insert is replaced with a twisted tape that is identical except that $y = 2.5$?

13.3 Convective condensation of R-22 vapor occurs in a horizontal tube having an aluminum twisted-tape insert. The tube inside diameter is 1.2 cm and the twist pitch-to-diameter ratio y for the insert is 2.0. The tape thickness is 0.4 mm. At a particular location, the pressure is 958 kPa and the quality is 0.7. The total mass flux through the tube is 200 kg/m^2 s. Estimate the condensation heat transfer coefficient at this location. Using a suitable correlation from Chapter Eleven for a plain round tube, also compute the heat transfer coefficient for the round tube without the insert. Repeat the above calculations for a quality of 0.3 at the same pressure.

13.4 R-12 condenses in a horizontal tube having an aluminum twisted-tape inset. The tube inside diameter is 8 mm, the twist–pitch diameter ratio y for the inset is 1.8, and the tape thickness is 0.5 mm. For a local pressure of 1602 kPa and a quality of 0.8, determine the heat transfer coefficient for total mass flux values of 100, 200, 400, and 800 kg/m^2 s. Compare the variation of h with G for this twisted tape with that predicted by an appropriate correlation from Chapter Eleven for a plain round tube.

13.5 Flow boiling of saturated R-12 occurs in a vertical channel containing the offset strip fin matrix described in Example 13.5. The total mass flux through the channel is 200 kg/m^2 s. Assuming that the pressure is essentially constant at 333 kPa over the length of the passage and that the prime surface is at a constant and uniform temperature of 20°C, use the method of Mandrusiak and Carey (1989) to estimate the heat transfer coefficient at locations where the quality is 0.2, 0.45, and 0.7.

13.6 Flow boiling of saturated ammonia occurs in a vertical channel containing the offset strip-fin matrix described in Example 13.5. The total mass flux through the channel is 150 kg/m^2 s. At a particular location, the pressure is 165 kPa, the quality is 0.4, and the prime surface wall temperature is 0°C. Use the method of Mandrusiak and Carey [13.81] to estimate the heat transfer coefficient at this location. Also use an appropriate correlation from Chapter Twelve to determine the convective boiling heat transfer coefficient at the same quality and imposed flow conditions for a plain round tube with the same wall temperature and hydraulic diameter. How do the values of h for the two geometries compare?

13.7 Flow boiling of nitrogen occurs in a vertical channel containing the offset strip-fin matrix described in Example 13.5. For a pressure of 229 kPa, a quality of 0.6, and a prime surface wall temperature of 90 K, use the method of Mandrusiak and Carey [13.81] to estimate the heat transfer coefficient for mass flux values of 50, 100, 200, and 400 kg/m^2 s. Using an appropriate correlation from Chapter Twelve, compare the variation of h with G for this offset strip-fin surface with that obtained at the same wall superheat and flow conditions for a plain round tube with the same hydraulic diameter.

APPENDIX
ONE

BASIC ELEMENTS OF THE KINETIC THEORY OF GASES

The success of the kinetic theory of gases is a direct consequence of the relatively simple nature of most gases. The kinetic theory of gases is based on the following idealizations of the structure of gases:

1. The gas is composed of a very large number of moving particles (atoms or molecules).
2. The particles collide with one another infrequently.
3. Collisions between particles are elastic; i.e., the total kinetic energy of the particles before and after the collision is constant.
4. Between collisions, the particles move in straight lines, in the absence of any force field, and hence obey Newton's first law of motion.
5. The motion of the particles is completely random or chaotic.

The relationship between the motion of molecules in the gas and the pressure P can be understood by considering the rectangular box of length L and cross-sectional area A shown in Fig. AI.1.

If a single molecule with a velocity u in the x direction hits the end of the box, it is reflected and travels in the opposite direction with velocity $-u$. Newton's second law of motion requires that the force F exerted on the wall during the collision is equal to the rate of change of momentum:

$$F = \frac{d}{dt}(mu) \tag{I.1}$$

where m is the mass of the molecule. The momentum of the molecule before

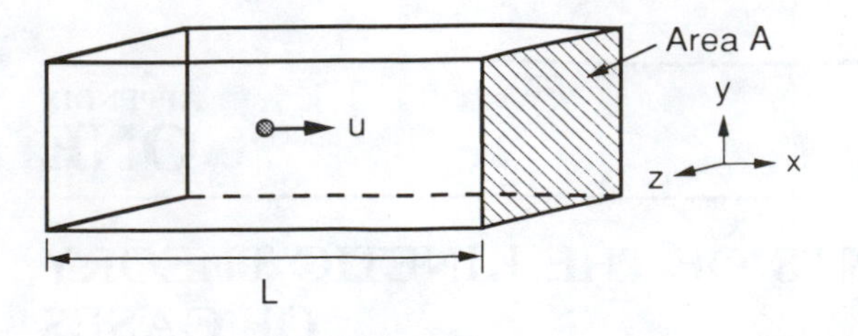

Figure AI.1 A single particle in a box moving between two parallel walls.

collision is mu, whereas after the collision it is $-mu$. The change in the momentum is $-2mu$.

The total change in momentum per unit time due to collisions of this molecule with the wall is equal to the change in momentum per collision multiplied by the number of collisions per unit time that this molecule makes with the wall. The time between collisions is equal to the time it takes the molecule to travel to the other end of the box and back again, $2L/u$. Thus the total change in momentum per unit time is $-2mu(u/2L)$, which must equal the force F on the molecule. The force on the wall, $F_w = -F$, must therefore equal mu^2/L. Since the pressure (normal stress) P' exerted by the molecule must equal F_w/A,

$$P' = \frac{mu^2}{AL} = \frac{mu^2}{V} \tag{I.2}$$

where V is the volume of the box.

If we have a collection of particles all moving in the x direction with various speeds $u_1, u_2, \ldots, u_n$, then the total force, and consequently the total pressure, will equal the sum of the contributions from all particles:

$$P = \frac{m}{V} \sum_{i=1}^{n} (u_i)^2 \tag{I.3}$$

Defining the average of the squares of the velocities, $\langle u^2 \rangle$, as

$$\langle u^2 \rangle = \frac{1}{n} \sum_{i=1}^{n} (u_i)^2 \tag{I.4}$$

where n is the total number of particles in the box, the expression for the pressure becomes

$$P = \frac{nm \langle u^2 \rangle}{V} \tag{I.5}$$

Because the molecules in this model can move in only one dimension, this can be thought of as being the equation for the pressure of a *one-dimensional* gas. Note that, so far, we have ignored the effect of molecular collisions. Since the molecules can move only in the x direction, collisions in this one-dimensional gas can occur only if two molecules travel in the exact same linear path. If we consider two such molecules, as shown in Fig. AI.2, it is clear that the leftmost molecule will never hit the right wall and the rightmost molecule will never hit the left wall.

However, because of collisions in the center of the box, the leftmost molecule will strike the left wall twice as often as it would if the two molecules traveled parallel paths and did not collide. Similarly, the rightmost molecule in Fig. AI.2 will strike the right wall twice as often as it would if the molecules traveled parallel paths. On the average, then, the net effect of the molecules traveling the same or parallel paths is identical. This implies that Eq. (I.5) is valid for a one-dimensional gas even if some of the molecules collide.

In a real gas, of course, molecules can move in three dimensions, and the velocity vector of a molecule has three components as shown in Fig. AI.3. The square of the velocity vector is related to the square of its components as

$$c^2 = u^2 + v^2 + w^2 \tag{I.6}$$

For any given molecule, the velocity components u, v, and w may generally all be different. However, averaging both sides of Eq. (I.6) over all the molecules yields

$$\langle c^2 \rangle = \langle u^2 \rangle + \langle v^2 \rangle + \langle w^2 \rangle \tag{I.7}$$

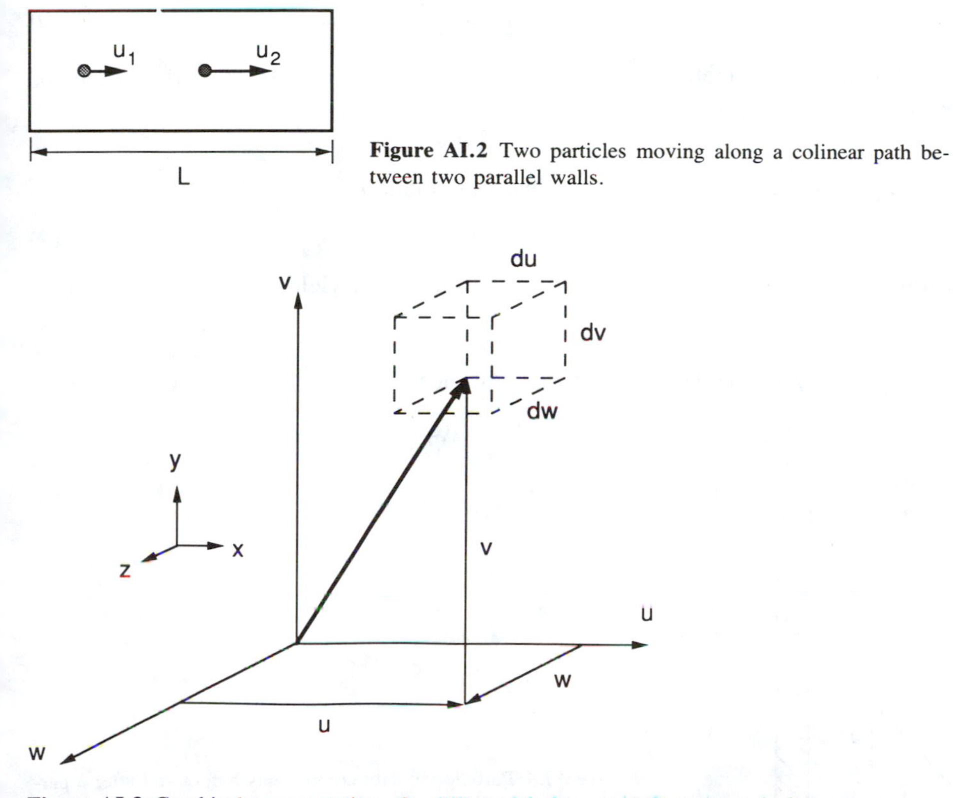

Figure AI.2 Two particles moving along a colinear path between two parallel walls.

Figure AI.3 Graphical representation of a differential element in Cartesian velocity space.

Since there is no reason to believe that any of the directions x, y, or z is preferred, it is expected that $\langle u^2 \rangle = \langle v^2 \rangle = \langle w^2 \rangle$. This conclusion, together with Eq. (I.7), implies that

$$\langle u^2 \rangle = \langle v^2 \rangle = \langle w^2 \rangle = \tfrac{1}{3} \langle c^2 \rangle \tag{I.8}$$

If we now consider a molecule that strikes a wall in a plane normal to the x axis, as shown in Fig. AI.4, it can be seen that only the normal velocity u is reversed in the collision. The tangential components v and w (not shown in Fig. AI.4) have the same direction and magnitude before and after the collision. Hence, only reversal of the normal velocity component is relevant to the momentum interaction with the wall. The frequency with which molecules strike the wall also depends on the normal velocity component only. The change in momentum per collision with the wall is $-2mu$, and the number of impacts per second is $u/2L$. By the same reasoning used for the one-dimensional gas, the pressure on the wall must equal

$$P = nm \frac{\langle u^2 \rangle}{V} \tag{I.9}$$

Using the expression (I.8) for $\langle u^2 \rangle$, Eq. (I.9) becomes

$$P = \tfrac{1}{3} nm \frac{\langle c^2 \rangle}{V} \tag{I.10}$$

If we denote the kinetic energy of a molecule as $\epsilon = \frac{1}{2}mc^2$, Eq. (I.10) can be written as

$$PV = \tfrac{2}{3} n \langle \epsilon \rangle \tag{I.11}$$

Equation (I.11) resembles the ideal gas relation,

$$PV = N\overline{R}T \tag{I.12}$$

Eliminating PV by combining Eqs. (I.11) and (I.12) yields

$$N\overline{R}T = \tfrac{2}{3} n \langle \epsilon \rangle \tag{I.13}$$

Noting that n/N is equal to Avogadro's number N_A, Eq. (I.13) can be written as

$$\overline{R}T = \tfrac{2}{3} N_A \langle \epsilon \rangle \tag{I.14}$$

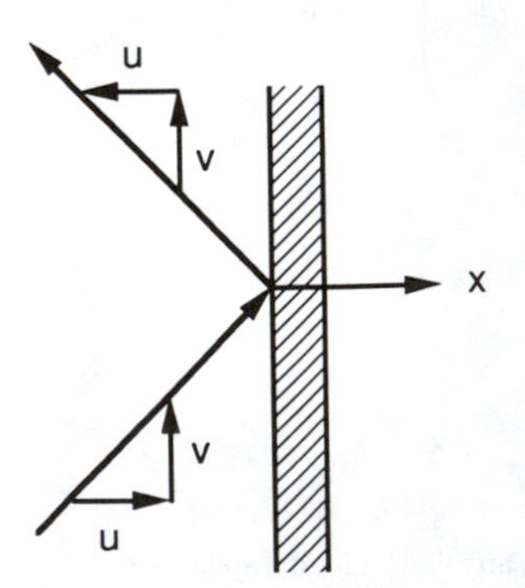

Figure AI.4 Particle velocity components before and after a perfectly elastic collision with a wall.

If we denote the total kinetic energy associated with the random motion of the molecules in one mole of gas as U, then U must be given by

$$U = \frac{1}{2} N_A m \langle c^2 \rangle = \frac{1}{2} N_A \langle \epsilon \rangle \tag{I.15}$$

Combining Eqs. (I.14) and (I.15), we obtain

$$U = \frac{3}{2} \overline{R} T \tag{I.16}$$

Equation (I.16) is one of the most interesting results of the kinetic theory. This relation links the temperature of the gas directly to the kinetic energy associated with the random motion of the molecules. For this reason, the random or chaotic motion of the molecules is sometimes referred to as the *thermal motion* of the molecules. This also implies that at absolute zero this thermal motion ceases completely.

If we define a root-mean-square (RMS) speed of the molecules, c_{RMS}, as

$$c_{RMS} = (\langle c^2 \rangle)^{1/2} \tag{I.17}$$

then Eqs. (I.15) and (I.16) can be combined to obtain

$$c_{RMS} = \left(\frac{3\overline{R}T}{\overline{M}} \right)^{1/2} \tag{I.18}$$

where $\overline{M} = N_A m$ is the molecular weight of the gas. Thus kinetic theory also provides a link between the mean speed of the gas molecules and the temperature of the gas.

To develop the kinetic theory of gases further, we must first consider the concept of distribution functions. So far, kinetic theory has provided a relation between the RMS speed of the molecules and the temperature of the system. However, individual molecules in the gas are traveling in various directions at different speeds. In general, each molecule will have a different combination of u, v, and w velocity components. Let dn_u be the number of molecules having a component of velocity in the x direction in the range between u and $u + du$. The probability of finding such a molecule in a box containing n molecules is dn_u/n.

If the width of the interval du is small, it is reasonable to expect that the number of molecules in the interval will be proportional to its size du (i.e., doubling its size would double the number of molecules in the interval). It is therefore expected that dn_u/n is proportional to du. Because the probability dn_u/n will depend on the velocity component u itself, the variation of dn_u/n with u and du is postulated as being of the form

$$\frac{dn_u}{n} = f(u^2)\, du \tag{I.19}$$

where $f(u^2)$ is an as-yet-undetermined function.

Note that f is postulated to be a function of u^2 rather than u. Because the molecular motion is completely random, motion in all directions is equally probable. Consequently, the probability of finding a molecule with velocity u between u and $u + du$ is equal to that of finding one with a velocity between $-u$ and $-u$

$-du$. By assuming that f is a function of u^2, this symmetry condition is automatically satisfied. Note also that it is implicitly assumed that the probability dn_u/n is independent of the velocity components v and w in the y and z directions, respectively. Although arguments can be presented to justify this assumption, here we will simply adopt it as a plausible idealization that is consistent with the expected symmetry of the probability distribution.

Because the arguments described above apply equally well to the velocity components in the y and z directions, we must also have

$$\frac{dn_v}{n} = f(v^2)\, dv \tag{I.20}$$

$$\frac{dn_w}{n} = f(w^2)\, dw \tag{I.21}$$

From basic probability theory, it is known that the probability of three independent things occurring simultaneously must equal the product of the probabilities of each occurring individually. It follows, therefore, that the probability of a molecule simultaneously having velocity components in the intervals u to $u + du$, v to $v + dv$, and w to $w + dw$ is equal to the product $(dn_u/n)(dn_v/n)(dn_w/n)$. Denoting this probability as dn_{uvw}/n, we thus obtain

$$\frac{dn_{uvw}}{n} = f(u^2)\, f(v^2)\, f(w^2)\, du\, dv\, dw \tag{I.22}$$

Noting that $dn_{uvw}/(du\, dv\, dw)$ is the *number density* of molecules at a particular location in velocity space coordinates, u, v, w, shown in Fig. AI.3, the above equation can be written as

$$\frac{dn_{uvw}}{du\, dv\, dw} = n\, f(u^2)\, f(v^2)\, f(w^2) \tag{I.23}$$

Suppose now that the above analysis was repeated with a different set of coordinate axes that had the same origin as our original set but were rotated relative to the original ones. If we denote these new coordinates as x', y', and z' and the respective velocities in each direction as u', v', and w', the line of reason previously described implies that the number density of molecules must be given by

$$\frac{dn_{u'v'w'}}{du'\, dv'\, dw'} = n\, f(u'^2)\, f(v'^2)\, f(w'^2) \tag{I.24}$$

Because the number density at a given location in velocity space must be independent of the coordinate system used, we conclude from Eqs. (I.23) and (I.24) that

$$f(u^2)\, f(v^2)\, f(w^2) = f(u'^2)\, f(v'^2)\, f(w'^2) \tag{I.25}$$

Also, since (u, v, w) and (u', v', w') represent the same point,

$$c^2 = u^2 + v^2 + w^2 = u'^2 + v'^2 + w'^2 \tag{I.26}$$

Furthermore, because the rotation of the primed axes is arbitrary, for convenience, we establish them by rotating the original axes in Fig. AI.3 (about the original v and w axes) so that in the primed coordinate system the point (u, v, w) corresponds to $(u', 0, 0)$. Using the fact that $v' = w' = 0$ and Eqs. (I.25) and (I.26), it is easily shown that

$$f(u^2)\, f(v^2)\, f(w^2) = A^2 f(u^2 + v^2 + w^2) \tag{I.27}$$

where

$$A = f(0) \tag{I.28}$$

If both sides of Eq. (I.27) are differentiated in turn with respect to u^2, v^2, and w^2, three relations are obtained, which indicate that

$$f'(u^2)\, f(v^2)\, f(w^2) = f(u^2)\, f'(v^2)\, f(w^2) = f(u^2)\, f(v^2)\, f'(w^2) \tag{I.29}$$

where f' denotes the derivative of f with respect to its argument. These relations can be rearranged to show that

$$\frac{f'(u^2)}{f(u^2)} = \frac{f'(v^2)}{f(v^2)} = \frac{f'(w^2)}{f(w^2)} \tag{I.30}$$

Because the expressions given above are each functions of a different independent variable, the equalities can hold only if they all equal a constant, which will designate as $-\beta$. If each one of these expressions is set equal to β and the differential equations are solved, it is found that

$$f(u^2) = Ae^{-\beta u^2} \tag{I.31a}$$

$$f(v^2) = Ae^{-\beta v^2} \tag{I.31b}$$

$$f(w^2) = Ae^{-\beta w^2} \tag{I.31c}$$

Note that either positive or negative values of β would satisfy Eqs. (I.30). However, if β were negative, f would become infinite as u^2 approaches infinity. This would mean that the probability of finding molecules with infinite kinetic energy is infinite, which is impossible for a gas having finite internal energy. We therefore conclude that only positive values of β are physically realistic. The variation of f for positive β values is the well-known Gaussian distribution associated with many random processes.

Substituting Eqs. (I.31a) through (I.31c) into Eq. (I.23) yields the following relation for the number density of molecules in velocity space:

$$\frac{dn_{uvw}}{du\, dv\, dw} = nA^3 e^{-\beta(u^2+v^2+w^3)} \tag{I.32}$$

It is also useful to determine the number density of molecules with speed c in the interval c to $c + dc$, regardless of direction. In the velocity space shown in Fig. AI.3, this corresponds to the number of points inside a spherical shell of radius c and thickness dc. If the velocity vector is transformed into the spherical coordinates shown in Fig. AI.5 using the relations

$$u = c \sin\theta \cos\phi \qquad v = c \sin\theta \sin\phi \qquad w = c \cos\theta \tag{I.33a}$$

$$c = (u^2 + v^2 + w^2)^{1/2} \tag{I.33b}$$

the velocity distribution can be written as

$$dn_{c\theta\phi} = nA^3c^2e^{-\beta c^2} \sin\theta \, d\theta \, d\phi \, dc \tag{I.34}$$

The left side of Eq. (I.34) is the number of molecules with speed c and directional angles in the ranges c to $c + dc$, θ to $\theta + d\theta$, ϕ to $\phi + d\phi$. The number of molecules with speeds in the range c to $c + dc$, regardless of direction, can be found by integrating Eq. (I.34) over all values of ϕ and θ:

$$dn_c = \int_{\phi=0}^{\phi=2\pi} \int_{\theta=0}^{\theta=\pi} dn_{c\theta\phi} = 4\pi nA^3c^2e^{-\beta c^2} \, dc \tag{I.35}$$

Although this analysis has suggested the form of the relation for the velocity distribution, the constants A and β are, as yet, undetermined. These constants can be determined, however, by requiring that the distribution be consistent with the total number of molecules and the total energy of the system. The total number of molecules n must equal the integral of dn_c over all possible speeds c from zero to infinity:

$$n = \int_{c=0}^{c=\infty} dn_c \tag{I.36}$$

Substituting Eq. (I.35) for dn_c into Eq. (I.36) yields

$$n = \int_{c=0}^{c=\infty} 4\pi nA^3c^2e^{-\beta c^2} \, dc \tag{I.37}$$

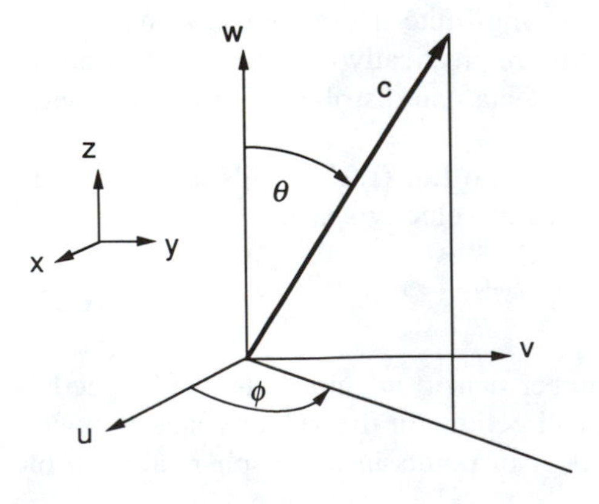

Figure AI.5 Velocity vector in Cartesian and spherical coordinates.

It is a straightforward manipulation to evaluate this integral. Upon doing so and solving for A, the following relation is obtained:

$$A = \left(\frac{\beta}{\pi}\right)^{1/2} \tag{I.38}$$

The average kinetic energy $\langle\epsilon\rangle$ per molecule is obtained by multiplying $(^1/_2)mc^2$ by the number dn_c of molecules (in each differential speed range), integrating over all values of c, and dividing by n:

$$\langle\epsilon\rangle = \frac{1}{n}\int_{c=0}^{c=\infty} \tfrac{1}{2}\, mc^2\, dn_c \tag{I.39}$$

Substituting Eq. (I.35) for dn_c into Eq. (I.39) yields

$$\langle\epsilon\rangle = \frac{1}{n}\int_{c=0}^{c=\infty} \tfrac{1}{2}\, mc^2(4\pi n)A^3c^2e^{-\beta c^2}\, dc \tag{I.40}$$

Canceling n on the right-hand side, substituting Eq. (I.38) to eliminate A, and evaluating the integral yields the following expression for $\langle\epsilon\rangle$:

$$\langle\epsilon\rangle = \frac{3m}{4\beta} \tag{I.41}$$

It has been shown above, using the kinetic theory, that the average energy per molecule $\langle\epsilon\rangle$ is related to the temperature of the gas as

$$\langle\epsilon\rangle = \frac{3}{2}\left(\frac{\overline{R}}{N_A}\right)T = \frac{3}{2}\, k_BT \tag{I.42}$$

where k_B is the Boltzmann constant, defined as

$$k_B = \frac{\overline{R}}{N_A} \tag{I.43}$$

Combining Eqs. (I.41) and (I.42) yields the following relation for β:

$$\beta = \frac{m}{2k_BT} \tag{I.44}$$

It follows directly using Eqs. (I.38) and (I.44) that

$$A = \left(\frac{m}{2\pi k_BT}\right)^{1/2} \tag{I.45}$$

Replacing A and β in Eq. (I.35) with the relations given in Eqs. (I.44) and (I.45) yields

$$dn_c = 4\pi n\left(\frac{m}{2\pi k_BT}\right)^{3/2} c^2e^{-mc^2/2k_BT}\, dc \tag{I.46}$$

The number distribution of molecules given by the above equation is the *Maxwell distribution*.

If the relations (I.44) and (I.45) are substituted into Eqs. (I.31*a*) through (I.31*c*), it can be shown using Eqs. (I.19) through (I.21) that the number distributions for the individual velocity components are given by

$$\frac{dn_u}{n} = \left(\frac{m}{2\pi k_B T}\right)^{1/2} e^{-mu^2/2k_BT}\, du \tag{I.47a}$$

$$\frac{dn_v}{n} = \left(\frac{m}{2\pi k_B T}\right)^{1/2} e^{-mv^2/2k_BT}\, dv \tag{I.47b}$$

$$\frac{dn_w}{n} = \left(\frac{m}{2k_B T}\right)^{1/2} e^{-mw^2/2k_BT}\, dw \tag{I.47c}$$

Substitution of Eqs. (I.44) and (I.45) into Eq. (I.32) also yields the following form of the Maxwell distribution:

$$dn_{uvw} = n\left(\frac{m}{2\pi k_B T}\right)^{3/2} e^{-m(u^2+v^2+w^2)/2k_BT}\, du\, dv\, dw \tag{I.48}$$

Associated with the Maxwell distribution of molecular velocities, one can consider the most probable speed c_{mp}, the root-mean-square speed $\langle c^2\rangle^{1/2}$, and the mean speed $\langle c\rangle$. These quantities can be determined directly from Eq. (I.46). The most probable speed is the value of c that maximizes dn_c/dc. Solving Eq. (I.46) for this quantity, differentiating, and setting the result equal to zero yields

$$c_{mp} = \left(\frac{2k_BT}{m}\right)^{1/2} \tag{I.49}$$

The mean speed $\langle c\rangle$ and the root-mean-square speed $\langle c^2\rangle^{1/2}$, each of which is weighted with the fraction of the molecules in each speed interval, are calculated using Eq. (I.46) as

$$\langle c\rangle = \int_0^\infty c\left(\frac{dn_c}{n\, dc}\right) dc = \left(\frac{8k_BT}{\pi m}\right)^{1/2} \tag{I.50}$$

$$\langle c^2\rangle^{1/2} = \left[\int_0^\infty c^2\left(\frac{dn_c}{n\, dc}\right) dc\right]^{1/2} = \left(\frac{3k_BT}{m}\right)^{1/2}$$

Thus the three representative molecular velocities defined above are all proportional to $(k_BT/m)^{1/2}$, with the proportionality constant ranging from $2^{1/2}$ to $3^{1/2}$.

The kinetic theory of gases provides an important conceptual foundation for understanding the behavior of real gases. Further discussion of the kinetic theory and its implications can be found in references [4.10–4.12].

APPENDIX
TWO

SATURATION PROPERTIES OF SELECTED FLUIDS*

Acetone

Chemical formula: CH_3COCH_3
Molecular weight: 58.1

Critical temperature: 508.15 K
Critical pressure: 4761 kPa
Critical density: 273 kg/m^3

T_{sat} (K)	329.25	340	360	380	400	420	440	460	480	508.15
P_{sat} (kPa)	101.3	152	274	452	731	1082	1637	2279	3252	4761
ρ_l (kg/m^3)	750	736	710	683	655	625	590	553	504	273
ρ_v (kg/m^3)	2.23	3.11	5.49	9.13	14.5	22.3	33.6	50.3	77.2	273
h_{lv} (kJ/kg)	506	494	465	439	414	382	344	300	242	
c_{pl} (kJ/kg K)	2.28	2.32	2.42	2.53	2.65	2.83	3.03	3.29	3.76	
c_{pv} (kJ/kg K)	1.41	1.46	1.55	1.66	1.79	1.95	2.18	2.54	3.38	
μ_l (μN s/m^2)	235	213	188	165	141	119	99	80	64	49
μ_v (μN s/m^2)	9.4	9.8	10.4	11.1	11.8	12.6	13.5	14.4	15.8	49
k_l (mW/m K)	142	137	129	121	112	104	96	87	77	58
k_v (mW/m K)	12.7	14.1	16.1	18.5	21.2	24.2	27.2	31.0	36.0	58
Pr_l	3.77	3.61	3.53	3.49	3.34	3.24	3.12	3.03	3.13	
Pr_v	1.04	1.01	1.00	1.00	1.00	1.02	1.08	1.18	1.48	
σ (mN/m)	18.4	17.0	14.5	12.1	9.6	7.1	4.6	3.1	1.6	

*Adapted from the *Heat Exchanger Design Handbook*, with permission, copyright © Hemisphere, New York, 1983.

Ammonia

Chemical formula: NH_3
Molecular weight: 17.032

Critical temperature: 405.55 K
Critical pressure: 11290 kPa
Critical density: 235 kg/m^3

T_{sat} (K)	239.75	250	270	290	310	330	350	370	390	400
P_{sat} (kPa)	101.3	165.4	381.9	775.3	1424.9	2422	3870	5891	8606	10280
ρ_l (kg/m^3)	682	669	643	615	584	551	512	466	400	344
ρ_v (kg/m^3)	0.86	1.41	3.09	6.08	11.0	18.9	31.5	52.6	93.3	137
h_{lv} (kJ/kg)	1368	1338	1273	1200	1115	1019	899	744	508	307
c_{pl} (kJ/kg K)	4.472	4.513	4.585	4.649	4.857	5.066	5.401	5.861	7.74	
c_{pv} (kJ/kg K)	2.12	2.32	2.69	3.04	3.44	3.90	4.62	6.21	8.07	
μ_l (μN s/m^2)	285	246	190	152	125	105	88.5	70.2	50.7	39.5
μ_v (μN s/m^2)	9.25	9.59	10.30	11.05	11.86	12.74	13.75	15.06	17.15	19.5
k_l (mW/m K)	614	592	569	501	456	411	365	320	275	252
k_v (mW/m K)	18.8	19.8	22.7	25.2	28.9	34.3	39.5	50.4	69.2	79.4
Pr_l	2.06	1.88	1.58	1.39	1.36	1.32	1.34	1.41	1.43	
Pr_v	1.04	1.11	1.17	1.25	1.31	1.34	1.49	1.70	1.86	
σ (mN/m)	33.9	31.5	26.9	22.4	18.0	13.7	9.60	5.74	2.21	0.68

n-Butanol

Chemical formula: $C_2H_5CH_2CH_2OH$
Molecular weight: 74.12

Critical temperature: 561.15 K
Critical pressure: 4960 kPa
Critical density: 270.5 kg/m^3

T_{sat} (K)	390.65	410.2	429.2	446.5	469.5	485.2	508.3	530.2	545.5	558.9
P_{sat} (kPa)	101.3	182	327	482	759	1190	1830	2530	3210	4030
ρ_l (kg/m^3)	712	688	664	640	606	581	538	487	440	364
ρ_v (kg/m^3)	2.30	4.10	7.9	12.5	23.8	27.8	48.2	74.0	102.3	240.2
h_{lv} (kJ/kg)	591.3	565.0	537.3	509.7	468.8	437.2	382.5	315.1	248.4	143.0
c_{pl} (kJ/kg K)	3.20	3.54	3.95	4.42	5.15	5.74	6.71	7.76		
c_{pv} (kJ/kg K)	1.87	1.95	2.03	2.14	2.24	2.37	2.69	3.05	3.97	
μ_l (μN s/m^2)	403.8	346.1	278.8	230.8	188.5	144.2	130.8	115.4	111.5	105.8
μ_v (μN s/m^2)	9.29	10.3	10.7	11.4	12.1	12.7	13.9	15.4	17.1	28.3
k_l (mW/m K)	127.1	122.3	117.5	112.6	105.4	101.4	91.7	82.9	74.0	62.8
k_v (mW/m K)	21.7	24.2	26.7	28.2	31.3	33.1	36.9	40.2	43.6	51.5
Pr_l	10.3	9.86	9.17	8.64	10.2	8.10	8.67	9.08		
Pr_v	0.81	0.83	0.81	0.86	0.87	0.91	1.01	1.17	1.56	
σ (mN/m)	17.1	15.6	13.9	12.3	10.2	7.50	6.44	4.23	2.11	0.96

Carbon Tetrachloride

Chemical formula: CCl_4
Molecular weight: 153.8

Critical temperature: 556.35 K
Critical pressure: 4560 kPa
Critical density: 588 kg/m^3

T_{sat} (K)	349.95	370	390	410	430	450	470	495	525	556.35
P_{sat} (kPa)	101.3	184	307	473	701	1020	1390	2020	3160	4560
ρ_l (kg/m^3)	1484	1442	1397	1351	1303	1250	1199	1107	989	588
ρ_v (kg/m^3)	5.44	9.40	15.2	23.4	34.8	50.3	71.2	108.5	184.5	588
h_{lv} (kJ/kg)	195	188	180	172	159	152	140	126	98	
c_{pl} (kJ/kg K)	0.92	0.94	0.97	1.01	1.06	1.14	1.24	1.36	1.57	
c_{pv} (kJ/kg K)	0.58	0.60	0.62	0.65	0.68	0.73	0.80	0.91	1.30	
μ_l (μN s/m^2)	494	407	352	309	274	241	205	154	98	63
μ_v (μN s/m^2)	11.9	12.5	13.3	14.1	14.9	15.7	16.7	18.9	21.0	63
k_l (mW/m K)	92	87	83	78	74	70	65	57	45	25
k_v (mW/m K)	8.6	9.3	10.0	10.7	11.5	12.3	13.2	14.3	16.3	25
Pr_l	4.94	4.40	4.16	4.08	3.93	3.92	3.91	3.67	3.42	
Pr_v	0.80	0.81	0.82	0.85	0.88	0.93	1.01	1.20	1.67	
σ (mN/m)	20.2	17.6	15.4	13.1	10.9	8.8	6.9	4.4	2.0	

Ethanol

Chemical formula: CH_3CH_2OH
Molecular weight: 46.1

Critical temperature: 516.25 K
Critical pressure: 6390 kPa
Critical density: 280 kg/m^3

T_{sat} (K)	351.45	373	393	413	433	453	473	483	503	513
P_{sat} (kPa)	101.3	226	429	753	1256	1960	2940	3560	5100	6020
ρ_l (kg/m^3)	757.0	733.7	709.0	680.3	648.5	610.5	564.0	537.6	466.2	420.3
ρ_v (kg/m^3)	1.435	3.175	5.841	10.25	17.15	27.65	44.40	56.85	101.1	160.2
h_{lv} (kJ/kg)	963.0	927.0	885.5	834.0	772.9	698.9	598.3	536.7	387.3	280.5
c_{pl} (kJ/kg K)	3.00	3.30	3.61	3.96	4.65	5.51	6.16	6.61		
c_{pv} (kJ/kg K)	1.83	1.92	2.02	2.11	2.31	2.80	3.18	3.78	6.55	
μ_l (μN s/m^2)	428.7	314.3	240.0	185.5	144.6	113.6	89.6	79.7	63.2	56.3
μ_v (μN s/m^2)	10.4	11.1	11.7	12.3	12.9	13.7	14.5	15.1	16.7	18.5
k_l (mW/m K)	153.6	150.7	146.5	141.9	137.2	134.8	129.1	125.6	108.0	79.11
k_v (mW/m K)	19.9	22.4	24.5	26.8	29.3	32.1	35.3	37.8	43.9	50.7
Pr_l	8.37	6.88	5.91	5.18	4.90	4.64	4.28	4.19		
Pr_v	0.96	0.95	0.96	0.97	1.02	1.20	1.31	1.51	2.49	
σ (mN/m)	17.7	15.7	13.6	11.5	9.3	6.9	4.5	3.3	0.9	0.34

Mercury

Chemical formula: Hg
Molecular weight: 200.51

Critical temperature: 1763.2 K
Critical pressure: 151,000 kPa
Critical density: 5500 kg/m^3

T_{sat} (K)	630.1	650	700	750	800	850	900	950	1000	1050
P_{sat} (kPa)	101.3	145	316	620	1120	1880	2990	4530	6580	9230
ρ_l (kg/m^3)	12737	12688	12567	12444	12318	12190	12059	11927	11791	11650
ρ_v (kg/m^3)	3.91	5.37	10.9	20.1	34.2	54.6	82.7	119.9	167.7	227.3
h_{lv} (kJ/kg)	294.9	294.2	292.3	290.2	287.8	285.1	282.1	278.6	274.7	269.2
c_{pl} (kJ/kg K)	0.136	0.136	0.137	0.138	0.140	0.142	0.144	0.146	0.149	0.153
c_{pv} (kJ/kg K)	0.104	0.104	0.105	0.106	0.107	0.108	0.109	0.111	0.113	0.116
μ_l (μN s/m^2)	884	870	841	816	794	776	760	746	736	723
μ_v (μN s/m^2)	61.7	63.5	68.6	73.5	78.4	83.5	88.4	93.2	98.0	103.0
k_l (mW/m K)	121.9	123.6	128.0	131.9	135.1	137.8	141.8	144.5	146.9	147.9
k_v (mW/m K)	10.4	10.8	11.7	12.6	13.5	14.4	15.3	16.2	17.2	18.1
Pr_l	0.987	0.957	0.900	0.854	0.823	0.800	0.772	0.754	0.744	0.748
Pr_v	0.617	0.612	0.616	0.618	0.621	0.626	0.630	0.637	0.644	0.660
σ (mN/m)	417	413	403	393	383	372	362	352	341	331

Methanol

Chemical formula: CH_3OH
Molecular weight: 32.00

Critical temperature: 513.15 K
Critical pressure: 7950 kPa
Critical density: 275 kg/m^3

T_{sat} (K)	337.85	353.2	373.2	393.2	413.2	433.2	453.2	473.2	493.2	511.7
P_{sat} (kPa)	101.3	178.4	349.4	633.3	1076	1736	2678	3970	5675	7775
ρ_l (kg/m^3)	751.0	735.5	714.0	690.0	664.0	634.0	598.0	553.0	490.0	363.5
ρ_v (kg/m^3)	1.222	2.084	3.984	7.142	12.16	19.94	31.86	50.75	86.35	178.9
h_{lv} (kJ/kg)	1101	1070	1022	968	922	843	756	645	482	
c_{pl} (kJ/kg K)	2.88	3.03	3.26	3.52	3.80	4.11	4.45	4.81		
c_{pv} (kJ/kg K)	1.55	1.61	1.69	1.83	1.99	2.20	2.56	3.65	5.40	
μ_l (μN s/m^2)	326	271	214	170	136	109	88.3	71.6	58.3	41.6
μ_v (μN s/m^2)	11.1	11.6	12.4	13.1	14.0	14.9	16.0	17.4	20.1	26.0
k_l (mW/m K)	191.4	187.0	181.3	178.5	170.0	164.0	158.7	153.0	147.3	142.0
k_v (mW/m K)	18.3	20.6	23.2	26.2	29.7	33.8	39.4	46.9	60.0	98.7
Pr_l	5.13	4.67	4.15	3.61	3.34	2.82	2.56	2.42		
Pr_v	0.94	0.91	0.90	0.92	0.94	0.97	1.04	1.35	1.81	
σ (mN/m)	18.75	17.5	15.7	13.6	11.5	9.3	6.9	4.5	2.1	0.09

Nitrogen

Chemical formula: N_2
Molecular weight: 28.016

Critical temperature: 126.25 K
Critical pressure: 3396 kPa
Critical density: 304 kg/m^3

T_{sat} (K)	77.35	85	90	95	100	105	110	115	120	126
P_{sat} (kPa)	101.3	229	360	540	778	1083	1467	1940	2515	3357
ρ_l (kg/m^3)	807.10	771.01	746.27	719.42	691.08	660.5	626.17	583.43	528.54	379.22
ρ_v (kg/m^3)	4.621	9.833	15.087	22.286	31.989	44.984	62.578	87.184	124.517	237.925
h_{lv} (kJ/kg)	197.6	188.0	180.5	172.2	162.2	150.7	137.0	119.9	95.7	32.1
c_{pl} (kJ/kg K)	2.064	2.096	2.140	2.211	2.311	2.467	2.711	3.180	4.347	
c_{pv} (kJ/kg K)	1.123	1.192	1.258	1.350	1.474	1.666	1.975	2.586	4.136	
μ_l (μN s/m^2)	163	127	110	97.2	86.9	78.5	70.8	59.9	48.4	19.1
μ_v (μN s/m^2)	5.41	5.60	6.36	6.80	7.28	7.82	8.42	9.25	10.68	19.1
k_l (mW/m K)	136.7	122.9	112.0	104.0	95.5	88.0	80.2	70.4	62.8	52.8
k_v (mW/m K)	7.54	8.18	9.04	9.77	10.60	11.69	14.50	20.76	30.91	51.11
Pr_l	2.46	2.17	2.10	2.07	2.10	2.20	2.39	2.71	3.35	
Pr_v	0.81	0.82	0.89	0.94	1.01	1.11	1.15	1.16	1.43	
σ (mN/m)	8.85	7.20	6.16	4.59	3.67	2.79	1.98	1.18	0.52	0.01

Oxygen

Chemical formula: O_2
Molecular weight: 32.00

Critical temperature: 154.77 K
Critical pressure: 5090 kPa
Critical density: 405 kg/m^3

T_{sat} (K)	90.18	97	104	111	118	125	132	140	146	154
P_{sat} (kPa)	101.3	196	352	583	908	1348	1924	2782	3591	3939
ρ_l (kg/m^3)	1135.72	1102.05	1065.07	1025.64	982.32	934.58	880.28	808.41	737.56	557.10
ρ_v (kg/m^3)	4.48	8.23	14.14	22.79	35.03	52.05	75.81	116.12	163.34	304.41
h_{lv} (kJ/kg)	212.3	205.9	198.3	189.4	178.7	165.7	150.1	127.3	104.6	46.1
c_{pl} (kJ/kg K)	1.63	1.66	1.70	1.76	1.86	2.00	2.22	2.63	3.28	
c_{pv} (kJ/kg K)	0.96	1.00	1.05	1.12	1.23	1.36	1.68	2.27	3.63	
μ_l (μN s/m^2)	195.83	161.75	136.55	116.80	101.20	89.00	80.15	69.66	60.65	42.48
μ_v (μN s/m^2)	6.85	7.50	8.35	9.36	10.6	11.24	13.35	15.8	18.5	26.9
k_l (mW/m K)	148	139	130	121	111	102	92.5	82.0	71.2	
k_v (mW/m K)	8.5	9.5	10.5	11.7	13.4	14.8	16.9	20.1	23.6	35.2
Pr_l	2.16	1.93	1.79	1.70	1.70	1.75	1.92	2.23	2.79	
Pr_v	0.77	0.79	0.84	0.90	0.97	1.03	1.33	1.78	2.85	19.93
σ (mN/m)	13.19	11.53	9.88	8.27	6.71	5.20	3.77	2.23	1.18	0.40

1-Propanol

Chemical formula: $CH_3CH_2CH_2OH$
Molecular weight: 60.1

Critical temperature: 536.85 K
Critical pressure: 5050 kPa
Critical density: 273 kg/m^3

T_{sat} (K)	373.2	393.2	413.2	433.2	453.2	473.2	493.2	513.2	523.2	533.1
P_{sat} (kPa)	109.4	218.5	399.2	683.6	1089	1662	2426	3402	3998	4689
ρ_l (kg/m^3)	732.5	711	687.5	660	628.5	592.0	548.5	492.0	452.5	390.5
ρ_v (kg/m^3)	2.26	4.43	8.05	13.8	22.5	35.3	55.6	90.4	118.0	161.0
h_{lv} (kJ/kg)	687	645	594	544	486	427	356	264	209	138
c_{pl} (kJ/kg K)	3.21	3.47	3.86	4.36	5.02	5.90	6.78	7.79		
c_{pv} (kJ/kg K)	1.65	1.82	1.93	2.05	2.20	2.36	2.97	3.94		
μ_l (μN s/m^2)	447	337	250	188	148	119	90.6	70.0	61.4	53.9
μ_v (μN s/m^2)	9.61	10.3	10.9	11.5	12.2	12.9	14.2	15.7	17.0	19.3
k_l (mW/m K)	142.4	139.2	138.4	133.5	127.9	120.7	111.8	100.6	94.1	89.3
k_v (mW/m K)	20.9	23.0	26.2	28.9	31.4	34.7	38.0	43.9	47.5	53.5
Pr_l	10.1	8.40	6.97	5.14	5.81	5.82	5.50	5.42		
Pr_v	0.76	0.82	0.80	0.82	0.85	0.88	1.11	1.41		
σ (mN/m)	17.6	16.15	14.42	12.7	10.77	8.85	6.35	4.04	2.6	0.96

Refrigerant-12

Chemical formula: CCl_2F_2
Molecular weight: 120.92

Critical temperature: 384.8 K
Critical pressure: 4132 kPa
Critical density: 561.8 kg/m³

T_{sat} (K)	243.2	260	275	290	305	320	335	350	365	384.8
P_{sat} (kPa)	101.3	200	333	528	793	1145	1602	2183	2907	4132
ρ_l (kg/m³)	1486	1436	1388	1338	1284	1225	1157	1075	969.7	561.8
ρ_v (kg/m³)	6.33	11.8	19.2	29.9	44.8	65.4	94.6	136.4	203.2	561.8
h_{lv} (kJ/kg)	168.3	161.5	154.7	146.6	137.7	127.2	114.0	97.6	75.8	
c_{pl} (kJ/kg K)	0.896	0.911	0.932	0.957	0.990	1.03	1.08	1.13	1.22	
c_{pv} (kJ/kg K)	0.569	0.614	0.646	0.689	0.746	0.825	0.920	1.22	1.68	
μ_l (μN s/m²)	373	303	262	231	208	187	167	144	119	
μ_v (μN s/m²)	10.3	11.0	11.7	12.5	13.3	14.2	15.2	16.5	18.1	
k_l (mW/m K)	95.1	87.4	80.5	73.3	66.8	59.8	53.0	46.2	39.2	15.4
k_v (mW/m K)	6.9	7.7	8.4	9.2	10.0	10.8	11.6	12.3	13.4	15.4
Pr_l	3.51	3.16	3.03	3.02	3.14	3.22	3.40	3.52	3.70	
Pr_v	0.85	0.88	0.90	0.94	0.99	1.08	1.21	1.64	2.27	
σ (mN/m)	15.5	13.5	11.4	9.4	7.7	5.9	4.2	2.8	1.3	

Refrigerant-22

Chemical formula: $CHClF_2$
Molecular weight: 86.48

Critical temperature: 369.3 K
Critical pressure: 4986 kPa
Critical density: 513 kg/m^3

T_{sat} (K)	242.4	250	265	280	295	310	325	340	355	369.3
P_{sat} (kPa)	101.3	218	376	619	958	1420	2020	2800	3800	4986
ρ_l (kg/m^3)	1413	1360	1313	1260	1206	1146	1076	991	877	513
ρ_v (kg/m^3)	4.70	9.59	16.1	26.3	40.6	60.9	90.2	134	208	513
h_{lv} (kJ/kg)	233.4	225.6	210.8	198.6	185.2	169.8	151.6	128.7	95.7	
c_{pl} (kJ/kg K)	1.10	1.13	1.16	1.19	1.24	1.30	1.41	1.65	2.43	
c_{pv} (kJ/kg K)	0.599	0.646	0.691	0.747	0.820	0.930	1.09	1.40	2.31	
μ_l (μN s/m^2)	332	282	251	225	204	187	172	150	119	
μ_v (μN s/m^2)	10.1	10.9	11.7	12.3	13.2	14.2	15.7	16.4	18.8	
k_l (mW/m K)	119	109	101	94.2	86.6	78.8	70.2	59.2	44.0	31.9
k_v (mW/m K)	7.15	8.22	9.10	10.1	11.2	12.4	14.0	16.0	18.8	31.9
Pr_l	3.07	2.92	2.88	2.84	2.92	3.09	3.45	4.18	6.89	
Pr_v	0.85	0.86	0.89	0.91	0.97	1.07	1.22	1.69	2.31	
σ (mN/m)	18.3	15.5	13.0	10.6	8.4	6.2	4.3	2.5	1.0	

Water

Chemical formula: H_2O
Molecular weight: 18.0156

Critical temperature: 647.3 K
Critical pressure: 22,129 kPa
Critical density: 351 kg/m^3

T_{sat} (K)	373.15	400	430	460	490	520	550	580	610	647.3
P_{sat} (kPa)	101.3	247	571	1172	2185	3773	6124	9460	14044	22129
ρ_l (kg/m^3)	958.3	937.5	910.3	879.4	844.3	803.8	756.1	697.2	619.5	315
ρ_v (kg/m^3)	0.597	1.370	3.020	5.975	10.95	18.90	31.52	51.85	87.5	315
h_{lv} (kJ/kg)	2256.7	2183	2092.8	1990.4	1871.5	1731.0	1562.6	1350.3	1064.2	0.0
c_{pl} (kJ/kg K)	4.22	4.24	4.28	4.45	4.60	4.84	5.07	5.70	8.12	
c_{pv} (kJ/kg K)	2.03	2.16	2.35	2.70	3.17	3.84	4.87	6.71	11.2	
μ_l (μN s/m^2)	277.53	218.9	175.73	147.24	126.6	111.05	99.21	89.40	78.60	23.1
μ_v (μN s/m^2)	12.55	13.57	14.716	15.86	17.00	18.14	19.33	20.51	21.68	23.1
k_l (mW/m K)	679.0	685.7	683.3	671.3	646.0	618.3	580.9	536.6	464.0	914
k_v (mW/m K)	25.0	28.1	31.6	36.6	42.3	50.1	60.2	77.3	111.4	914
Pr_l	1.72	1.35	1.10	0.98	0.90	0.87	0.87	0.950	1.38	
Pr_v	1.02	1.04	1.09	1.17	1.27	1.39	1.56	1.78	2.17	
σ (mN/m)	58.91	53.50	47.16	40.66	33.90	26.96	19.66	12.71	6.26	0.0

INDEX